Advances in Intelligent Systems and Computing

Volume 281

Series editor

Janusz Kacprzyk, Warsaw, Poland

For further volumes:
http://www.springer.com/series/11156

About this Series

The series "Advances in Intelligent Systems and Computing" contains publications on theory, applications, and design methods of Intelligent Systems and Intelligent Computing. Virtually all disciplines such as engineering, natural sciences, computer and information science, ICT, economics, business, e-commerce, environment, healthcare, life science are covered. The list of topics spans all the areas of modern intelligent systems and computing.

The publications within "Advances in Intelligent Systems and Computing" are primarily textbooks and proceedings of important conferences, symposia and congresses. They cover significant recent developments in the field, both of a foundational and applicable character. An important characteristic feature of the series is the short publication time and world-wide distribution. This permits a rapid and broad dissemination of research results.

Jiuping Xu · Virgílio António Cruz-Machado
Benjamin Lev · Stefan Nickel
Editors

Proceedings of the Eighth International Conference on Management Science and Engineering Management

Focused on Computing and Engineering Management

 Springer

Editors
Jiuping Xu
Business School
Sichuan University
Chengdu
China

Virgílio António Cruz-Machado
Faculdade de Ciências e Tecnologia
Departamento de Engenharia
  Mecânica e Industrial (DEMI)
Universidade Nova de
  Lisboa (FCT/UNL)
Caparica
Portugal

Benjamin Lev
Department of Decision Sciences
LeBow College of Business
Drexel University
Philadelphia, PA
USA

Stefan Nickel
Institute of Operations Research
Karlsruhe Institute of Technology (KIT)
Karlsruhe
Germany

ISSN 2194-5357
ISBN 978-3-642-55121-5
DOI 10.1007/978-3-642-55122-2
Springer Heidelberg New York Dordrecht London

ISSN 2194-5365   (electronic)
ISBN 978-3-642-55122-2   (eBook)

Library of Congress Control Number: 2014937306

Printed on acid-free paper

Springer is part of Springer Science+Business Media (www.springer.com)

Preface

Welcome to the *Proceedings of the Eighth International Conference on Management Science and Engineering Management* (ICMSEM2014) held from July 25 to 27, 2014 at Universidade Nova de Lisboa, Lisbon, Portugal.

The International Conference on Management Science and Engineering Management is the annual conference organized by the International Society of Management Science and Engineering Management (ISMSEM). The goals of the Conference are to foster international research collaborations in Management Science and Engineering Management as well as to provide a forum to present current research results in the forms of technical sessions, round table discussions during the conference period in a relaxed and an enjoyable atmosphere. This year, 1,337 papers from 37 countries were received and 138 papers from 14 countries were accepted for a presentation or poster display at the conference after a rigorous review. These papers are from countries including Spain, Australia, Germany, France, Canada, Pakistan, China, The USA, Japan, Portugal, Iran, The Netherlands, Korea, and Azerbaijan. They are classified into eight parts in the proceedings which are Intelligent Systems, Decision-Making Systems, Manufacturing, Supply Chain Management, Computing Methodology, Project Management, Industrial Engineering, and Information Technology. The key issues of the eighth ICMSEM cover various areas in MSEM, such as Decision-Making Methods, Computational Mathematics, Information Systems, Logistics and Supply Chain Management, Relationship Management, Scheduling and Control, Data Warehousing and Data Mining, Electronic Commerce, Neural Networks, Stochastic models and Simulation, Heuristics Algorithms, and Risk Control. In order to further encourage the state-of-the-art research in the field of Management Science and Engineering Management, the ISMSEM Advancement Prize for MSEM will be awarded at the conference to these researchers.

A total of 138 papers were accepted and divided into 2 proceedings, with 69 papers in each proceeding. In order to find out the research topics among the accepted papers, the NodeXL was applied. To begin with, key words from 69 papers were excerpted as follows: Computing methodology, Particle swarm optimization (PSO), Binary particle swarm optimization, Industrial engineering, Flexible job-shop scheduling problem (FJSP), Project management, Information technology, Knowledge network, Synergy effect, Self-organizing maps, Entropy, Black–Scholes model, Risk management, Customer churn prediction, Data mining, Classifiers, Demographic characteristics, Regression analysis, Dynamic programming, Optimization, Multiobjective Optimization, Scheduling, Assembly line balancing models (ALB), Electronic commerce, Vector Space Model, Information retrieval, Maintenance, KPI, Decision making, Autocorrelation, Default distance, Psychological capital, Turnover intention, Organizational support, Ethnic regions, Public utilities, Interpretative structural modeling, Grey topologic prediction, Relationship capital, Demand-pull absorptive ability, Innovation, International travels, Data envelopment analysis, Safety engineering, Clustering algorithms, RFID, Healthcare knowledge, Disease control, Process development, Bayesian networks, Quantitative test, Fuzzy sets, Virtual corporation, Cultural identity, Supply chain management, Random variables, Expected objective, Resource allocation, Behavioral research, Artificial neural network, Environment management, Decision-making systems.

The significance of the keywords not only lies in its frequency, but the connection between the keywords is also very important in our study of how these papers revolve around the theme of Engineering Management (EM). The field of EM provides a set of concepts and metrics to systematically study the relationships between the key words. The methods of information visualization have also become valuable in helping us to discover patterns, trends, clusters, and outliers, even in complex social networks. In the preface, the open source software tool NodeXL was designed especially to facilitate learning the concepts and methods of EM as a key component.

Using the NodeXL, all of the 506 keywords involved in the 69 papers were analyzed. To begin with, the preliminary processing was executed on all the key words. Except for a unified expression of words, all the key words with the same meaning and the words including the meaning of similar key words have been unitized. For example, "multiobjective problems," "multiobjetive models," and "multiobjective optimization" have finally been unified to "multiobjective optimization." Through the preliminary processing, the keywords have reduced to 453, making it possible to constitute network efficiently.

These processed keywords, represented as the vertexes in NodeXL will be visualized in a network diagram. In the network diagram, the vertexes' sizes have been set to depend on the number of other vertexes associated with it. The more the vertex connects with other vertexes, the higher centrality it would be, which reflects the keyword's important status in the field of EM. In other words, this key word is likely to represent an important issue in EM. At the same time, the vertexes' shapes have been set to depend on their betweenness and closeness centrality. When the degree of a vertex's betweenness and closeness centrality is beyond a certain value, the shape of this vertex would be square. The goal is to find out some key concepts in the field of EM. These key concepts are likely to be the important nodes that connect with other research topics.

Through the above steps, a network constituted by the keywords representing the relationship between them is demonstrated in Fig. 1.

Figure 1 shows that computing methodology, industrial engineering, project management, and information technology are key concepts, which are the

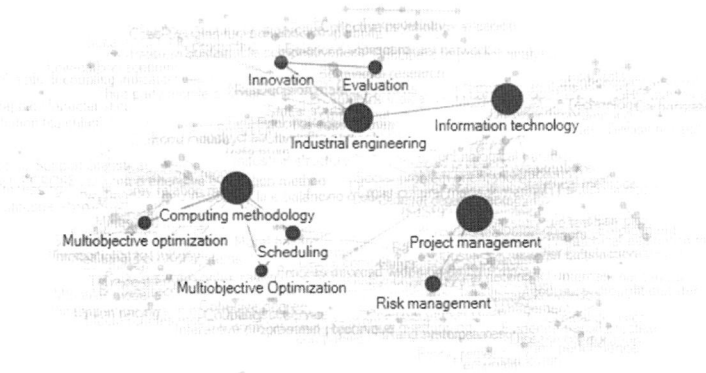

Fig. 1 Research topics in EM for the Eighth ICMSEM

important nodes connected with other research topics. In other words, they are key issues about EM in the accepted 69 papers in this volume.

In this volume, the proceedings concentrate on computing methodology, industrial engineering, project management, and information technology. To begin with, computing methodology is the theoretical foundation of solving the problems in management science and engineering management. In this part, Mehrbod et al. develop a vector space model to measure the similarity ratio of providers' e-catalogs with a buyer's e-catalog. Gen et al. concern with the design of multi-objective genetic algorithms (MOGAs) to solve a variety of manufacturing scheduling problems. He et al. propose an improved particle swarm optimization algorithm (IPSO) to prevent premature convergence and applied the IPSO to estimate the implied volatility for European option, which is a critical parameter in option pricing. Xiao et al. combine multiple classifiers ensemble technique, self-organizing data mining with cost-sensitive learning, and proposed one-step classifier ensemble model for imbalance data (OCEMI). Deng and Sun study the multiobjective dynamic programming in its investment system to improve the biogas energy development structure.

Project Management, Part VI, is the discipline of planning, organizing, securing, and managing resources to bring about the successful completion of specific project goals and objectives. Scholars in this section tend to focus on the accomplishment of desired goals and objectives by using restricted resources efficiently and effectively. Reis et al. present a methodology developed at LNEC for forecasting and early warning of wave overtopping in ports/coastal areas to prevent emergency situations and support their management and the long-term planning of interventions in the study area. Nazam et al. develop an evaluation model based on fuzzy set theory, analytical hierarchy process (AHP), and the technique for order performance by similarity to ideal solution (TOPSIS) methods for project bidding selection. Li et al. come up with the project management and technical scheme of the reinforced widening embankment with no extra land acquisition by keeping in line with the framework of conservative traffic and introducing the integrated management approach. Liang et al. propose a classified and graded safety management method for elevator users by combing both probabilities and consequences of accidents based on the coordination theory.

Part VII is Industrial Engineering. Industrial engineering is the branch of engineering which deals with the optimization of complex processes or systems. In this part, Maleki and Machado employ Bayesian Network (BN) and Analytic Network Process (ANP) to quantify mutual correlations between supply chain practices and customer values, and this approach is applied to a case study in the food industry to present its application in practice. Molina et al. present the analysis of this problem (contradictions) and the design of a noise filtration system based on Theory of Inventive Problem Solving (TRIZ). TRIZ has the potential to aid in the creation of innovative systems. Li et al. examine how human capital investment (including education and training investment) affects the firm performance. The results show that employees' educational level has a significantly positive impact on firm performance. Chen and Zhang develop a two-stage model to obtain a proactive and reactive schedule in resource-constrained project scheduling problems (RCPSP) under uncertainty, by testing the example, the effectiveness of the proposed model and approach is validated by the computation results.

Information technology (IT), the last part, is an appropriate technical platform for solving practical management problems, and is defined as "the study, design, development, implementation, support or management of computer-based information systems, particularly software applications and computer hardware," according to the Information Technology Association of America. IT is playing an increasingly dominant part in modern society. Azevedo et al. present a case study about the experience of three hospitals and one RFID technology provider to highlight the main architectural characteristics, functionality, and advantages associated with the RFID deployment. Cavaco and Machado provide a model for the creation of competitive advantage that integrates the principles of sustainability (triple bottom line) and the concepts of resilience and innovation. Luo and Amberg identify a set of dimensions of culture from the most known culture theories through quantitative and inductive analysis to form a research framework of culture and a classification of cultural impacts on information technology, which provides an agenda for systematically researching the cultural impacts on information technology. Xiang et al. develop guiding thoughts of all round evaluation and principles, and then established a fuzzy synthetic performance evaluation model based on realizing the performances achieved by the strategic alliances of technological innovation. Xu and Wu propose a large group of decision-making method, which is based on fuzzy preference dynamic information interaction given that the traditional fuzzy group decision-making model does not take the process of information interaction into consideration.

Except for high-quality proceedings, the conference also provides a suitable environment for discussions and exchanges of research ideas among participants during its well-organized conference. Although we present our research results in technical sessions and participate in round table discussions during the conference period, we will have extra and fruitful occasions to exchange research ideas with colleagues in this relaxed and enjoyable atmosphere of sightseeing.

We would like to take this opportunity to thank all the participants who have worked hard to make this conference a success. We appreciate the help from Universidade Nova de Lisboa and Sichuan University in conference organization. We also appreciate Springer-Verlag London for the wonderful publication of the proceedings. We are also grateful to Rector António Manuel Bensabat Rendas for being the General Chair and Prof. Fernando Santana for being the Local Arrangement Committee Chair. Besides, we appreciate the great support from all members of the Organizing Committee, Local Arrangement Committee, and Program Committee as well as all participants who have worked hard to make this conference a success. Finally, we also want to appreciate all the authors for their excellent papers in this conference. Due to these excellent papers, ISMSEM Advancement Prize for MSEM will be awarded again at the conference for the papers that describe a practical application of Management Science and Engineering Management. The Ninth International Conference on Management Science and Engineering Management will be hosted by Karlsruhe Service Research Institute (KSRI), Karlsruhe Institute of Technology, Germany in July, 2015. Prof. Dr. Stefan Nickel will be the Organizing Committee Chair for 2015 ICMSEM. We sincerely hope that you can submit your new findings on MSEM and share your ideas in Germany.

Lisbon, Portugal, 20 May 2014   Jiuping Xu
                                                     Virgílio António Cruz-Machado
                                                                        Benjamin Lev
                                                                        Stefan Nickel

Organization

ICMSEM 2014 was organized by International Society of Management Science and Engineering Management, Sichuan University (Chengdu, China), Universidade Nova de Lisboa (Caparica Portugal). It was held in cooperation with Advances in Intelligent Systems and Computing (AISC) of Springer.

Executive Committee

General Chairs	Prof. Jiuping Xu, Sichuan University, China
	Rector António Manuel Bensabat Rendas, Universidade Nova de Lisboa, Portugal
Program Committee Chairs	Prof. Hajiyev Asaf, National Academy of Sciences of Azerbaijan, Azerbaijan
	Prof. Shavkat Ayupov, Academy of Sciences of Uzbekistan, Uzbekistan
	Prof. Le Dung Trang, Abdus Salam ICTP, Vietnamese
	Prof. Benjamin Lev, Drexel University, Philadelphia, USA
	Prof. José Antonio de la Peña Mena, Mexican Academy of Sciences, Mexico
	Prof. Alberto Pignotti, University of Buenos Aires, Argentina
	Prof. Baldev Raj, Indian National Academy of Engineering, India
	Prof. Roshdi Rashed, National Center of Scientific Research, France
Organizing Committee Chair	Prof. V. A. Cruz-Machado, Universidade Nova de Lisboa, Portugal
Local Arrangement Committee Chair	Prof. Fernando Santana, Universidade Nova de Lisboa, Portugal

Local Arrangement Committee

Helena Carvalho, Universidade Nova de Lisboa, Portugal
Isabel Nunes, Universidade Nova de Lisboa, Portugal
Susana Duarte, Universidade Nova de Lisboa, Portugal

Program Committee

Prof. Mohammad Z. Abu-Sbeih, King Fahd University of Petroleum and Minerals, Saudi Arabia
Prof. Joseph G. Aguayo, University of Concepcion, Chile
Prof. Basem S. Attili, United Arab Emirates University, United Arab Emirates
Prof. Alain Billionnet, Ecole National Superieure Informatics for Industry and Enterprise, France
Prof. Borut Buchmeister, University of Maribor, Slovenia
Prof. Daria Bugajewska, Adam Mickiewicz University, Poland
Prof. Saibal Chattopadhyay, Indian Institute of Management, India
Prof. Edwin Cheng, Hong Kong Polytechnic University, Hong Kong
Prof. Anthony Shun Fung Chiu, De La Salle University, Philippines
Prof. Jeong-Whan Choi, Department of Mathematics, Republic of Korea
Prof. Kaharudin Dimyati, University of Malaya, Malaysia
Prof. Behloul Djilali, University of Sciences and Technology Houari Boumediene, Algeria
Prof. Eid Hassan Doha, Cairo University, Giza, Egypt
Prof. O'Regan Donal, National University of Ireland, Ireland
Dr. Siham El-Kafafi, Manukau Institute of Technology, New Zealand
Prof. Christodoulos A. Floudas, Princeton University, USA

Prof. Masao Fukushima, Kyoto University, Japan

Prof. Oleg Granichin, Sankt-Petersburg State University, Russia

Prof. Bernard Han, Western Michigan University, USA

Dr. Rene Henrion, Humboldt University, Germany

Prof. Voratas Kachitvichyanukul, Asian Institute of Technology, Thailand

Prof. Arne Løkketangen, Molde University College, Norway

Dr. Andres Medaglia, University of the Andes, Colombia

Prof. Venkat Murali, Rhodes University, South Africa

Prof. Shmuel S. Oren, University of California Berkeley, USA

Prof. Turgut Öziş, Ege University, Turkey

Prof. Panos M. Pardalos, University of Florida, USA

Prof. Gianni Di Pillo, Sapienza University of Rome, Italy

Prof. Nasrudin Abd Rahim, University of Malaya, Malaysia

Prof. Celso Ribeiro, Fluminense Federal University, Brazil

Prof. Hsin Rau, Chung Yuan Christian University, Taiwan

Prof. Jan Joachim Ruckmann, University of Birmingham, UK

Prof. Martin Skitmore, Queensland University of Technology, Australia

Prof. Frits C. R. Spieksma, Katholieke University Leuven, Belgium

Prof. Yong Tan, University of Washington, USA

Prof. Albert P. M. Wagelmans, Erasmus University Rotterdam, The Netherlands

Prof. Desheng Dash Wu, University of Toronto, Canada

Prof. Hong Yan, Hong Kong Polytechnic University, Hong Kong

Secretary-General	Prof. Zhineng Hu, Sichuan University, China
Under-Secretary General	Dr. Xiaoling Song, Sichuan University, China
	Dr. Zongmin Li, Sichuan University, China
	Dr. Cuiying Feng, Sichuan University, China
Secretaries	Yan Tu, Rui Qiu, Yusheng Wang, Xin Yang, Siwei Zhao

Contents

Part VI Project Management

Part VII   Industrial Engineering

Part V
Computing Methodology

Chapter 70
Recent Advances in Multiobjective Genetic Algorithms for Manufacturing Scheduling Problems

Mitsuo Gen, Wenqiang Zhang, Lin Lin and Jungbok Jo

Abstract Manufacturing scheduling is one of the important and complex combinatorial optimization problems in manufacturing system, where it can have a major impact on the productivity of a production process. Moreover, most of scheduling problems fall into the class of NP-hard combinatorial problems. In this paper, we concern with the design of multiobjective genetic algorithms (MOGAs) to solve a variety of manufacturing scheduling problems. In particularly, the fitness assignment mechanism and evolutionary representations as well as the hybrid evolutionary operations are introduced. Also, several applications of EAs to the different types of manufacturing scheduling problems are illustrated. Through a variety of numerical experiments, the effectiveness of these hybrid genetic algorithms (HGAs) in the widely applications of manufacturing scheduling problems are demonstrated. This paper also summarizes a classification of scheduling problems and the design way of GAs for the different types of manufacturing scheduling problems in which we apply GAs to a multiobjective flexible job-shop scheduling problem (MoFJSP; operation sequencing with resources assignment) and multiobjective assembly line balancing models (MoALB; shipments grouping and assignment). It is useful to guide how to investigate an effective GA for the practical manufacturing scheduling problems.

M. Gen (✉)
Fuzzy Logic Systems Institute, Lizuka, Japan
e-mail: gen@flsi.or.jp

W. Zhang
College of Information Science and Engineering,
Henan University of Technology, Zhengzhou 450000, People's Republic of China

L. Lin
School of Software Technology, Dalian University of Technology,
Dalian 116000, People's Republic of China

J. Jo
Division of Computer and Information Engineering,
Dongseo University, Busan, Korea

J. Xu et al. (eds.), *Proceedings of the Eighth International Conference on Management Science and Engineering Management*, Advances in Intelligent Systems and Computing 281, DOI: 10.1007/978-3-642-55122-2_70, © Springer-Verlag Berlin Heidelberg 2014

Keywords Hybrid genetic algorithms (HGA) · Multiobjective HGA (Mo-HGA) · Manufacturing scheduling · Flexible job-shop scheduling problem (FJSP) · Assembly line balancing models (ALB)

70.1 Introduction

In many real world applications in engineering design problems and information systems, a scheduling problem imposes on more complex issues, such as complex structure, complex constraints,and multiple objectives to be handled simultaneously and make the problem intractable to the traditional approaches. Network models and optimization for various scheduling and/or routing problems in manufacturing and logistics systems also provide a useful way as one of case studies in real world problems and are extensively used in practice [17]. Basically a scheduling problem is to determine the allocation of plant resources in manufacturing optimization. Tasks must be assigned to the process units, and the duration and amount of processed material related to those assigned tasks must be determined [34]. For a more extensive explanation of the various aspects of the scheduling model, the reader is directed to the reviews of [9]. Bidot et al. [1] gave detail definitions to avoid ambiguity of terms commonly used by different communities: complete schedule, flexible schedule, conditional schedule, predictive schedule,executable schedule, adaptive scheduling system, robust predictive schedule and table predictive schedule.However, to find the optimal solutions of manufacturing scheduling gives rise to complex combinatorial optimization, unfortunately, most of them fall into the class of NP-hard combinatorial problems.

Since the 1960s, there has been being an increasing interest in imitating living beings to solve the hard optimization problems. An evolutionary algorithm (EA) such as agenetic algorithm (GA) is a generic population-based meta-heuristic optimization algorithm [39]. An EA uses some mechanisms inspired by biological evolution: reproduction, mutation, recombination, and selection. Handa et al. [20] gave a comprehensive overview of recent advances of evolutionary computation (EC) studies. EAs has attracted significantly attention with respect to complexity scheduling, which is referred to evolutionary scheduling, it is vital research domain at interface of two important sciences–artificial intelligence and operational research [6]. Furthermore, many researches are focusing on the multi-objectives manufacturing scheduling problems. Li and Huo [28] proposed a GA for multi-objective flexible job-shop scheduling problem (MoFJSP) with consideration of maintenance planning, intermediate inventory, and machines in parallel, which had a background of practical scheduling problem in seamless steel tube production. Geiger [12] proposed a heuristic search, intensification through variable neighborhoods, and diversification through perturbations and successive iterations in favorable regions of the search space, and successfully tested on permutation flow shop scheduling problems under multiple objectives. Karimi-Nasab et al. [27] introduced a multi-product multi-period production planning problems. A novel multi-objective model for the production smoothing problem on a single stage facility that some of the operating times could be determined in a time interval for. The proposed model was solved by the GA, using a novel achievement function for exploring the solution space, based on LP-metric concepts. Gholami et al. [19] integrated simulation into GA to the dynamic scheduling of a flexible job shop with machines that suffer stochastic breakdowns. Zandieh et al. [41] considered a multi-objective group scheduling problem in hybrid flexible flow-shop with sequence-dependent setup times (SDST) by minimizing total weighted tardiness and maximum completion time simultaneously. Kachitvichyanukul et al. [26] proposed a two-stage genetic algorithm (2S-GA) for multi-objective Job Shop scheduling problems (MoJSP). The 2S-GA is proposed with three criteria: Minimize makespan, Minimize total weighted earliness, and Minimize total weighted tardiness.

Even if EAs have attracted significantly attention with respect to above complexity scheduling problems, it has a disadvantage: we have to design a specialized EA for each practical scheduling problem with the problem's specificity. So that means each class of EAs doesn't have a wide range of applications on manufacturing scheduling. Recently Gen and Lin [15] surveyed multiobjective evolutionary algorithm for manufacturing scheduling Problems. In order to design an effective EA with the problem's specificity, we have to consider (1) how to design a representation and a way of population initialization; (2) how to evaluate an individual by a fitness function; (3) how to improve population by evolutionary operators. In this paper, we focus on the effective multiobjective EA design for various applications on manufacturing scheduling. We will discuss the representation design of potential solutions to the problems, the fitness assignment mechanisms to the multiobjective scheduling problems, and the evolutionary operation design to improve solution simultaneously.

The rest of this paper is organized as follows: Sect. 70.2 introduces a hybrid genetic algorithms (HGA), multiobjective GA, and give fitness assignment mechanism for multiobjective scheduling problems. How to present a solution of the scheduling problem into a chromosome is given in Sect. 70.3. Section 70.4 presents the multiobjective flexible job-shop scheduling problem (MoFJSP) by Mo-HGA, as one of typical scheduling problems and multiobjective assembly line balancing models (MoALB) with worker capability by Mo-HGAis presented in Sect. 70.5. Finally, the conclusion of this paper and future researches are drawn in Sect. 70.6.

70.2 Multiobjective Hybrid Genetic Algorithms

1. Hybrid Genetic Algorithms with Fuzzy Logic

Genetic algorithms have proved to be a versatile and effective approach for solving NP-hard combinatorial optimization problems. Nevertheless, there are many situations in which the basic GA does not perform particularly well, and various methods of have been proposed. With the hybrid approach, local optimization such as a hill-climbing or neighborhood search is applied to each newly generated offspring to move it to a local optimum before injecting it into the population. GAs are used to perform global exploration among the population while heuristic methods are used to perform local exploitation around chromosomes. Because of the complementary properties of GAs and conventional heuristics, the hybrid approach often outperforms either method operating alone. The main idea is to use a fuzzy logic controller (FLC) to compute new strategy parameter values that will be used by the GAs. A FLC is comprised of four principal components: (1) a knowledge base, (2) a fuzzification interface, (3) an inference system, and (4) a defuzzification interface.

The experts' knowledge is stored in the knowledge base in the form of linguistic control rules. The inference system is the kernel of the controller, which provides an approximate reasoning based on the knowledge base. A hybrid genetic algorithms (HGA) combined with FLC routines proposed by Yun and Gen [40] and Lin and Gen [29] and the general structure of HGA in the pseudo code is described in Gen et al. [10].

2. Multiobjective Hybrid Genetic Algorithms

The multiple objective optimization problems (MOP) have been receiving growing interest from researchers with various backgrounds since early 1960 [21]. There are a number of scholars who have made significant contributions to the problem. Among them, Pareto is perhaps one of the most recognized pioneers in the field [32]. Recently, EAs have been received considerable attention as a novel approach to MOPs, resulting in a fresh body of research and applications known as evolutionary multiobjective optimization (EMO). The basic feature of EAs is the multiple directional and global searches by maintaining a population of potential solutions from generation to generation. The population-to-population approach is hopeful to explore all Pareto solutions. EAs are essentially a kind of meta-strategy methods. When applying the EAs to solve a given problem, it is necessary to refine upon each of the major components of EAs, such as encoding methods, recombination operators, fitness assignment, selection operators, constraints handling, and so on, in order to obtain a best solution to the given problem. Because the MOPs are the natural extensions of constrained and combinatorial optimization problems, so many useful methods based on EAs developed during the past two decades. One of special issues in the MOPs is fitness assignment mechanism. Since the 1980s, several fitness assignment mechanisms have been proposed and applied in MOPs [17]. Although most fitness assignment mechanisms are just different approach and suitable to different cases of multiobjective optimization problems, in order to understanding the development of multiobjective EAs (MOEAs), we classify algorithms according to proposed years of different approaches.

MOPs arise in the design, modeling,and planning of many real complex systems in the areas of industrial production, urban transportation, capital budgeting, forest management, reservoir management, layout and landscaping of new cities, energy distribution, etc. Since the 1990s, EAs have been received considerable attention as a novel approach to multiobjective optimization problems, resulting in a fresh body of research and applications known as EMO. Without loss of generality, a MOP with q objective functions conflicting each other and m constraints can be formally represented as follows:

$$\max \left\{ z_1 = f_1(x), z_2 = f_2(x), \cdots, z_q = f_q(x) \right\}$$
$$\text{s.t. } g_i(x) \leq 0, i = 1, 2, \cdots, m, x \in \mathbf{R}^n. \tag{70.1}$$

We sometimes graph the MOP problem in both decision space and criterion space. S is used to denote the feasible region in the decision space and Z is used to denote the feasible region in the criterion space respectively as follows: $S = \{x \in \mathbf{R}^n | g_i(x) \leq 0, \ i = 1, 2, \cdots, m\}$, $Z = \{z \in \mathbf{R}^q | z_1 = f_1(x), z_2 = f_2(x), \cdots, z_q = f_q(x), x \in S\}$, where $x \in \mathbf{R}^n$ is a vector of values of q objective functions. In the other words, Z is the set of images of all points in S. Although S is confined to the nonnegative region of $\mathbf{R}^n$ and Z is not necessarily confined to the nonnegative region of $\mathbf{R}^q$.

There usually exists a set of solutions for the multiple objective cases which cannot be simply compared with each other. Such kind of solutions are called nondominated solutions or Pareto optimal solutions, for which no improvement in any objective function is possible without sacrificing on at least one of other objectives.

Definition 70.1 For a given point $z_0 \in Z$, it is nondominated if and only if there does not exist another point $z \in Z$ such that for the maximization case, $z_k > z_k^0$, for some $k \in \{1, 2, \cdots, q\}$, $z_l \geq z_l^0$, for some $1 \neq k$, where z_0 is a dominated point in the criterion space Z with q objective functions.

3. Fitness Assignment Mechanisms
When applying the GAs to solve a given MOP problem, it is necessary to refine upon each of the major components of GAs, such as encoding methods, recombination operators, fitness assignment, selection operators, and constraints handling, and so on, in order to obtain a best solution to the given problem. One of special issues in the multiobjective optimization problems is fitness assignment mechanism. Although most fitness assignment mechanisms are just different approach and suitable to different cases of multiobjective optimization problems, in order to understanding the development of multiobjective EAs (MOEAs), we classify algorithms according to proposed years of different approaches.

Nondominated Sorting Genetic Algorithm II (NSGA II: [7]): Srinivas and Deb developed a Pareto ranking-based fitness assignment and it called NSGA. In each method, the nondominated solutions constituting a nondominated front are assigned the same dummy fitness value. These solutions are shared with their dummy fitness values (phenotypic sharing on the decision vectors) and ignored in the further classification process. Finally, the dummy fitness is set to a value less than the smallest shared fitness value in the current nondominated front. Then the next front is extracted. The procedure of NSGA II is repeated until all individuals in the population are classified [7].

Random-weight Genetic Algorithm (RWGA: [22]): Ishibuchi et al. proposed a weighted-sum based fitness assignment method. Weighted-sum approach can be viewed as an extension of methods used in the multiobjective optimizations to GAs. It assigns weights to each objective function and combines the weighted objectives into a single objective function. To search for multiple solutions in parallel, the weights are not fixed and able to uniformly the sample area towards to the whole frontier.

Adaptive Weight Genetic Algorithm (AWGA: [14]): Gen and Cheng utilized some useful information from the current population to readjust weights to obtain a search pressure toward a positive ideal point. For the examined solutions at each generation, we define two extreme points for the kth objective (maximum: z^+, minimum: z^-) as follows: $z_k^{max} = \max\{f_k(\mathbf{x})|x \in P\}, k = 1, 2, \cdots, q, z_k^{min} = \min\{f_k(x)|x \in P\}, k = 1, 2, \cdots, q$.

The weighted-sum objective function for a given chromosome x is given by the following equation:

$$eval(x) = \sum_{k=1}^{q} w_k \left(z_k - z_k^{min}\right) = \sum_{k=1}^{q} \frac{z_k - z_k^{min}}{z_k^{max} - z_k^{min}} = \sum_{k=1}^{q} \frac{f_k(x) - z_k^{min}}{z_k^{max} - z_k^{min}}, \quad (70.2)$$

where w_k is adaptive weight for the kth objective function as shown in the following equation: $w_k = \frac{1}{z_k^{max} - z_k^{min}}, k = 1, 2, \cdots, q$.

The Eq. (70.2) driven above is a hyperplane defined by the following extreme points in current solutions:

Strength Pareto Evolutionary Algorithm 2 (SPEA 2: [44]): Zitzler and Thiele proposed strength Pareto Evolutionary Algorithm (SPEA: [43]) and an extended version SPEA 2 [44] that combines several features of previous MOGA in a unique manner. The fitness assignment procedure is a two-stage process. The individuals in the external nondominated set P' are ranked.

Interactive Adaptive-weight Genetic Algorithm (i-AWGA: [29]): Lin and Gen proposed an interactive AWGA, which is an improved adaptive-weight fitness assignment approach with the consideration of the disadvantages of weighted-sum approach and Pareto ranking-based approach. They combined a penalty term to the fitness value for all of dominated solutions. Firstly, we calculate the adaptive weight $w_i = 1/(z_i^{max} - z_i^{min})$ for each objective by using AWGA. Afterwards, we calculate the penalty term $p(v_k) = 0$, if v_k is nondominated solution in the nondominated set P. Otherwise $p(v_k') = 1$ for dominated solution v_k'. Lastly, we calculate the fitness value of each chromosome by combining the i-AWGA method: $eval(v_k) = \sum_{i=1}^{q} w_i(z_i^k - z_i^{min}) + p(v_k), \forall k \in popSize$.

Hybrid Sampling Strategy-based EA (HSS-EA: [42]): Zhang et al. proposed a hybrid sampling strategy-based evolutionary algorithm (HSS-EA). A Pareto dominating and dominated relationship-based fitness function (PDDR-FF) is proposed to evaluate the individuals. The PDDR-FF of an individual S_i is calculated by the following function: $eval(S_i) = q(S_i) + \frac{1}{p(S_i+1)}, i = 1, 2, \cdots, popSize$, where $p(\cdot)$ is the number of individuals which can be dominated by the individual S. $q()$ is the number of individuals which can dominate the individual S. The PDDR-FF can set the obvious difference values between the nondominated and dominated individuals. The general structure in the pseudo code of multiobjective hybrid genetic algorithms (Mo-HGA) is described as shown in Fig. 70.1.

```
procedure: Multi objective Hybrid GA with Preserving Pareto
input: problem data, GA parameters
output: Pareto optimal solutions E(P,C)
begin
    t← 0;
    initialize P(t) by encoding routine;              // P(t): population
    calculate objectives f(P) by decoding routine;
    create Pareto optimal solution E(P) by nondominated routine;
    evaluate P(t) by fitness assignment routine and keep the best Pareto solution; //eval(P)
    while (not terminating condition) do
create C(t) from P(t) by crossover routine;       // C(t): offspring
create C(t) from P(t) by mutation routine;
improve C(t) by local search routine;
calculate objectivesf(C) by decoding routine;
update Pareto optimal solution E(P,C) by nondominated routine;
evaluate C(t) by fitness assignment routine and update the best Pareto; //eval(C)
select P(t+1) from P(t) and C(t) by selection routine;
tune parameters by fuzzy logic controller routine;
t     ← t+1;
    end
    output Pareto optimal solutions E(P,C);
end;
```

Fig. 70.1 The general structure of multiobjective hybridgenetic algorithms

70.3 Evolutionary Representation

How to present a solution of the scheduling problem into a chromosome is a key issue for EAs. For evaluating the effectiveness of the different chromosome representation, there are several critical issues are summarized by Gen and Lin [17].

- Space: Chromosomes should not require extravagant amounts of memory.
- Time: The time complexities of evaluating, recombining, and mutating chromosomes should be small.
- Feasibility: All chromosomes, particularly those generated by simple crossover (i.e., one-cut point crossover) and mutation, should represent feasible solutions.
- Uniqueness: The mapping from chromosomes to solutions (decoding) may belong to one of the following three cases: 1-to-1 mapping, n-to-1 mapping and 1-to-n mapping. The 1-to-1 mapping is the best one among three cases and 1-to-n mapping is the most undesired one.
- Heritability: Offspring of simple crossover (i.e., one-cut point crossover) should represent solutions that combine substructures of their parental solutions.
- Locality: A mutated chromosome should usually represent a solution similar to that of its parent.

We need to consider these critical issues carefully when designing an appropriate representation so as to build an effective EA. As known, scheduling problem is the implement of production plan, with considering production processes, lot-size, amount and customer requirements etc. And scheduling problem is how to decide the resources assignment to the production, with considering constrains of resources capabilities and capacities. There are two decision making parts for scheduling optimization: (1) operation sequencing and (2) resources assignment.

1. Representation for Operation Sequencing

In the past few decades, the following 6 representations for job-shop scheduling problem (JSP, an operation sequencing problem with considering the precedence constraints of operations) have been proposed: • Operation-based representation [23]; • Job-based representation; • Preference list-based representation [5]; • Priority rule-based representation [8]; • Completion time-based representation [37]; • Random key-based representation [30].

The flexible job-shop scheduling problem (FJSP) is expanded from the traditional JSP, which possesses wider availability of machines for all the operations (a combinatorial optimization problem considering both of the operation sequence and the resource assignment). The following 4 representations for FJSP have been proposed: • Parallel machine-based representation [13]; • Parallel jobs representation [13]; • Operations machines-based representation [24]; • Multistage operation-based representation [16].

Permutation-based representation is perhaps the most natural representation of operation sequences. Unfortunately because of the existence of precedence constraints, not all the permutations of the operations define feasible sequences. For job shop scheduling problem, Cheng et al. [3, 4] applied job-based representation: they name all operations for a job with the same symbol and then interpret them according to the order of occurrence in the sequence of a given chromosome. Gen and Zhang [16] also applied this representation to advanced scheduling problem. The job-based representation can also be used to represent the operation sequences for the FJSP problem. However, if the operation precedence is more complex than JSP or extend JSP problems, the job-based representation cannot be used directed.

Cheng and Gen [2] proposed a priority-based representation firstly for solving Resource-constrained Project Scheduling Problem (RcPSP). This representation encodes a schedule as a sequence of operations and each gene stands for one operation. As known, a gene in a chromosome is characterized by two factors: locus, i.e., the position of the gene located within the structure of chromosome, and allele, i.e., the value the gene takes. In this encoding method, the position of a gene is used to represent operation ID and its value is used to represent the priority of the operation for constructing a schedule among candidates. A schedule can be uniquely determined from this encoding. However, the nature of the priority-based representation is a kind of permutation representations. Generally, this representation will yield illegal offspring when using one-cut point crossover or other simple crossover operators. That means some node's priority may be duplicated in the offspring. There are several crossover operators proposed for permutation representation, such as partial-mapped crossover (PMX), order crossover (OX), position-based crossover (PX), heuristic crossover, and so on [17]. Norman and Bean [30] proposed random key-based representation for JSP.

2. Representation for Resources Assignment

After the operation sequence is fixed, the resources assignment can be formulated as a multi-stage graph problem. For each stage (operation), we decided the state number (which resource should be assigned). This multi-stage graph problem can be solved by dynamic programming. Yang [38] proposed a GA-based discrete dynamic programming approach for scheduling in FMS environment. However, the IPPS problem is the combination of the operation sequencing (NP-hard problem) and resources assignment. Considering the computation times of the algorithm, and the most of practical IPPS problems are multi-resources assignment, the most of researches combined a state permutation representation into the chromosome [10, 18, 31], called multi-stage representation.

70.4 Multiobjective Flexible Job-shop Scheduling Models

1. Background of MoFJSP Model

As discussed above, manufacturing scheduling problem is how to decide the resources assignment to the production, considering constrains of resources capabilities and capacities; and is the implement of production plan, with considering production processes, lot-size, amount and customer requirements etc. Flexible job-shop scheduling problem (FJSP) is a generalization of the job-shop scheduling problem (JSP) [17] and the parallel machine environment, which provides a closer approximation to a wide range of real manufacturing systems. In particular, there are a set of parallel machines with possibly different efficiency. The FJSP allows an operation to be performed by any machine in a work center. The FJSP model is NP-hard since it is an extension of the JSP [11] and it is a combined assignment and scheduling decision.

- Every machine processes only one operation at a time and the operation sequence of a job is prespecified.
- The execution of each operation requires one machine selected from available machines for the operation.
- The operations are not preemptible, i.e., once an operation has started it cannot be stopped until it has finished.
- The set-up times for the operations are sequence-independent and are included in the processing times.

The problem is to assign each operation to an available machine and sequence the operations assigned on each machine in order to minimize the make span, that is, the time required to complete all jobs. The multiobjective FJSP model (MoFJSP) will be formulated as a 0-1 mixed integer programming (0-1MIP) model as follows:

$$\min t_M = \max_{i,k} \{c_{ik}\} \tag{70.3}$$

$$\min W_M = \max_{j} \{W_j\} \tag{70.4}$$

$$\min W_T = \sum_{j=1}^{m} W_j \tag{70.5}$$

$$\text{s.t. } c_{ik} - t_{ikj}x_{ikj} - c_{i(k-1)} \geq 0, k = 2, \cdots, K_i; \forall i, j \tag{70.6}$$

$$\sum_{j=1}^{m} x_{ikj} = 1, \forall k, i \tag{70.7}$$

$$x_{ikj} \in \{0, 1\}, \forall j, k, i \tag{70.8}$$

$$c_{ik} \geq 0, \forall k, i. \tag{70.9}$$

The objective functions accounts Eq. (70.3) is to minimize the make span, Eq. (70.4) is to minimize the maximal machine workload (i.e., the maximum working time spent at any machine), Eq. (70.5) is to minimize the total workload (i.e., the total working time over all machines). Equation (70.6) states that the successive operation has to be started after the completion of its precedent operation of the same job, which represents the operation precedence constraints. Equation (70.7) states that one machine must be selected for each operation.

2. Hybrid GA for MoFJSP

The $P(t)$ and $C(t)$ are parents and offspring respectively in current generation t, the implementation structure of HGA for scheduling is described as shown in Fig. 70.2.

3. Numerical Experiments

In order to test the effectiveness and performance of EAs, three representative instances (represented by problem $n \times m$) were selected for simulation. The works by Kacem et al. [24], Xia and Wu [36], and Gen et al. [17] are among the most recent progresses made in the area of FJSP. All the simulation experiments were performed with Delphi on Pentium 4 processor (2.6-GHz clock). Table 70.1 gives the performance of EAs. "Approach by Localization" and "AL + CGA" are two algorithms by Kacem et al. [24, 25]. "PSO + SA" is the algorithm by Xia and Wu [36], and "HGA" is proposed by Gen et al. [18].

70.5 Assembly Line Balancing Problem with Worker Capability

1. Background of ALB-WC Model

The assembly line balancing (ALB) problem determines the assignment of various tasks to an ordered sequence of stations, while optimizing one or more objectives without violating restrictions imposed on the line in a manufacturing system. How to allocate the proper workers to proper stations to obtain the best efficiency of the

```
procedure: hybrid GA for MoFJSP
input: scheduling data, HGA parameters
output: Pareto optimal solutions E
begin
        t←0;
        initialize P₁(t)operation sequence section by priority-based encoding routine;
        initialize P₂(t) resources assignment sections by permutation encoding routine;
        calculatethree objective functions f(P) by decoding routine; //P(t) = [P₁(t),P₂(t)]
        create Pareto optimal solution E(P) by nodominated routine;
        evaluate P(t) by fitness assignment routineand keep the best Pareto solution; //eval(P)
        while (not terminating condition) do
            create C(t) from P(t) by exchange crossover routine;
            create C(t) from P(t) by allele-based mutation routine;
            improve C(t) by bottleneck shifting routine;
            calculate three objective functions f(C) by decoding routine;
            update Pare to optimal solution E(P,C)by no dominated routine;
            evaluate C(t) by fitness assignment routine and update the best Pareto; //eval(C)
            select P(t+1) from P(t) and C(t) by mixed sampling routine;
            tune parameters by fuzzy logic controller routine;
            t←t+1;
        end
        output Pare to optimal solutions E;
end;
```

Fig. 70.2 The general structure of multiobjective hybridgenetic algorithms for MoFJSP

Table 70.1 Performance of multiobjective HGAs for the 3-FJSP problems

Problem		Classical GA	AL+CGA		PSO+SA		Mo-HGA
8×8	t_M	16	15	16	15	16	15
	w_M	–	–	–	12	13	12
	w_T	77	79	75	75	73	75
10×10	t_M	7	7		7		7
	w_M	7	5		6		5
	w_T	53	45		44		43
15×10	t_M	23	24		12		11
	w_M	11	11		11		11
	w_T	95	91		91		91

line and reduce the total cost is also a problem in multiobjective ALB with worker capability (MoALB-WC). The MoALB-WC problem concerns with the assignment of the tasks to stations and the allocation of the available workers for each station in order to minimize the cycle time and minimize the total cost under the constraint of precedence relationships. The notation used in the model can be summarized as follows:

Indices:

j, k : indices of task $(j, k = 1, 2, \cdots, n)$;
i : index of station $(i = 1, 2, \cdots, m)$;
w : index of worker $(w = 1, 2, \cdots, m)$.

Parameters:

n : number of tasks;
m : number of stations/workers;
d_{jw} : worker cost of worker $w$process task j;
t_{jw} : processing time of task j by worker w;
$\mathrm{Suc}(j)$: set of direct successors of task j;
$\mathrm{Pre}(j)$: set of direct predecessors of task j;
S_i : set of tasks assigned to station i;
$t(S_i)$: processing time at station i, $t(S_i) = \sum_{j=1}^{n} \sum_{w=1}^{m} t_{jw} x_{ij} y_{iw}, \forall i$;
u_i : utilization of the station S_i, $u_i = t(S_i)/\max_{1 \le i \le m} \{t(S_i)\}$;
u : average utilization of all stations, $u = \frac{1}{m} \sum_{i=1}^{m} u_i$.

Decision Variables:

$$x_{ij} = \begin{cases} 1, & \text{if task } j \text{ is assigned to station } i \\ 0, & \text{otherwise,} \end{cases}$$

$$y_{iw} = \begin{cases} 1, & \text{if worker } w \text{ is working in station } i \\ 0, & \text{otherwise.} \end{cases}$$

Mathematical Model:

$$\min c_T = \max_{1 \le i \le m} \left\{ \sum_{j=1}^{n} \sum_{w=1}^{m} t_{jw} x_{ij} y_{iw} \right\} \tag{70.10}$$

$$\min d_T = \sum_{i=1}^{m} \sum_{j \in S_i} \sum_{w=1}^{m} d_{jw} y_{iw} \tag{70.11}$$

$$\text{s.t.} \sum_{i=1}^{m} i x_{ij} \ge \sum_{i=1}^{m} i x_{ik}, \forall k \in Pre(j), \forall j \tag{70.12}$$

$$\sum_{i=1}^{m} x_{ij} = 1, \forall j \tag{70.13}$$

$$\sum_{w=1}^{m} y_{iw} = 1, \forall i \tag{70.14}$$

$$\sum_{i=1}^{m} y_{iw} = 1, \forall w \tag{70.15}$$

$$x_{ij}, y_{iw} \in \{0, 1\}, \forall i, j, w. \tag{70.16}$$

The first Eq. (70.10) of the model is to minimize the cycle time of the assembly line. The second Eq. (70.11) is to minimize the total worker cost. Inequity Eq. (70.12) states that all predecessor of task j must be assign to a station, which is in front of or the same as the station that task j is assigned in Eqs. (70.13) to (70.14) ensures that task j must be assigned to only one station and only one worker can be allocated to station respectively. Equation (70.15) ensures worker w can be allocated to only one station and Eq. (70.16) represents the nonnegative restrictions.

2. Hybrid EA for MoALB-WC

The detailed genetic representation consists of three steps:

Step 1. Creating a task sequence

(1) Generating a real number based on a random key for each task as task priority in task priority vector.

(2) Creating a task sequence by decoding method

Step 2. Assigning worker to each station

(1) Encoding worker allocation vector by randomly assign the worker to each station.

Step 3. Assigning tasks to each station

(1) Calculating the lower bound and the upper bound cycle time.

(2) Finding an optimal cycle time by bisection searching.

(3) Dividing the task sequence to form the breakpoint vector.

In this study, we consider the MoALB-WC and propose a new multiobjective evolutionary algorithm with strong convergence of multi-area (MOEA-SCM) based on hybrid sampling strategy-based EA (HSS-EA). The MOEA-SCM could converge to the center and two edges areas of Pareto front strongly and could both preserve the convergence rate and guarantee the better distribution performance.

The solution procedure of one generation includes 3 phases [42] as follows:

Phase 1: Generating Mating Pool

Step 1. Selecting individuals into subpopulations 1 and 2 by VEGA (good for the edges area of Pareto front).

Step 2. Combining the subpopulations and archive $A(t)$ to form the mating pool (good for the central area of Pareto front).

Phase 2: Reproducing the new population

Arithmetical crossover (for task priority vector), weight mapped crossover (WMX) (for worker allocation vector), and swap mutation are used to reproduce new individuals.

Phase 3: Updating the Archive

The individuals of $A(t)$ and $P(t+1)$ are combined to form a temporary archive $A'(t)$. Thereafter, the PDDR-FF values of all individuals in $A'(t)$ are calculated and sorted. The smallest $|A(t)|$ individuals in $A'(t)$ are copied to form $A(t+1)$.

This archive updating mechanism likes an elitist sampling strategy to keep the better individuals with better PDDR-FF values.

3. Experimental Comparisons

We employed Gunther's problem data with 35 tasks and 6 stations, which is widely used in the ALB problem literature [33]. The number of workers is the same as the number of stations, 6 and the worker cost data are generated randomly according to work level (see [35] for details). All the simulation are performed on Pentium Dual-Core processor (2.70 GHz clock) and 2 GB memory. The adopted parameters

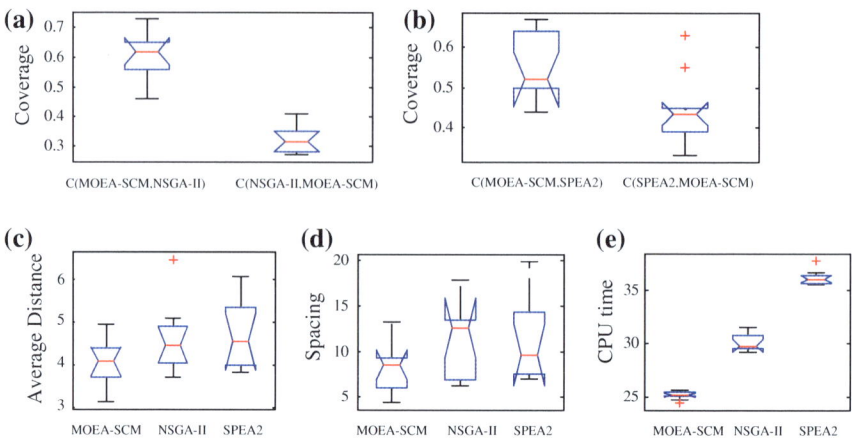

Fig. 70.3 C, GD, SP and CPU times by 3 methods for ALB-WC. **a** Coverage for MOEA-SCM and NSGA-II; **b** Coverage for MOEA-SCM and SPEA2; **c** Average distance, **d** Spacing, **e** CPU time

are listed as follows: population size, 100; maximum generation, 500; archive size, 50; crossover probability, 0.80 and 0.3; mutation probability, 0.40 and 0.1. MOEA-SCM, NSGA-II, and SPEA2 are run 30 times to compare the results with each other. It should be noted that the parameters of all 3 methods are the same, except for the size of archive. The archive sizes of MOEA-SCM is set to be half the population size, 50, while of NSGA-II and SPEA2 are set to be the same as the population size, 100.

Let S_j be a solution set for each method. PF^* is a known reference Pareto solutions. In this study, PF^* comes from combining all of the obtained Pareto set with 30 runs by 3 methods. The following three performance measures are considered:

Coverage $C(S1, S2)$ is the percent of the individuals in $S2$ which are weakly dominated by $S1$ [43]. The larger $C(S1, S2)$ means that $S1$ outperforms $S2$ in convergence.

Generational distance $GD(Sj)$ finds an average minimum distance of the solutions of S_j from PF^* [17]. The smaller GD of Sj means better Sj in approaching $PF*$.

Spacing $SP(Sj)$ is the standard deviation of the closest distances of individuals by Sj [42]. Smaller SP means better distribution performance.

The C, GD are used to verify convergence performance while SP is used to check the distribution performance.

The Fig. 70.3 shows the numerical comparison of the box-and-whisker plots for C, GD and SP and the CPU times by three methods. From Fig. 70.3 we can see that the convergence and distribution performance of MOEA-SCM is better than famous NSGA-II and SPEA2, and the efficiency is better.

70.6 Conclusions

Recently, manufacturing companies are faced with global market demands for a variety of low cost products with a high quality. For responding rapidly to demand fluctuations and reducing costs related to manufacturing scheduling and logistics networks, hybrid genetic algorithms (HGA) and multiobjective HGA (Mo-HGA) have received considerable attention regarding their potential for solving various complex manufacturing and logistics problems as NP-hard combinatorial multiobjective optimization problems (MOPs).

In this paper, firstly we summarized recent advances in hybrid genetic algorithms and multiobjective HGA (Mo-HGA) with several fitness assignment mechanisms for solving the different types of multiobjective scheduling problems. Then we introduced how to design a representation for different types of scheduling problems. We also introduced the design ways to apply evolutionary algorithms (EA) to the different typical manufacturing scheduling problems, including multiobjective flexible job-shop scheduling problem (MoFJSP; operation sequencing with resources assignment) and multiobjective assembly line balancing problem (MoALB; shipments grouping and assignment) with their mathematical programming models respectively. Through the numerical experiments, we demonstrated the effectiveness of these Mo-HGAs in the applications of manufacturing scheduling problems. We will expand the widely applications to various multiobjective optimization problems (MOPs) byMo-HGA including a hybrid sampling strategy-based EA.

Acknowledgments This work is partly supported by JSPS: Grant-in-Aid for Scientific Research (C) (No.24510219), Taiwan NSF (NSC 101-2811-E-007-004, NSC 102-2811-E-007-005), the National NSF of China (No. U1304609), the Education Dept. of Henan Province: Basic Research Program of Sci. and Tech. Key Project (No. 13A520203), the Plan of Nature Science Fundamental Research in Henan Univ. of Tech. (No. 2012JCYJ04), the Fundamental Research Funds (Software+X) of Dalian Univ. of Tech. (No.DUT12JR05, No.DUT12JR12) and also the Dongseo Frontier Project Research Fund of Dongseo University.

References

1. Bidot J, Vidal T et al (2009) A theoretic and practical framework for scheduling in a stochastic environment. J Sched 12(3):315–344
2. Cheng R, Gen M (1994) Evolution program for resource constrained project scheduling problem. In: Proceedings of IEEE international conference of evolutionary computation, pp 736–741
3. Cheng R, Gen M, Tsujimura Y (1996) A tutorial survey of job-shop scheduling problems using genetic algorithms, part i: representation. Comput Ind Eng 30(4):983–997
4. Cheng R, Gen M, Tsujimura Y (1999) A tutorial survey of job-shop scheduling problems using genetic algorithms, part ii: hybrid genetic search strategies. Comput Ind Eng 36(2):343–364
5. Croce F, Tadei R, Volta G (1995) A genetic algorithm for the job shop problem. Comput Oper Res 22:15–24
6. Dahal K, Tan K, Cowling P (2007) Evolutionary scheduling. Springer, Berlin
7. Deb K (2001) Multiobjective optimization using evolutionary algorithms. Wiley, Chichester

8. Dorndorf W, Pesch E (1995) Evolution based learning in a job shop scheduling environment. Comput Oper Res 22:25–40
9. Floudas C, Lin X (2005) Mixed integer linear programming in process scheduling: modeling, algorithms, and applications. Ann Oper Res 139:131
10. Gao J, Sun L, Gen M (2008) A hybrid genetic and variable neighborhood descent algorithm for flexible job shop scheduling problems. Comput Oper Res 35(9):2892–2907
11. Garey M, Johnson D, Sethi R (1976) The complexity of flowshop and jobshop scheduling. Math Oper Res 1:117–129
12. Geiger M (2011) Decision support for multi-objective flow shop scheduling by the pareto iterated local search methodology. Comput Ind Eng 61:805–812
13. Gen M, Cheng R (1997) Genetic algorithms and engineering design. Wiley, New York
14. Gen M, Cheng R (2000) Genetic algorithms and engineering optimization. Wiley, New York
15. Gen M, Lin L (2013) Multiobjective evolutionary algorithm for manufacturing scheduling problems: State-of-the-art survey. J Intell Manuf. doi:10.1007/s10845-013-0804-4:1-18
16. Gen M, Zhang H (2006) Effective designing chromosome for optimizing advanced planning and scheduling. Intelligent engineering systems through artificial neural networks, vol 16, ASME Press, pp 61–66
17. Gen M, Cheng R, Lin L (2008) Network models and optimization: multiobjective genetic algorithm approach. Springer, London
18. Gen M, Lin L, Zhang H (2009) Evolutionary techniques for optimization problems in integrated manufacturing system: state-of-the-art survey. Comput Ind Eng 56(3):779–808
19. Gholami M, Zandieh M (2009) Integrating simulation and genetic algorithm to schedule a dynamic flexible job shop. J Intell Manuf 20(4):481–498
20. Handa H, Kawakami H, Katai O (2008) Recent advances in evolutionary computation. IEEE J Trans Electron Info Syst 128(3):334–339
21. Hwang C, Yoon K (1981) Multiple attribute decision making: methods and applications. Springer, Berlin
22. Ishibuchi H, Murata T (1998) A multiobjective genetic local search algorithm and its application to flowshop scheduling. IEEE Trans Syst Man Cybern 28(3):392–403
23. Jong KD (2005) Genetic algorithms: a 30 year perspective. Perspect Adapt Nat Artif Syst 11:125–134
24. Kacem I, Borne SHP (2002) Approach by localization and multiobjective evolutionary optimization for flexible job-shop scheduling problems. IEEE Trans Syst Man Cybern Part C 32(1):1–13
25. Kacem I, Hammadi S, Borne P (2002) Pareto-optimality approach for flexible job-shop scheduling problems: hybridization of evolutionary algorithms and fuzzy logic. Math Comput Simulat 60:245–276
26. Kachitvichyanukul V, Siriwan S (2011) A two-stage genetic algorithm for multi-objective job shop scheduling problems. J Intell Manuf 22(3):355–365
27. Karimi-Nasab M, Aryanezhad M (2011) A multi-objective production smoothing model with compressible operating times. Appl Math Model 35:3596–3610
28. Li X, Gao L, Li W (2012) Application of game theory based hybrid algorithm for multi-objective integrated process planning and scheduling. Expert Syst Appl 39:288–297
29. Lin L, Gen M (2009) Auto-tuning strategy for evolutionary algorithms: balancing between exploration and exploitation. Soft Comput 13(2):157–168
30. Norman B, Bean J (1995) Random keys genetic algorithm for job-shop scheduling: unabridged version. Technical report, University of Michigan
31. Okamoto A, Gen M, Sugawara M (2009) Integrated scheduling using genetic algorithm with quasi-random sequences. Int J Manuf Technol Manage 16(1/2):147–165
32. Pareto V (1906) Manuale di economica politica, vol 13. Societa Editrice Libraria, Milan
33. Scholl A (1995) Data of assembly line balancing problems. Technical report, Darmstadt Technical University, Department of Business Administration, Economics and Law, Institute for Business Studies (BWL)

34. Verderame P, Christodoulos A (2008) Integrated operational planning and medium—term scheduling for large-scale industrial batch plants. Ind Eng Chem Res 47(14):4845–4860
35. Wang D, Liu H (2008) Fms schedule based on hybrid swarm optimization. In: Proceedings of 7th world congress on control and automation, pp 7054–7059
36. Xia W, Wu Z (2005) An effective hybrid optimization approach for multi-objective flexible job-shop scheduling problem. Comput Ind Eng 48(2):409–425
37. Yamada T, Nakano R (1992) A genetic algorithm applicable to large-scale job-shop problems Parallel Problem Solving from Nature: PPSN II, pp 281–290
38. Yang J (2001) Ga-based discrete dynamic programming approach for scheduling in fms environments. IEEE Trans Syst Man and Cybern Part B 31(5):824–835
39. Yu X, Gen M (2010) Introduction to evolutionary algorithms. Springer, London
40. Yun Y, Gen M (2003) Performance analysis of adaptive genetic algorithms with fuzzy logic and heuristics. Fuzzy Optim Decis Making 2(2):161–175
41. Zandieh M, Karimi N (2011) An adaptive multi-population genetic algorithm to solve the multi-objective group scheduling problem in hybrid flexible flowshop with sequence-dependent setup times. J Intell Manuf 22(6):979–989
42. Zhang Q, Gen M, Jo J (2013) Hybrid sampling strategy-based multiobjective evolutionary algorithm for process planning and scheduling problem. J Intell Manuf. doi:10.1007/s10845-013-0814-2:1-17
43. Zitzler E, Thiele L (1999) Multiobjective evolutionary algorithms: a comparative case study and the strength pareto approach. IEEE Trans Evolut Comput 3(4):257–271
44. Zitzler E, Thiele L (2001) SPEA2: improving the strength pareto evolutionary algorithm. Technical report, Computer Engineering and Communication Networks Lab (TIK)

Chapter 71
A Vector Space Model Approach for Searching and Matching Product E-Catalogues

Ahmad Mehrbod, Aneesh Zutshi and António Grilo

Abstract In e-procurement, companies use e-catalogues to exchange product information with business partners. The large variety of e-catalogue formats which are used by various companies make it difficult to match a product request from a buyer (buyer e-catalogue) with products e-catalogues. While, there are too many different standards for e-catalogues in use, often companies do not follow standard formats. Hence we often encounter a plethora of catalogue formats ranging from unstructured text to well-structured XML documents. One traditional approach to solve this problem is to convert different formats to a general common structure. But within this heterogeneous set of known or even unknown structures achieving a global structure is impractical. In this paper, vector space model has been used to measure the similarity ratio of providers' e-catalogues with a buyer's e-catalogue. Attributes of known structures and their values have been used as terms and their weights in the vectors to find the correlation of e-catalogues based on relationship of common tags. In order to associate the structures in calculating similarity, levels of attributes in xml documents are also included in the terms. Natural language processing is used to extract the same attributes from unstructured or unknown structured documents.

Keywords E-procurement · E-catalogue · Vector space model · Information retrieval

A. Mehrbod (✉) · A. Zutshi · A. Grilo
Faculdade de Ciências e Tecnologia, Universidade Nova de Lisboa, Campus de Caparica, 2829-516 Caparica, Portugal
e-mail: a.mehrbod@campus.fct.unl.pt

A. Mehrbod · A. Zutshi · A. Grilo
Unidade de Investigação em Engenharia Mecânica e Industrial (UNIDEMI), FCT/UNL, Campus de Caparica, 2829-516 Caparica, Portugal

J. Xu et al. (eds.), *Proceedings of the Eighth International Conference on Management Science and Engineering Management*, Advances in Intelligent Systems and Computing 281, DOI: 10.1007/978-3-642-55122-2_71, © Springer-Verlag Berlin Heidelberg 2014

71.1 Introduction

E-catalogues play a critical role in e-procurement marketplaces. They can be used in both the tendering (pre-award) and the purchasing (post-award) processes. Companies use e-catalogues to exchange product information with business partners. Suppliers use e-catalogues to describe goods or services that they offer for sale. Meanwhile buyers may use e-catalogues to specify the items that they want to buy [5, 6].

Matching a product request from a buyer with products e-catalogs that have been provided by the suppliers, helps companies to reduce the efforts needed to find partners in e-marketplaces [12, 16].

The large variety of e-catalogue formats [21] which are used by various companies is one of the major challenges in the matching process. Since each business actor may use a different structure, classification and identification code for describing e-catalogues, it is not easy to match a product with the e-catalog requested by another partner [16]. This heterogeneity makes it difficult and time-consuming to the integrate and query e-catalogues [3].

While, there are too many different standards for e-catalogues and product classifications in use, often companies do not follow standard formats and prefer to have their individual structures [4]. Hence we often encounter a plethora of catalog formats ranging from unstructured text to well-structured XML documents.

One traditional approach to solve this problem is to transform different formats into a uniform catalogue model [3, 6, 12]. But within this heterogeneous set of known or even unknown structures achieving a uniform structure for e-catalogues is usually not practical. Development of a uniform e-catalogue model requires precise and detailed understanding of each of the various formats of catalogues [1]. There is always a chance to encounter with a new format which may cause difficulties in its interpretation. Furthermore for transformation to a uniform model, e-catalogues must be completely validated and in conformance to the expected format with no tolerance from format deviations.

This paper proposes to exploit Vector Space Model [18] which has been used by information retrieval systems to measure the similarity ratio of documents to match providers' e-catalogues with a buyer's e-catalogue. VSM uses occurrence of terms in documents to produce a table of vectors. Having a vector model of documents, mathematical vector operations can be applied to determine the similarity of a document with another one or with a search query. The simplest example is to use the deviation angle between vectors of frequent terms to calculate the relevance between text documents. While it is used to deal with flat textual data (i.e. classical free text documents), IR is being extended, since the last two decades, so as to treat complex structured and semi-structured data [22].

The rest of this paper is structured as follows: Sect. 71.2 has a deeper view on heterogeneity of e-catalogues and reviews related works. In the Sect. 71.3 the proposed approach has been described in details. Experimental results are described in Sect. 71.4. We conclude with future work in Sect. 71.5.

71.2 Related Works

Regarding to the usage of e-catalogues in e-commerce, interoperability of e-catalogues (Catalogue integration) and personalization of e-catalogues are two main challenges which have been studied in the literatures. Although these challenges are related and many researchers studied both together, former is to match a search query with product e-catalogues and the latter is more focused on customizing e-catalogue selection based on user profile.

As mentioned, the heterogeneity of e-catalogues which come from various sources causes [8] difficulty in the matching process. Generally we encounter with two aspects of heterogeneity in e-catalogues which are semantic and syntactic diversity. Syntactic heterogeneity is the result of different document structures and catalogue formats. Semantic heterogeneity refers to the different meanings of the words in various contents [16, 17].

In order to avoid this, diversity classification systems such as CPV[1], UNSPSC[2] and eCl@ss[3] try to standardize the terms used for describing goods and services which are the subject of procurement. Additionally, e-catalogue standards such as PEPPOL[4], BMEcat[5] and ePRIOR[6] recommend using of these classification systems and furthermore propose common data structures for unifying e-catalogue schemas usually for exchanging purposes.

But catalogue standards are not sufficient to the meet all requirements of data exchange [17]. Consequently, often enterprises do not follow standard formats and prefer to have their individual structures [4]. Not to mention variety of standards makes it impractical to reach the classification and structure unification goal. These standards differ in addressed markets, capabilities to represent product information, market acceptance, and standardization processes [21]. This problem is more visible in multi-source e-marketplaces [6–9]. There are at least 25 standards relating to e-catalogue and product classification, and thousands of enterprise products database and e-commerce sites [4, 14, 21].

Based on two aspects of heterogeneity, syntactic integration and semantic integration of multi-source electronic catalogues have been attended to make e-catalogues interoperable. Both of the two require integration of international product classification standards, enterprise product database and product e-catalogue standards [4, 14]. Some researches such as [6] and [5]more considered syntactic integration and other such as [13, 14] and [11] more focused on semantic integration. But regardless of semantic or syntactic aspect of the problem, general solution in e-catalogue integration is to define a global model and convert e-catalogues to this uniform model or simply interpret them based on this reference model. For example [3] tries to formalize e-catalogues by offering an ontological model of e-catalogues. The main work in this kind of solutions is to introduce generic attributes to design e-catalogue ontology model [4]. Due to different e-catalogue ontology being generated from different data sources are heterogeneous, the key of semantic integration of e-catalogue turns out to be the mapping and integration of catalogue ontologies [4, 10].

Therefore these traditional solutions either for semantic integration or syntactic integration are dependent on universal formal models. As previously mentioned creating such general models has the following problems:

- Requires proper knowledge of the underlying catalogues' structures. But individual formats which are used by companies are usually unknown and usually there is chance to always encounter new formats. Lee et al. [16] provides a search index to match e-catalogues regardless of structures. But usually the structures contain valuable information. Our proposed approach exploits the structures as much as their details is known whilst is independent of structure. Structure independency provides the ability to match unstructured information as well as unknown structures. Information existing in structures is valuable and can be helpful in matching process.
- E-catalogues must be completely validated for conformance to their formats with no tolerance for format deviations. Since usually each structure is transformed to the general model, it has to be completely compatible with the structure which the convertor expects. Furthermore development of such convertors is crucial and time-consuming task.
- All the various matching cases in graphs or models must be predefined in matching algorithms. For example, Kwon et al. [15] tried to cover all the possible conditions in matching two structures. But within a heterogeneous set of structures always there is chance to encounter a new unconsidered condition.

71.3 Document Similarity in E-Catalogues

Since in the area of e-catalogue we often encounter a plethora of formats and developing an integrated model is crucial, this paper proposes to apply a more flexible model to e-catalogues search problem. Vector Space Model which is the base of many search techniques and document similarity methods can be applied to both semantic [19, 24] and syntactic [2, 18] aspects of search problem. Although semantic issues are also considered recently in document similarity techniques [19, 24], we will apply it to our approach in a future work. In this paper we will focus on overcoming syntactic diversity of e-catalogues including unstructured, unknown structured and known structured documents.

In an e-procurement e-marketplace at least two scenarios for finding similar documents are possible [5]. First a buyer who makes a call for tender needs to select some suppliers based on their e-catalogues in order to send the invitation. Second scenario is when a supplier searches to find opportunities in e-marketplace. Supplier may upload a product e-catalogue to e-marketplace in order to find similar call for tenders.

Three general cases may be considered in syntactic interoperability of e-catalogues. First, unstructured text such as pdf files which are common in online commerce. Second, structured or semi-structured documents which are unknown for the system such as individual formats. Third structured standard documents which are known for the system such as PEPPOL e-catalogues. XML is one of the most common formats for exchanging structured and semi-structured data and also standard e-catalogues in B2B e-commerce [20]. Among 25 e-catalogue standards, 16 of them are based on XML [21]. Hence in the following sub-sections we will apply Vector Space Model to three groups of documents at the same time: unstructured text documents, XML documents and standard e-catalogues.

71.4 Unstructured Documents

In VSM, sets of keywords or terms have been extracted from documents and user queries. Then, a vector has been made to represent occurrence of terms in each document. If a term occurs in a document, its value or weight in the vector of the document is non-zero. Documents that are similar to a given query can be calculated by comparing deviation of angle between the vector of each document and that of the query.

Depending on the application, several methods have been proposed to define the weights. Keywords are commonly weighted in order to reflect their relative importance in the query or document at hand. The underlying idea is that terms that are of more importance in describing a given query or document are assigned a higher weight [18, 22]. For example one well-known method of weighting the terms is TF-IDF that takes into consideration both document and collection statistics [22].

Usually natural language processing techniques are utilized to extract important terms automatically form the documents and queries. Among various types of processing which can be applied to text documents, we used a Natural Language Processing tool to tokenize, lemmatize and remove stop words. Tokenization is to decide what constitutes a term and how to extract terms from raw text. Since our application is to match e-catalogues and term matching is more valuable than phrase matching, we filtered the stop words. Stop words are some of most common words such as the, is, at and so on. Lemmatisation is to convert the different inflected forms of a word to the lemma form so they can be analyzed as a single item [23].

71.5 Structured Documents

XML documents is widely used to represent structured information. Any structured or semi-structured document can be shown using XML files. Hence XML-based similarity becomes a central issue in the structured information retrieval. Since in conventional information retrieval, documents are unstructured data, Vector Space Model has been extended towards XML information retrieval [22]. Using these extensions we can represent structured and unstructured queries and documents in vector space model and compute matching between them.

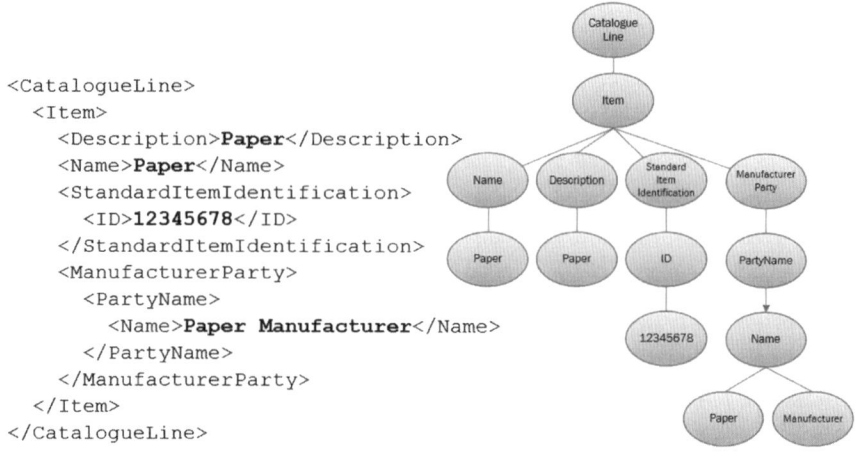

```
<CatalogueLine>
  <Item>
    <Description>Paper</Description>
    <Name>Paper</Name>
    <StandardItemIdentification>
      <ID>12345678</ID>
    </StandardItemIdentification>
    <ManufacturerParty>
      <PartyName>
        <Name>Paper Manufacturer</Name>
      </PartyName>
    </ManufacturerParty>
  </Item>
</CatalogueLine>
```

Fig. 71.1 A part of a standard e-catalogue (D1)

Hierarchical structure of XML documents are generally modelled as trees. In a traditional model, nodes of tree represent XML elements and are labelled with corresponding element tag names. Since content is distributed at different levels of the tree, location of a term in the tree is effective on the value of the term [22] and should be considered in term extraction.

In the comparison process generally values of attributes are disregarded. This approach is useful for structure-only comparing of XML documents [22]. But in the context of product features, similarity measure is more sensitive to the values which have been saved in the e-catalogue structures. Therefore in matching process of e-catalogues, values are crucial and are even more important than structures. Consequently, we used a structure-and-content tokenization process [18] to define the terms.

As an example, Fig. 71.1 shows a portion of a standard e-catalogue D1 which is used in PEPPOL [5]. In the matching process this e-catalogue should be similar to e-catalogue D2 in Fig. reffig:lurong04Ahamdfig2 that has the word paper. Moreover it should have a higher similarity ratio with document D3 in Fig. 71.2 that has the word paper in attribute name and even higher to document D4 in Fig. 71.2, that has paper in hierarchy of name and item and so on.

One way of doing this is to define a term as a value together with its position within the XML tree. Figure 71.2 illustrates this representation. We use all the sub-trees of a document that contain at least one value as terms [2, 18]. In other words, we first take each value (paper) as a term. This help the matching process to match this documents with unstructured documents or structured document such as D2 that have same values but in different structure. Next we add values with last level of their position (name/paper) to the terms. It helps matching process to increase similarity of D1 with structured documents such as D3. Then we continue adding levels of positions (item/name/paper) to the terms to root of tree. It keep increasing

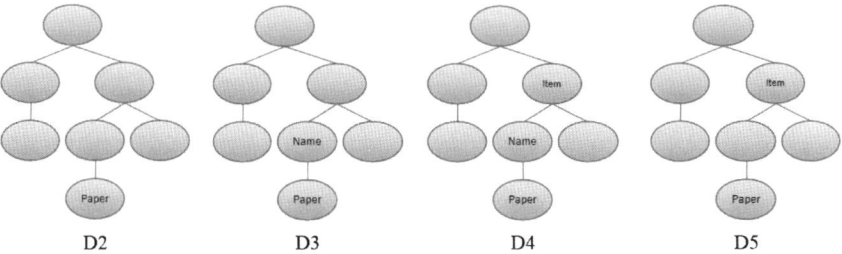

D2   D3   D4   D5

Fig. 71.2 Similar e-catalogues to D1

similarity of D1 with structures documents such as D4. Table 71.1 shows all possible terms for the tree of Fig. 71.1. Therefore D1 will have one common term with D2, two common term with D3 and tree common term with D4 which guarantees more ration of similarity for documents with resembling structures. Note that having value attached to all the terms helps search process to avoid matching documents with same structure but different products.

Documents such as D5 should have a lower matching ratio with D1 as compared to its matching ratio with D3 and D4, because the same value for an attribute (item) exists in both documents but not necessarily in same path order. Therefore the terms of Table 71.2 have also been added to the terms of D1 to cover this type of similarity. In order to decrease the similarity ratio for documents that match D1 using the terms of Table 71.2 instead of the terms of Table 71.1, we divide the weight of a term by twice the number of nodes between the value and the attribute. With this simple approach we don't have to change the similarity formula as proposed in [2] and [18].

71.6 Standard E-Catalogues

Standard e-catalogues are source of diverse types of information. For example a PEPPOL e-catalogue includes general document data, product data and partners' data. This extra information can mislead product search process. Furthermore various attributes of product data can have different value in the matching process. For example classification code of a product has more value than a description in matching process. Hence we used a table of coefficients to adjust the impact of each attribute in similarity of known structures. These coefficients are values between 0 and 1 which are multiplied to the weight of terms. Undesired information such as partners' data can be simply excluded from matching process by putting 0 coefficients. Using this simple mechanism a new known structures can be easily added to the search system. Default value for all coefficients are 1 which reduces the status of an e-catalogue to an unknown structure for the matching process.

Table 71.1 All possible terms for D1

Value	Terms
Paper manufacturer	Manufacturer
	Name/manufacturer partyname/name/manufacturer
	Manufacturerparty/partyname/name/manufacturer
	Item/manufacturerparty/partyname/name/manufacturer
	Catalogueline/item/manufacturerparty/partyname/name/manufacturer
	Paper name/paper partyname/name/paper
	Manufacturerparty/partyname/name/paper
	Item/manufacturerparty/partyname/name/paper
	Catalogueline/item/manufacturerparty/partyname/name/paper
12345678	12345678
	ID/12345678
	Standarditemidentification/ID/12345678
	Item/Standarditemidentification/ID/12345678
	Catalogueline/item/Standarditemidentification/ID/12345678
Paper	Paper
	Description/paper
	Item/description/paper
	Catalogueline/item/description/paper
Paper	Paper
	Name/paper
	Item/name/paper
	Catalogueline/item/name/paper

Table 71.2 Additional terms for the last entry of Table 71.1

Value	Terms	Weight ratio
Paper	Item/paper	1/2
	Catalogueline/item/paper	1/2
	Catalogueline/name/paper	1/2
	Catalogueline/paper	1/4

71.7 Conclusion and Future Works

Based on many standards and data resource, query and search of e-catalogues is affected by integration problems [4]. Since IR-based methods are applicable to wide range of structured and unstructured documents which we encounter in matching e-catalogues, this paper proposes a vector space model approach to search e-catalogues. Furthermore these methods target loosely structured data, thus useful and generally exploited for fast simple structured search and retrieval [22].

Combinations of values and attributes of structured documents have been used to find the correlation of documents based on relationship of common tags. Then we have proposed a simple table of coefficients to specify the matching model for standard e-catalogues. This mechanism increases the search precision by removing unrelated information from the matching process and boosting weights of important tags. In future these tables can be customized for users using a learning mechanism based on their profiles and search interests.

In order to test matching process we used a set of e-catalogues in various formats. First we utilized an open source full text search tool and a natural language analyser to extract terms from flat text files. Then we extended the search tool to consider the locational values of words in term extraction process when such information is available. Having tokenized all types of e-catalogues, we made term vectors for them and applied the coefficient tables to known structures.

With this tool, users will be able to search within the available files or simply upload their e-catalogues to find similar documents. This search mechanism allows users who prefer to specify the tag relations while searching [22] to get rid of using content-and-structure queries [19]. The experimental results show the matching process is capable to match diverse formats of catalogues from various sources.

Acknowledgments The research of this work has been partially funded by project VortalSocialApps, co-financed by VORTAL and IAPMEI and the European Funds QREN COMPETE, and also would like to thank Fundação da Ciência e Tecnologia for supporting the research center UNIDEMI through the grant PEst-OE/EME/UI0667/2011.

References

1. Benatallah B, Hacid MS et al (2006) Towards semantic-driven, flexible and scalable framework for peering and querying e-catalog communities. Info Syst 31(4–5):266–294
2. Carmel D, Maarek Y et al (2002) An extension of the vector space model for querying xml documents via xml fragments. SIGIR Forum
3. Chen D, Li X et al (2010) A semantic query approach to personalized e-catalogs service system. J Theor Appl Electron Commer Res 5(3):39–54
4. Chen D, Li X, Zhang J (2010b) User-oriented intelligent service of e-catalog based on semantic web. In: 2nd IEEE international conference on information management and engineering (ICIME), pp 449–453
5. Ghimire S, Jardim-Goncalves R et al (2013a) Framework for inter-operative e-procurement marketplace. In: 17th IEEE international conference on computer supported cooperative work in design (CSCWD 2013), pp 459–464
6. Ghimire S, Jardim-Goncalves R, Grilo A (2013b) Framework for catalogues matching in procurement e-marketplaces. Iber Conf Inf Syst Technol Cist.
7. Grilo A, Jardim-Goncalves R (2013) Cloud-marketplaces: distributed e-procurement for the aec sector. Adv Eng Info 27(2):160–172
8. Grilo A, Jardim-Goncalves R, Ghimire S (2013a) Cloud-marketplace: new paradigm for e-marketplaces. In: Technology management in the IT-driven services (PICMET), 2013 proceedings of PICMET, vol 13, pp 555–561
9. Grilo A, Jardim-Goncalves R, Ghimire S (2013b) E-procurement in the era of cloud computing. In: 4th international conference on IS management and evaluation (ICIME 2013), pp 104–110
10. Guo J (2009) Collaborative conceptualisation: towards a conceptual foundation of interoperable electronic product catalogue system design. Enterp Info Syst 3(1):59–94
11. Huang JZ, Huang G (2005) Ontology-based e-catalog matching for integration of gdsn and epcglobal network. In: IEEE international conference on e-business engineering, pp 212–215

12. Kim D, Kim J, Lee S (2002a) Catalog integration for electronic commerce through category-hierarchy merging technique. In: Proceedings of twelfth int work res issues data eng eng E-commerce/e-bus Syst RIDE-2EC 2002, pp 28–33
13. Kim D, Kim J, Lee S (2002b) Catalog integration for electronic commerce through category-hierarchy merging technique. In: Proceedings twelfth int work res issues data eng eng e-commerce/e-bus Syst RIDE-2EC 2002, pp 28–33
14. Kim W, Choi DW, Park S (2007) Agent based intelligent search framework for product information using ontology mapping. J Intell Info Syst 30(3):227–247
15. Kwon IH, Kim CO, KPK et al (2008) Recommendation of e-commerce sites by matching category-based buyer query and product e-catalogs. Comput Ind 59(4):380–394
16. Lee J, Lee T, et al. (2007) Massive catalog index based search for e-catalog matching. In: 9th IEEE intetrnational conference on e-commerce technology 4th IEEE international conference on enterprise computing e-commerce e-services (CEC-EEE 2007), pp 341–348
17. Leukel J, Schmitz V, Dorloff F (2002) Exchange of catalog data in B2B relationships. Analysis and improvement 2002(ICWI), pp 403–410
18. Manning CD, Prabhakar R, Schutze H (2008) Introduction to information retrieval. Cambridge University Press, Cambridge
19. Mukerjee K, Porter T, Gherman S (2011) Linear scale semantic mining algorithms in microsoft sql server's semantic platform. In: Proceedings of 17th ACM SIGKDD international conference on knowledge discovery and data mining—KDD'11, p 213
20. Schmitz V, Leukel J, Dorloff F (2003) Xml data modeling concepts in B2B catalog. In: Proceedings of IADIS international conference e-Society 2003 (ES 2003) 2003(Es), pp 227–234
21. Schmitz V, Leukel J, Dorloff F (2005) Do e-catalog standards support advanced processes in B2B e-commerce? The CEN/ISSS Workshop ECAT 00(C), pp 1–10
22. Tekli J, Chbeir R, Yetongnon K (2009) An overview on XML similarity: background, current trends and future directions. Comput Sci Rev 3(3):151–173
23. Turney P, Pantel P (2010) From frequency to meaning: vector space models of semantics. J Artif Intell Res 37:141–188
24. Widdows D (2008) Semantic vector products: some initial investigations. In: Second AAAI symposium on quantum interaction, March

Chapter 72
One-Step Classifier Ensemble Model for Customer Churn Prediction with Imbalanced Class

Jin Xiao, Geer Teng, Changzheng He and Bing Zhu

Abstract In customer churn prediction, an important yet challenging problem is the class imbalance of data distribution. After analyzing the disadvantages of the commonly used "two-step" methods, this study combines multiple classifiers ensemble technique, self-organizing data mining with cost-sensitive learning, and proposes one-step classifier ensemble model for imbalance data (OCEMI). For each test customer, it can adaptively select out the more appropriate one from the two kinds of dynamic ensemble approach: dynamic classifier selection (DCS) and dynamic ensemble selection (DES). Meanwhile, new cost-sensitive selection criteria for DCS and DES are constructed respectively to improve the classification ability for imbalanced data. The empirical results show that this strategy can be used to predict customer churn more effectively.

Keywords Customer churn prediction · One-step ensemble model · Self-organizing data mining · Multiple classifiers ensemble

72.1 Introduction

With the aggravation of market competition, customer churn affects the profits of enterprises amazingly in many industries such as telecommunications, banking, etc. The survey data of 9 industries in U.S. show that the industry average profit will increase by from 25 to 85 % when the customer churn rate reduces by 5 % [1]. Therefore, it is important to predict the customer churn correctly and implement

J. Xiao · C. He · B. Zhu (✉)
Business School, Sichuan University, Chengdu 610064, People's Republic of China
e-mail: zhubing1866@hotmail.com

G. Teng
The Faculty of Social Development and Western China Development Studies, Sichuan University, Chengdu 610064, People's Republic of China

J. Xu et al. (eds.), *Proceedings of the Eighth International Conference on Management Science and Engineering Management*, Advances in Intelligent Systems and Computing 281, DOI: 10.1007/978-3-642-55122-2_72, © Springer-Verlag Berlin Heidelberg 2014

the customer retention strategies in time for enhancing the core competitiveness of enterprise.

The customer churn prediction can be considered as a binary classification issue, i.e. all customers are classified into two types: churn and non-churn. It is found that the class distribution is usually imbalanced, i.e. the number of customer with different classes varies greatly, and the ratio of churn and non-churn customer reaches to 1:100, even over 1:10 000 [2]. In this case, although the classification accuracy in the total customer samples is very high, the classification accuracy in the churn customer is very poor. However, the correct identification for the churn customer is what we focus on most, because it can bring more profits for enterprise [2].

To solve the above issues, this study combines self-organizing data mining (SODM) suited to the characteristic of CRM data, multiple classifiers ensemble technology with cost-sensitive learning, and proposes one-step classifier ensemble model for imbalance data (OCEMI) and applies it to customer churn prediction. The empirical study shows that the proposed method can achieve better customer churn prediction performance than some existing methods.

The remainder of this study is organized as follows. We first discuss the related work on customer churn prediction in Sect. 72.2, and introduce the basic principle and modeling process of SODM in Sect. 72.3. Next, we explain details of the one-step classifier ensemble model for class imbalance data in Sect. 72.4. In Sect. 72.5, we show the empirical study to analyze the churn prediction performance of the proposed model. Finally, the conclusions and future research directions are contained in Sect. 72.6.

72.2 Literature Review

In recent years, the imbalance class distribution of customer churn prediction data have gotten more and more attention. In order to predict customer churn effectively, Kim et al. [3] proposed a framework with three phases:

1. Data collection and preprocessing, which were to collect the transaction data and questionnaire data for customer churn prediction, and preprocess these data, for example, how to handle noise, missing values, sample data with class imbalance, etc;
2. Customer churn prediction modeling. It found the factors that affected customer churn from customer information, and constructed the classification model according to these factors to predict customer churn;
3. Marketing strategies formulation. In this phase it developed appropriate strategies for different customers to maximize their values to the enterprise according to the results of prediction in Phase 2 (see Fig. 72.1).

In the above framework, most of the existing researches often preprocess the class imbalanced data by resampling techniques first, and then implement prediction modeling. For instance, Verbeke et al. [4] utilized over-sampling to handle the class

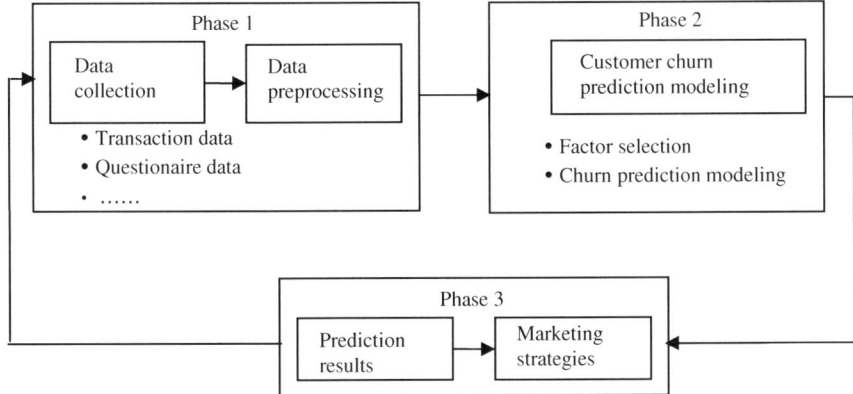

Fig. 72.1 The commonly used customer churn prediction research framework

imbalance data, and then constructed customer churn prediction model based on rule induction. The data preprocessing in Phase 1 and customer churn prediction modeling in Phase 2 are implemented independently, and we call it "two-step" customer churn prediction strategy. However, the effects of classification models are usually related to data preprocessing methods, which will affect the classification performance of "two-step" strategies. Crone et al. [5] showed that the classification models were sensitive to data preprocessing, Van Hulse and Khoshgoftaar [6] further pointed out that the classifiers were sensitive to the sampling preprocessing methods under the situation of class imbalance. Thus, the "two-step" customer churn prediction should be improved.

In order to improve the performance of prediction model further, some scholars attempt to introduce ensemble learning that has received extensive attention in recent years to customer churn prediction modeling. For example, Lemmens and Croux [7] applied the two important ensemble learning algorithms Bagging and Boosting to customer churn prediction, to handle the data with class imbalance, they combined ensemble learning with re-sampling technique. They found that Bagging and Boosting could improve the prediction accuracy to a large extent. Burez and Van den Poel [2] introduced two sampling techniques (random down-sampling and advanced down-sampling) and two ensemble methods (gradient Boosting method and weighted random forest) to customer churn prediction. Their experimental results showed that the prediction performance with down-sampling was better than that without sampling. The above researches have improved the accuracy of customer churn prediction to some extent, while they still belong to "two-step" customer churn prediction strategies.

New issue emerges from the practice, so we must look for new solution for it. SODM is an inductive modeling technique and proposed by Ivakhnenko [8], which has been applied to many areas such as macro economic analysis, financial and ecology modeling. However, the application of GMDH in CRM is rare. Different from the existing researches, this study combines ensemble learning with SODM, begins from the characteristics of customer data, proposes one-step ensemble research framework for customer churn prediction.

72.3 Self-Organizing Data Mining

SODM is a heuristic, automatic modeling and identification method for complex system, and group method of data handling (GMDH) is its core technology. Given a system modeling issue, multilayer GMDH neural network builds the general relationship between output and input variables in the form of mathematical description, which is also called reference. Generally, the description can be considered as a discrete form of the Volterra functional series or Kolmogorov-Gabor polynomial:

$$Y = f(X_1, X_2, \cdots, X_n) = a_0 + \sum_{i=1}^{n} a_i X_i + \sum_{i=1}^{n} \sum_{j=1}^{n} a_{ij} X_i X_j \qquad (72.1)$$

$$+ \sum_{i=1}^{n} \sum_{j=1}^{n} \sum_{k=1}^{n} a_{ijk} X_i X_j X_k + \cdots$$

Eq. (72.1) is also known as K-G polynomial. Here Y is the output, $X = (X_1, X_2, \cdots, X_n)$ is the input vector and a is the vector of coefficients or weights. In particular, the form of the first order (linear) K-G polynomial including n variables (neurons) is as follows:

$$f(X_1, X_1, \cdots, X_n) = a_0 + a_1 X_1 + a_2 X_2 + \cdots + a_n X_n. \qquad (72.2)$$

If the linear reference function like the form of Eq. (72.2) is chosen, we need to regard all sub-sections of Eq. (72.2) as $n + 1$ initial models of the multilayer GMDH neural network, that is $v_1 = a_0, v_2 = a_1 X_1, \cdots, v_{n+1} = a_n X_n$. The specific modeling process is as follows: Combine every two initial models to form a unit according to the transfer function $y = f(v_i, v_j) = a_1 + a_2 v_i + a_3 v_j$, then there are $n_1 = C_{n_0}^2$ ($n_0 = n + 1$) partial models in total in the first layer (see Fig. 72.2). Their forms are as follows: $w_k = a_1^k + a_2^k v_i + a_3^k v_j, i, j = 1, 2, \cdots, n_0, i \neq j, k = 1, 2, \cdots, n_1$, where y_k^1 is estimated output, $a_1^k, a_2^k and a_3^k$ ($k = 1, 2, \cdots, n_1$) are the parameters estimated by least squares (LS) in the model learning set. Then outputs of $F_1(n_1)$ partial models are selected as per the threshold measure to pass on to the second layer as inputs in pairs. In the second layer, we check the functions of the form $z_k = b_1^k + b_2^k w_i + b_3^k w_j, i, j = 1, 2, \cdots, F_1; i \neq j; k = 1, 2, \cdots, n_2$. The number of such partial models is $n_2 = C_{F_1}^2$. The process continues until we find the optimal complexity model by the termination principle (also called optimal complexity theory): along with the increase of model complexity, the value of external criterion will decrease first and then increase, and the global extreme value corresponds to the

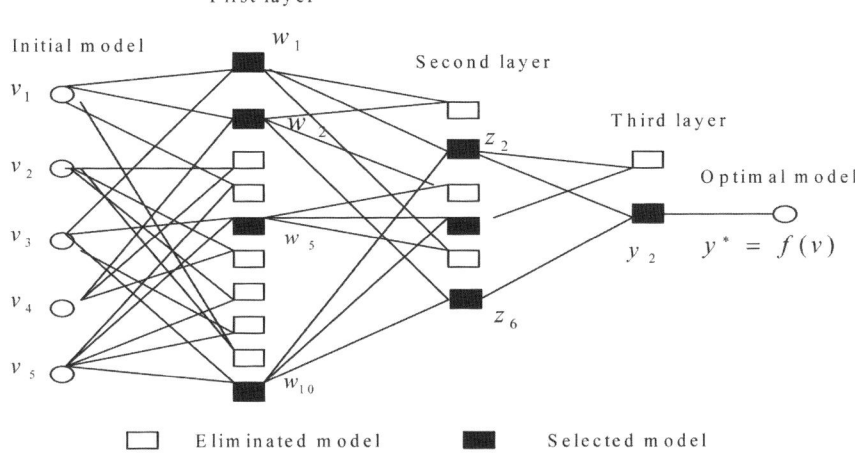

Fig. 72.2 The diagram of optimal model generation process by GMDH

optimal complexity model [9]. In this way, algorithm can determine the variables of getting into model, structure and parameters automatically, realize the process of self-organizing modeling, and also can avoid over-fitting [10].

GMDH divides the training set into two parts: model learning set A and model selecting set B. It utilizes the inner criterion (LS) to estimate the parameters of middle candidate models in set A, and uses the external criterion to evaluate and select the middle candidate models in set B. Thus, if we can apply GMDH to classifier ensemble selection, it can automatically select the member classifiers of the ensemble, identify the parameters (weights) and structure of the optimal complexity combination model, and then accomplish the self-organizing modeling process. For the specific modeling steps of GMDH, please refer to the reference by Xiao et al. [10].

72.4 One-Step Classifier Ensemble Model for Class Imbalance Data

1. The Basic Idea of Model

In general, the dynamic classifier ensemble strategies include two basic ideas: dynamic classifier selection (DCS) and dynamic ensemble selection (DES) [11, 12]. The former is to select an optimal base classifier for each candidate classification sample to classify it, and the latter is to select an optimal classifier ensemble for each candidate classification sample to classify it. Both DCS and DES have advantages and disadvantages: given a candidate classification sample, if the performance of a base classifier is significantly better than that of other classifiers, DCS should be a better choice; while if there is no significant difference among all base classifiers, the performance of DES strategy may be better. Therefore, in this study, OCEMI model chooses the most suitable ensemble strategy for each candidate classification customer from the DCS method DCS-LA [12] and dynamic classifier ensemble selection method GDES [10], which achieves adaptive switch between DCS and dynamic classifier ensemble selection. Further, OCEMI assigns different misclassification costs to different class samples, adopts cost sensitive learning to reconstruct the external criterion of GDES, and then builds cost sensitive GDES strategy as well as cost sensitive DCS-LA.

2. The External Valuation Criteria of Cost Sensitive

The GDES strategy proposed by Xiao and He [13] regards symmetrical regularization criteria (SRC) as external criteria, and its form is as follows:

$$d^2(T_{rain}) = \Delta^2(A) + \Delta^2(B) = \sum_{t \in A} (y_t - y_t^m(B))^2 + \sum_{t \in B} (y_t - y_t^m(A))^2, \quad (72.3)$$

where, y_t is the true class label in the corresponding subset, $y_t^m(B)$ is the predicted value of subset A by the model trained in subset B. Therefore, $\Delta^2(A)$ means the classification error in set A of the model trained in set B, $\Delta^2(B)$ means the opposite. It can be seen from Eq. (72.3) that all the customer samples have the same misclassification cost when the criterion SRC is adopted. However, the misclassification costs of customers with different class are often very different in customer churn prediction with class distribution imbalance. Therefore, this study combines cost sensitive learning with SRC and constructs a new cost sensitive symmetric regularity criterion (CS-SRC) for GDES:

$$S = Cost(A) + Cost(B), \quad (72.4)$$

$$Cost(A) = \sum_{t=1}^{n_{11}} k_1(y_t - y_t^m(B))^2 + \sum_{t=1}^{n_{12}} k_2(y_t - y_t^m(B))^2, \quad (72.5)$$

$$Cost(B) = \sum_{t=1}^{n_{21}} k_1(y_t - y_t^m(A))^2 + \sum_{t=1}^{n_{22}} k_2(y_t - y_t^m(A))^2. \quad (72.6)$$

Here, k_1 means the misclassification cost of churn customer (minority sample); k_2 means the misclassification cost of non-churn customer (majority sample); n_{11} and n_{12} mean the number of minority sample and majority sample respectively in subset A; n_{21} and n_{22} mean the number of minority sample and majority sample respectively in subset B. Therefore, $Cost(A)$ describes the total misclassification cost in subset A by the model trained in subset B; $Cost(B)$ means the opposite.

3. Model Constructing

Let T_{rain}, V_a, T_{est} be the training set, validation set and test set for a customer churn prediction issue with class imbalance, and the basic steps of one-step ensemble model OCEMI are as follows:

Step 1. Train N base classifiers $C_1, C_2, \cdots, C_N$ in training set to compose the base classifier pool (BCP);

Step 2. For each candidate classification sample $x^* \in T_{est}$,

Step 2.1. Find K nearest neighbors of x^* from validation set V_a with K-nearest neighbors method to construct its local area R_K;

Step 2.2. Divide R_K into minority $D_{\min}$ (churn class) and majority D_{maj} (non-churn class), and calculate the total misclassification cost of each base classifier C_i $(i = 1, 2, \cdots, N)$ in local area R_K: $\varepsilon_i = \sum_{t \in D_{\min}} k_1(y_t - y(C_i))^2 + \sum_{t \in D_{\max}} k_2(y_t - y(C_i))^2$, find the single classifier C_{opt} with the lowest misclassification cost;

Step 2.3. According to the external criterion CS-SRC in Eqs. (72.4) $\sim$ (72.6) select a classifier ensemble E_{opt} with the lowest misclassification cost in local area R_K with GDES strategy from BCP;

Step 2.4. Utilize the best single classifier C_{opt} and the optimal classifier ensemble E_{opt} to classify R_K, and calculate their misclassification costs in R_K (denote them as err_1 and err_2 respectively);

Step 2.5. Compare err_1 and err_2, if $err_1 \leq err_2$, then adopt the optimal single classifier to classify x^*; otherwise, adopt ensemble E_{opt} to classify x^*.

72.5 Empirical Analysis

We conducted experiments in the credit card customer churn prediction "China churn" dataset of one commercial bank in Sichuan province China. Further, we compared OCEMI with four commonly used "two-step" ensemble strategies. The four strategies first adopted resampling technique to balance the class distribution of the training set (make the ratio of non-churn and churn samples be 1 : 1), and then utilized genetic algorithm based static ensemble (GASE) [14], neural network based static ensemble (NNSE) [15], dynamic classifier selection (DCS-LA) [12], and GMDH based dynamic classifier ensemble selection GDES [13] to classify the test sample respectively, here we abbreviate them as R-GASE, R-NNSE, R-DCS, R-GDES. At the same time, we also compared OCEMI with the single cost sensitive support vector machine (CSSVM) method proposed by Zou et al. [16].

1. The Data Description

The dataset is about churn prediction of credit card customer from one commercial bank in Sichuan province China ("China churn"). The data interval is 2010.5 $\sim$ 2010.12. According to the basic principle of churn prediction variable selection and the availability of data, we selected 25 prediction variables (see Table 72.1). For the

Table 72.1 Attribute description of the "China churn" dataset

Feature	Name	Feature	Name
x_1	Total consumption times	x_{14}	Cash times in the last 1 month
x_2	Total consumption amount	x_{15}	Cash times in the last 2 months
x_3	Total cash times	x_{16}	Cash times in the last 3 months
x_4	Customer survival time	x_{17}	Cash times in the last 6 months
x_5	Total contributions	x_{18}	Months of transaction times reducing continuously
x_6	Valid survival time	x_{19}	If overdue in the last 1 month
x_7	Average amount ratio	x_{20}	If overdue in the last 2 months
x_8	Whether associated charge	x_{21}	Amount usage ratio in the last 1 month /historical average usage ratio
x_9	Consumption times in the last 1 month	x_{22}	Sex
x_{10}	Consumption times in the last 2 months	x_{23}	Annual income
x_{11}	Consumption times in the last 3 months	x_{24}	Nature of work industry
x_{12}	Consumption times in the last 4 months	x_{25}	Education
x_{13}	Consumption times in the last 5 months		

customer class label, we defined the churn customer as someone who canceled card from May 2010 to October 2010 or did not consume for 3 months. After simple data cleaning, we obtain 3255 customer instances from the database finally, in which there are 302 churn customers and 2953 non-churn customers. The churn rate is 9.28 % and it belongs to class imbalanced dataset.

2. Experimental Setup

For static ensemble strategies R-GASE, R-NNSE as well as the single CSSVM model, this study selected 70 % customer data randomly as training set, the remaining 30 % as test set. For dynamic classifier ensemble strategies OCEMI, R-GDES and R-DCS, the study randomly divided the training set into two subsets further, among them 60 % samples were used to train models, and the validation set was composed of the remaining ones. We chose support vector machine as the basic classification algorithm. We chose kernel function with the form of radial basis function, because we found that the SVM could obtain the best performance. Let the size of base classifier pool (BCP) be 20. Then we generated the base classifiers with RSS method as the following steps: first selected 20 feature subsets randomly from the original feature space (the number of features was half of the original feature space), got 20 training subsets by mapping, and finally trained a SVM classifier in each subset at last. As for the parameters of OCEMI algorithm, we let $K = 10$ through repeated experiments; for k_1 and k_2, we let $k_1 = 5$, $k_2 = 1$ according to [6]. Finally, all the experiments were performed on the MATLAB 6.5 platform with a dual-processor 3.0 Ghz Pentium 4 Windows computer, and in each case the final classification result was the average of 10 experiments.

Table 72.2 The confusion matrix of customer churn prediction

Feature	Name	Feature	Name
	Predicted negative	Predicted positive	Total
Actual negative (non-churn)	D_1	D_2	$D_1 + D_2$
Actual positive (churn)	D_3	D_4	$D_3 + D_4$
Total	$D_1 + D_3$	$D_2 + D_4$	$D_1 + D_2 + D_3 + D_4$

Note D_1 is the number of true negative, D_2 is the number of false positive, D_3 is the number of false negative and D_4 is the number of true positive

3. The Evaluation Criteria

To evaluate the performance of the models referred in this study, we introduced the confusion matrix in Table 72.2. On this basis, six commonly used evaluation criteria were adopted [4, 17].

- Total accuracy $= \frac{D_1+D_4}{D_1+D_2+D_3+D_4} \times 100\,\%$;
- The receiver operating characteristic (ROC) curve and the area under the receiver operating characteristic curve (AUC) value. ROC curve is an important evaluation criterion of classification model in the data with class distribution imbalance. For a issue of two classes, the ROC graph is a true positive rate-false negative rate graph, where Y-axis is true positive rate $(D4/(D3 + D4) \times 100\,\%)$ and the X-axis is false negative rate $(D2/(D1 + D2) \times 100\,\%)$. However, sometimes it is difficult to compare ROC curves of different models directly, so AUC is more convenient and popular;
- Livelihood and survival mobility are oftentimes coutcomes of uneven socioeconomic development.
- Type I accuracy $= \frac{D_4}{D_3+D_4} \times 100\,\%$;
- Type II accuracy $= \frac{D_1}{D_1+D_2} \times 100\,\%$;
- Hit rate $= \frac{D_4}{D_2+D_4} \times 100\,\%$;
- Lift coefficient $= \frac{\text{Hit rate}}{\text{the churn rate of test set}}$.

Figure 72.3 shows the ROC curves of six strategies on the "China-churn" dataset, in which the higher the curve is, the better the performance of the corresponding strategy is. It can be seen that the ROC curve of OCEMI is at the top and those of R-NNSE and R-GASE are the lowest. Therefore, we can roughly get the conclusion that the customer churn prediction performance of OCEMI is the best. At the same time, it can also be seen from the figure that the ROC curves of six classification strategies can be divided into three groups roughly:

- OCEMI, R-GDES and R-DCS show the best performance, and they are dynamic classifier ensemble strategies;
- CSSVM is single one-step model and in the middle;
- R-GASE and R-NNSE are static classifier ensemble strategies, and the performance of them is the worst.

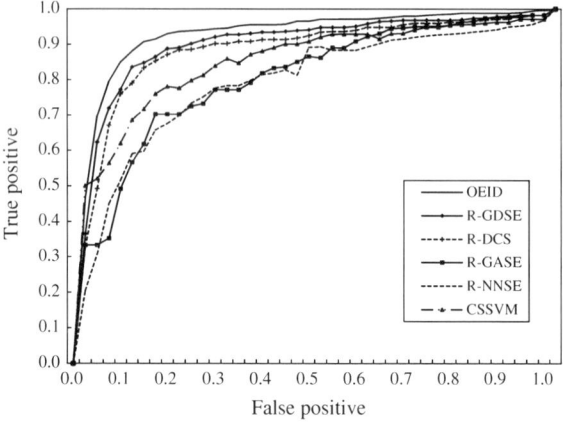

Fig. 72.3 ROC curves of six strategies in "China-churn" dataset

Table 72.3 The comparison of performance for six models in "China-churn" dataset

Models	AUC	Total accuracy (%)	Type I accuracy (%)	Type II accuracy (%)	Hit rate (%)	Lift coefficient
OCEMI	0.9542	94.98	93.55	95.13	66.92	7.21
R-DCS	0.9344	94.26	91.40	94.56	63.91	6.89
R-GDES	0.9442	94.57	90.32	95.02	65.63	7.07
R-GASE	0.8767	94.16	83.87	95.24	65.00	7.00
R-NNSE	0.8629	93.24	80.65	94.56	60.98	6.57
CSSVM	0.9269	94.33	88.95	94.60	64.64	6.92

Therefore, it can be concluded that the performance of dynamic classifier ensemble strategies (whether it is "two-step" or one-step) is better than that of static classifier ensemble methods, and it is consistent with the analysis of Woods et al. [12]. On the other hand, the performance of "two-step" static ensemble strategies is no better than that of single one-step model, which validates further that "two-step" strategies need to be improved really.

Table 72.3 shows the AUC values, Total accuracy, Type I accuracy, Type II accuracy, Hit rate and Lift coefficient of six models in "China-churn" dataset, and the bold face shows the maximum of each column. It can be seen from the table that OCEMI strategy outperforms the other five strategies on AUC value, Total accuracy, Type I accuracy, Hit rate and Lift coefficient, and its Type II accuracy is only lower than that of R-GASE. Thus, it can be concluded that the customer churn prediction performance of the proposed one-step ensemble strategy for class imbalance data is better than that of the commonly used "two-step" strategies R-GDES, R-DCS, R-GASE, and R-NNSE, and also better than that of single one-step model CSSVM.

72.6 Conclusions

Customer churn prediction is very important for CRM. This paper proposes a one-step classifier ensemble customer churn prediction ensemble framework for data characteristics combining ensemble learning with SODM. It can overcome thedisadvantages of commonly used "two-step" customer churn prediction strategies to some extent. Further, we take the customer churn prediction with class imbalance data as an example, and present the corresponding one-step ensemble strategy OCEMI. The empirical results show that the customer churn prediction performance of OCEMI is better than that of the commonly used "two-step" strategies R-GASE, R-NNSE, R-DCS, R-GDES and single cost sensitive classifier CSSVM.

Acknowledgments This research is partly supported by the Natural Science Foundation of China under Grant Nos. 71101100, 71273036 and 71211130018, New Teachers' Fund for Doctor Stations, MOE (Ministry of Education) under Grant No. 20110181120047, Excellent Youth fund of Sichuan University under Grant No. 2013SCU04A08, China Postdoctoral Science Foundation under Grant Nos. 2011M500418 and 2012T50148, Frontier and Cross-innovation Foundation of Sichuan University under Grant No. skqy201352, Soft Science Foundation of Sichuan Province under Grant No. 2013ZR0016, Youth Project of Humanities and Social Sciences, MOE under Grant No. 13YJC630249, Scientific Research Starting Foundation for Young Teachers of Sichuan University under Grant No. 2012SCU11013.

References

1. Bhattacharya CB (1998) When customers are members: customer retention in paid membership contexts. J Acad Mark Sci 26(1):31–44
2. Burez J, Van den Poel D (2009) Handling class imbalance in customer churn prediction. Expert Syst Appl 36(3):4626–4636
3. Kim SY, Jung TS et al (2006) Customer segmentation and strategy development based on customer lifetime value: a case study. Expert Syst Appl 31(1):101–107
4. Verbeke W, Martens D et al (2011) Building comprehensible customer churn prediction models with advanced rule induction techniques. Expert Syst Appl 38(3):2354–2364
5. Crone SF, Lessmann S, Stahlbock R (2006) The impact of preprocessing on data mining: an evaluation of classifier sensitivity in direct marketing. Eur J Oper Res 173(3):781–800
6. Van Hulse J, Khoshgoftaar T (2009) Knowledge discovery from imbalanced and noisy data. Data Knowl Eng 68(12):1513–1542
7. Lemmens A, Croux C (2006) Bagging and boosting classification trees to predict churn. J Mark Res 43:276–286
8. Ivakhnenko AG (1976) The group method of data handling in prediction problems. Sov Autom Control 9(6):21–30
9. Xiao J, He C, Jiang X (2009) Structure identification of bayesian classifiers based on gmdh. Knowl -Based Syst 22(6):461–470
10. Xiao J, He C et al (2010) A dynamic classifier ensemble selection approach for noise data. Inf Sci 180(18):3402–3421
11. Ko AH, Sabourin R, Britto AS Jr (2008) From dynamic classifier selection to dynamic ensemble selection. Pattern Recognit 41(5):1718–1731
12. Woods K, Kegelmeyer WP Jr, Bowyer K (1997) Combination of multiple classifiers using local accuracy estimates. IEEE Trans Pattern Anal Mach Intell 19(4):405–410

13. Xiao J, He C (2009) Dynamic classifier ensemble selection based on gmdh. computational sciences and optimization, 2009. cso 2009. IEEE Int Joint Conf 1:731–734
14. Zhou ZH, Wu J, Tang W (2002) Ensembling neural networks: many could be better than all. Artif Intell 137(1):239–263
15. El-Melegy MT, Ahmed SM (2007) Neural networks in multiple classifier systems for remote-sensing image classification In Soft Computing in Image Processing. Springer, Berlin
16. Zou P, Hao YY, Li YJ (2010) Customer value segmentation based on cost-sensitive learning support vector machine. Int J Serv Technol Manage 14(1):126–137
17. Coussement K, Van den Poel D (2008) Churn prediction in subscription services: an application of support vector machines while comparing two parameter-selection techniques. Expert Syst Appl 34(1):313–327

Chapter 73
Synergy Effect of Knowledge Network and Its Self-Organization

Yue Wu, Xin Gu and Tao Wang

Abstract Knowledge network is formed by interchain coupling of knowledge chains, there is a non-linear structural link formed among the knowledge chains. The set of synergy effect of knowledge network is a complex system that stems from its self-organization. The relationship between network topology entropy and structure of knowledge networks was studied in this paper, which derived that the topology entropy of such a complex network is between $\frac{1}{2}\ln 4(n-1) \sim \ln n$. Different types of knowledge networks have different entropy distribution, but all of them follow power law distribution.

Keywords Knowledge network · Synergy effect · Self-organization · Entropy

73.1 Introduction

The national innovation systems, which is complex and large systems, is not a closed subsystem, but a open system. In national innovation systems, different components can exchange information, technology, knowledge, energy, resources with each other. Any innovative organization cannot develop alone, it depends on other innovation system. In the increasingly complex national innovation large systems, any innovative organization is not a closed subsystem, but a open system that continuously exchanging information, technology, knowledge, energy, resources with national

Y. Wu · X. Gu (✉) · T. Wang
Busniss School, Sichuan University, Chengdu 610064, People's Republic of China
e-mail: guxin@scu.edu.cn

X. Gu
Institute for Soft Science, Sichuan University, Chengdu 610064, People's Republic of China

T. Wang
Institute for Innovation and Entrepreneurial Management, Sichuan University,
Chengdu 610064, People's Republic of China

J. Xu et al. (eds.), *Proceedings of the Eighth International Conference on Management Science and Engineering Management*, Advances in Intelligent Systems and Computing 281, DOI: 10.1007/978-3-642-55122-2_73, © Springer-Verlag Berlin Heidelberg 2014

innovation system with, and any innovative organization cannot develop individually independent of a particular innovation system [3]. Knowledge network is an important operational mode of collaborative innovation, it is such an innovation network system composed of businesses, universities and research institutions by means of contractual relations, cooperation networks and social relations. Owing to the in-depth cooperation and integration of resources from knowledge creators and innovative subjects, a nonlinear interaction '1 + 1 + 1 > 3' is formed.

'Synergy effect' comes from the synergy theory (Synergetics) which describes a synergy between elements and system, system and system, system and environment, even among the various elements. The whole system is connected via cooperation, synchronization, coordination and complementarity, it refers to the nonlinear interaction among various elements within the complex system, when external control parameter reaches a certain threshold value, interaction and interrelation among various elements occupy a leading position by replacing their relative independence and mutual competition, and thus demonstrates coordination or cooperation, the overall effect was enhanced, and the system goes from a disordered state to an ordered state, e.g. 'synergy generates the order'. For knowledge networks, synergy effect is just such an overall effect arising from the synergies among various innovative subjects, which mainly manifests collaborative share of resources such as information, knowledge, technology, talent and financial to reduce the cost of innovation, and to share the innovation risk and improve the innovation performance, so that the overall innovation performance is greater than simple sum of independent innovation performances, to achieve real '1 + 1 + 1 > 3' at the level of inter-organization under the influence of synergy effect. As professor R. Moss Kanter from Harvard University pointed out: the only reason for existence of diversified organizations is to obtain synergy effect.

73.2 Reviews and Proposition

Synergetic innovation is the result from integration of different organizational knowledge and technology, and inter-organizational relationships will affect innovative interaction and complementarity effects [6], it is also related to a full integration of knowledge, resources, behavior and performance [10].

In the pattern type and its construction, from a study on Spain's manufacturing companies, Nieto and Santamaria [8] found that the establishment of technological innovation network is the key to achieve product innovation, collaboration with suppliers, customers and research institutions is helpful to innovation, yet collaboration with competitors is not conducive to innovation.

In relations between the structure and the subjects, the structure evolution of participants with the network determines structure and performance of collaboration network, and the enterprise-scale, the new production value brought about by new customers and their partners further explaining the structure evolution of innovative subjects in the context of networks [5]. Eschenbaecher and Graser [4] also built up a structural relation management and optimization model. The ability of members in the innovation network to related to the ability of entire innovation network, the cultural constitution and organizational structure are also prominent issues, successful implementation of synergetic innovation is inseparable from a interactive learning in the knowledge network [7]. Using an agent-based model SKIN (simulation dynamic knowledge innovation network), Petra et al [1] made an analysis on the links between universities and enterprises, and stated that general collaboration improves the overall knowledge innovation capacity of the university, and increases the company's knowledge of species, and also promotes the knowledge diffusion of innovation network.

In the outputs of knowledge network, Tsai [11] has studied the relationship between the absorptive capacity on technology or knowledge of collaborative members and innovation performance. The results showed as follow: first, absorptive capacity positively moderates the impact of vertical collaboration on the performance of technologically new or improved products; second, the effect of absorptive capacity on the relationship between supplier collaboration and the performance of new products with marginal changes varies based on firm size and industry type; third, absorptive capacity negatively affects the relationship between customer collaboration and the performance of marginally changed products; fourth, absorptive capacity positively affects the relationship between competitor collaboration and the performance of new products with marginal changes for large firms; fifth, absorptive capacity negatively affects the relationship between collaboration with research organizations and the performance of technologically new or improved products.

What we state, knowledge network is formed by interchain coupling of knowledge chains, in this case, there is a non-linear structural link formed between the knowledge chains, including coordination, conflict links, and finally promotes the formation of knowledge networks with functions as knowledge sharing and knowledge creation, as shown in Fig. 73.1. In this structure, knowledge is the core resource, a transferable production factor, and the source of sustainable competitive advantage of business entities. Core activities of knowledge network include the acquisition of knowledge, transfer, exchange, learning, creation and application etc., the performance of the knowledge vertex forms, can either be business organizations, or universities, research institutes, service agencies and other organizations.

73.3 Formation of Synergy Effect for KN: Self-Organization

Thermodynamic theory points out that only an open system can have development potential, and can organize itself to a more orderly state of development. Most systems in real world are open, the corresponding networks are open too, which constitutes a large and complex network by constantly adding new elements to the system, the increase of system elements is throughout the system lifecycle. That means, complex network system continues to grow by obtaining new vertex, and is constantly developing [9].

particle of the swarm is denoted by $P_i = (P_{i1}, P_{i2}, \cdots, P_{iD})$, and the global best position (gbest) is denoted by $P_g = (P_{g1}, P_{g2}, \cdots, P_{gD})$. At each iteration, the velocity of particle and its new position will be updated according to the following equations,

$$V_{id}(t+1) = wV_{id}(t) + c_1 r_{1i,d}(t)(P_{id}(t) - X_{id}(t)) + c_2 r_{2i,d}(t)(P_{gd}(t) - X_{id}(t)),$$
(74.2)

$$X_{id}(t+1) = X_{id}(t) + V_{id}(t+1),$$
(74.3)

where $i = 1, 2, \cdots, n$, $d = 1, 2, \cdots, D$, w is the inertia weight, c_1 and c_2 are cognitive and social scaling parameters respectively, $r_{1i,d}$ and $r_{2i,d}$ are uniformly distributed random numbers in the interval $(0, 1)$.

The binary particle swarm optimization algorithm (BPSO) was proposed by Kennedy and Eberhart [19], which used the concept of velocity as a probability that a bit takes on one or zero. In the BPSO system, the Eq. (74.2) of updating particle's velocity keeps the same, but the Eq. (74.3) of updating particle's position is replaced by the following rule:

$$X_{id}(t + 1) = \begin{cases} 0, & \text{if } rand() \geq S(v_{id}(t + 1)) \\ 1, & \text{if } rand() \leq S(v_{id}(t + 1)), \end{cases}$$
(74.4)

where $rand()$ is a quasi-random number selected from a uniform distribution in $(0, 1)$, and $S(\cdot)$ is a sigmoid function for transforming the velocity to the probability constrained to the interval $(0, 1)$ which is defined as follows: $S(v_{id}(t + 1)) = \frac{1}{1+e^{-v_{id}(t+1)}}$.

74.3 An Improved Particle Swarm Optimization Algorithm

The PSO algorithm suffers a critical weakness that a solution for an optimization problem converged too early which resulting in being suboptimal. In order to prevent premature convergence, based on the PSO, we proposes an improved PSO algorithm in this section. Firstly, we consider some modifications on the global best position (gbest).

1. Substitute mbest for gbest where mbest is described as the mean of the gbest positions during the past iteration period, i.e.,

$$\text{mbest} = \frac{1}{n} \sum_{i=1}^{n} g_i(t).$$
(74.5)

2. Do mutation operation on gbest. The process is as follows:

$$\text{if } r_0 \leq P_0$$
$$\text{for } k = 1 \text{ to } D_0$$
$$d = [r_{k1} * D]; \ P_{gd} = L + r_{k2} * (U - L); \qquad (74.6)$$

end

end,

where r_0, r_{k1} and r_{k2} are random numbers in the interval $(0,1)$, P_0 and D_0 are mutation probability and mutation dimension respectively, $[\cdot]$ rounds the element to the nearest integers greater than or equal to the element, and U and L are the maximal and the minimal value of P_{gd} respectively.

The IPSO's procedure flow is as follows:
Step 1 Initialize all particles, and let iteration count $t = 0$;
Step 2 Compute the fitness values of current particles;
Step 3 Replace the current position with pbest once particle's current fitness value is worse than its pbest value;
Step 4 Replace the current particle's position with gbest if the current fitness value is worse than the swarm's gbest value;
Step 5 Renew gbest by Eqs. (74.5) and (74.6);
Step 6 Renew iteration velocity and position of each particle by Eqs. (74.2) and (74.3), and let $t = t + 1$;
Step 7 Turn to Step 2 when t is less than the maximal iteration count, unless end the process.

In the IPSO system, let $c_1 = c_2 = 2$, then we have: $w(t) = w_{min} + \frac{T-t}{T-1}(w_{max} - w_{min})$, where T and t are maximal and present iteration counts respectively, and w_{max} and w_{min} are maximal and minimal values of inertia weight, respectively. We set $w_{max} = 1$ and $w_{min} = 0.4$. Then let $D_0 = 5$ and P_0 be as follows: $P_0 = P_{min} + \frac{T-t}{T-1}(P_{max} - P_{min})$, where P_{max} and P_{min} are maximal and minimal values of P_0, respectively.

74.4 Numerical Experiments and Comparison Analyses

In this section, we apply the IPSO to estimate the volatility of a European call option. All data in numerical experiments are collected from [20]. For finding better estimation of the volatility in Black-Scholes model, we expect that the approximate price of the call option is the same as its actual option price. Thus, to minimize the difference between estimated and actual call option value, we choose the sum of absolute deviation as a fitness function. In our experiments, a 20 dimensional binary code is used in both the BPSO1 and the BPSO2, a 1 dimensional real value code is considered in the IPSO. Four different scenarios with different number of particle

Table 74.1 Minimum fitness values of three algorithms

Scenario	n	T	BPSO1	BPSO2	IPSO
S1	10	50	0.1719	0.1424	4.6674E-004
S2	10	100	0.0887	0.0778	1.8849E-005
S3	20	50	0.0564	0.0987	1.3658E-004
S4	20	100	0.0362	0.0326	5.2370E-006

Table 74.2 Fitness comparison between BPSO2 and IPSO

Scenario	Algorithm	Min	Max	Mean	Std.Dev.
S1	BPSO2	0.0019	0.3821	0.1424	0.0149
	IPSO	4.1339E-007	0.0025	4.6674E-004	2.9676E-007
S2	BPSO2	0.0058	0.2446	0.0778	0.0035
	IPSO	4.5125E-007	8.2782E-005	1.8849E-005	7.0702E-010
S3	BPSO2	0.0058	0.3438	0.0987	0.0080
	IPSO	1.3528E-006	3.5266E-004	1.3658E-004	1.0450E-008
S4	BPSO2	0.0048	0.1195	0.0326	0.0011
	IPSO	3.0509E-007	2.2859E-005	5.2370E-006	3.4502E-011

swarm n and maximal iteration count T are considered. All tests are repeated twenty times and relative results are showed as average values.

In Table 74.1, minimum fitness values of three algorithms under different scenarios are listed. From Table 74.1, it is clear that IPSO has a better results and shows a stronger searching ability than other two algorithms in all scenarios. The main reason is that IPSO does not only concern about the information of global optimal position during iteration period, but also uses mutation operation to explore and exploit the particle's search ability. The better performance of BPSO2 than that of BPSO1 should attribute to the bit change mutation.

Further, fitness comparison between the BPSO2 and the IPSO are given in Table 74.2. It is clear that both the BPSO2 and the IPSO obtain best fitness values when $n = 20$ and $T = 100$. Compared with the BPSO2, the IPSO shows a better computational stability and need less iterations. Note that the fitness value is related to the difference between estimate and actual option value. Thus, the IPSO have a better estimation for call option than other algorithms. Moreover, the comparison results between the BPSO2 and the IPSO are shown in Fig. 74.1. From Fig. 74.1, we have that the IPSO could escape the local optimum effectively. Moreover, the IPSO converges faster and need less iterations than the BPSO2. Further, the approximate solutions of the BPSO2 and the IPSO are shown in Fig. 74.2. From Fig. 74.2, we also have that the IPSO converges much faster than the BPSO2. Table 74.3 shows the estimation of volatility from different algorithms. From Table 74.3, it is clear that the IPSO outperforms both the BPSO1 and the BPSO2 on the premise that the difference between estimated and actual call option value is minimized.

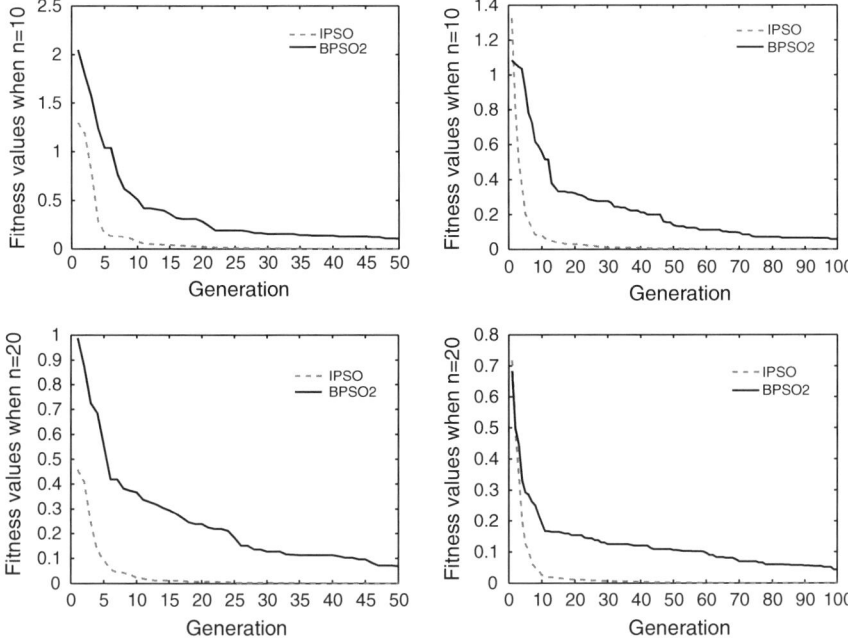

Fig. 74.1   Fitness values of the BPSO2 and the IPSO respectively

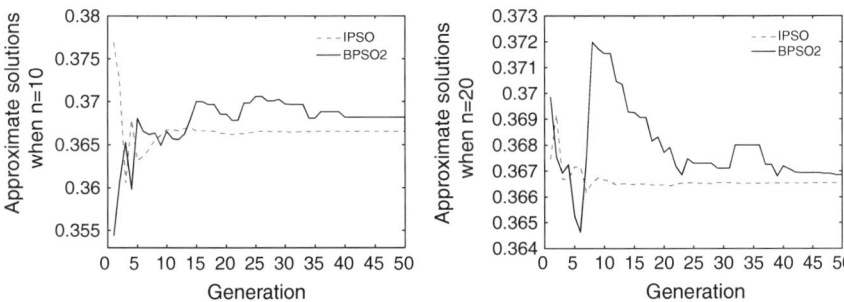

Fig. 74.2   Approximate solutions of the BPSO2 and the IPSO respectively

Table 74.3 Volatility estimate values of three algorithms	Scenario	BPSO1-σ	BPSO2-σ	IPSO-σ
	S1	0.3638	0.3673	0.3665
	S2	0.3652	0.3663	0.3665
	S3	0.3658	0.3671	0.3665
	S4	0.3666	0.3667	0.3665

74.5 Conclusions

In this paper, we propose an improved particle swarm optimization algorithm (IPSO), which is used to find volatility estimate value for option pricing. After comparing with both the BPSO1 and the BPSO2, the IPSO displays simpler implementation and faster convergence on global optimal value. Moreover, the IPSO could avoid premature convergence more effectively and gets more accurate solution in numerical experiments with fewer fluctuation. Thus, the IPSO could find more reliable solution of fitness function than binary particle swarm optimization algorithms which implies the IPSO is a more suitable algorithm for volatility estimation in option pricing.

Acknowledgments This work was supported by the National Natural Science Foundation of China (70831005, 11171237, 11301359, 71101099) and by the Construction Foundation of Southwest University for Nationalities for the subject of Applied Economics (2011XWD-S0202).

References

1. Black FS, Scholes MS (1973) The pricing of options and corporate liabilities. J Polit Econ 81:637–654
2. Merton RC (1973) Theory of rational option pricing. Bell J Econ Manag Sci 4:141–183
3. Hofmann N, Platen E, Schweizer M (1992) Option pricing under incompleteness and stochastic volatility. Math Finance 2:153–187
4. Avellaneda M, Paras ALA (1995) Pricing and hedging derivative securities in markets with uncertain volatilities. Appl Math Finance 1:73–88
5. Schweizer M (1995) Variance-optimal hedging in discrete time. Mathematics of Operations Research 20:1–32
6. Dindar ZA, Marwala T (2004) Option pricing using a committee of neural networks and optimized networks. In: Proceedings of IEEE international conference on syetem, man and cybernetics, pp 434–438
7. Kumar S, Thulasiram RK, Thulasiraman P (2008) A bioinspired algorithm to price options. In Proceedings of conference of computer science and software engineering, pp 11–22
8. Sharma B, Thulasiram R, Thulasiraman P (2013) Normalized particle swarm optimization for complex chooser option pricing on graphics processing unit. J Supercomputing 66:170–192
9. Manaster S, Koehler G (1982) The calculation of implied variances from the black-scholes model: a note. J Finance 37:227–230
10. Bruce K (2000) Black-scholes option pricing via genetic algorithms. Appl Econ Lett 7:129–132
11. Kennedy J, Eberhart R (1995) Particle swarm optimization. In: Proceedings of IEEE international conference on neural networks, pp 1942–1948
12. Liu Y, Zhao Q, Niu B (2011) Constrain particle swarm optimizer for solving self-fiancing portfolio model. Math Pract Theor 41:78–84
13. Deng W, Chen R et al (2012) A novel two-stage hybrid swarm intelligence optimization algorithm and application. Soft Comput 16:1707–1722
14. He G, Huang N (2012) A modified particle swarm optimization algorithm with applications. Appl Math Comput 219:1053–1060
15. Lee S, Lee J et al (2007) Binary particle swarm optimization for black-scholes option pricing. Lect Notes Artif Intell 4692:85–92
16. Zhao X, Sun J, Xu W (2010) Applicaion of quantum-behaved particle swarm optimization in parameter estimation of option pricing. In: Proceedings of international symposium on distributed computing and applications to business, engineering and science, pp 10–12

17. Mo Y, Liu F, Ma Y (2013), Application of gso algorithm to the parameter estimation of option pricing model. In: Proceedings of the 9th international conference on intelligent, computing, pp 199–206
18. He G, Lu J (2013) An improved multi-objective particle swarm optimization algorithm with an application. Trans Math Prog Appl 1(9):31–42
19. Kennedy J, Eberhart R (1997) A discrete binary version of the particle swarm algorithm. In: Proceedings of systems, man and cybernetics, IEEE international conference on, pp 4104–4108
20. Stampfli J, Goodman V (2003) The mathematics of finance: modeling and hedging. China Machine Press, Beijing

Chapter 75
How Venture Capital Institutions Affect the Structure of Startups' Board of Directors

Yingkai Tang, Maomin Wu, Qilin Cao and Jing Zhou

Abstract Venture Capital institutions is different from the general investors, they provide enterprises with not only funds but also improvements on the aspect of corporate. This paper took 285 companies listed before December 31, 2011 as samples; mainly studies VCI affect the board independence and professional structure. Results showed that the VCs can significantly improve the professionalism and independence of the board of directors. Research provides new empirical evidence for the inner mechanism of the venture capital to improve startups performance.

Keywords Venture capital · Startup enterprise · Structure of board of director

75.1 Introduction

China's Venture Capital (VCs) started from the end of last century, and developed rapidly after the setting of small and medium-sized enterprises board in 2004 and GEM (Growth Enterprises Market) in 2009. The profound influence it had on the development of Chinese enterprises is paid great attention by academia. China's earlier researches focuses on the positive role it played in the improvement of startup enterprises' operating performance, including, for example, improvement of startup enterprises' listed value by VCs [11], the effects of VCs for enterprise long-term performance on the aspect of time to market, issue expenses, short-term excess share price after listing, long-term excess share return, etc [16], VCs effects for startup enterprises' IPO premium [11]. Other researches studied VCs effects for enterprises

Y. Tang · M. Wu · Q. Cao
Business School, Sichuan University, Chengdu 610064, People's Republic of China

J. Zhou (✉)
International Business School, Southwestern University of Finance and Economics, Chengdu 611130, People's Republic of China
e-mail: jing.zhou@swufe.edu.cn

J. Xu et al. (eds.), *Proceedings of the Eighth International Conference on Management Science and Engineering Management*, Advances in Intelligent Systems and Computing 281, DOI: 10.1007/978-3-642-55122-2_75, © Springer-Verlag Berlin Heidelberg 2014

operating performance. Recently, scholars began to study the effects of VCs for startups' management behavior, for example, VCs reducing earnings management of startups [2], VCs effects for investment and financing behavior of listed enterprises [20], he thinks that VCs can not only inhibit the excessive investment of free cash flow, but also increase the company's short-term interest-bearing debt financing and external equity financing. He also agrees on VCs function of easing inadequate investment problem for lack of cash flow to some extent. Another analytical perspective of this field was the mechanism of VCs effects for startups' management behavior and operating performance. A representative example was the empirical research of VCs role in making startups governance structure more reasonable [22].

To sum up, China's research of VCs effects for startups went from market performance to concrete behavior and to inner mechanism finally, while study for the last step was not so convincing relatively. This paper is an empirical research of VCs effects for startups' governance structure in the perspective of professionalism and independence. It aims to offer empirical evidence for the inner mechanism of VCs improving startups' performance.

75.2 Review

Venture capital institution's (VCI) participation in corporate governance is not rare around the world, and it arouses the attention of many scholars. By analyzing the USA data, it is proved that VCI participation can significantly improve the corporate governance situation 1986 [18]. It is also believed that VCI is the only way to establish governance system for newly setup companies [8]. Among all the VCI in Europe, 66 % was entitled as the board of directors, and in USA, where VCs is the most developed, 41.4 % of board of directors was venture capitalists and a quarter of the enterprises were controlled by VCI [14].

What we can learn from international experiences is that by affecting the structure of the board of directors, VCI affect enterprise in two ways: the first is by way of regular inspection, guidance, etc in the role of shareholders; the second is to change the structure directly by sending directors. Starting from the two points, domestic scholars did a series of researches about the relation between VCs and corporate governance. In theories, most of the scholars agrees on the idea that for the advantages of professional management and human resource, VCI participation can greatly improve the corporate governance situation; whereas, for the different investment environments in different countries VCI facing with as well as the different strength of participation, scholars reached to various of results by using different data.

In 1988, by analyzing limited data, Rosenstein creatively studied the VCI effects for the board of directors. He found that the board of directors participated with VCI are more independent and professional. In 2000, Hellman collected the related data of 170 startups companies of Silicon Valley and studied the VCI effects for startups. He drew the following two results about the relation between VCI and corporate governance: first, participation of VCI can affect the personnel selection system and make the enterprise to hire more professional staff and independent CEO; second, enterprise participated with VCI have a better incentive system. The enterprise and the managers are combined closely through the form of stock option, which will reduce the moral hazard.

In 2002, some scholars widen the sample and drew the same result [1]. Besides, Bouresli also took the VCI effects for ownership concentration into consideration, but it is proved that these two variable has no obvious relation. In 2003, based on the previous study, Baker and Gompers found that different VCI can improve the independence of the board of directors in different degree; and the better reputation, the greater improvement.

In 2009, by analyzing the data of small and medium-sized listed enterprises in China, Liu Guofeng drew a completely different conclusion. In his paper, there was no obvious difference in percentage of independent directors, times of board meeting or executive compensation between the enterprises with and without VCI. At the same time, in the perspective of mean value, governance level of enterprise participated with VCI is lower than those without VCI. Thus, the author thinks that VCI participation in enterprise was insufficient. In 2010, by analyzing the same data but different governance index, Jin Ming and Wang Juan drew the same result. These two authors also think that the main reason for insufficient participation was not only the weak consciousness but also the underdeveloped financial system as well as the culture of small and medium-sized enterprises.

75.3 Theoretical Framework and Research Analysis

Separation of ownership and management leads to two agency problems: first, principal-agent problem resulting from inconsistent interests between shareholders and managers; second, major shareholders occupying the interests of minority shareholders by using their control power. To protect the interests of all investors, modern enterprises supervise the behavior of managers and shareholders by introducing the board mechanism [5]. More studies show that the board of directors abandon their role of being supervisor gradually; and become the instrument for major shareholders and managers to deprive the interests of external investors. Thus, Optimization of the board mechanism becomes the key of corporate governance.

Traditional financing institution, such as banks, emphasis on safe and steady operation, whereas, VCI, which is distinguished for its high risk and high return, commits to invest in risky areas like high technology enterprise and new company. Serious information asymmetry in these enterprises make the investors face greater risks. Limited by the high supervise cost, minority shareholders, who wants to be "free rider", dare not to compete with managers and major shareholders. VCI like investing in large scale and long term project, so they are more willing to supervise the enterprises. This is the original reason for VCI participation in corporate governance. The idea of regulatory efficiency is that efficiency and cost were positively correlated. By playing the role of supervisor, VCI participation makes it easier to find problems. Besides, most staffs in VCI, which is a professional institutional investor, are knowledgeable in economy, finance, law and related areas, are more professional in supervising. Based on the above analysis, we believe that VCI can improve corporate governance structure; while how VCI participation affects corporate governance is closely related with shareholding ratio. It is generally accepted that the bigger ratio, the stronger incentive of VCI to participate and the greater improvement. In addition, improvement of corporate structure by VCI participation is related to the number of investment institution. During the actual operation, most of co-investment of from VCI. So we think that joint investment is conductive to improve corporate governance structure. This result can not be reached without supervision; so it is reasonable that improvement of corporate governance structure mainly resulted from VCI role of being the supervisor.

The board of directors has two main functions: first, supervision, i.e. replacing shareholders to supervise managers [5]; second, giving opinions for strategy making, which can reduce cost and enhance value. Therefor, the board structure is improved in two ways: enhance of independence for supervision and professionalism for service.

Many scholars hold that enhance of board independence was directly reflected by the proportion of independent directors. The bigger proportion, the greater independence [7, 19]. The independent board system appeared lately in China, enterprises did not pay enough attention to it. Most of the Independent directors are set to meet the requirement of CSRC, not to enhance independence. What's worse, many of the independent directors serve for several other companies and care little about the invesed enterprise. For all of these reasons, the present proportion of independent directors in China can not reflect the real situation of the board of directors. There is another saying that board independence can not be reflected by the proportion, the rights to contend managers instead. If the board is controlled by managers, its supervision function will disappear. Chairman and CEO being the same person is an important point of judging if the board is controlled by the managers [6, 9]. This paper holds that the above point can only reflect the independence of chairman, not the whole board. To measure the independence of the board, this paper considers the proportion of external directors[1] as a main index, and the bigger proportion, the greater Independence.

To make the supervision function of board into full play, maintaining of professionalism is as important as that of independence [7, 17]. There are two ways to understand professionalism: first, it is specialized knowledge about the field the enterprise is in; we define this the first kind of professional director; second, the background knowledge for economy, finance, law, account, etc; and this is the second kind of professional director. For high-tech enterprises or newly setup companies, they focus on product development and market development. Technician takes a big proportion of the whole staff, while professional managers is insufficient. This makes the internal structure a chaos. VCI participation is a great way to solve this problem.

---

[1] External director in this paper refer to the directors who do not take position in the enterprise or the parent company.

Therefor, we think that VCI can improve board professionalism by increasing the proportion of the second kind of professional director. While increasing of the first kind of professional director is not obvious in short term.

In addition to analysis for relation between VCI participation and board governance structure, this paper did a further research about if different kind of VCI (state-owned or not) affects board structure in the same way.

For the parallelized economy orientation and policy orientation, it is generally accepted that the efficiency of state-owned enterprise is relatively low.

75.4 Research Design

Based on the above analysis, this paper define the proportion of external directors and the second kind of professional directors as explained variable. By reading the previous document, we learned that there are many key factors affecting board independence and professionalism, for instance, enterprise performance, ownership structure, corporate scale and other variables which can be seen as the reflection of internal structure like CEO and chairman being the same person as well as board scale [1, 4, 12]. Thus, we take the related index as control variable in the econometric-mode. To study VCI effects for board governance structure, we define the following variables as indicators of risky-investment bearing: VCI participation, VCI share-holding ratio, holding time of VCI and property of VCI (see Table 75.1).

75.5 Data Sources and Descriptive Statistics

The author of this paper collected the prospectus of 285 enterprises listed before December 31, 2011 and establish a database for GEM enterprise board governance. In terms of the information of VC, it mainly came from Zdatabase, except for these that supplemented by prospectus. The prospectus were downloaded on the Shenzhen stock exchange website. The softwares used in this paper are eviews 3 and SPSS 17.0.

Table 75.2 is the information about the 285 enterprises. We can learn that average asset value of the 285 enterprises 1 year before listing is 360.1323 million $. The standard deviation is 32503.41, which means that asset size varies widely between different enterprises. The maximum of shareholding ratio of managers is as high as 100 and both mean and median are over 50 %, which means that managers are not so willing to seek personal gain at the expense of damaging corporate interest. Seen from the perspective of board scale, mean scale of GEM enterprise is 8.42; among these, only 5 of the 17 directors are independent.[2] This proportion is the minimum

---

[2] This data is not included in Table 75.3, it comes from the original data of this paper.

Table 75.1 The related index

Type	Name	Description
Explained variable	OUTDIR PROF	Proportion of external director
		The proportion of the second kind of professional director
Explaining variable	$D1$	VCs participation or not. Dummy variable. If so, $D1 = 1$, otherwise, $D1 = 0$
	VCSHARE	Shareholding ratio of VCI
	VCTIME	Holding time of VCI (the longest holding time of VC before the prospectus date)
	CHA	Property of VCI, for state-owned enterprise, CHA $= 1$, otherwise, CHA $= 0$
Control variable	FIRST	Shareholding ratio of dominant shareholder
	ROE	Return on equity
	ASSET	Asset size (take its logarithm)
	MSHARE	Shareholding of managers (sum of all the top management's)
	BSIZE	Board scale (number of the board of directors)
	CEOCHA	CEO and chairman being the same person or no, if so, CEOCHA $= 1$, otherwise, CEOCHA $= 0$

Table 75.2 Descriptive statistics

	Mean	Median	Max	Min	Sd
ASSET	36013.23	27437.59	321932.36	1942.58	32503.41
BSIZE	8.42	9	13	5	1.438
CEOCHA	0.54	1	1	0	0.499
FIRST	48.7077	47.91	96.27	11.7	18.37107
MSHARE	51.32	54.73	100	0	28.719
ROE	30.3214	27.85	81.79	5.21	12.96624
OUTDIR	65.887	66.67	100	22.22	12.22634
PROF	41.2166	40	100	11.11	17.8439

requirement of Corporate Law. The mean proportion of external directors is 65.89 %, and that of professional directors is 41.22 %.

Since this paper aims to study VCI participation effects for board governance structure, we classified the 285 enterprises into two samples according to whether they were joined with VCI. Then we compute the statistics respectively and do T test analysis in order to see if the average of the two sets of data has significant difference. The results are showed in Table 75.3. The table says four points. First, board scales is roughly the same between enterprises participated with VCI and the without, while the professional director ratio of the former is bigger than the latter at 1 % level. In addition, external director ratio of the former is significantly bigger than the latter

Table 75.3 Descriptive statistics of subsample indicators

	Samle ($N = 285$)		WithVC ($N = 182$)		Without VC ($N = 103$)		t statistics (Prob)
	Mean	Sd	Mean	Sd	Mean	Sd	
ASSET	36013.23	32503.41	38240.08	34923.52	32078.41	27436.84	1.541(0.124)
BSIZE	8.42	1.438	8.51	1.36	8.25	1.56	1.461(0.145)
CEOCHA	0.54	0.499	0.59	0.493	0.45	0.5	2.405(0.017 **)
FIRST	48.7077	18.37107	50.1	18	46.27	18.85	1.700(0.09*)
MSHARE	51.32	28.719	53.72	27.52	46.06	30.01	2.183(0.030 **)
ROE	30.3214	12.96624	30.06	13.27	30.78	12.47	−0.448(0.655)
OUTDIR	65.887	12.22634	66.89	12.2	64.13	12.13	0.413(0.068*)
PROF	41.2166	17.8439	42.74	18.65	35.76	15.42	3.227(0.001 ***)

Notes *, **, *** means it is significant at 10, 5, and 1 % level respectively. It is the same in the rest part of this paper

Table 75.4 Indicator of VCI participation

Indicator	Mean	Median	Max	Min	Sd
VCNUM	2.4	2	12	1	1.57
VCSHARE	22.14	15	43.2	0.34	21.16
VCTIME	27.33	22	125	4	22.28

at 10 % level. Second, the possibility of CEO and chairman being the same person of the former is significantly bigger than the latter at 5 % level. Third, shareholding ratio of controlling stockholders of the former is significantly bigger than the latter at 10 % level, and that of managers is at 5 % level. Fourth, asset scale, ROE, board scale and other indicators has no significant difference between the two sets of data.

To study VCI effects for enterprise investment, we take the 182 enterprises joined with VCI as sample and have the statistics showed in Table 75.4. Fromuyu Table 75.3, we can learn that among the 182 enterprises, each enterprise is joined with 2.4 VCI in average; and the median 2 is smaller than the mean, which results from the situation that some enterprises are joined with several VCI. For example, FPI Ltd. alone is joined with 12 VCI. In the perspective of shareholding ratio, the average shareholding ratio of VCI is 22.14 %. The average holding time of VCI is about 27 months; but this data is overestimated for we take the longest holding time of VCI as our data. Thus, we just consider it as descriptive statistics, but did not put it into the regression model.

75.6 Empirical Analysis

To test VCI participation effects for board governance, this paper modelling the variables in Table 75.1. After correlation analysis, we found that three indicators (joined with VCI or not, shareholding ratio of VCI and number of VCI) about VCI

Table 75.5 Regression results of the relation between VC and board independence and professionalism

	Model 1 OUTDIR			Model 2 PROF		
C	58.306	56.14	59.644	2.973	4.124	7.083
	(0.000 ***)	(0.000 ***)	(0.000 ***)	(0.051*)	(0.001 ***)	(0.000 ***)
ASSET	2.677	3.092	2.665	5.094	5.383	4.881
	(0.000 ***)	(0.000 ***)	(0.000 ***)	(0.000 ***)	(0.000 **)	(0.000 ***)
ROE	−0.094	−0.0852	−0.069	−0.158	−0.17	−0.129
	(0.000 ***)	(0.000*)	−0.1013	(0.000 ***)	(0.000 ***)	(0.000 ***)
MSHARE	−0.076	−0.077	−0.076			
	(0.000 ***)	0.000 ***	(0.000 ***)			
BSIZE	−9.059	−8.255	−8.986	−1.403	−1.42	−1.666
	(0.000 ***)	(0.000 ***)	(0.000 ***)	(0.000 **)	(0.000 ***)	(0.000 ***)
CEOCHA	−1.314	−1.301	−1.418			
	(0.000 ***)	(0.000 ***)	(0.000 ***)			
$D1$	4.474			6.406		
	(0.000 ***)			(0.000 ***)		
VCSHARE		0.005			−0.004	
		(0.000 ***)			−0.3156	
VCNUM			1.091			1.861
			(0.000 ***)			(0.000 ***)
FIRST				−0.054	−0.035	−0.0404
				(0.000 ***)	(0.0001 ***)	(0.000 ***)

Notes values in the bracket refers to Prob; and there are no heteroscedasticity in the two models

participation are of serious multicollinearity, and can not be put into one regression model. So we decided to put them into different models since they are crucial to this paper.

To test VCI participation effects for board independence, we build the following regression model:

$$OUTDIR = \alpha_0 + \alpha_1 ASSET + \alpha_2 ROE + \alpha_4 MSHARE + \alpha_5 BSIZE + \alpha_6 CEOCHA + \alpha_7 VC + \varepsilon. \tag{75.1}$$

To test VCI participation effects for board professionalism, we build the model (75.2):

$$PROF = \alpha_0 + \alpha_1 ASSET + \alpha_2 ROE + \alpha_3 BSIZE + \alpha_4 FIRST + \alpha_5 VC + \varepsilon. \tag{75.2}$$

In the model, variable VC refers to variable related to VCI ($D1$, VCSHARE, VCNUM), the result is showed in Table 75.5.

If considering external director ratio as the standard for judging how independent the board is, the regression results say the following points: VCI participation and external director ratio is of significant positive correlation at 1 % level, which means that VCI participation is conductive to increase external director ratio. Improvement of independence by VCI participation and shareholding ratio as well as number of joined VCI is significantly correlated at 1 % level; the bigger ratio and the bigger number, the greater independence. In model (75.1), asset scale, the controlling variable and external director ratio are positively correlated and significant at 1 % level. This means that board independence of giant enterprise is greater than small company in China. Shareholding ratio of managers and external director ratio are negatively correlated and significant at 1 % level. There are two explanations: first, if shareholding ratio of managers increase, their interests are consistent with the enterprise and the company is relatively independent of external director supervision. Second, if shareholding ratio of managers increase, the managers will have more rights in the company and tend to hire less external directors in order to gain more personal interests. In addition, increasing of board scale will reduce external director ratio in some degree; and the board independence will decrease either. This result is consistent with the conclusion of researches done by some scholars [3, 15]. Whether CEO and chairman being the same person and external director ratio are negatively correlated at 1 % level, i.e. if CEO and chairman are the same person, the ratio will decrease and the independence decrease either. Corporate performance and external director ratio is negatively correlated at 1 % level. This result is partly consistent with the research done by some scholars [10, 13, 21]. There are also two explanations for this: first, poor performance of enterprises with directors of greater independence results from the "un-professional" independent director missing the best invest opportunity. Second, the poorer performance, the more likely to hire independent director; because hiring independent director during the appropriate accounting period can make the enterprise be more confident for market. Number of independent director is a passive response of corporate performance.

VCI effects for board professionalism is tested in model (75.2). The result is showed in Table 75.5. We can learn that VCI participation contributes to the increasing of the ratio of the second kind of director; and the bigger number, the faster rise. Those results are all significant at 1 % level. So we can confirm that the relation between VC and board professionalism is consistent with hypothesis 2 and 4. In addition, asset scale, corporate performance, board scale and shareholding ratio of controlling directors are all key factors to board professionalism.

75.7 Further Test

To test whether the property of VCI is consistent with improvement of board independence and professionalism, this paper takes 182 enterprises joined with VCI as sample and did further research. First of all, the 182 enterprises were classified into two subsamples according to if they are joined with state-owned VCI. Sample 1 is enterprises joined with state-owned VCI, and sample 2 is those without. We did independent sample T tests. The descriptive statistics are showed in Table 75.6.

Table 75.6 Descriptive statistics of related indicators

	Subsample 1 ($N = 24$)		Subsample 2 ($N = 158$)		TVALUE (Prob)
	Mean	Sd	Mean	Sd	
ASSET	32289.65	15984.33	39143.94	36911.13	−0.895(0.372)
ROE	29.63	16.58	30.13	12.76	−0.17(0.866)
FIRST	50.02	17.91	50.11	18.08	−0.022(0.982)
MSHARE	42.24	27.92	55.46	27.12	−2.217(0.028 **)
VCSHARE	11.4	7.79	23.77	22.06	0.002(0.007 ***)
VCNUM	1.75	0.94	2.49	1.63	−2.181(0.03 **)
VCTIME	25.88	20.97	27.83	22.37	−0.42(0.689)
PROF	42.49	20.87	42.79	18.36	−0.07(0.944)
OURDIR	66.3	14.56	66.97	11.85	−0.252(0.801)

**Significant at the 5 % level, ***at the 1 % level

Table 75.7 Empirical results of VCI of different property

	OUTDIR	PROF
C	54.66(0.000 ***)	0.82(0.962)
ASSET	3.12(0.002 ***)	5.63(0.001 ***)
ROE	−0.09(0.081*)	−0.16(0.045 **)
FIRST		−0.043(0.466)
MSHARE	−0.07(0.007 ***)	
BSIZE	−8.30(0.000 ***)	−1.37(0.075*)
CEOCHA	−1.20(0.011 **)	
CHA	0.008(0.997)	2.73(0.4645)

Notes Values in the bracket are Prob, and there is no heteroscedasticity

We can learn from Table 75.6 that only 24 of the 182 enterprises are joined with VCI. Seen from the perspective of the number of state-owned invest enterprise, VCI investment is not very positive; and seen from the property of invested enterprise, there is no obvious difference between the two samples. Though all the indicators of subsample 1 is smaller than subsample 2, only three of them (shareholding ratio of managers, shareholding of VC and number of VCI) are significant at 5, 1 and 5 % level respectively. Mean of subsample 2 is significantly smaller than subsample 1.

This paper also aims to study board independence and professionalism. Though means of the two subsamples are roughly the same and not significant, we still can not confirm that property of VCI has no effects for board independence and professionalism based on the univariate analysis alone. Therefor, we have variable VC in the previous model replaced by dummy variable CHA which symbolize the property of VCI, and the regression results are showed in Table 75.7.

After we add a variable symbolize property of VCI into the model, we can find that this variable did not significantly affect board independence and professionalism. In addition, this new variable reduces R2 of the model and make the variables significantly affect explained variable not so significant as before. We can judge from this point that there is no obvious difference between state-owned VCI investment and the rest VCI investment. So state-owned VCI has no low-efficincy problem as it is said.

75.8 Conclusion

Based on the collected information of 285 GEM enterprises listed before December 31, 2011, this paper studed the relation between VCI and board governance structure, especially board independence and professionalism. The regression results show that VCI participation improves GEM enterprises independence and professionalism and this kind of improvement is positively correlated with shareholding ratio of VCI. At the same time, this paper reflects that VCI usually co-investment to supervise the enterprise. This measure is conductive to improve board governance situation. The above empirical results prove that China VCI is roughly the same as VCI abroad, and they paly the paralleled roles of capital providers and strategy maker. This paper also shows that state-owned VCI and non-state VCI improve board independence and professionalism in the same degree by and large. The defect of this paper is that we did not take VCI affects for board efficiency into account, nor did we widen the governance mechanism effects to effects for cooperate performance. These two points will be further studied in the future.

Acknowledgments This research was supported by the National Natural Science Fund of China (71072066, 71302183) and the Distinguished Young Scholars Fund of Sichuan University (SKJC201007, 2013JCPT006, SKQY201225).

References

1. Bouresli AK, Davidson III, Abdulsalam FA (2002) Role of venture capitalists in IPO corporate governance and operating performance. Q J Bus Econ 41(3/4):71–82
2. Chen X (2010) Study of venture capital and earnings management of IPO companies. Res Financ Econ Issues 1:64–69
3. Chen Y, Wu Z (2008) Factors influencing board size and structure of listed companies. Securities Mark Herald 4:70–77
4. Dalton DR, Kesner IF (1987) Composition and CEO duality in boards of directors: An international perspective. J Int Bus 18(3):33–42
5. Fama EF, Jensen MC (1983) Separation of ownership and control. J Law Econ 2B:301–325
6. Gong CL, Yang C (2009) Equity structure, board property and external security of listed company. Friends Acc 6:84–89
7. Haung J, Pan M (2010) Effectiveness of independence director and related transactions of controlling shareholders. Theor Prac Financa Econ 163:47–51
8. Jeng LA, Wells PC (2000) The determinants of venture capital funding: evidence across countries. J Corp Finance 6(3): 241–289
9. Jensen MC, Meckling WH (1976) Theory of the firm: managerial behavior agency costs and ownership structure. J Fin Econ 3:306–360
10. Li C, Lai J (2004) Did property of board affects corporate performance. J Finance Res 5:64–77
11. Li K, Tang Y (2011) Can venture capital increase listed company's value. Reform Econ Syst 1:55–59
12. Mak YT, Yuan L (2001) Deteminants of corporate ownship and board structure: evidence from singapore. J Corp Finance 7:235–256
13. Shivdasani A, Yermack D (1999) CEO involvement in the selection of new board members: an empirical analysis. J Finance 54:1829–1854

14. Steven K, Strömberg P (2001) Financial contracting theory meets the real world:an empirical analysis of venture capital contracts. University of Chicago, Chicago
15. Song YX, Zhang Z (2008) Toward director compensation, board independence, and corporate governance empirical evidence from China liseted companies. Mod Econ Sci 2:95–128
16. Tan Y, Yang Y (2011) The effect of venture capitalist participation on long-term performance. Shanghai Econ Rev 5:72–96
17. Tang Q, Zhang D (2005) Which is more important, idependence and compensation, or knowledge and information. Contemp Econ Manag 6:32–35
18. Timmons JA, Bygrave WD (1986) Venture capital's role in financing innovation for economic growth. J Bus Ventur 1:161–176
19. Weisbach MS (1988) Outside directors and ceo turn over. J Financ Econ 20:431–460
20. Wu C, Wu S et al (2012) The role of venture capital in the investment and financing behavior of listed companies: Evidence from China. Econ Res J 2:105–160
21. Yu D, Wang H (2003) Independent directors and corporate governance theory and experience. Acc Res Study 8:8–14
22. Zhang X, li L (2011) VCs' backgrounds, IPO underpricing and post-IPO performance. Econ Res J 2:118–132

Chapter 76
Customer-Value Based Segmentation and Characteristics Analysis of Credit Card Revolver

Changzheng He, Mingzhu Zhang, Jianguo Zheng, Xiaoyu Li and Dongyue Du

Abstract The purpose of this study is trying to find high-value revolvers and analyze their demographic characteristics using credit card data collected from a real Chinese bank instead of a survey. Due to the unique character of credit card, we develop RFM model to establish a new model, called RFMCT. The SOM neural network clusters the revolvers based on RFMCT and the revolvers are divided into high-value, potential-value and low-value based on the clustering results. In addition, demographic characteristics are analyzed by logistic regression. The results show education has negative relationship with high-value revolver and we find female, younger, high income, works in non-government organizations and non-state-owned enterprises have a higher probability of being high-value revolver.

Keywords Revolver · Value differentiation · Demographic characteristics · RFMCT · Logistic regression

76.1 Introduction

A credit card is both a payment tool and a convenient source of credit [4]. It is one of the profitable financial products for credit card issuers. Credit card holding has increased steadily over the past 30 years. According to the Survey of Consumer Finances (SCF), in 1998 more than two-thirds of U.S. households had a bank-type credit card, compared to only 43 % in the 1983 Survey. This also happens in Asian region. According to MasterCard, they expect that the total number of credit cards in circulation in China in the next 15 years would increase at an average annual rate of 11 %, with the total number of cards reaching 1.1 billion in 2025 (i.e. 0.75 cards a

C. He · M. Zhang · J. Zheng (✉) · D. Du
Business School, Sichuan University, Chengdu 610065, People's Republic of China
e-mail: zhengjianguo@scu.edu.cn

X. Li
Deparment of Experiment Education, Guizhou University of Finance and Economics, Guiyang 550004, People's Republic of China

J. Xu et al. (eds.), *Proceedings of the Eighth International Conference on Management Science and Engineering Management*, Advances in Intelligent Systems and Computing 281, DOI: 10.1007/978-3-642-55122-2_76, © Springer-Verlag Berlin Heidelberg 2014

person) and the total amount of credit card spending to increase at an average annual rate of 14 % reaching US $ 2.5 trillion by 2025.

Over time, the number of credit card revolvers has increased in US credit card market. Revolver is a main profit source for credit card issuer. In American market, approximately 60 % credit card users are revolvers and they contribute about 70 % of card revenue in 2005 [10]. According to the researches on customer management and lifetime value, if a firm intends to be profitable in the long run, it must either convert unprofitable customers to a profitable status or "fire" them [2]. However, a better customer management strategy would require the firm to identify the more profitable customers from the less desirable ones. The credit card industry's best customers are among the revolvers. It is quite important, therefore, to understand the behaviors and characteristics of revolvers, especially high-value ones. The credit card issuers who know the revolvers better are more likely to win in the intense competition.

Chinese banks have recognized that credit card is profitable and with brilliant prospect, and many resources had injected since 2002. Up to the end of November 2011, the Chinese financial institutions issued about 260 million credit cards. Unlike US mature credit card market, there are only about 14 % revolving credit card users in Chinese credit card market and they contribute about 47 % profits. As a McKinsey's survey report, most Chinese people use credit cards as a convenient way to make purchases (or accumulate points toward a small gift), not as a credit facility. Chinese credit card market is still in its infancy.

The main purpose of this study is not to provide a new segmentation algorithm and characteristics analysis algorithm, but to focus on finding high-value consumers by the results of revolving credit behavior segmentation and finding which kind of people are more likely to be high-value revolvers. Hoping to provide a framework of understanding the knowledge of revolvers, especially high-value revolvers, uses the data of a Chinese bank. In this study, we also develop the RFM model and establish RFMCT model to evaluate the revolvers' behavior more appropriately.

The rest of the paper is organized as follows. The RFMCT model and the SOM segmentation based on it is in Sect. 76.2 and the profiling by logistic regression is presented in Sect. 76.3. In Sect. 76.4, we conclude the whole paper.

76.2 Customer Segmentation

1. RFMCT Model

Recency, Frequency, and Monetary (RFM) model which was first proposed by Stone and Bob [7] is a behavior-based model, which is used to analyze and predict customers' behavior in the database. R is the time period since last purchase; F is the number of purchases in a certain period of time and M is the total or average monetary amount spent during the certain time period. RFM model has continued to evolve during the past 20 years. Yeh et al [9] establishes RFMTC model by adding two parameters: time since first purchase (T) and churn probability (C), and applies it in direct marketing. In addition, Hsieh applies RFM model to segment bank credit card customers, where (R) value measures the average time distance between the day of makes a charge and the day pays the bill, frequency (F) value measures the average number of credit card purchases made, and monetary (M) value measures the amount of consumption spent during a yearly time period. On this basis, there is an extra parameter-the Repay Ability (RA) for segmenting customers, where RA is number of months without delayed pay off divided by number of months holding the card.

RFM performs well when segments normal customer. However, when it is applied in revolvers, something special should be considered. Because revolver is a kind of cardholder who uses the revolving credit to overdraw, we need to pay more attention to the overdraft besides the traditional R, F, M variables. For different customers, in the case of equal monthly consumption amount, the higher the overdraft proportion, the higher value he/she belongs to. Therefore, this study employs 'Average Ratio of Credit Used' (C) to model card usage behavior. It is the mean of ratio of used credit since holding the card. On one hand, it reflects the consumers' general degree of overconsuming, on the other hand, it also reflects whether the consumers take full advantage of credit limit. The formula is as follows:

$$C = \left(\sum_{i=1}^{T} \frac{\alpha_i}{\beta} \right) / T, \tag{76.1}$$

where α_i is overdraft in the i month, β is credit limit, T is available time of holding credit card which is the time period since the first purchase.

We also add T—available time of holding credit card—into the model as Yeh et al. [9]. The RFMCT model is ultimately established for the revolvers. Where R measures the days since last purchase, F measures the monthly average number of credit card purchases made since first purchase, and M is the monthly average consumption amounts since first purchase. The cardholders who are high-value to the credit card issuer are with high F, M, C, T and low R.

2. The Data Source

An anonymous China commercial bank provides the data for this study. We select the data from 2008 when the bank began to issue the credit cards to January, 2011. The samples used in the study must have at least one revolving credit behavior experience and 1,047 samples are selected at this period. The data set includes 84,856 transaction records, 1,047 revolvers' demographic information and credit card information. Before analysis, we preprocess the data as follows: (1) select the customers who hold the card more than half a year; (2) if customer's latest consumption did not occur in the previous six months, then we regard them as churned, and filter them. We get the 838 samples finally.

We present summary statistics of the RFMCT variables in Table 76.1. The mean value of indicator R (29.12 days) is so long that the issuer may be unsatisfactory, while the mean of F is also much less than mature credit card markets. The mean of C catches our attention. The mean of Average ratio of used credit is about 20 % indicating even if they are revolver, their attitudes toward overdraft is cautious.

Table 76.1 Descriptive statistics of behavioral data

	Minimum	Maximum	Mean	Standard deviation
R (day)	0.0	178.00	29.12	38.1
F (times)	0.03	25.00	3.18	2.69
M (CNY)	11.56	48981.94	2879.67	4041.20
C (%)	0.01	100.00	21.89	25.68
T (month)	6.00	32.00	18.86	7.00

Table 76.2 Customer clustering results

	R	F	M	C	T	Number of customers
Cluster one	20.74	2.75	3727.15	79.44	22.68	96
Cluster two	11.70	4.09	20531.39	50.61	20.18	22
Cluster three	23.71	2.02	1394.59	8.27	10.95	153
Cluster four	16.48	2.83	2262.96	15.38	22.97	161
Cluster five	4.87	10.30	3633.51	21.30	17.36	55
Cluster six	101.26	1.73	1196.55	11.47	18.58	44
Cluster seven	129.78	1.35	1672.18	15.27	20.56	53
Cluster eight	12.93	4.63	3953.68	17.50	11.17	117
Cluster nine	20.13	2.11	1622.36	9.71	26.60	137
Mean value	29.12	3.18	2879.67	21.89	18.86	

3. SOM Clustering Analysis

Before the clustering, we normalized the data as follows:

$$x_{ij} = (X_{ij} - \overline{X}_j)/DS_j, \tag{76.2}$$

where X_{ij} is the value of the ith sample on the jth indicator, $\overline{X}_j$ and DS_j are the mean value and standard deviation of all the samples on the jth indicator, respectively.

Using SOM to cluster the credit card users based on the assumption that the historical behavior patterns and the future behavior patterns are similar. We use the self-organizing map network to cluster customer behavior automatically by Matlab toolbox. After several times of parametric fitting, customers are clustered into nine clusters ultimately. The mean value of R, F, M, C and T of each cluster and the number of customers in each cluster are shown in Table 76.2.

The indicators of each cluster are compared with the mean value using the signal '↑' and '↓'. Where '↑' indicates that the indicator is greater than the mean value and otherwise is '↓'. According to the analysis, we found some clusters are similar. We combine the clusters and get six revolver groups.

Group 1 (Cluster one): $R \downarrow F \downarrow M \uparrow C \uparrow T \uparrow$. This type of revolvers are with shorter consumption time interval, lower consumption frequency, larger consumption amount, higher average ratio of used credit and longer available time of holding credit card. Such kind of customers' single transaction amount is large and therefore they may have a strong consumer demand for big-ticket items and luxury goods. Due to the average ratio of used credit is the highest one, indicating that this type of customers have strong willingness to overdraw. In addition, longer available time of holding credit card suggests they have long relationship with the card issuer. However, their behaviors of using the card are not very stable which may cause some potential risk in the transaction (e.g. cash out risk). Thus more attention should be paid on their financial situation. Such customers, therefore, are the ones with risk but high return and issuers want to attract. For these customers, the issuer can introduce shopping installments to them. The small installments business not only converts the uncertain interest income into a stable fee-based interest income, but also avoids potential cash out risk. And some promotions also should be introduced to them to increase their loyalty.

Group 2 (Cluster two): $R \downarrow F \uparrow M \uparrow C \uparrow T \uparrow$. This type of customers are with shorter consumption time interval, higher consumption frequency, larger amount of consumption, higher average ratio of used credit and longer available time of holding credit card. This group of customers has stable behavior as well as good overdraft utilization. They are perfect customers with high value for the issuers. Further analysis shows that this group of customers is better than Group 1 in terms of consumption frequency and amount of consumption. But average ratio of used credit is less than Group 1. Similar business can be introduced to this group as Group 1 to increase profits and avoid churn.

Group 3 (Cluster three): $R \downarrow F \downarrow M \downarrow C \downarrow T \downarrow$. The five indicators are all lower than mean value. The unsatisfactory of credit card usage and overdraft made them cannot bring high interchange fees neither can provide high interests to the issuers. However, shorter available time of holding credit card suggests that they accept and use the credit card for a short time. The issuer needs to introduce more promotions to help them recognize the advantage of credit card.

Group 4 (Cluster four and nine): $R \downarrow F \downarrow M \downarrow C \downarrow T \uparrow$. All the indicators are lower than the mean value except for the available time of holding credit card. They hold the credit card for a long time but with low activities and overdraft. They cannot contribute high profits to the issuer, but taking the resources of the issuer for a long time. Some targeted promotions should be taken to these cardholders to increase the frequency of using card. Meanwhile these cardholders may have low loyalty with the issuer, so the issuer should pay more attention to their risk.

Group 5 (Cluster five and eight): $R \downarrow F \uparrow M \uparrow C \downarrow T \downarrow$. For this group, customers have shorter consumption time interval, higher consumption frequency, and larger consumption amount, lower average ratio of used credit and shorter available time of holding credit card. Although they are better than group 3 and 4 for the issuer in terms of higher consumption frequency and larger consumption amount, the interest is low because the overdraft proportion is low. From another angle, the short available time of holding credit card means the behavior of the cardholder can develop better. The potential overdraft needs, therefore, can be further tapped. For such customers, the issuers should adopt a targeted marketing strategy to increase their acceptance with the credit card.

Group 6 (Cluster six and seven): $R \uparrow F \downarrow M \downarrow C \downarrow T \uparrow$. For this group, customers have longer consumption time interval, lower consumption frequency, and smaller consumption amount, lower average ratio of used credit and longer available time of holding credit card. The more longer consumption time interval (101 and 130 days) suggests that this group may more likely to churn.

4. Customers Segmentation

In order to help out the issuer in customer management and marketing, we integrate the customers in three dimensions based on the SOM clustering results.

- High-value revolvers ($F \uparrow M \uparrow C \uparrow$ and $F \downarrow M \uparrow C \uparrow$). Among the revolvers, customers in Group 2 are prominent no matter the consumption frequency, the consumption amount or average ratio of used credit. They can bring high profits for the credit card business. The Group 1's consumption frequency is lower, but the average consumption amount is high. They are able to create a substantial interchange fees cash advance fees. Moreover, the high proportion of overdraft may contribute high interest profits. Both groups of customers, therefore, are key accounts for the issuers that they must pay more attention and maintain well relationship. Targeted strategies should be provided to them, such as shopping installments, cash installment and billing installments and other installment business, to increase the profits. These strategies may also prevent them from going to the competitors. Meanwhile, high return means high risk almost the time. The risk management should be done well. The high—value revolvers consist of Cluster one and two, accounting for about 14.1 % of samples.
- Potential value revolvers ($F \uparrow M \uparrow C \downarrow$). Such consumers have high trading activities. They contribute to the issuer by interchange fees and cash advance fees mostly, but little interest. However, interests are the main income in the credit card business for credit card issuers. These customers do not have a strong consumer demand for revolving credit. Thus the key point is to improve their consumption habits, increase the proportion of the using overdraft. The revolving credit interests interest eventually. This category consists of Cluster five and eight, accounting for 20.5 % of the samples.
- Low-value revolvers ($F \downarrow M \downarrow C \downarrow$). Such consumers have low trading activities, consumption amount and low average ratio of used credit. Moreover, this type of customer may have churn tendency. One reason they use revolving credit may be their poor financial situation which forces them to use a credit card to obtain financing payments. They bring high risk instead of providing many profits. Another reason may be that for some reasons they did not repay in time and be regarded as revolver. All these reasons indicate they are low-value customers at present. The issuers should focus on the liabilities of such customer. In addition, issuers need to reduce the number of these customers gradually. This category includes Cluster three, four, six, seven and nine, accounting for 65.4 % of the samples.

76.3 Logistic Regression Analysis

1. Modelling

Only if consumers use credit cards, the database records their behaviors. However, the marketing, especially attracting new customers, we observe their basic information only. The analysis of demographic characteristics, therefore, is particularly important. Since the high-value revolving credit are the target of the issuers, and the low-value revolvers are who they want to reduce. We analyze the demographic characteristics of these two types of revolvers by binary logistic regression analysis in this section.

Kicking out of the missing samples, a total of 596 samples are remaining. In the first step, the dependent variable is equal to 1 if the cardholder is high-value and is equal to 0 if low-value. The demographic characteristics include gender, age, education level, annual revenue and industry. Sample characteristics are presented in Table 76.3. There are almost as many women as men among the samples. Yong and middle-aged people are dominant in the samples. The majority of the revolvers are well-educated. Most samples' annual revenue is less than 100,000.

Kinsey [6] pointed out that the demographic characteristics such as age, income and industry are highly correlated. Thus these variables are added in the regression modelling respectively. We selected men, 36–50 years old, 50,000–100,000 work in government agency and state run enterprises, and undergraduate degree as based group of gender, age, annual income, industry and education degree. The regression results are shown in Table 76.4.

2. Results Analysis

In terms of gender, women have higher probability to be high-value revolvers in all the equations, indicating that women are more likely to use credit cards and overdraw. This is not the same as the view of Western scholars who find the women are more self-control and cautious consumers [3]. Women in China have same social position with men and earned as much as men. Meanwhile many women are head of the families in China, which made them have more chances to consume.

There is a negative correlation between the education degree and being a high-value revolver. However, many Western researches show that revolvers with more education are likely to have higher credit worthiness and be confident about their future financial status, thus they increase the current consumption and lead to liability ultimately [5]. Similar conclusion also appears in the research of Chinese credit card market. Wang and Ren [8] find education is positive related to the overdraft behaviors. The above researches use the survey data, while the data used in this study is collected from a real Chinese bank which is more objective. Future analysis is done by dividing the data into three groups randomly. All the groups show education is negative related to high-value revolver in terms of statistical analysis. Why this happens? May be credit card stills in its infancy in China and Chinese traditional consumption habits result in the difference.

Customers between 21 and 35 years old have a higher probability of being high-value revolvers compared with consumers between 36 and 50 years old, but customers

Table 76.3 Descriptive statistics of demographic characteristics

Variables	Value	Frequency	Percentage	Cumulative percentage
Gender	Male	297	49.9	49.9
	Female	299	50.1	100
Age	21–35 years old	242	40.7	40.7
	36–50 years old	291	48.8	89.5
	Older than 50	63	10.5	100
Education level	High school and below	63	10.5	10.5
	College	178	29.9	40.4
	Undergraduate	239	40	80.4
	Graduate and above	117	19.6	100
Annual revenue (CNY)	Less than 30,000	150	25.2	25.2
	30,00–50,00	226	37.9	63.1
	50,000–100,000	166	27.9	90.9
	100,000–250,000	44	7.3	98.3
	More than 250,000	10	1.7	100
Industry	State-owned	404	67.7	67.7
	Sole investment or joint ventures	88	14.8	82.5
	Private or non-fixed work	104	17.5	100

older than 50 years old are less likely to be high-value revolvers compared with consumers between 36 and 50 years old. This conclusion supports the view of Berthoud and Kempson [1] "young people are not only more likely to use credit card, but also have credit liabilities", suggesting that the issuers should pay more attention to younger revolvers.

Income is positively related to be a high-value revolver. Even though the low-income groups need to use the credit more, their credit limit may lower and bad financial situation made them cannot consume very often. Customers with high income may have higher credit limit reflecting buying power. This buying power might lead high-income revolvers to have more consumption which brings high profits.

Consumer works in non-government organizations and non-state-owned enterprises has a higher probability of being high-value revolver. The Chinese have a common idea that one has a job in government agency means he has a lifelong job. If one works in a state run company, then the job is also more stable than working in other kinds of companies. Wang and Ren [8] also find card users work in industry such as: bank, insurance and foreign-funded company, are more likely to overdraw. The two similar research results mean that more stable job using credit card less.

Table 76.4 Comparative results of the high value and low value customers

Basis set	Variable	Coefficient	Coefficient	Coefficient	Coefficient
Male	Female	1.277**	1.205**	1.105**	1.257**
Undergraduate	High school and below	3.239**	2.123**	2.581**	2.011**
	College	2.585**	1.860**	2.328**	1.779**
	Graduate and above	−1.555 **	−1.247 **	−1.481 **	−1.177 **
36–50 years old	21–35 years old	0.565**			
	More than 50 years old	−1.209 **			
Annual revenue: 50,000–100,000	Less than 30,000			−0.869 **	
	30,000–50,000			−0.697 **	
	100,000–250,000			0.920**	
	More than 250,000			1.310**	
Institutions or state-owned enterprises	Sole investment, joint ventures				1.025**
	Private or non-fixed work				1.326**
Likelihood ratio		150.076	190.001	204.432	133.831
Chi-square		126.994	144.044	144.383	136.82
The model significance level		0	0	0	0

**Significant at the 5 % level

76.4 Conclusions

Using a dataset obtained from a bank in China, this study recognizes high-value consumers by the results of SOM clustering results based on RFMCT model and analyzes the demographic characteristics of revolvers by logistic regression.

We introduce Average Ratio of Used Credit (C) and available time of holding credit card (T) into the RFM to establish a new model RFMCT. Then, we use of SOM neural network clustering model to divide revolvers based on RFMCT. According to the analysis we have divided the revolvers in to three categories: (1) high-value revolvers who are with large consumption amount and high average ratio of used credit, accounting for 14.1 % of the overall samples; (2) potential value revolvers who have high consumption frequency and large consumption amount, accounting for 20.5 % of the overall samples; (3) low-value revolvers who have low trading activities, low consumption amount, and low average ratio of used credit., accounting for 65.4 % of the samples.

The logistic regression analysis shows that women have higher possibility to be high-value revolver. This reminds us to mine the consuming habits of the female cardholders deeply, and provide products more suitable for female. The education degree is negatively related to be high-value revolver which is different from some researches. Credit card still in its infancy in China and Chinese traditional consumption habits may result in the difference. Consumer works in non-government organizations and non-state-owned enterprises has a higher probability of being high-value revolver. Income is positively related to be a high-value revolver. These results help the marketers to find new high-value customers in target market and provide more suitable services for this group to increase their loyalty.

Acknowledgments This work is supported by National Natural Science Foundation of China (No. 71071101). Ministry of Education in China Youth Project of Humanities and Social Sciences (No. 13YJC630249), Research Fund for the Doctoral Program of Education of the Ministry of Education of China (No. 0120181120074).

References

1. Berthoud R, Kempson E (1992) Credit and debt: the PSI report. Technical report, Policy Studies Institute, London
2. Blattberg R, Gary G, Jacquelyn S (2001) Customer equity: building and managing relationships as valuable assets. Harvard Business School Press, Boston
3. Brown S (2005) Debt and distress: evaluating the psychological cost of credit. J Econ Psychol 26(5):642–663
4. Garman E, Forgue R (1997) Personal finance. Houghton Mifflin Company, Boston
5. Kim H, DeVaney S (2001) The determinants of outstanding balances among credit card revolvers. Financ Couns Plann 12(1):67–79
6. Kinsey J (1981) Determinants of credit card accounts: an application of tobit analysis. J Consum Res 8(2):172–182
7. Stone B (1994) Successful direct marketing methods. NTC Business Books, Lincolnwood
8. Wang M, Ren X (2004) The relationship between the demographic characteristic of credit card holder and overdraft. J Financ Res 286(4):106–117
9. Yeh I, Yang K, Ting T (2008) Knowledge discovery on rfm model using bernoulli' the sequence. Expert Syst Appl 36:5866–5871
10. Zhao Y, Zhao Y, Song I (2009) Predicting new customers' risk type in the credit card market. J Mark Res XLVI:506–517

Chapter 77
Applying Dynamic Programming Model to Biogas Investment Problem: Case Study in Sichuan

Yanfei Deng and Caiyu Sun

Abstract To improve the biogas energy development structure, this paper studies the multi objective dynamic programming in its investment system. Limited resource has bandaged the ideal of investors. Variety of stages in the systems and in object function us state diversion, stage decision and overall decision constitute optimization problem. This paper establish the math-model of having disagreement of amount, and the resource allocation problem. The decision makers need to make a decision assigning the different area condition and resource to invest different scales of biogas projects under exploring constraint. Due to the lack of historical data, some coefficients are considered as fuzzy numbers according to experts' advices. Therefore, a multi-objective dynamic optimization model with possibilistic constraints under the fuzzy environment is developed to control the pollution and realize the economic growth. Finally, a practical case is proposed to show the efficiency of the proposed model and algorithm.

Keywords Biogas plant development · Dynamic programming · Fuzzy optimization

77.1 Introduction

The development and biogas utilizations projects have not only solved the problems of environmental and ecological issues caused by excrement, reduced greenhouse gases emissions, but alsooped a new pattern of energy consumption to

Y. Deng (✉)
School of Management, Southwest University for Nationalities, Chengdu 610064,
People's Republic of China
e-mail: dengyanfei921@163.com

C. Sun
Finance Department, Southwest University for Nationalities, Chengdu 610064,
People's Republic of China

J. Xu et al. (eds.), *Proceedings of the Eighth International Conference on Management Science and Engineering Management*, Advances in Intelligent Systems and Computing 281, DOI: 10.1007/978-3-642-55122-2_77, © Springer-Verlag Berlin Heidelberg 2014

effectively integrate environment protection with ecological protection, and created conditions to fundamentally solve the problem of energy, resources and environment [3]. By contrast to the traditional economic analysis, biophysical analysis as a measure of both quantity and quality of energy and materials is preferred for the evaluation of complex systems instead of using the money value of goods, services and resources [6]. Analyzing the spatial distribution of biomass byproducts resources can help define the biomass resource endowment of China and suggest reasonable government policy, which plays an important role in guiding the distribution of development of biomass energy industry and market [12].

For quantitatively assessing different social-economic or ecological systems, holistic evaluation based on bio-physics can be added to the body of knowledge on the poor coherence between economic profitability and ecological sustainability [9]. The emissions and energy input were determined for the operation of different scale biogas system at mesophilic temperatures, such as farm-scale and large-scale. Therefore, it is vitally necessary to work out varied and target-oriented investment based on the biogas projects differentiations. As the limitation of resources the decision-makers should determine a suitable investment distribution for dynamically allocating them among different biogas systems. Investment analysis, integrating economic and ecological processes in a common unit, is suitable for evaluating sustain ability of the ecological engineering and may help to identify the appropriate biogas production and consumption mode.

Choosing how to invest local resources in China is the responsibility of community leaders. Several decision-making tools have been developed to aid in local policy development. In the past, artificial intelligence methods, including artificial neural networks (ANN), ant colony optimization (ACO), genetic algorithms (GA), and particle swarm optimization (PSO) algorithms have been used extensively for energy demand projections. Sozen et al employed ANN to determine the sectoral energy consumption in Turkey [10]. Assareh et al applied GA techniques to estimate oil demand in Iran [2]. A PSO method was proposed to investigate long term electricity load forecasting [1]. The advantage of artificial intelligence models is that they are elective techniques with the capability to find the global optimal solution although with long computational time generally.

In this paper, the optimal control model under fuzzy environment is considered and the fuzzy theory is used to deal with such uncertain information in modeling the concrete investment system of the different biogas system projects. The other sections of this paper are organized as follows. In Sect. 77.2, the basic problem why we use the optimal control model to optimize the biogas investment is explained. A multi-objective programming model is developed and its crisp equivalent model is obtained in Sects. 77.3 and 77.4. Then the fuzzy simulation-based improved simulated algorithm is proposed to solve the multi-objective programming model with fuzzy coefficients. In Sect. 77.5, a practical case is given to show the significance of the proposed models and algorithms. Finally, some conclusions are made in Sect. 77.6.

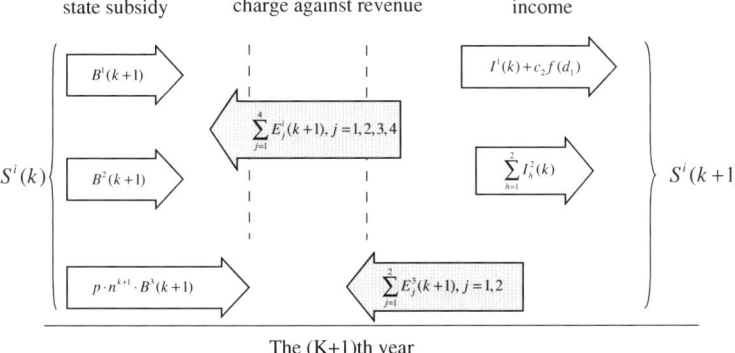

Fig. 77.1 Dynamic allocation process of capital

77.2 Problem Statement

Although different scale and components of biogas projects system had different economic effects. It is important to note the difference between the different kinds of digesters regarding economic feasibility. It is difficult to evaluate the economic benefits of utilizing biogas. As for the peasants, it can be categorized into direct and indirect economic benefits. Many different designs of biogas digesters are available both in small and large scale operations throughout the world. In the context of China, there are three major categories of the biogas project constructions: Rural household biogas digesters, urban sewage purification biogas digesters, large and medium-scaled biogas projects.

According to the operation of small biogas projects, biogas is used primarily for household cooking, ranch heating and as foliar fertilizer for vegetables. For large and medium-scaled biogas projects, biogas is used to generate electricity and waste water treatment, mostly.

77.3 The Proposed Optimization Model

The dynamic programming for optimizing the biogas industry can be mathematically formulated as follows. As a developing country, China has focused its development strategy on economic construction. For a regional economy, economic growth will create stability and stimulate trade and investments in the whole region. Thus, maximizing economic benefit is the objective of the local government. The dynamic allocation process of capital was shown in Fig. 77.1.

Table 77.1 Main variables related to modeling

Sort	Notation	Variable
1	X^1_{ik}	total investment of the ith class new biogas project in k year, $i = 1, 2, 3$
2	X^2_{ik}	total investment of the ith class renovation biogas project in k year
3	C^1_{ik}	cost of one ith class new biogas project in k year
4	C^2_{ik}	cost of one ith class renovation biogas project in k year
5	$P_{i(k-1)}$	the number of all ith class biogas projects in $k - 1$ year
6	$h_{i(k-1)}$	the scrapped rate of the ith class biogas projects in $k - 1$ year
7	θ_1	the output value of one full benefits project
8	θ_2	the output value of one non-full benefits project
9	q_1	the amount of national funding for one full benefits biogas project
10	q_2	the amount of national funding for one non-full benefits biogas project
11	B_k	the local financial budget in k year
12	$d_{i(k-1)}$	the scrap amount of the ith biogas project in $k - 1$ year

$$S^1(k+1) = S^1(k) - \sum_{j=1}^{4} E^i_j(k+1) + B^1(K+1) + I^1(k) + c_2 f(d_1),$$

$$S^2(k+1) = S^2(k) - \sum_{j=1}^{4} E^i_j(k+1) + B^2(k+1) + \sum_{h=1}^{2} I^2_n(k),$$

$$S^3(k+1) = S^3(k) - \sum_{j=1}^{2} E^3_j(k+1) + p \cdot n^{k+1} \cdot B^3(k+1).$$

The total economic benefit is calculated as the sum of full benefits and non-full benefits. Full benefits represent the economic benefits of new projects and renovation projects. Non-full benefits represent the economic benefits of the old projects without renovation. The main variables related to modeling was shown in Table 77.1. The number of biogas projects is given by: $X^i_1(k)/C^i_1(k) + X^i_2(k)/C^i_2(k)$, where $X^i_1(k)$ is the total investment of the ith class new biogas project in k year. $C^i_1(k)$ is the cost of one ith class new biogas project in k year. $X^i_2(k)$ is the total investment of the ith class renovation biogas project in k year. $C^i_2(k)$ is the cost of one ith class renovation biogas project in k year.

The number of old biogas projects is calculated by $P^i(k-1) - h^i(k-1) \times P^i(k-1)$, where $P^i(k-1)$ is the number of all ith class biogas projects in $k-1$ year. $h^i(k-1)$ is the scrapped rate of the ith class biogas projects in $k-1$ year.

The scrapped rate of the ith class biogas projects in $k-1$ year is given by:

$$h^i(k-1) = \frac{\sum_{i=1}^{3} \sum_{k=1}^{10} d^i(k-1)}{\sum_{i=1}^{3} \sum_{k=1}^{10} P^i(k-1)},$$

where $d^i(k-1)$ is the scrap amount of the ith biogas project in $k-1$ year. Then the objective can be expressed, it follows that:

$$P^i(k) = \theta_1 \frac{X_1^i(k)}{C_1^i(k)} + \frac{X_2^i(k)}{C_2^i(k)} + \theta_2 \left(P^i(k-1) - h^i(k-1) \times P^i(k-1) - \frac{X_2^i(k)}{C_2^i(k)} \right).$$

Therefore, we have:

$$P^i(k) = \theta_1 \frac{X_2^i(k)}{C_2^i(k)} + \theta_1 \frac{X_1^i(k)}{C_1^i(k)} - \theta_2 \times h^i(k-1) \times P^i(k-1) - \theta \frac{X_2^i(k)}{C_2^i(k)}$$

$$= (\theta_1 - \theta_2)\theta_1 \frac{X_2^i(k)}{C_2^i(k)} + \theta_1 \frac{X_1^i(k)}{C_1^i(k)} + \theta_2 P^i(k-1)(1 - h^i(k-1))$$

$$= \theta_2(1 - h^i(k-1)) P^i(k-1) + \frac{(\theta_1 - \theta_2)X_2^i(k)}{C_2^i(k)} + \frac{X_1^i(k)}{C_1^i(k)}. \qquad (77.1)$$

Then the constraint can be expressed as:

$$X^i(k) - q_1 \left(\frac{X_2^i(k)}{C_2^i(k)} + \frac{X_1^i(k)}{C_1^i(k)} \right) - q_2(P^i(k-1) - h^i(k-1) \times P^i(k-1) - \frac{X_2^i(k)}{C_2^i(k)}) \le B_k,$$

where B_k is the local financial budget in k year.

Finally, the dynamic programming model for the investment and development of biogas project can be formulated as follows:

$$\max F = \sum_{k=1}^n \sum_{i=1}^3 \theta_1 \frac{X_1^i(k)}{C_1^i(k)} + \frac{X_2^i(k)}{C_2^i(k)} + \theta_2(P^i(k-1) - h^i(k-1)) \times P^i(k-1) - \frac{X_2^i(k)}{C_2^i(k)}$$

$$\text{s.t.} \begin{cases} h^i(k-1) = \dfrac{\sum\limits_{i=1}^3 \sum\limits_{k=1}^{10} d^i(k-1)}{\sum\limits_{i=1}^3 \sum\limits_{k=1}^{10} P^i(k-1)} \\[2mm] X^i(k) - q_1 \left(\frac{X_2^i(k)}{C_2^i(k)} + \frac{X_1^i(k)}{C_1^i(k)} \right) - q_2(P^i(k-1) - h^i(k-1) \times P^i(k-1) - \frac{X_2^i(k)}{C_2^i(k)}) \le B_k \\[2mm] S^1(k+1) = S^1(k) - \sum\limits_{j=1}^4 E_j^i(k+1) + B^1(K+1) + I^1(k) + c_2 f(d_1) \\[2mm] S^2(k+1) = S^2(k) - \sum\limits_{j=1}^4 E_j^i(k+1) + B^2(k+1) + \sum\limits_{h=1}^2 I_n^2(k) \\[2mm] S^3(k+1) = S^3(k) - \sum\limits_{j=1}^2 E_j^3(k+1) + p \cdot n^{k+1} \cdot B^3(k+1). \end{cases}$$

77.4 Optimization Algorithm

Since there exist objective function and constraints (i.e., $p_i^0(k)$ and $p_i^j(k)$) that are not linear in the model, and the number of summation items in the objective function and constraints will change depending on the decision variables, the above equivalent crisp model is in fact an NP-hard problem, which could not be solved by general linear programming method [8]. In solving our expected value model, a particle swarm optimization (PSO) algorithm is proposed (Fig. 77.2). The advantage of using a PSO algorithm over other techniques is that it is computationally tractable, easy to implement, and does not require any gradient information of an objective function except for its value. Therefore, a PSO algorithm is developed to solve the previously mentioned problem. Key features of the algorithm are explained in detail, including the general concept of PSO, solution representation, initializing method, adjusting method, decoding method, fitness value function, and parameter selection, followed by the framework of the algorithm.

Solution representation of the DIAP is one of the key elements for effective implementation of PSO. An indirect representation is proposed here. In this study, every particle consists of $4 \times (M_i - 1)$ dimensions i.e., $H = 4 \times (M_i - 1)$, and is divided into four parts, which are expressed as: $P_I^i(\tau) = P_{I1}^i(\tau), P_{I2}^i(\tau), \cdots, P_{I[4 \times (M_i-1)]}^i(\tau) = [Y_{I1}^i(\tau), Y_{I2}^i(\tau), Y_{I3}^i(\tau), Y_{I4}^i(\tau)]$, where $Y_{I(k+1)(\tau)}^{i\theta}$ = the $(k+1)$th part of the lth particle in the τth generation for biogas project $i, k = 1, 2, \cdots, n$. Note that every part of a particle is $a(Mi - 1)$ dimensional vector, and can be denoted as:

$$Y_{I(k+1)}^i(\tau) = \left[y_{I(k+1)}^{i^1}(\tau), y_{I(k+1)}^{i^2}(\tau), \cdots, y_{I(k+1)}^{i^{M_i-1}}(\tau) \right],$$

where $y_{I(k+1)}^{i^\theta}(\tau)$ the θth dimension of $Y_{I(k+1)}^i(\tau)$ for the lth particle in the τth generation; $k = 1, 2, \cdots, n; \theta = 1, 2, \cdots, M_i - 1$. In order to be in accordance with the expression $P_I^i(\tau) = [P_{I1}^i(\tau), P_{I2}^i(\tau), \cdots, P_{IH}^i(\tau)]$, we have $y_{I(k+1)}^{i^\theta}(\tau) = P_{I[k \times (M_i-1+\theta)]}^i(\tau)$, there $H = 4 \times (M_i - 1)$.

77.5 Case Study

1. Presentation of the Case Problem

Sichuan Province has the largest number of biogas plants, with 2.94 million running. There are several fermentation techniques mainly including "Up-Flow Solids Reactor (USR)", "Continuous Stirred Tank Reactor (CSTR)", "Plug Flow Reactor (PFR)", and etc. Currently, among the established biogas projects in Sichuan, USR and CSTR are mainly adopted fermentation techniques, since these techniques are more mature and have provided concrete technical support for the construction of large and medium biogas projects in Sichuan. Regional assignments of the three biogas system projects was shown in Table 77.2. Due to special weather conditions, it is very important to choose appropriate heating instruments, especially fermentation tanks, for biogas projects. As for biogas cogeneration projects, waste heat from cogeneration is mostly used as heating for fermentation tanks. For projects with insufficient capacity of power generation, some alternate heating boilers (such as coal-burning steam boilers or biogas boilers) are needed for fermentation tanks heating.

Table 77.2 Regional assignments of the three biogas system projects

Type index	Biogas system	Initial number	Regional index	Regional assignments	Throughput (expected value)
i	Biogas project (volume; maximum)	α_i (unit)	i^θ		$E[Q_{i^\theta}]$ (m^3/year)
1	Family-scale rural biogas project	32	1^a	Hilly areas and Panxi region	3470
			1^b	Northwest ethnic regions	3510
			1^c	Chengdu plain region	2980
			1^d	Mountainous regions surrounding basin	3900
2	Sewage purification biogas project	4 8	2^a	Hilly areas and Panxi region	5100
			2^b	Northwest ethnic regions	5020
			2^c	Chengdu plain region	4680
			2^d	Mountainous regions surrounding basin	3900
3	Biogas power generation project	86	3^a	Hilly areas and Panxi region	6180
			3^b	Northwest ethnic regions	6090
			3^c	Chengdu plain region	8400
			3^d	Mountainous regions surrounding basin	5900

To begin investigating energy issues in China, we used the representative, prosperous Sichuan province for case studies. This section presents two parts: the first covers the urban households' appliances nationwide; the second one includes rural households nationwide. The biogas potential of Sichuan is large and at present almost unexploited. Interest in biogas technology in Sichuan began in the late 1930s but it was not until the middle 1980s did biogas technology receive the needed attention from government. Sichuan Province is the nation's top biogas state which has the largest number of biogas plants.

By the end of 2009, there were over 2.76 million rural household biogas digesters (account for more than 25 % of the national total biogas digesters), 47,000 urban sewage purification biogas plants and 117 large and medium-sized biogas plants in Sichuan. which is equivalent to 19.0 million t of standard coal (tce). In recent years, at the same time the number of household-size biogas digesters is gradually increasing, the number of medium and large-scale biogas plants is rising rapidly.

It is estimated that 2.76 million rural household biogas digesters in Sichuan have produced biogas 980 million m^3, equivalent to 734,000 tons of standard coal. The 2.76 million biogas digesters can increase income 13.8–1.66 billion RMB each year.

A survey of 47,000 sewage purification biogas digesters were conducted in Sichuan at present, which the total volumes reached 2.24 million m^3 and could dispose 250 million domestic sewerages every year.

Biogas projects for large scale in Sichuan began to construct in 2003, and the first biogas project was put into operation with a 150 m^3 fermentation tank in 2004. Despite late start of biogas projects, the State has invested heavily in Sichuan, in recent years. From 2005 to 2007, the number of new biogas projects for each year was 3, 1, and 3, respectively.

By the end of 2009, there were 117 large biogas projects in Sichuan, where total capacity of biogas pool is 12,400 m^3, annual biogas production is 4.35 million m^3, and average daily biogas

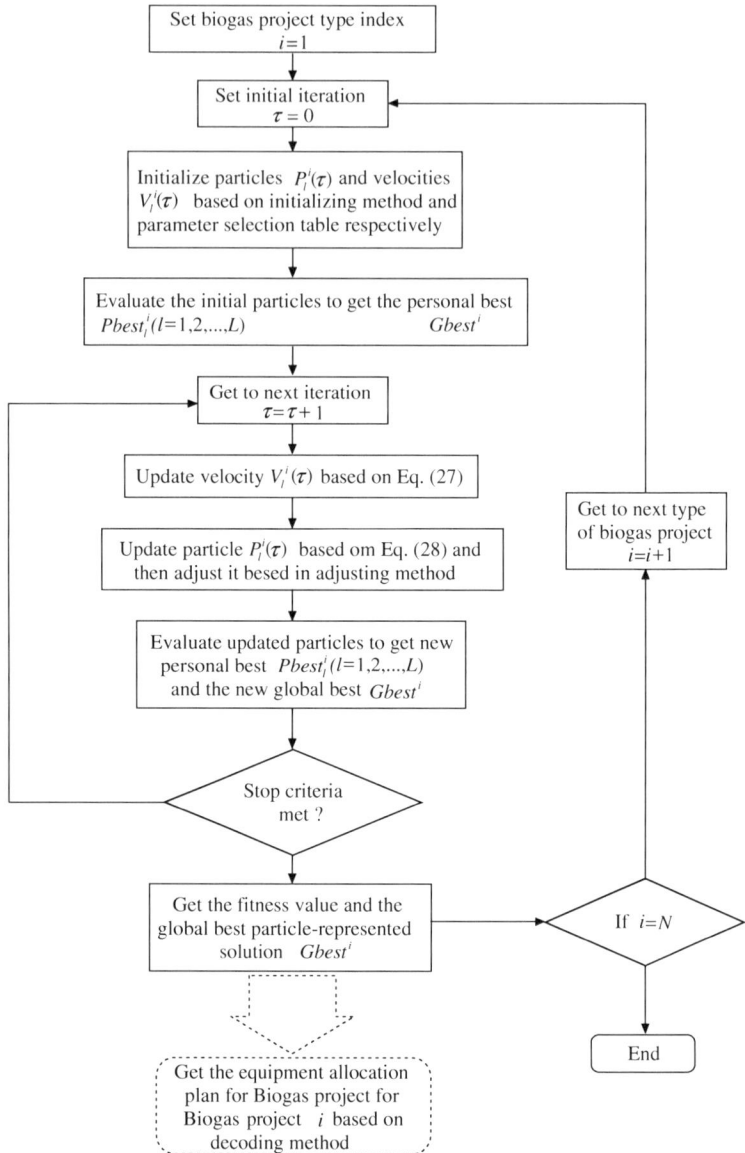

Fig. 77.2 Procedure of the PSO framework

production is more than $1300\,m^3$ for each project. By the end of 2008, 14 large and medium biogas projects were completed, 6 biogas projects were still under construction and 36 new biogas projects were granted. Currently in Sichuan, biogas comprehensive utilizations is less than 1 % of full potential, and there is much room for development.

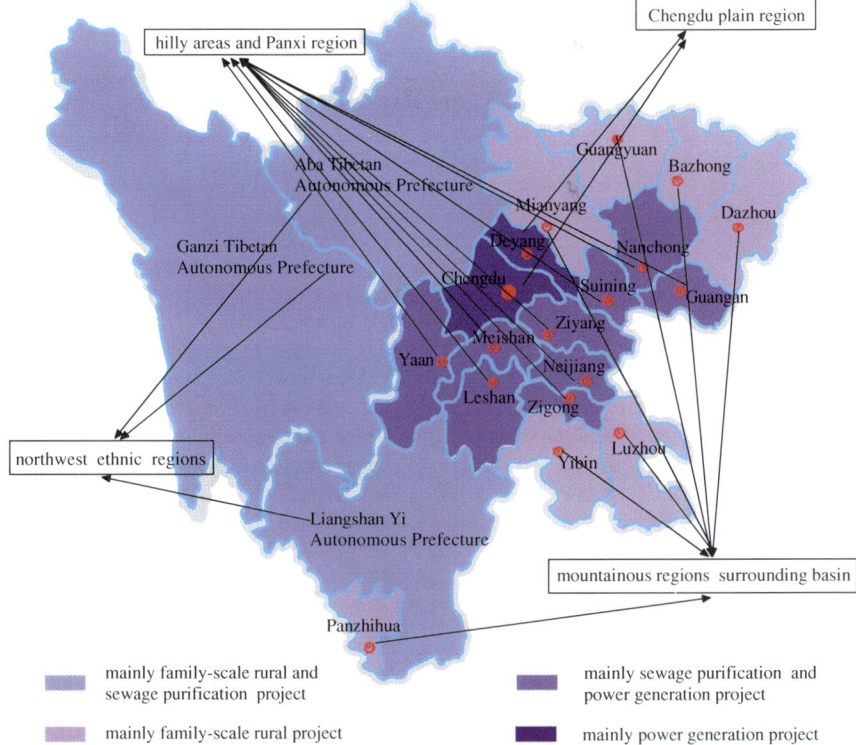

Fig. 77.3 Regional assignments of three biogas system projects in sichuan

2. Results and Analysis

However, because the aim of our study is to provide an alternative and effective method for optimizing equipment allocation in construction projects by utilizing the PSO algorithm, our solution method and computational results are sufficient and significant, which demonstrates the practicality and efficiency of our optimization method. Investing in the biogas electricity project has been estimated at the total amount of 958,950. This amount consists of the investments into fixed and current assets. Anaerobic digestion systems for fermentation of organic matters used widely with commercial digesters of 70–5000 m^3, small units are used mainly for heating, while large units for generation of electricity. The regional assignments of three biogas system projects in Sichuan was shown in Fig. 77.3.

Enlargement of the renewable energy production is clearly an imperative, but only economically viable construction and operation can result in long-term attainability, which is initially the goal when deciding upon such investments. This article thought that the large-scale biogas power generation construction project has static large investment, time limit is long, construction is high in intensity, participate and significant effect in unit's complicated characteristic, so it suitable for the region of high level in the economy development and the urbanization and abundant biogas resources. Sewage purification biogas project could provides an eco-friendly solution to the problem of sewage disposal with moderate investment. As the byproduct of anaerobic sludge treatment, the biogas is a kind of potential utilizable resource in municipal waste water treatment plant. So it suitable for small town. The construction of a family-scale rural biogas project is relatively simple and has less operation staff due to its small scale and minor investment, it could provide one effective method to improve ecological agriculture, rural economy, and life quality of farmers with advantages of lower cost, small individual investment, low running expenses, less field, etc. So it suitable for the rural areas and less well-off places.

77.6 Conclusion

In this study we have developed a multi-objective optimization model with possibilistic constraints under the fuzzy environment to simulate the investment of biogas plant. The aim of this study is to provide an alternative and effective method for optimizing investment structure in biogas project. The paper also proposed an algorithm called fuzzy simulation-based to solve the model. Our computational results do show that the proposed optimization method has made noticeable differences as compared to actual total construction throughput collected from the biogas project in China. Finally, a practical case proved that the proposed model and algorithm was efficient. Note that our mathematical model is formulated with some assumptions and, so, it may not represent the exact construction environment. Although the model proposed in this paper should be helpful for solving some real-world problems, it is only dealt with by the possibilistic constraints. In further research to be under taken, a detailed analysis will be given.

Acknowledgments This research was supported by the National Science Foundation for Distinguished Young Scholars (Grant No. 70425005) and the Key Program of National Natural Science Foundation of China (NSFC) (Grant No. 70831005), People's Republic of China. The writers would like to thank various people for their helpful and constructive comments and suggestions.

References

1. AlRashidi MR, EL-Naggar KM (2010) Long term electric load forecasting based on particle swarm optimization. Appl Energy 87:320–326
2. Assareh E, Behrang MA et al (2010) Application of PSO (particle swarm optimization) and GA (genetic algorithm) techniques on demand estimation of oil in Iran. Energy 35:5223–5229
3. Chang S, Zhao J (2011) Comprehensive utilizations of biogas in Inner Mongolia, China. Renew Sustain Energy Rev 15:1442–1453
4. Clerc M, Kennedy J (2002) The particle swarm: explosion, stability, and convergence in a multidimensional complex space. IEEE Trans Evol Comput 6(1):58–73
5. Eberhart RC, Shi Y (2001) Tracking and optimizing dynamic systems with particle swarms. In: Proceedings of the IEEE congress on evolutionary computation, pp. 94–97
6. Hau JL, Bakshi BR (2004) Promise and problems of emergy analysis. Ecol Model 178:215–225
7. Kennedy J, Eberhart RC (1995) Particle swarm optimization. In: Proceedings of the IEEE international conference on neural networks, pp. 1942–1948
8. Mekler VA (1993) Setup cost reduction in the dynamic lot-size model. J Oper Manage 11(1):35–43
9. Sciubba E, Ulgiati S (2005) Emergy and exergy analyses: complementary methods or irreducible ideological options? Energy 30:1953–1988
10. Sozen A, Ulseven Z, Arcaklioglu E (2007) Forecasting based on sectoral energy consumption of GHGs in Turkey and mitigation policies. Energy Policy 35:6491–6505

11. Xu J, Zeng Q (2011) Applying optimal control model to dynamic equipment allocation problem: case study of concrete-faced rockfill dam construction project. J Constr Eng Manage-ASCE 137(7):536–550
12. Yang YL, Zhang PD, Guangquan L (2012) Regional differentiation of biogas industrial development in China. Renew Sustain Energy Rev 16:6686–6693
13. Yapicioglu H, Smith AE, Dozier G (2007) Solving the semi-desirable facility location problem using bi-objective particle swarm. Eur J Oper Res 177:733–749
14. Zhang H, Li H, Tam CM (2006) Permutation-based particle swarm optimization for resource-constrained project scheduling. J Comput Civil Eng 20(2):141–149

Chapter 78
Determination of the Control Chart CUSUM-ln(S^2)'s Parameters: Using a Computational Tool to Support Statistical Control

José Gomes Requeijo, Ricardo Costa Afonso, Ricardo Barros Cardoso and José Pedro Borrego

Abstract The increasing attention devoted to the control of process variance has stimulated the development of a set of control charts, which includes the CUSUM-ln(S^2) control chart that is considered one of the most effective tools to control the variance and consequently one of the most used whenever a SPC is implemented. However, a common problem which arises with the application of CUSUM-ln(S^2) control charts is the limited set of tables of results for consultation. In general, only the most common situations are available. Therefore, these resources become ineffective when one need to cover other less common cases, but equally important and considered target of interest. On the other hand, there are not available the abacuses correspondent to the numerical tables. This paper intends to present a new approach (methodology), based on a computational tool (FCSCE), which provides us, not only with the abacuses, but also with the respective tables. This software tool, implemented in MATLAB environment, is a key instrument to deal with generic SPC case studies involving the CUSUM-ln(S^2) control chart, responding effectively to the previously listed impairments.

Keywords Statistical Process Control (SPC) · Control chart CUSUM-ln(S^2) · Abacuses · Tables of results · Computational tool to support statistical control (FCSCE)

J. G. Requeijo (✉)
UNIDEMI, Departamento de Engenharia Mecânica e Industrial, Faculdade de Ciências e Tecnologia, Universidade Nova de Lisboa, 2829-516 Caparica, Portugal
e-mail: jfgr@fct.unl.pt

R. C. Afonso · R. B. Cardoso
Departamento de Engenharia Mecânica e Industrial, Faculdade de Ciências e Tecnologia, Universidade Nova de Lisboa, 2829-516 Caparica, Portugal

J. P. Borrego
Departamento de Eletrónica, Telecomunicações e Informática, Instituto de Telecomunicações, Universidade de Aveiro, Campus Universitário de Santiago, 3810-193 Aveiro, Portugal

J. Xu et al. (eds.), *Proceedings of the Eighth International Conference on Management Science and Engineering Management*, Advances in Intelligent Systems and Computing 281, DOI: 10.1007/978-3-642-55122-2_78, © Springer-Verlag Berlin Heidelberg 2014

78.1 Introduction

In the context of the current market environment, increasingly complex and competitive, competitiveness has become a major concern of the players and simultaneously an intrinsic need of the companies. In order to raise the standards of industrial competitive edge, the organizations have to take into account several factors to improve the overall performance of their business model.

Because of this, the quality control has been receiving an increasing importance in the management domain, by any industrial company, aiming the optimization of the production processes and thereby the productivity levels.

The Statistical Process Control (SPC) offers a set of tools and techniques to perform this control of production processes, and the control charts are the most well known in this area, due to its proven efficiency. This area of studies has been identified as fundamental, receiving the most notable contributions, from Quesenberry [7], Oakland [5] and Montgomery [4].

The control charts were created and introduced by Shewhart [9], resulting in an important contribution to the scientific community in general and to the field of quality in particular. These charts are applied to processes, in which, there is a huge amount of data, being possible in some instances their application in a real production environment.

The control charts can be classified according to the number of features, the type of feature and the type of memory. Therefore depending on the subject of study and characteristics present in the production process, it's possible to choose the most adequate control chart to use.

Since the days of Walter Shewhart several control charts were developed, being various the contexts in which they should be applied. Most of the control charts are focused on monitoring the mean of the process. However in recent years the control of process variance has become an area of growing attention by the researchers.

Indeed, since the mid-1980s, there has been increasing attention to monitoring variance of a given manufacturing process. The main reason for the growing attention in this specific area of Statistical Process Control has been the increasing demand of the markets and the consequent need for companies to submit products of high technical consistency.

The control charts of cumulative sums are considered special control charts, having been developed in order to be better suited to the current realities of production and market. This specific control charts has been first introduced by Page (1961) and subsequently been subject of several studies such as Woodall [12], Gan [3], Sparks [11] and Ryu and Wan [8]. As a result there are now several charts of cumulative sums developed along various assumptions and intended for different realities.

It is in this context that appear control charts of cumulated sums specially designed for the detection of small and moderate shifts in the variance of the process, being the CUSUM-ln(S^2) one of these specific charts. Over the years some researchers have focused their attention on the efficiency of this type of control charts, among

Table 78.1 Summary of variables and parameters related to the control chart CUSUM-ln(S^2)

Control chart	CUSM-ln(S^2)	
Statistic	$Y_t = \ln(S_t^2/\sigma_0^2)$	
Cumulative sum	$C_t = \max(0; C_{t-1} + Y_t - k_C)$	$D_t = \min(0; D_{t-1} + Y_t - k_D)$
Initial values	$C_0 = u$ $D_0 = v$	
Rules to detect shifts	$C_t > h_C$ $D_t > -h_D$	

which are the studies of Chang and Gan [2] and more recently Shu et al. [10] who advocate an adaptive approach of several control charts existing.

In this case the focus is confined to the CUSUM-ln(S^2) as reference when attempting to detect small and moderate shifts in the variance of the process, considering that the data are independent and normally distributed. It was from these assumptions that was developed the methodology inherent to the creation of the Computational Tool to Support Statistical Control (FCSCE).

This article aims to shown an alternative approach to the limitations regarded in the determination of the parameters of the control chart CUSUM-ln(S^2). The application software inherent to this new approach was developed according to four sequential phases. By setting the input values in each phase it's possible to reach a valid set of results in the form of abacuses and tables.

78.2 Control Chart CUSUM-ln(S^2)

In order to build the control chart CUSUM-ln(S^2) are defined two variables C and D, which are determined according to ln(S^2) [6].

It is through the graphical representation of the ordered pairs (t, C_t) and (t, D_t) that it's designed the two half-lines essential to control the dispersion process.

The ray C is intended to detect small and moderate shifts in terms of increases in the variance of the process, while the ray D is responsible for detecting decreases in the variance of the process.

The variables C and D are set, at time t, by: $C_t = \max(0; C_{t-1} + Y_t - k_C)$, $D_t = \min(0; D_{t-1} + Y_t - k_D)$, $C_0 = u$, $D_0 = v$, where $0 \leq u < h_C$ and $-h_D < v \leq 0$. The values of h_C and h_D match to the control limits for the variables C and D, respectively. Consider u and v equal to zero for this specific chart.

The variable Y is given at time t by: $Y_t = \ln(S_t^2/\sigma_0^2)$, where ln($S^2$) is the natural logarithm of the variance of sample t (S_t^2).

Is detected an increase or a decrease of the variance of the process when one of these two following conditions is verified: $C_t > h_C$, $D_t > -h_D$, where h_C and h_D are, respectively, the Upper Control Limit and the Lower Control Limit.

The Table 78.1, shown below, summarizes the statistics and variables considered in the construction of a control chart CUSUM-ln(S^2).

The previous table adopts the following notation:

S_t : Standard deviation of the sample t,
S_t^2 : Variance of sample t,
σ_0 : Initial value of the process standard deviation,
k_C : Reference value to detect an increase of δ_C in the process standard deviation,
k_D : Reference value to detect a decrease of δ_D in the process standard deviation,
h_C : Control limit to detect an increase of δ_C in the process standard deviation,
h_D : Control limit to detect a decrease of δ_D in the process standard deviation.

The design of CUSUM-ln(S^2) and its graphical representation requires the determination of the parameter value k, which is optimal for detecting an increase or a decrease in the standard deviation of the process. The decision interval h is then determined so that the control board has a specific [2].

The CUSUM-ln(S^2) charts are used considering that data is organized in the form of samples.

The process of design a CUSUM-ln(S^2) chart implies a logical sequence of steps. According to Pereira and Requeijo [6] the procedure to be followed for the construction of CUSUM-ln(S^2) when data are organized in samples is as follows:

1. Choosing the sample size n.
2. Select the lowest acceptable value for $ARL_{\text{In Control}}$.
3. Choosing the value of standard deviation shift σ_1, which may be different for the decrease and the increase detecting.
4. Determine the shift $\delta = \sigma_1/\sigma_0$.
5. Determine the best values for the parameters k and h, knowing the values of δ, n and $ARL_{\text{Out of Control}}$.

Considering this particular chart and in order to determine the parameter values of k_C, k_D, h_C and h_D, Chang and Gan [2] suggest the use of the tables elaborated by both in sequence of their studies in 1995 for situations considered to be the most common.

78.3 Computational Tool (FCSCE): Methodology

The tables elaborated by Chang and Gan in their studies in 1995 become ineffective whenever is needed to access the parameters of this control chart for situations not covered in them. This is one of the main reasons that led to the development of this computational tool. Also there aren't available the abacuses related to the numerical tables.

The development of this Computational Tool to Support Statistical Control aims to provide an important contribution to the scientific community in the area of Quality. At the same time aims to answer the limitations referred above, by designing a tool that allows to create a set of tables and abacuses for the control chart CUSUM-ln(S^2). Therefore the main goal of this tool is to flexible the working of engineers and/or technicians who are dedicated to the Statistical Process Control.

The development approach of this computational tool is based on one of the procedures considered in the dissertation work of Alves [1] referring to a different control chart.

This tool (FCSCE) was developed by designing a set of algorithms which represent the basis for this particular tool.

These algorithms were designed in the form of a programming language named M-code, which is the programming language used in software MATLAB. The algorithms developed under the procedure of the FCSCE are specific, which makes this tool unique and unlike any other.

The procedure inherent to the development of the computational tool to support statistical control (FCSCE) is sequential, having been divided into the following phases for both studies (detection of increases and detection of decreases in the variance of the process):

Phase I. Initially it starts by obtaining multiple tables of results regarding the ARL values for different values of h as a function of k and δ.

Phase II. Using the results obtained in the previous Phase, the goal is to create an intermediate auxiliary table of results regarding $ARL_{\text{In Control}}$, as function of h and k for $\delta = 1$. Therefore this Phase summarizes the values obtained previously, but now only for $\delta = 1$, which represents a shift zero, since $\sigma_0 = \sigma_1$.

Phase III. After the construction of the auxiliary table and based on the results of that, is constructed by computer simulation and by the numerical method of successive approximations, a table of h as a function of k and $ARL_{\text{In Control}}$. The computer simulation and the use of the numerical method of successive approximations related to this phase, allows obtaining the first set of abacuses.

Phase IV. This last Phase consists in a sensitivity analysis in order to find the optimal values of k and h that are able to minimize the value of $ARL_{\text{Out of Control}}$ for a given value of δ. This sensitivity analysis is performed to the parameter δ and as well to all combinations of (k, h) obtained in the previous phase. Thus obtains finally the second set of abacuses and the final table of results.

In Fig. 78.1 is shown a flowchart that represents all the phases inherent to the procedure for the development of this Computational Tool to Support Statistical Control.

It's important to note that the processes involved in each phase of the development procedure of the FCSCE are considered in duplicate, taking into account the two studies of detection of small and moderates shifts in the variance of the process.

However the logical reasoning is very similar for the detection of increases and decreases in the variance of the process.

78.4 Results of the Application of FCSCE

The authors of this article conducted the development of the FCSCE as described in Sect. 78.3. Simulations were run for all of the phases of the current methodology considering a sample size of 5. Below are presented the results obtained for this methodology, being shown in the form of abacuses and tables.

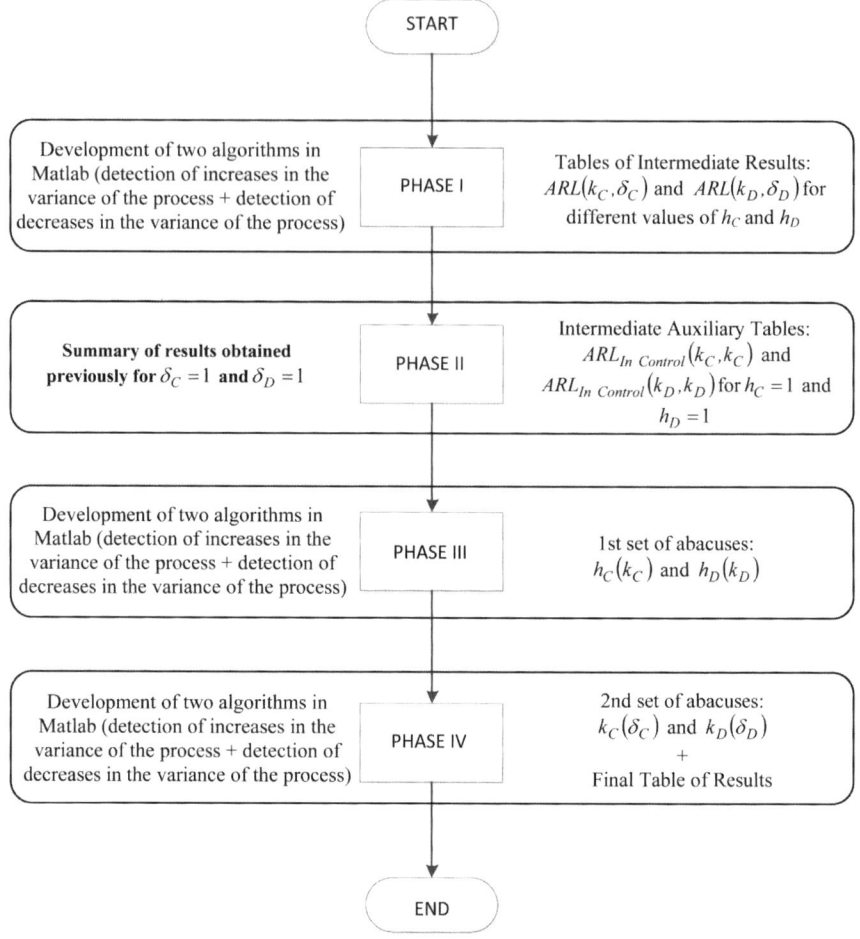

Fig. 78.1 Flowchart related to all the phases inherent in the development of FCSCE

It's also important to note that the algorithms related to these simulations were very complex which resulted in a very long time of simulation. Indeed the computational effort involved in creating the first set of abacuses (Phase I, II and III) resulted in a continuous period of time of approximately 3 months of computer processing. As to the second set of abacuses and tables (Phase IV) the computational effort took 3 and half months.

1. Abacuses

To build the best CUSUM-ln(S^2) chart, which is the one that has the lowest $ARL_{\text{Out of Control}}$ for a given $ARL_{\text{In Control}}$, it were determined the parameters of this control chart for two different studies of detection.

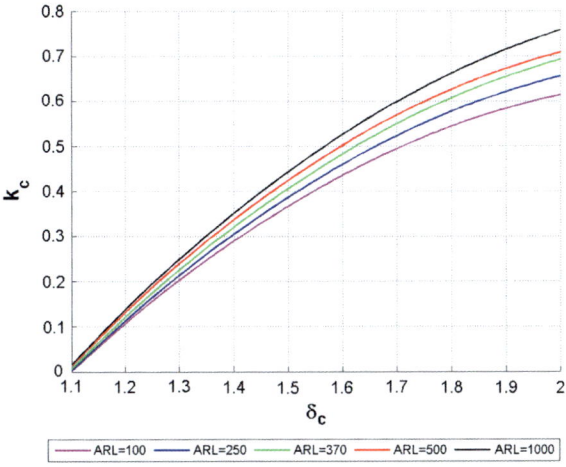

Fig. 78.2 Values of k_C in order to δ_C, related to the control chart CUSUM-ln(S^2), for several desired $ARL_{InControl}$ ($n = 5$)

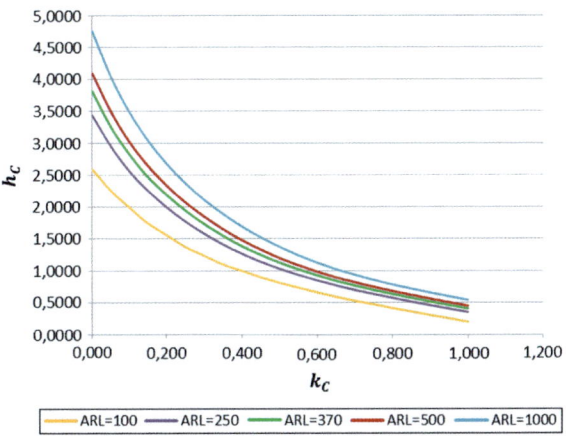

Fig. 78.3 Values of k_C and h_C related to the control chart CUSUM-ln(S^2), for several $ARL_{In\ Control}$ ($n = 5$)

Detect an increase in the variance of the process—Figs. 78.2 and 78.3; Detect a decrease in the variance of the process—Figs. 78.4 and 78.5.

2. Tables

The values of the parameters considered in the CUSUM-ln(S^2) for both studies of detection are presented in the Tables 78.2 and 78.3.

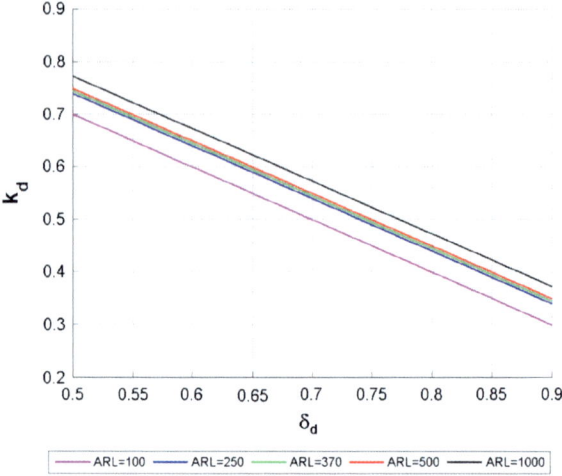

Fig. 78.4 Values of k_D and δ_D related to the control chart CUSUM-ln(S^2), for several $ARL_{\text{In Control}}$ ($n = 5$)

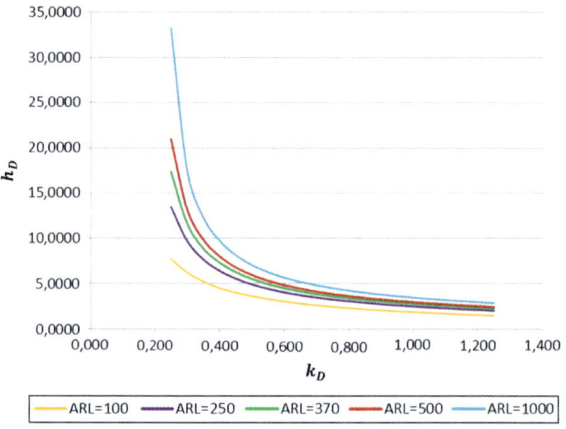

Fig. 78.5 Values of k_D and h_D related to the control chart CUSUM-ln(S^2), for several $ARL_{\text{In Control}}$ ($n = 5$)

78.5 Conclusions

This article is focusing on the development of a computational tool which was given the name of Computational Tool to Support Statistical Control (FCSCE) and that was specifically and specially developed for determining the parameters of the control chart CUSUM-ln(S^2).

The methodology in which is based consists in four sequential phases that generate a set of results in the form of abacuses and tables. These could be consulted if implementing this particular chart is the goal.

Table 78.2 Final table of results related to the detection of increases in process variance ($n = 5$)

			δ_C					
			1.10	1.20	1.30	1.40	1.50	1.75	2.00
ARL	100	$k_{C\ optimal\ (int.)}$	0.0011	0.1065	0.2026	0.2893	0.3668	0.5197	0.6144
		$h_{C\ optimal}$	2.5938	1.7500	1.3750	1.1000	0.9000	0.7375	0.6625
		$ARL_{Out\ of\ Control}$min	28.152	13.456	8.093	5.544	4.197	2.577	1.919
	250	$k_{C\ optimal\ (int.)}$	0.0023	0.1124	0.2131	0.3045	0.3865	0.5505	0.6559
		$h_{C\ optimal}$	3.4375	2.5781	2.0000	1.4125	1.0313	0.8500	0.7000
		$ARL_{Out\ of\ Control}$min	47.357	19.647	11.201	7.388	5.423	3.069	2.185
	370	$k_{C\ optimal\ (int.)}$	0.0066	0.1207	0.2252	0.3203	0.406	0.5788	0.6925
		$h_{C\ optimal}$	3.8125	3.8125	2.4844	1.5488	1.1266	0.9281	0.7000
		$ARL_{Out\ of\ Control}$min	58.120	22.682	12.582	8.211	5.862	3.290	2.311
	500	$k_{C\ optimal\ (int.)}$	0.0129	0.131	0.239	0.3367	0.4242	0.5982	0.7083
		$h_{Coptimal}$	4.0938	4.0938	2.6602	1.8516	1.1984	0.9867	0.8219
		$ARL_{Out\ of\ Control}$min	67.121	24.933	13.675	8.831	6.265	3.455	2.406
	1000	$k_{C\ optimal\ (int.)}$	0.0158	0.1373	0.249	0.3509	0.4431	0.6307	0.7572
		$h_{C\ optimal}$	4.7609	4.7609	3.0598	2.1152	1.5234	1.0250	0.8573
		$ARL_{Out\ of\ Control}$min	93.390	30.462	16.353	10.352	7.303	3.888	2.632

Table 78.3 Final table of results related to the detection of decreases in process variance ($n = 5$)

			δ_D				
			0.9	0.8	0.7	0.6	0.5
ARL	100	$k_{D\ optimal\ (int.)}$	0.3000	0.4000	0.5000	0.6000	0.7000
		$h_{D\ optimal}$	6.2500	4.5000	3.6250	3.0250	2.6000
		$ARL_{Out\ of\ Control}$min	31.0440	14.4836	8.4102	5.2798	3.5093
	250	$k_{D\ optimal\ (int.)}$	0.3400	0.4400	0.5400	0.6400	0.7400
		$h_{D\ optimal}$	7.6875	5.5469	4.9063	3.7188	3.2375
		$ARL_{Out\ of\ Control}$min	49.1603	20.5780	11.0330	6.7059	4.3827
	370	$k_{D\ optimal\ (int.)}$	0.3450	0.4450	0.5450	0.6450	0.7450
		$h_{D\ optimal}$	8.8750	6.2500	4.9297	4.1406	3.5898
		$ARL_{Out\ of\ Control}$min	57.9621	23.1185	12.2293	7.3787	4.7742
	500	$k_{D\ optimal\ (int.)}$	0.3500	0.4500	0.5500	0.6500	0.7500
		$h_{D\ optimal}$	9.8750	6.7813	5.3203	4.4531	3.8593
		$ARL_{Out\ of\ Control}$min	65.4505	25.0294	13.0950	7.8588	5.0694
	1000	$k_{D\ optimal\ (int.)}$	0.3730	0.4730	0.5730	0.6730	0.7730
		$h_{D\ optimal}$	12.3125	8.1094	6.2500	5.1836	4.2168
		$ARL_{Out\ of\ Control}$min	83.9032	30.0851	15.2196	8.9888	5.7424

As is known the tables of Chang and Gan [2] become ineffective whenever is needed to access the parameters of this control chart for situations not covered in them, and this was one of the main reasons to support the development of this computational tool. Also there aren't available the abacuses related to the numerical tables.

It is largely due to this limitation that was considered necessary to develop an alternative approach to allowing the generation of results for many different situations not covered in the studies of Chang and Gan [2]. By using the algorithms related to this tool it's only necessary to define in advance a set of values for the input variables of each phase.

Through small changes in the input values of the algorithms considered in the development of this tool, it's possible to generate a set of valid results for this particular chart. Upon completion of this study came to the conclusion that the tool has developed valid results. Therefore it'possible to reach a large set of results by defining the desired input variables.

Thus, it is considered that FCSCE is a significant contribution to the scientific community linked to the Quality area in general and in particular to the Statistical Process Control area.

78.6 Additional Conclusions

Considering the whole procedure of development the FCSCE it was concluded that the existing capabilities in most PCs for common use are still insufficient for the procedure of this tool perform more quickly. Naturally it's expected that the constant changes that occur in the computer industry soon result in the development of personal computers with better and more efficient processing skills.

References

1. Alves YM (2009) Determination of parameters of statistical control charts using specific computational tool. Master's Thesis, Faculty of Science and Technology, New University of Lisbon (In Portuguese)
2. Chang TC, Gan FF (1995) A cumulative sum control chart for monitoring process variance. J Qual Technol 27:109–119 (In Chinese)
3. Gan F (1991) An optimal design of cusum quality control charts. J Qual Technol 23(4):279–286
4. Montgomery DC (2012) Statistical quality control, 8th edn. Wiley Global Education, Hoboken
5. Oakland JS (2003) Statistical process control, 5th edn. Butterworth& Heinemann, Amsterdam
6. Pereira ZL, Requeijo JG (2012) Quality: statistical process control and planning, 2nd edn. Foundation FCT/UNL Publisher, Lisboa (In Portuguese)
7. Quesenberry CP (1997) SPC methods for quality improvement. Wiley, New York
8. Ryu JH, Wan H, Kim S (2010) Optimal design of a cusum chart for a mean shift of unknown size. J Qual Technol 42(3):311–326
9. Shewhart WA (1931) Economic control of quality of manufactured product. D. Van Nostrand Company, New York, p 501
10. Shu L, Yeung HF, Jiang W (2010) An adaptive cusum procedure for signaling process variance changes of unknown sizes. J Qual Technol 42(1):69–85
11. Sparks RS (2000) Cusum charts for signalling varying location shifts. J Qual Technol 32(2):157–171
12. Woodall WH (1986) The design of cusum quality control charts. J Qual Technol 18(2):99–102

Chapter 79
Credit Risk Measurement of the Listed Company Based on Modified KMV Model

Liang Dong and Junchao Wang

Abstract In this paper, the non-ferrous metal industry has been employed to build a credit risk measurement model. By modifying the model parameters and setting five different default points, we conformed that the predicted results of original KMV model was invalid, while the revised model has a better recognition ability between the blue chips and low quality stocks, under the redefining the default distance. It's best to set the default point to the short-term debt. The results showed that the revised KMV model was able to improve the validity of the model and monitor the change of the credit risk of listed companies more accurately.

Keywords Credit risk · KMV model · Default distance · Listed company

79.1 Introduction

Credit risk, which is the main types of financial risk, is always the core content of risk control by financial institutions and regulators. The traditional expert grading method, discrimination analysis and Logit model cannot meet the requirements of modern risk management. A lot of research scholars at home and abroad, put forward many measures of credit risk models, including the KMV model (with the analysis of stock price changes, the credit risk monitoring model) which is relatively mature and most widely used. But in the aspect of credit risk measurement in China is still in the stage of traditional method. So the research about applying KMV model more effectively to the Chinese financial market to improve the level of credit risk management of financial institutions is necessary.

L. Dong · J. Wang (✉)
School of Economics and Management, Jiangsu University of Science and Technology,
Zhenjiang 212003, People's Republic of China
e-mail: wjclifegood@126.com

J. Xu et al. (eds.), *Proceedings of the Eighth International Conference on Management Science and Engineering Management*, Advances in Intelligent Systems and Computing 281, DOI: 10.1007/978-3-642-55122-2_79, © Springer-Verlag Berlin Heidelberg 2014

As is known to all, different industries in the external environment and the characteristics of capital structure have little in common, neither the default trigger conditions. In this paper, the non-ferrous metal industry listed company which is highlighted in credit risk has been taken as an example, to make the comparative analysis of prediction effect before and after the correction of KMV model, which provides the reference for commercial Banks and strengthen the control of credit risk.

79.2 KMV Model and Literature Review

1. The Principle and Calculation Steps of KMV Model

KMV model established by KMV company (purchased by Moody) is a modern credit risk measurement model which reverses the bank lending problem on the basis of Black [1] research and Merton models [6] based on option pricing theory. Assuming the company's assets consisting of equities, bonds and other debt, including bank loans, bank loans must be repaid within a specified time, or be treated as default. If the asset value after 1 year is greater than the value of its debt, the company would not break the contract, and the company asset value at this level is defined as the default point, conversely, the company will select default. That means similarly when the shareholders of the company borrow an amount of debt, it is the equivalent of buying a call option, a company's asset value as the underlying asset, default point as the strike price. What's more, KMV company thinks corporate default point is between current liabilities and total liabilities. Computation process of this model is usually divided into the following three steps:

(1) Estimation of the value of asset and its volatility value

According to the theory of equity value in Merton model, using the BSM option pricing formula, the structural relationship between the asset value and enterprise equity market can be got as follows:

$$V_E = V_A N(d_1) - De^{-r\tau} N(d_2),$$
$$d_1 = \frac{\ln(V_A/D) + (r + \sigma_A^2/2)\tau}{\sigma_A \sqrt{\tau}},$$
$$d_2 = d_1 - \sigma_A \sqrt{\tau},$$

where V_E refers to market value of firms equity, V_A refers to market value of firms asset, D refers to nominal value of firm's debt, r refers to risk free rate, τ refers to the rest of the debt repayment period, N refers to cumulative standard normal distribution function.

Volatility and equity are related under the Merton model as follows: $\sigma_E = \frac{V_A}{V_E} N(d_1)\sigma_A$, then, V_A and σ_A can be calculated.

(2) Estimation of the default distance (DD)

KMV company takes default point as the value of short-term debt plus half the book value of all long term debt: $DP = STD + \frac{1}{2}STD$.

As a commercial secret, KMV company did not explain how to get the relationship.

By KMV company default distance is defined as multiples that a year later the distance between the default point and the expected value of assets relative to the standard deviation of the future asset returns, its computation formula is as follows: $DD = \frac{E(V_A) - DP}{E(V_A)\sigma_A\sqrt{\tau}}$. This is a standardized index of default risk measurement, so different listed companies also can use this index to compare. The greater the distance to default, it shows that the value of assets, the farther away from the point of default expiration and the smaller the possibility of default.

(3) Estimation of the Expected Default Frequency (EDF)

The theoretical EDF is rigorous mathematical derivation and $EDF = 1 - N(DD)$. But the assumption of that is still in a large academic debate. KMV company set up a relationship between the default distance and the probability of default, based on its large sample information database. But at present there is no a similar database in our country, therefore, this article uses the default distance directly as indicators of credit risk.

2. Literature Review

Since the KMV model creation, domestic scholars have done a lot of research, hoping to apply the model to the financial market in our country, the research mainly focus on the following several aspects. The first is the emphasis on the principle of KMV model, and comparison with several other credit risk models. The second is directly applying original KMV model to our country stock market. The empirical results by Chen [2] showed that the KMV model had good applicability in our country. Another major is to modify the model parameters, enable it to meet our country's economy background. Peng [8], Zhang [11] through correction of individual parameters of the model, found that the revised model has better prediction ability. What's more, many scholars applied KMV model to other aspects. Li [4] introduced the model to the local government bond credit risk research field. At present, many scholars are making further research on credit risk. Ni et al. [10] amended the model by using uncertain interest rate to compare the uncertain KMV model to traditional KMV model Dierkes [3] found that business credit information sharing substantially improves the accuracy of aggregate and firm-specific default predictions. da Silva [7] showed that credit risk is procyclical and default risk depends on structural features and the banking regulator is able to set up a policy to promote financial stability and efficiently reduce fluctuations in the output.

More and more scholars are trying to perfect the KMV model in the aspect of credit risk measurement in China, but mainly in the model parameter optimization, not the optimization of default distance correction. And there is lack of study aiming at a specific industry, as we all know that there are significant differences in asset structure, debt characteristics, in different industries. In order to improve the level of credit risk management in China, it's better to modify KMV model for a certain industry, which is the main purpose and new contributions of this paper.

79.3 Sample Selection and Model Parameters Correction

1. Sample Selection
This article sets the first day of 2011 as basis points, to predict the future change of credit risk in 1 year. First of all, the non-ferrous metal industry listed companies can be divided into blue chip stocks and low quality stocks. Then 20 listed companies were selected as research samples in their respective categories, the selected 40 listed companies must have listed before 2008, and in the forecasting point shareholding reform completed and non-current liabilities existed.

2. Parameters Correction
(1) Demarcation of the value of equity
Because our country has basically completed the shareholding reform, the calculation of the value of equity will no longer consider the problem of non-tradable shares, considering the possible malicious speculation on the fluctuation of stock price in Chinese stock market, this paper multiplied average closing price in 20 trading days before the basis points by the shares outstanding to calculate the value of the equity of the day.

(2) The volatility of equity value
Assuming that share prices obey the lognormal distribution, through the historical data of stock return standard deviation to estimate the annual standard deviation of the share price volatility, calculation method is as follows:

$$\sigma_E = \sigma_E' \sqrt{n}, \ \ \sigma_E' = \sqrt{\frac{1}{n-1} \sum_{t=1}^{n} (\mu_t - \mu_{t-1})^2}, \ \ \mu_t = \ln\left(\frac{S_t}{S_{t-1}}\right),$$

where n refers to trading days in 1 year, S_t refers to t day's closing price, μ_t refers to t day's Logarithmic yield, σ_E refers to annual volatility, σ_E' refers to daily volatility.

(3) The risk-free rate
Because our country is lack of the risk-free rate of data in a strict sense, according to the 1-year term lump sum time deposit interest rates published by the people's bank in 2010, this paper calculated the risk-free rate on a daily basis of the weighted average, $r = 0.0230$.

(4) The growth rate of assets value
KMV company used the return on assets deducted company bonuses and payment of interest to estimate the expected growth rate of assets, but in China listed company's dividend is mainly in the form of stock dividend and transfer which does not affect the company's equity structure, and some scholar's study found that China's stock dividends do not affect the change of the value of equity. What's more, few in number, cash dividends is very few, so this paper ignored the impact of these factors on the value of equity. Many scholars assumed the company's predicted growth rate of assets value to zero during the forecast period, but this does not conform to the actual situation. Yang et al. [9] found that using the financial statement data to calculate the

growth rate was more effective. So this paper take the arithmetic average growth rate of the total assets in 3 years as growth rate of asset value in the prediction period.

(5) Selection of default point
For Chinese capital market is still in the development and non-ferrous metal industry is different from others in the degree of funds utilization, debt characteristics also different, default point chose by KMV company is not necessarily suitable for non-ferrous metals industry in China. We set the default point of the non-ferrous metal industry listed companies as $DP = STD + aLTD$, where "a" is respectively defined $0, 0.25, 0.5, 0.75, 1$ to determine the best default point of prediction effect by comparison.

79.4 Empirical Research

1.Sample Data Calculation
According to the nonlinear equations mentioned above and model parameter setting, by Matlab 7.0 programming iteration to calculate the value of the assets and its volatility in each default point, then calculates the default distance of two types of sample companies under each default point DD1, DD2, DD3, DD4, DD5. Calculation results are shown in Table 79.1.

Obviously from the above-mentioned, default distances of two kinds of sample companies are not significant differences, even there is on obvious differences between the ST (special treatment) companies and not ST companies, which cannot achieve good prediction effect. For example, 600338 (ST Zhufeng) in 2010 whose annual report showed that its net profit, earnings per share, operating cash flow per share are negative and asset-liability ratio reached 100.76 %, the financial situation has seriously deteriorated, but relative to other sample firms have not shown a very low default distance, apparently do not tally with the actual situation.

One of the main reasons for the failure of the model application is that default distance was not set unreasonably. Firstly the assets value calculated according to the stock price volatility is always positive and the original formula did not reflect the direction of the asset value volatility. Secondly the original formula did not reflected solvency under certain default point. Therefore this paper uses the redefined default distance put forward by Liu [5]:

$$DD = \begin{cases} \frac{E(V_A)+E(V_A)^*\sigma_A-DP}{DP}, & \text{Net profit is positive} \\ \frac{E(V_A)-E(V_A)^*\sigma_A-DP}{DP}, & \text{Net profit is negative.} \end{cases} \qquad (79.1)$$

This is also a standardized index which can be used in comparison between the different listed companies. To calculate the default distance using the same sample data again, and the results are shown in Table 79.2.

Table 79.1 The default distance of 40 sample companies

Stock code	DD1	DD2	DD3	DD4	DD5	Stock code	DD1	DD2	DD3	DD4	DD5
600547	1.32	1.33	1.33	1.33	1.33	000762	1.86	1.86	1.86	1.86	1.86
600595	1.44	1.45	1.47	1.48	1.49	600311	1.86	1.86	1.86	1.86	1.86
600489	1.44	1.45	1.45	1.46	1.46	600456	1.92	1.92	1.92	1.92	1.92
600111	1.87	1.87	1.87	1.87	1.88	000807	2.28	2.32	2.35	2.38	2.41
002155	1.82	1.82	1.82	1.82	1.82	000426	1.87	1.87	1.87	1.87	1.87
600366	1.69	1.69	1.70	1.70	1.70	600711	2.62	2.62	2.62	2.62	2.62
600888	1.92	1.93	1.94	1.95	1.95	000697	2.26	2.26	2.26	2.26	2.26
600497	2.28	2.28	2.29	2.29	2.30	600961	1.87	1.87	1.87	1.87	1.87
000060	1.41	1.42	1.42	1.43	1.43	600338	2.77	2.77	2.77	2.77	2.77
000960	1.66	1.66	1.66	1.67	1.67	002114	2.01	2.01	2.01	2.01	2.01
600219	3.28	3.29	3.29	3.30	3.30	600331	2.12	2.12	2.12	2.13	2.13
601168	2.16	2.16	2.16	2.16	2.16	000751	2.16	2.16	2.16	2.16	2.17
601958	1.95	1.95	1.95	1.95	1.95	000831	1.43	1.19	0.96	0.73	0.50
002237	2.50	2.51	2.52	2.54	2.55	002167	1.15	1.15	1.15	1.15	1.15
600139	2.08	2.08	2.08	2.08	2.08	600259	1.64	1.64	1.64	1.64	1.64
600193	2.38	2.38	2.39	2.39	2.39	600459	1.91	1.91	1.91	1.91	1.91
002171	2.17	2.17	2.17	2.17	2.17	000758	1.73	1.74	1.74	1.74	1.74
00612	2.08	2.08	2.08	2.08	2.09	600255	2.01	2.01	2.01	2.01	2.01
000962	1.89	1.89	1.90	1.91	1.92	002149	2.04	2.05	2.05	2.06	2.06
600432	2.16	2.17	2.17	2.18	2.19	002160	2.15	2.15	2.15	2.15	2.15
Mean	1.97	1.98	1.98	1.99	1.99	Mean	1.98	1.97	1.96	1.96	1.95

2. The Empirical Result Analysis

Testing whether there was a significant difference in redefined default distance between the two kinds of sample companies, and selecting the most appropriate default point to improve the applicability of the model is needed. Using statistical software SPSS 16.0, the results of independent sample t-test are shown in Table 79.3.

The above test results showed that all the five F value concomitant probability of default distance are less than 0.05 significant level, so reject the assumption of equal variances, there are significant differences between the two types of companies variance; In the range of variance t test results showed that only the DD1 concomitant probability is 0.038 that is less than significance level of 0.05, and refused to t test null hypothesis, which indicated that two types of the average default distance DD1 exist significant differences. In addition, the sample mean difference of 95 % confidence interval, only DD1, does not contain 0, which also suggests that two types of the average default distance DD1 exists significant differences.

In the sample default distance sorting, the last eight enterprises are 000697, 002160, 600691 (ST), 600338 (ST), 002114 (ST), 600331, 000751 (ST), 000831 (ST), respectively, and five of which had been special treatment (ST). All their default distance are less than 7, and the financial statements of the eight companies in 2011 showed that in addition to 002160, 600331, 000831, the other five companies' net profits are negative. Though 000831 (ST Guan lv) surplus, its operating cash

Table 79.2 The refined default distance of 40 sample companies

Stock code	DD1	DD2	DD3	DD4	DD5	Stock code	DD1	DD2	DD3	DD4	DD5
600547	65.17	56.98	50.62	45.55	41.42	000762	59.78	54.4	49.91	46.11	42.84
600595	52.14	25.39	16.86	12.66	10.17	600311	40.44	40.34	40.24	40.14	40.03
600489	35.28	27.81	22.97	19.58	17.07	600456	15.44	14.94	14.47	14.03	13.61
600111	34.7	31.94	29.6	27.57	25.81	000807	8.27	6.28	5.08	4.28	3.71
002155	33.21	32.42	31.67	30.95	30.27	000426	7.74	7.29	6.88	6.52	6.2
600366	23	22	21.07	20.23	19.45	600711	7.13	6.85	6.59	6.35	6.13
600888	19.35	15.7	13.22	11.42	10.06	000697	6.67	6.65	6.63	6.6	6.58
600497	14.9	13.94	13.09	12.34	11.68	600961	4.63	4.25	3.93	3.65	3.41
000060	11.77	10.66	9.75	8.98	8.32	600338	2.94	2.94	2.93	2.92	2.92
000960	11.16	10.4	9.74	9.15	8.64	002114	2.29	2.23	2.17	2.11	2.06
600219	10.44	10.35	10.27	10.18	10.09	600331	1.56	1.44	1.34	1.26	1.18
601168	8.2	7.82	7.48	7.17	6.88	000751	0.66	0.65	0.65	0.65	0.66
601958	207.61	192.11	178.76	167.15	156.95	000831	0.23	0.18	0.15	0.13	0.11
002237	8.19	7.95	7.73	7.52	7.32	002167	34.34	33.85	33.38	32.92	32.47
600139	142.29	142.14	141.99	141.85	141.71	600259	26.25	25.93	25.62	25.32	25.02
600193	59.98	28.14	18.39	13.66	10.87	600459	12.79	12.22	11.7	11.21	10.77
002171	12.58	12.42	12.26	12.1	11.95	000758	8.96	7.78	6.87	6.16	5.58
000612	12.53	11.22	10.15	9.28	8.54	600255	8.95	8.85	8.76	8.66	8.57
000962	12.01	10.6	9.5	8.61	7.87	002149	8.08	7.39	6.81	6.31	5.89
600432	9.52	8.78	8.16	7.62	7.15	002160	4.7	4.67	4.64	4.61	4.57
Mean	39.2	33.94	31.16	29.18	27.61	Mean	13.09	12.46	11.94	11.5	11.12

flow per share was −0.09, asset-liability ratio is 98.76 %, the credit risk was still very serious. Easy to see, calculation result of the adjusted KMV model is in line with the actual situation, and 1 year ahead of the forecast effect is good.

Visibly, the different settings of default points have an impact on listed companies to identify two types of credit situation, above studies show that setting the non-ferrous metal industry listed companies default point to short-term debt and using the redefined default distance can get a better prediction of the changes of credit risk. As to the reasons, further study found that the majority of listed companies in the non-ferrous metals industry liabilities are mainly short term, and long term liabilities will not expire during the forecast period. When the assets value is less than the short-term liabilities, companies would be faced with enormous repayment pressure and shareholders may choose to default, which is characterized by a greater relationship with the certain industry.

Table 79.3 The refined default distance of 40 sample companies

	F	Sig.	t	df	Sig. 2- tailed	Mean difference	Std. error difference	95 % Confidence interval of the difference	
								Lower	Upper
DD1	6.348	0.016	2.201	38	0.034	26.1047850	11.8626083	2.0901900	50.1193800
			2.201	22.519	0.038	26.1047850	11.8626083	1.5356873	50.6738827
DD2	4.427	0.042	1.920	38	0.062	21.4797750	11.1872858	−1.1677011	44.1272511
			1.920	22.561	0.068	21.4797750	11.1872858	−1.6878492	44.6483992
DD3	4.449	0.042	1.794	38	0.081	19.2247600	10.7189134	−2.4745458	40.9240658
			1.794	22.550	0.086	19.2247600	10.7189134	−2.9735184	41.4230384
DD4	4.459	0.041	1.713	38	0.095	17.6810000	10.3208292	3.2124263	38.5744263
			1.713	22.551	0.100	17.6810000	10.3208292	−3.6957918	39.0547918
DD5	4.426	0.042	1.653	38	0.107	16.4951200	9.9787283	−3.7057593	36.6959993
			1.653	22.561	0.112	16.4951200	9.9787283	−4.1697119	37.1599519

79.5 Conclusions

In order to make better use of original KMV model to Chinese financial market, first select the serious credit risk of listed companies in non-ferrous metals industry as the research object, the parameters of the original KMV model have been improved, including calculation methods of equity value, risk free rate, the growth rate of asset value. Then set up five different default points to build a more accurate KMV model and measure the credit risk of listed enterprises. However, the results show that the definition of original default distance do not well distinguish the level of credit risk among different types of companies, and there is even contrary phenomenon to the actual situation. The use of the redefinition of the default distance can effectively distinguish blue chip stocks and underperformance, significantly improving the ability of credit risk measurement model. And for non-ferrous metals industry, above study shows that setting the default point as the amount of short-term is more reasonable. The adjusted KMV model has good prediction ability of predicting changes in the credit risk of listed companies, which is accord with the debt characteristics of non-ferrous metal industry listed companies.

Certainly, the above study is conducted under certain assumptions, and there is no breach of the default distance into the probability of default, which is the inadequacies of this article, and further research is needed.

References

1. Black F, Scholes M (1973) The pricing of options and corporate liabilities. J Polit Econ 8:637–654
2. Chen H, Xia H (2012) The credit risk measurement of listed companies in China and the empirical research of its influence factors. Res Finance Educ 25(1):28–34
3. Dierkes M, Erner C et al (2013) Business credit information sharing and default risk of private firms. J Bank Finance 37:2867–2878
4. Li J, Wang J (2011) Analysis on credit risk of local government bonds based on the KMV model. J Finance Econ 26(5):23–31
5. Liu Z, Liu X, Zheng G (2011) Credit default risk management of listed companies based on IKMV model. Chin J Manage Sci 19:367–373
6. Merton R (1974) On the pricing of corporate debt: the risk structure of interest rates. J Finance 29(2):449–470
7. da Silva M, Divino J (2013) The role of banking regulation in an economy under credit risk and liquidity shock. North Am J Econ Finance 26:266–281
8. Wei P (2012) Study on credit risk of listed small and medium size enterprises based on the analysis of KMV model. South Financ 3:6
9. Yang Y, Zhou Z (2010) Growth rate of asset value in corporate credit risk assessment. Inq Econ Issues 7:93–98
10. Zhan N, Lin L, Lou T (2013) Research on credit risk measurement based on uncertain kmv model. J Appl Math Phys 1:12–17
11. Zhang P, Cao Y (2012) Credit risk measurement study of listed companies. Res Financ Econ Issues 3:11

Chapter 80
Why Do They Leave? The Mechanism Linking Employees' Psychological Capital to Their Turnover Intention

Chen Zhao and Zhonghua Gao

Abstract In this study, the influence mechanism of psychological capital on employees' turnover intention has been revealed by a moderated mediation model based on conservation of resource theory. 600 employees from seven branches of a high-tech corporation group located at Beijing and Hangzhou were involved in this survey. Regression analysis and moderated path analysis were adopted and the results show that: (a) psychological capital is negatively associated with employees' turnover intention; (b) this association is fully mediated by role stress; and (c) organizational support moderates not only the influence of psychological capital on role stress, but also the mediating effect of role stress on the association between psychological capital and employees' turnover intention. The results implicate that that employees' psychological capital can decrease turnover intention by alleviating their role stress; meanwhile, organizational support positively exacerbates the decrease of psychological capital on turnover intention. Our results can provide guidance to improve human resource management practices for Chinese enterprises in future.

Keywords Psychological capital · Turnover intention · Organizational support · Moderated mediation model

80.1 Introduction

Recently, psychological capital has attracted much attention from the scholars in both fields of positive psychology and positive organizational behavior. Prior studies demonstrated that individuals with higher psychological capital are more likely

C. Zhao
School of Management, Capital Normal University, Beijing 100089, People's Republic of China

Z. Gao (✉)
College of Business Administration, Capital University of Economics and Business, Beijing 100070, People's Republic of China
e-mail: gzhruc@gmail.com

J. Xu et al. (eds.), *Proceedings of the Eighth International Conference on Management Science and Engineering Management*, Advances in Intelligent Systems and Computing 281, DOI: 10.1007/978-3-642-55122-2_80, © Springer-Verlag Berlin Heidelberg 2014

to think about the problems and evaluate the environments by more positive and optimistic way, which leads to higher level of subjective well-being, and job performance [2]. In addition to these positive results, there are some negative influences of psychological capital on work attitude and behavior [3], especially on turnover intention [1, 4]. However, few studies revealed the in-depth influence mechanism of psychological capital on turnover intention.

In this study, we take role stress as a mediator to explain the influence of psychological capital on the turnover intention. In reality, everyone faces multiple role expectations or requests from various constituencies, as a result when people cannot meet role expectations or requirements using their psychological resources, they might experience some tension that can be described as role stress [15]. In other words, people with lower psychological capital often feel higher role stress. Consequently, role stress will lead to higher turnover intention which could be regarded as a very typical role stress reaction [11, 24, 35], since results from previous studies indicates that role stress has significant negative impact on staff's working attitude [26]. Therefore, the mediating effect of role stress between psychological capital and turnover intention will be tested in this study.

In addition, many studies showed that, as a state characteristic, psychological capital can be developed through certain procedures and its formation and function will be influenced by training developments and supportive interventions, such as creating supportive atmosphere [23]. Thus, we assume that the psychological capital can predict outcome variables more significantly through interacting with organizational support. Therefore, the main effects of psychological capital on turnover intention, the mediating effect of role stress and the moderating effect on this mediation will be analyzed in this study.

80.2 Theoretical Background and Hypotheses

1. Conservation of Resource Theory

More specifically, the resource conservation theory, a branch of the psychological resources theories, can be used to better explain the psychological capital formation and its mechanism. According to this theory, people are always trying to obtain and maintain valuable resources, such as work control and decision, autonomy, self-efficacy, self-esteem and so on [15, 16], which can be used to inspire people effectively ponder and deal with problems in work environment. However, various role expectations people often face would lead to role stress [18]. In order to alleviate their stress feeling, people need to seek opportunities to obtain new resources, for example, by making investment at the cost of other resources or accepting training, thereby to satisfy the existing demands or avoid potential loss of resources for the future. However, sometimes, acquiring new resources requires people to give up or consume some existing resources. In this case, people would carry on some assessments or evaluations, and thus people can decide whether to maintain existing resources. If not, they would reduce resources investments on meeting other role requirements. If people want to give up roles related to work, intention to leave the organization would be stirred up and strengthened.

According to the resource conservation theory, people's psychological capital will enhance their ability to meet multiple roles, and thus reduce their intention to leave the organization. The higher the psychological capital, the more psychological resources people will possess. Consequently, they will have t higher ability to decrease the pressure caused by their multiple roles. Besides, the requirements and expectations from the multiple roles will be effectively satisfied. These results might lead to the decrease of the intention to leave current position and organization caused by role stress. In the workplace, various interventions, such as some training developments, employee assistant programs and etc, are main approaches to help individuals get more psychological resources and increase their psychological capital. Supports provided by organization, for example, prompt incentives, supportive environment atmospheres, and psychological counseling, would promote the formation and function of employees' psychological capital [21].

2. Main Effects of Psychological Capital on Employees' Turnover Intention

Due to huge potential cost and destruction to organization, the phenomenon of voluntary turnover has been widely studied since its initial inception [7, 12]. In previous studies, the factors that lead to employees' voluntary turnover have drawn more attention [14, 17]. According to Lee et al.'s unfolding model of turnover, there are five psychological paths that can be used to explain employees' voluntary turnover decision, and the source of these paths is the "system shock" that is widely experienced by employees, such as organizational change and downsizing; however, because of individual differences, people have different levels of ability to cope with these "system shock" [20].

According to the resource conservation theory, employees with higher psychological capital often possess more psychological resources (self-efficacy, hope, optimism, and resiliency) and higher self-motivation, thus they are always confident in their competency to cope with the changes at work, take positive efforts to integrate various resources, and overcome various obstacles and barriers to their success. When faced with setbacks, employees with higher psychological capital often persevere in trying all possible ways to seek the breakthrough, and never give up the current work [34]. Therefore, we assume that:

Hypothesis 1. psychological capital is significantly associated with employees' turnover intention such that employees with higher psychological capital often have lower turnover intention.

3. Mediating Effect of Role Stress

According to the role theory, people play multiple roles in social interactions which compel them to meet various requirements and expectations [18]. When people are unable to meet all role requirements and expectations, they would experience a kind of psychological tension which is proposed as role stress. In literature, there are three kinds of role stress: role ambiguity, role conflict and role overload [8]. Role ambiguity refers to the situation that people cannot discern the specific role requirements and expectations they should deal with [19]; role conflict refers to the incompatibility of the requirements from multiple roles, which means that the satisfaction of one role would lead to the unfulfilling of other role [33]; role overload means that the requirements from various roles exceed the people's coping ability [38].

It was proved by previous studies that both employees' work attitude and behavior were negatively and destructively influenced by role stress [26]. For instance, role stress would significantly enhance employees' turnover intention [27, 35, 37]. It was revealed that turnover intention was negatively related to role ambiguity, role conflict and role overload based on a survey conducted among 887 full-time pastors in Hong Kong [25]. Besides, the effects of role stress on other outcomes have also been researched, such as job burnout, psychological retreat, etc. However, it has not been answered that which factors lead to higher role stress.

According to the resource conservation theory, psychological capital can be considered one of the major factors predicting and alleviating role stress. If people have lower psychological capital, they tend to be lack of confidence in satisfying multiple role expectations; when faced with multiple roles, people tend to lose the sense of direction and cues, which leads to role ambiguity; when their limited resources were invested into the limited roles, they would have no time to answer for other role calls, which leads to role conflict; when people don't have enough energy to deal with various role requirements and expectations, they would easily feel exhausted, which leads to role overload. Instead, people with higher psychological capital can meet all role requirements and expectations breezily. Thus, psychological capital can exert an alleviation on role stress by enhancing employees' ability to cope with role ambiguity, role conflict and role overload. Therefore, we assume that:

Hypothesis 2. psychological capital is negatively and significantly associated with role stress such that role stress plays as a mediator between psychological capital and turnover intention.

4. Moderating Effects of Organizational Support

Organizational support refers to employees' subjective perception that organization values their contribution and care about their individual well-being [10]. With respect to social exchange theory, there are mutual identification and commitment between organizations and employees [5]. According to the principle of reciprocity [13], when employees perceive higher organizational support, such as rewards and recognition, their recognition and commitment to the organization would be enhanced by reducing the unexpected outcomes such as psychological retreatment and turnover intention, and improving the expected outcomes such as job involvement and organizational citizenship behavior. In addition, organizational support can also offer psychological security to employees. When they encounter problems and confusions in work, organization would provide them with the necessary help and support, which can improve the job satisfaction and reduce work pressure [32]. Besides, organizational support can also buffer the influence of negative factors, for instance, the negative influence of work family conflict on job performance [36]; and enhance the influence of positive factors, such as the positive influence of trust on help behaviors.

In this study, we assume that organizational support can moderate the influence of psychological capital on role stress. According to the resource conservation theory, higher organizational support means providing more clear job requirements, and thus people with higher psychological capital will be much easier to see the directions, goals and expectations, and feel less role ambiguity; higher organizational support also means providing more assistance and guidance, and thus people with higher psychological capital will be more confident and optimistic, and believe that their persistence can help them meet multiple role requirements and expectations. In this case, people feel less role conflict and overload. If they are lack of both clear job requirements and necessary help and guidance, employees may feel stress caused by role ambiguity, conflict and overload even if they have high psychological capital. Therefore, we assume that:

Hypothesis 3. organizational support positively moderates the association between psychological capital and role stress such that when employees feel less organizational support, psychological capital would not alleviate role stress to a smaller scale; when they perceive more organizational support, their psychological capital would more significantly alleviate role stress.

In addition, as an important assistant means to reduce employees' role stress, organizational support may moderate the relationship between role stress and turnover intention. In other words, assistance provided by organizations can not only decrease the role stress perceived by employees, but also reduce the negative influences caused by role stress. These ideas have been proved both directly and indirectly [28]. Combined with the moderating effect described above, we assume that:

Hypothesis 4. organizational support positively moderates the mediation of role stress between psychological capital and turnover intention such that when employees feel less organizational support, the mediation of role stress will be lower; when they perceive more organizational support, the mediation of role stress will be higher.

80.3 Method

1. Procedures

In this study, surveys were conducted in seven companies located at Beijing and Hangzhou, branches of a domestic automatic control system manufacturer. With the coordination provided by human resources department, researchers went to each branch to collect data. During the surveys, complete confidential has been confirmed to all subjects. Researchers gave prompt answers to questions raised by subjects. Meanwhile, guidance and manual were provided to subjects before surveys. Questionnaires were collected after the completion as soon as possible, thus higher return rate has been assured in each survey. Totally, 592 questionnaires have been returned with 600 distributed and 545 questionnaires can be used in the study, which means a return rate of 90.8 %.

2. Measurements

Psychological capital. Psychological capital was measured by 24 items in 4 dimensions adopted from Luthans et al. [22]. Sample items were like "I believe I can solve long-term problems, and figure out solutions" and "at present, I'm in full energy to achieve my job objectives", and three of them were recorded, for example, "when I face setbacks at work, I can hardly recover and move on". Reliability of (Cronbach-α) of psychological capital is 0.88. Five point Likert scale was adopted, the same below.

Role stress. Role stress was measured by 13 items in 3 dimensions adopted from Peterson et al. [29]. Sample items were like "I'm used to dealing with some conflict situations" and "at work, I feel too much burden", and four of them were recorded, for example, "I always make clear plans and goals". Reliability of (Cronbach-α) of role stress is 0.79.

Organizational support. Organizational support was measured by 8 items abstracted from Eisenberger et al. [10]. Sample items were like "our unit attaches great importance to my goals and values" and "our unit really care about whether I have a good life", and two of them were recoded, for example, "my interest cannot be guaranteed in this unit". Reliability of (Cronbach-α) of organizational support is 0.76.

Turnover intention. Turnover intention was measured by 5 items adapted from Bluedorn [6]. Sample items were like, "I probably change my job next year" and "I'm not going to stay in this unit to continue my career." Reliability of (Cronbach-α) of turnover intention is 0.92.

Control variables. In order to exclude unexpected influence of company difference, we introduced the sixth company as the reference and other companies as dummy variable into regression because samples in this study are from seven branches. Meanwhile, we introduced sex, age, education, tenure, position and income as control variables. In this study, 52.84 % of subjects are male, and 47.16 % are female; 70.28 % are from 20 to 30 years old; 70.09 % hold bachelor degree or above; 87.71 % have five years or less working experience; 75.96 % are ordinary employees; and most employees' income are distributed at two intervals: 2000–5000 (45.32 %) and 5000–8000 (29.36 %).

80.4 Results

1. Descriptive Analysis

Table 80.1 shows mean, standard deviation, reliability (Cronbach-α) and correlation coefficient matrix of all variables. Reliabilities of psychological capital, role stress, organizational support and turnover intention are all >0.7, which represents high internal consistency of all constructs. Psychological capital is negatively correlated to turnover intention ($\beta = 0.25$, $p < 0.01$), which preliminarily support hypothesis 1.

Table 80.1 Mean, standard deviation, reliability, and correlation[a]

Variables	1	2	3	4	5	6	7	8	9	10
Sex										
Age	0.09**									
Education	0.13**	0.02	S							
Tenure	−0.02	0.49**	−0.05							
Position	0.01	−0.24**	−0.03	−0.26**						
Income	0.18**	0.30**	0.43**	0.20**	−0.30**					
Psychological capital	0.02	0.05	0.03	0.04	−0.08*	0.08*	0.88			
Role stress	0.06	0.10**	0.02	0.10**	0.00	−0.05	−0.34**	0.79		
Organizational support	−0.04	−0.06	0.09**	−0.09**	−0.05	0.13**	0.15**	−0.40	0.76	
Turnover intention	−0.01	0.02	−0.09**	0.03	−0.03	−0.07*	−0.25**	0.54**	−0.50**	0.92
Mean	0.53	1.33	2.97	1.66	3.67	2.44	3.61	2.78	2.96	2.76
S.D.	0.5	0.55	0.89	0.8	0.68	1.02	0.44	0.48	0.55	0.93

[a] $N = 545$; **$p < 0.01$; *$p < 0.05$; bold numbers represent reliabilities of all scales (Cronbach-α)

Table 80.2 Results of regression[a]

Dependent variables	Model 1 Turnover intention	Model 2 Role stress	Model 3 Turnover intention	Model 4 Role stress	Model 5 Turnover intention
Sex	0.03	0.08	−0.04	0.04	−0.07
Age	0.04	0.08	−0.04	0.07	−0.05
Education	−0.06	0.04	−0.10*	0.05*	−0.07
Tenure	0.02	0.06	−0.04	0.03	−0.07
Position	−0.07	0.00	−0.07	−0.01	−0.09
Income	−0.05	−0.05*	0.00	−0.03	0.03
Branch 1	0.70	0.23	0.47	0.33	0.75
Branch 2	0.52	0.19	0.33	0.31	0.61
Branch 3	0.63	0.28	0.34	0.44	0.72
Branch 4	0.91	0.49	0.41	0.54	0.67
Branch 5	0.40	0.26	0.14	0.20	0.17
Branch 7	0.59	0.10	0.48	0.27	0.84
Psychological capital (PC)	−0.52**	−0.37**	−0.14	−0.34**	−0.11
Role stress (RS)			1.02**		0.77**
Organizational support (OS)				−0.27**	−0.59**
PC × OS				−0.22**	−0.07
RS × OS					0.10
Intercept	−0.11	−0.42	0.32	−0.50	0.05
R^2	0.08	0.16	0.32	0.28	0.42
Adjusted R^2	0.06	0.14	0.30	0.26	0.40
F	3.67	7.74	17.70	13.88	22.34

[a] $N = 545$; $**p < 0.01$, $*p < 0.05$

2. Mediating Effect of Role Stress Between Psychological Capital and Turnover Intention

Hypothesis 2 describes the mediating effect of role stress between psychological capital and turnover intention, just as model 1–3 in Table 80.2 show. According to results of model 1, psychological capital has significant negative influence on turnover intention ($\beta = -0.52$, $p < 0.01$), and hypothesis 1 and the first condition of mediation have been supported; Results of model 2 indicate that psychological capital has significant negative influence on role stress ($\beta = -0.37$, $p < 0.01$), and the second condition of mediation has been supported; Results of model 3 show that when regress independent variable and mediator on dependent variable in the same model, the influence of independent variable- psychological capital on dependent variable-turnover intention decreased significantly ($\beta = -0.14$, $p > 0.01$), and the mediator-role stress still has significant and positive influence on dependent variable turnover intention ($\beta = 1.02$, $p < 0.01$), and hypothesis 2 has been totally supported.

3. Moderating Effects of Organizational Support on the Relationship Between Psychological Capital and Role Stress

Hypothesis 3 predicts the positive moderation of organizational support on the relationship between psychological capital and role stress. Centralized data of

Table 80.3 Results of moderated path analysis[a]

Moderator	Stage		Effects	
	One	Two	Direct	Indirect
Low organizational support (−1 S.D.)	−0.22**	0.71**	−0.07	−0.15**
High organizational support (+1 S.D.)	−0.46**	0.82**	−0.15	0.37**
Difference between low and high level	0.24**	−0.11	0.07	0.22**

[a] $N = 545$; ** $p < 0.01$, * $p < 0.05$

psychological capital, organizational support and the interaction of these two variables have been entered into regression. As model 4 in Table 80.2 shows, the interaction negatively influenced the role stress ($\beta = -0.22$, $p < 0.01$). In other words, this negative relationship has been strengthened by organizational support. Hypothesis 3 has been verified.

4. Moderated Mediation Model: Moderating Effect of Organizational Support on the Mediation of Role Stress

As mentioned above, we conducted moderated path analysis proposed by Edwards and Lambert to test the moderated model in this study [9]. Simple effects showed in Table 80.3 have been abstracted by combining results from both model 4 and model 5.

Stage one means the path from psychological capital to role stress, while stage two refers to the path from role stress to turnover intention. Direct effect refers to the immediate path from psychological capital to turnover intention, and indirect effect equals to the product of stage one and stage two. Coefficient significance of each path (stage one, stage two and direct effect) follows the procedure used to test simple slope rate, and significance test of difference equals to the significance test of interaction.

Bootstrap has been adopted to test the product (indirect effect) coefficient and the significance. According to Table 80.3, negative effects are all significant whenever organizational support is low ($\beta = -0.22$, $p < 0.05$) or high ($\beta = -0.46$, $p < 0.01$), but the difference between low and high organizational support is significant at stage one ($\beta = 0.24$, $p < 0.05$). Although effects at stage two are all positively significant, there is no significant difference. In addition, there is no significant difference of all direct effects, and there is significant difference of all indirect difference ($\beta = 0.22$, $p < 0.05$). Hypothesis 4 has been proved.

80.5 Discussion and Conclusion

Above all, the mediating effect of role stress on the relationship between psychological capital and turnover intention has been fully verified. According to the results of regression, full mediation of role stress exists: After entering role stress into the regression, significance of the influence on turnover intention of psychological capital became insignificant any more. The finding supports the mediation mechanism proposed according to the resource conservation theory; in other words, role stress can fully explain the negative influence of psychological capital on turnover intention. The stress aroused by role stressors (role ambiguity, role conflict and role overload) leads to higher turnover intention. Besides, the moderating effects of organizational support have been fully supported. With respect to the results of two-stage moderated path analysis, the moderating effects of organizational support mostly exist and strengthen the relationship between psychological capital and role stress.

There are some insights to human resource management practices with respect to the findings in this study. Firstly, the main effect of psychological capital on employees' turnover intention indicates that voluntary turnover can be decreased by developing employees' psychological capital, so as to avoid potential loss of talents caused by turnover. Secondly, the mediating effect of role stress demonstrates that managers should pay attention to the multiple roles of employees, and help them analyze various role requirements objectively, and thus they can enhance the self-confidence to meet multiple roles. Thirdly, the moderating mediation effect shows that supportive assistance provided to employees can strengthen the negative influence of psychological capital on role stress. All these findings provide evidences to the implementations of Employee Assistant Program (EPA for short) which has been widely recognized and accepted by Chinese enterprises. Additionally, our findings in this study can enrich the content of EPA by adding psychological capital into the supportive assistance and providing them with guidance to meet various requirements and expectations from multiple roles.

Of course, there are still some limitations in this study. Firstly, this study has been threatened by same source bias. In fact, we have taken measures to minimize the influence of same source bias. The data were collected at two times with an interval of one month: At the beginning of the month, we collect the data of psychological capital and organizational support; at the end of the month, we collect the data of role stress and turnover intention. To test whether this avoidance method makes the difference or not, we conduct Harman single factor test [30, 31], and the results show that same source bias don't distort our research findings significantly. In addition, although respondents were from seven companies located in Beijing, and Hangzhou, results in this study may be vulnerable to the corporate culture because these companies belong to the same high-tech group. In future, control variables should be included in models to exclude the influence of cultural factors.

Acknowledgments This work was supported by National Natural Science Foundation of China (Grant No. 71302170; 71302119), and MOE (Minstry of Education in China) Project of Humanities and Social Sciences (Project No. 13YJC630036).

References

1. Avey JB, Luthans F, Jensen SM (2009) Psychological capital: a positive resource for combating employee stress and turnover. Hum Res Manage 48(5):677–693
2. Avey JB, Luthans F et al (2010) Impact of positive psychological capital on employee well-being over time. J Organ Behav 15(1):17–28
3. Avey JB, Reichard RJ et al (2011) Meta-analysis of the impact of positive psychological capital on employee attitudes, behaviors, and performance. Hum Res Dev Q 22(2):127–152
4. Avey JB, Luthans F, Youssef CM (2010) The additive value of positive psychological capital in predicting work attitudes and behaviors. J Manage 36(2):430–452
5. Blau PM (1964) Exchange and power in social life. Transaction Publishers, New Brunswick
6. Bluedorn AC (1982) A unified model of turnover from organizations. Hum Relat 35(2):135–153
7. Chang WA, Wang Y, Huang T (2013) Work design related antecedents of turnover intention: a multilevel approach. Hum Res Manage 52(1):1–26
8. Eatough EM, Chang CH et al (2011) Relationships of role stressors with organizational citizenship behavior: a meta-analysis. J Appl Psychol 96(3):619–632
9. Edwards JR, Lambert LS (2007) Work role stressors and turnover intentions: a study of professional clergy in hong kong. Psychol Methods 12(1):1–22
10. Eisenberger R, Huntington R et al (1986) Perceived organizational support. J Appl Psychol 73(3):500–507
11. Glazer S, Beehr TA (2005) Consistency of implications of three role stressors across four countries. J Organ Behav 26(5):467–487
12. Glebbeek AC, Bax EH (2004) Is high employee turnover really harmful? an empirical test using company records. Acad Manag J 47(2):277–286
13. Gouldner AW (1960) The norm of reciprocity: a preliminary statement. Am Sociol Rev 25:161–178
14. Griffeth RW, Hom PW, Gaertner S (2000) A meta-analysis of antecedents and correlates of employee turnover: update, moderator tests, and research implications for the next millennium. J Manage 26(3):463–488
15. Hobfoll S (2002) Social and psychological resources and adaptation. Rev Gen Psychol 6(4):307–324
16. Hobfoll SE (1989) Conservation of resources: a new attempt at conceptualizing stress. J Manage 44(3):513–524
17. Holtom BC, Mitchell TR et al (2008) Turnover and retention research: a glance at the past, a closer review of the present, and a venture into the future. Acad Manag Ann 2(1):231–274
18. Kahn RL, Wolfe DM et al (1964) Organizational stress: studies in role conflict and ambiguity. Wiley, New York
19. Katz D, Kahn RL (1978) The social psychology of organizations. Wiley, New York
20. Lee TW, Mitchell TR et al (1999) The unfolding model of voluntary turnover: a replication and extension. Acad Manag J 42(4):450–462
21. Luthans F, Luthans KW, Luthans BC (2004) Positive psychological capital: beyond human and social capital. Bus Horiz 47(1):45–50
22. Luthans F, Youssef CM, Avolio BJ (2007b) Psychological capital. Oxford University Press, Oxford
23. Luthans F, Norman SM et al (2008) The mediating role of psychological capital in the supportive organizational climate employee performance relationship. J Organ Behav 29(2):219–238
24. Ngo H, Foley S, Loi R (2005a) Work role stressors and turnover intentions: a study of professional clergy in hong kong. Int J Hum Res Manage 16(11):2133–2146
25. Ngo H, Foley S, Loi R (2005b) Work role stressors and turnover intentions: a study of professional clergy in hong kong. Int J Hum Res Manage 16(11):2133–2146
26. O'Driscoll MP, Beehr TA (1994) Supervisor behaviors, role stressors and uncertainty as predictors of personal outcomes for subordinates. J Organ Behav 15(2):141–155

27. Panaccio A, Vandenberghe C (2011) The relationships of role clarity and organization-based self-esteem to commitment to supervisors and organizations and turnover intentions. J Appl Soc Psychol 41(6):1455–1485

28. Parasuraman S, Greenhaus JH, Granrose CS (1992) Role stressors, social support, and well-being among two-career couples. J Organ Behav 13(4):339–356

29. Peterson MF, Smith PB et al (1995) Role conflict, ambiguity, and overload: a 21-nation study. Acad Manag J 38(2):429–452

30. Podsakoff PM, Organ DW (1986) Self reports in organizational research: problems and prospects. J Manage 12(4):531–544

31. Podsakoff PM, MacKenzie SB et al (2003) Common method biases in behavioral research: a critical review of the literature and recommended remedies. J Appl Psychol 88(5):879–903

32. Rhoades L, Eisenberge R (2002) Perceived organizational support: a review of the literature. J Appl Psychol 87(4):698–714

33. Rizzo JR, House RJ, Lirtzman SI (1970) Role conflict and ambiguity in complex organizations. Adm Sci Q 15:150–163

34. Stajkovic AD, Luthans F (1998a) Self-efficacy and work-related performance: a meta-analysis. Psychol Bull 124(2):240–261

35. Vandenberghe C, Panaccio A et al (2011) Assessing longitudinal change of and dynamic relationships among role stressors, job attitudes, turnover intention, and well-being in neophyte newcomers. J Organ Behav 32(4):652–671

36. Witt LA, Carlson DS (2006) The work-family interface and job performance: moderating effects of conscientiousness and perceived organizational support. Int J Hum Res Manage 11(4):343–357

37. Wittmer JLS, Martin JE (2011) Work and personal role involvement of part-time employees: implications for attitudes and turnover intentions. J Organ Behav 32(32):767–787

38. Zickar MJ, Balze WK et al (2008) The moderating role of social support between role stressors and job attitudes among roman catholic priests. J Appl Soc Psychol 38(12):2903–2923

Chapter 81
Interpretative Structural Modeling of Public Service Supply in Ethnic Regions

Qian Fang

Abstract Taking system theory as the guide and using systematic analysis method—interpretative structural modeling, this paper focuses on the elements of public service supply system in ethnic regions, analyzes the element relationship, and sets up an "interpretative structural model of public service supply in ethnic regions". Based on this model, this paper gives in-depth discussion about the structural features of the public service supply system in ethnic regions, and finally points out three problems influencing the normal operation of China's public service supply system.

Keywords Ethnic regions · Public service supply · Interpretative structural modeling

81.1 Introduction

As an important content of people's livelihood project, public service has attracted much attention of academic circles and political circles, and its research achievements are relatively abundant with different focuses. Some focuses on the management model of public service, believing that we should pay more attention to the results of public management from the aspects such as efficiency, effectiveness and service quality [6]. Some focuses on the practice of the government public service, thinking that the public service practice has transformed from emphasizing on the efficiency of management process to greatly emphasizing on the management results and the personal responsibility of the managers [3]; Some focuses on the channel strategies of public service [2, 8, 13]; Some focuses on the performance evaluation, believing that the Chinese government performance evaluation pays more and more attention to the public service [1]; And some promotes from the perspective of democratic political consultation [11]. Of course, most of the researches have more or less analyzed the

Q. Fang (✉)
Economic Research Institute of Sichuan Academy of Social Sciences, Chengdu
610072, People's Republic of China
e-mail: lily_lily009@163.com

J. Xu et al. (eds.), *Proceedings of the Eighth International Conference on Management Science and Engineering Management*, Advances in Intelligent Systems and Computing 281, DOI: 10.1007/978-3-642-55122-2_81, © Springer-Verlag Berlin Heidelberg 2014

public service elements, but there are still rare systematic and in-depth analyses. But seen from the literature, the existing researches are still insufficient: first, the discussion about the elements of public service supply is too casual, lack of in-depth analysis of the element relationship; second, the research in ethnic regions is not enough, failing to reflect the regional characteristics; third, there is lack of systematic research. Based on the above problems, the author sets up a concept model on the basis of system theory and using the systematic analysis method—interpretative structural modeling (ISM), translating the unclear thoughts and views existed in the current research of public service supply into intuitive and well-structured model, so as to provide clear thinking for public service management in ethnic regions, and finally to explore a path for improving the public service supply capacity in ethnic regions.

81.2 Interpretative Structural Modeling

Interpretative structural modeling (ISM) is a system analysis method to analyze the complex social and economic system, developed by John Nelson Warfield, an American systematic scientist. The ISM method is characterized by dividing the complex system into several subsystems (elements), making use of people's practical experience and knowledge as well as the help of electronic computers, eventually constructing a multilevel hierarchical structural model [12].

This study chooses this method of modeling for the following reasons:

- Visual intuition of the structure model can help the managers to understand the system. But what should be emphasized is that we cannot get structure model by subjective judgment, and we need to find out the relationship between the elements by modeling technology;
- It is a kind of concept model, which can translate the unclear thoughts and views into intuitive and well-structured model;
- This study requires to construct a multilevel hierarchical structural model based on prior knowledge and by using of computers.

1. Elements Selection

First of all, set the research topic as "to determine the elements and the element relationship of public service supply system in ethnic regions". Secondly, extract the key elements of the system from public service theories and literatures, and make a list of elements, as shown in Table 81.1. In order to simplify the modeling calculation, we collectively call the education department, the health department, the human resources and social security department, the family planning department, the urban and rural construction department, the culture department, the transportation department, the environmental protection department and the financial department as "functional departments".

Table 81.1 Element codes and names

Element code	Element name	Element code	Element name
S_1	Capital	S_{18}	Employment service
S_2	Human resource	S_{19}	Social security
S_3	Facilities	S_{20}	Medical treatment and public health
S_4	Information	S_{21}	Family planning
S_5	Government	S_{22}	Housing security
S_6	Central government	S_{23}	Public culture
S_7	Local government	S_{24}	Infrastructure
S_8	Functional departments	S_{25}	Environmental protection
S_9	Supply object	S_{26}	Minority language and character
S_{10}	Supply content	S_{27}	Special requirement of minorities
S_{11}	Supply standard	S_{28}	Traditional medicines of minorities
S_{12}	Supply process	S_{29}	Traditional culture of minorities
S_{13}	Supply policy	S_{30}	Housing of farmers and herdsmen
S_{14}	Public service product	S_{31}	Citizens
S_{15}	General public service	S_{32}	Public service demand
S_{16}	Special public service	S_{33}	Systems
S_{17}	Public education	S_{34}	Resources

2. Elements Relationship

Referring to a flood of literature, we determine the relationship between elements. It is important to note that the lower-level elements and the upper-level elements have direct relationship (classification relationship or function relationship), for instance, the "central government" and the "local government" are both the lower-level elements of "government". The "functional departments" decide the "supply object" of public service, and the "supply object" is the lower-level element of the "functional departments".

3. Modeling Process

It is to make location for the basic relationship between elements according to the existing research results. Study the relationship between S_i and S_j $(i, j = 1, 2, \cdots, n)$, and set up upper triangular relationship matrix. Then based on the upper triangular relationship matrix, draw up the adjacency matrix A. After element selection and establishing the adjacency matrix, we need to make calculation on adjacency matrix, so as to get reachable matrix. By adding unit matrix to the adjacency matrix, through not more than $(n - 1)$ times of calculation, we can get reachable matrix R. The evolving characteristic of reachable matrix is as follows:

- $A_1 \neq A_2 \neq \cdots \neq A_{r-1}, r \leq n - 1$.
- In this equation, n is as the matrix order, $A_{r-1} = (A + I)^{r-1} = R$.
- Using computer for matrix operation, we acquire: $A_1 \neq A_2 \neq \cdots \neq A_7 = A_8$.
- Get reachable matrix R, namely, $A_7 = A_8 = R$. It indicates the reachable degree when the passing length between each node is not more than 7.

4. Regional Division

It is to make regional division. Divide the relationships between elements into accessible and inaccessible; decide which elements are connected, that is to divide the system into several related parts or subdivisions.

First of all, through operation of predecessor set and the reachable set, we determine the bottom elements, and then judge the connectivity of these elements. If the intersection of reachable sets of elements is null, the elements belong to different connected domains, otherwise belong to the same connected domain.

The lowest level of elements obtained: $T = \{S_1, S_2, S_3, S_4\}$. Because, $R(s_1 \bigcap s_2) = \{s_5, s_6, \cdots, s_{33}, s_{34}\} \neq \Phi$, so s_1, s_2 belong to the same connected domain. Similarly, s_1, s_2, s_3, s_4 belong to the same connected domain. Then make interlevel classification. All the elements in the system are divided into different levels (layer) based on the criterion of reachable matrix. If n_i is the highest level unit, it must satisfy that:

$$R(n_i) = R(n_i) \bigcap A(n_i), \quad \Pi(n) = [L_1, L_2, \cdots, L_k], \tag{81.1}$$

$$L_k = \{n_i \in N - L_0 - \cdots - L_k | R_{k-1}(n_i) = R_{k-1}(n_i) \bigcap A_{i-1}(n_i)\}, \tag{81.2}$$

$$R_{j-1}(n_i) = \{n_j \in N - \cdots - L_{J-1} | m_{ij} = 1\}, \tag{81.3}$$

$$A_{j-1}(n_i) = \{n_j \in N - \cdots - L_{J-1} | m_{ji} = 1\}. \tag{81.4}$$

Successively, elements of different levels are acquired as follows:

- Bottom level: $T = \{S_5, S_{31}\}$.
- First level: $L_1 = \{S_6, S_7, S_{32}\}$.
- Second level: $L_2 = \{S_8\}$.
- Third level: $L_3 = \{S_1, S_2, S_3, S_4, S_9, S_{10}, S_{11}, S_{12}, S_{13}\}$.
- Fourth level: $L_4 = \{S_{33}, S_{34}\}$.
- Fifth level: $L_5 = \{S_{17}, S_{18}, S_{19}, S_{20}, S_{21}, S_{22}, S_{23}, S_{24}, S_{25}, S_{26}, S_{27}, S_{28}, S_{29}, S_{30}\}$.
- Sixth level: $L_6 = \{S_{15}, S_{16}\}$.
- Seventh level: $L_7 = \{S_{14}\}$.

According to the classification result, draw the system hierarchical structural model (omit). Then based on the system hierarchical structural model, construct the "interpretative structural model of public service supply in ethnic regions", as shown in Fig. 81.1. Considering that it goes with feedback mechanism in the input and output process, we add two feedback lines (dashed line); one is from "public service products" to "government sectors", and the other is from "public service products" to "citizens".

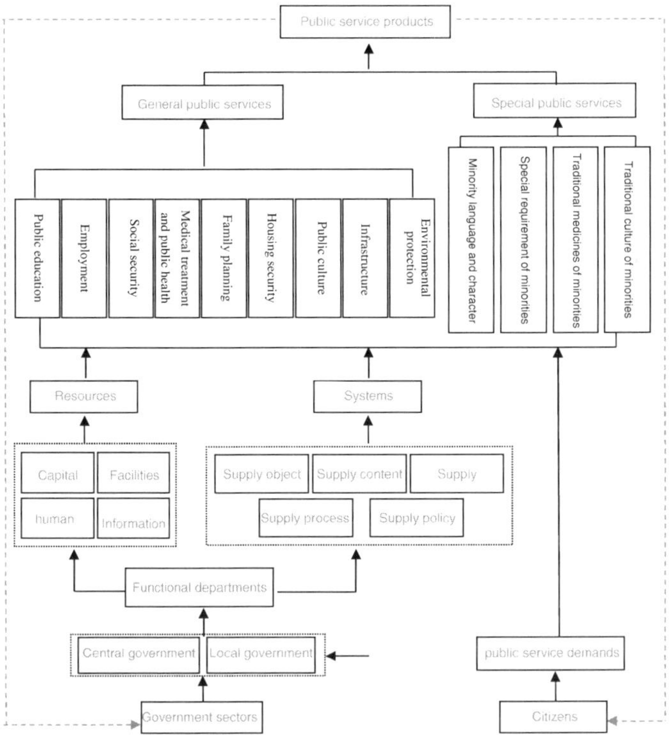

Fig. 81.1 Interpretative structural model of public service supply in ethnic regions

81.3 The Structure Characteristics of Public Service Supply System in Ethnic Regions

1. The Hierarchical Structure of the Model

The first level is "subject and object element level" for the public service supply, including "government sectors" and "citizens" elements. The basic goal of public service supply is to meet people's increasing material, cultural and spiritual needs. As the subjective party of the supply, the government should regulate the supply content according to the changing requirements of citizens, providing different supply content according to the different demand of groups, providing different degree of supply and support according to the importance and emergency degree of demand. As the "supply object", citizens have the responsibility and obligation to feedback their own demands to the government, reminding the government to timely adjust the supply content, supply priority and supply degree.

The second level is "detailing element level" of the public service supply. The government sectors are detailed into "central government" and "local government". At present in China, the administration authority of public service supply belongs to more than 20 departments. These departments shall be divided into local and central in accordance with the geopolitical relation as well as the function emphasis. Obviously, the local government is the main planner and implementer of public service supply. Local government has direct and in-depth understanding about the demand of local citizens, and it will be more accurate and precise on the supply content, supply group and supply degree. Therefore, the local government is the mainstream of public service supply. In service supply, the "central government" mainly functions on the fiscal balance affairs, such as transfer payments. By fiscal balance, the central government provides more fund guarantees to the key and difficult regions (groups) of public service supply, extending in the form of project, which is one of the major functions of the central government. The second major function of the central government is to implement management from top to bottom, supervise and ensure the service supply quality, avoiding social problems due to the local government's supply "preference".

At the second level, citizens are detailed into "public service demand". This detailing process has very important significance, affecting the mass satisfaction on public service supply. Citizens' demand for public services is changing with the economic and social development over time. For a single individual, his demand for public services is also changing. People pay attention to public education at young age, pay attention to employment service at middle age, and pay attention to health in old age. Citizens of different groups have different demand for public services. For a child in the compulsory education stage, the public education should be listed as the emphasis of supply; for a migrant worker coming from rural area, who has no fixed residence, the housing security is more urgent for him; for a retired worker, he pays more attention on social security and medical care. As a result, the difficulties of public service supply not only lie in reasonably and effectively allocating the limited resources to the people, but also lie in how to timely and effectively follow up people's different demand and improve people's satisfaction.

The third level is "functional department element level" of public service supply. The main reason of further decomposing the government element to the functional departments is that, we found there are so many departments involved in the public service supply, and generally each function department acts on parallel tracks in its own way. There is no entity department to exercise unified coordination for these functional departments. Seen from the current situation, "the 12th five-year plan outline of basic public service system" of our country is enacted by the State Council, while "the basic public service system plan" in some provinces and cities is issued by the local governments. Although these departments have the authority to rule over the subordinate departments, it is not easy to implement the unification and coordination of the nine public services. The fourth level is the "supply level" of public service supply, which is divided into the "supply resource element group" and the "supply decision element group". The "supply resource element group" includes four elements, which are capital, human resources, facilities and information; the "supply decision element group" includes five elements, which are supply object, supply content, supply standard, supply process and supply policy. This level contains the most important element of public service supply. Take resource elements for instance, capital, information, human resources and facilities are undoubtedly the basic resources and guarantees of the public service supply. In the economic less developed areas, even though the country gave money and took the facilities into account, but if without enough human resources to participate in the public service, the quality of public service supply is still difficult to be promoted. The "supply decision element group" is the core element for our country to improve the effectiveness of public service supply. The country's resources are limited, and how to make people get better service and more satisfied under such limited resources? This requires us to understand the object of the public service supply, analyze the general demands and special demands of different supply objects, control the supply standard, guarantee the fairness of the supply process, and give corresponding prior- ity on the supply policies. The fifth level is the "process source level" of public service supply, including resources, systems and demands. Although the demand element is in the second level, it together with resource element and system element constitute the sources of public service supply. In the supply processing, the government has to guarantee abundant resources, and guarantee the high efficiency and fairness of supply with sound system, as well as guide according to people's demands.

The sixth level is the "product content level" of the public service supply, including 13 elements, among which 9 elements are the content of the basic public service system construction of our country, and the other 4 elements are the core content of public service supply for ethnic regions.

The seventh level is the "product classification level" of public service supply, including general public services and special public services, which are differentiated services proposed against the demand difference of people in ethnic regions.

The top level shows the public service products, which are the final results of all inputs.

2. Subsystem and Function Division

The model can be simplified as shown in Fig. 81.2, being applied to analyze the sys- tem function. By Fig. 81.2, we know the system consists of five subsystems, namely the government subsystem, the citizen subsystem, the supply resource subsystem, the supply decision subsystem and the public service product subsystem. Specific as follows: In the government subsystem, the government is responsible for resource input and system input, as well as determining the final outputs (public service prod- ucts) according to the citizens' demand; in the citizen subsystem, the citizens are responsible for accepting outputs (public service products) and feedback informa- tion; the supply resource subsystem embodies the base element support of public service supply; the supply decision subsystem embodies the government's manage- ment condition on the public service supply; the public service product subsystem is the final product of the government's input, which is directly supplied to citizens.

In the five different functional divisions, we need to pay special attention to the management class. China has implemented the public service for a short time, so the refinement management of public service is a hard work for us to think seriously about and actively break through in the coming decades. Under the condition of splitted administrative divisions, how to achieve harmony among the different service

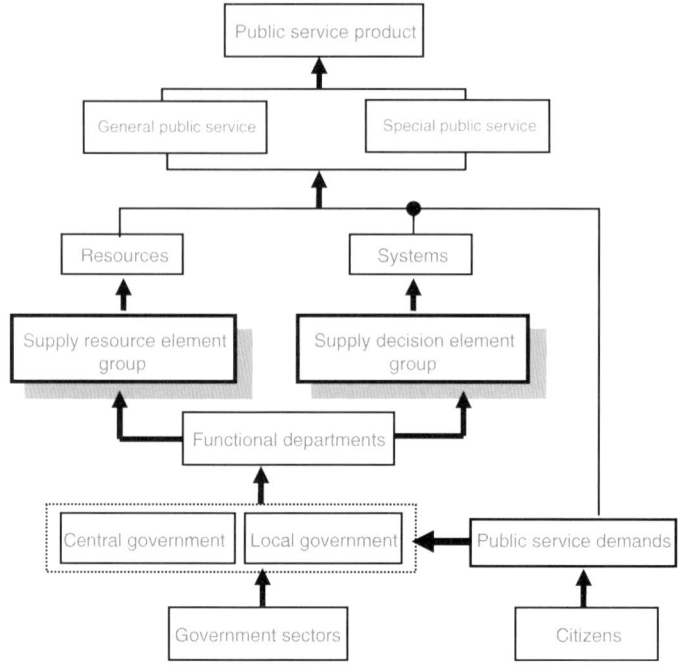

Fig. 81.2 Function division of structural model system

contents? How to avoid resource battle and repeating waste? To a large extent, solving these problems depends on the promotion of management ability.

3. Regulation of Performance and Sensitivity

Every system has its purpose. The purpose of public service supply system is to meet the public demands and realize the goal of the government through rational allocation of input resources. There are two feedback lines (dashed line) in Fig. 81.1; it means that the public service products may affect the government sectors' understanding of product supply, and may affect the citizens' judgement on demands. The government departments will regulate the supply of public service products, and citizens will adjust their demand content according to the demand fulfillment condition; the two lines simultaneously affect the internal management structure of the supply system, finally to improve the outputs accordingly.

If there existed large difference between output and goal, the organization will first check whether the inputs are in accordance with the current management. Secondly, we will consider whether the internal management flows smoothly and whether there exist problems such as poor execution or lack of supervision. Finally, we will rectify the deviation of output difference through input improvement and management control. The merits of regulating performance depend on the corresponding relationship between demand and supply, therefore, increasing the regulating performance of the supply system is an important step to improve the efficiency and quality of supply.

The standards of measuring the regulating performance lie in the regulating speed and the regulating strength, etc. Here, information system, evaluation system and supervision system together play a crucial role to promote the regulating function of the supply system.

Sensitivity is an important indicator for measuring the regulating function; the sensitivity of the supply system is determined by the feedback function of the supply system. From the current situation, our country's public service supply system is a top-down system dominated by the government, lack of feedback mechanism from bottom to top. Therefore, it is effective for the running of the whole system to increase the feedback lines and cut short the feedback cycle of the system.

81.4 Major Problems Affecting the Public Service Supply in Ethnic Regions

From the model, there are three problems affecting the normal running of the public service supply system in ethnic regions, namely the supply resources, the supply methods and the supply solutions.

1. Supply Resources
In economic less developed areas, ethnic regions and remote mountainous areas, it is more likely to be in short supply of resources, which is the reason of these areas having the priority to get more public service supply. Seen from resources classification, hard resources (capital and facilities) are easier to get among all the resources, while soft resources (human resources and information) are more difficult to obtain or difficult to be promoted in short time. Take demand information for example: in order to collect the demand information of citizens, the government needs to collect information by means of investigation and research, and extract effective information through analysis by professionals, and then reflect the analyzed results into the next supply solution. This is a dynamic process needing both professional team and professional technology, as well as time to wait. The human resource in ethnic regions is insufficient, being seriously lack of professional team and professional technology for extracting information.

2. Supply Methods
For supply methods, there are two basic categories based on whether the government is as a producer? The first category is that the government is not only the producer but also the provider of the services; the second category is that the government only plays the provider's role, and the government buys the producer from other professional organizations or individuals in the form of purchase. The first category of supply method may cause large size, inefficiency and high costs to the government organization to a certain extent. If we take the supervision problem into consideration, this monopoly pattern of proprietary sales may greatly influence the quality of service.

In the late 1960s, western scholars began to focus on introducing competition mechanism into the government public service supply system. Landau believed that "the government internal competition will not necessarily cause waste; on the contrary, it may enhance the reliability of the government organization and its services to a certain extent" [4]. Lerner thought, "while maintaining the government internal competition, it is necessary to make use of corresponding control strategies to provide better quality public services" [5]. Savas believed that, "appointing part of the public services to a third sector or a private sector can better improve the quality of public services than just relying on government internal competition, which is helpful to reduce the size of the government" [9]. Osborne and Plastrik made classification of all the steps and measures adopted by western countries in improving the public service quality, and summed up 12 kinds of "management tools" that can be used to improve the public service quality, namely, performance budgeting, flexible performance framework, bidding, companization management, enterprise funds, internal enterprise management, competitive public choice system, voucher and compensation plan, comprehensive quality management, organization process reengineering, franchise and community governance structure [7]. Savas confirmed the different institutional arrangements of public services, summarizing 10 forms of public-private partnership that are helpful to public service supply [10]. These viewpoints provide sufficient theoretical basis for the government to buy public services.

In recent years, Chinese government has gradually transformed from the "patriarchal" supply mode that the government directly involves in the production, to a production mode that multiple subjects compete and cooperate with each other. Non-governmental organizations involve in more and more service production fields, and the government also slowly adjust its role in public service supply, transforming from grasping "production+supply" to provider rather than producer. Admittedly, these developments are good, but obviously not enough in breadth and depth. Compared with the public service developed countries, our country's public service promotion time is short, and lack of experience, so many of the existing practices are learned from the preceding areas. Some experiences will be effective and some will not. Especially for ethnic regions, it is not wise to fully copy the experience of developed areas.

3. Supply Solution

If the supply method emphasizes the difference between producer and provider, which is deconstruction from the perspective of the supply main body—the government, then the supply solution is a plan aimed at the supply object - the people. The supply solution concerns about: the public service supply object; supply quantity; who for more; who for less and who for zero; who should the supply lean to; what should be supplied in the current period; what should be supplied in the next period, and some other problems. To sum up, the supply solution is a supply task goal made by the government according to previous investigation.

At present, the problem of supply solution in ethnic regions mainly shows that the supply content is seriously disconnected from the people's demand. Take rural resident training for example, residents can accept training from educational department, agricultural department and other departments, but these training contents sometimes are not corresponding to people's demand. On the one hand, the training providers (functional departments) carry out work according to their own projects (poverty alleviation project, special transfer payments), neglecting people's demand. On the other hand, there is lack of information communication between departments, so the training content is often found to be repeated. Therefore, a supply solution based on people's demand is an important guarantee to improve the effectiveness of the public service supply in ethnic regions.

References

1. Burns J, Zhiren Z (2010) Performance management in the government of the people's republic of china: accountability and control in the implementation of public policy. OECD J Budgeting 2(1):7–34
2. Ebbers WE, Pieterson WJ, Noordman HN (2008) Electronic govermant: rethinking channel management strategies. Gov Inf Q 25(2):181–201
3. Hughes OE (1994) Public management and administration. The Macmillan Press LTD, Indiana
4. Landau M (1969) Redundaney rationality, and the problem of duplication and overlap. Public Adm Rev 29(4):346–358
5. Lerner A (1986) There is more than one way to be redundant. Adm Soc 18(2):334–359
6. Mathiasen DG (1999) The new public management and its critics. Int Public Manag J 2(1):90–111
7. Osborne D, Plastrik P (2002) Abandon the bureaucracy: five strategies for reinventing government. China Renmin University Press, Beijing
8. Reddick CG, Turner M (2012) Channel choice and public service delivery in canada: comparing government to traditional service delivery. Gov Inf Q 29(1):1–11
9. Savas E (1987) Privatization: the key to better government. Chatham House, Chatham
10. Savas ES, Zhou Z (2002) People first and public-private partnership. China Renmin University Press, China
11. Stephen E (2010) The third generation of deliberative democracy. Polit Stud Rev 8(3):291–307
12. Wang Y (2001) System engineering theory, method and application. Higher Education Press, China
13. Wijingaert LV, Pieterson W, Teerling ML (2011) Influencing citizen behavior: experiences from multichannel marketing pilot projects. Int J Inf Manage 31(5):415–419

Chapter 82
Road Map to the Statistical Process Control

José Gomes Requeijo, Rogério Puga-Leal and Ana Sofia Matos

Abstract Since 1920s, when Walter Shewhart first introduced the foundations for control charts, that several developments regarding new implementations of the statistical process control (SPC) have been presented in order to suit different situations that can be found in several processes. It is noted, among others, the Short Run SPC, data non-normality, the presence of auto-correlation in process data, detection of small and moderate shifts in process parameters and the simultaneous control of various quality characteristics. This great diversity of situations is crucial for academic researchers and quality managers in making decisions regarding the choice of the best technique to implement the statistical processes control. For answering this diversity of situations in production systems, this paper presents a road map that allows the decision maker choosing the best technique for implementation. Various techniques are shown, such as the traditional Shewhart control charts, cumulative sums (CUSUM) charts, exponential weighted moving average (EWMA) charts, dimensionless Z/W and Q charts, residuals/forecast errors charts to processes with a significant autocorrelation and multivariate control charts.

Keywords Statistical process control (SPC) · Process capability · Short runs · Auto-correlation

82.1 Introduction

The implementation of traditional SPC is basically performed using the control charts developed by Shewhart [16]. Other important references are Woodall [17] and Montgomery [9]. The Shewhart charts have shown to be less sensitive in detecting

J. G. Requeijo (✉) · R. Puga-Leal · A. S. Matos
UNIDEMI, Departamento de Engenharia Mecénica e Industrial, Faculdade de Ciências
e Tecnologia, Universidade Nova de Lisboa, 2829-516 Caparica, Portugal
e-mail: jfgr@fct.unl.pt

J. Xu et al. (eds.), *Proceedings of the Eighth International Conference on Management Science and Engineering Management*, Advances in Intelligent Systems and Computing 281, DOI: 10.1007/978-3-642-55122-2_82, © Springer-Verlag Berlin Heidelberg 2014

assignable causes of variation, particularly when the process shifts are small or moderate. For solving this situation, Page [11] developed the cumulative sums (CUSUM charts) and Roberts [15] the exponentially weighted moving average (EWMA charts). The developments of Gan [6] for the CUSUM and the developments of Crowder [5] for the EWMA must be considered. There are many developments of these charts, like the studies of Maravelakis [8].

Nowadays, production systems are characterized by a great diversity of products and low volume production. The classical SPC is difficult to implement, since Shewhart charts are designed for high production volume and few products to control. The most effective way to overcome this limitation is to apply dimensionless charts. Pereira and Requeijo [12] suggest the use of Z and W charts when it is possible to estimate the process mean and variance, and Q charts when the number of data is too small, not allowing this estimation. These developments constitute an important contribution to widespread the implementation of SPC to situations with great diversity of products and limited data. These developments are usually known as Short Run SPC, and are studied by many researchers like Del Castilloet et al. [3] and Crowder and Halblieb [4].

The aforementioned developments are based on the assumption that the values for the quality characteristic are independent and Normally distributed. Alwan and Roberts [1] were the first researchers to study the effect of significant autocorrelation in processes. Several approaches have been made since then, suggesting the application of Shewhart charts for residuals and prediction errors, as shown in Reynolds and Lu [13], Pereira and Requeijo [12] and Requeijo and Cordeiro [14].

Despite there is often an interest for controlling several variables simultaneously, scientific work carried out so far typically addresses single variable studies. In these circumstances, process monitoring should apply multivariate techniques. Many developments were made, among which we highlight the contribution of Alt [2], Nedumaran and Pignatiello [10], Mason et al. [7]. Independence and normality of data are required for applying multivariate SPC. Since T^2 control charts are based on dimensionless statistics, they can be used for the controlling short production runs when there is sufficient data to estimate process parameters. For a restricted amount of data, the MQ charts are an effective way to overcome the impossibility of implementing traditional multivariate SPC [12].

82.2 Methodology

In order to help identifying the best technique for implementation in a specific process, this paper suggests a comprehensive methodology covering the various situations that may occur in a production process, as presented in Fig. 82.1. This road map aims to be a valuable decision making tool to scientific researchers and quality managers.

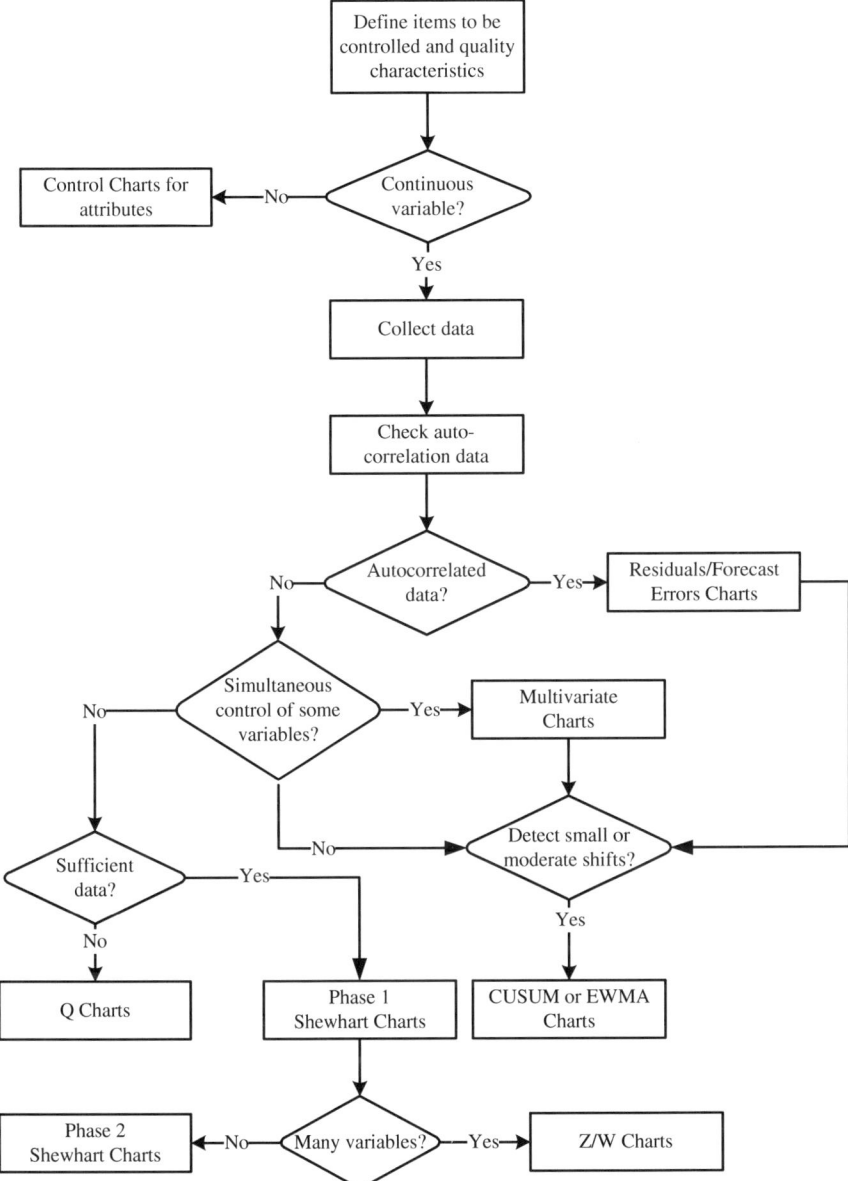

Fig. 82.1 Methodology for implementation of SPC-road map

82.3 SPC Techniques

1. Traditional Univariate SP The implementation of Shewhart charts is based on the following principles:
(1) samples should be homogeneous;

Table 82.1 Shewhart control chartse

Chart	Phase 1			Phase 2			Estimation	
	LCL	CL	UCL	LCL	CL	UCL	$\hat{\mu}$	$\hat{\sigma}$
X	$\overline{X} - 3\frac{MR}{d_2}$	$\overline{X}$	$\overline{X} + 3\frac{MR}{d_2}$	$\mu - 3\sigma$	μ	$\mu + 3\sigma$	$\overline{X}$	–
$\overline{X}$	$\overline{\overline{X}} - A_2\overline{R}$	$\overline{\overline{X}}$	$\overline{\overline{X}} + A_2\overline{R}$				$\overline{\overline{X}}$	
(Average)	$\overline{\overline{X}} - A_3\overline{S}$		$\overline{\overline{X}} + A_3\overline{S}$	$\mu - A\sigma$	μ	$\mu + A\sigma$	$\overline{\overline{X}}$	–
S (Standard deviation)	$B_3\overline{S}$	S	$B_4\overline{S}$	$B_5\sigma$	$C_4\sigma$	$B_6\sigma$		$\frac{\overline{S}}{c_4}$
R (Range)	$D_3\overline{R}$	$\overline{R}$	$D_4\overline{R}$	$D_1\sigma$	$d_2\sigma$	$D_2\sigma$		$\frac{\overline{R}}{d_2}$
MR (Moving range)	$D_3\overline{MR}$	$\overline{MR}$	$D_4\overline{MR}$					$\frac{\overline{MR}}{d_2}$

(2) the sampling frequency is defined according to process characteristics, expecting to maximize the opportunity of change between samples;

(3) the collected data should be independent ($x_i = \mu + \varepsilon_i$, where $\varepsilon_i \sim N(0, \sigma^2)$ is a random variable designated by white noise);

(4) the collected data should follow a Normal distribution ($X \sim N(\mu, \sigma^2)$);

(5) the control limits are located at ± 3 standard deviations from the average (Center Line) of the statistical distribution of the sample under examination (significance level is 0.27 %).

When there is sufficient data to estimate the processes parameters, for continuous variables, the $\overline{X} - R$, $\overline{X} - S$, $X - MR$ charts are implemented. Two stages are usually considered in SPC: Phase 1 (preliminary) and Phase 2 (monitoring). In Phase 1 the process mean and variance are not known and the purpose of this phase is to estimate these parameters. Phase 2 proceeds for monitoring the process in real time, and is based upon the parameter estimates made in Phase 1. The control limits (upper UCL and lower LCL) and the center line (CL) are determined using the formulas shown in Table 82.1.

The Table 82.1 adopts the following notation:

LCL : lower control limit,
CL : central line,
UCL : upper control limit,
$\overline{X}$: sample average,
R : sample range,
S : sample standard deviation,
MR : moving range,
$\overline{\overline{X}}$: overall average of the sample averages,
$\overline{R}$: average of the sample ranges,
$\overline{S}$: average of the sample standard deviations,
$\overline{MR}$: average of the moving ranges,
μ : process mean,
σ : process standard deviation,

Table 82.2 Statistics and control Limits of the traditional multivariate charts

Data	Chart	Estatística	LCL	UCL
Samples	T^2 Chart	$(T^2)_k = n(\overline{X}_k - \overline{\overline{X}})' S^{-1} (\overline{X}_k - \overline{\overline{X}})$	0	$\dfrac{p(m-1)(n-1)}{mn-m-p+1} F_{\alpha;p;mn-m-p+1}$
	(Phase 1)			
	T^2 Chart		0	$\dfrac{p(m+1)(n-1)}{mn-m-p+1} F_{\alpha;p;mn-m-p+1}$
	(Phase 2)			
	χ^2 Chart	$(\chi^2)_k = n(\overline{X}_k - \mu)' \Sigma^{-1} (\overline{X}_k - \mu)$	0	$\chi^2_{\alpha;p}$
	(Phase 2)			
	T^2 Chart	$(T^2)_k = n(X_k - \overline{X})' S^{-1} (X_k - \overline{X})$	0	$\dfrac{(m-1)^2}{m} \beta_{\alpha;p/2;(m-m-p-1)/2}$
	(Phase 1)			
Individual	T^2 Chart		0	$\dfrac{p(m+1)(m-1)}{m(m-p)} F_{\alpha;p;m-p}$
data	(Phase 2)			
	χ^2 Chart	$(\chi^2)_k = n(X_k - \mu)' \Sigma^{-1} (X_k - \mu)$	0	$\chi^2_{\alpha;p}$
	(Phase 2)			

$A, A_2, A_3, B_3,$
$B_4, B_5, B_6, D_1,$: constants that depend on the sample size n.
D_2, D_3, D_4, d_2, c_4

When the quality characteristic is Normally distributed, the study of process capability is carried out through the classical capability indices C_p and C_{pk}, defined by:

$$C_p = \frac{USL - LSL}{6\sigma}, \tag{82.1}$$

$$C_{pk} = \min(C_{pkL}, C_{pkS}), \tag{82.2}$$

$$C_{pkL} = \frac{\mu - LSL}{3\sigma}, \tag{82.3}$$

$$C_{pkS} = \frac{USL - \mu}{3\sigma}, \tag{82.4}$$

where LSL denotes lower specification limit, USL denotes upper specification limit.

2. Traditional Multivariate SPC

The multivariate SPC also includes two phases, Phase 1 (retrospective study) and Phase 2 (monitoring). The process mean vector, μ, is controlled using T^2 charts in Phase 1 and T^2 or χ^2 charts in Phase 2, depending whether the process parameters are estimates or are known. The statistics of multivariate SPC will depend on whether the data are grouped into samples or individual observations. Table 82.2 presents these statistics and corresponding limits.

Table 82.3 CUSUM and EWMA control charts

Chart	Statistic	LCL	CL	UCL
CUSUM	$T_t = \min(0, T_{t-1} + (Z_t + k)); T_0 = 0$ $C_t = \max(0, C_{t-1} + (Z_t - k)); C_0 = 0$	$-h$	$-$	$+h$
EWMA	$E_t = (1 - \lambda)E_{t-1} + \lambda \overline{X}_t, t = 1, 2, \cdots$ $\sigma_E^2 = \frac{\sigma^2}{n}(\frac{\lambda}{2-\lambda})(1 - (1 - \lambda)^{2t}), 0 < \lambda \le 1$	$E_0 - K\sigma_E$	E_0	$E_0 + K\sigma_E$

The Table 82.2 adopts the following notation:

n	:	sample size,
$(T^2)_k$	:	T^2 statistic in instant k,
$(\chi^2)_k$	:	χ^2 statistic in instant k,
X_k	:	individual data vector in instant k,
$\overline{X}_k$	:	samples average vector in instant k,
$\overline{\overline{X}}$	:	overall average of the samples average vector,
S	:	sample covariance matrix,
μ	:	overall average of the sample averages,
Σ	:	process covariance matrix,
$F_{\theta;v_1,v_2}$	:	percentile of the right-sided probability α for the Fisher, distribution with v_1 and v_2 degrees of freedom,
$\beta_{\alpha;p/2,(m-p-1)/2}$	:	percentile of the right-sided probability α for the Beta, distribution with parameters $p/2$ and $(m - p - 1)/2$.

3. CUSUM and EWMA Charts

These control charts are techniques with memory, i.e. the value of the statistic in each moment is an average of values for the sample's statistic, considering that instant as well as previous periods. The CUSUM and EWMA statistics, for controlling process mean and corresponding control limits, are shown in Table 82.3. The parameters of CUSUM and EWMA charts are given by Gan [6] and Crowder [5], respectively.

The Table 82.3 the following notation:

T_t	:	CUSUM statistic to detect an increase in process mean,
C_t	:	CUSUM statistic to detect a decrease in process mean,
k	:	reference value,
h	:	decision interval (control limit),
E_t	:	EWMA statistic to detect a shift in process mean,
E_0	:	initial value of E_t,
λ	:	constant smoothing,
σ_E^2	:	variance of E_t,
K	:	decision interval (control limit).

Table 82.4 Z and W Control charts

Chart	Transformation	LCL	CL	UCL
Z	$(Z_i)_j = (\frac{\overline{X}_i - \mu}{\sigma_{\overline{X}}})_j$ or $(Z_i)_j = (\frac{X_i - \mu}{\sigma})_j$	-3	0	$+3$
W_S	$(W)_j = (\frac{S_i}{S})_j$	B_3	1	B_4
W_R	$(W)_j = (\frac{R_i}{R})_j$	D_3	1	D_4
W_{MR}	$(W)_j = (\frac{MR_i}{MR})_j$			

Table 82.5 Q control charts

Chart	Transformation
$Q(X)$	$Q_r(X_r) = \Phi^{-1}(G_{r-2}(\sqrt{\frac{r-1}{r}}(\frac{X_r - \overline{X}_{r-1}}{S_{r-1}}))), \ r = 3, 4, \cdots$
$Q(MR)$	$Q_r(MR_r) = \Phi^{-1}(F_{1,v}(\frac{v(MR)_r^2}{(MR)_2^2 + (MR)_4^2 + \cdots + (MR)_{r-2}^2})), \ r = 4, 6, \cdots$
$Q(\overline{X})$	$Q_i(\overline{X}_i) = \Phi^{-1}(G_{n_1 + \cdots + n_i - i}(\omega_i)) = \Phi^{-1}(G_{v_1 + \cdots + v_i}(\omega_i)), \ i = 2, 3, \cdots$
	$\omega_i = \sqrt{\frac{n_i(n_1 + \cdots + n_{i-1})}{n_1 + \cdots + n_i}}(\frac{\overline{X}_i - \overline{\overline{X}}_{i-1}}{S_{p,i-1}}), \ i = 2, 3, \cdots; \ S_{p,i}^2 = \frac{v_1 S_1^2 + \cdots + v_i S_i^2}{v_1 + \cdots + v_i}$
$Q(S^2)$	$Q_i(S_i^2) = \Phi^{-1}(F_{n_i - 1, n_1 + \cdots + n_{i-1} - i + 1}(\theta_i)) = \Phi^{-1}(F_{v_i, v_1 + \cdots + v_{i-1}}(\theta_i)), \ i = 2, 3 \cdots$
	$\theta_i = \frac{(n_1 + \cdots + n_{i-1} - i + 1)S_i^2}{(n_1 - 1)S_1^2 + \cdots + (n_{i-1} - 1)S_{i-1}^2} = \frac{S_i^2}{S^2 p, i-1}, \ i = 2, 3, \cdots$

4. Short Run SPC

(1) SPC for a Significant Amount of Data

Phase 1 uses the traditional Shewhart charts for assessing process stability (in-control) and for estimating the parameters of all processes. When there are enough data, it is suggested in Phase 2 the application of dimensionless Z and W control charts. The transformation of the sample statistics and the control limits of these charts are presented in Table 82.4.

The notation used in Table 82.4 is as follows:

$(Z_i)_j$:   statistic to control de process mean,
$(W_i)_j$:   statistic to control de process variation,
$\sigma_{\overline{X}}$:   standard deviation of $\overline{X}$.

(2) SPC for a Limited Amount of Data

When there is a restrict number of data, this piece of research proposes the implementation of dimensionless Q control charts. The transformation of sample statistics for these charts is presented in Table 82.5. The control limits and the center line for all these charts are $UC = +3$, $CL = 0$, $LL = -3$.

The equations in Table 82.5 consider that:

X_r : observation at time r,
$\overline{X}_{r-1}$: average of $r - 1$ observations,
S_{r-1} : standard deviation of $r - 1$ observations,
MR_r : moving range calculated at time r,
$\Phi^{-1}(\cdot)$: inverse of the Normal Distribution Function,
$G_v(\cdot)$: T-student distribution Function with v degrees of freedom,
$F_{v_1,v_2}(\cdot)$: Fisher distribution Function with v_1 e v_2 degrees of freedom,
n_i : size of the sample i,
v_i : degrees of freedom of the sample i ($v_i = n_i - 1$),
$\overline{X}_i$: average of the sample i,
$\overline{\overline{X}}_i$: sequencial average of i samples,
S_i^2 : variance of the sample i,
$S_{p,i}^2$: pooled variance of i samples.

5. SPC for Processes with Significant Autocorrelation

When processes have autocorrelated data, the following procedure is suggested for implementing SPC:
(1) Utilize the autocorrelation function (ACF) and partial autocorrelation function (PACF); for checking what kind of autocorrelation exists in the process;
(2) adjust the best autoregressive integrated moving average (ARIMA) model;
(3) in Phase 1 compute the residuals $e_t = X_t - \hat{X}_t$;
(4) in Phase 2 compute the prediction errors $e_\tau = X_{T+\tau}(T) - \hat{X}_{T+\tau}(T)$;
(5) apply the control charts mentioned above for the residuals/prediction errors (one considers that the distribution of residuals and prediction errors follow a Normal distribution with zero mean and variance) σ_ε^2 (residuals) or $\sigma_\varepsilon^2(1 + \sum_{j=1}^{\tau-1}(\psi_j^2))$ (prediction errors).

The aim of ARIMA methodology is determining and adjusting the better mathematical model to the collected process observations, thus eliminating the existing autocorrelation and obtaining a good prediction for each observation. An ARIMA model is defined by: $\Phi_p(B)\nabla^d X_t = \Theta_q(B)\varepsilon_t$, where

$$B = \frac{X_{t-1}}{X_t}, \nabla = \frac{X_t - X_{t-1}}{X_t} = 1 - B,$$

$$\Phi_p(B) = (1 - \phi_1 B - \phi_2 B^2 - \cdots - \phi_p B^p), \theta_q(B) = (1 - \theta_1 B - \theta_2 B^2 - \cdots - \theta_q B^q).$$

B : backward shift operator,
∇ : backward difference operator,
d : order of the differencing to make the process stationary,
$\Phi_p(B)$: autoregressive polynomial of order p,
$\theta_q(B)$: moving average polynomial of order q,
X_t : observation in period t,
ε_t : "white noise" in period t, ($\varepsilon_t \sim N(0, \sigma_\varepsilon^2)$).

82.4 Conclusions

Process monitoring promotes production stability, checking and analyzing the capability to produce products in accordance to their technical specification. The most effective way for monitoring processes is through techniques such as statistical control charts. The analysis of process capability is obtained by indices like C_p and C_{pk}.

This piece of research presents a methodology for adequate monitoring of processes, considering the various situations that may occur. This methodology, when properly applied, is an asset to the implementation of SPC and helps organizations in the implementation of the most appropriate technique to specific circumstances of processes. The great advantage associated to the proposed methodology when compared to some SPC developments, lies on presenting a global approach that includes most important situations. In literature, SPC developments focus only a special situation, such as univariate or multivariate.

A brief presentation on how to implement the techniques is also mentioned in the methodology. These techniques were further developed in Pereira and Requeijo [12], being advised a consultation of such work for a better implementation of SPC.

References

1. Alwan LC, Roberts NV (1988) Time-series modeling for statistical process control. J Bus Econ Stat 6:87–95
2. Alt FB (1985) Multivariate quality control. Encyclopaedia Stat Sci 6:110–122
3. Castillo ED, Grayson JM et al (1996) A review of statistical process control for short run manufacturing systems. Comm Stat Theor Methods 25:2723–2737
4. Crowder SV, Halblieb LL (2001) Small sample properties of an adaptive filter applied to low volume spc. J Qual Technol 23:29–45
5. Crowder SV (1989) Design of exponentially weighted moving averages schemes. J Qual Technol 21:155–162
6. Gan FF (1991) An optimal design of cusum quality control charts. J Qual Technol 23:279–286
7. Mason RL, Chou Y, Young JC (2001) Applying hotelling's t^2 statistics to batch processes. J Qual Technol 33:466–479
8. Maravelakis PE (2012) Measurement error effect on the cusum control chart. J Appl Stat 39:323–336
9. Montgomery DC (2012) Introduction to statistical quality control. Wiley, New York
10. Nedumaran G, Pignatiello JJ (1999) On constructing t^2 control charts for on-line process monitoring. IIE Trans 31:529–536
11. Page ES (1961) Cumulative sum charts. Technometrics 3:1–9
12. Pereira ZL, Requeijo JG (2012) Quality: statistical process control and planning. Foundation of FCT/UNL, Lisbon (in Portuguese)
13. Reynolds MR, Lu C (1997) Control charts for monitoring processes with autocorrelated data. Non Linear Anal Theor Methods Appl 30:4059–4067
14. Requeijo JG, Cordeiro J (2013) Implementation of the statistical process control with autocorrelated data in an automotive manufacturer. Int J Ind Syst Eng 13:325–344
15. Roberts SW (1959) Control chart tests based on geometric moving averages. Technometrics 1:239–250
16. Shewhart WA (1931) Economic control of quality of manufactured product. D. Van Nostrand Company, New York
17. Woodall WH (2000) Controversies and contradictions in statistical process control. J Qual Technol 32:341–350

Chapter 83
Temporal Distribution and Trend Prediction of Agricultural Drought Disasters in China

Zongtang Xie and Hongxia Liu

Abstract China is one of the countries where agricultural drought disasters occur particularly frequently. The prediction of agricultural drought disasters plays an important role in the drought disaster defense and reduction. This paper, based on the agricultural drought disaster statistics between 1978 and 2011 years in China, analyzed the temporal distribution characters of agricultural drought disasters during the period from 1978 to 2011 in China. It can be seen from the analysis that agricultural natural disasters fluctuated in the research period. Three phases can be evidently indicated: the drought disasters had lower harm to agriculture in the 1978–1984; great disaster influenced area and losses increased evidently in the 1985–2000; the drought disasters had lower harm to agriculture in the period of 2001–2011, except for the year of 2001. Based upon the topologic predicting method of grey system theory, this paper established predicting models. Through applying these models in fact, it is testified that topologic prediction is an effective method that can predict the sequences with wide fluctuating range. This paper also predicted the changing trend of agricultural drought disasters in 15 years in China. It found that heavy agricultural drought disaster will occur in all the 15 years except for 2012 and 2022. The trend of disaster-suffering will be over the average levels in the past 34 years and will appear 4 year of great drought respectively 2015, 2017, 2020 and 2025. At the same time, the interval time of the great drought is between 3 and 5 years.

Keywords Agricultural drought disaster · Grey topologic prediction · Predicting model · China

Z. Xie
School of Management, Northwest University for Nationalities, Gansu 730030, People's Republic of China

H. Liu (✉)
School of Economics, Northwest University for Nationalities, Gansu 730030, People's Republic of China
e-mail: shirleygnd@163.com

J. Xu et al. (eds.), *Proceedings of the Eighth International Conference on Management Science and Engineering Management*, Advances in Intelligent Systems and Computing 281, DOI: 10.1007/978-3-642-55122-2_83, © Springer-Verlag Berlin Heidelberg 2014

83.1 Introduction

Drought disaster is one of the most seriously natural disasters and its impacts on agriculture are enormous. The drought events also have huge harm to economies, societies and environments [1]. It has an important impact on the development of world instability and becomes bottleneck of the sustainable development of national economy. In recent decades, the impacts of drought have escalated in response to population increase, environmental degradation, industry development, and fragmented government authority management [1]. China is one of the countries where drought disasters occur particularly frequently, and China is a great agricultural country. A basic industry of the national economy, agriculture is easily affected by agro-meteorological hazards (particularly drought). The frequent occurrence of drought, coupled with the impact of global warming, poses an increasingly severe threat to the Chinese agricultural production [2]. According to statistics, the tremendous drought has occurred 1056 times and severe drought occurred averagely every two years within the 2155 years from 206 BC to 1949 [3, 4]. Among the five climatic disasters (drought, flood, typhoon, cold damage, heat wave) in the statistics from 1949 to 2005, the frequency of droughts is most frequently and about one-third of the total frequency of disasters [5]. Especially in the recent years, drought disasters continuously happen and cause serious impact to people's production and life. For example, 2009–2010, in the southwest of China, the five regions (Yunnan, Guangxi, Guizhou, Sichuan and Chongqing) happened serious droughts. Therefore, it is significant to predict the area of drought-suffering by drought assessment, and adopt effective disaster prevention measures. In economic literature, there are a wide variety of forecasting techniques available, ranging from subjective judgment to sophisticated models [6–8] and the search for appropriate modeling techniques to accurately forecast future trends has been of great interest to many researchers. In recent studies, many researchers applied grey topologic prediction to decision-making problems [9, 10]. In this paper, a changing trend of agricultural drought disaster prediction, based on grey topologic prediction model, was put forward and applied to assessment trend prediction of agricultural drought disasters in China.

83.2 Grey System Theory and Grey Prediction Models

In the past half-century, system science is developed and growing prosperity [11]. The overall effect and internal relations among different parts of systems has been revealed profoundly [12]. Considering the information people obtain is always uncertain and limited, a variety of uncertainty theories emerged, such as fuzzy mathematics proposed by Zadeh [13] and rough sets theory by Pawlak [14, 15]. However, because of limited information and knowledge, only part of system structure could be fully recognized. To solve the problem, in 1982, Professor Deng Julong, a Chinese scholar, initiated the grey system theory which is a branch of uncertainty systems science, under the assumption that circumstantial information obtained by decision makers or researchers may be partially unknown, uncertain or incomplete [16, 17]. The grey forecasting model adopts the essential part of the grey system theory, and it has been successfully used in finance, physical control, engineering and economics [18, 19]. Thus, the main purpose of this research is to forecast the trend prediction of agricultural drought disasters in China. The findings of the study would be encouraging for general economists and China's government authorities policy makers with more accurate estimates for the future.

83.3 Methodology

1. The Data
Agricultural drought disaster index generally have the drought-affected area, drought-suffering area, quantity of grain damage, etc., and various derived indicators, each indicator reflects the intensity of drought disasters from different angles on the harm of agricultural production system. This paper selected the area of drought disaster affected and suffering (Table 83.1) and the total sown area of crops in the China statistical yearbook from 1978 to 2011.

As each year's crop sown area is different, even the drought-affected area and drought-suffering area are equal in different years, the relative loss and the harm caused by disaster are distinct. Therefore, only using the absolute value of the drought-affected area and drought-suffering area to analysis time of dynamics of disaster cannot really reflect the characteristics of disaster. For ease of comparison time distribution characteristics of agricultural drought disaster, it is necessary to carry out data processing appropriately.

Defined as the ratio of the drought affected area with the total sown area of crops is drought-affected rate, it can reflect the effects of drought disasters basic scope and scale. Similarly, the ratio of the drought-suffering area with the total sown area of crops is drought-suffering rate, it reflects the degree of hazard-formative drought disasters.

According to definition, the paper calculated the calendar year ratios of drought-affected area and drought-suffering area; and on the basis of the calendar year drought-affected area ratio and drought-suffering area ratio and Eq. (83.1), the paper calculated anomaly indices of drought-affected area ratio and anomaly indices of drought-suffering area ratio.

$$\psi_i = \frac{M_i - \bar{M}}{\delta}, \tag{83.1}$$

in the formula, ψ_i is the anomaly indices of drought-affected area ratio and anomaly indices of drought-suffering area ratio of the corresponding year; M_i is the drought-affected area ratio sand drought-suffering area ratio of the corresponding year; $\bar{M}$ is the mean of drought-affected area ratio sand drought-suffering area ratio of the corresponding year; δ is the mean square error.

Table 83.1 The area of drought-affected and drought-suffering in the year 1978 ~ 2011

Year	Drought-affected area (ha)	Drought-suffering area (ha)	Year	Drought-affected area (ha)	Drought-suffering area (ha)
1978	4017.00	1797.00	1995	2345.50	1040.20
1979	2465.00	932.00	1996	2015.20	624.70
1980	2611.10	1417.40	1997	3351.60	2001.20
1981	2569.27	1213.40	1998	1423.60	506.90
1982	2069.73	997.20	1999	3015.60	1661.43
1983	1608.87	758.60	2000	4054.10	2678.40
1984	1581.87	701.47	2001	3847.20	2369.80
1985	2298.90	1006.30	2002	2212.40	1317.40
1986	3104.20	1476.50	2003	2485.20	1447.00
1987	2492.00	1303.30	2004	1725.34	848.16
1988	3290.40	1530.30	2005	1602.81	847.92
1989	2935.80	1526.20	2006	2073.79	1341.13
1990	1817.50	780.50	2007	2938.57	1616.99
1991	2491.40	1055.87	2008	1213.68	679.76
1992	3298.10	1704.70	2009	2925.87	1319.71
1993	2109.70	865.60	2010	1325.90	898.65
1994	3042.30	1705.00	2011	1630.42	659.85

When $\psi < 0$ is small; when $0 \le \psi < 0.5$ is light; when $0.5 \le \psi < 1.0$ is light; when $\psi \ge 1.0$ is most suffered.

2. Grey Topologic Prediction Model

The most common and important procedure in the grey generating approach is the grey topologic prediction model, which is used to discover the potential regular pattern or trend of data series through the accumulation of data in order to change the data variations.

Step 1. Taking a series of numbers arranged according to a certain $x^{(0)} = \{x^{(0)}(k)|k = 1, 2, \cdots, n\}$.

Step 2. Depicting the line chart of $x^{(0)}$ according to the point $(k, x^{(0)}(k))$.

Step 3. Taking some threshold ξ_i $(i = 1, 2, \cdots, m)$.

Letting max $x^{(0)} = \max\{x^{(0)}(1), x^{(0)}(2), \cdots, x^{(0)}(n)\}$,
   min $x^{(0)} = \min\{x^{(0)}(1), x^{(0)}(2), \cdots, x^{(0)}(n)\}$,
   min $x^{(0)} \le \xi_i \le max x^{(0)}$ $(i = 1, 2, \cdots, m)$.

Step 4. On the line chart, making the threshold line according to the height of ξ_i. Assuming that the threshold line and the fold line of $x^{(0)}$ have p intersection points, and abscissa point is respectively.

Assume $x_i^{(0)}$ to be the original data sequence, meaning:

$$x_{\xi_i}^{(0)} = \left(x_{\xi_i}^{(0)}(1), x_{\xi_i}^{(0)}(2), \cdots, x_{\xi_i}^{(0)}(n) \right), \tag{83.2}$$

and AGO is:

$$x_{\xi_i}^{(1)} = \left(\sum_{k=1}^{1} x_{\xi_i}^{(0)}(k), \sum_{k=1}^{2} x_{\xi_i}^{(0)}(k), \cdots, \sum_{k=1}^{n} x_{\xi_i}^{(0)}(k) \right). \tag{83.3}$$

The approximate differential equation differs from the differential equation in ways that the differential equation is used to handle a continuous and differential object, but the grey system is able to use data series to build a model when the object is not continuous and differential. In addition, the differential equation normally is an infinite information space, but the grey data series belongs to finite information space. The GM (1, 1) in grey theory is defined below [17].

$$\sum_{i=0}^{k} a_i \frac{d^{((i))} x_i^{(1)}}{dt^{(i)}} = \sum_{j=2}^{N} b_j x_j^{(1)}(k). \tag{83.4}$$

Let $h = 1$ and $N = 2$, then we have:

$$\frac{dx_1^{(1)}}{dt} + a_1 x_1^{(1)} = b_2 \Longrightarrow \frac{dx_1^{(1)}}{dt} + a_1 x_1^{(1)} = b, \tag{83.5}$$

as a result, the grey difference equation of GM (1, 1) is obtained:

$$x_{\xi_i}^{(0)}(k) + a z_{\xi_i}^{(1)}(k) = b_2 \Longrightarrow x_{\xi_i}^{(0)}(k) + a z_{\xi_i}^{(1)}(k) = b, \tag{83.6}$$

where a is called the development coefficient and b is called grey input.

The grey differential whitening equation can be generated by those two values:

$$\frac{dx_{\xi_i}^{(1)}}{dt} + a x_{\xi_i}^{(1)}(t) = b. \tag{83.7}$$

The relationship between the grey differential equation and its whitening equation is presented below: $x_{\xi_i}^{(0)}(k) = x_{\xi_i}^{(1)}(k) - x_{\xi_i}^{(1)}(k-1)$, $z_{\xi_i}^{(1)}(k) = x_{\xi_i}^{(1)}(t)$, for the purpose of solving a and b, by putting the original series and mean series into the grey differential equation, the $n - 1$ linear equation is obtained.

$$x_{\xi_i}^{(0)}(2) + a z_{\xi_i}^{(1)}(2) = b,$$
$$x_{\xi_i}^{(0)}(3) + a z_{\xi_i}^{(1)}(3) = b,$$
$$\cdots$$
$$x_{\xi_i}^{(0)}(n) + a z_{\xi_i}^{(1)}(n) = b. \tag{83.8}$$

The linear equations can be converted into the form of a matrix of $B\hat{\theta} = Y$, where,

$$
Y = \begin{bmatrix} x_{\xi_i}^{(0)}(2) \\ x_{\xi_i}^{(0)}(3) \\ \cdots \\ x_{\xi_i}^{(0)}(n) \end{bmatrix}, \quad B = \begin{bmatrix} x_{\xi_i}^{(0)}(2) \ 1 \\ x_{\xi_i}^{(0)}(3) \ 1 \\ \cdots \\ x_{\xi_i}^{(0)}(n) \ 1 \end{bmatrix} \quad \text{and} \quad \hat{\theta} = \begin{bmatrix} \alpha \\ \mu \end{bmatrix}.
$$

Thus, the least square method is used to solve α and μ:

$$
\hat{\theta} = \begin{bmatrix} \alpha \\ \mu \end{bmatrix} = (B^T B)^{-1} B^T Y, \tag{83.9}
$$

when α and μ are determined, α and μ can be used as substations of the whitening equation:

$$
\frac{d_{\xi_i}^{(1)}(t)}{dt} + ax_{\xi_i}^{(1)}(t) = b
$$

to obtain the results:

$$
\frac{d_{\xi_i}^{(1)}(t)}{dt} = -\alpha\left(x_{\xi_i}^{(1)}(t) - \frac{\mu}{\alpha}\right), \tag{83.10}
$$

$$
\frac{d(x_{\xi_i}^{(1)}(t) - \frac{\mu}{\alpha})}{dt} = -\alpha\left(x_{\xi_i}^{(1)}(t) - \frac{\mu}{\alpha}\right). \tag{83.11}
$$

Let $x_{\xi_i}^{(1)}(t) - \frac{\mu}{\alpha} = w$, and to integrate both sides. Because $x_{\xi_i}^{(1)}(t) = x_{\xi_i}^{(0)}(t)$, let $t = 1$, then the prediction formula can be presented as:

$$
\hat{x}_{\xi_i}^{(1)}(n + p) = \left(\hat{x}_{\xi_i}^{(1)} - \frac{\mu}{\alpha}\right)e^{-\alpha(n+p+1)} + \frac{\mu}{\alpha}. \tag{83.12}
$$

The Eq. (83.11) is equivalent to:

$$
\hat{x}_{\xi_i}^{(1)}(t + 1) = \left(\hat{x}_{\xi_i}^{(1)} - \frac{\mu}{\alpha}\right)e^{-\alpha(t)} + \frac{\mu}{\alpha}, \tag{83.13}
$$

where $\hat{x}_{\xi_i}^{(1)}(t + 1)$ is the predicted value of $x_{\xi_i}^{(1)}(t + 1)$ at time $(t + 1)$. After the completion of an inverse-accumulated generating operation on (83.13), $\hat{x}_{\xi_i}^{(0)}(t + 1)$ the predicted value of $x_{\xi_i}^{(0)}(t + 1)$ at time $(t + 1)$ becomes available and therefore:

$$
\hat{x}_{\xi_i}^{(0)}(t + 1) = \hat{x}_{\xi_i}^{(1)}(t + 1) - \hat{x}_{\xi_i}^{(1)}(t), \tag{83.14}
$$

where $t = 1, 2, 3, \cdots, n$.

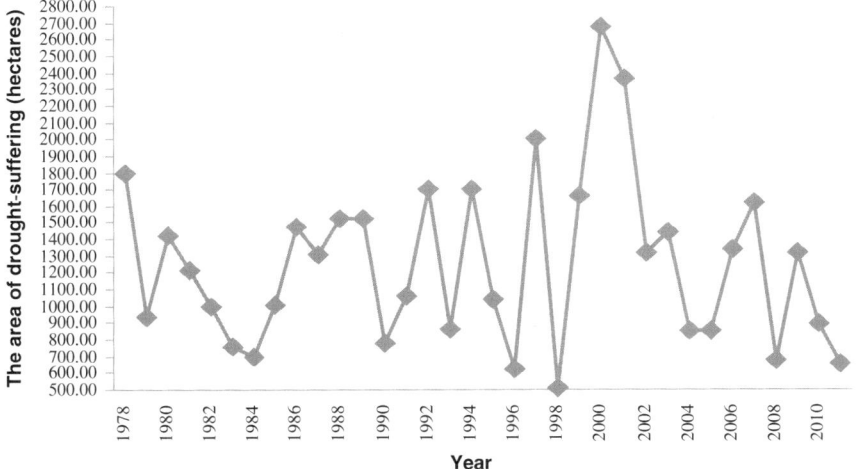

Fig. 83.1 The line chart of the threshold

Step 5. After-test residue checking. Assuming that S_2 is the mean square error of original data sequence; that S_1 is the mean square error of the residual sequence between the original sequence and the simulation sequence. Therefore, the posterior difference ratio for C is: $C = S_1/S_2$, there into:

$$S_1 = \sqrt{\frac{1}{m-1}\sum_{t=2}^{n}(e_{\xi_i}^{(0)}(k) - \bar{e})}, \quad S_2 = \sqrt{\frac{1}{m}\sum_{t=1}^{n}(e_{\xi_i}^{(0)}(k) - \bar{x})}.$$

Small error probability (p) is: $p = p\{|e_{\xi_i}^{(0)}(k) - \bar{e}| < 0.6745 \times S_2\}$.

83.4 Grey Topologic Prediction Model Applications and Drought Disaster Forecasting

1. Fixing the Threshold (ξ_i) and Making the Threshold Line on the Line Chart On the basis of topologic predicting method of grey system theory and the changing trend of agricultural drought disaster in the years of 1978–2011 in China, the paper established predicting models. The max area of drought-suffering is 1678.40 ha; the min area of drought-suffering is 506 ha; therefore, the threshold are between 700 and 2600 ha. The total of threshold are 20 (Fig. 83.1).

The 14 threshold lines are: $\xi_1 = 700$ ha, $\xi_2 = 800$ ha, $\xi_3 = 900$ ha, $\xi_4 = 1000$ ha, $\xi_5 = 1100$ ha, $\xi_6 = 1200$ ha, $\xi_7 = 1300$ ha, $\xi_8 = 1400$ ha, $\xi_9 = 1500$ ha, $\xi_{10} = 1600$ ha, $\xi_{11} = 1700$ ha, $\xi_{12} = 1800$ ha, $\xi_{13} = 1900$ ha, $\xi_{14} = 2000$ ha.

Table 83.2 The GM (1,1) model of threshold series

Threshold ($\times 10^4$ ha)	GM (1,1) model
$\xi_1 = 700$	$x_{\xi_1}^{(1)}(t+1) = 141.7938e^{0.1241t} - 122.9750$
$\xi_2 = 800$	$x_{\xi_2}^{(1)}(t+1) = 100.4427e^{0.1222t} - 93.1195$
$\xi_3 = 900$	$x_{\xi_3}^{(1)}(t+1) = 681.8717e^{0.0245t} - 674.2204$
$\xi_4 = 1000$	$x_{\xi_4}^{(1)}(t+1) = 175.3260e^{0.0571t} - 173.4046$
$\xi_5 = 1100$	$x_{\xi_5}^{(1)}(t+1) = 176.0725e^{0.0.0570t} - 174.2267$
$\xi_6 = 1200$	$x_{\xi_6}^{(1)}(t+1) = 176.6867e^{0.0569t} - 174.9965$
$\xi_7 = 1300$	$x_{\xi_7}^{(1)}(t+1) = 207.8302e^{0.0532t} - 206.2556$
$\xi_8 = 1400$	$x_{\xi_8}^{(1)}(t+1) = 590.1442e^{0.0197t} - 588.6853$
$\xi_9 = 1500$	$x_{\xi_9}^{(1)}(t+1) = 120.3879e^{0.0897t} - 119.0446$
$\xi_{10} = 1600$	$x_{\xi_{10}}^{(1)}(t+1) = 149.3473e^{0.0878t} - 119.0446$
$\xi_{11} = 1700$	$x_{\xi_{11}}^{(1)}(t+1) = 186.0015e^{0.0741t} - 184.8893$
$\xi_{12} = 1800$	$x_{\xi_{12}}^{(1)}(t+1) = 192.5887e^{0.0992t} - 172.7349$
$\xi_{13} = 1900$	$x_{\xi_{13}}^{(1)}(t+1) = 194.4265e^{0.0984t} - 172.7349$
$\xi_{14} = 2000$	$x_{\xi_{14}}^{(1)}(t+1) = 172.4084e^{0.1102t} - 152.4093$

2. Calculating the Sequence of Abscissa Point

$$x_{\xi_1}^{(0)}(t) = \{18.8188, 19.0547, 20.8703, 21.1680, 30.9784, 31.0316, 33.8319\},$$
$$x_{\xi_2}^{(0)}(t) = \{7.3232, 12.9739, 13.0708, 18.5781, 19.1274, 20.8034, 21.2540,$$
$$30.8717, 31.1879, 33.4131\},$$

$\cdots$

$$x_{\xi_n}^{(0)}(t) = \{22.9230, 23.2540\}.$$

3. Establishing the GM (1, 1) Model

Establish the equidistance GM (1, 1) group according to above threshold sequence. For $\xi_1 = 700$, the corresponding GM (1, 1) model is: $x_{\xi_1}^{(1)}(t+1) = 141.7938e^{0.1241t} - 122.9750$. For $\xi_2 = 800$, the corresponding GM (1, 1) model is: $x_{\xi_2}^{(1)}(t+1) = 100.4427e^{0.1222t} - 93.1195$, and so forth, the GM (1, 1) model of other threshold series is in Table 83.2.

Table 83.3 Long-term prediction of great drought disasters in China

The year	Emerging year	Serial number	Predicted value	Residual	Relative error
	2015	38	38.9090	−0.9090	0.0023
	2017	40	40.0350	−0.0350	0.0087
	2020	43	43.4410	−0.4410	0.0101
	2025	48	48.5020	−0.5020	0.0102

83.5 Conclusions and Discussion

1. Results

Gray modeling results indicate that: To the year of 2026, heavy agricultural drought disaster will occur in all the 15 years except for 2012 and 2022. The disaster-suffering area will be over the average levels in the past 34 years and will appear 4 year of great drought respectively 2015, 2017, 2020 and 2025 (Table 83.3). At the same time, the interval time of the great drought is between 3 and 5 years.

2. Discussion

Temporal distribution and trend prediction of agricultural drought disasters provides an important reference for the effective work of disaster reduction, but this is only one small step for a whole system of the work of disaster reduction. The work of agricultural drought disaster reduction is a complicated system engineering, which involved the coordination and cooperation of the government functional departments and agricultural scientific research institution. In order to effectively deal with the next 15 years tendency of increasing drought in China, this paper put forward the measures of strengthening the prevention of agricultural drought disaster and the reduction of agricultural drought disaster.

- Establishing and improving the network system of the agricultural drought disaster integrated monitoring and early warning prediction, strengthening the agricultural drought disaster monitoring and prediction, and improving the overall of agricultural drought disaster prevention function.
- Improving and perfecting the management system of agricultural drought disaster reduction, strengthening the society ability of agricultural drought disaster reduction.
- Strengthening regional and seasonal agricultural drought disaster research, improving the efficiency work of agricultural drought disaster prevention and mitigation.

Acknowledgments This research was supported by the National Social Science Foundation of P. R. China (No. 11XMZ069), Scientific and Technological Project of Gansu Province (No. 1205ZCRA193, 1405ZCRA167) and Northwest University for Nationalities "Young and middle-aged project".

References

1. Pei F, Li X et al (2013) Assessing the impacts of droughts on net primary productivity in china. J Environ Manage 114:362–371
2. Zhu G, Li D, Hu Y (2004) The extreme dry/wet events in northern china during recent 100 years. J Geogr Sci 3:275–281
3. Ni S, Gu Y, Wang H (2005) Study on frangibility zoning of agricultural drought in china. Adv Water Sci 5:705–709
4. Du X, Huang S (2010) Comprehensive assessment and zoning of vulnerability to agricultural drought in tianjin. J Nat Disaster 5:138–145 (In Chinese)
5. Huang H (2008) Analysis of statistical characteristics of droughts in china from 1949 to 2005. Meteorol Sci Technol 5:552–555 (In Chinese)
6. Tang S, Selvanathan E, Selvanathan S (2008) Foreign direct investment, domestic investment and economic growth in china: a time series analysis. World Econ 10:1292–1309
7. Holz C (2008) China's economic growth 1978–2025: what we know today about china's economic growth tomorrow. World Dev 10:1665–1691
8. Liu Z (2003) The economic impact and determinants of investment in human and political capital in china. Econ Dev Cult Change 4:823–849
9. He B, Wu J, Lv A (2010) New advances in agricultural drought risk study. Prog Geogr 5:557–564 (In Chinese)
10. Hamed R, Masoud R (2009) Project selection using fuzzy group analytic network process. World Acad Sci Eng Technol 58:457–461
11. Liu S, Forrest J (1997) The role and position of grey systems theory in science development. J Grey Syst 4:51–356 (In Chinese)
12. Liu S, Lin Y (2006) Grey information: theory and practical application. Springer, London
13. Zadeh L (1965) Fuzzy sets. Inf Control 3:338–353
14. Pawlak Z (1982) Rough sets. Int J Comput Inf Sci 5:341–356
15. Pawlak Z, Grzymala-Busse J et al (1995) Rough sets. Comm ACM 11:88–95
16. Deng J (1989) Introduction to grey system theory. J Grey Syst 1:1–24 (In Chinese)
17. Liu S (2010) Grey systems: theory and applications. Springer, New York
18. Zhao Z, Wang J et al (2012) Using a grey model optimized by differential evolution algorithm to forecast the per capita annual net income of rural households in china. Omega 40:525–532
19. Yin M, Wen H, Tang V (2013) On the fit and forecasting performance of grey prediction models for china's labor formation. Math Comput Model 57:357–365

Chapter 84
The Research of Relationship Capital and Demand-Pull Absorptive Capacity Impact on Innovation Performance

Man Yang, Wei Liu and Huiying Zhang

Abstract This paper provides new perspective to study the impact of relationship capital and absorptive ability on the innovation performance. Considering the character of Chinese enterprise, absorptive ability discussed in this paper is demand-pull absorptive capacity, which can be divided into potential demand-pull absorptive capacity and realized demand-pull absorptive ability. On the basis of questionnaire survey, we confirm the relationship among relationship capital, demand-pull absorptive capacity and innovation performance. We probe into the intermediary roles of the potential demand-pull absorptive capacity and the realized demand-pull absorptive capacity under the positive effects of relationship capital to innovation performance. In addition, we also implement an empirical research on the inner relation between potential absorptive ability and realized absorptive capacity.

Keywords Relationship capital · Demand-pull absorptive ability · Innovation performance

84.1 Introduction

Under the fierce competitive situation, innovation plays an increasingly significant role in process of globalization. Then innovative capability can be core competitiveness power for enterprise sustained development. In fact, the incessant innovation is motive to improve enterprise situation continuously. However, innovation and knowledge are always intertwined that innovation occurs typically based on new knowledge while the application of new knowledge often bring about innovation [6]. Therefore, knowledge plays an irreplaceable role in the innovation process of enterprise.

M. Yang · W. Liu · H. Zhang (✉)
College of Management and Economics, Tianjin University, Tianjin 300072,
People's Republic of China
e-mail: hyzhang@tju.edu.cn

J. Xu et al. (eds.), *Proceedings of the Eighth International Conference on Management Science and Engineering Management*, Advances in Intelligent Systems and Computing 281, DOI: 10.1007/978-3-642-55122-2_84, © Springer-Verlag Berlin Heidelberg 2014

Innovation becomes complicated under the changing and competitive environment, increasingly severe technical innovation environment, shorter lifecycle of product and more and more diversified of the market demand. Conducting an innovation activity solely rely on knowledge in the technical field is insufficient. It is necessary for enterprise to have various areas of knowledge, such as market knowledge, suppliers and customer knowledge, the other competitor information, etc Therefore, the enterprise needs to absorb profitable information and knowledge from various sources to establish a vast and solid knowledge foundation [1]. To timely obtain the information and knowledge at a low cost, the enterprise must establish and maintain a good relational network with its stakeholders, in which the relationship capital in the network will become the key innovation resources for enterprise. Especially in China, with such a "heavy relations" traditional culture background, relationship capital plays a more prominent role. In such complex and dynamic environment, relationship capital has become the sources for enterprise to maintain competitive advantage for sustainable development [7]. Relationship capital is beneficial to the enterprise in obtaining useful knowledge and information from relative sources [12], which can promote the knowledge transformation through effective communication between enterprises and reduce the opportunity cost in the process of cooperation. However, obtaining knowledge does not mean using knowledge. The enterprise must have good absorptive ability for the sake of better apply knowledge. The so-called absorptive ability is that the enterprise can get important knowledge from the external sources and digest, transform and eventually commercially apply them [4]. Cohen and Levinthal in 1990 put forward the absorptive ability is very important to the innovation of enterprise [3]. Through the existing research we found absorptive ability can be divided into two kinds, technology-driven absorptive capacity and demand-pull absorptive ability. In China, due to the lack of the original enterprises low labor costs, a lot of foreign enterprises are attracted to our country. They have advanced technology and our local business makes profit by being OEM, which makes the market information and knowledge become particularly important. In this paper, absorptive ability discussed here is demand-pull absorptive ability. In existing researches, despite there are a small amount related research about the relation among absorptive ability, innovation and relationship capital set in the western developed economic background, they all focus on the impact of relationship capital and absorptive capacity on innovation performance [9], but mechanism of action of them is yet to seek. Yu [13] find that by increasing absorptive capability, it will systematically increase the slope and amplitude of the positive effects for firm innovation performance. Some current research assumes that the relationship between relationship capital and innovation performance is linear [10], but there are also people who hold a different view. In order to make up this deficiency and to be different from the previous studies, this paper aims at:

1. Exploring and proving the effects of Chinese enterprise relationship capital and demand-pull absorptive capacity on innovation performance;
2. Refining the absorptive capacity to potential ones and realized ones, this paper probes the potential absorptive capacity and the realized ones' intermediary role

in the positive effects of relationship capital to innovation performance, simultaneously, exploits mediated effect model based Chinese enterprise date;
3. Confirming promotion effect of potential absorptive capacity to realized absorptive capacity.

The rest of this paper is organized as follows. Section 84.2 presents a literature review of previous related research, and derives several hypotheses. Section 84.3 is devoted to constructing modeling methodology. Section 84.4 provides an empirical analysis. Finally, Sect. 84.5 offers concluding remarks.

84.2 Literature Review and Hypothesis

1. Literature Review
(1) Concept and measure of relationship capital
Relationship capital is derived from the social capital relational dimension, and it was initially based on mutual trust and friendly relations in the personal level. Relationship capital in this paper is the longitudinal relationship capital, namely the relational network established through the supply chain, mainly including relations with suppliers and customers. There is a common point for the measurement of relationship capital referred to the existing literature [2], as in Table 84.1.

(2) Concept and measure of demand-pull absorptive capacity
As the study about the absorptive ability starts later, the absorbency were falt into potential demand-pull absorptive capacity and realized demand-pull absorptive ability [14]. The former includes knowledge acquisition and digestion ability, and the latter includes knowledge transformation and the application ability. In addition, Nika and Lgor [8] proposes two classes of absorption ability: technology-driven absorption capacity and demand-pull absorption ability. This paper mainly studies the relationship capital of the enterprise in the longitudinal relational network, so it is about demand-pull absorptive ability. With reference to Jansen [5], potential demand-pull absorptive capacity and realized demand-pull absorptive capacity were measured [11] as in Table 84.1.

(3) Concept and measure of innovation performance
In terms of innovation performance, the academia has still not a unified concept. Researchers defined the innovation performance from the perspective of knowledge. Some Chinese scholars regarded innovation performance as the concept similar to the concept for the assessment of efficiency and effect of business operations. We put forward, innovation performance is the measurable output based on enterprise innovation, measured as in Table 84.1.

2. Hypothesis
Some hypothesis was proposed:
H1: Relationship capital has a positive impact on innovation performance.
H2a: relationship capital has the positive impact on potential demand-pull absorptive ability;

Table 84.1 Measurement sheet

Variable	Number	Item
Relationship capital	RC1	We cherish and active in the relations with customers and suppliers
	RC2	Customers and suppliers will consider this enterprise's interests when making great decisions
	RC3	Customers, suppliers and us trust each other
	RC4	Customers, suppliers and us often contact
Potential demand-pull absorptive ability	PA1	We have procedures and methods to acquire information of customers and suppliers
	PA2	We regularly hold special meetings with customers and suppliers
	PA3	We have special plans to train the staff mastering new knowledge
	PA4	We have special mechanisms to solve the conflict due to employees' different understanding and explanation of the new knowledge
Realized demand-pull absorptive ability	RA1	We have specific procedures and practices to help digest new knowledge and combine them with the existing knowledge
	RA2	We have special practices and structures to store and record new knowledge
	RA3	We put forward the improvement suggestions of products and processes according to the new knowledge
	RA4	We have systematic programs to develop new products by applying new knowledge
Enterprise innovation performance	IP1	Introduce into more new products
	IP2	The improvement of new product has good market reaction
	IP3	Introduce into more new production process
	IP4	The input-output efficiency is very high in new product development

H2b: relationship capital has the positive impact on realized demand-pull absorptive ability;

H3: potential absorptive ability has the positive impact on realized absorptive ability;

H4a: potential demand-pull absorptive ability has positive impact on innovation performance;

H4b: realized demand-pull absorptive ability has positive impact on innovation performance;

H5: relationship capital has positive impact on innovation performance through the intermediary role of demand-pull absorptive ability;

H6: relationship capital has positive impact on realized demand-pull absorptive ability through the intermediary role of potential demand-pull absorptive ability;

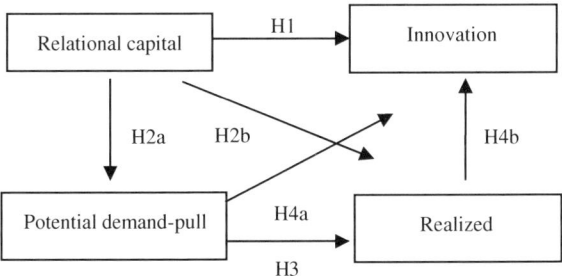

Fig. 84.1 Conceptual model

H7: potential demand-pull absorptive ability has positive impact on innovation performance through the intermediary role of realized demand-pull absorptive ability (Fig. 84.1).

84.3 Research Method

1. Research Design and Sample

This study tested the hypotheses using samples of manufacturing industries, including household electrical appliances, the automobile, the textile and electronic communication industries, which belong to the export-oriented manufacturing to a large extent, so the demand-pull absorptive ability is important for them. The selected firms were contacted by telephone calls, E-mails, faxes and letters. In order to insure the validity of the questionnaires, we asked the man who filling in the questionnaire to be the manager level or above and to leave their name and contact way. Data collection occurred from 2011 May to 2011 October. In total, 174 questionnaires were feed back, and 157 were valid, corresponding to an effective rate of recovery is 68.26 %.

2. Measurements

The measurements of all variables are referred to previously used measurement scales. Measuring construct includes 4 latent variables, namely relationship capital, potential demand-pull absorptive ability, realized demand-pull absorptive ability and innovation performance. Each latent variable contains a set of indicators (see Table 84.1). A 7-point interval rating scale system was used in the survey, with 7 equaling the highest extent of agreement.

84.4 Empirical Assessment

An exploratory factor analysis firstly was applied to examine the reliability and validity of these measures. Cronbach's reliability estimate test was used to proving the reliability of the scales and within-scale factor analysis was used to measure the

Table 84.2 Exploratory factor analysis

Indicators	ID α	α	Loading scores	Contribution rate
RC1	0.686	0.729	0.720	0.594
RC2	0.696		0.707	
RC3	0.637		0.781	
RC4	0.652		0.771	
PA1	0.715	0.724	0.701	0.589
PA2	0.662		0.738	
PA3	0.649		0.766	
PA4	0.617		0.804	
RA1	0.707	0.791	0.832	0.616
RA2	0.726		0.806	
RA3	0.783		0.705	
RA4	0.736		0.790	
IP1	0.841	0.869	0.833	0.719
IP2	0.806		0.891	
IP3	0.847		0.820	
IP4	0.834		0.845	

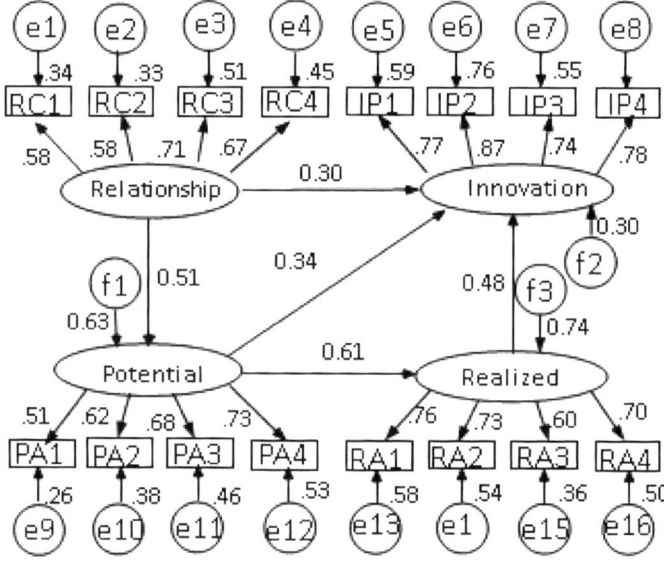

Fig. 84.2 Path diagram

extent to which the indicators measure the same latent variable. Table 84.2 presents the summary of the exploratory factor analysis.

This study uses SEM method to test the conceptual model. $2/\mathrm{df} = 1.687 < 2$, $CFI = 0.932 > 0.90$, $IFI = 0.941 > 0.90$, $GFI = 0.926 > 0.90$, $RMSEA =$

Table 84.3 Overview of hypotheses and findings

Hypotheses	Path	S.C	p	Findings
H1	RC→IP	0.30	0.007	Supported
H2a	RC→PA	0.51	0.001	Supported
H2b	RC→RA	0.06	0.347	Rejected
H3	PA→RA	0.61	0.008	Supported
H4a	PA→IP	0.34	0.006	Supported
H4b	RA→IP	0.48	0.005	Supported

$0.04 < 0.05$, it shows the data fits the model well. The final results are presented in Fig. 84.2.

The data showed that relationship capital has a positive impact on innovation performance and potential absorptive capacity, supporting hypotheses H1 and H2a, but the effect of relationship capital on realized absorptive capacity (H2b) is not significant. The weak result maybe explained: realized absorptive capacity comprises conversion ability and application ability. Conversion ability is the ability to combine the enterprise's original knowledge with the acquired knowledge to come out new knowledge that is useful for enterprise. Application ability is ability to apply the internalized knowledge to acquire commercial profits. Both of the abilities seem to have less relation with the external resource. It is the firm itself matter the most. The results also support H3, H4a and H4b. See Table 84.3. Hypothesis 2 is rejected through the above analysis, so the H6 is rejected. But an indirect effect exists between relationship capital and realized demand-pull absorptive ability through potential ones.

84.5 Conclusions

This research analyzes the impact of relationship capital and absorptive ability on the innovation performance. We refined the absorptive capacity to potential ones and realized ones, probing the potential absorptive capacity and the realized ones' intermediary roles in the positive effects of relationship capital to innovation performance. This study provides interesting results. The first conclusion is that relationship capital exerts a positive impact on innovation performance, and the demand-pull absorptive capacity plays a partial intermediary role in the positive impact. Innovation is a complex process and it is unwise to depend on the enterprise itself solely. Especially in China, with such a "heavy relations" traditional culture background, relationship capital is more important. So relationship capital is helpful for enterprise to acquire valuable information from relative sources and reduce the opportunity cost in the process of cooperation. However, obtaining knowledge does not mean using knowledge. The enterprise must have good absorptive capacity for the sake of better apply information.

The results also show that relationship capital exerts positive impact on potential absorptive capacity, but not on realized ones. It may be because that knowledge identification and knowledge acquisition are close to the relations of enterprise with external resources such as customers and suppliers. Keeping good relations with them can be beneficial for enterprise to obtain useful knowledge and information in time, but it is mainly depend on the enterprise itself in the transformation and application because most of the work about transformation and application is completed inside the enterprise.

The third conclusion is that potential demand-pull absorptive capacity has a positive impact on realized ones and both of them exert positive impacts on innovation performance. The potential absorptive capacity helps enterprise learn and master knowledge obtained from external resources while the realized absorptive capacity helps to strengthen the combination of existing knowledge with obtained knowledge, and further develop, utilize, and even create new knowledge. Obviously, the development and utilization of knowledge is set up on the basis of the existing knowledge. So a strong potential absorptive capacity will positively affects the realized ones. Innovation is a process of creating new knowledge. Considering the innovation value chain, from the produce of new idea to the final success of innovation, every phase is closely correlated with information and knowledge. Both potential absorptive capacity and realized ones positively affect the innovation performance.

Further, this study shows that the realized absorptive capacity plays an intermediary role in the positive effect of potential absorptive capacity on innovation performance. Potential absorptive capacity concludes knowledge acquisition capacity and knowledge digestion capacity. Enterprise can master useful knowledge through these two abilities, but conducting an innovation activity depends on them solely is not enough. Enterprise must combine the existing knowledge with digested knowledge and then it can develop and apply new knowledge to an innovation activity.

Like all studies, this paper has some limitations that further study should over come. First, relationship capital in this study mainly includes two kinds of relationship, that is, the relation with customers and the one with suppliers. Although the result is appropriate in this study, it can't be generalize the findings to other relations such as relation with competitors, government, colleges. Future study should fully consider these relations. Second, we use cross-sectional data in this study. The variables are measured at the same moment, which may course problems. We should have a longitudinal study in future research to come to a better conclusion. And the one with suppliers, not including other relationships.

Acknowledgments The research was funded by the special project of Science and Technology Development Strategy in Tianjin in 2012 (12ZLZLZF10200), and in 2013 (13ZLZLZF08900).

References

1. Aurélie B (2009) Meaningful distinctions within a concept: relational, collective, and generalized social capital. Soc Sci Res 38(2):251–265
2. Chen YS, Lin MJ, Chang CH (2009) The positive effects of relationship and absorptive capacity on innovation performance and competitive advantage in industrial market. Ind Mark Manage 38:152–158

3. Cohen W, Levinthal D (1994) Fortune favors the prepared firm. Manage Sci 42(2):227–251
4. Escribano A, Fosfuri A, Tribó JA (2009) Managing external knowledge flow: the moderating role of absorptive capacity. Res Policy 39:96–105
5. Jansen JP, Bosch FAJ, Volberda HW (2005) Managing potential and realized absorptive capacity: how do organizational antecedents matter. Acad Manage J 48(6):999–1015
6. Jensen MB, Johnson B et al (2007) Forms of knowledge and modes of innovation. Res Policy 36:680–693
7. Konstantin's K, Alexandra's P et al (2011) Absorptive capacity, innovation, and financial performance. J Bus Res 64:1335–1343
8. Nika M, Lgor P (2009) Absorptive capacity, its determinants, and influence on innovation output: Cross-cultural validation of the structural model. Technovation 29(12):859–872
9. Rass M, Dumbach M et al (2013) Open innovation and firm performance: the mediating role of social capital 22(2):177–194
10. Steinfield C, Scupola A, Lopez-Nicolas C (2010) Social capital, ICT use and company performance: findings from the medicon valley biotech cluster. Technol Forecast Soc Chang 77:1156–1166
11. Tessa CF, Andreas E et al (2011) A measure of absorptive capacity: scale development and validation. Eur Manage J 29(2):98–116
12. Torodova G, Durisin B (2007) Absorptive capacity: valuing a reconceptualization. Acad Manage Rev 32(3):774–786
13. Yu SH (2013) Social capital, absorptive capability, and firm innovation. Technol Forecast Soc Chang 80:1261–1270
14. Zahra SA, George G (2002) Absorptive capacity: a review, reconceptualization, and extension. Acad Manage Rev 27(2):185–203

Chapter 85
Modeling and Forecasting International Arrivals from China to Australia

Zheng Yong and Robert Brook

Abstract China has risen to second ranking source of Australia's international tourists by the end of 2012. In term of market share, Chinese represented about 7.69 % international arrivals to Australia in 2012. This paper undertake a statistical analysis of inbound tourism arrivals to Australia from China. In particular, ARIMA model will be estimated by tourist arrivals quarter data from $Q1$ 1995 to $Q4$ 2009 and validated by $Q1$ 2010 to $Q2$ 2013. The expected outcomes is the determination of a statistically optimal time series model of Australia tourism demand, which will provide a better understanding of the importance of the Chinese tourist market for Australia, and provide a practice quarterly forecast with confidence 95 % on international tourist arrivals from China to Australia between Quarter3 2013 and Qutarter4 2016. Such knowledge will aid in tourism planning by private and public sectors to achieve efficient development and management of tourism facilities and infrastructure.

Keywords International travels · ARIMA · Modeling and forecasting

85.1 Introduction

Tourism demand has long been an intriguing issue for academic researchers and industry practitioners. Because of its importance in the economy and society, it has been constantly under scrutiny by various parties in different field.

Z. Yong (✉)
International Business, Beijing Normal University, Zhuhai 519087,
People's Republic of China
e-mail: zhengyong2011@bnuz.edu.cn

R. Brook
Business and Economics, Monash university, Melbourne 3000, Australia

J. Xu et al. (eds.), *Proceedings of the Eighth International Conference on Management Science and Engineering Management*, Advances in Intelligent Systems and Computing 281, DOI: 10.1007/978-3-642-55122-2_85, © Springer-Verlag Berlin Heidelberg 2014

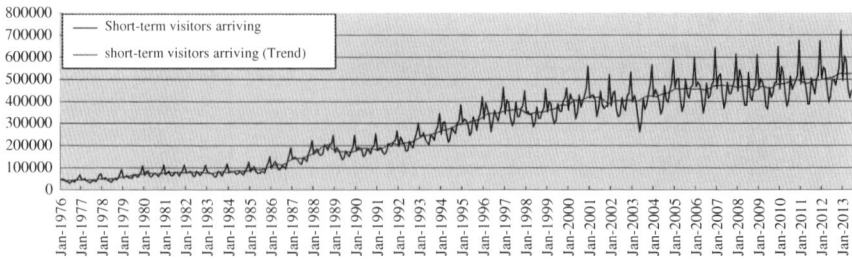

Fig. 85.1 Short-term international arrivals Australia (Jan 1976–Dec 2012) (Month)

The growing interest in this area is related to the fast expansion of the international tourism industry in both developed and developing economies. According to the statistics compiled from the Yearbook of Tourism Statistics, World Tourism Organization (WTO) (various issues, up to 2012), Global visitor arrivals grew at an average of 6.9%, increased from 25.3 million arrivals to 1035 million arrivals between 1950 and 2012, while visitor receipts grew at an average growth rate of 11.5% per annual, an increase from US 2.1 billion to US 1,070 billion over the same period. Over the past three decades, Australia's international tourism demand has grown at an rate faster than the worldwide rate from 1976 to 2012. Global international visitor arrivals grew at a average of 6.9% between 1976 and 2012, and Australia international visitor grew at an average of 16.90%, or about 0.16 million out of the worldwide visitors (of total worldwide tourist arrivals) selected as their destination. By 2012, this figure increased over 13 million arrival out of 1.035 billion the worldwide visitors (0.13% of total worldwide tourist arrivals). Placing demands on its tourist services, and spending almost US 27.6 Billion (around 1.77% of Gross Domestic Product GDP). In the same year, Australia was ranked ninth in terms of international receipts in the world. As shown in Table 85.1, Australia achieved obviously long-term growth in term of tourism demand over the period from 1976 to 2012 (Fig. 85.1).

Many countries have increasingly been interesting in the Chinese outbound markets because of their high disposable income. Chinese tourists are the top tourism spenders in the Asia pacific, and were ranked as the second highest international tourism spender in 2012 (World Tourism Organization). Australia was one of Chinese top ten most popular oversea touring destination and hottest spot of immigration in the world. China is, therefore, one of the Australia major source of international visitors. The average annual compound growth rate of short-term (namely, no more 12 months) tourist arrivals from China to Australia from 1991 to 2012 up to 18%, or the annual number of short-term arrivals increased almost 37-fold from 16,500 to 62,6400. The double-digit growth are 17 years during the same period, the highest growth rate achieved at 43.92% in 1995, the negative growth rate only appeared in 2003 and 2009, the reason is SARS and financial crisis respectively (as showing in Fig. 85.2).

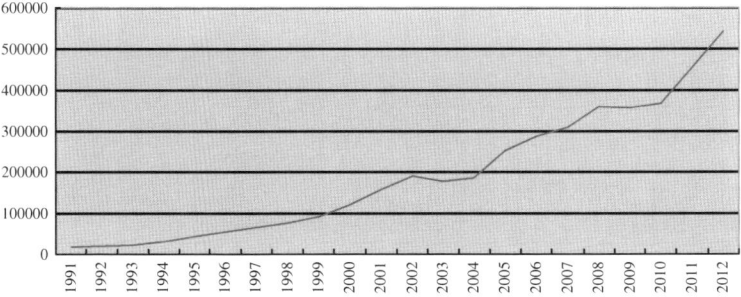

Fig. 85.2   Short-term international arrives from China to Australia (year)

It imperative for destinations such as Australia to examine market trends. The characteristics of Chinese tourists and the specific travel objectives and activities they wise to persue. Many studies have examined the consumer behaviour characteristics unique to Chinese tourists, such as their main reason of travel, motivation [1], their concerned about price [3] and what was the reason Chinese like to choose organized tour [4]. For the Chinese, they like to spent their time on shopping on which they spend high budget, and for westerner, shopping is not necessary which their expenditure is relatively low, beside buying some local products. Chinese tourists would like to stay with their own group (family group or the group the small region where they are from) ; for the Chinese organized tour, in all the case the package is a mixture of all the elements such as cultural, natural, architecture etc. the Chinese tourists visit the artificial or man-made sights such as museum, square, monument. Despite the limited opportunities for the group tours, a new group of tourists has emerged in china-FIT (free and independent traveler) [7].

85.2 Methodology and Modeling of China Tourist Arrivals to Australia

The purpose of this section is to examine international tourism demand by China for Australia, as measured by tourist arrivals from $Q1$ 1995 to $Q4$ 2009. This study focus on the demand for tourism in Australia from China and forecast 24 quarters ahead, which is $1Q$ 2010 to 4 2016. Forecast result (ex ante point forecasts and their 95 % prediction intervals on a 14-quarter horizon for the SARIMA model) are provided in table. The foretasted date points fitted by the final models are potted along with the actual arrival observations in corresponding figure. In addition, the the author indicates three periods in the plot:

- "M" represents model fitting period,which contains actual "in sample" data points used for building a forecasting model,

- "v" represents validation period, which contains out-of-sample(holdout) data points for measuring a forecasting model's "*expost*" predicting ability on holdout data, and
- "F" represents forecasting period, which shows "*exante*" predictions of data points in the forecasting horizon.

In this study, we used seasonal ARIMA model to forecast one-period ahead by applying ARIMA. This model is generally referred to as an ARIMA (p, d, q) 4 model, where p, d, q are integers greater than or equal to zero and refers to the order of the auto regressive, integrated and moving average aspects. In this study, the ACF and PACF are use to identify the stationary of time sires; Z-test, SSE (Sum of square error) and BIC (Schwarz Bayesian Information Criterion) were employed by model estimation and selection; MAPE (Mean Absolute Percentage Error) and Thiel'U (Theil's inequality coefficient) were used to measure the accuracy of forecast model. Many papers and textbooks have detailed discussion on the equation of ARIMA & SARIMA and the expression of statistical magnitude mentioned above, which are not supplied due to the space limitation [2, 5, 6].

85.3 Empirical Result

Data series profiles:

- Total Q China to Australia series $= 1995.Q1 - 2013.Q2$ (74 observations).
- Model building(81 % of data) $= 1995.Q1 - 2009.Q4$ (60 observation).
- Ex post forecasting(19 % as holdout sample) $= 2010.Q1 - 2013.Q2$ (observation).
- Ex ante forecasting(14 quarters ahead) $= 2013.Q3 - 2016.Q4$.

A graphical analysis of the unadjusted quarterly data from 1995(1) to 2009(4) suggest that the original tourist arrivals from China to Australia are likely nonstationary (See Fig. 85.3). This result is supported by correlogram, which display the estimated autocorrelation and partial autocorrelation function of the residuals (Fig. 85.4). We can see from the Fig. 85.4, the ACF (K) & PACF (K) ($K = 1, 2, \cdots, 36$) is not significant as zero respectively, and ACF (1) & PACF (1) is positive respectively, this implies that the series is non stationary (see Fig. 85.4).

Taking natural logarithm for each observation, a logarithmic time series can be seen as Fig. 85.5, looking intuitively, this series is much more stable than the original series. Need to explain,in order to ensure the accuracy of forecasting model, this research use a average value between $Q1$ 2003 and $Q3$ 2003 observation to replace the actual value of $Q2$ 2003, because SARS epidemic sweeping the world cause a rapid reduction in Australia visitor arrivals from China, and the visitor arrivals returned to the normal after the epidemic was brought under control.

For the quarterly nature logarithm series, the final SARIMA model was identified as SARIMA (3, 1, 0) (0, 0, 0). The step identifying as following: Stationary identifying. Using first difference at length 0, it is found that the ACF (K) & PACF (K)

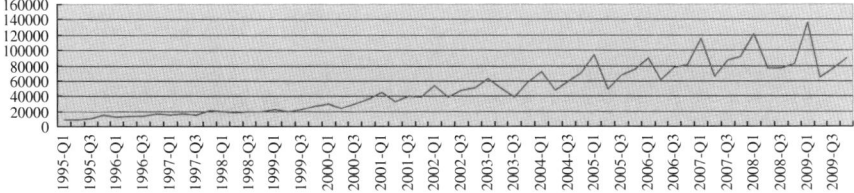

Fig. 85.3 Number of movements: China (exclude SAR and Taiwan) original quarterly

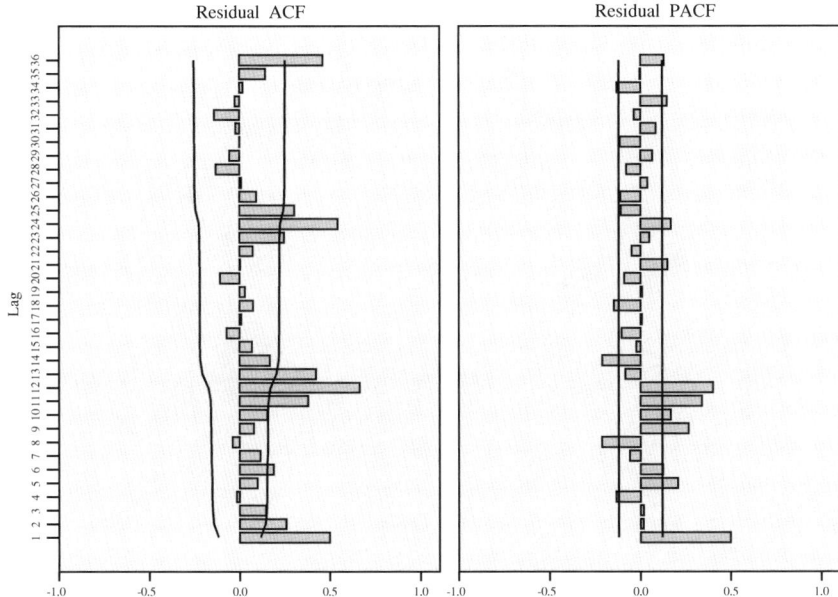

Fig. 85.4 Persons China to Australia exclued Taiwan and SAR Hong Kong residual ACF and PACF (original)

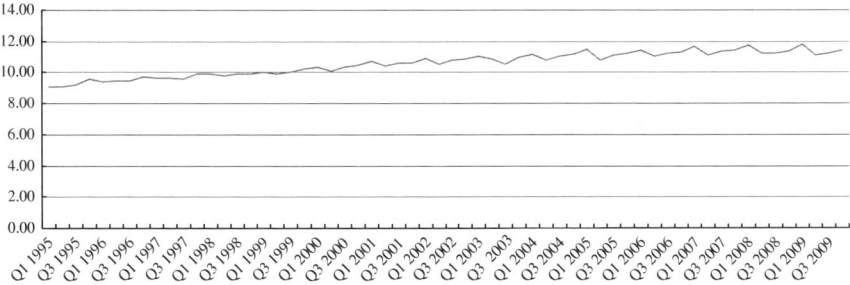

Fig. 85.5 Number of movement: China (exclude SAR and Taiwan) original (logarithm nature)

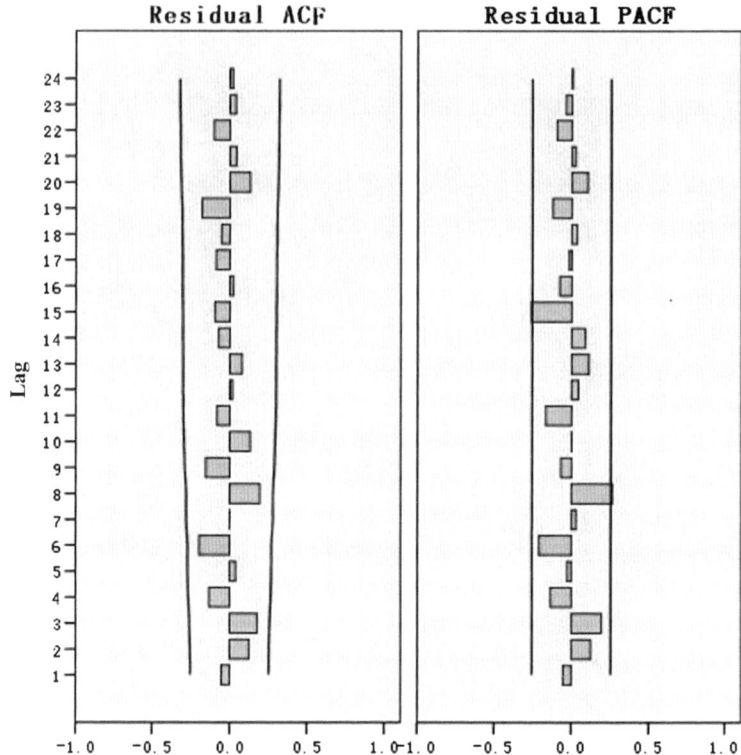

Fig. 85.6 Persons China to Australia exclued Taiwan and SAR Hong Kong residual ACF and PACF (seasonally)

all lie within the confidence interval (less than |0.5|), and the ACF (1) is closed to zero and PACF (K) are all small display a random pattern. Thus the tourist arrivals form China to Australia are integrated of order one, whereas the first difference of tourist arrivals series are integrated of zero, I(0), are and hence are stationary (see Fig. 85.6).

85.4 Model Estimated and Selected

Both estimated parameters achieved significant t-values at $a = 0.05$ level. Model selection were BIC $= -3.568$. Adjusted sum of $squares = 1.04$. The in-sample error measure for SARIMA model were: MAD $= 0.103$, MSE $= 0.02$, SMAPE $= 0.987$, RMSPE $= 1.35$, Theil's U $= 0.55$ (Indicating this final model is better than its naive 1 model) (Tables 85.1 and 85.2).

Table 85.1 A best-fitting models according to normalized BIC (smaller values indicate better fit)

		Variable	Estimate	SE	t	Sig.
Natural log	Constant	Lag 0	0.153	0.020	7.677	0.000
		Difference	1			
	AR	Lag 1	−0.890	0.082	−10.893	0.000
		Lag 2	−0.870	0.086	−10.149	0.000
		Lag 3	−0.843	0.080	−10.597	0.000

Table 85.2 Seasonal ARIMA model statistics for tourist arrivals from China to Australia

Statistics		Model
Number of predictors		1
Model fit statistics	Stationary R-squared	0.720
	TR-squared	0.961
	RMSE	0.146
	RMSE	0.146
	MAPE	0.978
	MAE	0.103
	MAD	0.103
	MaxAPE	4.901
	MaxAE	0.471
	Normalized BIC	−3.568
	Theil's U	0.55
Ljung-Box Q(18)	Statistics	19.179
	DF	15
	Sig.	0.206
Number of outliers		0

A sensible strategy for estimate seasonable simple autoregressive integrated moving average ARMIA (p, q) models, with values for p, q from 3 to 1, is to start with small p & q and work up to larger values. Only models with statistic significant (at 5 %) AR and MA coefficient are selected, ten appropriate seasonal ARIMA models have been selected. For the quarterly China to Australia series, the final SARIMA model was identified as SARIMA $(3, 1, 0) (0, 0, 0)_4$ in the follow equations below by the backshift notation, Which is the "best-fitting" model selected, with the lowest BIC namely(with absolute t ratios in parentheses): $(1 - \phi_1 B - \phi_2 B^2 - \phi_1 B^3)(1 - B)Y_t = a + e_t$, $\ln y_t = \ln a + e_t - \ln(1 - \phi_1 B - \phi_2 B^2 - \phi_1 B^3) - \ln(1 - B)$.

It is clear that the seasonality effects do not give any clear evidence for Australia tourist arrivals from China. Therefore, in order to forecast Australia tourist arrivals from China, this study used only ARIMA $(3, 1, 0)$ model without any seasonal effect to identify the autoregressive (AR) and moving average (MA) effects.

Table 85.3 Quarterly China to Australia ex ante forecast

Horizon	Forecast	95 % UCL	95 % LCL
Q3 2013	147267	233281	92967
Q4 2013	161135	255250	101722
Q1 2014	221904	362217	137310
Q2 2014	179872	293608	110194
Q3 2014	174556	284930	105873
Q4 2014	185350	305590	113550
Q1 2015	252711	420837	150242
Q2 2015	213203	358613	128027
Q3 2015	204843	344552	121783
Q4 2015	217510	365858	128027
Q1 2016	284930	488942	167711
Q2 2016	252711	433653	147267
Q3 2016	242802	416649	140084
Q4 2016	252711	438011	145801

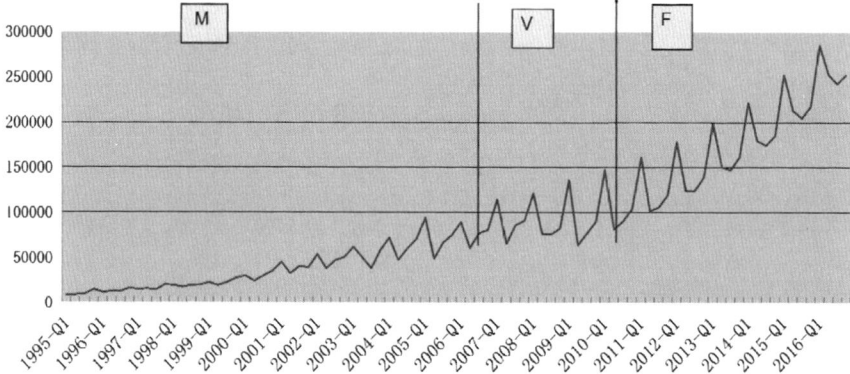

Fig. 85.7 Quarterly international arrivals China to Australia historical and forecasting chart

85.5 Forecasting Tourist Arrivals from China to Australia

During the holdout sample validation, error measure for ARIMA model were: MAD = 0.11, MSE = 0.02, MAPE = 0.97 %, RMSPE = 1.14 %, and Theil's U statistic = 0.33.

The evidence suggests that the ARIMA model be a good one. Using this final model, ex ante forecasts and their 95 % prediction intervals were made for an 14-quarter horizon staring from 2013 $Q3$ to 2016 $Q4$. These quarterly forecasts with their forecasting intervals are provided in Table 85.3 and plotted with actual historical observations in Fig. 85.7.

85.6 Conclusion

In this paper, the moving trend of China tourists to Australia have been discussed briefly. Univariate time series modeling using ARIMA processes has been used to explain the pattern of tourist arrivals from China to Australia. The residual ACF and PACF was applied to identify the original tourist arrivals from China to Australia from $Q1$ 1995 to $Q4$ 2009. Visual inspection of the correlograms also suggested that the series are nonstationary, which led to the use of nature logarithm. According to Normalized BIC, the SPSS identified ARIMA (3, 1, 0) as a Best-Fitting Models, the author used this model to provide a practice forecasting results (ex ante point forecasts and their 95 % prediction intervals on a 14-Quarterly horizon for ARIMA model). This study found the seasonality does not have an effect on the numbers of China tourist arrivals to Australia.

References

1. Ankomah PK, Crompton JL (1996) Influence of cognitive distance in vaction choices. Annu Tourism Res 23:138–150
2. Box GEP, Jenkins G, Reinsel G (1994) Time series analysis: forecasting & control. Prentice Hall Inc, Upper Saddle River
3. Chen N, Clark C et al VC Ethnographies of the urban in later twentieth century China. Duke University Press, Durham
4. Graburn N (1995) Tourism, modernity and nostalgia. In Ahmed AS, Shore C (eds) The future of anthropology: its relevance in the contemporary world, The Athlone Press
5. Jones RH (1975) Fitting auto regressions. J Am Stat Assoc 70:590–592
6. Shibata R (1976) Selection of the order of an auto regressive model by Akaike's information criterion. Biometrika 63(1):117–126
7. Sun Y (2006) Understanding the Chinese tourist-intrinsic and extrinsic reasons of travelling to Europe, pp 76–85

Chapter 86
A New Algorithm for Dynamic Scheduling with Partially Substitutability Constrained Resource on Critical Chain

Qin Yang, Yushi Wang, Si Wu, Wenke Wang and Tingting Wang

Abstract A dynamic scheduling for the constrained resource with partially substitutability based on the critical chain can not only improve the work efficiency of the system, but also achieve the replacement of resources and reduce the utilization of the core resources. This problem is influenced by operating time and setup time of jobs and aims at minimizing total tardiness. Plant growth simulation algorithm is inspired by the photostrophism mechanism and mainly be utilized to solve the global optimization of integer programming problem in the beginning. It possesses the convenience to determine the parameters, stability and accuracy to solve the problem and excellent global optimization ability which make it get extensive and effective application and promotion in many engineering and technical fields. Firstly, describe the problem as a dynamic scheduling problem with machine eligibility restrictions who has machines in parallel with different speeds and jobs with different release date and due date. Secondly model the scheduling problem and design the rescheduling rule according to the time series of jobs. Thirdly, adopt and optimize the new algorithm-plant growth simulation algorithm to get optimization by iterating. Finally, conduct simulation experiment and test the feasibility and superiority of the method by comparing with PSO and GA.

Keywords Plant growth simulation algorithm · Fungibility · RCPSP · Dynamic scheduling · Critical chain

Q. Yang (✉) · Y. Wang · S. Wu · W. Wang · T. Wang
Business School, Sichuan Normal University, Chengdu 610064, People's Republic of China
e-mail: yyuu2006@aliyun.com

J. Xu et al. (eds.), *Proceedings of the Eighth International Conference on Management Science and Engineering Management*, Advances in Intelligent Systems and Computing 281, DOI: 10.1007/978-3-642-55122-2_86, © Springer-Verlag Berlin Heidelberg 2014

86.1 Introduction

Critical chain is a schedule project method based on TOC, which can simplify the large-scale complex scheduling problem by optimizing the bottleneck of the project. Critical chain project management (CCPM) has attracted public attention since its origin [6] in 1997 and has been widely applied to all areas. Available Scheduling for Partially Substitutability Constrained Resource on critical chain can improve the work efficiency of the system considerably. Resource-constrained project scheduling problem (RCPSP) exists widely in manufacturing enterprises. RCPSP is an important combinatorial optimization NP-hard problem. On condition that the resources constraints and project precedence constraints are satisfied, it can achieve the goal of minimizing the total due date by arranging the start and terminal time of all jobs rationally. At present, scholars have adopted many heuristic algorithms [3, 5, 7], such as genetic algorithm, ant colony algorithm and scatter search to solve problem. In daily life, compared with shortening the work period, reducing the requirement level of project resources is also important [4]. But the study of requirement level of project resources in constrained resource mainly concerned whether the resource can be updated or not, namely, the acquisition and consumption of a particular resource is for the limit of one stage. Only by completing one job, can the other job be executed which may neglect the fungibility of resources. Namely, though resources have dissimilitudes on different level, there only exists time difference in the process of completing jobs.

Plant growth simulation algorithm (PGSA) was put forward by Li in 2005 when she was inspired by the photostrophism mechanism. It was mainly used to solve the global optimization [11] of integer programming problem at first, now this algorithm is applied to solve the static problem information which related to the problem are well known at initial time. Relying on the convenience to determine the parameter, PGSA has got extensive and effective application and promotion [8–10, 15] in many engineering and technical fields. Based on pre-existing method to identify critical chain which has alternative resources, this thesis combines the research results [1, 2, 12, 14] of bottleneck resources scheduling and dynamic scheduling in manufacturing and adopts PGSA to solve Dynamic Scheduling for Partially Substitutability Constrained Resource on the critical chain.

This thesis tries to solve the problem which considered alternative resources, resource restriction and dynamic scheduling. At present, the discussions of such issue among scholars domestic and overseas are restricted in comprehension theoretically, meanwhile, there are not detailed studies which relevant to such issue. But this is a typical multi-NP-hard problem, it is complicated to solve this problem by considering optimization algorithm in hand, and cannot even reflect the dynamic natural and so on. The improvement in this thesis embodies in two ways. Firstly, by designing mix rescheduling tactics to achieve dynamic scheduling. But this tactics will turn a NP-hard problem into a multi-NP-hard problem which will increase difficulty of problem. Secondly, in order to simplify the procedure, it needs rule algorithm to work out initial solutions specifically.

86.2 Problem Statement and Model Design

Based on the typical RCPSP problem, this thesis focuses on the applicability and expansibility of the model and describes the research question as a dynamic scheduling problem with machine eligibility restrictions who has machine in parallel with different speeds and jobs possesses different release date and due date. This problem is influenced by operating time and setup time of the jobs and aims at minimizing the total tardiness. In this scheduling, superscript i indicates constrained resources and alternative resources ($i = 1, 2, \cdots, m$), subscript j denotes job ($j = 1, 2, \cdots, n$) and subscript t denotes points in time.

86.2.1 Hypothesis of the Problem

Hypothesis 1. Multiple jobs form one project, the number of jobs is limited and each job's operating time can be considered as certain;

Hypothesis 2. A specific resource can at most be utilized for one job at any moment;

Hypothesis 3. A Job can only utilize constrained resources or alternative resources for once which means it can't be interrupted to proceed the next job until current job is finished;

Hypothesis 4. n numbers of jobs are applied to m' numbers of constrained resources, and m' is much smaller than $n (m' \ll n)$;

Hypothesis 5. Other resources can partially replace constrained resources, but the constrained resources have applicable restrictions for different jobs.

86.2.2 Problem Description

Dynamic scheduling for Partially Substitutability Constrained Resource on the critical chain can be described by a triple $\alpha/\beta/\gamma$.

(1) Environment of Resources (α field)

The α field can be equivalent to machines in parallel with different speeds (Q_m). m_u denotes the quantum of constrained resources. On the basis of category, $m_1, m_2, \cdots, m_L$ denotes alternative resources. m denotes the quantum of constrained and alternative resources.

(2) Characteristic and Restraint of Jobs (β field)

Job j starts to operate from release date r_j on constrained or alternative resources, go through operating time p_j^i (which is co-determined by quantities of jobs Q and the operating speed of resources μ^i) on resource, then the work can be completed. But during the period between the job k and job j, there exists a setup time S_{jk}^i. Jobs on critical chain often have a due date d_j, but the release date r_j had principle of uncertainty, thus jobs have dynamic nature.

Job j has two constraints, the first is temporal constraints, which means some specific precedence relationships indwell between jobs. The second is resource constraints. m kinds of resources are in requirement to complete the project, the quantum of resources i is m_j.

(3) Goal of Scheduling (γ field)
The primary goal of schedule is to minimize the delay time. The delay time of jobs can be measured by $T_j^i = \max(C_j^i - d_j, 0)$ (C_j^i represents the actual make-span of job j on resource i).

86.2.3 Modeling

The model can be designed as follows:

$$\min \sum_{j=1}^{n} T_j \tag{86.1}$$

$$\text{s.t. } x_{jt}^i \in \{0, 1\}, \quad j \in N, \ i \in M, \tag{86.2}$$

$$\sum_{t=0}^{C_{max}} \sum_{i=1}^{m} \sum_{j=1}^{n} x_{jt}^i = n, \tag{86.3}$$

$$\sum_{t=0}^{C_{max}} \sum_{i=1}^{m} x_{jt}^i = 1, \quad j \in N, \tag{86.4}$$

$$T_j = \max(C_j^i - d_j, 0), \quad j \in N, \ i \in M, \tag{86.5}$$

$$C_j^i - p_j^i \geq r_j, \quad j \in N, \ i \in M, \tag{86.6}$$

$$C_j^i = (C_k^i + s_{jk}^i + p_j^i)x_{jt}^i, \quad (i, k \to i, j) \in A, \tag{86.7}$$

$$p_j^i = Q_j/\mu^i, \quad j \in N, \tag{86.8}$$

$$d_j \geq 0, \ r_j \geq 0, \quad j \in N. \tag{86.9}$$

Decision variable in the model has two cases:

$$x_{jt}^i = \begin{cases} 1, & \text{at time } t, \text{resource } i \text{ begins to be utilized by job } j \\ 0, & \text{otherwise.} \end{cases}$$

M denotes the set of constrained and alternative resources; N denotes the set of jobs; A is the set of line constraints $i, k \to i, j$, means that job k utilized resource i first, then job j utilize it.

Equation (86.1) is the objective function to minimize the total tardiness of jobs. Equation (86.2) restrains the evaluation of decision variables; Eq. (86.3) ensures all the jobs can be completed, C_{max} represents the completion time of the last job; Eq. (86.4) guarantees the establishment of Eq. (86.3); Eq. (86.5) can calculate the tardiness time of jobs; Eq. (86.6) guarantees job j with ample operating time on resource i; Eq. (86.7) measures the completion time of jobs; Eq. (86.8) confirms the actual operating time; Eq. (86.9) restricts expected completion time and release date of job in value ranges.

86.3 Designment of Improved PGSA

First, devising composite rescheduling tactics, then choose related composite dispatching rules; Next, design and determine the method of growing points and morphactin concentrations. Finally, put PGSA into utilizing and get the goal of algorithm optimization.

86.3.1 Strategy of Rescheduling

There are three kinds of scheduling: event-driven rescheduling, periodical rescheduling and hybrid rescheduling. The thesis chooses the hybrid rescheduling strategy, which the strategy has the following flow chart in Fig. 86.1.

There are two conditions: when specific events appear, update resource information and conduct rescheduling; at the end of the rescheduling period, shift the completed jobs out of job window, moreover, shift the jobs to be processed into job window to ensure the conformance of jobs' quantities in the job window. Meanwhile, check idle resources then reschedule jobs in the job window. Repeat the aforesaid processes until complete all the jobs.

86.3.2 Stage 1: Getting Initial Solutions with Composite Dispatching Rules

The goal of LIST rule can distribute jobs to all of the resources and minimize the total operating time C_{max} of the jobs to be processed on each machine at the same time.

According to Apparent Tardiness Cost with Setups rule, sequencing jobs of which attributed to each machine, the goal of schedule is to minimize $\sum w_j T_j$. Calculating the index I_j^i of jobs on each machine,

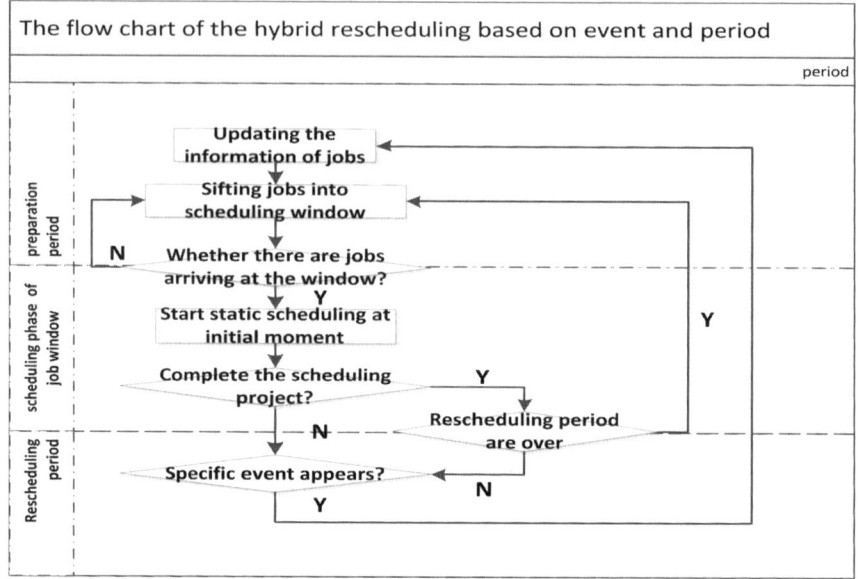

Fig. 86.1 Flow chart of the hybrid rescheduling strategy based on event and period

$$I_j^i(t,k) = \frac{w_j}{p_j^i} \exp\left(-\frac{\max(d_j - p_j^i - t, 0)}{K_1 \bar{p}}\right) \exp\left(-\frac{s_{jk}^i}{K_2 \bar{s}}\right). \qquad (86.10)$$

$\bar{p}$ denotes average operating time of surplus jobs, $\bar{s}$ denotes average setup time, K_1 is proportion parameter in small rules related to due date. K_2 denotes the regular proportion parameter related to setup time. Choose jobs with biggest I_j^i to be scheduled preferentially.

86.3.3 Stage 2: Iterating and Solving by PGSA

PGSA [13] takes the plants' whole growing space as feasible region of solutions, illuminant as global optimal solution and phototropism decides the dynamic growth mechanism.

1. Growth mechanism

There are four necessary concepts in PGSA: Roots, trunks, branches and growing points, above all, the growing point is the most important one, namely, the plant growth cell.

The first procedure of plant growth:
There are two key issues need to be settled: choose one of the growing points which are able to branch out; how to make sure the branches can get close to the illuminant.

Morphactin concentration decides whether the growth function works or not and the possible speed of growth. Branching happens in two *a* dimensional plane, suppose that it grows in unit length each time.

The second procedure of plant growth:
Finding out the sit *a* with maximum morphactin concentration. After the following growing points coming into being, morphactin will be reallocated to each growing point adapting to the change of the environment. Finally the optimal solutions will be found.

2. Abstract of variables and mathematic simulation in PGSA
Abstract all resources: alternative resources M_L ($L = 1, 2, \cdots, L$), m_L denotes quantum of each alternative resources, quantum of constrained resources M_U is m_U. Here are the agreements:

(1) The length of the each growth is one unit, which can put PGSA into practical application and make it tally with the unit change of real tasks.
(2) Each job has a job coefficient y_i, $yj \in (0, 1)$, as for the coefficient set y_{jx} of each job on a certain machine, the smaller the coefficient y_j, the earlier the job to be processed.
(3) Any growing point is denoted by S_{mx} with the information of jobs and resources M_L, M_U and the job set Y_{jm} which is composed of y_j: $S_{mx} = (y_{jL}, M_L|y_{jU}, M_U|\cdots)$.
(4) Change of operating jobs or sequences of jobs will generate different growing points.

Actually, the scheduling here is with regard to resources and jobs separately:

a. Transfer contiguous jobs on other resources to the certain idle resources to be operated, so it will increase a unit of job on the idle resource.
b. When jobs exist on one resource, jobs on this resource will decrease a unit. Operation diagram of a and b is as Fig. 86.2.
   In formulation of growing points, it considers the last job on resource connects to the next which means there can be an adjustment about job1 between resource 1 and resource 2 growing point $S_{mx} = (y_{1x}, y_{11}, M_1|y_{2x}, \cdots, M_2|y_{3x}, \cdots, M_3|\cdots)$. After adjustment, $S_{mx} = (y_{1x}, M_1|y_{11}, y_{2x}, \cdots, M_2|y_{3x}, \cdots, M_3|\cdots)$. As for the resource M_1, the job on it has decreased one unit, so it has a corresponding growing length of -1, and the growing length of resource M_2 is 1.
   Generally speaking, if we count all the possible growing conditions, there will be $[2(m-1)-1]$ kinds of them which can represent the number of new growing points.

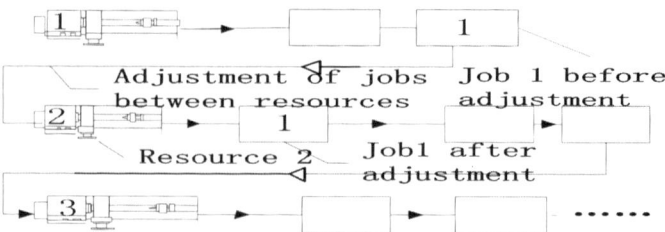

Fig. 86.2 Adjustment of jobs based on difference resources

c. To make the completed jobs differ from the uncompleted jobs on one certain resource, the coefficient y_i of completed jobs will be added one unit. For instance, the job set $Y_{jm} = (y_{j1}, y_{j2}, y_{j3}, \cdots)$ on one resource, if y_{jx} has been operated firstly, then $y_{jx} = y_{jx} + 1$, reordering the sequence, $Y_{jm} = (y_{j2}, y_{j3}, \cdots, y_{jl} + 1)$. There will be n numbers of growing points if n jobs are growing respectively. Here are conditions of it: Coefficient y_j of a certain job is already the largest one in all coefficients. It happens only once on each resource in a certain scheduling moment, so the real number of new growing points is $(n - m)$. Growing points denote all possible sequence conditions on certain resource.

Completed the procedure of a, b and c, then preliminary scheduling scheme is formed.

(5) Whether the new growing point is superior to the primary one depends on the relative size of morphactin concentration.

(6) If the two growing points appears, merge the two growing points, then the number of growing points will decrease correspondingly.

Set of growing points in the model is $S_{ML} = (S_{m_1}, S_{m_2}, \cdots, S_{m_l})$ for example, as for every growing point on branch M_U, the auxin concentration is denoted by P_{r_k}, $P_{r_h} = (pr_1, pr_2, \cdots, pr_l)$. The formation mechanism of auxin concentration has been introduced before, so the model is as follows:

$$P_{R_l} = \frac{f(x_0) - f(S_{mx})}{\sum_{x=1}^{m_x}(f(x_0) - f(S_{mx}))}. \tag{86.11}$$

x_0 is the initial cardinal point which is the initial solution generated in phase one, $f(*)$ is the located environment information function which can be reflected by optimal function $f(S_{mx}) = \sum \omega_j T_j(S_{mx})$. Smaller value of the function means a better environment conditions of the corresponding points, and if the auxin concentration is bigger, it is more possible to branch out new branches. From analysis of the formula, $P_{R_l} \in (0, 1)$ and $\sum_{x=1}^{m_x} P_{R_L} = 1$; This closed system will only distributes materials within system and the sum of morphactin concentration is of value 1 constantly.

86.4 Algorithm Steps of Dynamic Scheduling for Partially Substitutability Constrained Resource on the Critical Chain

Step 1. Generate growing points randomly. Due to this thesis has already illustrated about the growing points, it can be considered that the growth of plants only connect with cells at the growing points. Owing to this thesis is aimed at dynamic scheduling for partially substitutability constrained resource on the critical chain, it needs to be emphasized that every growing point is represented by the resource index and job index.

Step 2. Make sure the morphactin concentration P_{M_x} of each growing point ($X = 1, 2, \cdots, x$).

Step 3. Choose the growing point. Based on the calculation consequences in *Step 2*, establish the probability interspace between 0 and 1 of each growing point, then choose the iterative growing point S_{mx} by adopting random numbers. It can be comprehended from Fig. 86.3. The system will generate random numbers in the interval of [0, 1], these numbers are located in one of the homologous probability interval $pr_1, pr_2, \cdots, pr_l$, then the corresponding random growing point will have prior right to grow. When the former new finishes growing, all the morphactin concentration will be divided again, which is in the terms of Eqs. (86.1) and (86.2), besides, the newly branching branches should replace the origin growing points.

Step 4. Make decision of the growing length $\lambda(\lambda = 0, 1)$.

Step 5. Choose initial growing point a_h in accordance with random initial data. Keep all parameters growing separately while other parameters remaining unchanged. Then there will be $(n + m - 3)$ growing points when new growing points are superior than original ones, replace the former ones by utilizing the newer which the result of target function is minimum.

Step 6. When new growing points are superior than the original ones, replace the original one with the new one S that has minimum corresponding objective function value.

Step 7. Repeat the aforesaid steps until there is no new one superior than the optimal one.

86.5 Case Study

Sign each tranche with $n \times m[M_I, M_U, \cdots]$, in which n represents number of activities, m denotes the quantum of resources. Generate the test data randomly and adopt the hybrid rescheduling tactic then solve the problem by using PGSA, what is more, there are 30 iterations. Finally compared the solutions with particle swarm optimization and genetic algorithm.

Fig. 86.3 Chart of
concentration section that
random number has located

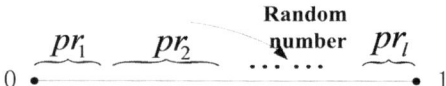

$$0 \bullet \hspace{8cm} \bullet 1$$

Table 86.1 The comparison of scheduling performance

Questions $n \times m[M_L, M_U, \cdots]$	The parameter of scheduling performance $\rho(H, S)$		
	Improved PGSA	PSO	GA
$45 \times 3[M_L(2), M_U(4)]$	1.000	1.014	1.023
$95 \times 3[M_L(2), M_U(4)]$	1.000	1.029	1.038
$95 \times 6[M_L(4), M_U(4)]$	1.000	1.037	1.046
$150 \times 6[M_L(4), M_U(4)]$	1.000	1.051	1.067
$150 \times 10[M_L(4), M_U(6)]$	1.000	1.060	1.078

- Scope of total resources is located in [3, 10] and the scope of jobs is located in [45, 150];
- In each job, the quantum of task Q_i can be evenly distributed in [10, 20] and the release date r_j can be evenly distributed in [0, 8], moreover, the operating date p_j^i should be in [2, 8].
- The setup time of resources for transformation in different tasks is in associated with the execution sequence of activities. a_j indicates the circumstance of activity j which on the basis of resource. While resource i is being utilized, activity k proceed after activity j, moreover, the setup time can be described as $s_{jk}^i = |a_k^i - a_j^i|$, and a_j is evenly distributed in the interval [0, 1], besides, the original setup time is 0.
- To avoid setting the estimated due time a_j too loose or tight, it can be measured by formula $d_j = r_j + kp$. p denotes average operating time of tasks in different equipment, meanwhile, k is evenly distributed in the interval [0, 2].

Here are the assessment criteria [31], taking $I (I \in S)$ as an example, $T(H, I)$ denotes the objective function value of example I which based on algorithm H. Optimal solution can be described as $BEST(I) = \min\{T(H, I)\}$. For each group of question S, computing the following performance index:

$$\rho(H, S) = \frac{\sum_{I \in S} T(H, I)}{\sum_{I \in S} BEST(I)}. \tag{86.12}$$

Here comes to the following conclusions from the data by Table 86.1.

- With satisfied of the same resources and activities, the scheduling performance parameter which adopted the improved PGSA takes prior compared with PSO and the GA.
- With enlargement of the issues scale, larger performance parameter of the PSO and GA means a wider the discreteness. Meanwhile, the performance parameter value of the improved PGSA remains 1.000, which takes prominent role to show superiority of improved GPSA.

- On condition of the same number of activities, the more the quantum of resources is, the more prominent superiority of improved GPSA will be; If the quantum of resources is same, more activities means more prominent role to show superiority compared with the PSO and GA.

86.6 Conclusion

A dynamic scheduling for the constrained resource with partially substitutability based on the critical chain has vital values. This thesis analyses the characteristics of this scheduling model and sets up corresponding optimal scheduling model. This algorithm can arrange the start and terminal time of all jobs rationally and it has simple and stable demands for corresponding parameters. Therefore, the combines of mixed rescheduling tactics and improved PGSA can guarantee the solving with efficiency and accuracy.

Acknowledgments Project supported by National Natural Science Foundation of China 71202166; MOE (Ministry of Education in China) Project of Humanities and Social Sciences (Project No. 13YJC630202); The Department of Education Project of Sichuan Province (14ZA0026, 14SB0022).

References

1. Betterton CE, Cox III, James F (2009) Espoused drum-buffer-rope flow control in serial lines: a comparative study of simulation models. Int J Prod Econ 117(1):66–79
2. Chen C, Chen C (2009) A bottleneck-based heuristic for minimizing makespan in a flexible flow line with unrelated parallel machines. Comput Oper Res 36(11):3073–3081
3. Chen R, Wu C et al (2010) Using novel particle swarm optimization scheme to solve resource-constrained scheduling problem in psplib. Expert Syst Appl 37(3):1899–1910
4. Fang C, Wang L (2010) Survey on resource-constrained project scheduling. Control Decis 25(5):641–650
5. Fin A, Homberger J (2013) An ant-based coordination mechanism for resource-constrained project scheduling with multiple agents and cash flow objectives. Flex Serv Manuf J 25(1–2):94–121
6. Goldratt EM (1997) Critical chain: a business novel. North River Press, Great Barrington
7. Gonçalves JF, Resende MGC, Jorge JJM (2011) A biased random-key genetic algorithm with forward-backward improvement for the resource constrained project scheduling proble. J Heuristics 17(5):467–486
8. Guo G (2011) Study on rescheduling method for flow shop problem with uncertainty. Huazhong University of Science and Technology (In Chinese)
9. Haibo T, Chunming Y (2010) Application of plant growth simulation algorithm to solving job shop scheduling problem. Mech Sci Technol Aerosp Eng 11:35
10. Li T, Wang Z (2008) Application of plant growth simulation algorithm on solving facility location problem. Syst Eng Theor Pract 28(12):107–115
11. Li T, Wang C et al (2005) A global optimization bionics algorithm for solving integer programming-plant growth simulation algorithm. Syst Eng Theor Pract 25(1):76–85

12. Liu S, Song J (2011) Combination of constraint programming and mathematical programming for solving resources-constrained project-scheduling problems. Control Theor Appl 28(8):1113–1120
13. Pinedo M (2012) Scheduling: theory, algorithms, and systems. Springer, Boston
14. Vanhoucke M (2012) Project management with dynamic scheduling. Springer, Berlin
15. Yang Q, Zhou G, Lin J (2012) Scheduling of machines in parallel with different speeds based on planted growth simulation algorithm—taking an example of the bottleneck in 4s auto dealership maintenance shop. Syst Eng Theor Pract 32(11):2433–2438

Chapter 87
Decision Methodology for Maintenance KPI Selection: Based on ELECTRE I

César Duarte Freitas Gonçalves, José António Mendonça Dias
and Virgílio António Cruz-Machado

Abstract Measure the performance of maintenance service and the influence of implemented practices in maintenance activities is essential to the success of companies. Maintenance managers deal with the complex tasks of find out the best maintenance performance indicators which can help them in achieve goals. In this chapter a new approach is introduced for maintenance Key Performance Indicators (KPIs) selection using the original ELECTRE I, which is a Multi Criteria Decision Making method. This chapter describes the methodology of the decision model and a case study to exemplify its use is presented. The proposed methodology determines a ranking of possible alternatives after its evaluations according important criteria. The proposed methodology that involves decision maker's preference information makes the decision process more explicit, rational and efficient. Light to the results, this chapter proves that the new methodology is an effective tool to assist maintenance managers in accurate KPIs selection tasks in line with maintenance objectives and strategies.

Keywords Maintenance · KPI · Multi criteria decision making · MCDM · SRF · ELECTRE

C. D. F. Gonçalves (✉)
Departamento de Engenharia Mecanica (DEM),
Instituto Superior de Engenharia, Universidade do Algarve (ISE/UALG),
8005-139 Faro, Portugal
e-mail: cgoncal@ualg.pt

J. A. M. Dias · V. A. Cruz-Machado
Faculdade de Ciências e Tecnologia,
Departamento de Engenharia Mecanica e Industrial (DEMI),
Universidade Nova de Lisboa (FCT/UNL), 2829-516 Caparica, Portugal

J. Xu et al. (eds.), *Proceedings of the Eighth International Conference on Management Science and Engineering Management*, Advances in Intelligent Systems and Computing 281, DOI: 10.1007/978-3-642-55122-2_87, © Springer-Verlag Berlin Heidelberg 2014

87.1 Introduction

Maintenance management is an important activity within companies. The maintenance management cares about ensuring the smooth operation of the facilities, systems and equipment, ensuring that good operating conditions with maximum availability are achieved, all at an optimized global cost.

The maintenance managers need relevant indicators to measure important elements of maintenance performance. Establish a usable set of maintenance Key Performance Indicators (KPIs) depends mainly on maintenance objectives and these on the company's goals.

Maintenance performance measurement is in fact a multidisciplinary process once taking into account multiple aspects of maintenance activities and, for this reason, the selection of the best maintenance performance indicators is a complex task that can be formulated as a Multi Criteria Decision Making (MCDM) problem. In this chapter, we propose a methodology by using MCDM methods that deal with such kind of problems in order to assist the maintenance managers.

87.2 Maintenance Key Performance Indicators

Typically, companies use Key Performance Indicators (KPIs) to measure the successes of the activities in which they are engaged. Maintenance can serve itself a management system of appropriate indicators to measure its performance.

Performance indicator is defined by the European Standard EN 15341 [1] as a measured characteristic (or set of characteristics) of a phenomenon, according to a specific formula that evaluates its evolution.

In the EN 15341 [1] are provided three categories and three levels of maintenance performance indicators to appraise and improve efficiency and effectiveness, in order to achieve excellence in the maintenance of technical assets. In order to select relevant KPIs, the standard EN 15341 [1], refers that the first step is to define the objectives to be reached at each level of the enterprise, whose requirement is to identify the proper maintenance management model in order to improve global performance. This standard mentions that in the search for relevant KPIs a first approach can be made, by choosing from among the list of existing indicators, which after analysis, fulfill the requirements. Furthermore, the KPIs can be developed whenever we need to measure a fact, to analyze it and to allow its optimization.

A relevant indicator shall be a key element in the decision making, which means that its evaluation and its data should be related to the performance parameter to be measured and with the defined objective. Some authors provide guidelines for choosing maintenance KPIs that seek to align maintenance objectives with production objectives [2, 3], which relationship is an important factor for the success of maintenance performance measurement and to reach customer's needs [4].

The number of indicators used for each department, in an organization, should be limited by identifying key features or key factors [5]. A large number of indicators to measure each maintenance aspect hinder comprehension and the work for which they are developed.

Different categories of KPIs and distinct frameworks have extensively been discussed and proposed in the literature to monitor and control maintenance activities. However, it was observed that few publications propose methodologies for selecting relevant KPIs, especially in the field of maintenance.

87.3 Multiple Criteria Decision Making

Multiple Criteria Decision Making (MCDM) or Multi Criteria Decision Analysis (MCDA) are different designations for the same concept. These well-known terms are concerned with the analysis and solving decision problems involving multiple criteria and are considered a sub-discipline within Operations Research. The purpose is to aid decision makers to make preference decisions when multiple alternatives are evaluated by multiple criteria, which in most cases are conflicting.

The acronym ELECTRE stands for "ELimination Et Choix Traduisant la REalité (Elimination and Choice Expressing the Reality)" and is a well-known MCDM method presented by Bernard Roy in the late 1960s. The ELECTRE I was the first decision aid method using the concept of outranking relation [6] and was originally designed to lead to "choice-type" results. The ELECTRE approach has evolved into a number of variants based on the same concept covering the different types of decision problems, see [7]. A complementary analysis to the results of the ELECTRE I method by introducing the calculation of aggregate differential concordance and discordance values (net concordance and net discordance values) [8] was also introduced, which allow the ranking of alternatives from the best to the worst.

Some approaches in literature dealing with the original ELECTRE I method which deserved attention in our research. For instance, we have found authors proposing methodologies of decision making in suppliers selection [9, 10], evaluation of transport sustainability [11], bioinformatics investigation of human genes [12], environmental impact assessment [13] and tourism management [14]. In another approach, the original ELECTRE I and a complementary analysis through the net concordance and net discordance values were used for selecting appropriate materials [15].

Still emerge two recent approaches of MCDM methods application that called attention relatively to the subject of our research. Both using the Analytic Network Process (ANP), an extension of the Analytic Hierarchy Process (AHP) method: an, for choose the most appropriate practices and KPIs in a supply chain [16]; and another, in order to assist the maintenance manager in the definition and selection of the relevant KPIs [17].

87.4 The Proposed Methodology

We propose a methodology which makes use of the original ELECTRE I method to select the most significant KPIs for a maintenance management framework. Within the ELECTRE family methods, the decision maker is asked to assign numerical values to the weights of the criteria. For assessing those weights we advise the revised Simos' procedure aided by the SRF (Simos–Roy–Figueira) software.

Ranking is significantly different from choice or sorting and therefore requires the use of specific methods. ELECTRE I is a simple method for a choice problems, nevertheless, their outputs can be worked upon to obtain a ranking of alternatives. Towards that goal, the net concordance and net discordance values calculation is proposed. These methods (used in sequence) make it possible to select and rank the most relevant maintenance KPIs after their evaluation by a decision maker.

The conceptual framework (Fig. 87.1) intends to aid decision makers conducting the methodology.

In Table 87.1, a assessment "Likert" scale is presented, with which the decision makers can specify their level of agreement with each alternative in relation to different criteria.

In the two following subsections the procedures to obtain the weights of criteria by the SRF software and the mathematical formulation of the original ELECTRE I method are both described.

1. The Revised Simos' Procedure—SRF Software

The proposed multi criteria approach is aided by the SRF (Simos–Roy–Figueira) software in order to obtain the relative importance of the criteria, which is often referred as weights of the criteria. The SRF software is an implementation of the revised Simos' procedure [18] and was developed with an algorithm for assigning numerical values to the weights of the criteria. The higher the weight of a criterion is, the more important it is.

In revised Simos' procedure, the names of the criteria are written in a set of cards which the decision maker ranks from the least important criterion to the most important one. Thus, the ranking will be in ascending order and some criteria may take the same rank if are viewed as having the same importance. Then, blank cards may be introduced between successive criteria taking into account the smaller or bigger difference of their importance. Using the SRF software, these inputs are introduced through a simple questionnaire where the decision maker is also called to state how many times the most important criterion is considered to be more important than the least important one in the ranking (Z value), see [18]. In addition, different values concerning the ratio Z can be introduced since the decision maker did not feel confident enough expressing this ratio using a single constant value. Finally, the weights of criteria are determined and displayed as non-normalized weights and normalized weights.

2. The ELECTRE I Method

The original ELECTRE I method is presented in the literature [8, 15, 19] and consists of the following steps:

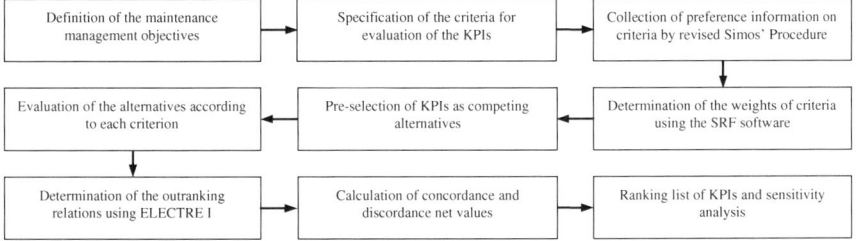

Fig. 87.1 Conceptual framework of the proposed methodology

Table 87.1 Likert scale of scores

Score	Definition
1	Unimportant
3	Little importance
5	Important
7	Very important
9	Absolutely important
2, 4, 6, 8	Intermediate values

Step 1. Consider a finite set of m possible alternatives $A = (A_1, A_2, \cdots, A_m)$, evaluated on a coherent set of n decision criteria, denoted by $C = (C_1, C_2, \cdots, C_n)$. The assignment values of each alternative A_i according to each criterion C_j are defined as x_{ij} and form a performance matrix $X = (x_{ij})_{m \times n}$. Let w_n be the relative importance coefficient (or weight) attached to criterion C_n and so, $W = (w_1, w_2, \cdots, w_n)$ represent the set of all criteria weights, satisfying $\sum_{i=1}^{n} w_i = 1$.

Step 2. The normalization of the performance matrix X to a comparable scale with the maximization objective is done using the Eq. (87.1) and a normalized performance matrix $R = (r_{ij})_{m \times n}$ is obtained,

$$r_{ij} = \frac{x_{ij}}{\sqrt{\sum_{i=1}^{m} x_{ij}^2}}, \quad i = 1, 2, \cdots, m, \ j = 1, 2, \cdots, n. \qquad (87.1)$$

Step 3. A weighted and normalized performance matrix $V = (v_{ij})_{m \times n}$ is obtained by multiplying the columns of normalized performance matrix R by the associated weights w_j of the criterion C_j, such as Eq. (87.2) provides below.

$$v_{ij} = r_{ij} w_i, \quad i = 1, 2, \cdots, m, j = 1, 2, \cdots, n. \qquad (87.2)$$

Step 4. For each pair of alternatives A_p and A_q ($p, q = 1, 2, \cdots, m$ and $p \neq q$) the set of criteria is divided into two distinct subsets. If alternative A_p is preferred to alternative A_q for all criteria, the concordance set $C(p, q)$ is composed and can be determined as follows:

$$C(p, q) = \{j | v_{pj} > v_{qj}\}, \quad j = 1, 2, \cdots, n. \tag{87.3}$$

The complementary of $C(p, q)$, called the discordance set $D(p, q)$, contains all criteria for which A_p is worse than A_q and is determined by:

$$D(p, q) = \{j | v_{pj} < v_{qj}\} = J - C(p, q), \quad j = 1, 2, \cdots, n, \quad J = \{j | j = 1, 2, \cdots, n\}. \tag{87.4}$$

Step 5. The concordance index c_{pq} of the set $C(p, q)$ express in what measure the alternative "A_p outranks A_q", according to the decision maker preferences and is obtained using the Eq. (87.5):

$$c_{pq} = \sum_{j \in C(p,q)} w_j, \quad j = 1, 2, \cdots, n. \tag{87.5}$$

All concordance indices make up the concordance matrix $C = (c_{pq})_{m \times m}$.

The discordance index d_{pq} of the set $D(p, q)$ indicates in what measure the decision maker preferences oppose the assertion of "A_p outranks A_q" and is obtained using the Eq. (87.6):

$$d_{pq} = \frac{\max_{j \in C(p,q)} |v_{pj} - v_{qj}|}{\max_{j \in J} |v_{pj} - v_{qj}|}, \quad j = 1, 2, \cdots, n, \quad J = \{j | j = 1, 2, \cdots, n\}. \tag{87.6}$$

All discordance indices make up the discordance matrix $D = (d_{pq})_{m \times m}$.

Step 6. The net concordance value NC_p measures the dominance degree that each alternative A_p has competing with the others and can be calculated by Eq. (87.7). Here, the objective is evaluating alternative to know how strong than others it is in concordance (the higher the better). Thus,

$$NC_p = \sum_{q=1}^{m} c_{pq} - \sum_{q=1}^{m} c_{pq}, \tag{87.7}$$

In its turn, the net discordance value ND_p measures the relative weakness of alternative A_p with respect to the other alternatives. However, a lower net discordance value is required for a greater preference (the lower the better). The net discordance value can be calculated by Eq. (87.8):

$$ND_p = \sum_{q=1}^{m} d_{pq} - \sum_{q=1}^{m} d_{pq}. \tag{87.8}$$

Hence, the final selection should satisfy the condition that its net concordance value should be at a maximum and its net discordance value at a minimum [19].

The presented mathematical procedure was programmed in "Excel VBA" (Visual Basic for Applications) to provide computational results.

87.5 Case Study

Suppose a maintenance department of an important airport realized that there should be a relevant set of indicators for maintenance service quality measurement on their facilities. Those responsible for maintenance management find it difficult to establish strict preference judgements, due to the vast number of alternatives.

In general, the organization's strategic goals are continuous improvement of efficient management of airport infrastructure, productivity levels and quality of service to passengers and airlines. The goal is also striving towards optimal maintenance with motivated employees.

A first way to approach the problem is to generate appropriate criteria as constraints in evaluating the set of alternatives. Thus, we started using the literature survey and a panel of maintenance experts to identify the criteria, in order to select relevant KPIs for maintenance. In a subsequent analysis with airport maintenance managers, we have focused on the matter of determining the most important criteria in the context of maintenance service quality measurement. As a result, we met a set of five decision criteria, described as follows:

- *Maintenance quality* (C_1): to evaluate how the KPI quantitatively measures the quality aspects of maintenance service.
- *Maintenance objectives* (C_2): to evaluate how the KPI measures the maintenance state in fulfilment of established objectives for the maintenance and according to company goals.
- *KPI influence* (C_3): to evaluate how the result of the KPI can influence decision making in maintenance.
- *Understanding and interpretation of KPI* (C_4): to evaluate how the KPI expresses a measurement that easily translates an obtained or expected real result.
- *Sharing results* (C_5): to evaluate how the results of the KPI has interest in being shared with others company sectors, employees or customers.

Then the maintenance manager was called upon to elicit the relative importance of criteria in an indirect way by using the SRF software. The required data for the revised Simos' procedure are given in Table 87.2. Through this software the appropriate values to the weights (W_n) of the five criteria (C_n) were generated, which are also presented in Table 87.2 as normalized weights.A set of competing alternatives, assuming which are all possible best choices, was pre-selected. Table 87.3 presents a set of 20 KPIs with possible interest to measure the performance of maintenance in terms of service quality. Regarding sources of KPIs, we research the own maintenance department of airport, the standard EN 15341 [1], articles from several authors in the scope of maintenance measurement and other KPIs were developed by us for this specific case.

Table 87.2 Weights of criteria generated by RSF software

Criteria code	Criteria rank	Blank cards	Non-normalized weights	Normalized weights
C_5	1	2	1	2.24
C_2	2		8	17.91
C_3	2	1	8	17.91
C_1	3	0	12.67	28.36
C_4	4		15	33.58
Total	–	3	Z value = 15	100

The performance of the 20 alternatives (KPIs) evaluated according to the 5 criteria by the decision maker is shown in Table 87.4. Making use of the informatics application built in VBA, the 20 rankings of the alternatives were provided. The preference rank based on the net concordance is given in Table 87.5, where the KPIs are presented by descending order of the net concordance value.

87.6 Results and Discussion

The six first ranked alternatives are classified with clear evidence, where "Rank C" equals "Rank D". For the subsequent ranked alternatives, although both classifications do not establish the same order, there are no very significant differences. Thus, it is establishes a balanced pattern of preferences in the selection of the main KPIs for measurement of maintenance service quality at the airport.

Table 87.6 gives a list of KPIs ranked in order of concordance and considered important by maintenance managers.

The human brain can only handle four to eight measurements intended to quantify one aspect [5], which suggests that it is reasonable to use six performance indicators for measurement of each maintenance aspect. However, it is noted that with only the six highest-ranked alternatives, the measurement of some important aspects of maintenance quality will not be considered. Therefore, we selected the eight highest-ranked KPIs with respect to the concordance and which discordance shows no dispersion in the set. The set of KPIs shown in Table 87.6 covers the main aspects to ensure the desired measurements, such as, the management and execution of work orders, the control of events, the quality of planning and scheduling of maintenance activities, the use of time and the contribution of maintenance for customer satisfaction.

The methodology allowed quantitatively compares the interest of KPIs for maintenance, through concordance values and also by checking the discordance, according to the preference and experience of managers.

The methodology is not difficult to apply by maintenance managers using the developed software, since it is not necessary master the mathematical calculation but instead use the technique. We realized facility and consistency in the procedures for the evaluation of the criteria and alternatives according preference information of the

Table 87.3 List of propose KPIs (competing alternatives)

KPIs	KPIs description	Calculation formula	Source
1	Number of incidents per 1000 passengers	N^o of incidents/(N^o of passengers/1000)	Airport
2	Average response time to emergency calls (incidents)	Total response time to emergency calls/N^o of emergency calls	Airport
3	Average time to replacement of functionality	Total replacement time /N^o of incidents	Airport
4	Availability related to maintenance	Total operating time/(Total operating time + Downtime due to maintenance)	[1]
5	Availability due to planned and scheduled maintenance	Total operating time/(Total operating time + Downtime related to the planned and scheduled maintenance)	[1]
6	Mean time between work orders that cause downtime	Total operating time/N^o of maintenance work orders causing downtime	[1]
7	Percentage of systems covered by criticality analysis	N^o of systems covered by a critical analysis/Total N^o of systems	[1]
8	Mean time to repair (MTTR)	Total time to restore/ N^o of failures	[1]
9	Percentage of working time lost due to accidents (severity rate)	Man hours lost due to injuries for maintenance personnel/Total man-hours worked by maintenance personnel	[1]
10	Fulfilment rate of work orders	No of work orders performed as scheduled/Total N^o of scheduled work orders	[1]
11	Percentage of work orders in backlog	N^o of overdue work orders/N^o of received work orders	[3]
12	Percentage of available man hours used for proactive work	Man-hours used for proactive work/Total man-hours available	[3]
13	Percentage of work orders requiring rework (quality of execution)	N^o of work orders to rework/Total N^o of work orders	[3]
14	Rate of planned work (planning intensity)	Planned work/Total work done	[3]
15	Quality of planning	N^o of work orders requiring rework due to planning/Total N^o of work orders	[3]
16	Quality of scheduling	N^o of work orders with delayed execution due to material or man-power/Total N^o of work orders	[3]
17	Execution Index	Hours of work orders planned and executed/Expected hours of work orders planned	
18	Rate of customer complaints due to causes imputed to the maintenance (per 1000 passengers)	N^o of complains with cause imputed to maintenance/(N^o of passengers/1000)	
19	Maintenance quality index	Total hours of maintenance/Rework hours	
20	Maintenance performance index	Expected hours of maintenance/Hours used for maintenance	

Table 87.4 KPIs evaluation against criteria (performance of alternatives)

Alternatives A_m	Criteria				
	C_1	C_2	C_3	C_4	C_5
1	8	9	9	8	7
2	7	8	8	7	7
3	9	8	7	9	5
4	8	7	8	8	8
5	9	8	7	7	7
6	8	5	1	5	7
7	7	4	4	7	6
8	8	6	8	7	8
9	5	7	8	7	9
10	9	8	9	8	9
11	9	9	9	9	9
12	7	8	7	4	7
13	9	9	9	8	9
14	8	3	4	8	8
15	9	8	8	8	9
16	9	8	9	7	6
17	7	9	8	7	7
18	8	7	9	8	9
19	8	9	8	5	8
20	3	7	6	5	5

decision maker. It allows obtaining quick results and the decision process becomes more explicit, rational and efficient.

87.7 Conclusions

A new approach is presented in order to aid maintenance managers in the hard task of selecting relevant KPIs to measure the performance of maintenance service. For such, a simple ranking MCDM methodology based on the original ELECTRE I method is proposed, which proved to be a suitable and efficient tool to deal with that problems type, such as the one presented in the case study.

In the case study of the airport, used for exploratory demonstration, the proposed methodology was applied to ranking and selecting a set of KPIs with the preference of the maintenance manager and the results were well accepted. A ranking list of alternatives was generated, from the best to the worst, which was considered acceptable and reliable by the maintenance managers of the airport. The KPIs ranked in the first eight positions meet the needs of intended measurement, covering the main aspects to be ensured for maintenance service quality in airport.

Table 87.5 Ranking of KPIs by concordance ascending order

Alternatives A_m	Net values		Ranking	
	NC_m	ND_m	Rank C	Rank D
11	15.2387	−19.0000	1	1
13	12.2165	−16.0161	2	2
10	10.0673	−12.8535	3	3
3	8.2534	−10.1294	4	4
1	7.7981	−9.4813	5	5
15	7.7390	−9.1803	6	6
16	4.7389	−5.0338	7	8
18	4.1268	−5.2597	8	7
4	1.5745	−1.9931	9	9
5	0.7988	−1.4539	10	10
17	−2.3510	−1.1079	11	11
19	−2.3654	6.3972	12	14
14	−3.4403	8.8398	13	15
8	−4.0222	4.7772	14	13
2	−4.5002	3.1863	15	12
9	−7.4403	10.3287	16	16
12	−10.9924	13.1710	17	18
7	−11.3061	12.4794	18	17
6	−11.5444	15.4671	19	19
20	−14.5897	16.8622	20	20

Table 87.6 Selected KPIs for measurement of maintenance service quality at the airport

Rank	KPIs (A_m)	KPIs description
1	11	Percentage of work orders in backlog
2	13	Percentage of work orders requiring rework (quality of execution)
3	10	Fulfilment rate of work orders
4	3	Average time to replacement of functionality
5	1	Number of incidents per 1000 passengers
6	15	Quality of planning
7	16	Quality of scheduling
8	18	Rate of customer complaints due to causes imputed to the maintenance (per 1000 passengers)

Acknowledgments This work is supported by a cooperation protocol between Instituto Superior de Engenharia of Universidade do Algarve (ISE/UALG) and Faculdade de Cincias e Tecnologia of Universidade Nova de Lisboa (FCT/UNL). The authors acknowledge the support of UNIDEMI - R&D Unit in Mechanical and Industrial Engineering of FCT/UNL.

References

1. EN 15341 (2007) Maintenance key performance indicators. European Committee for Standardization (CEN), Brussels
2. Kumar J, Soni V, Agnihotri G (2013) Maintenance performance metrics for manufacturing industry. Int J Res Eng Technol 2(2):136–142
3. Muchiri P, Pintelon L et al (2011) Development of maintenance function performance measurement framework and indicators. Int J Prod Econ 131(1):295–302
4. Pacaiova H, Nagyova A et al (2013) Systematic approach in maintenance management improvement. Int J Strateg Eng Asset Manage 1(3):228–237
5. Kumar U, Galar D et al (2013) Maintenance performance metrics: a state-of-the-art review. J Qual Maintenance Eng 19(3):233–277
6. Roy B (1991) The outranking approach and the foundations of ELECTRE methods. Theor Decis 31(1):49–73
7. Figueira J, Greco S et al (2013) An overview of ELECTRE methods and their recent extensions. J Multi-Criteria Decis Anal 20:61–85
8. Nijkamp P (1976) A multi-objective decision model for regional development, environmental quality control, and industrial land use. Papers Reg Sci 36(1):35–58
9. Hatami-Marbini A, Tavana M (2011) An extension of the ELECTRE I method for group decision-making under a fuzzy environment. Omega 39(4):373–386
10. Sevkli M (2010) An application of the fuzzy ELECTRE method for supplier selection. Int J Prod Res 48(12):3393–3405
11. Bojković N, Anić I, Pejčić-Tarle S (2010) One solution for cross-country transport-sustainability evaluation using a modified ELECTRE method. Ecol Econ 69(5):1176–1186
12. Hartati S, Wardoyo R, Harjoko A (2011) ELECTRE methods in solving group decision support system bioinformatics on gene mutation detection simulation. Int J Comput Sci Inf Technol 3(1):40–52
13. Kaya T, Kahraman C (2011) An integrated fuzzy AHP–ELECTRE methodology for environmental impact assessment. Expert Syst Appl 38(7):8553–8562
14. Botti L, Peypoch N (2013) Multi-criteria ELECTRE method and destination competitiveness. Tourism Manage Perspect 6:108–113
15. Shanian A, Savadogo O (2006) ELECTRE I decision support model for material selection of bipolar plates for polymer electrolyte fuel cells applications. J New Mater Electrochem Syst 9(3):191
16. Cabral I, Grilo A, Cruz-Machado V (2012) A decision-making model for lean, agile, resilient and green supply chain management. Int J Prod Res 50(17):4830–4845
17. Horenbeek AV, Pintelon L (2014) Development of a maintenance performance measurement framework-using the analytic network process (ANP) for maintenance performance indicator selection. Omega 42(1):33–46
18. Figueira J, Roy B (2002) Determining the weights of criteria in the ELECTRE type methods with a revised Simos' procedure. Eur J Oper Res 139(2):317–326
19. Yoon KP, Hwang CL (1995) Multiple attribute decision making: an introduction. Sage University Paper Series on Quantitative Applications in the Social Sciences, Sage, Thousand Oaks, pp 07–104

Part VI
Project Management

Chapter 88
Process Developments in FSW

Telmo G. Santos, Jorge Martins, Luis Mendes and Rosa M. Miranda

Abstract FSW and processing have largely evolved in recent years in terms of process knowledge, materials processed and industrial applications. Considering process developments friction stir has been studied in an hybrid welding process associated with electric current for welding Al based alloys. In this variant a temperature increase due to Joule effect was seen to improve the viscoplasticity of base material without affecting metallurgical characteristics, reducing lack of penetration, which is detrimental for fatigue resistance. Diffusion welding triggered by FSW, named friction stir diffusion welding (FSDW) has also been investigated to join materials with very distinctive properties as copper to stainless steel preventing the formation of brittle intermetallics and improving joining efficiency using downward force of 5,500 N, rotation speed of 1,800 rpm and travel speed of 90 mm/min. The same variant was tested to produce sandwich, or composite, materials as Al reinforced with NiTi ribbons taking advantage of the high conductive soft Al material and the damping capacity and high mechanical resistance shown by NiTi. A good embedment of the reinforcing material under appropriate processing conditions was seen due to the viscoplastic material flow around the NiTi. This paper aims at presenting some applications involving FSW and its variants in distinct industrial applications.

Keywords FSW · Diffusion joining assisted by FSW · Process development · Composites

T. G. Santos · J. Martins · L. Mendes · R. M. Miranda (✉)
UNIDEMI, Faculdade de Ciências e Tecnologia, Universidade Nova de Lisboa,
2829-516 Caparica, Portugal
e-mail: rmiranda@fct.unl.pt

J. Xu et al. (eds.), *Proceedings of the Eighth International Conference on Management Science and Engineering Management*, Advances in Intelligent Systems and Computing 281, DOI: 10.1007/978-3-642-55122-2_88, © Springer-Verlag Berlin Heidelberg 2014

88.1 Introduction

Friction Stir Welding (FSW) is a recent solid state welding process that presents a set of very attractive advantages, mostly related to the absence of melting of base materials but it still struggles with several technical and productivity issues which contribute to a limited range of industrial applications. Some major drawbacks are the tool lifetime and cost and the existence of defects, specially oxides alignments and lack of penetration (LoP) in the weld root. Intensive and extensive research work has been devoted to FSW, mostly to improve the process performance, to develop new variants, including the association of FSW to other processes in hybrid or assisted FSW processes and to identify new potential industrial applications. Results are presented in three distinct areas: (1) Development of an hybrid friction stir welding process with electric current. (2) Use of FSW to trigger diffusion welding of dissimilar materials. (3) A new application of FSW to produce composite material based on Al reinforced with shape memory alloys.

88.2 Hybrid FSW with Electric Current

Very common defects in FSW are oxide alignment is the root defect or lack of penetration which results from an insipientviscoplastic material flow in the weld root that leads to a poor stirring in the bottom part of the weld and is insufficient to fully recrystallize the material in this region [1, 2, 5]. Additionally, the viscoplasticity of this zone is also affected by the heat loss into the back plate. So, an improvement of the material viscoplasticity in the weld root would be beneficial and, for this, an external energy source is proposed. The concept relies on the use of an external electrical energy source delivering a high intensity current passing through a thin layer of material between the back plate and the probe lower tip, generating heat by Joule effect. Figure 88.1 depicts the overall setup.

The heat generated by Joule effect raises the material temperature, increasing its ductility and viscoplastic behavior, while reducing the mechanical strength. This facilitates the local material stirring and the dynamic recrystallization, preventing oxide alignment and lack of penetration defects in the weld root.

The concept was validated in Aluminium alloys performing butt welds in 4 mm thick plates with a travel speed of 180 mm/min, a rotation speed of 1,120 rev/min and a tilt angle of 1.5° with a probe plunging depth of 3.3 mm, to deliberately produce a lack of penetration (LOP) since the plate has 4 mm thickness. FSW was continuously done along the plate with and without electrical current in order to analysis the effect of the current on the dimension of the root defect, and on the extension of the different zones observed in FSW. Electric currents of 700 and 800 A were applied. Samples were removed for metallographic analysis and as depicted in Fig. 88.2. The aperture of the LoP defects was seen to be reduced from about 15 to 3.3 micron in the areas where electric current was applied.

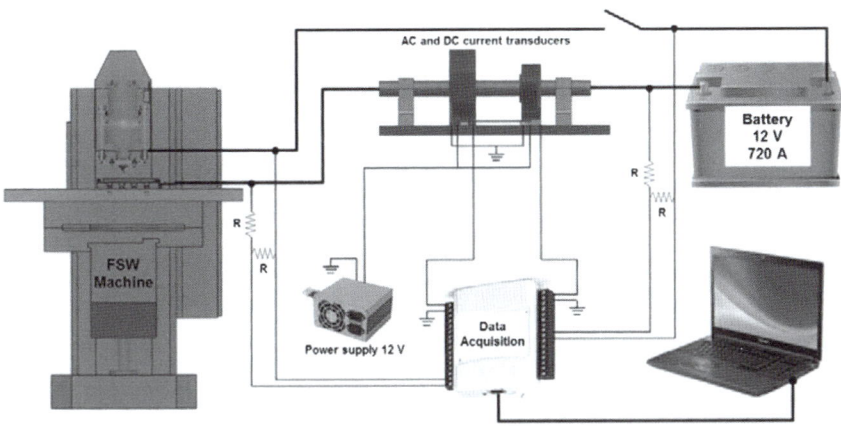

Fig. 88.1   Schematic representation of the overall installation

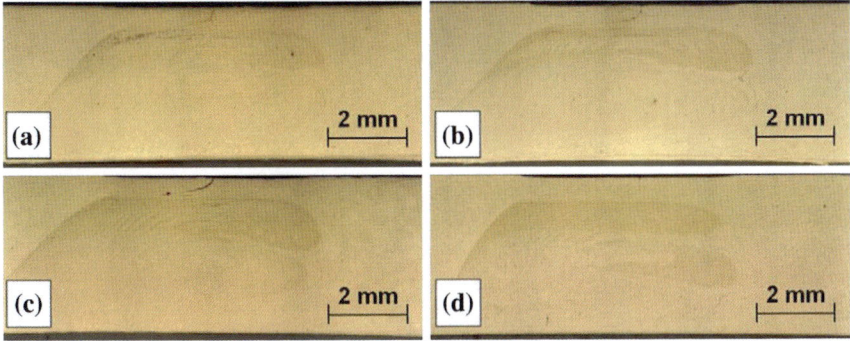

Fig. 88.2   Macrographs of butt welds. **a** and **b** with electrical current of 800 A at different positions on the plate, **c** without electrical current and **d** with electrical current of 700 A

A concept of hybrid FSW with electrical current was developed, tested and validated aiming at minimizing weld root defects in FSW. The concept relies on forcing an electrical current to pass in the weld root to increase the local temperature, improving the material viscoplasticity. An increase in temperature of about $200-300\,°C$ was computed by a simple analytical model to validate the concept for the aluminium alloy under study. Experimental results showed that passing an electrical current through the weld root, LoP defects reduced in size from a width of 20 to $3\,\mu m$, under the conditions tested. Electrical current seems to increase the viscoplastic material flow creating the conditions to suppress defects.

Fig. 88.3 Schema of tool
positioning in FSDW

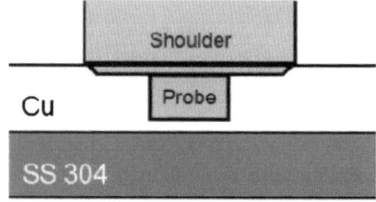

88.3 Friction Stir Diffusion Welding

One of the major trends in welding and joining technology is to join dissimilar materials taking advantage of individual materials properties. Among these, copper to stainless steel joining has significant industrial applications where e.g. the highest electrical and thermal conductivities are required to engineering materials associated to good corrosion resistance. However, joining these materials is difficult due to their very different mechanical and thermo-physical properties [3, 6]. Fusion based welding processes have limited application and for the FSW was tested. However, the high hardness of stainless steel reduces the tool lifetime.

So a variant of FSW was tested aiming to use the process for heat generating to promote diffusion joining between both materials.

Friction stir diffusion process (FSDP) promotes joining only by the rapid diffusion of heat generated between the tool and one of the metals, unlike the FSW that joints both metals by diffusion and plastic strain. Thus, the tool processes preferably the materials with the lowest melting point and hardness, minimizing the detrimental effect on the tool wear and on the interface of materials to join. Multi pass technique has been used mostly in aluminum, for the processing of larger surface areas composed of several overlapping passes. To ensure homogenization of the properties is needed nugget interpenetration, thereby producing fine grain microstructure. Gandraet [4] proposed the overlapping on the advancing side (AS), once the hardness is higher than that in the overlapping on the retreating side (RS).

Figure 88.3 depicts the relative positioning of the tool relative to the join interface. Lap joints were performed varying the most relevant processing parameters, namely, rotation and travel speeds and downward force. The effect of processing parameters on the width of effective joining was studied and the joints were characterized for mechanical resistance properties and microstructural features at the interface (Figs. 88.4 and 88.5). The thermo-mechanical conditions and time during the FSDP resulted in an interface with diffusion between both materials below 3 μm. The shear strength of the lap joints depends on the material thickness involved, but joining efficiencies up to 73.8 % were achieved.

An analysis was conducted with EDS under SEM along a line across the interface, as shown in Fig. 88.5. Cu, Fe, Ni and Cr were inspected along this line and depicted in (b). An IMC layer of 2–3 μm thick was identified, however, the resolution of BSE is of 1–3 μm thus, there is an incertitude in this IMC identification.

Fig. 88.4 Cross section view of Cu/SS2 sample. **a** Joined width. Parameters: Df = 5,500 N, Ω = 1,400 rev/min, v = 120 mm/min, **b** detail of the previous

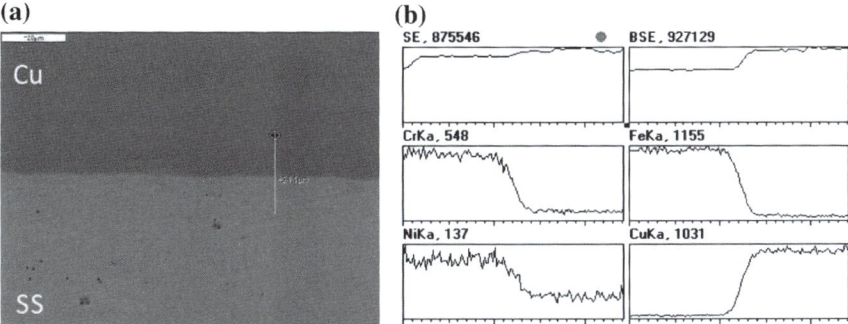

Fig. 88.5 EDS line scanning analysis across the sample interface. **a** BSE image at the interface under center processing, **b** EDS profiles along the line

FSW was used to trigger diffusion at the interface of distinct materials promoting joining of very dissimilar materials as copper to stainless steel. The pin rotated at a close distance from the stainless steel surface and the produced heat was sufficient to promote diffusion at the interface together with the shoulder pressure. The subsequent viscoplastic material flow breaks surface oxide layers, exposing metal-to-metal contact and enabling bonding. Bonded area showed no defects and minimal dilution without distortion.

88.4   Production of Composite Material by FSW Based on Al Reinforced with Shape Memory Alloys

Several solid state technologies have been widely used for the fabrication of metal-matrix composites. Amongst these is diffusion bonding technology and hot pressing. Research on the use of FSW to create composite materials has being performed [8, 9] due to its capability to join materials below melting point. In this study, NiTi ribbons

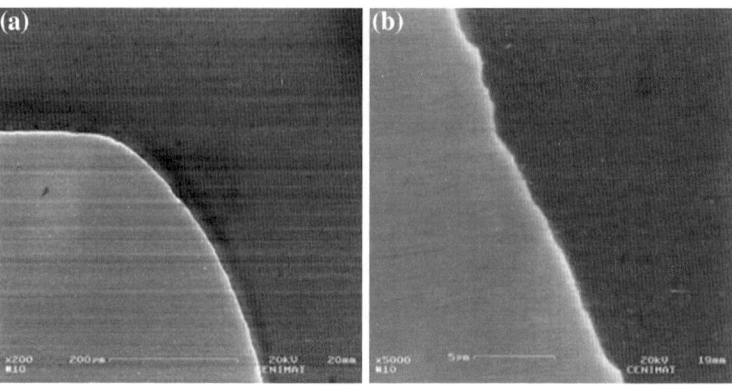

Fig. 88.6 Interface cohesion between Al1100 (*darker*) and NiTi (*lighter*) observed under SEM **a** general view, **b** detail of (**a**)

were used as reinforcements of Al 1100 substrate by placing the NiTi between two Al plates in pre machined grooves and friction stir welding the lap joints studying the optimal embedment of the NiTi reinforcements, under a viscoplastic state without destroying its functional properties of the reinforcing material. Three main variables were studied, and these were the tool position relatively to NiTi ribbons, tool plunge depth and multi-pass configurations.

Multi-pass overlapping by the advancing greatly increased the AA1100 embedment, when compared to overlapping by the retreating side or full overlapping. Multi-pass overlapping successfully minimizes the material flow defects around the NiTi ribbon. Volume defects were minimised with the offset of the tool in relation to the NiTi ribbons, in multi-pass the counter material flow of the subsequent steps reduces the previous, therefore being able to completely embed the NiTi [7].

To study the interface between matrix and reinforcing materials, scanning electron microscope (SEM) was performed in composite cross sections. Figure 88.6 shows a good embedment of the reinforcing material. The NiTi rough waved shape has been totally covered by the intensively deformed matrix materials. This interface was subjected to high compressive forces, resulting from the stirred material flow of the FSW and the disturbing 'object' counter pressure. The interface present no voids or cavities, whereas most common liquid state joining techniques would rather present more irregular micro structures at this range. The overall material embedment is good, although there is clear physical interface suggesting that there is no inter diffusion between materials.

This investigation aimed at producing and characterizing composites consisting of two very distinctive materials, by friction stir welding. The process proved to be viable to produce Al/NiTi composites in a lap joint configuration where the NiTi ribbons were mechanically bonded to the Al matrix. A good embedment of the NiTi was achieved by the intensively deformed matrix aluminium. This interface was subjected to high compressive forces, resulting from the stirred material flow of the FSW process.

Acknowledgments TS and RM acknowledge Pest OE/EME/UI0667/2011 from the Portuguese Fundação para a Ciência e a Tecnologia (FCT-MEC).

References

1. Cui L, Yang X et al (2012) Characteristics of defects and tensile behaviors on friction stir welded AA6061-T4 T-joints. Mater SciEngA 543:58–68
2. Ferrando WA (2008) The concept of Electrically Assisted Friction Stir Welding (EAFSW) and application to the processing of various metals. Technical Report NSWCCD-61-TR-2008/13, Naval Surface Warfare CenterCarderock Division
3. Gandra J (2010) Preliminary study on the production of functionally graded materials by friction stir processing. Master's thesis, Mechanical Engineering, Departamento de Engenharia Mecânica e Industrial, Universidade Nova de Lisboa-Faculdade de Ciências e tecnologia, Lisboa
4. Gandra J, Miranda RM, Vilaça P (2011) Effect of overlapping direction in multipass friction stir processing. Mater Sci Eng A 528:5592–5599
5. Lammlein D, Trepal N, Posada MA (2011) Defect significance and detection in aluminum friction stir welds: a literature search. Technical Report NSWCCD-61-TR-2011/20, Naval Surface Warfare Center, Carderock Division, West Bethesda
6. Ma ZY (2008) Friction stir processing technology: a review. Metall Mater Trans A Phys Metall Mater Sci 39:642–658
7. Mendes L, Miranda RM, Santos T (2013) Production of al based composites reinforced with embedded niti by fsw. LAP, Saarbrücken
8. Nascimento F, Santos T et al (2009) Microstructural modification and ductility enhancement of surfaces modified by FSP in aluminium alloys. Mater Sci Eng A Struct Mater Prop Microstruct Process 506:16–22
9. Yadava MK, Mishra RS et al (2010) Study of friction stir joining of thin aluminium sheets in lap joint configuration. Sci Technol Weld Joining 15:70–75

Chapter 89
Human Resource Practices: A Case Study of Karachi Port Trust (KPT)

Nadeem A. Syed, Asif Kamran, Sulaiman Basravi and S. Faheem H. Bukhari

Abstract Global changes and technological advancements have affected the process leads to innovation, new forms of work organization, changes in the nature of work and new occupational categories. With these changes, human resources development takes the position of significance. The purpose of this research paper is to understand the HR policies being conducted at Karachi Port Trust (KPT). For this purpose, positivism approach has been used and survey was conducted with the sample size of 100 (employees of KPT). The data obtained from survey are analyzed by applying both descriptive and inferential statistics to determine the proper HR practices in KPT. The result shows that the factors highlighted in this research have a strong relationship with employee's performance. Other than that, there is a strong relationship between compensation and employees promotion towards employee performance. This study is one of its kinds pertaining to government institution in Pakistan where there is a room for improvement in employee's performance and this research will act as an instrument in achieving the strategic HR goals of KPT.

Keywords Human resource management (HRM) · Karachi port trust (KPT) · Performance appraisals (PA) · Organization (ORG)

N. A. Syed · S. F. H. Bukhari
Management Science Department, Shaheed Zulfikar Ali Bhutto Institute of Science and Technology (SZABIST), Karachi, Pakistan

A. Kamran (✉)
School of Management and Economics, University of Electronic Science and Technology of China, Chengdu 610064, People's Republic of China
e-mail: asifkamrankhan@gmail.com

S. Basravi
Management Science Department, Bahria University, Karachi 752400, Pakistan

J. Xu et al. (eds.), *Proceedings of the Eighth International Conference on Management Science and Engineering Management*, Advances in Intelligent Systems and Computing 281, DOI: 10.1007/978-3-642-55122-2_89, © Springer-Verlag Berlin Heidelberg 2014

89.1 Introduction

Employers in the department are the implementers of Human Resource Management policy, and as such, vital to successful management practices. They hold key to performance management of employees and get things done from them. To get the best output managers have to identify individual objectives for staff and assigned specified task to the individuals [2]. They also provide employees with on-going guidance and supervision, including regular performance feedback. They have to conduct appraisals and counsel to initiate appropriate action where necessary to address poor performance or if any misconduct happens in the organization. Employers have the responsibility to identify the training and development needs for employees and implement the training plan when appropriate. Available Communication also plays a vital role in the performance of the employees. Managers have to regularly communicate with the employees on subjects that affect them and should consider them while taking decisions [7].

89.2 Literature Review

1. The Organizational Context of Human Resource Management
Alvares [1] argues on the organisational context in which HRM at present is one of the rapid change and considerable uncertainty. As various views of HRM were placed on the table and discussed, it became apparent that this sub-field of management is in a phase of transition. In the context of traditional organisational structures, HRM was placed as a function within a "silo", as were other organisational functions; for example finance, production and marketing. However, a turbulent environment has brought a concurrent change in organisational structures and the nature of HRM and its functions are in the process of change. Many organisations are now structured around multi-disciplinary project teams with the HRM professional as one member of the team or as consultant to the team, and where line managers take on various HRM roles [6].

2. The Nature of Human Resource Management Roles
Beatty et al. [3] argues that the human resource management is also in the process of change with regard to the nature of the role performed. In the past many functions were performed by HRM professionals themselves, the role they are taking on, is one of consultant to line management, where line managers perform many of the functions traditionally handled by HRM professionals. To do this, those issues which are going to shape the future for HRM practices (termed transformation and issues) need to be identified and analyzed, especially in relation to current roles that will still be required of HRM practitioners. These issues are central to the activity of generating unit standards (e.g. outsourcing, societal responsibility) and their impact on HRM roles (e.g. staffing, performance management). In addition, supportive roles or functions required by HRM practitioners will also have to be identified in order

to complete the HRM practitioners qualifications design package (e.g. Financial, IT) [5].

3. Statement of the Problem

Human Resource proper practices play an important role in organizational development. At KPT different HR practices like job description, appraisal, compensation, supervision and other HR policies needs to be studied so that proper feedback and analysis must be provided to the top management of KPT to improvise in their HR practices.

4. Objectives of the Study

The objective of the study is to analyze the current procedures being followed by the HR department of KPT. The purpose of this research is to analyse the HR function at KPT in terms of its proper management by qualified HR professionals who contribute towards bringing efficiency in the organization performance.

5. Justification

This case study illustrates how KPT tailors its human resource practices to get specific behaviours from employees in order to be productive in the competitive corporate market. The author has chosen Karachi Port Trust to fulfil the purpose of this research and the reason is the importance of sea port in development of the Pakistan's economy, furthermore, 90 % of Pakistan's international trade takes place through marine transport.

6. Limitations

The management was not supportive in providing the confidential data related to KPT employees to the researcher.

Top management was not ready to co-operate in providing the data about their HR policies.

7. Scope

This sort of research will be beneficial for the top management of KPT as the analysis will add value in terms of knowing the HR practices being implemented by KPT management and most importantly changes can be implemented in lights of the research results.

8. Sample and Sampling Method

The questionnaire was distributed among 100 employees of KPT to extract the information. Convenience sampling method has been used in this study.

9. Research Methodology

(1) Research design

The research design, which involves a series of rational decision-making choices, was originally presented in a simple manner. Specifically, in this research the changes that are made in Karachi Port will be highlighted. The analysis of data from HR records to determine the effectiveness of past and present HR practices.

Table 89.1 Gender wise analysis

Gender	Frequency	Percent
Male	92	92
Female	8	8
Total	100	100

Table 89.2 Age wise analysis

Age	Frequency	Percent
20–30	4	4
30–40	27	27
40–50	55	55
50–60	14	14
Total	100	100

(2) Procedure

- Data from secondary sources was collected and analyzed. Management and Government consultants as well as reports were consulted and relevant data was compiled.
- Interview questions from Supervisors and Managers would be collected and based on that data survey questionnaire will be collected from employees and the complete analysis will be done.

(3) Variables
HR Systems and Procedures Evaluation; Performance Appraisals; Hiring; Orientation and Training; Compensation and Benefits; Discipline and Termination.

89.3 Analysis and Interpretation

1. Frequencies
The Tables 89.1, 89.2 and 89.3 shows frequency statistics of gender, age and Grade.

The Table 89.1 shows the total population with respect to gender. The number of people considered in KPT is 100 out of which 92 are males with a valid percent of 92 % and there are only 8 females having a valid percent of 8 %. Due to the nature and requirement of the work, weight age of male employees is far greater than female on Port area.

The Table 89.2 shows the analysis in terms of age of the employees present in KPT. 4 % of people lie between the ages 20–30 have a frequency of 4. 27 % people who lie between the ages of 30–40 have a frequency of 27. 55 % people lie between the ages 40–50 have a frequency of 55. 14 % people lie between the ages 50–60 have a frequency of 14. From the last 8–9 years appointment are very less in number. They usually prefer the more experienced employees. The manpower has been reduced in large number due to outsourcing/privatization of major activities of Port operation.

Table 89.3 Grade wise analysis

Cgrade/bps	Frequency	Percent
Non-Respondent	3	3
16–18	38	38
9–22	11	11
8-Jan	48	48
Total	100	100

Table 89.4 Reliability statistics

Cronbach's Alpha	N of items
0.856	6

Table 89.5 Reliability statistics

Cronbach's Alpha	N of items
0.895	45

The Table 89.3 depicts the analysis of employee's in terms of their Grade or Basic pay Scale. According to which the people lie between the grades of 16–18 with a frequency of 38 and the people who lie between the grades of 19–22 with a frequency of 11 because officers are less in number. The people having the grades or basic scale of 1–8 has a frequency of 48 are due to the large number of lower level employees on the Port operation. A reliability of 0.856 (Table 89.4) shows that the data gathered in measuring the six factors of organizational HR practices (HR system and Procedures Evaluation, Performance Appraisal, Orientation and Training, Hiring, Compensation and Benefits, Discipline and Termination) is reliable and consistent.

A reliability of 0.895 (Table 89.5) shows that the data gathered in measuring the 45 factors of organizational HR practices is reliable and consistent. It shows that the respondents have answered all the questions correctly and the respondents are fair and unbiased in providing their response to the questions asked in the questionnaire.

The Table 89.6 shows the descriptive analysis contain the population of 100 employees. The Maximum experience of the employees working in KPT is 35 with a Mean of 20.55 and a Standard Deviation of 8.508 because the employees are more experienced and they are not willing to leave their position due to the better Financial benefits provided by the organization as compare to other similar organizations. Employees have a maximum 35 years of Association with the organization having a mean of 18.24 and a standard deviation of 8.507. The employees are more associated with KPT due to other incentives like Medical, pension, etc.

The Table 89.6 shows the descriptive analysis of the factors (HR system and Procedures Evaluation, Performance Appraisal, Orientation and Training, Hiring, Compensation and Benefits, Discipline and Termination are considered for the analysis.

- The HR systems and Procedures are followed in the organization. The analysis shows the mean of 2.18 and a standard deviation of 0.404. It means that company has to make their rules and procedures more adoptable by the employees for the better outcome in the future.

Table 89.6 Descriptive statistics

	N	Minimum	Maximum	Mean	Std. deviation
HR systems and procedure evaluations	100	1	3	2.18	0.404
Performance appraisal	100	1	3	2.11	0.467
Orientation and training	100	1	4	2.17	0.52
Hiring	100	1	3	1.49	0.364
Compensation and benefits	99	1	3	2.06	0.407
Discipline and termination	99	1	3	1.49	0.446

- KPT is providing Performance Appraisal to their employees according to their performance level and the received outcome with a mean of 2.11 and a standard deviation of 0.467 or 46.7 %.
- The organization is also offering new Training programs to their employees. They send their employees on training mostly according to the requirement of the position of the employee and their Grade level. The analysis shows a mean of 2.17 and a standard deviation of 0.520 or 52.0 %.
- From the last 8–9 years, KPT is in the process of rightsizing. The manpower has been reduced in large number due to outsourcing/privatization of major activities of Port operation. Hiring of new employees was not done on a large scale. The analysis shows the mean of 1.49 and a standard deviation of 0.364 or 36.4 %.
- As compare to other similar organizations, KPT is offering more financial benefits and compensation to their employees, that is why there are more experienced employees present in KPT and they are not willing to leave their job. The analysis shows the mean of 2.06 and a standard deviation of 0.407.
- Termination of the employees had been done if the employee's performances are not up to the mark. The analysis shows a mean of 1.49 and a standard deviation of 0.446 or 44.6 %.

2. Correlations

Correlation is significant at the 0.01 level. The correlation between performance appraisal and hiring is 0.502 means that are is a positive relation between the two variables. Performance appraisal will be highly based on the annual performance of the employees. There is a high degree of correlation between Orientation programs and Performance appraisal i.e. 0.843, which shows that employees performance will by highly affected with these training and orientation programs. Employees who attend these training and orientation programs will result in better performance. The correlation between Hiring and compensation is 0.216 showing that there exists a relation but on a lower extent as hiring of the employees are not done from the past several years. Performance Appraisal and HR systems and procedures also have a high correlation of around 0.640; therefore performance appraisal is linked with the systems and procedures, the employees get the performance appraisal based on their job evaluation programs conducted by the HR managers. The employees who performed well according the job descriptions and will result in more outcome of

Table 89.7 Correlations

		a	b	c	d	e	f
HR systems and procedure evaluations	P*	1	0.640**	0.491**	0.497**	0.328**	0.471**
	N	100	100	100	100	99	99
Performance appraisal	P*	0.640**	1	0.843**	0.502**	0.549**	0.484**
	N	100	100	100	100	99	99
Orientation and training	P*	0.491**	0.843**	1	0.417**	0.540**	0.528**
	N	100	100	100	100	99	99
Hiring	P*	0.497**	0.502**	0.417**	1	0.216*	0.287**
	N	100	100	100	100	99	99
Compensation and benefits	P*	0.328**	0.549**	0.540**	0.216*	1	0.520**
	N	99	99	99	99	99	99
Discipline and termination	P*	0.471**	0.484**	0.528**	0.287**	0.520**	1
	N	99	99	99	99	99	99

a HR systems and procedure evaluations; *b* Performance appraisal; *c* Orientation and training; *d* Hiring; *e* Compensation and benefits; *f* Discipline and termination.
P* Pearson correlation; **: Significant at 5 %

Table 89.8 Gender * age cross tabulation

		Age				Total
		20–30	30–40	40–50	50–60	
Gender	Male	4	23	52	13	92
	Female	0	4	3	1	8
Total		4	27	55	14	100

the company, they then received the more incentive and benefits as compare to other similar organizations. It shows the high correlation between each other of about 54.9 % (Table 89.7).

3. Crosstabs

Table 89.8 shows the gender analysis with respect to age. Total population considered for analysis is 100 out of which 92 are males and only 8 are females.

- In KPT, the number of male employees in the age group of 20–30 is 4 and there is no female working in this age group as the work requirement is more preferable to the male employees as compare to the female employees of this age group. Due to the nature of the work on Port management hired the male employees. But they are also not in greater number due to their policy of rightsizing. But KPT should encourage hiring the new employees as they are more energetic and adaptable according to the changing needs and environments of the market as compare to the older one. The young employees have the better skills to produce the more outcomes on the Port.
- The number of male employees in the age group of 30–40 is 23 and there are 4 female employees working in this age group. The organization has to take some measures to provide employment to the female employees as well. They have to

Table 89.9 Gender grade/ BPS cross tabulation

		Grade/ BPS				Total
		0	16–18	19–22	8-Jan	
Gender	Male	3	33	11	45	92
	Female	0	5	0	3	8
Total		3	38	11	48	100

design their operations in a more competitive manner so that female have also get the equal chance to work there as well as to get benefited from their policies and financial benefits.

- In KPT there are more experienced and old officers present as compare to the new ones. The management feels that they require more experienced worker as the nature of the work on PORT is based more on operations. The number of male employees in the age group of 40–50 is 52 and there are 3 female working in this age group.
- The number of male employees in the age group of 50–60 is 13 and there is only 1 female working in this age group. Because at this age group level, mostly the positions are holder to the higher grade employees.

The Table 89.9 shows the gender analysis with respect to grade/basic pay scale. Total population considered for analysis is 100 out of which 92 are males and only 8 are females.

- The number of male employees in the grade of 16–18 is 33 showing the large number of male present in this grade as the requirement of the work and there are only 5 female employees having this grade. They are less in number due to the nature of the work as well as the management preferred the male as compare to female.
- The number of male employees in the grade of 19–22 is 11 and there is no female employee having this grade.
- KPT requires more labor force, as their major work are done on the Port that is why they have the higher number of workers at this grade level. The number of male employees in the grade of 1–8 is 45 and there are only 3 female employees having this grade.

4. Descriptive Analysis of Factors

The Table 89.10 shows the descriptive analysis of all the factors considered with respect to age. In the analysis of in HR systems and Procedure evaluation shows that the employees who are in the age group of 20–30 have a mean of 2.20 and standard deviation of 16.3 % as they are very less in number. The performance appraisal of this age group has a mean of 2.20 and a standard deviation of 28.3 %, their performance at the job is not satisfactory. Orientation and training program has a mean of 2.19 and a standard deviation of 59.1 % which a strong deviation in this factor. KPT is involved more in the training and orientation of the new employees according to

Table 89.10 Age wise analysis

Age		Mean	Std. deviation
20–30	HR systems and Procedure Evaluations	2.2	0.163
	Performance appraisal	2.2	0.283
	Orientation and training	2.19	0.591
	Hiring	1.56	0.171
	Compensation and benefits	2.25	0.228
	Discipline and termination	2.43	0.429
30–40	HR systems and procedure evaluations	2.1	0.462
	Performance appraisal	2.1	0.485
	Orientation and training	2.14	0.474
	Hiring	1.53	0.42
	Compensation and benefits	2.1	0.381
	Discipline and termination	2.6	0.445
40–50	HR systems and procedure evaluations	2.19	0.405
	Performance appraisal	2.1	0.465
	Orientation and training	2.17	0.519
	Hiring	1.49	0.36
	Compensation and benefits	2.01	0.425
	Discipline and termination	2.4	0.463
50–60	HR systems and procedure evaluations	2.3	0.311
	Performance appraisal	2.13	0.518
	Orientation and training	2.17	0.64
	Hiring	1.36	0.293
	Compensation and benefits	2.16	0.414
	Discipline and termination	2.52	0.357

their work requirement hiring of the employees has a mean of 1.56 and a standard deviation of 17.1 % which shows that hiring is very less in number. The management is not hiring the new employees due to their policy of rightsizing. Compensation and benefits program of the employees has a mean of 2.25 and a standard deviation of 22.8 % means that the employees of this age group are not benefited more from the financial benefits of the company. They provide benefits to those employees who are more experienced and associated with KPT from the longer time. Discipline and Termination program of the employees has a mean of 2.43 and a standard deviation of 42.9 %. The employees who are in the age group of 30–40 have a mean of 2.10 and standard deviation of 46.2 % in HR systems and evaluation, the policies and procedures are followed by the employees. The performance appraisal of the employees was given according to the performance of the employees and it shows a mean of 2.10 and a standard deviation of 48.5 %. KPT is highly in a process of Orientation of the new hired employees, so that they easily understand their work and they also organized training program for their existing employees so that they will organize their work in a more competitive manner which shows a mean of 2.14 and a standard deviation of 47.4 %. From the last several years hiring of the employees in KPT has been equal to none due to the reason of rightsizing which has a mean of 1.53 and a standard deviation of 42.0 %. KPT provides the Compensation and benefits program to those employees who are more experienced and associated with the organization as compare to the lower experienced one which shows a mean of 2.10 and a standard deviation of 38.1 %. Discipline and Termination program of the employees has a mean of 2.60 and a standard deviation of 44.5 % which shows a high deviation in this factor.

The employees who are in the age group of 40–50 have a mean of 2.19 and standard deviation of 40.5 % deviation present in HR systems and Procedure evaluation. The performance appraisal of this age group has a mean of 2.10 and a standard deviation of 46.5 %, means that the employees of this age group received more appraisals. Orientation and training program has a mean of 2.17 and a standard deviation of 51.9 %, refers that the employees of this age group were sent to more training programs so that they clearly understand the level of their work and organize their work in alignment with the organizations goals and mission. Hiring of the employees has a mean of 1.49 and a standard deviation of 36.0 % which shows a deviation in this factor. Compensation and benefits program of the employees has a mean of 2.01 and a standard deviation of 42.5 %, means that the KPT is providing more benefits and incentives as compare to other similar organization and it will be in the form of Medical, pension, etc. Discipline and Termination program of the employees has a mean of 2.40 and a standard deviation of 46.3 % which shows a high deviation in this factor. The employees who are in the age group of 50–60 have a mean of 2.30 and standard deviation of 31.1 % deviation present in HR systems and Procedure evaluation, because at this age group level mostly employees are at the upper officer level. The performance appraisal of this age group has a mean of 2.13 and a standard deviation of 51.8 % refers that they get more appraisal due to the level of their work. Orientation and training program has a mean of 2.17 and a standard deviation of 64.0 %, means that the employees of this age group are more send to the training programs so that they will more learn the management techniques according to the changing needs and the requirements of their position and job. Hiring of the employees has a mean of 1.36 and a standard deviation of 29.3 % which shows a deviation in this factor. Compensation and benefits program of the employees has a mean of 2.16 and a standard deviation of 41.1 % which shows a deviation in this factor. Discipline and Termination program of the employees has a mean of 2.52 and a standard deviation of 35.7 %, means that mostly these employees were not terminated from their position.

The Table 89.11 shows the descriptive analysis of all the factors considered with respect to Grade or Basic Pay Scale.

The employees who have the grade level between 16–18 have a mean of 2.26 and standard deviation of 26.6 % deviation present in HR systems and Procedure evaluation, shows that employees of this level are not directed towards to the HR policies and procedures described by the management. KPT is appreciated the employees who perform well in their job as described by the upper management and provide them more appraisals to these employees. It shows a mean of 2.08 and a standard deviation of 50.6 %. The employees of this grade level are more send to the Orienta-

Table 89.11 Grade wise analysis

Grade/ BPS		Mean	Std. deviation
16–18	HR systems and procedure evaluations	2.26	0.366
	Performance appraisal	2.08	0.506
	Orientation and training	2.08	0.602
	Hiring	1.53	0.447
	Compensation and benefits	1.98	0.483
	Discipline and termination	2.47	0.498
19–22	HR systems and procedure evaluations	2.13	0.539
	Performance appraisal	1.85	0.559
	Orientation and training	1.9	0.629
	Hiring	1.35	0.237
	Compensation and benefits	2.01	0.489
	Discipline and termination	2.22	0.516
8-Jan	HR systems and procedure evaluations	2.15	0.402
	Performance appraisal	2.17	0.408
	Orientation and training	2.29	0.4
	Hiring	1.47	0.319
	Compensation and benefits	2.13	0.303
	Discipline and termination	2.53	0.379

tion and training program because they are more capable of adopting and changing the activities at the work level and it has a mean of 2.08 and a standard deviation of 60.2 %. Hiring of the employees has a mean of 1.53 and a standard deviation of 44.7 % which shows a deviation in this factor. Compensation and benefits program of the employees has a mean of 1.98 and a standard deviation of 48.3 % which shows a deviation in this factor. Discipline and Termination program of the employees has a mean of 2.47 and a standard deviation of 49.8 %.

The employees who have the grade level between 19 and 22 have a mean of 2.13 and standard deviation of 53.9 % deviation present in HR systems and Procedure evaluation, the policies and procedures are followed by the employees. The performance appraisal of these employees has a mean of 1.85 and a standard deviation of 55.9 %. Orientation and training program has a mean of 1.90 and a standard deviation of 62.9 % which shows a deviation in this factor. Hiring of the employees has a mean of 1.35 and a standard deviation of 23.7 %. Compensation and benefits program of the employees has a mean of 2.01 and a standard deviation of 48.9 % which shows a deviation in this factor. Discipline and Termination program of the employees has a mean of 2.22 and a standard deviation of 51.6 %.

The employees who have the grade level lie between 1 and 8 have a mean of 2.15 and standard deviation of 40.2 % deviation present in HR systems and Procedure evaluation. The performance appraisal of these employees has a mean of 2.17 and a standard deviation of 40.8 %. Orientation and training program has a mean of 2.29 and a standard deviation of 40.0% which shows a deviation in this factor. Hiring of the employees has a mean of 1.47 and a standard deviation of 31.9 %. Compensation

Table 89.12 Gender wise analysis

Gender		Mean	Std. deviation
Male	HR systems and procedure evaluations	2.18	0.413
	Performance appraisal	2.1	0.476
	Orientation and training	2.14	0.524
	Hiring	1.48	0.368
	Compensation and benefits	2.06	0.404
	Discipline and termination	2.46	0.463
Female	HR systems and procedure evaluations	2.2	0.302
	Performance appraisal	2.2	0.355
	Orientation and training	2.42	0.415
	Hiring	1.6	0.312
	Compensation and benefits	2.12	0.463
	Discipline and termination	2.58	0.128

and benefits program of the employees has a mean of 2.13 and a standard deviation of 30.3% which shows a deviation in this factor. Discipline and Termination program of the employees has a mean of 2.53 and a standard deviation of 37.9%.

The above analysis (Table 89.12) shows the descriptive analysis of all the factors considered with respect to Gender.

According to which the Male employees have a mean of 2.18 and standard deviation of 41.3% deviation present in HR systems and Procedure evaluation. The performance appraisal of these employees has a mean of 2.10 and a standard deviation of 47.6%. Orientation and training program has a mean of 2.14 and a standard deviation of 52.4% which shows a deviation in this factor. Hiring of the employees has a mean of 1.48 and a standard deviation of 36.8% which shows a deviation in this factor. Compensation and benefits program of the employees has a mean of 2.06 and a standard deviation of 40.4% which shows a deviation in this factor. Discipline and Termination program of the employees has a mean of 2.46 and a standard deviation of 46.3% which shows a deviation in this factor.

The Female employees have a mean of 2.20 and standard deviation of 30.2% deviation present in HR systems and Procedure evaluation. The performance appraisal of these employees has a mean of 2.20 and a standard deviation of 35.5%. Orientation and training program has a mean of 2.42 and a standard deviation of 41.5% which shows a deviation in this factor. Hiring of the employees has a mean of 1.60 and a standard deviation of 31.2% which shows a deviation in this factor. Compensation and benefits program of the employees has a mean of 2.12 and a standard deviation of 46.3% which shows a deviation in this factor. Discipline and Termination program of the employees has a mean of 2.58 and a standard deviation of 12.8%. which shows a deviation in this factor.

89.4 Conclusions

It has been concluded from the research that HR Practices plays a great role in the success of the KPT. The factors that have been analyzed in this research have a strong relation with the employees and its performance.

The effective performance evaluation practices exist in KPT which help the HR department and upper management to better analyze the employees work and relate their performance to the goals of the organization. The evaluation system of KPT is in such a manner that it is linked with promotion and compensation of the employees. As far Compensation Practices are concerned they have direct impact on the employees' performance. Currently KPT is engaged in providing compensation practices to their employees and they are always in a process of announcing attractive salary and other bonus packages to their employees which are a very positive step to enhance employees' performance. Promotion practices not only help employees to grow in organizational hierarchy but also serve as mode for professional development.

Since this study proves the relationship of promotion and compensation practices with employees' performance, it is up to the government and the management to devise career development programs for their employees which should allow them to grow in their careers as well as there should be opportunities to grow professionally.

Human capital management (HCM) is an essential component of KPT management system. It refers to the systems, policies, procedures, and practices of managing human capital within the organizations.

89.5 Recommendations

- KPT should regularly conduct and updates a thorough analysis of its human capital needs. This goal hinges on the extent to which a public organization is aware of and addresses its human capital capacity y over time.
- The organization acquires the employees and its needs in a more competitive manner and according to the changing needs of the environment. This goal addresses the extent to which the KPT is able to obtain needed employees and determine the quality of its new hires.
- KPT should retain a skilled workforce; the employees who are highly skilled and capable should retain in the organization and should not become the process of rightsizing.
- KPT have to develop its workforce and should develop the career development program. This goal captures an organization's commitment to training and developing its employees and its future leaders, especially in areas that are critical to its mission.
- The public organization mainly has to manage its workforce performance programs effectively. This goal focuses on whether a public organization is able to encourage employees to perform effectively in support of its goals, discipline poor performers,

and terminate employees who cannot or will not meet performance and behavioral standards.

- Effective motivation typically rests on the use of appropriate monetary and non-monetary rewards and incentives, an effective performance appraisal system linking individual and KPT's goals, and sound mechanisms that both facilitate and use employee feedback are also needed.

References

1. Alvares KM (1997) The business of human resource management. Hum Resour Manage 36(1):9–16
2. Bahrami H, Evans S (1997) Human resource leadership in knowledge-based entities: shaping the context of work. Hum Resour Manage 36(1):23–28
3. Beatty RW, Schneier CE (1997) New hr roles to impact organizational performance: from "partners" to "players". Hum Resour Manage 36(1):29–37
4. Domsch ME, Hristozova E (2006) Human resource management in consulting firms. Springer, Berlin
5. Fritzel I, Vaterrodt JC (2002) Flexible auszeit für berater. Manage Training 2:14–15
6. Kubr M (2002) Management Consulting: a guide to the profession. 4th edn. International Labour Organization, Geneva
7. Norman C, Powell A (2004) Transforming HR to deliver innovation at Accenture. Strateg HR Rev 3(3):32–35

Chapter 90
Female Director Characteristics of Audit Committee and Corporate Transparency

Rui Xiang, Xiaojuan He and Yun Cheng

Abstract This paper empirically examines whether the characteristics of female directors of audit committee affect the corporate transparency based on the data of China's Shenzhen A-share listed companies from 2004 to 2007. The results from our sample show that, overall, the academic professional background of female directors has significantly positive impact on corporate transparency; the age of female directors has significantly negative impact on corporate transparency; the accounting expertise and education of female directors have no significant effects on corporate transparency. Further analysis also show that the significant impact relationship between the academic professional background or age of female directors and corporate transparency only in the lower legal level region. The result shows that the different characteristics of female directors pay different roles in affecting corporate transparency. This chapter provides some empirical evidences for perfecting the China's audit committee system.

Keywords Audit committee · Female directors · Corporate transparency

90.1 Introduction

As Enron, WorldCom, Xerox, and other serious accounting fraud occurred, the United States congress and government in June 2002 passed the Sarbanes–Oxley Act (SOX), and stressed the importance of the audit committee in corporate governance. There are also accounting frauds in Chinese listed companies, such as Zheng BaiWen, Yin GuangXia. In order to solve these problems, at the beginning of 2002, China securities regulatory commission and state economic and trade commission (SETC) jointly issued the governance guidelines of the listed corporate, and required that

R. Xiang (✉) · X. He · Y. Cheng
Business School, Sichuan University, Chengdu 610064, People's Republic of China
e-mail: xiangrui@scu.edu.cn

J. Xu et al. (eds.), *Proceedings of the Eighth International Conference on Management Science and Engineering Management*, Advances in Intelligent Systems and Computing 281, DOI: 10.1007/978-3-642-55122-2_90, © Springer-Verlag Berlin Heidelberg 2014

the board of directors of the listed company can establish the audit committee in accordance with the resolutions of the shareholders' meeting. So, the audit committee characteristics and governance efficiency caused extensive concern of the academia.

This chapter used the data of China's Shenzhen A-share listed companies from 2004 to 2007, for the first time from the perspective of corporate transparency, investigates the impact characteristics of audit committee female directors having on corporate governance efficiency. Research results show that academic professional background of the female directors is beneficial to the improvement of company's transparency, older female directors are not beneficial to the improvement of corporate transparency, the accounting expertise and education level of female directors have no significant effects on corporate transparency. In addition, our further study found that the above relationship exists only in listed companies where the legalization level is low, but this significant influence disappears in the region where the legalization level is high.

The main contributions in this chapter are as follows: firstly, we study the impact of audit committee female directors on corporate transparency. Secondly, we study the impact of audit committee personal characteristics on corporate governance in China. Therefore, this chapter provides some empirical evidences for perfecting the China's audit committee system.

The reminder of this chapter is organized as follows. The next section develops the theory analysis and research hypotheses whilst Sect. 90.3 presents the research design. Section 90.4 provides research analyses and findings of the study. Finally, Sect. 90.5 concludes the study.

90.2 Theory Analysis and the Research Hypothesis

Gender theory is that women made a significant contribution to the success of economic [28]. Psychology and management thought, there are significant gender differences between women and men in cognitive function, communication skills, decision-making and leadership style [5, 7, 20]. Women's economic theory is that in the process of value judgment, male decisions are biased, and women's decisions tend to be more neutral [18]. Therefore, this chapter studies that the female directors' characteristics affect corporate transparency based on the relating theory about female, along the logical thinking of "female directors' characteristics, the efficiency of audit committee, management decision-making, and corporate transparency".

1. Accounting Expertise and Corporate Transparency

The famous "reputation hypothesis" that independent directors have the motivation to improve its reputation as experts of decision-making and supervision [8]. Bedard et al. [1] find that there is a negative correlation between the audit committee financial experts and earnings management. Ge [10] found that CFO with CPA qualification has a significant negative impact on the company earnings management. Emilia and Sami found women CFO leads to less income accrued profit manipulation, and illustrates that the female CFO prefers to conservative earnings management strategy. Bill [2] find that the company's financial report is more conservative with female CFO. Women with accounting expertise thought that the company earnings management will have a higher level of legal and professional risk. Daniela and Lukas [6] find that women with financial expertise prefer to more risk-averse behavior and less overconfidence. Singh [23] find that in Britain, 35% of women directors hold accounting vocational qualification, and 25% of women directors have the CFO experience. Based on the above analysis, we believe that in the audit committee, female directors with accounting expertise may help improve the transparency of the corporate. So, this chapter puts forward the following hypothesis:

Hypothesis 1. The more accounting expertise the female directors have, the higher the corporate transparency is.

2. The Academic Professional Background and Corporate Transparency

Professional background is one of the important considered factors when the companies appoint independent directors [26]. Kaplan and David [13] regard the academic profession as a symbol of external links. Academic directors can cultivate more independent thought and create knowledge-based value for the company [15]. Hilleman et al. [11] argue that female directors can bring professional expertise and knowledge to the board of directors. Sealy et al. [21] find that British female directors often have more appellations or titles than male directors, these titles include the academic titles, such as "Dr.", "professor". In the USA, the company prefers to appointing the so-called "high-end lady" as directorship of the board, and send a positive signal on company's strength to the market [16]. Singh et al. [23] find that in Britain, 12.5% of female directors have academic professional background. In our country, the female from the institutions of higher education are appointed as directors of audit committee mainly because they have received a good education, good theoretical background and good social reputation, all of these will improve the supervision efficiency of audit committee, which will affect the information disclosure decisions of company management. Based on the above analysis, we believe that in the audit committee, female directors from universities will help to improve the transparency of the corporate. So, this chapter puts forward the following hypothesis:

Hypothesis 2. The more academic professional background female directors have, the higher the corporate transparency is.

3. Education Level and Corporate Transparency

Education level reflects the level of female directors' cognitive ability and speciality, and can provide important resources support for the company's information disclosure. Hilleman et al. [11] argue that female directors can bring a higher level of education to the board of directors. Burgess and Tharenou [3] find that 69% of Australian female directors have a bachelor's degree. Smith [24] find that when female directors received the higher education level, the positive effect on corporate performance is more significant, whereas less or no significant effect. Roberts et al. [19] think that a higher level of education is one of important factors that non-executive directors can effectively perform their duties, because the higher education can help to foster their independent thinking. Jiang et al. [12] find that there is a certain negative correlation relationship between management education level and the enter prise excessive investment. Tang and Ma [25] find that the market reaction is more negative when independent directors with high degree resign. In our country, female directors in the audit committee generally have a higher degree, with solid theoretical knowledge and rich practical experience, they can provide more supervision and consultation advises in the process of corporate information disclosure. Based on the above analysis, we believe that in the audit committee, the higher education level female directors have, the more helpful to improve the quality of information disclosure. So, this chapter puts forward the following hypothesis:

Hypothesis 3. The higher level of the female directors' education is, the higher the corporate transparency is.

4. Age and Corporate Transparency

The requirements of the retirement age in China is that Female employees' retirement age is 50 years old and female cadres is 55. If approaching retirement, their social relations, knowledge and business experience will face the possibility of a rapidly aging. Some retired people have no appetite to learn business, they are satisfied with the dawdle, often become "nice guy" on the job. Burgess and Tharenou [3] find that the average age of Australian female directors is 47 years old, the age median is in 45–49 years old. Sheridan and Milgate [22] find the average age of the female directors is lower than the average age of the male directors. Waelchli and Zeller [27] find that there is a significant negative correlation relationship between the age of chairman of board and corporate performance. Tang and Ma [25] find that older independent directors' resignation is a "good news" to outside investors, and the extent of the enterprise value decreasing is smaller. Xiang [29] find that the older financial independent directors are, the lower the accounting conservatism is. Based on the above analysis, we believe that in the audit committee, older female directors may not have enough energy and power to perform the regulatory responsibilities of the audit committee, which results in a lower level of corporate transparency. So, this chapter puts forward the following hypothesis:

Hypothesis 4. The older women directors are, the lower the corporate transparency is.

90.3 Research Design and Descriptive Statistics

1. Sample Selection and Data Sources

This chapter takes Shenzhen A-share listed companies as the research object, the sample period is between 2007 and 2009. We screen the samples according to the following principles:

• Exclude the listed companies which do not disclose the audit committee information;
• Eliminate the listed companies that there is no female directors in audit committee;
• Exclude financial listed companies, as they are subject to different disclosure requirements in China;

Table 90.1 Distribution of the information disclosure score of Shenzhen listed company

Year	Excellent 4		Good 3		Qualification 2		Disqualification 1		Total	
	Quantity	Ratio (%)	Quantity	Ratio (%)	Quantity	Ratio (%)	Quantity	Ratio (%)	Quantity	Ratio (%)
2004	4	10.5	24	63.2	10	26.3	0	0	38	100
2005	5	8.3	40	66.7	11	18.3	4	6.7	60	100
2006	11	16.4	34	50.8	22	32.8	0	0	67	100
2007	10	11.2	40	44.9	36	40.5	3	3.4	89	100

- Exclude the listed companies whose data is missed.

The final sample was composed of 254 firm-years observations. The data about the female directors in audit committee come from the annual report of listed companies, www.cninfo.com.cn, www.jrj.com.cn, and completed by hand-collection. The data about information disclosure score of listed companies come from integrity records of Shenzhen stock exchange website directly, the rest data comes from CSMAR.

2. Test Model and Variable Definition

To check the influence of the female directors' characteristics on corporate transparency, we build the following model Eq. (90.1):

$$\text{TRANSP} = \beta_0 + \beta_1\text{ACCOUNT} + \beta_2\text{ACAD} + \beta_3\text{EDU} + \beta_4\text{AGE} + \beta_5\text{ANUM}$$
$$+ \beta_6\text{WNUM} + \beta_7\text{WDIR} + \beta_8\text{SIZE} + \beta_9\text{LOSS} + \beta_{10}\text{LEV} + \beta_{11}\text{GROW} + \varepsilon. \tag{90.1}$$

Dependent variable: TRANSP is the corporate transparency, ie information disclosure score. The specific definition: 1 stands for disqualification, 2 stands for qualification, 3 stands for good, and 4 stands for excellent. Table 90.1 is the distribution of information disclosure score of sample companies from 2004 to 2007.

Variable definitions: corporate transparency, unqualified for 1, eligible for 2, good for 3, excellent for 4.

Independent variables: ACCOUNT-number of female directors who have accounting professional background divided by total number of female directors; ACAD-number of female directors from universities divided by total number of female directors; EDU-average degree score of female directors in audit committee, when a female director obtained a Ph.D. degree is 5 scores, Master's degree is 4 scores, Bachelor's degree for 3 scores College degree is 2 scores, and secondary education and below is 1 score; The AGE-average age of female directors in audit committee.

Control variable: ANUM-total number of audit committee; WNUM-total number of female directors in the audit committee; WDIR-dummy variable, 1 if the gender of the director in audit committee is female, 0 otherwise; SIZE-natural logarithm of the total amount of the company's assets; LEV-total liabilities divided by total asserts; LOSS-dummy variable, 1 if the company has a loss, 0 otherwise; GROW-the growth rate of operating income this period compared with the previous period.

Table 90.2 Descriptive statistics

Variable	Minimum	Maximum	Mean value	Median	Standard deviation
TRANSP	1.000	4.000	2.752	3.000	0.693
ACCOUNT	0.000	1.000	0.710	1.000	0.710
ACAD	0.000	1.000	0.379	0.000	0.457
EDU	1.000	5.000	2.930	3.000	0.953
AGE	28.000	71.000	47.076	46.000	7.429
ANUM	1.000	7.000	3.772	3.000	1.191
WNUM	1.000	3.000	1.236	1.000	0.486
WDIR	0.000	1.000	0.358	0.000	0.480
SIZE	18.028	24.126	21.279	21.176	1.087
LOSS	0.000	1.000	0.126	0.000	0.332
LEV	0.033	5.970	0.565	0.544	0.412
GROW	−1.799	190.219	1.058	0.153	12.075

3. Descriptive Statistics

Table 90.2 is the descriptive statistics of the main variables. The average value of TRANSP is 2.752, indicating that the information disclosure of most listed companies in our country is qualified and good. The average value of ACCOUNT is 0.710, indicating that most of female directors have accounting specialty in audit committee. The average value of ACAD is 0.379, indicating that nearly 37.9 % of female directors have academic profession background in audit committee. The average value of EDU is 2.93, indicating that the education level of female directors is close to undergraduate in audit committee. The maximum, minimum and average value of AGE is 71, 28 and 47.076 years old, respectively, indicating that the age distribution gap of female directors in audit committee is larger.

90.4 The Empirical Results and Analysis

90.4.1 Regression Results

The dependent variable TRANSP in model Eq. (90.1), and its value is between 1 and 4 showing a non-normal distribution. Therefore, we use the OLS and Tobit methods to analyze the model Eq. (90.1) respectively, regression results are shown in Table 90.3.

In Table 90.3, coefficient of ACCOUNT in OLS and Tobit regression results are all positive, but not significant, showing that there is no significant relationship between accounting specialty of female directors in the audit committee and corporate transparency. Therefore, hypothesis 1 is not supported by empirical evidence. Coefficient of *ACDA* in OLS and Tobit regression results are significant positive in 10 and 5 % respectively. The result shows Academic profession background of female

Table 90.3 The regression results between characteristics of female directors and corporate transparency

| Independent variable | Dependent variable: TRANSP | | | | | |
| | OLS regression | | | Tobit regression | | |
	Coefficient	t value	p value	Coefficient	t value	p value
Constant	-1.901^b	-2.308	0.022	-2.465^c	-2.64	0.009
ACCOUNT	0.043	0.395	0.693	0.038	0.31	0.758
ACAD	0.193^a	1.897	0.062	0.241^b	2.04	0.042
EDU	-0.062	-1.220	0.223	-0.078	-1.36	0.176
AGE	-0.009^a	-1.659	0.098	-0.010^a	-1.70	0.090
ANUM	0.028	0.776	0.439	0.030	0.73	0.465
WNUM	0.018	0.217	0.828	-0.004	-0.05	0.964
WDIR	-0.121	-1.368	0.173	-0.146	-1.45	0.147
SIZE	0.245^c	6.493	0.000	0.283^c	6.58	0.000
LOSS	-0.488^c	-3.995	0.000	-0.512^c	-3.67	0.000
LEV	-0.149	-1.481	0.140	-0.300^a	-1.79	0.075
GROW	-0.001	-0.347	0.729	-0.001	-0.26	0.795
N		254			254	
Adj.R2		24.9%				
Pseudo.R2					13.7%	
F value		8.635				
(p value)		(0.00)				
Chi2 value					83.95	
(p value)					(0.00)	

Note [a] indicates 10% significance level, [b] 5% significance level, [c] 1% significance level

directors in audit committee could significantly promote the improvement of corporate transparency, supporting hypothesis 2. Coefficient of EDU in OLS and Tobit regression results are both negative, but not significant. The result shows that there is no significant relationship between education degree of female directors in the audit committee and corporate transparency. Therefore, hypothesis 3 is not supported by empirical evidence. Coefficient of AGE in OLS and Tobit regression results are significant negative in 10% respectively, suggesting that the older female directors may not have enough energy and power to perform supervisory duties in audit committee, resulting in a decline in corporate transparency. The result supports hypothesis 4.

For the other control variables, coefficient of SIZE in OLS and Tobit regression results are both significantly positive; coefficient of LOSS in OLS and Tobit regression results are both significantly negative; coefficient of LEV in Tobit regression results is significantly negative, but the coefficient can't pass the test of significance in OLS regression results.

Table 90.4 Female directors characteristic, legal environment and corporate transparency (OLS regression)

| Independent variable | Dependent variable: TRANSP | | | | | |
| | High Legalization level | | | Low Legalization level | | |
	Coefficient	t value	p value	Coefficient	t value	p value
Constant	−1.431	−1.325	0.187	−2.087	−1.539	0.127
ACCOUNT	0.186	1.338	0.183	−0.149	−0.787	0.433
ACAD	0.139	1.004	0.317	0.360[b]	2.240	0.028
EDU	−0.048	−0.778	0.438	−0.127	−1.393	0.167
AGE	−0.009	−1.212	0.227	−0.016[b]	−2.027	0.046
ANUM	−0.010	−0.214	0.931	0.077	1.064	0.290
WNUM	0.116	1.059	0.291	−0.138	−0.930	0.355
WDIR	−0.144	−1.263	0.209	0.047	0.300	0.765
SIZE	0.222[c]	4.464	0.000	0.263[c]	4.396	0.000
LOSS	−0.329[b]	−2.074	0.040	−0.789[c]	−3.807	0.000
LEV	−0.244[b]	−2.240	0.027	0.531	1.544	0.126
GROW	−0.002	−0.724	0.470	0.031	1.581	0.117
N		154			100	
Adj.R2		18.7%			37.1%	
F value		4.198			6.319	
(p value)		(0.00)			(0.00)	

Note [a] indicates 10% significance level, [b] 5% significance level, [c] 1% significance level

90.4.2 Further Analysis

Emperical research indicates that a country's legal environment has important influence on corporate's disclosure quality [4, 14, 17]. In our country, although the national legal environment of differently listed company is the same, their region's legalization level are comparatively large different, and enforcement and implementation effect of the same law are also different in indifferent regions. Therefore, on the basis of above analysis, this chapter divide sample firms into high level and low level legalization according to the overall median legalization index in year 2004 which reported by Fan et al. [9], then use OLS and Tobit methods to make the further analysis on the relationship between characteristics of female directors in audit committee and the corporate transparency respectively. Regression results are shown in Tables 90.4 and 90.5.

From Tables 90.4 and 90.5, in high legalization region, coefficient of female directors' characteristics variables (ACCOUNT, ACAD, EDU and AGE) in audit committee are not significant whether in OLS regression results or Tobit regression results, suggests that in the region where legalization level is high, the legal environment plays a monitoring role in corporate information disclosure, and characteristics of female directors in audit committee has no impact on corporate transparency, which indicates that there is a certain substitution effect between legal environment and audit committee in corporate governance. In low legalization region, coefficient

Table 90.5 Female directors characteristic, legal environment and corporate transparency (Tobit regression)

| Independent variable | Dependent variable: TRANSP | | | | | |
| | High Legalization level | | | Low Legalization level | | |
	Coefficient	t value	p value	Coefficient	t value	p value
Constant	−1.948	−1.62	0.108	−2.874[a]	−1.90	0.061
ACCOUNT	0.218	1.40	0.163	−0.214	−1.03	0.306
ACAD	0.214	1.34	0.182	0.410[c]	2.35	0.021
EDU	−0.071	−1.01	0.314	−0.153	−1.55	0.124
AGE	−0.012	−1.47	0.143	−0.017[a]	−1.96	0.053
ANUM	−0.023	−0.46	0.646	0.088	1.12	0.265
WNUM	0.121	0.99	0.323	−0.183	−1.15	0.255
WDIR	−0.178	−1.40	0.164	0.049	0.29	0.771
SIZE	0.266[c]	4.73	0.000	0.303[c]	4.63	0.000
LOSS	−0.321[a]	−1.80	0.074	−0.848[c]	−3.77	0.000
LEV	−0.560[b]	−2.30	0.023	0.584	1.58	0.117
GROW	−0.002	−0.63	0.528	0.033	−1.54	0.128
N		154			100	
Pseudo.R2		12.3%			23.4%	
Chi2 value		45.41			57.13	
(p value)		(0.00)			(0.00)	

Note [a] indicates 10% significance level, [b] 5% significance level, [c] 1% significance level

of ACAD are significant positive whether in OLS regression results or Tobit regression results, but significant negative for the coefficient of AGE, this suggests that in the region where legalization level is low, the legal environment doesn't plays a monitoring role in corporate information disclosure, and academic professional background and age of female directors have significant impact on corporate transparency, further stating that there is a certain substitution effect between the legal environment and the audit committee in the company's corporate governance. At the same time, the result also illustrates that the significant impact of the academic professional background, age of female directors and the corporate transparency mainly exist in the listed companies located in low legalization area.

90.4.3 Robustness Analysis

In order to further investigate the robustness, we made the following robustness analysis for the above results:

1. Extreme value test. In order to eliminate the possible effect of extreme value, we use winsorization to process all the continuous variable value which is at 1 and 99% in model Eq. (90.1), reusing the method of OLS and Tobit tomake regres-

sion analysis for model Eq. (90.1) respectively, and regression results are almost consistent with Tables 90.3, 90.4 and 90.5, supporting the research conclusion above.
2. Multicollinearity test. In order to control the effect of multicollinearity, we calculate the variance inflation factor (VIF value) to each regression equation in model Eq. (90.1), and find that VIF value are all less than 2, indicating that there is no multicollinearity problem among variables.

90.5 The Research Conclusion

As one of the most important internal governance mechanism, the audit committee efficiency has been the core theme of corporate governance research. Previous study on the audit committee often tend to focus on investigating the effect that the structure characteristics of audit committee (such as size, independence, activity and financial experts, ownership, etc.) having on the efficiency of corporate governance.

This chapter extend the previous research category on the audit committee, from the Angle of characteristics of female directors in the audit committee, empirically examines whether the characteristics of female directors in audit committee affect corporate transparency based on the data of China's Shenzhen A-share listed companies from 2004 to 2007. The results show that, overall, the academic professional background of female directors has significantly positive impact on corporate transparency; the age of female directors has significantly negative impact on corporate transparency; the accounting expertise and education level of female directors have no significant effects on corporate transparency. Further analysis also show that the significant impact relationship between the academic professional background or age of female directors and corporate transparency only exist in the lower legal level region, however, in the listed companies located in the area where the legalization level is high, the impact of female directors' characteristics on the corporate transparency disappeared completely. Therefore, in order to further improve the corporate transparency in our country, on the one hand, we should increase the degree of gender diversity in the audit committee based on the characteristics of professional background and age of female directors; on the other hand, improve and perfect the legalization degree in different areas of our country.

References

1. Bedard J, Chtorou SM, Courteau L (2004) The effect of audit committee expertise, independence, and activity on aggressive earnings management. Auditing J Pract Theor 23(2):13–25 NULL
2. Bill BF, Hasan I et al (2009) Gender differences in financial reporting decision-making: Evidence from accounting conservatism. Available at SSRN

3. Burgess Z, Tharenou P (2002) Women board directors: characteristics of the few. J Bus Ethics 37:39–49
4. Bushman RJP, Smith A (2004) What determines corporate transparency. J Account Res 42:207–252
5. Dallas L (2002) The new managerialism and diversity on corporate boards of directors. Tulane Law Rev 76:1363–1405
6. Daniela B, Lukas M (2008) Will women bewomen? analyzing the gender difference among financial experts. KYKLOS 61(3):364–384
7. Eagly A, Carli L (2003) The female leadership advantage: an evaluation of the evidence. Leadersh Q 14:34–807
8. Fama FE, Jensen MC (1983) Separation of ownership and control. J Law Econ 26:301–326
9. Fan G, Wang XL, Zhu HP (2007) The report on the relative process of marketization of each region in China (2006). Economic Science Press, Beijing
10. Ge W, Dawn M, Zhang JL (2011) Do cfos have style? an empirical investigation of the effect of individual cfos on accounting practices. Contemp Account Res 28(4):1141–1179
11. Hillman A, Cannella A et al (2002) Women and racial minorities in the boardroom: how do directors differ? J Manage 28(6):747–7632
12. Jiang F, Yi X et al (2009) Background characteristics of the manager and excessive investment behavior of enterprises. Manage World China 1:30–39
13. Kaplan SN, David R (1990) Outside directorships and corporate performance. J Financ Econ 27(2):389–410
14. La Porta R, Shleifer A et al (1998) Law and finance. J Polit Econ 106:1113–1155
15. Maher P, Munro M (2000) Today's board and the academic option. Ivey Bus J 64(6):8–11
16. Mattis MC (2000) Women corporate directors in the united states. In: Burke R, Mattis M (eds) Women on corporate boards of directors. Kluwer Academic, Netherlands, pp 43–56
17. Nenova T (2003) The value of corporate voting rights and control: a cross-country analysis. J Financ Econ 68:325–351
18. Nelson J (1996) Feminism. Objectivity and economics. Routledge, London
19. Roberts J, McNulty T, Stiles P (2005) Beyond agency conceptions of the work of the non-executive director: creating accountability in the boardroom. Brit J Manage 16:S5–S26
20. Schubert R (2006) Analyzing and managing risks—on the importance of gender difference in risk attitudes. Manag Financ 32:706–715
21. Sealy R, Singh V, Vinnicombe S (2007) The female FTSE report 2007. Cranfield
22. Sheridan A, Milgate G (2005) Accessing board positions: a comparison of female and male board members' views. Corp Governance Int Rev 13(6):847–855
23. Singh V, Terjesen S, Vinnicombe S (2008) Newly appointed directors in the boardroom: how do women and men differ? Eur Manage J 26(1):48–58
24. Smith N, Smith V, Verner M (2005) Do women in top management affect firm performance? A panel study of 2 500 danish firms. IZA Discussion Paper No. 1:708
25. Tang XS, Ma C (2012) Background characteristics, resignation behavior of the independent directors and enterprise value. Account Econ Stud Chin 4:3–13
26. Trautman L (2012) The matrix: the board's responsibility for director selection and recruitment. Fla State Univ Bus Rev 11(1):1–66
27. Waelchli U, Zeller J (2013) Old captains at the helm: chairman age and firm performance. J Banking Financ 37:1612–1628
28. Waring M (1988) If women counted: a new feminist economics. Harper and Row, San Francisco
29. Xiang R (2013) The characteristics of financial independent directors and the accounting conservatism. In: The 19th Chinese finance annual conference

Chapter 91
Empirical Analysis on the Relation Between Capital Structure and Corporate Performance of Listed Companies in SME Boards in China

Chaojin Xiang, Xiaofeng Wu and Aoyang Miao

Abstract Small and medium-sized board market in China was established in 2004 and developing very fast. By the end of 2011, there are 648 listed SMEs in China. Small and medium-sized listed enterprises (SMEs) play an important role in economic development, but financing problems and lack of capital structure flexibility are also difficult issues they need to deal with. This article makes regression analysis on the data of small and medium-sized listed enterprises from 2009 to 2011 to study the relationship between capital structure and corporate performance. It is concluded that different performance indicators lead to different relationships. There was a significantly negative correlation between the Tobin Q and capital debt ratio, while comprehensive accounting indicator and capital debt ratio are significantly positive correlation. The relation between short-term debt ratio and performance is significantly mutual negative correlation, no matter what indicators are used. Long-term debt ratio does not present a significantly positive impact on Tobin Q. According to empirical conclusions, it is suggested to help improve the capital structure and corporate performance of the SMEs.

Keywords Capital structure · SMEs · Corporate performance · Financial indicators

91.1 Introduction

As an important part of social economy, small and medium-sized enterprises (SMEs) are playing a momentous role in the economic and social development. Figure 91.1 is the statics in 2011, it is shown in the figure that whatever employment, innovation, and economic development, SMEs can't be replaced.

C. Xiang (✉) · X. Wu · A. Miao
Business School, Sichuan University, Chengdu 610064, People's Republic of China
e-mail: 122402961@qq.com

J. Xu et al. (eds.), *Proceedings of the Eighth International Conference on Management Science and Engineering Management*, Advances in Intelligent Systems and Computing 281, DOI: 10.1007/978-3-642-55122-2_91, © Springer-Verlag Berlin Heidelberg 2014

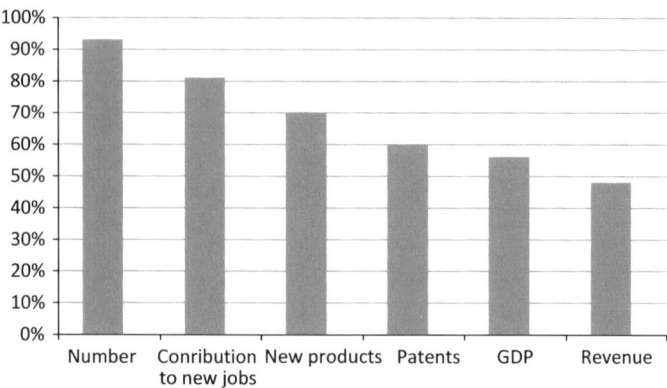

Fig. 91.1 Contribution rate of SMEs in 2011

However, default in SMEs does exist, such as poor ability to resist risk, less competition and short life circle, etc. After the global financial crisis in 2008, a large number of SMEs are affected deeply into trouble or even bankrupt. Some important reasons are financing difficulties and lack of capital structure elasticity in SMEs. The following European debt crisis even makes SMEs in China face a bigger challenge. SMEs enhance the innovation ability and realize the upgrading of industrial structure, as well as find a reasonable capital structure, which improve corporate performance and resist risks.

Therefore, this article targeted listed companies in small and medium-sized boards, and tries to find out the relationship between capital structure and corporate performance of listed company in small and medium-sized board. Different with the earlier articles, two kinds of indicators, financial index and market index, are used in this article. Through the comprehensive analysis, more comprehensive and accurate results are concluded. It is beneficial to make a suggestion on optimizing capital structure and improving corporate performance of SMEs.

91.2 Relevant Literatures on Capital Structure and Corporate Performance

Based on the assumptions of perfect capital markets, Modigliani and Miller proposed MM theory, which is called the beginning of the modern capital structure. Since then, a large number of theoretical and empirical research on capital structure and corporate performance are emergent.

1. Capital Structure and Corporate Performance are Positively Correlative

Lv and Wang [7] pointed that, according to their empirical research conclusions, there is a significantly positive correlation between corporate performance and capital structure, and the same conclusion was found by Zhang etc, who selected listed companies in electric power industry as samples. Wang et al. study [7] shows that the interaction between capital structure and agency costs of listed companies can be explained by the agency cost hypothesis and efficiency risk hypothesis. It also means that the capital structure and corporate performance are positively related, but the speed of interaction is different.

2. Capital Structure and Corporate Performance are Negatively Correlative

Booth et al. [1] studied ten developing countries, found that corporate performance is the most important factors influencing the capital structure, and corporate performance and capital structure in developing countries are significantly negative correlation, in addition to Zimbabwe. Tang and Huang [4] considered EBIT as the representative of company performance and thought the debt level is significantly affected by the corporate performance. Mao and Zhao researched on data of 648 listed companies in six categories and chose EVA as a representative of the company's performance, then found that corporate performance is negatively related to the debt levels in most of the industries, except some utilities industry. Wang and Li [6] found that there is a significant negative correlation between profitability and Debt-to-assets ratio (the representative of capital structure) through the empirical analysis of listed companies in Jiangsu province.

3. Capital Structure and Corporate Performance are Non-linearly Correlative or Irrelevant

Zhang's research showed that the relationship between capital structure and corporate performance is a inverted "U" type on the data of 39 listed companies in Anhui. Song and Zhang selected the state-owned holding company of liaoning province as sample for analysis, and it is concluded that capital structure and corporate performance present positive correlation when the Debt-to-assets ratio in 0–40 %; otherwise, a negative correlation is set up. Yang et al. [8] selected listed companies in energy industry from 1999 to 2008 as samples. Empirical results show that when Debt-to-assets ratio is under 26.4 %, there is significantly positive correlation between the two; on the other side, they has significantly negative correlation.

4. Relation Between Capital Structure and Corporate Performance in SMEs

Ying [9] found that corporate performance and its Debt-to-assets ratio has a negative-linear correlation but doesn't exist linear relations to its long-term debt by the research on SMEs listed in ShenZhen small and medium-sized board market. Zhang and Zhang [10] based on the panel data of 381 listed SMEs from 2003 to 2009, analyzes influence on enterprise performance from capital structure and corporate governance and found that debt constraint and the corporate performance of listed SMEs is significantly negative correlation.

In addition, Wang and Li [6] took listed companies in small and medium-sized boards as samples and the results have shown that SMEs do exist the optimal capital structure, which contains that the short-term debt ratio and interest-free debt ratio are positively correlated with corporate performance, and long-term capital debt ratio and interest-bearing debt ratio are negatively related with corporate performance.

Relationship between capital structure and corporate performance has been a hot topic of financial world. Scholars do a lot of theoretical and empirical studies, but it comes out different conclusions. It can be seen that most researches on the relationship of capital structure and corporate performance are focused on the major board market or concentrated in listed companies of certain industry. However, there is few studies on relationship between capital structure and corporate performance of listed companies in small and medium-sized boards, while small and medium-sized board market is developing very fast with strong profitability and great development potential since established in 2004.

91.3 Data Sources and Model Design

According to main motivation of this article, data on SME board market are collected from year 2009 to 2011.

1. Sample Selection and Model Design

Since small and medium-sized enterprise board market was officially opened on June 25, 2004, there have been 648 SMEs successfully listed till December 31, 2011. Considering the missing problems on financial insurance industry and the particularity of ST companies' financial information, this article will exclude ST companies and financial insurance industry company. There are totally 318 listed companies selected in 2009 except 8 ST companies and 2 missing problem companies. In the same way, there are 533 listed companies selected in 2010 and 648 listed companies selected in 2011. We can get a graph below about the industry distribution of all the companies selected (Fig. 91.2).

Data comes from RESSET and CSMAR. Sample data used by empirical analysis are relevant financial indicators from RESSET database about capital structure and corporate performance of small and medium-sized board listed company, while tobin Q value is derived from the CSMAR database. Tools used for analysis are statistical description, the least squares (OLS) and other analytic tools.

2. Research Hypothesis and Model Design

This article analyze the performance of listed companies in small and medium-sized boards from the perspective of financial index and market index, selecting returns on equity (ROE) and tobin Q value as represent of company performance and Debt-to-assets ratio for capital structure.

(1) Hypothesis

Hypothesis 1. listed companies in small and medium-sized board has a reasonable capital structure range.

Hypothesis 2. short-term debt ratio of small and medium-sized enterprises is negatively related with corporate performance.

Hypothesis 3. long-term debt ratio of small and medium-sized enterprise is positively related with corporate performance.

Hypothesis 4. corporate performance of small and medium-sized enterprises is negatively related with capital structure.

Fig. 91.2 Industry distribution ratio of selected companies

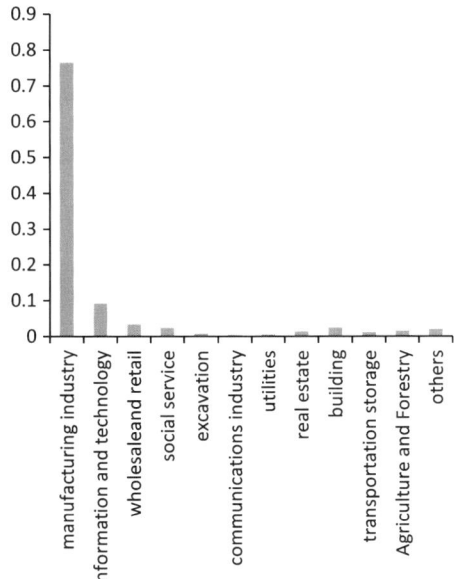

(2) Model Assumption

Model In this paper is divided into two parts: one is the model of influence that capital structure has on corporate performance, and the other one is the model of influence that corporate performance has on capital structure.

Firstly, model of influence that capital structure has on corporate performance: According to Hypothesis 1,

Model 1: $TQ_1 = a1 + b1DAR + c1DAR^2 + \varepsilon1$,

Model 2: $TQ_2 = a2 + b2DAR + c2DAR^2 + d2ROA + e2SIZE + \varepsilon2$,

Model 3: $COM_1 = a3 + b3DAR + c3GROW + \varepsilon3$.

According to Hypothesis 2,

Model 4: $TQ_3 = a4 + b4SFB + c4ROA + d4SIZE + \varepsilon4$,

Model 5: $COM_2 = a5 + b5SFB + c5GROW + \varepsilon5$.

According to Hypothesis 3,

Model 6: $TQ_4 = a6 + b6LFB + c6ROA + d6SIZE + \varepsilon6$,

Model 7: $COM_3 = a7 + b7LFB + c7GROW + \varepsilon7$.

secondly, model of influence that corporate performance has on capital structure: According to Hypothesis 4,

Model 8: $Y = a8 + b8TQ + c8TQ_{il1} + \varepsilon8$,

Model 9: $Y = a9 + b9COM + c9COM_{il1} + \varepsilon9$.

3. Definition of the Variables

According to the model mentioned above, we define the following variables:

(1) Company Performance Variable

In this paper, we take the the combination of financial indicators and market indicators (tobin Q) as the corporate performance indicators.

Table 91.1 Financial indicators

Profitability	Gross profits margin, ROE, Cash return on total assets
Operating efficiency	Assets turnover ratio, ratio of sales to cost
Growth	TAGR
Asset management capability	Inventory turnover ratio, accounts receivable turnover
Cash flow ability	Cash income of operating levels = NCFO/operating income

Table 91.2 Capital structure variable

Debt-to-assets ratio (DAR)	Total liabilities/General assets
Short-term debt rate	Short-term debt/Total liabilities
Long-term debt rate	Long-term debt/Total liabilities

Financial indicators contain profitability, operation ability, asset management capability, cash flow ability and development ability and other comprehensive index (Table 91.1)

Tobin $Q(TQ)$ represents the market performance of listed companies. Tobin $Q =$ market value/reset value of total assets (Table 91.2).

This article selects nine financial indicators. It is hard to avoid has the correlation and multicollinearity between indicators and can be solved by factor analysis, which could reduce the workload.

(2) Capital Structure Variable

We choose Debt-to-assets ratio, short-term debt ratio and long-term debt ratio as the indicators of capital structure.

(3) Control Variables

Profitability, firm size and growth ability are selected as Control variable. It is obvious that the stronger the profitability, the better the corporate performance. Similarly, the bigger the company size and the stronger the growth ability, the better the company performance will be. This paper use ROE on behalf of profitability and the logarithm of total assets represent the company size, total assets growth rate represents the company's growth ability.

91.4 Analysis of Empirical Results

It is proved that most of the indicators and comprehensive performance value have significant correlation through factor analysis and variable correlation analysis, so we can build models of capital structure and performance indicators. Through the multicollinearity test, we can also prove that all the models are valid for model 1 to model 9.

91.4.1 The Impact of the Capital Structure on the Corporate Performance

Hypothesis 1. listed companies in small and medium-sized board has a reasonable capital structure range.

Model 1: $TQ_1 = a1 + b1DAR + c1DAR^2 + \varepsilon1$,

Model 2: $TQ_2 = a2 + b2DAR + c2DAR^2 + d2ROA + e2SIZE + \varepsilon2$,

Model 3: $COM_1 = a3 + b3DAR + c3GROW + \varepsilon3$.

For model 1, according to regression analysis of the relevant data from 2009 to 2011, we found that a significant negative correlation between corporate performance and debt ratio.

For model 2, 3 years' regression analysis results show that control variable ROA was significantly positively correlated with corporate performance, which explains the tobin Q value is positively affected by the rate of return on total assets. At the same time, the company size and corporate performance was significantly negative, which means that SMEs does not exist economies of scale.

For model 3, there is significantly positive correlation between DAR and corporate performance. This conclusion stands at the opposite of model 1 and model 2 and reject the hypothesis 1, which means that different corporate performance index used leads to empirical results of the relation.

Hypothesis 2. short-term debt ratio of small and medium-sized enterprises is negatively related with corporate performance.

Model 4: $TQ_3 = a4 + b4SFB + c4ROA + d4SIZE + \varepsilon4$,

Model 5: $COM_2 = a5 + b5SFB + c5GROW + \varepsilon5$.

For model 4, 3 years' results of the regression analysis turns out that short-term debt ratio was significantly negative correlation with corporate performance. For model 5, it has the same conclusion with model 4. This conclusion verifies the hypothesis 2, that is, with the increase of current liabilities ratio, corporate performance decreases.

Hypothesis 3. long-term debt ratio of small and medium-sized enterprise is positively related with corporate performance.

Model 6: $TQ_4 = a6 + b6LFB + c6ROA + d6SIZE + \varepsilon6$,

Model 7: $COM_3 = a7 + b7LFB + c7GROW + \varepsilon7$.

For model 6, regression analysis concluded that the long-term debt ratio were positively correlated with corporate performance, but the correlation isn't significant so much. This conclusion verifies the hypothesis 3, that is, with long-term debt ratio increase, the company performance increase.

For the model 7, thar the regression analysis of 2009 show the long-term debt didn't pass the test of significance means the explain ability of corporate performance ability is not strong. Long-term debt ratio of 2010 pass the significance test but fail of 2011. Therefore, long-term debt ratio is significantly positive correlation with corporate performance, but the significance level is uncertain.

This conclusion is the same with model 6 and verifies the hypothesis 3 though the significance level is not sure, generally speaking, corporate performance will increase with the increase of long-term debt ratio.

91.4.2 The Impact of Corporate Performance on the Capital Structure

Hypothesis 4. corporate performance of small and medium-sized enterprises is negatively related with capital structure.

Model 8: $Y = a8 + b8TQ + c8TQ_{il1} + \varepsilon 8$,

Model 9: $Y = a9 + b9COM + c9COM_{il1} + \varepsilon 9$.

For model 8, regression analysis result turns out that the company performance and asset-liability ratio has significantly negative correlation relationship. This conclusion verifies the hypothesis 4, which means with the increase of corporate performance of last two years, asset-liability ratio will decline. There is not a significant impact on the long-term debt ratio by corporate performance and the relationship between them is not sure, but the influence on short-term debt ratio is significant, it will decline with the increase of company performance of the last two years.

For model 9, we can see that corporate performance was significantly positively correlated with the asset-liability ratio, which rejected hypothesis 4 and is obviously different to model 8. Besides, corporate performance does not have significant influence the short and long-term debt ratio.

91.5 Conclusions and Recommendations

According to the empirical analysis, results and conclusions can be summarized.

1. Results Analysis

   Through the empirical analysis, results analysis can be concluded in Tables 91.3 and 91.4.

 2. Conclusions and Recommendations

   At the end of this article, conclusions and recommendations are suggested, based on the empirical analysis results.

(1) Conclusions

   Firstly, while corporate performance indicators of listed companies in small and medium-sized boards is different, the relationship between corporate performance and capital structure is also different. If we take tobin Q on behalf of company performance, then capital structure and corporate performance are significantly negative correlation relationship. Instead, if comprehensive performance value calculated by nine financial indicators presents company performance, then capital structure and corporate performance are significantly positive correlation.

Table 91.3 Influence on capital structure by corporate performance

	Tobin Q	Comprehensive performance value
DAR	Negatively correlated	Positively correlated
Short-term debt ratio	Negatively correlated	Negatively correlated
Long-term debt ratio	Positively correlated	Positively correlated

Table 91.4 Influence on corporate performance by capital structure

	DAR	Short-term debt ratio	Long-term debt ratio
Tobin Q	Negatively correlated	Negatively correlated	Uncertain
Comprehensive performance value	Positively correlated	Uncertain	Uncertain

Secondly, debt financing level of listed companies in small and medium-sized boards in china is low and optimal capital structure is uncertain. There is much more space for the development of the capital structure and financial leverage should be made full use of to improve company performance.

Thirdly, capital structure of small and medium-sized enterprises in our country is unreasonable. short-term debt ratio is very high, some can even reach 100% level. This situation relates to the characteristics-financing difficulties and small size in small and medium-sized enterprise.

Fourthly, long-term debt ratio has a positive impact on corporate performance. Generally speaking, although the empirical results is not significant, the increase long-term debt ratio levels of small and medium-sized companies will promote the corporate performance. It is not certain about the impact on long-term debt ratio with corporate performance changes. Besides, long-term liabilities also help balance the debt term structure as well as increase their own financial status.

Fifthly, we can see that the increase of the rate of return and growth rate of total assets can increase the corporate performance through the analysis of the empirical model while company size does exactly different effect. Small and medium-sized enterprises lead to low efficiency of management, information asymmetry and increasing agent cost with expending the companies' scale, and does not improve company performance.

(2) Recommendations

Factors affecting capital structure and corporate performance of listed companies in small and medium-sized board are so many that can't be discussed totally. This article emphasizes on the relationship between them and would try to make some suggestions about optimizing capital structure and improving corporate performance.

Firstly, SMEs should improve capital structure and increase long-term debt. Despite that different indicators of listed companies in small and medium-sized boards lead to different results, we can also get the conclusion that there is significant correlation between them. Capital structure of listed company in small and medium-sized boards is unreasonable and depends on equity finance and current liabilities too much. Enterprises should seek a reasonable capital structure, increase the long-term debt ratio, make a buffer space for capital structure and avoid financial trouble.

Secondly, SMEs should improve their own credit level and profitability. More than 75 % of the listed companies in small and medium-sized boards focused on the manufacturing industry, which has low technology and profitability. SMEs should enhance innovation capability, improve the scientific and technological content to achieve the optimization and upgrading of industrial structure. Devoting to improve the overall level of profitability and solvency can also improve the enterprises' credit level, alleviate the financing trouble.

Thirdly, government should create the environment to develop various methods of financing services. Government can create some financial support policies and encourage the development of some banks, credit guarantee situations or micro-finance companies specifically for SMEs due to that loan is still the mainly platform for SMEs to finance. It can also bring guidance in to force, strengthen supervision and protect the legitimate rights and interests of SMEs.

References

1. Booth L, Aivazian V et al (2001) Capital structures in developing countries. J Financ 56(1):87–130
2. Ma QG (2002) Management statistics. Science Press (in Chinese)
3. Qin NN (2012) Empirical study on capital structure and corporate performance of listed companies. Shandong Text Econ 9:21–22 (in Chinese)
4. Tang HR, Huang D (2005) Analysis of factors changing capital structure of listed companies—evidence from listed companies in China. Manag Rev 4:3–8 (in Chinese)
5. The Chinese Institute of Certified Public Accountants (2011) Financial cost management, 1st edn. Perking University Press, pp 274–277 (in Chinese)
6. Wang CJ, Li C (2011) Relevant research on corporate profitability and capital structure—based on the factor analysis of listed companies in Jiangsu province. J SE Univ 13:6 (in Chinese)
7. Wang CF, Zhou M, Fang Z (2008) Empirical research on interaction impact of capital structure and corporate performance—based on the analysis of stochastic frontier approach method. J Shanxi Financ Econ Univ 30(4):77–83 (in Chinese)
8. Yang H, Chen X, Tian HG (2011) Non-linear relationship study on capital structure and corporate performance—empirical evidence from listed companies in China energy industry. Financ Econ 156:101–106 (in Chinese)
9. Ying J (2009) Empirical analysis on capital structure and corporate performance of listed companies in sme board in China. Master's Thesis, Jiangxi finance University (in Chinese)
10. Zhang YM, Zhang ZH (2011) Capital structure, corporate governance and performance in listed SMEs. J Shanxi Financ Econ Univ 33:11 (in Chinese)

Chapter 92
RFID Application Infant Security Systems of Healthcare Organizations

Susana G. Azevedo, Helena Carvalho and Virgílio António Cruz-Machado

Abstract The Radio Frequency Identification (RFID) technology is actually considered a hot topic in all scientific areas and has been described as a major enabling technology for the automation of many processes. Although it is not a new technology it has only recently come to the awareness of the public and widely used in many sectors and particularly in the Healthcare. This paper aims to illustrate the RFID technology deployment in Healthcare organisations, more precisely in infant security systems inside the hospitals. To attain this objective a case study about the experience of three hospitals and one RFID technology provider is presented to highlight the main architectural characteristics, functionality, and advantages associated with the RFID deployment.

Keywords RFID · Healthcare · Case study · Infant security systems · Hospital security

92.1 Introduction

The RFID is a technological application that uses the radio waves for automatic identification of objects, positions or persons through electromagnet answers and at considerable distances [24]. This technology application is growing in various sectors and is being used for very different purposes. The RFID has been object of academics' attention from widespread fields of knowledge. We can find researches

S. G. Azevedo (✉)
UNIDEMI, Department of Management and Economics, Faculty of Social Sciences,
University of Beira Interior, Covilhã, Portugal
e-mail: sazevedo@ubi.pt

H. Carvalho · V. A. Cruz-Machado
UNIDEMI, Department of Mechanical and Industrial Engineering, Faculdade de Ciências
e Tecnologia da Universidade Nova de Lisboa, Caparica, Portugal

J. Xu et al. (eds.), *Proceedings of the Eighth International Conference on Management Science and Engineering Management*, Advances in Intelligent Systems and Computing 281, DOI: 10.1007/978-3-642-55122-2_92, © Springer-Verlag Berlin Heidelberg 2014

on RFID in the following knowledge' areas: logistics [6]; supply chain management [26]; innovation [10]; marketing [21]; and manufacturing operations [12].

In the Healthcare sector have been also some experiences with this kind of information technologies showing the promising applications of them [11]. Kumar and Shim [15] had extended the deployment of RFID to the whole Healthcare supply chain as a tool of remotely tracking supplies, equipment, and even people as they move through the supply chain from manufacturers to suppliers, wholesalers, hospitals, pharmacies, and intermediaries. Monitoring and controlling effectively (and efficiently) the hospital operations, although difficult, it is vital to patient care. The errors could have harmful effects, in extreme the patient death, and lead to the increase of service cost [19]. Information technologies in Healthcare organisations are used mainly for collect integrate and share information, for control costs and for medication safety [9]. Therefore, an important feature in RFID Healthcare system is the ability to integrate and exchange information with others systems improving the hospital information system visibility [28]. More recently, the RFID has also reached an important role in Healthcare organisations because of its enormous potentialities mainly to track personnel, patients and equipment, to admit and to register patients, to support a set of tasks associated directly with patients such as bill payment, dosage and disposal of medicines, and to update medical records [19]. One potential application of people tracking is the infant abduction protection systems [11]. Since Healthcare facilities are the facilities where more than two thirds of the infant abductions occur [22], hospitals managers are looking for solutions to improve the infant security. One important solution to overcome these problems inside the hospitals is the deployment of RFID-based security systems. However, despite the potentials of RFID-based security systems in Healthcare industry their applications have not been extensively studied in this area [32]. Fisher and Monahan [7] also referred the lack of empirical evidence on how to implement these systems effectively.

This paper aims to illustrate the RFID technology deployment in infant security systems inside the hospitals. Adopting a case study methodology, this study seeks to emphasize this technology deployment in Healthcare organizations highlighting the main advantages that hospitals reach with it. Also, the main characteristics of the infant security system, functionality and advantages are presented.

The paper is structured as follows; first, we focus on the main architectural characteristics of the RFID System in terms of the elements that constitute it (readers, tags, and software). Next, it focuses on the RFID deployment specifically in Healthcare organizations highlighting its main areas of deployment, advantages and disadvantages. In the following section, the research methodology is discussed, followed by a case study about the application of the RFID in infant security system within three hospitals and also the experience of a RFID-based systems provider. Finally, the main conclusions are presented.

92.2 RFID Technology in Healthcare

The enormous advantages associated with this technology, has justified its large application in several functional areas. We can find the RFID technology in different contexts namely in: (1) electronic keys; (2) warehouses [16]; (3) distribution centres [2]; (4) points of sales; (5) security applications in the transport [14]; (6) demotic [13]; (7) retailing [1]; (8) e-business [30]; (9) supply chain execution applications [16]; (10) Healthcare [23].

Since RFID can remotely identify and track tagged objects as they moved around the hospital area, it can provide solutions to these challenges. In Healthcare the RFID has been deployed supporting several tasks and with different objectives. One of the primary applications of the RFID is improving patient safety and security [23] including reducing medical errors and improving effectiveness of services. Tzeng et al. [27] referred the RFID utilization during the 2003 SARS epidemic. This technology was used in Taiwan hospitals to track patients in order to determine the paths of infection sources monitoring and avoiding contaminations.

Additionally, this technology has been used to scan prescriptions and transmit them to the pharmacy to eliminate hand-written prescriptions and reduce prescriptions fill-rate errors [25]. In this case, the tag is used to identify out-of-date products contributing to reduce the possibility of a fatal or ineffective dose. Another interesting RFID application consists of embedding RFID tags into blister packaging systems to monitor electronically the date and time a patient opens a medicine package and takes out a pill. In outpatient settings, the patient would return the used packaging to the clinic, the package would be scanned and patient usage patterns plotted [18]. This provides a more effective evaluation of patient compliance with prescription medication therapy. The RFID deployment can be also used to support drug anti-counterfeiting and to track recall counterfeit or contaminated medication and supplies [4]. Beyond this, the RFID has been used to help improving the management of blood distribution. Hospitals and laboratories are dealing with a highly perishable, hugely sensitive product that is always in short supply and is always difficult to procure. So, temperature sensitive tags can provide accurate tracking to ensure that blood stored at less than optimal temperatures would not be distributed to a patient [20]. Yao et al. [31] also stress the value the RFID integration with other sensors for monitoring and management; for example wireless detection of patients temperature and for monitoring.

Brooke [3] identified several aspects of the Healthcare services where RFID can be beneficial, including the ability to trace high value assets in the hospital and the ability to track assets over time, thus verifying that certain procedures have been completed (in this case, decontamination of surgical instruments). As state by Qu et al. [19] an effective equipment management in hospitals is critical to deliver high quality care as well as reducing Healthcare cost. The surgical instruments management is a major problem for most Healthcare facilities. In addition to the loss issue (ranging from simply lost or misplaced instruments to outright theft), there has been a need to track both the instruments themselves and the entire process associated with them,

Table 92.1 Areas of RFID deployment in healthcare organizations

RFID	Deployment authors
Tracking/identifying patients	[4, 8, 11, 19, 23, 31]
Tracking bags of blood, recording transfusions and ensuring the right match patient-blood	[8, 11, 20]
Tracking paths of infection sources	[27]
Tracking equipment, staff and documents	[11, 19, 23, 31]
Managing surgical instruments and associated process	[3, 11, 19]
Developing security systems	[5]
Tracking/monitoring medication therapy	[4, 11, 18]

Table 92.2 Infant abduction from healthcare facilities from 1983 to 2011, *Source* [17]

Abductions	Numbers (%)
From Mother's room	74 (58)
From nursery	17 (13)
From pediatrics	17 (13)
From "on premises"	20 (16)

aiming at optimizing instrument inventory, and patient safety. The surgical instrument cycle includes procurement, assembly, packaging, sterilization, storage, distribution, utilization in the surgical suite and other clinical settings, and the decontamination process [8]. The RFID also can be used to track the real composition of a sterile surgical kit, prior to the start of operations, allowing checking if there is any missing item after surgery [11]. Some of the main areas of RFID technology deployment in healthcare organizations can be found in Table 92.1.

Summing up, most of the Healthcare organizations deploy RFID systems in the following contexts: (1) to track/identify patients; (2) to track bags of blood to record transfusions and ensure that correct blood is given to each patient; (3) to track equipment, staff and documents; (4) to avoid thefts of medical equipment; and (5) to infant security systems. These applications are in line with search for improving healthcare safety. The application of RFID in infant security systems deserves a special highlight. According to the National Center for Missing and Exploited Children [17] as of this date, there have been 128 identified infants abducted from hospitals between 1983-April 2011 (Table 92.2). Of this number, five are still missing. It is interesting to note also that among these cases, 58 percent of the children were abducted from mother's room. Thus, it is obvious that the mother's room is where increased security efforts are needed. Infant tracking with RFID is becoming more and more common as hospitals in today's competitive environment realize the benefits [29].

92.3 Methodology

The main objective of this research is to illustrate the RFID technology deployment in infant security systems inside the hospitals, more precisely its main architectural characteristics, the way it works, and advantages.To attain the research objective, was used a two stage research methodology similar to the one used by Kumar and Shim [15]. In a first stage, three illustrative case studies were conducted based on secondary data using external sources, namely books, journals, business magazines and websites. Despite, the study was limited to the selected case studies and to the available data in external sources, this research design helps to define issues specifically before undertaking a primary study like an in-depth case study. Next, in a second stage, one case study in an RFID-based infant security system provider is conducted to obtain primary data related to the RFID deployment in Healthcare organizations. Using a methodology similar to Kumar and Shim [15] the secondary data for this research was gathered from the analysis of published literature based on a broad range of sources from Healthcare and RFID experts including newspapers, conference proceedings, industry reports, white papers, press releases and books. In addition, was used the specialized magazine on RFID: the RFID Journal. Selected articles describing case studies were analyzed and, finally, picked aspects are briefly described with a special focus on the main characteristics of the RFID-based infant security system, operational issues and advantages. The objective is not to offer further insight into the single cases, but to bring them together to get a wider picture and learn from the cross-case analyses. For the collection of primary data and to limit expert bias, the data concerned to personal judgment of the participants were obtained through semi-structured interviews.

92.4 Case Studies Findings

The John H. Stroger Jr. Hospital serves as one of 10 Illinois hospitals designated as prenatal centers for high-risk maternal and infant services. It has 460 beds, of which 322 are adult and/or pediatric care, 8 are burn intensive care, 34 are intensive care, and 24 are surgical intensive care. This hospital has thousands of authorized transfer of babies per week, between its general maternity wards and its intensive care, pediatrics and obstetrics facilities. To protect babies from abduction a parental/baby mismatching safety solution based on RFID technology is used particularly when babies are transferred between departments. The RFID-based infant security solution deployed by these hospital is the BabyMatch and consists of 150 active, long-range, supervised RFID baby and 150 corresponding mother tags that transmit safe radio frequency and coded messages. These transmissions are received by strategically placed readers, and then automatically passed onto the BabyMatch host computer. A standard Windows based touch-screen interface enables medical and security staff to monitor alarms; personalize tags, discharge babies, temporarily deactivate tags perform searches or follow the movements of a particular infant. BabyMatch's open standard architecture enables seamless integration with security, CCTV (Closed Circuit Television) access control systems as well as patient care and billing platforms.

At the time of birth both the infant and the mother are issued preconfigured personalized tags that are only removed upon discharge from the hospital. The baby's tag (attached to the infant's ankle) is paired to the mother's wrist tag, which enables mothers to confirm that babies they are with are their own.

The strategically placed radio frequency readers (wall or ceiling mounted) throughout the hospital are used to determine the exact room location of the infants in real-time. Additionally, low frequency exciters are mounted at entry and exit points inside the protected areas so when an unauthorized badged newborn is physically near the protected exit/entrance an alert notification is immediately generated by BabyMatch. This means that staff and family can move babies freely within the protected zones, but no infants can be removed without prior authorization.

The system ensures full supervision of each tag (including low battery conditions, device tampering, or if the location of the tag is unknown) from the time of birth through discharge. Also, many supplemental baby and mother tags can be added to the initial installation without risk to infant safety.

Babies requiring medical treatment from outside departments can be escorted through the protected exit by authorized staff members.

Baby tags are easily attached to the infant's ankle. If the tag is tampered with or removed from, a tamper 'State' alert will automatically be generated.

Prevents accidental baby switching by permitting a mother to confirm that the baby she is with is her own. Match tests can be performed by both parents and supports twins, triplets and other multiple births.

92.4.1 Case 2: New Hampshire Hospital

New Hampshire Hospital (NHH) provides psychiatric services to the people of Concord district. This hospital has 212 beds and deploys the infant security systems supplied by the Accutech entitled "Cuddles". New Hampshire Hospital is a state operated, publicly funded hospital providing a range of specialized psychiatric services.

The infant abduction prevention solution, based on active RFID technology, deployed in this is the Cuddles. This system works on 418 MHz frequency and has a portable STAD units control tag functions and a windows-based software with password protection. This software enables to activate tags, admit and discharge patients and generate reports. No enrolment is necessary for instant protection, multi-floor monitoring capabilities are available, and the software can be updated for free to ensure that it is always operating with the latest features and benefits.

The Cuddles Soft Bracelet is a light-weight, non-allergenic, self-adjusting band that fits snugly and comfortably around the ankle or wrist. Made of an ultra-soft polyester blend, the bracelet won't cut or chafe the skin, and won't fall off due to movement or changes in weight. In the event of removal or cutting, the Soft Bracelet immediately activates an alarm—preventing abductions and ensuring the continued safety and security of the infant wearing it.

A red, pulsating LED serves as a continuous visual indication that the tag is active. The tag can be turned on or off at any time to conserve battery power. If the tag is cut off or tampered with, it immediately locks down the perimeter and/or activates an alarm to alert staff. The bracelet design means greater comfort for the infant and easier cleaning for the nurse. The tag's small size and light-weight construction mean it will never hinder the infant's movement. The device provides universal activation capabilities, secured with unique user codes.

This infant security system is strapped through a band (usually around the ankle) that incorporates skin sensing technology in case the band is removed.The soft bracelet worn by infants is self-adjusting, preventing fall-off due to any post-birth weight loss.When the band is removed or cut from the baby's body, antennas placed throughout the facility pick up the alarm signal and relay it to a centralized alarm at the nursing station and on computer software. Usually the facility incorporates locks, and the hospital unit will go into a "lockdown mode" when a band alarm or attempted unauthorized exit occurs.

Some infant security solutions can also interface with existing systems in the hospital. For example, with the Cuddles system deployed in the New Hampshire Hospital, when the tagged infant's cradle approaches a door, a surveillance camera near the exit can be triggered. In addition to this, the alarm can be exported from the Cuddles software to nurse call pagers. The type of information that can be shown on the pagers is where the alarm is occurring, which baby is creating the alarm, and the type of alarm. The recommended code to activate the hospital emergency response to an infant abduction (or suspected abduction) is code pink. When this code is announced, there are specific actions of specific staff to control various access points and perform searches. This code, announced overhead, alerts all staff to watch for an infant being openly carried or signs of possible concealment (in back packs, bags, etc.). This code can be used to include pediatric patients, as well as a newborn, but generally carries an age identifier and/or abduction location.

The main advantages associated with the security system deployed in the New Hampshire Hospital are mainly the following ones:self-adjusting, soft bracelet, simple operation, quick patient assignment, no enrolment needed for instant protection, easy report generation, free on-site training, easy to use windows-based software, reusable, and easy-clean tags.

92.4.2 Case 3: Centro Hospitalar do Médio AVE

The Centro Hospitalar do Mdio Ave is a Portuguese Hospital which has 301 beds of which 101 are internal medicine, 71 are surgery, 10 are gynecology, 21 are obstetrics, 45 are orthopedics, 45 are pediatric/neonatology, and 8 intensive care.

The infant security system deployed in the Centro Hospitalar do Médio Ave is named Hugs and Kisses. It is constituted by hugs tags, receiver, exciters, security server software, and system manager software. It works in a standard Windows based PC environment. The hug tags are small radios that are attached to the infant with the tamper-proof strap. Receivers are radio frequency reception devices installed at regular intervals throughout the monitored area of the facility. They are installed in ceilings, usually out of view. Exciters monitor the exits from the safe area. Each exciter also includes two relays, which can be used to control a variety of devices, including magnetic door locks or audio and visual alarm devices. Like the receivers, exciters are continually monitored by the server software, and a warming message is automatically displayed if there is a problem. The security server software is installed on a server PC that is connected to the device network. As regards the system manager software, it is installed on the server and all client PCs. Also, it can be installed on any other computer with an Ethernet connection to the server PC.

The security system is based on RFID technology and has the following as main components: exciters, tags, receivers and controller PC. Every infant wears a HUGs tags on the ankle, and every exit point of the obstetrics unit is electronically monitored to detect the tags. This means staff and family can move infants freely within the protected Zone, but no one can remove an infant without the system alerting hospital staff. The Hugs tag contains a tiny radio transmitter. Once activated, the tag emits a special signal every 10 seconds. These signals are picked up by reception devices through the monitored area and relayed to the server PC via network. If a tag in close proximity to an open exit is detected by door monitors, an alarm occurs. The Hugs application software shows the tag ID number and indicates the exact location on a floor plan map of the facility. In addition, with the integrated CCTV option, the Hugs system automatically displays images from the exact CCTV camera when an alarm occurs, so that staff can respond with full knowledge of the situation. The Hugs systems can also support magnetic door locks, and can be interfaced with other hospital security systems such as pagers and alarm devices. The optional mother/infant matching component provides automatic matching of mothers and infants. Each time mother and baby are brought together, an audible signal will alert staff of a mismatch. The main advantages of this system are:

Any unauthorized person trying to take an infant from the bed nursery will set off an alarm 10 feet before they hit the exit door since if the alarm goes off everyone comes running; In operational terms, it is easy to attach the tag since it is done automatic enrolment; No manual checks is needed given that the system software continually monitors the status of all devices, and will generate an alarm if something goes wrong;

It also allows an automatic mother/infant matching; the system immediately confirms that the right baby is with the right mother.

There are no buttons to push, no numbers to match and no wall-mounted lamps to check.

It is user friendly given that the users only see the menus and commands they need, all in a standard Windows based PC environment.

92.4.3 Case 4: RFID-Based Infant Security System Provider

The experience of the Portuguese hospitals with the RFID technology is relatively recent and reduced in its scope. Among the seventy tree (73) hospitals that constitute the Healthcare system in the country in 2010 the following Healthcare organizations are using RFID-based infant security systems: Hospital da Luz, Hospital do Barreiro (Lisboa), Hospital S. Teotónio (Viseu), Hospital de S. João de Deus (V.N. Famalicão), Casa de Saúde da Boavista (Porto), Hospital de S. Marcos (Braga), Hospital dos Lusíadas (Lisboa), Centro Hospitalar de Trás-os-Montes e Alto Douro (Vila Real), Centro Hospitalar de Trás-os-Montes e Alto Douro (Chaves), Centro Hospitalar do Nordeste (Bragança). Beyond these hospitals other ones are actually using the same system, the technology supplier refers: (⋯)*"More three hospitals are in the introduction phase of this technology: Hospital Infante D. Pedro (Aveiro), Hospital da Cruz Vermelha Portuguesa (Lisboa) and Hospital Conde S. Bento (Santo Tirso)"*(⋯)

When asked to the system' supplier about the hospitals' past experience on this kind of technology, he says: (⋯)*"All the thirteen hospitals where the infant security system was deployed did not have any previous experience with this kind of systems"*(⋯)

As regards the time spend to install the RFID-based infant security system, he refers: (⋯)*"In average the installation process leaves one week. He said also that the system were requested by hospitals and not proposed by his firm"*(⋯)

How this infant security system works? The security system is based on RFID technology and has as main components the following: exciters, tags, receivers and controller PC. The exciters monitor the exits from the safe area. The hug tags incorporate a tamper mechanism that is enabled as soon as the tag is attached with the tamper proof strap. The receivers receive hugs tag transmissions, time stamp them and relay them to the controller PC. The controller PC contains the Hugs system software and controls the operation of the entire system. Besides these components, he says: (⋯)*"The infant security system has beyond these components also electromagnets, sirens with flash; strobe) and doors' magnetic contacts"*(⋯)

Unlike bar codes, an RFID chip can be sensed from many feet away and without human intervention. Sensors, for example, in the ceiling detect a chip that is embedded in a baby's wristband, triggering an alarm if the child is in an off-limits zone, or prompting a jingle when the baby comes close to its mother's pre-programmed RFID band.

Every infant in a medical unit wears a tag with a unique ID number on the ankle, and every exit point is electronically monitored to detect the tags. This means staff and family can move infants freely within the protected zone, but no one can remove an infant from the unit without the organisation' staff being alerted. Beyond this superior and active supervision of infants, the system monitors its own functionality and alerts staff of any problems.

In the event of an alarm, the system can automatically activate magnetic door locks or hold an elevator. It can also integrate with and activate other security and access control systems, such as alpha-numeric pagers and cameras. When asked the interviewed about the situations in which the alarm goes on, he said: (· · ·)*"The alarm goes on in the following situations: (i) someone tries to exit via a monitored door or elevator with a protected infant, without authorization; (ii) the strap has been cut or tampered with; (iii) the tag's signal has not been detected by the system for a specified time period, (iv) the tag's battery power is low; (v) an authorized exit has occurred but someone tries to "piggyback" through the protected exit with another infant; and (vi) an authorized exit has occurred but the infant has not been returned to the designated safe area in the specified time."*(· · ·)

For security purposes, all system transactions are password controlled, time and date stamped and logged into the database on the system controller. A permanent record is possible of who is admitted, signed out and discharged of all babies. Also have a record of when and where alarms occurred and who cleared them can be obtained. The system controller itself is equipped with a watchdog timer card to output an alarm signal in the unlikely event of a problem with the operating system, providing an extra level of security. When mother and baby are brought together with a correct match, a pleasant lullaby will sound from tag. An incorrect match generates a buzzing tone.

As regards the interoperability of the system, he said: (· · ·)*"The Hugs system's advanced radio frequency technology not affects or be affected by other electronic hospital equipment which represents an important advantage."*(· · ·)

Each tag has a unique code to ensure easier identification of every infant, the strap is easily "snagged-up" to accommodate weight loss and extremely durable and the tags are also reusable and waterproof which allows its permanence with the baby all over the time.Also, a full supervision is granted by the system. Beyond the advantages pointed out, he refers: (· · ·)*"The deployment of this system in hospitals allows these organisations to improve its image, customer service and babies' flow control, to decrease the number of babies' shrinkages and thefts and also to get more durable tags."*(· · ·)

92.4.4 Cross Case Findings

After the four within-case analyses it is important to do a cross-case analysis in order to identify similarities and differences in the RFID-based infant security systems deployed in the hospitals and systems provider. To illustrate how the RFID-based infant security system works the Fig. 92.1 is drawn from the evidences collected in the case studies.

First, every infant wears a tag on the ankle with a unique ID which is can be paired to the mother's wrist tag; this functionality is optional in Hospital of S. João de Deus and RFID provider. If the band is removed from the baby, an unauthorised exit occurs, or the tag's signal is not detected by the system, antennas placed throughout the facility pick up the alarm signal and relay it to a centralized alarm on computer software. After this, the tag emits a signal to the reception devices through the monitored area and an alarm occurs. Following, the system automatically displays

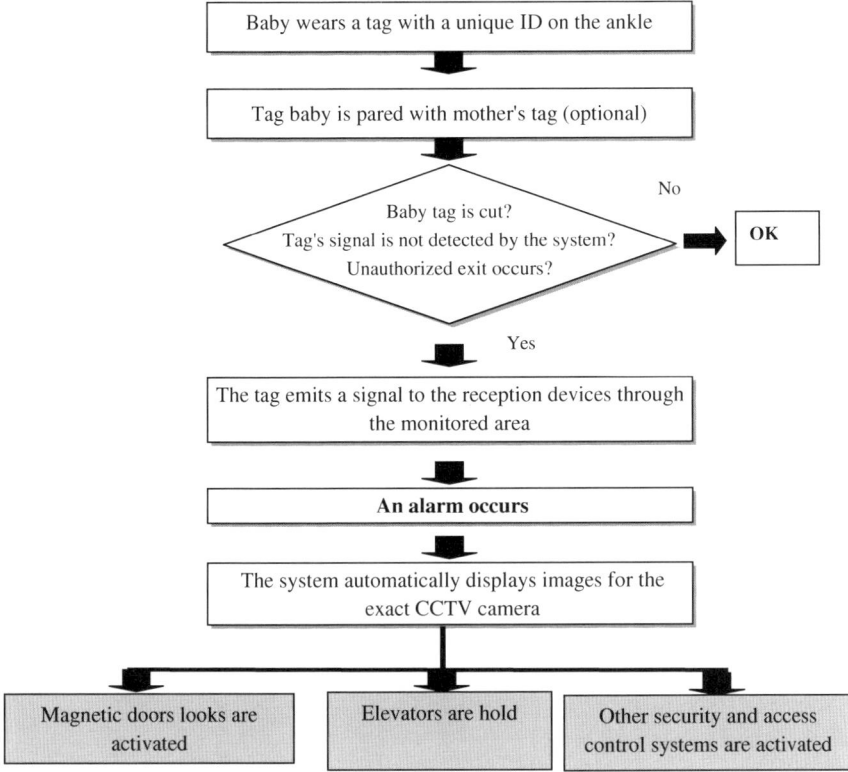

Fig. 92.1 RFID-based infant security system overview

images for the exact CCTV camera, activates magnetic door looks and other security and access control systems, and holds elevators. To identify the main architectural characteristics, functionality, and advantages associated with the RFID deployment, the evidences in four case studies were compiled in Table 92.3.

Summing up, doing a cross-case analysis it is possible to state that the RFID-based infant security systems deployed among the research hospitals and RFID provider are very similar. They have the following main components: exciters, tags, receivers, controller PC, CCTV and windows-based software. Also, the way they work it is quite similar. A unique identification tag is associated with each baby, with the ability of transmit and receive radio frequency signals. The RFID security system is supported by windows-based software being easily interfaced with others conventional security systems (e.g. CCTV camera or activating magnetic door looks). The system automatically triggers alarms when the tag is out of the safety/secured zone or if the tag is removed without authorization.

Table 92.3 Cross-case analysis of the infant security systems deployed in the four case studies

		John H. Stroger Jr. Hospital	New Hampshire Hospital	Centro Hospitalar do Médio AVE	RFID provider
Who?	Babies	✓	✓	✓	✓
	Mother	✓		✓(optional)	✓(optional)
How?	RFID tag in patients (transmit and reception)	✓	✓	✓	✓
	Strategic position of receivers (hospital departments exit and entry points)	✓		✓	
	Interface with others existent security system	✓	✓	✓	✓
	Automatic alarm if tag is removed without authorization	✓	✓	✓	✓
	Automatic alarm if the baby is out of the authorized zone	✓	✓	✓	✓
Advantages	Automatic supervision of each tag maintenance state	✓		✓	
	Easy and safe use of the tag in babies and mothers	✓	✓		
	Babies can be escorted by authorized staff members	✓			
	Avoid babies switching	✓		✓	
	Easy scalable for multiple births	✓			✓
	Quick patient assignment		✓		
	Easy to use the RFID without extensive training	✓	✓	✓	✓
	Reduced implementation time				✓
	Real time location of the patients			✓	
	It not interfere with other hospital equipment				
	Improve hospital image				✓
	Improve customer service and babies flow control				✓
	Improve babies safety				✓

Despite the enormous improvements in the children safety, one main advantage of the RFID-based infant security system recognized by the research case studies is the simplicity of operations associated with the tag attachment since it is done an automatic enrolment. This advantage reflects the perspective of the hospitals professionals which had to adapt their daily routines to the RFID system deployment.

92.5 Conclusion

In a business context the RFID technology has reached many adepts by the huge potentials that it presents for organisations. The RFID technology has received also considerable attention from academics and practitioners because of its potentialities and diverse fields of use in organisations such as: manufacturing, transportation, distribution, information systems, Healthcare, and others. The increased use of the RFID has been pointed out by several kinds of organisations because of the advantages reached with its use. It has been recognized that the adoption of this technology in Healthcare organizations allows a better patient flow management, improves the organisation' productivity, reduces human errors, speeds the data access and multiple item identification, allows the automation of some process activities. The main disadvantage signalled is its cost. The RFID-based security system deployed in Healthcare organizations beyond others applications intend to prevent infant abductions and inadvertent child mishandlings in hospitals.After the case studies analysis it is possible to state that the infant security systems deployed in Healthcare organizations with different characteristics and sited in diverse countries are not so different. In all case studies the infant security system involves RFID tagging patients. Also, the researched Healthcare organisations had implemented RFID solutions with an interface with others security systems and if the baby is out of the authorized zone an automatic alarm goes off. As regards the advantages pointed out, all of them highlight the RFID easy to use not requiring an extensive training. Besides the Healthcare organizations had already wake up to the potentialities of the RFID Technology in some specific applications there are however other medical services and valences that could be improved through the RFID technology. The Joint Commission (2007) has signed up some errors that must be avoided in any kind of Healthcare organization such as: (i) patient care hand-over errors; (ii) wrong site and procedures; (iii) wrong person surgical errors; (iv) medication errors; and (v) high concentration drug errors. Also to overcome these errors the RFID technology could be the answer. Being so, we propose to study the influence of the RFID application in these areas on the performance of Healthcare organizations. Another suggestion is to extend this investigation to other departments inside hospitals.

References

1. Azevedo S, Ferreira J (2008) The rfid as an innovative technology in retailing: a case study. J Bus Retail Manage Res 3:16–26
2. Borck J (2006) Tuning in to RFID. InfoWorld 28(16): 31–36
3. Brooke M (2005) RFID best practices, part 1: addressing problem unknown, Information management magazine. http://www.information-management.com/issues/20050801/1033579-1.html. Accessed Jan 2011
4. Cerlinca T, Turcu C, et al. (2010) RFID-based information system for patients and medical staff identification and tracking. In. In Turcu C (ed) Sustainable Radio Frequency Identification Solutions, InTech
5. Collins J (2005) RFID delivers newborn security. RFID J. http://www.RFIDjournal, com/article/purchase/1372. Accessed Jan 2011
6. Delen D, Hardgrave B, Sharda R (2007) Rfid for better supply-chain management through enhanced information visibility. Prod Oper Manage 16:613–624
7. Fisher JA, Monahan T, (2008) Tracking the social dimensions of RFID systems in hospitals. Int J Med Informatics 7:176–183
8. Fuhrer P, Guinard D (2006) Building a smart hospital using RFID technologies. In: 1st European conference on eHealth (ECEH06), Fribourg Switzerland
9. Furukawa M, Raghu T et al (2008) Adoption of health information technology for medication safety in us hospitals. Health Aff 27:865–875
10. Holmqvist M, Stefansson G (2006) Smart goods and mobile rfid: a case with innovation from volvo. J Bus Logistics 27:251–259
11. Iadanza E (2009) RFID technologies for the hospital, how to choose the right one and plan the right solution? In: Recent advances in biomedical engineering, pp 519–536
12. Jones E, Riley M et al (2007) Case study the engineering economics of rfid in specialised manufacturing. Eng Econ 52:285–303
13. Kelly E, Scott P (2005) RFID tags: commercial applications v privacy rights. Ind Manage & Data Syst 105:703–715
14. Kevan T (2004) Calculating rfid's benefits. Frontline Solutions 5:16–21
15. Kumar A, Shim SJ (2009) Centralization of intensive care units: process reengineering in a hospital. Int J Eng Bus Manage 1:49–54
16. Meyerson J (2007) Rfid in supply chain: a guide to selection and implementation new york study of the adoption, usage and impact of rfid. Inf Technol Manage 8:87–110
17. NCMEC (2001) Infant abduction by state, from 1983 to 2011. http://www.missingkidscom/en_US/documents/InfantAbductionStatspdf. Accessed Jan 2011
18. Parks L (2003) New microchip watchdog could boost patient compliance. Drug Store News 25(7): 26
19. Qu X, Simpson LT, Stanfield P (2011) A model for quantifying the value of rfid-enabled equipment tracking in hospitals. Adv Eng Inform 25:23–31
20. Roberts S (2004) When the supply chain becomes a matter of life and death. Frontline Solutions 12:14–16
21. Rundh B (2008) Radio frequency identification (rfid) invaluable technology or a new obstacle in the marketing process? Mark Intel & Plann 26:97–114
22. Saad MK, Ahamed SV (2007) Vulnerabilities of rfid systems in infant abduction protection and patient wander prevention. SIGCSE Bull 39:160–165
23. Sini E, Locatelli P, Restifo N (2008) Making the clinical process safe and efficient using rfid in healthcare. Eur J e Pract 2:1–18
24. So S, Liu J (2006) Securing rfid applications: issues, methods, and controls. Inf Syst Secur 15:43–56
25. Sun P, Wang B, Wu F (2008) A new method to guard inpatient medication safety by the implementation of rfid. J Med Syst 32:327–332
26. Turcu C, Graur A (2009) Improvement of supply chain performances using RFID technology. In: Huo Y, Jia F (ed) Supply chain the way to flat organisation, InTech, Rijeka

27. Tzeng S, Chen W, Pai F (2008) Evaluating the business value of rfid: evidence from five case studies. Int J Prod Econ 112:601–613
28. Vanany I, Shaharoun AB (2008) Barriers and critical success factors towards RFID technology adoption in south-east asian healthcare industry. In: Proceedings of the 9th Asia pasific industrial engineering and management systems conference, Bali, Indonesia
29. Wang K, Lin N, Liu C (2009) Infant management system based on RFID and internet technologies. In: Proceedings of the 2009 first IEEE international conference on information science and engineering ICISE '09, Nanjing, China
30. Want R, Fishkin K et al (1999) Bridging physical and virtual worlds with electronic tags. In: CHI '99 Proceedings of the SIGCHI conference on human factors in computing systems: the CHI is the limit, Pittsburgh, Pennsylvania, USA
31. Yao W, Chu C, Li Z (2010) The use of RFID in healthcare: benefits and barriers, RFID-technology and applications. In: 2010 IEEE international conference on RFID-technologies and applications (RFID-TA), Guangzhou, China
32. Zhou W, Piramuthu S (2010) Framework, strategy and evaluation of health care processes with rfid. Decis Support Syst 50:222–233

Chapter 93
Performance Evaluation of China's Performing Arts Groups from the Perspective of Totality and Subdivision

Xiaoyi Huang, Yongzhong Yang and Lin Zhong

Abstract Performing arts groups' performance is a focused issue. Use method of DEA-Malmquist productivity index to evaluate these groups' performance in 2003–2011. Results show that although with progressive performing technology, our performing arts groups' total factor productivity (TFP) has grown by 2.8 %, there are highlight problems of depending on governmental subsidy and support excessively, inadequate risk resistance capacity and unbalanced development. So these groups should emancipate and develop productivity, deepen internal system reform, enhance management level and self-financing rate and promote coordinated development of performing arts groups of diversified ownership, different levels and different types.

Keywords Performance evaluation · Performing arts groups · Data envelopment analysis (DEA) · Total factor productivity (TFP)

93.1 Introduction

1. Summarize

With the rapid development of productivity, people live and work in peace and contentment, and begin to pursue the cultural life of rich and colorful continuously. In recent years, China's performing arts groups at all levels which integrated with market have begun to develop artistic creation and production. They have created remarkable social and economic benefits, and played an important role in booming performing market, spreading advanced culture, promoting the quality of the whole people. They have been driving the development of the socialist cultural construction.

X. Huang · Y. Yang (✉)
Business School, Sichuan University, Chengdu 610064, People's Republic of China
e-mail: yangyongzhong116@163.com

X. Huang · L. Zhong
Sichuan College of Architectural Technology, Deyang 618000, People's Republic of China

J. Xu et al. (eds.), *Proceedings of the Eighth International Conference on Management Science and Engineering Management*, Advances in Intelligent Systems and Computing 281, DOI: 10.1007/978-3-642-55122-2_93, © Springer-Verlag Berlin Heidelberg 2014

In the field of performance evaluation in cultural industry, Lan and Han [4] had used data envelopment analysis (DEA) and Malmquist productivity index to analyze the business efficiency and its changes of China's performing arts groups from 2004–2009 in their paper "The Efficiency's Present Situation and the Evaluation of the Dynamic Efficiency of China's Performing Arts Groups". They found most groups were unbalanced and had big internal differences, serious overcrowding input and deficient output.

Guo and Zheng [3] had used data envelopment analysis (DEA) technology to evaluate developmental performance of six provinces in central China's cultural industry, and used structural equation model to simulate the input-output acting path of cultural industry in their paper "The Evaluation and Research of Developmental Performance in Six Provinces of Central China's Cultural Industry". Combining with the research conclusion, they offered proposals to six provinces in central China's cultural industry.

Wang and Zhang [11] found that cultural industry's technical efficiency and scale efficiency was low, the regional difference was obvious, the environmental factors had a significant influence on regional cultural industry's development in their paper "The Efficiency Study of Cultural Industry in 31 Provinces of China Based on Three Stages DEA Model".

2. Sources of Data and Innovation

Performing arts groups are cultural institutions which are sponsored by cultural department or managed by industry. They specialize in performing arts activities and contain Chinese opera, drama, song and dance drama, puppet, shadow puppet and many other art categories. They are an important part of China's cultural system. On April 1, 2004, "the Classification of cultural and related industry" which marked China's cultural industry had the first statistical standard was formally issued by the National Bureau of Statistics [9].

On the basis of looking up "China Statistical Yearbook" [8], from the perspective of totality and subdivision, this paper use art performance groups' total expenditure and the number of institutions as input variables, use the performance session and total income as output variables, use Out-orientated Malmquist DEA Model and DEAP 2.1 software developed by Coelli to analyze the total factor productivity change in different periods of our performing arts groups. Finally, evaluate their performance and put forward reasonable suggestions for their development.

93.2 Research Model and Method

1. Brief Introduction of DEA Model

In 1978, the DEA method was proposed by Charnes et al. for the first time [12]. Using mathematical programming model, this method is a typical nonparametric method to calculate decision making units' relative efficiency simultaneously which have different units of measurement and multiple inputs and multiple outputs.

C^2R model is a typical model in DEA model. This model assumes that there are n decision making units (DMU), each DMU_j $(j = 1, 2, \cdots, n)$ has m types of input and s types of output. If input vector is expressed as X_j, output vector is expressed as Y_j, the input vectors and output vectors are expressed respectively as $X_j = (X_{1j}, X_{2j}, \cdots, X_{mj})$, $Y_j = (Y_{1j}, Y_{2j}, \cdots, Y_{sj})$, $(j = 1, 2, \cdots, n)$. For a selected DMU_0 (input vector: X_0, output vector: Y_0), the C^2R model's dual program to determine the effectiveness is model (93.1) [10]:

$$\min \theta$$

$$\text{s.t.} \begin{cases} \sum_{j=1}^{n} X_j \lambda_j + S^- = \theta X_0 \\ \sum_{j=1}^{n} Y_j \lambda_j + S^+ = \theta Y_0 \\ \lambda \geq 0, \quad j = 1, 2, \cdots, n \\ S^+ \geq 0, \quad S^- \geq 0. \end{cases} \tag{93.1}$$

The decision making unit DMU_0's effective value is expressed as θ which means the degree of effective utilization of input relative to the output. Relative to DMU_0, λ_j is combination ratio which is the number j decision making unit in a new effective combination of decision making unit. $S^- = [S_1^-, S_2^-, \cdots, S_m^-]$ and $S^+ = [S_1^+, S_2^+, \cdots, S_m^+]$ are slack variables. The former is redundant input and the latter is insufficient output. When $\theta = 1$ and $S^- = S^+ = 0$, DMU_0 is effective DEA, that is, in the economic system composed of n decision making units, the output Y_0 has reached optimum on the basis of the original input X_0. When $\theta = 1$ and $S^- \neq 0$ or $S^+ \neq 0$, DMU_0 is weak effective DEA. It means that if reduce S^- in the input X_0, the original output Y_0 can still be got or keeping the original input X_0 remains the same, the original output Y_0 can increase S^+. When $\theta < 1$, DMU_0 is non effective DEA [2].

2. Brief Introduction of Malmquist DEA Model

In 1953, Malmquist index was put forward by Swedish economist and statistician Malmquist for the first time [7]. In 1982, based on distance function, Caves Christensen and Diewert constructed Malmquist productivity index which was used for measuring the TFP change and widely used in measuring the productivity's growth. In 1994, Fare proposed the DEA-Malmquist method [6] which opened up a new field for the performance evaluation method. In this method, (x_t, y_t) was input and output vector in period t. (x_{t+1}, y_{t+1}) was input and output vector in period $(t + 1)$. In the technical condition of period t, the distance function of period t production point was $d_o^t(x_t, y_t)$ and $d_o^t(x_{t+1}, y_{t+1})$ was the distance function of period $(t + 1)$ production point. The subscript "o" of $d_o^t(x_t, y_t)$ mean that the Malmquist productivity index was calculated based on the out-oriented DEA model. Therefore, in the technical condition of period t, the out-oriented Malmquist productivity index was expressed as $m_o^t(y_{t+1}, x_{t+1}, y_t, x_t) = d_o^t(x_{t+1}, y_{t+1})/d_o^t(x_t, y_t)$. Similarly, in the technical condition of period $(t + 1)$, the out-oriented Malmquist productivity index was expressed as $m_o^{t+1}(y_{t+1}, x_{t+1}, y_t, x_t) = d_o^{t+1}(x_{t+1}, y_{t+1})/d_o^{t+1}(x_t, y_t)$.

To avoid the difference due to the arbitrary period choosing, in 1982, according to the structure method of Fisher's ideal index, Caves, Christensen and Diewert used the geometric mean of two periods' Malmquist index as Malmquist productivity index: $m_o^t(y_{t+1}, x_{t+1}, y_t, x_t)$, which was used to measure the productivity change of period t to period $(t+1)$, and was also referred to as the TFP change (TFPch), Eq. (93.2) [1]:

$$\text{TFPch} = m_o^t(y_{t+1}, x_{t+1}, y_t, x_t) = \left[\frac{d_o^t(x_{t+1}, y_{t+1})}{d_o^t(x_t, y_t)} \times \frac{d_o^{t+1}(x_{t+1}, y_{t+1})}{d_o^{t+1}(x_t, y_t)} \right]^{1/2}. \tag{93.2}$$

How to calculate the Malmquist productivity index? Firstly, calculate $\max_{\phi, \lambda} \phi$ through Out-orientated VRS DEA Model: formula (93.3). Secondly, calculate four distance functions including four linear programs [1]. Lastly take these four distance functions into Eq. (93.2).

$$\max_{\phi, \lambda} \phi$$
$$\text{s.t.} \begin{cases} -\phi y_i + Y\lambda \geq 0 \\ x_i - X\lambda \geq 0 \\ N1'\lambda = 1 \\ \lambda \geq 0, \end{cases} \tag{93.3}$$

$$[d_o^t(x_t, y_t)]^{-1} \max_{\phi, \lambda} \phi$$
$$\text{s.t.} \begin{cases} -\phi y_{it} + Y_t\lambda \geq 0 \\ x_{it} - X_t\lambda \geq 0 \\ \lambda \geq 0, \end{cases} \tag{93.4}$$

$$[d_o^{t+1}(x_{t+1}, y_{t+1})]^{-1} \max_{\phi, \lambda} \phi$$
$$\text{s.t.} \begin{cases} -\phi y_{i,t+1} + Y_{t+1}\lambda \geq 0 \\ x_{i,t+1} - X_{t+1}\lambda \geq 0 \\ \lambda \geq 0, \end{cases} \tag{93.5}$$

$$[d_o(x_{t+1}, y_{t+1})]^{-1} \max_{\phi, \lambda} \phi$$
$$\text{s.t.} \begin{cases} -\phi y_{i,t+1} + Y_t\lambda \geq 0 \\ x_{i,t+1} - X_t\lambda \geq 0 \\ \lambda \geq 0, \end{cases} \tag{93.6}$$

$$[d_o^{t+1}(x_t, y_t)]^{-1} \max_{\phi, \lambda} \phi$$
$$\text{s.t.} \begin{cases} -\phi y_{i,t} + Y_{t+1}\lambda \geq 0 \\ x_{i,t} - X_{t+1}\lambda \geq 0 \\ \lambda \geq 0. \end{cases} \tag{93.7}$$

The change of the Malmquist productivity index was decomposed into technical change (TECHch) and efficiency change (EFFch) by Nishimizu and Page in 1982.

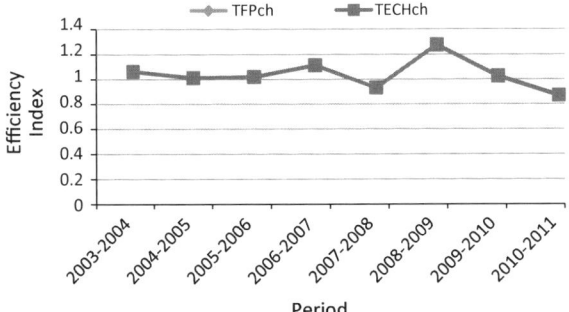

Fig. 93.1 Technical change and total factor productivity's change of performing arts groups of China (2003–2011)

In 1994, efficiency change was further decomposed into pure technical efficiency change (PEch) and scale efficiency change (SEch). Therefore, Eq. (93.2) could be further decomposed into Eq. (93.8) [1, 5]. "TFPch> 1" expresses that from period t to period $(t + 1)$, the TFP has improved. On the contrary, "TFPch< 1" expresses that the TFP has declined.

$$TEPch = TECHch \times EFFch = TECHch \times PEch \times SEch. \qquad (93.8)$$

93.3 Performance Evaluation

1. Overall Performance Evaluation of Performing Arts Groups of China
According to Fig. 93.1, affected by technical progress, the average growth rate of China's performing arts groups' TFP was 2.8 %. In 2008 and 2011, under the influence of technical retrogression, the year-on-year TFP declined by 6.9 and 12.8 % respectively. Of CNY 15.40263 billion grossed by our performing arts groups in 2011, CNY 8.971 billion was governmental subsidy which accounted for 58.24 % of the total income. CNY 1.164 billion was earned by other ways which accounted for 7.56 % of the total income. Performing income accounted for only 34.2 % of the total. At the same time, in 2011, our performing arts groups' total expenditure was CNY 14.59 billion which was an increase of 110.98 % over the previous year. The range of total expenditure growth was greater than the range of total income growth.

It is thus clear that the productivity growth of China's performing arts groups largely depend on national financial and technical support. As an important indicator to measure self-hematopoietic capacity, our performing arts groups' self-sufficiency rate of funds is still in a relatively low level.

Table 93.1 Total factor productivity's change and it's composing of China's performing arts groups of diversified ownership (2008–2011) (Unit, %)

Group category	EFFch	TECHch	PEch	SEch	TFPch
State-owned	0.932	1.058	1.000	0.932	0.986
Collective	0.944	1.058	1.013	0.932	0.999
The other	1.000	0.667	1.000	1.000	0.667
Mean	0.958	0.907	1.004	0.954	0.869

Note
The result data is calculated by DEAP v2.1 software and all the means of the Malmquist index are geometric means
Calculating data source: "China Statistical Yearbook" (2004–2012)
TFPch TFP Change, *TECHch* Technical change, *EFFch* Efficiency change, *PEch* Pure technical efficiency change, *SEch* Scale efficiency change

2. Performance Evaluation of China's Performing Arts Groups of Diversified Ownership

From Table 93.1, the technical progress rate of our state-owned performing arts groups was 5.8%. Affected by 6.8%'s decrease of the scale efficiency rate, TFP declined by 1.4%. Collective performing arts groups' technical progress rate was 5.8%, the pure technical efficiency rate increased by 1.3%, the scale efficiency rate declined by 5.6%, and the TFP declined by 0.1%. Other performing arts groups' scale efficiency was unchanged, only affected by technical change, the TFP declines by 33.3%.

In 2002, the "Implementing Rules for Management Regulations on Commercial Performances" which was revised by Ministry of Culture cancelled the ownership constraint of producer's right as principal. In 2005, Ministry of Culture, Ministry of Finance, Ministry of Personnel and State Administration for Taxation jointly issued a document "Opinions about Encouraging the Development of Private Literary and Artistic Performing Groups" which reflected loose market access policy and appeared a coexisting situation of performing arts groups of diversified ownership. By the end of 2010, in 2086 state-owned performing arts groups which undertook the task of reform in national cultural system, 461 groups had become enterprises [9]. The completion rate was 22.1%.

According to the requirements of "market oriented, vitality enhanced", our reformed performing arts groups explored and established a work mechanism under the new situation to promote the development of these groups. In the aspect of economic and social benefits, they have achieved new progress. Although the scales of state-owned and collective performing arts groups shrink in different degrees, compared with other performing arts groups, the state-owned and collective performing arts groups have stronger strength on technical renovation and progress. We should see that the fundamental reform of our performing market's operation and management system has not yet been completed. When the market environment changes, such as financial crisis occurs, the disadvantages of poor ability to resist risk will be revealed, especially the other performing arts groups of non-state-owned and non-collective ownership.

Table 93.2 Total factor productivity's change and it's composing of China's performing arts groups at different levels (2008–2011) (Unit, %)

Group category	EFFch	TECHch	PEch	SEch	TFPch
Period 2008–2009					
Central level	1.000	1.595	1.000	1.000	1.595
Provincial level	0.932	1.587	1.000	0.932	1.479
Prefectural level	0.903	1.566	0.981	0.921	1.415
County level	1.000	1.191	1.000	1.000	1.191
Mean: 2008–2009	0.958	1.474	0.995	0.962	1.412
Period 2009–2010					
Central level	0.808	0.975	1.000	0.808	0.789
Provincial level	1.104	0.982	1.000	1.104	1.085
Prefectural level	1.057	0.985	0.968	1.092	1.040
County level	1.000	1.025	1.000	1.000	1.025
Mean: 2009–2010	0.986	0.992	0.992	0.994	0.977
Period 2010–2011					
Central level	0.961	0.729	1.000	0.961	0.701
Provincial level	0.793	0.729	0.849	0.934	0.578
Prefectural level	0.900	0.737	0.860	1.046	0.664
County level	1.000	0.677	1.000	1.000	0.677
Mean: 2010–2011	0.910	0.718	0.925	0.984	0.653

Note
The result data is calculated by DEAP v2.1 software and all the means of the Malmquist index are geometric means
Calculating data source: "China Statistical Yearbook" (2004–2012)
TFPch TFP Change, *TECHch* Technical change, *EFFch* Efficiency change, *PEch* Pure technical efficiency change, *SEch* Scale efficiency change

3. Performance Evaluation of China's Performing Arts Groups at Different Levels

From Table 93.2, to abate the impact of the situation of external economy and natural disasters in 2008, the performing arts groups at different levels were supported greatly in 2009. The TFP which was driven by the technology improved greatly. The year-on-year TFP growth of performing arts groups at central, provincial, prefectural and county level was 59.5, 47.9, 41.5 and 19.1 %, respectively. But the situation of strong growth failed to get the continuation, especially the central performing arts groups. Under the influence of the structural reform of the national cultural system, the central performing arts groups' TFP which was affected by scale efficiency declined by 21.1 %. In 2011, this weak situation continued in performing arts groups of China at all levels, especially the performing arts groups at provincial level which affected greatly by the scale efficiency and technical efficiency (Table 93.3).

At the same time, we should see the unequal development of our performing arts groups at all levels. The central, provincial and prefectural performing arts groups' financial strength is relatively strong. They have excellent equipment, advanced technology and a galaxy of talents. Although the county troupes have outstanding subject status of the cultural activities of the grass-roots, such as the proportion which

Table 93.3 Total factor productivity's change and it's composing of China's performing arts groups of different types (2005–2011) (Unit, %)

Group category	EFFch	TECHch	PEch	SEch	TFPch
Drama, child play, comic troupe	1.020	1.043	1.000	1.020	1.064
Opera, ballet, song and dance drama troupe	1.006	1.033	1.000	1.006	1.039
Song and dance troupe, light music troupe	1.019	1.018	1.000	1.019	1.037
Orchestra and chorus	0.992	1.019	1.000	0.992	1.010
Art and cultural troupe, cultural propaganda team, A Nei Monggol revolutionary cultural troupe Mounted on Horseback	1.002	1.009	1.000	1.002	1.010
Chinese opera troupe	1.004	1.046	1.000	1.004	1.050
Beijing opera	1.000	1.050	1.000	1.000	1.050
Folk arts, acrobatics, puppetry, shadow play troupe	1.000	0.934	1.000	1.000	0.934
Comprehensive performing arts groups	1.017	1.054	1.010	1.007	1.071
Mean: 2005–2011	1.006	1.022	1.001	1.005	1.029

Note
The result data is calculated by DEAP v2.1 software and all the means of the Malmquist index are geometric means
Calculating data source: "China Statistical Yearbook" (2004–2012)
TFPch TFP Change, *TECHch* Technical change, *EFFch* Efficiency change, *PEch* Pure technical efficiency change, *SEch* Scale efficiency change

accounts for nationwide institutions rose from 74.2 % in 2007 to 84.9 % in 2010 [9], the majority of the county troupes have small scale, inadequate investment, talent and funds, backward technology and equipment, backward artistic production ability and management level. For lack of funds, part of the performing groups at the county level unable to rehearse large program and purchase advanced performing equipments. In addition, the existing venues cannot satisfy normal performance due to the informal performing venues and the existing obsolete facilities. It is thus clear that represented by performing arts groups at the county level, our country's small and medium-sized performing arts groups still rely on the support of national finance and technology. At the same time, the fare of the performing market in China which is much higher than the developed countries and make ordinary people especially the low-income people at the county level cannot afford also affect the development of performing arts groups.

4. Performance Evaluation of China's Performing Arts Groups of Different Types
The document "Opinions about Encouraging the Development of Private Literary and Artistic Performing Groups" in 2005 which has the significance of milepost is the first document to encourage the development of private performing arts groups since the establishment of People's Republic of China. According to the documental spirit, with flexible and varied ways of management, distinctive local style, energetic marketing mechanism and hard working spirit, our private performing arts groups perform a lot of lively plays. In 2011, in 7,055 nationwide performing arts groups, the number of non-cultural departmental groups was 4,639 [9] which account for

65.8 % of the total number of national performing arts groups. These groups cover drama, folk arts, song and dance drama, acrobatics, magic, circus, puppetry, shadow play and many other art categories.

The diversification of the culture promotes the structural adjustment of the masses' cultural consumption and enriches the cultural life. But the performing arts groups of different types also present unbalanced development. For example, traditional opera performing groups have redundant construction and descending scale efficiency, while new forms of performing arts are obviously insufficient in innovation and market cultivating.

93.4 Conclusions and Recommendations

Firstly, our performing arts groups should further emancipate and develop productivity, and improve the level of management.

Our performing arts groups should face the needs of the market, use the audience's demand as the starting point and ending point, exploit performing works to meeting the demands of market and introduce the achievements of modern science and technology into performing market and each link of the artistic production, management and service. They should also establish scientific system of organization and management, make great efforts to cultivate a large number of marketing team in which there are a lot of people who are good at management, carry out marketing activities actively, enhance the self-financing rate and the proportion of performing income and optimize the allocation of resources. Therefore, form a production and operation mechanism to adapt to the requirements of market economy and strengthen the ability to participate in market competition.

Secondly, our performing arts groups should deepen the reform of cultural system and increase the intensity of becoming enterprises.

The basic characteristic of modern enterprises system is clearly established ownership, well defined power and responsibility, separation of enterprise from administration, and scientific management. Therefore, separating enterprises from administration, making most of the performing arts groups become enterprises, deepening the internal reform of cultural system and making them become independent, self-financing legal entity and main body of market competition are reasonable choice. State-owned performing arts groups should expand the autonomy in management in order to enhance the market competitiveness. In addition, through asset reorganization, collective groups and many other systems of ownership should be developed coordinately.

Finally, our performing arts groups should encourage and support the coordinated and balanced development of the groups of different sizes and types.

Performing arts groups at central, provincial and prefectural levels which have relatively strong strength should centralize superior resources and strength, and according to the requirement of the intensive management, actively make a trans-regional, trans-departmental performing groups, such as "Northern Union" and "Western Union" which have taken a welcome step in operation of the performing industry collectivization [9]. On the basis of doing well in domestic performance, performing arts groups at central, provincial and prefectural levels should actively seize the moment, go out and introduce in, actively learn the advanced performing technology of international performing arts groups and introduce excellent foreign performing equipments and experience. While occupying the international market, improve their international competitiveness. Based on the current situation, performing arts groups at the county level should make effective use of resources and funds, actively cultivate and introduce technical personnel and management talent, capture the position of public consumption and make great efforts to develop the performing markets of small city and countryside. From the small theater stage to the grassroots and rural big society stage, performing arts groups at the county level should explore the new way of market-oriented operation and form an industrialization operation pattern which contains artistic originality, programming, performance and marketing. In addition, while promoting and carrying forward the traditional Chinese art, we should vigorously encourage the development of the different art types and achieve cultural boom fast and stably in China.

Acknowledgments We gratefully acknowledge the research support received from the National Natural Science Fund (71173150), the Key Project of the National Social Science Fund (12AZD018), Program for New Century Excellent Talents in University (NCET-12-0389) and the key project of System Science and Enterprise Development Center of Sichuan Province (XQ12A01).

References

1. Coelli T (1996) A guide to DEAP version 2.1: a data envelopment analysis (computer) program. CEPA Working Paper 96/08, University of New England, Australia
2. Deng H, Yan Z (2006) Application of dea method on analyzing the economic development of each city of hubei province. J Huaihua Univ 25(5):17–19 (In Chinese)
3. Guo G, Zheng S (2009) The evaluation and research of developmental performance in six provinces of central china's culture industry china industrial. Economics 12:76–85 (In Chinese)
4. Lan Y, Han X (2011) The efficiency's present situation and the evaluation of the dynamic efficiency of china's art performance groups value. Engineering 23:9–11 (In Chinese)
5. Liu Z, Ye Q (2006) Comparison analysis on TFP of high technology industry in our country's east-central-west regions—based on nonparametric Malmquist Index method. High-Technol Industrialization 4:22–24 (In Chinese)
6. Malmquist S (1953) Index numbers and indifference surfaces. Tra-bajos de Esladisica 4: 209–232
7. Meng L, Gu H (2001) Nonparametric method of measuring productivity change. J Quant Tech Econ 2:48–51 (In Chinese)
8. National Burea of Statistics of China. China Statistical Yearbook (EB/OL). http://www.stats.gov.cn/tjsj/ndsj/
9. Plan and Financial Secretary. The analysis of the developing situation of national performing arts groups in recent years (EB/OL). http://www.ccnt.gov.cn. Accessed 23 Aug 2011
10. Sheng Z (1996) DEA theory, method and application. Science Press, Beijing (In Chinese)
11. Wang J, Zhang R (2009) The efficiency study of cultural industry in 31 provinces of China based on three stages DEA model. China Soft Sci Mag 9:75–82 (In Chinese)
12. Wei Q (1988) DEA method to evaluate the relative effectiveness—new field of operational research. China Renmin University Press, Beijing (In Chinese)

Chapter 94
Graded and Classified Safety Management for Elevator Users Based on the Coordination Theory

Xuedong Liang, Xiaojian Wang, Feng Wang and Jian Zhang

Abstract The elevator safety management can be a matter of life safety, however, traditional research works focus on the single factor of equipment without integrating essential factors together. Based on the coordination theory, a classified and graded safety management method for elevator users is proposed by combing both probabilities and consequences of accidents. Synchronously, the evaluation indexes are defined. Accident probability emphasizes the comprehensive effect of four aspects, including management, environment, equipment and staff system, combined with the assessment of accident consequence, which makes the elevator risk evaluation more systematic and scientific. The safety evaluation index system is beneficial for elevator users to remove potential safety hazards and save resources of special equipment survey. Its validity and efficiency are proved by case study.

Keywords Elevator users · Safety management · Coordination theory · Grading and classification

94.1 Introduction

Elevator is important special equipment which closely related to people's life [2]. Traditional researches mainly focus on equipment's potential safety hazard, but less from a comprehensive perspective of the device users' management, environment, equipment, and staff to study the safety management problems. Ren et al. studied the main parameters effecting elevator' traction performance, and put forward the method

X. Liang (✉) · X. Wang · F. Wang
School of Business, Sichuan University, Chengdu 610064, People's Republic of China
e-mail: liangxuedong@scu.edu.cn

J. Zhang
Inspection Institute of Fujian Special Equipment, Fuzhou 350008, Fujian,
People's Republic of China

J. Xu et al. (eds.), *Proceedings of the Eighth International Conference on Management Science and Engineering Management*, Advances in Intelligent Systems and Computing 281, DOI: 10.1007/978-3-642-55122-2_94, © Springer-Verlag Berlin Heidelberg 2014

of optimizing elevator' traction performance by controlling the baking torque. Xia [12] introduced the main components associated with elevator' safe running, and discussed the method to reduce the risk of elevator running. Luo and Fan [5] made brief analysis about the main reasons of elevator safety running, from the view of elevator mechanical point. (Special equipment includes eight categories: boilers, pressure vessels (including gas cylinders), penstock, elevators, lifting machinery, passenger rope way, large recreation facilities and special vehicles in factory [4]).

This paper based on the coordination theory, resolve the safety risk management into four areas, management, environment, equipment and staff, and confirm the safety risk level from probabilities and consequences of accidents, and make distinguish management [14], thus put forward the graded and classified safety management model. This study is beneficial for elevator users to remove potential safety hazard, on the other hand, it helps saving resources for special equipment survey, and furthermore, it has important significance for promoting special equipment inspection departments' elevator safety management work.

94.2 Graded and Classified Safety Management Model for Elevator Users Based on the Coordination Theory

1. Connotation of the Model

Coordination theory, proposed by the coordinating interdisciplinary research center of Massachusetts institute of Technology, studied the widespread dependencies in cross-disciplinary, based on the analysis principles, put forward the method of how to coordinate these dependencies [6]. Coordination theory regard coordination means as process in dependencies between things, or management mechanism which coordinate dependencies among participants [1]. Elevator safety management not only has the close relation with the elevator equipment, but also is affected by management, environment and staff, and these four aspects have different degrees of interdependence.

This chapter based on the analysis of coordination theory, combined with the thought and principle of safety system engineering, resolve the safety risk management into four areas, management, environment, equipment and staff, and confirm the safety risk level for elevator users from probabilities and consequences of accidents, thus realize the graded and classified safety management (Fig. 94.1).

2. Safety Risk Classification

Safety risk classification for elevator includes two dimensions: probabilities and consequences of accidents, we get each dimension's score through grading the safety risk analysis index, thus determine the safety risk grade (safe, slight, less, general, serious, critical) which based on the accident probability-consequence severity, called the two-dimensional risk matrix [13].

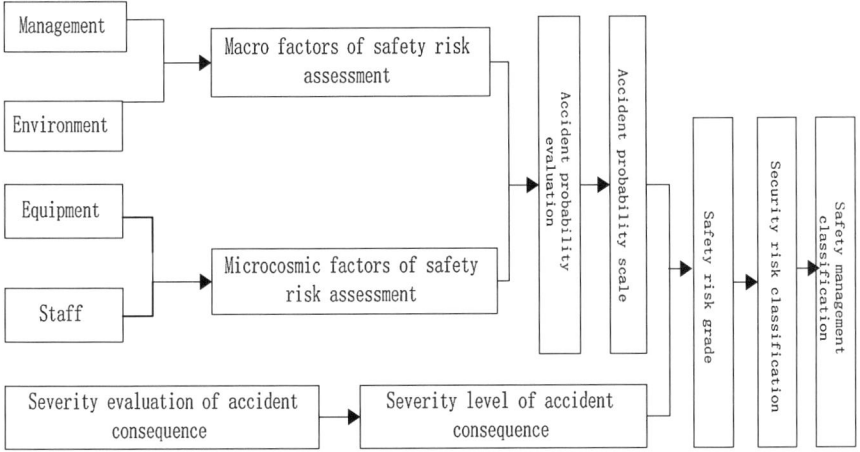

Fig. 94.1 Graded and classified safety management model

(1) Safety risk analysis index system
Elevator user's safety risk analysis includes two index systems: the probability of accident and the severity of accident consequence.

a. Index system of accident probability
    Index system of accident probability based on coordination theory, combines four factors: the macro management and environment, equipment and staff to the microscopic aspect, establish the comprehensive safety risk analysis index system (Fig. 94.2).

b. Index system of accident consequence severity
    Index system of accident consequence severity includes the severity of intrinsic consequence and reduction of the severity as well as its subordinate index (Fig. 94.3).

(2) Index weight
The index weight is determined by mathematical statistic analysis and AHP analytic hierarchy process method [7], including accident probability and accident consequence severity.
    The index weight of accident probability includes management, environment, equipment and staff, and each subordinate index weight.

a. Weight of management, environment, equipment and staff
    This paper collected 8 years (2005–2012 years) data of reasons for the special equipment accidents, divided it into four factors, environmental management, equipment and staff, and carried out statistical analysis (Table 94.1) from the perspective of coordination theory, then obtain the index weight above, and calculated the weight of the macroscopic and microscopic aspects (Table 94.2).

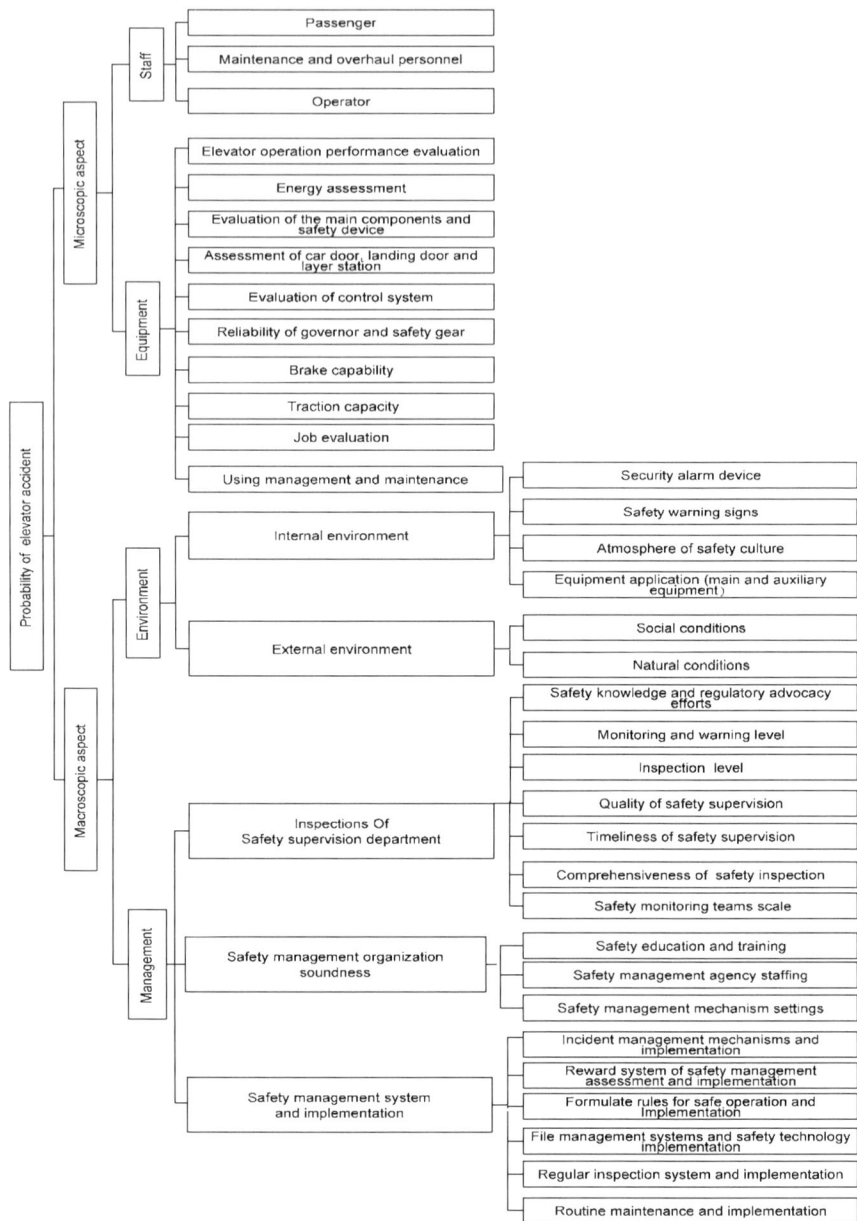

Fig. 94.2 Accident probability index system of safety risk analysis

b. Subordinate index weight of management, environment, equipment and staff

The subordinate index weight is determined by using AHP method [8]. This paper collected some index data graded by 10 elevator safety management experts, after

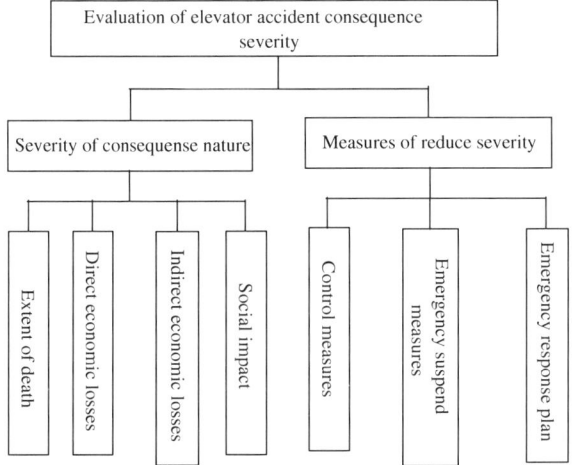

Fig. 94.3 Accident consequences severity index system of safety risk analysis

Table 94.1 Statistical analysis results of four factors (unit: number)

Classification statistics	Management	Environment	Equipment	Staff	Total number
Macro	440	40			480
Microcosmic			205	1393	1598

Table 94.2 Index weight of four factors

	Index	Weight	Macro index weight	Microscopic index weight
Macro index	Management	0.231	0.916	
	Environment		0.084	
Microcosmic index	Equipment	0.769		0.128
	Staff			0.872

dealing with it, reached the subordinate index weight (Tables 94.3, 94.4, 94.5 and 94.6).

The index weight of accident severity consequences is figured out by processing the data above graded by ten elevator safety management expert, using the AHP method (Table 94.7).

(3) Safety risk grade

First, determine the grade of an accident probability. Accident probability constitutes four areas: management, environment, equipment and staff, we can figure out the accident probability score and its grade, through scoring all aspects of elevator users in percentile.

- Management factors score M_{sp}, $M_{sp} = \sum_{i=1}^{m} \sum_{j=1}^{n} w_{spij} M_{spij}$.

Table 94.3 Subordinate index weight of management factors

Level 1 of subordinate index	Weight	Level 2 of subordinate index	Weight
Safety management organization soundness	0.143	Safety management institutions settings	0.2
		Safety management institutions staffing	0.2
		Safety education and training	0.6
Safety management system and implementation	0.429	Routine maintenance system and implementation	0.265
		Regular inspection system and implementation	0.183
		Safety records management system and implementation	0.073
		Formulate rules for safe operation and implementation	0.220
		Safety management assessment and reward system and implementation	0.106
		Incident management system and implementation	0.153
Safety inspections of safety supervision and management department	0.429	Safety monitoring teams scale	0.052
		Comprehensiveness of safety supervision	0.113
		Timeliness of safety supervision	0.132
		Quality of safety supervision	0.155
		Level of inspection and testing	0.181
		Level of monitoring and early warning	0.155
		Propaganda of safety knowledge and legislation	0.212

Table 94.4 Subordinate index weight of environmental factors

Level 1 of subordinate index	Weight	Level 2 of subordinate index	Weight
External environment	0.25	Natural conditions	0.75
		Social conditions	0.25
External environment	0.75	Equipment use (main and vice production equipment)	0.091
		Atmosphere of safety culture construction	0.139
		Safety warning signs	0.360
		Safety alarm device	0.409

- M_{c2i} is the score of the i-th index, W_{c2i} is the weight of the i-th index. The severity score of the accident result M_c is, $M_{ep} = \sum_{i=1}^{m} \sum_{j=1}^{n} w_{epij} M_{epij}$.
- M_{epij} is the score of the j-th index, W_{epij} is the weight of the j-th index. Macroscopic score M_{mac}, $M_{mac} = [w_s \quad w_e][M_{sp} \quad M_{ep}]^{\mathrm{T}}$.
- W_s is the weight of management, W_e is the weight of environment. Equipment score: M_{mp}, $M_{mp} = \sum_{i=1}^{n} w_{mpi} M_{mpi}$.

Table 94.5 Subordinate index weight of equipment factors

Level 1 of subordinate index	Weight
Management of using and maintenance	0.045
Assessment of working condition	0.029
Capacity of elevator traction	0.135
Capability of elevator brake	0.108
Reliability of speed limiter and safety clamp	0.168
Evaluation of elevator controlling system	0.108
Evaluation of car door, door and floor	0.135
Evaluation of the main parts of elevator with safety device	0.149
Evaluation of elevator energy consumption	0.045
Evaluation of elevator operation performance	0.078

Table 94.6 Subordinate index weight of staff factors

Staff	Level 1 of subordinate index	Weight
	Operator	0.258
	Maintenance workers	0.637
	Passenger	0.105

Table 94.7 Subordinate index weight of accident severity consequence

Severity of accident consequences	Level 1 of subordinate index	Weight	Level 1 of subordinate index	Weight
	Severity of the consequences intrinsic	0.75	Level of casualties	0.625
			Direct economic losses	0.125
			Indirect economic losses	0.125
			Social impact	0.125
	Measures of reduce accident severity	0.25	Monitoring measures	0.281
			Emergency stop measures	0.584
			Contingency plans	0.135

- M_{mpi} is the score of the i-th index, W_{mpi} is the weight of the i-th index. Staff score: $M_{mp}, M_{hp} = \sum_{i=1}^{n} w_{hpi} M_{hpi}$.
- M_{hpi} is the score of the i-th index, W_{hpi} is the weight of the i-th index. Microcosmic score: $M_{mic}, M_{mic} = [w_m \ \ w_h][M_{mp} \ \ M_{hp}]^{\mathrm{T}}$.
- W_m is the weight of equipment, W_h is the weight of staff. Accident probability score: $M_p, M_p = [w_{mac} \ \ w_{mic}][M_{mac} \ \ M_{mic}]^{\mathrm{T}}$.
- W_{mac} is the weight of macroscopic, W_{mic} is the weight of microcosmic. The safety risk grade of accident probability is, according to the Table 94.8.

Table 94.8 Comparison Table of safety risk grade

Score	[0, 10]	[11, 40]	[41, 60]	[61, 90]	[91, 100]
Safety risk grade	A level	B level	C level	D level	E level

Table 94.9 Matrix of accident probability-severity of result

Accident probability (P)	A level	B level	C level	D level	E level
A-level	Safe	Slight	General	Serious	Serious
B-level	Slight	Less	General	Serious	Serious
C-level	General	General	Serious	Serious	Serious
D-level	General	Serious	Serious	Serious	Critical
E-level	Serious	Serious	Serious	Critical	Critical

Next, determine the grade of an accident consequence severity.

- Score of intrinsic accident consequence severity: M_{c1}, $M_{c1} = \sum_{i=1}^{n} w_{c1i} M_{c1i}$.
- M_{c1i} is the score of the i-th index, W_{c1i} is the score of the i-th weight. Score of reducing accident consequence severity: M_{c2}, $M_{c2} = \sum_{i=1}^{n} w_{c2i} M_{c2i}$.
- M_{c2i} is the score of the i-th index, W_{c2i} is the score of the i-th weight. Score of reducing accident consequence severity: M_c, $M_c = [w_{c1} \quad w_{c2}][M_{c1} \quad M_{c2}]^T$.
- W_{c1} is the weight of intrinsic severity of the result, W_{c2} is the weight of measures to reduce the severity of the result.

According to the grade table of the risk of special equipment unit usage (Table 94.8), the grade of safety risk of accident result severity is R_c.

At last, confirm the grade of safety risk. According to the matrix of accident probability-severity of result (Table 94.9), the grade of safety risk can be reached.

3. Classification of Safety Management

According to the result of safety risk classification, combined with the actual situation of the users' safety management, it can be classified from low to high into A, B, C categories in accordance with the safety risks [9]. Specific criteria for the classification are as follows (Table 94.10).

Different types of elevator users use different safety management to achieve elevator category management (Table 94.11).

94.3 Applications

1. Application Background

Residential area is located in the southwest of Fuzhou city and adjacent to the Minjiang River. In May 2004, it was put into use. Around 2010, for special reasons, the property management company left the area, leading it into the state of non-residential

Table 94.10 Classification standards of elevator users

A	(1) Volunteer declare as required, all special equipment are regularly inspected and registered
	(2) Emphasis on safety, equipped with management institution of special equipment safety and full-time safety management staff
	(3) All workers take appointment with certificate
	(4) Timely safety education and training for special equipment operator
	(5) Establish and improve rules and regulations, and operate effectively
	(6) Strictly implement safety procedures
	(7) Timely maintain equipment, make safety checks as required, and make inspection records. Special equipment is allocated to persons who are responsible for safety management. Hazards identified are rectified timely
	(8) All relevant files are created
	(9) Have special equipment emergency plan and organize drills at least once a year
	(10) Workers for special equipment has no punishment for violations within 3 years
	(11) No above the average accident within 5 years
B	(1) Have dedicated institution for special equipment safety management or part-time safety management workers; Basically volunteer declare as required
	(2) Less emphasis on safety, have self-examination. Regular inspection and maintenance for equipment. Occasionally have hidden danger. Hazards identified are rectified after supervise and urge
	(3) Special equipment is allocated to persons who are responsible for safety management. The registration rate, regular inspection rate and rate of work with certificates of staff is 100 %
	(4) Basic safety education and training for special equipment workers
	(5) Establish and basically implement safety responsibility system, management systems and procedures standards
	(6) No records of punishment for violations in the use of special equipment within 2 years
	(7) Timely rectify after inspection
	(8) No above the average accident within 3 years
C	(1) Have special equipment with no regular inspection and registration
	(2) Equipped with no staff for special equipment management
	(3) Weak awareness and under-investment of safety. Equipped with no safety responsibility system, management systems and procedures standards
	(4) Workers with no certificates
	(5) Equipment has severe hidden dangers. No timely rectification after inspection
	(6) No special equipment emergency plan
	(7) Have the average accident within 1 year

property management. In the same year, the residents established the property owners' committee [3]. It reflects the lack of standardization and specialization on property management. It threatens safe use of elevator. In the following part, the author uses classification and grading methods to determine the level of the safety risk of the residential area and category management.

Table 94.11 Classification standards of elevator users

Category	Category management approach
Class A	Implement a relatively loose safety management approach. Users present report on elevator safe on a quarterly status. Special equipment inspection department makes comprehensive inspection every 2 years
Class B	In addition to the rectification for safety risks, users present report on elevator safe on a quarterly status. Special equipment inspection department makes comprehensive inspection every year
Class C	Implement a key regulation. Users' present report on elevator safe on a weekly status until the hidden danger is under control. Comprehensive inspection until it reaches the B and the above level

2. Calculations

(1) Accident probability Level

According to the actual situation of the residential area, one can grade the index of management, environment, equipment and staff (data in Table 94.12) to determine the grade of accident probability.

Comprehensive score of management factors:

$$\begin{bmatrix} [3.3\ 2.3\ 4.7\ 5.7\ 6.0\ 9.3]\,[0.265\ 0.183\ 0.073\ 0.220\ 0.106\ 0.153]^{\mathrm{T}} \\ [8.7\ 14.3\ 8.7]\,[0.2\ 0.2\ 0.6]^{\mathrm{T}} \\ [1.0\ 2.0\ 1.0\ 1.0\ 1.0\ 1.0\ 1.0]\,[0.052\ 0.113\ 0.132\ 0.155\ 0.182\ 0.155\ 0.212]^{\mathrm{T}} \end{bmatrix} \begin{bmatrix} 0.429 \\ 0.142 \\ 0.429 \end{bmatrix} = 3.94.$$

When each management factors take full score, the score is:

$$\begin{bmatrix} [6\ 9\ 5\ 9\ 6\ 10]\,[0.265\ 0.183\ 0.073\ 0.220\ 0.106\ 0.153]^{\mathrm{T}} \\ [10\ 15\ 10]\,[0.2\ 0.2\ 0.6]^{\mathrm{T}} \\ [3\ 4\ 4\ 3\ 2\ 2\ 2]\,[0.052\ 0.113\ 0.132\ 0.155\ 0.182\ 0.155\ 0.212]^{\mathrm{T}} \end{bmatrix} \begin{bmatrix} 0.429 \\ 0.142 \\ 0.429 \end{bmatrix} = 5.93.$$

Similarly, according to the environment factors score data we get the composite score for environmental factors, and the score is 6.69.

When each environment factors take full score, the score is 16.25.

Thus the score of macroscopic aspect is: $100[3.94/5.93\ \ 6.69/16.65]$ $[0.916\ 0.084]^{T} = 64.58$.

According to the equipment and staff factors score data, we can get the microscopic score is 32.43.

And then the accident probability score is: $[64.58\ 32.43][0.231\ 0.769]^{T} = 39.86$.

Table 94.12 Data for score on management

Factor	Index	Full score (100)	Actual score
Management	Routine maintenance and implementation of the system	6	3.3
	Regular inspection and enforcement system	9	2.3
	Safety records management system and implementation	5	4.7
	Formulate rules for safe operation and implementation	9	5.7
	Safety management assessment and reward system and implementation	6	6.0
	Incident management system and implementation	10	9.3
	Safety management institutions settings	10	8.7
	Safety management institutions staffing	15	14.3
	Safety education and training	10	8.7
	Safety monitoring teams scale	3	1.0
	Comprehensiveness of safety supervision	4	2.0
	Timeliness of safety supervision	4	1.0
	Quality of safety supervision	3	1.0
	Level of inspection and testing	2	1.0
	Level of monitoring and early warning	2	1.0
	Propaganda of safety knowledge and legislation	2	1.0

According to the data of accident consequence severity, the composite score for accident consequence severity is 2.24, and when items of accident consequences severity get full score, the score is 18.73.

Thus the score of accident consequences severity is $100 \times 2.24/18.73 = 11.96$.

The score matrix of accident probability and consequence severity (probability, consequence severity) is equal to (39.86, 11.96). According to the comparison table (Table 94.8) of the safety risk grade of special equipment building user, the safety risk grade is B.

In accordance with the score matrix of accident probability and consequence severity (Table 94.9), the safety risk grade of elevator building user is small.

3. The Strategy of Safety Management

The safety risk grade of elevator in this community is small. At the same time, in line with reality, there are some risks of the elevator:

- No professional or full-time managers of the use of elevators;
- Unsound safety management system of elevators;
- Ineffective problems of anti clamping protective devices, emergent lighting devices, and others.

According to the safety management classification methods of building users, it belongs to B. Special Inspection Department orders the building users to overhaul potential safety hazard, report the safe condition of the use of elevators monthly, and do a complete inspection yearly.

94.4 Conclusions

According to Coordination theory, this paper studies the probability of accident of the elevator unit from management, environment, equipment, and staff. It also evaluates the accident consequence severity and determines the safety risk grade for elevator users. Based on the research, classified management can be applied on elevator users. The research conclusions are as follows:

1. The safety risk of the elevator is not only influenced by equipment but also management, environment and staff. Those four factors are interdependent.
2. The effectiveness of evaluation is significantly enhanced if safety risk is studied from probability of accident and severity of consequence.
3. Safety management classification based on safety risk classification saves the resources of special inspection, which contributes to improving the elevator safety management level of special inspection department.

Further research direction of this paper is analysis of classified safety management towards special equipment users not only elevator.

Acknowledgments This research is funded by the National Intrinsic Science Foundation of China (71131006,71302134) [14], the science and technology planning project of general administration of quality supervision inspection and quarantine of PRC (2013QK317) [11], China Postdoctoral Science Foundation Funded Project (2012M521705) [15], Postdoctoral Science Special Foundation of Sichuan Province and the Fundamental Research Funds for the Central Universities (skzx2013-dz07) [10].

References

1. Bennetts I, Moinuddin KA et al (2005) Testing and factors relevant to the evaluation of the structural adequacy of steel members within fire-resistant elevator shafts. Fire Saf J 40(8):698–727
2. Du D, Qiao JX (2010) Analysis of cooperative project management platform based on collaborative management system. Project Manage Technol 10:262–276 (In Chinese)
3. Fan YX (2009) System safety engineering. Chemical Industry Press, Beijing (In Chinese)
4. Jing D (2010) Safety evaluation and analysis of special equipment in China, vol 04. Chemical Engineering & Machinery, China (In Chinese)
5. Luo Z, Fan W (2012) Analysis of the main reasons affecting the safe operation of elevators. China Sci Technol Inf 12:183–196 (In Chinese)
6. Malone TW, Crowston K, Herman GA (2003) Organizing business knowledge: the MIT process handbook. MIT press, Cambridge
7. Min Y, Yu Y (2013) Calculation of mixed evacuation of stair and elevator using evacnet 4. Procedia Eng 62:478–482
8. Ren TX, Wan JR (2007) Main parameters of effecting elevator traction capacity and its optimization. J Tianjin Univ 40(10):1247–1250 (In Chinese)
9. Wei ZQ, Men ZF (2006) Risk assessment of special equipment, vol 06. China Plant Engineering (In Chinese)
10. Wu ZZ, Gao JD (2001) Identification and control of major hazard. Metallurgical Industry Press, BeiJing (In Chinese)

11. Wu ZZ, Gao JD, Wei LJ (2001) Risk assessment method and application. Metallurgical Industry Press, BeiJing (In Chinese)
12. Xia RM (2010) Discussion of elevator safe operation. Metrol Measur Tech 03:169–175 (In Chinese)
13. Xia XY (1997) Causes and countermeasures of casualty accident in elevator operation. Labour Prot 12:38–39 (In Chinese)
14. Yang Z (2009) Research on risk management of special equipment. Ph.D. thesis. Tianjin University, Tianjin (In Chinese)
15. Yu Y (2009) Current situation and analysis of special equipment management problems of industrial enterprise. China Plant Eng 12:205–212 (In Chinese)

Chapter 95
Research on Comprehensive Risk Management System Construction Based on SPIC for Local State-Owned Investment Finance Companies

Yongjun Tang, Qinghua Qin and Yi Jiang

Abstract Local State-Owned Investment Finance Companies, as microcosmic bodies to intervene economic operation for the local government, play irreplaceable roles in guiding the social investment, promoting the adjustment of economic structure, strengthening controlling force of state-owned economy and increasing the value of state-owned assets. Jet more and more operational risk management has been accumulated. This article analyzes the comprehensive risk management theory at home and abroad, and summarizes the risk and countermeasures of local state-owned investment Finance Companies. Furthermore, based on the overall risk management framework (ERM) which is built by American national anti Fraudule Financial Reporting Council Initiative (COSO), and on the *Guidelines to comprehensive Risk Management of Central Enterprise* in China and other research results, this article puts forward the comprehensive risk management system for local state-owned investment and Finance Companies including organizational structure, business processes, information systems and management culture, which may provide references for the government to conduct a comprehensive risk management for local state-owned investment Finance Companies.

Keywords State owned investment finance companies · The comprehensive risk management · SPIC

95.1 Introduction

In recent years, local governments at all levels have set up a large number of local state-owned investment finance companies for financing, managing or operating of a large number of city infrastructure and public welfare projects construction.

Y. Tang · Q. Qin (✉) · Y. Jiang
Baise University, Baise 533099, Guangxi, People's Republic of China
e-mail: 120645952qq.com

J. Xu et al. (eds.), *Proceedings of the Eighth International Conference on Management Science and Engineering Management*, Advances in Intelligent Systems and Computing 281, DOI: 10.1007/978-3-642-55122-2_95, © Springer-Verlag Berlin Heidelberg 2014

These companies have played a positive role in financing, city infrastructure construction, and important industrial investment, and optimization of the economic layout and structure adjustment, and transformation of the pattern of economic development. But this kind of company is developing rapidly in recent years, the operation standard and rowing risk has caused widespread concern by the central government and the society. Therefore, it is vital important to establish comprehensive risk management system for local state-owned investment finance companies to strengthen its risk management, which would enhance the investment management capabilities of local state-owned investment finance companies and further consolidate their important roles in the national economy.

95.2 Research on Comprehensive Risk Management

At present, foreign enterprise risk management theory has been more mature. Enterprises have started to implement comprehensive risk management. The main point of current research on academic and business is how to make it more effective execution and implementation. The representative theories are below.

1. COSO internal control theory based on comprehensive risk management framework. In 2004, Enterprise Risk Management-Integrated Framework was released by USA National Anti Fraudulent Financial Reporting Committee of Sponsoring Organization (COSO). COSO-ERM proposes comprehensive risk management is the enterprise strategic decision management activities, which is covering decision layer, management layer and execution layer. The aim is to identify potential risk effectively and ensure the whole process of enterprise development goals.

2. Emphasize on quantitative analysis framework for comprehensive enterprises risk management to support business decisions. Stanford University emphasizes quantitative analysis framework can be used to control decision-making and risk on important business decisions for comprehensive Enterprises risk management. On the other hand, in order to help company to achieve management objectives, the company should establish enterprise risk management system including risk identification, risk assessment, risk response and risk control. Try to find out science method to identify and analyze all kinds of risk. Furthermore, risk management strategy should be research for implement the system, organization, processes and functions in order to help enterprises reach management objectives.

95.3 Domestic Researches on Comprehensive Risk Management

1. The Comprehensive Risk Management Theory

In China, comprehensive risk management research is still in the introduction stage, which results in the finance sector and project management. However, some scholars have already committed to a comprehensive enterprise risk management theory in certain industry field. Huai [10] proposed that ETRM three-dimensional fuzzy evaluation model can be use to analysis risk for the whole process dynamic risk management which starts from price, preference probability, risk identification, risk assessment, risk warning, risk control and risk response. Zhang and Chen [11] refer to enterprise risk management including risk management strategy and process, to support and ensure the ERM efficiency. According to Wei et al. [8], environmental risk information system (ERIS) provides a reference for enterprise risk management implementation. The total three aspects include the subject management, control elements and resources. There are five aspects content including the perfect support system, improve the ability of information collection, information processing, strengthen the decision to promote regional linkage and public participation. Zhao et al. [12] presents in COSO-ERM should integrate into the social responsibility, make a change to comprehensive risk management in the elements, in order to avoid and manage the enterprise risks.

2. The Central Enterprises Comprehensive Risk Management

In 2006, The State-Owned Assets Supervision and Administration Commission issued "Central Enterprise Wide Risk Management Guidelines". Guidelines point out the enterprise should focus on the overall business objectives. Establish risk management culture and comprehensive risk management system. To improve the risk management strategy, risk management measures and internal control system. Since then, the Shanghai stock exchange, Shenzhen stock exchange and the Ministry of finance, Securities Regulatory Commission, the Audit Commission, China Banking Regulatory Commission, the China Insurance Regulatory Commission and other departments have issued guidelines and regulations related to internal control in the enterprise, strengthen enterprise risk management.

3. Local State-owned Enterprises Risk Management

The central government state-owned enterprises have implemented comprehensive risk management. However, the local state-owned enterprises comprehensive risk management system theory is initial exploration stage. Wang [7] describes there are four parts of State-owned Invest-holding Company comprehensive risk management system including comprehensive organizational structure, the basic process of risk management, risk management and risk management information system. Lin [5] points out the main aim of local government financing platform risk management is to control the behavior, which set the local government standard to borrow and guarantee liabilities. It also can used to broaden the financing channels and reduce the proportion of indirect financing. On the other hand, the local government investment and financing platform should be standardized. Financial supervision departments and banks should reduce the credit risk. Yang et al. [9] focus on potential three debt risk management which should be strengthen the local government investment and financing platform debt scale expansion. The first is the non transparent operation and soft constraint mechanism easily lead to debt risk. Secondly, there are two aspects of moral hazard. Through the investment and financing platform, a large amount of funds can be obtained by the local government. It may lead to push up land cost, Real Estate cost and affect the fair market operation. The third risk is monetary policy adjustment to the market economy operation space.

95.4 The Necessity of Local State-owned Investment Finance Companies to Implement Comprehensive Risk Management

The necessity of comprehensive risk management in local state-owned investment Finance Companies can be explored from both the micro enterprise management internal dynamics and macroeconomic operation aspects.

1. Enterprise Management Demand
   At first, improve the enterprise management level and realize scientific development demand. Secondly, adapt the external environmental risk and enhance the level of market economy demand. Finally, improve the enterprise intrinsic value, the state-owned assets maintenance and appreciation demand. Improve and perfect comprehensive risk management system can effectively resolve and transfer local state-owned investment finance companies risks, reduce expense, indirectly enhance the enterprise intrinsic value.

2. The Operation Macro Economy Requirement
   In recent years, State-owned Enterprises and investment Finance Companies have occurred major crises frequently in China. It caused the difficult to expected losses, such as Cao Singapore Investment derivatives event, Sanlu incident, Huayuan company events. The reason is there are not enough preparing on business and financial risk in capital chain management, supply chain management and safety production process for the enterprises. Therefore, supply chain risk management (SCRM) has been a growing field of interest among researchers in the area of supply chain [3]. However, there is no good enough risk management strategies when the risk arrived.

95.5 Local State-owned Investment Finance Companies Risk Areas and Countermeasures

1. Local State-Owned Investment Finance Companies Risk Field
   Local state-owned investment Finance Companies has great advantages in the aspects of policy support, resource monopoly, funding. However, the company developing period is short, with the rapid investment and financing expansion, all kinds of problems and risks have increased. The main reason is the main strategic direction unclear, the comprehensive risk management system, enterprise internal control, the supervision and restraint mechanism have not been established. The financing channel is single. The loan guarantee is not standardized. On the other hand, other risk factors are debt growth, high debt ratio, lack of sinking fund arrangements, source of repayment is not stable, assets structure is not reasonable, some policy project asset profitability, solvency is weak. Mainly in the following nine risk areas: (1) Strategic Risk. The main fault is economic losses. (2) Decision Risk. Risk decision cannot achieve the expected purpose or cause losses to the enterprise. (3) Investment

Risk. The risk is investment project income uncertainty lead to income loss and the principal loss. (4) Financial Risk. It refers to the assets structure is not reasonable, financing, accounting and management or financial reporting errors and lost for the company. Raj and Sindhu [6] point out Financial Risks may be subdivided into Credit Risk and Market Risk. In fact, local state-owned investment Finance Companies debts could bring risk, including cash flow risk, financial report risk, the risk of accounts receivable report, commissioned by the financial risk, guarantee risk etc. (5) Financing risk. It is face risks in the process of raising money, the reason is the capital required for local state-owned investment Finance Companies investment are huge. That is also relates to many aspects., including the financing pressure, lack of company financing system and the government financing policy separated mutually etc. (6) Marketing Risk. The market changing leads to the local state-owned investment risk Finance Companies produce economic loss. (7) The Operation Risk. The main risk is the corporate governance imperfection, internal process is not standard, safety management is not in place, quality control is not strict, personnel structure is not reasonable system, person or external factors causing economic losses to the enterprise, including its risk control risk, the company safety production risk, business risk, the information system risk, external events. (8) The Legal Risk. The risk for the enterprise is due to external legal environment changes and various subjects, including enterprise itself is not in accordance with the law or contract negative legal consequences or other related legal affairs. (9) Integrity Risk. Significant financial and reputation losses for companies are due to education, system, supervision is not in place and the cadres, workers in the work life can not clean fingered self-discipline and caused possible corruption.

2. Local State-Owned Investment Finance Companies Risk Response Measures

In order to prevent and respond to the risk, the general enterprise risk management experience has taken the following measures to the local government, state-owned assets supervision and Administration Commission and in charge of the industry sector. (1) Make clear developing direction and restricted the high risk investment industry. Since 2009, the central government, local state-owned assets supervision and Administration Commission have issued documents to clear and strict restrictions high risk investment market for the central and local state-owned enterprise's main business, such as real estate, stock, futures, to ensure that the enterprises in the limited resources within the scope of the bigger and stronger industry, enhance the company's core competitiveness. (2) Establish Organizational Structure. The local government establishes enterprise legal advisor system, set specialized institutions with legal affairs, legal affairs personnel, and management to strengthen the major contract in order to improve the corporate governance structure, a risk management committee, audit committee, risk management departments and other measures, multi level common risk prevention and management. (3) Decision Making System and Business Process. The local government should formulate and implement "Three-Importance & One-Large" system to standardize the internal decision-making process, implementation of major investment projects auditing system, establish the investment committee, optimizing the investment and operation mode, strengthen project information collection, feasibility, track propulsion and post evaluation. (4) Improvement

Financial System. The local government should control enterprises behaviors on standard corporate loans, owing on the loan, the external guarantee. On the other hand, in order to reduce financial risks, the local government should optimize the asset structure, adjustment borrowing and owing on the loan scheme, the compression debt scale, strictly control the assets and liabilities rate and strengthening external security controls. (5) Broaden the Financing Channels. We encourage enterprises to direct financing through issuance of bonds, medium-term notes and short-term financing bonds, it also can reduce the financing cost. Meanwhile, we lead partners into innovative financing cooperation way and try to get external funding support. (6) Establish Enterprise Management Information Platform. We establish an information communication system to sound internal control mechanism. We also establish information collection, risk analysis and perfect the feedback, timely grasp of market dynamics and respond to change.

95.6 Local State-owned Investment Finance Companies' SPIC Overall Risk Management System

The research combines on "Overall Risk Management Framework" and "Central Enterprise Wide Risk Management Guidelines" [2], to build a comprehensive risk management system in four aspects, including the organizational structure, business processes, information systems and enterprise culture.

1. Comprehensive Risk Management Framework (Structure)

Local state-owned investment Finance Companies conduct comprehensive risk management, the first task is to establish risk management organizational structure. The basis for implementing the comprehensive risk management and an important guarantee are three levels management division into risk management responsibilities including strategy, execution and operation. The local state-owned investment Finance Companies can use the "1 + 3 + 3" pattern for comprehensive risk management organization structure, which is "a foundation, three line, three levels", as shown in Fig. 95.1.

The organization includes corporate governance, risk management committee, audit committee, risk management departments, legal audit department, headquarters functions, subordinate branch, sub branch.

A foundation based on the corporate governance structure. The Content includes establishing and perfecting the board of directors, board of supervisors and managers. Consider to balance the rights rational allocation, fair distribution of interests, define their duties, in order to establish an effective incentive, supervision and restriction mechanism, and the establishment of enterprise "Three-Importance and One-Large" system.

Three lines The first defense line is a business unit, refers to the enterprise functions, molecular branch business department and the functional departments, responsible for organization, risk management departments and legal audit department coordination, guidance and supervision, to fulfill their risk management

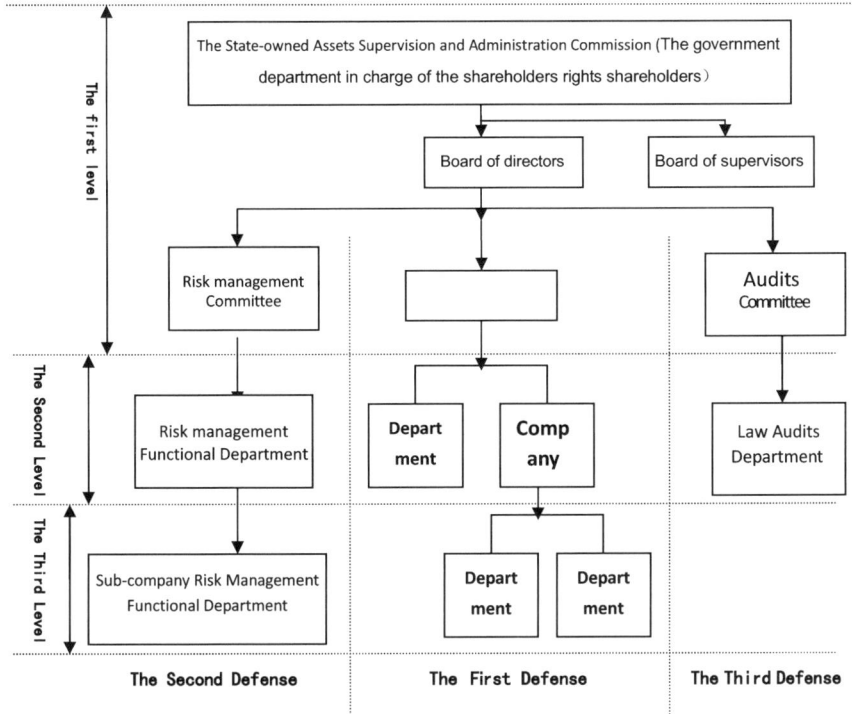

Fig. 95.1 The local State-owned investment finance companies comprehensive risk management organization structure

responsibilities. The second defense line is a risk management unit, including the risk management committee and the department risk management. The risk management committee is responsible for the approval and reviews the major risk criteria. Risk management departments are responsible for the proposed criteria and judgment mechanism. The third defense line is the audit unit, including the audit committee and the legal audit. According to Kochetova-Kozloski et al. [4], auditing standards require that auditors to use the knowledge about the identified entity-level business risks and their potential for misstatement to assess the risk of material misstatement at the process level. The audit committee set by the board of directors, and takes responsible for the enterprise internal and external audit communication, monitoring and verification work. The audit committee supervises and evaluates the comprehensive effectiveness risk management. Legal audit department is mainly responsible for the internal audit and related legal responsibilities of the audit committee.

Three levels The first level refers to the enterprise decision-making layer, such as the risk management committee, management and the Audit Committee. The second level is the group company departments and subsidiary companies, such as risk management function, enterprise related functional departments, subordinate units. The third level is sub-departments branch, such as risk management function, functional department and business department.

2. Comprehensive Risk Management Framework (Structure)

There are five major steps to local state-owned investment finance companies comprehensive risk management process. It can be divided into risk management initial information collection, risk assessment, risk management strategy, proposed the risk management to solve the supervision and improvement of risk management. (1) Collecting Initial Information. According to different enterprises and industry characteristics, functional departments can gather outside the initial information and related risk management, including historical data, data and predicted data. (2) Risk Assessment. Risk assessment should be organized by the enterprise organizations relevant functional departments and business units. At the same time, the enterprise should hire agencies with good reputation and professional quality to assistance implementation. (3) Establish a Risk Management Strategy. According to the enterprises own conditions and external environment, the development strategy, determine the risk preference, risk tolerance, the effectiveness of risk management standards, enterprises can choose risk avoidance, risk transfer, risk transfer, risk hedging, risk compensation and risk control. Furthermore, enterprises can formulate corresponding risk management strategy. (4) Risk Management Solutions. According to the risk management strategy, the enterprise can develop risk management solutions, including solve the risk target, personnel organization, business processes, resources qualification, risk coping strategies and tools, the enterprises development "internal risk control handbook", and guarantee the legitimate compliance requirements. (5) Risk Management Supervision and Improvement. The key point is to emphasize the major financing investment issues, major decisions and important business process. It can use to control the risk information collection, risk assessment, the measures implementation, key management and the solution. There are four aspects include decision management layers understand the goal realization degree, achieving the management goal level, the reliability of financial reporting and obey the law to ensure the risk management activities are effective whether or not.

3. Comprehensive Risk Management Information System (Information System)

Risk management information system is an important part of the comprehensive risk management system. It provides necessary technical support to the enterprises for implement comprehensive risk management in order to risk assessment, risk solutions, the implementation of risk management, internal control system.

1. Basic Elements and Internal Control System: According to "central enterprise wide risk management guidelines" on the establishment of risk management information system and the basic requirements of COSO "enterprise risk management framework" in three dimensions, the basic elements of local state-owned investment Finance Companies risk management information system and internal control system each link as shown in Fig. 95.2.

2. The Overall Framework. The comprehensive risk management information system includes information processing, risk identification, risk prediction, early warning and alarm, risk management quantification and analysis software, analysis and decision support system.

The local state-owned investment Finance Companies basic framework planning objectives are shown in Figs. 95.3 and 95.4.

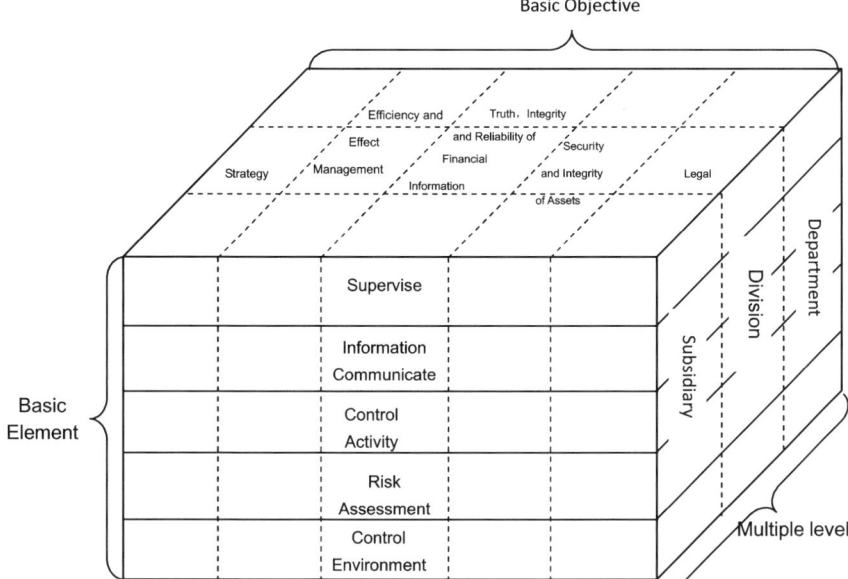

Fig. 95.2 Risk management information system basic elements and internal control system link

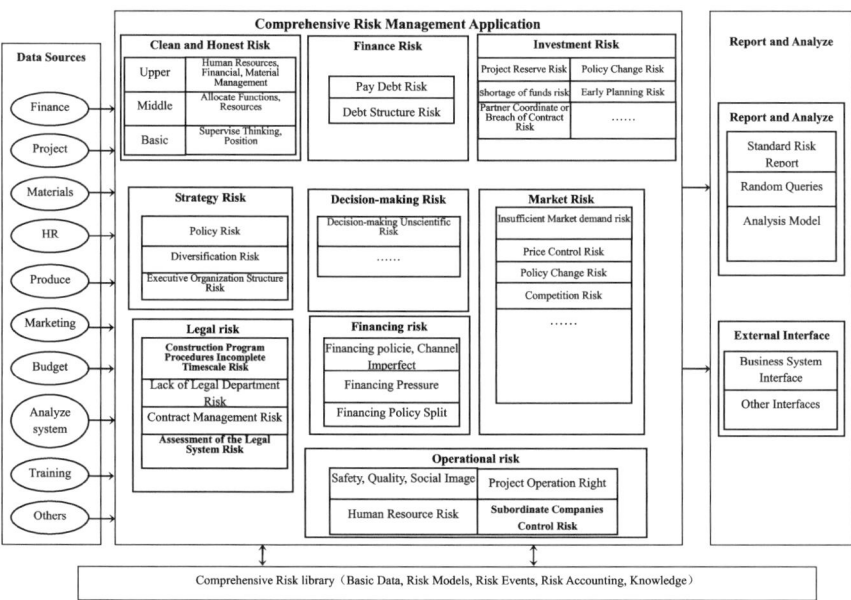

Fig. 95.3 Local state-owned investment and finance companies risk management information system overall application framework

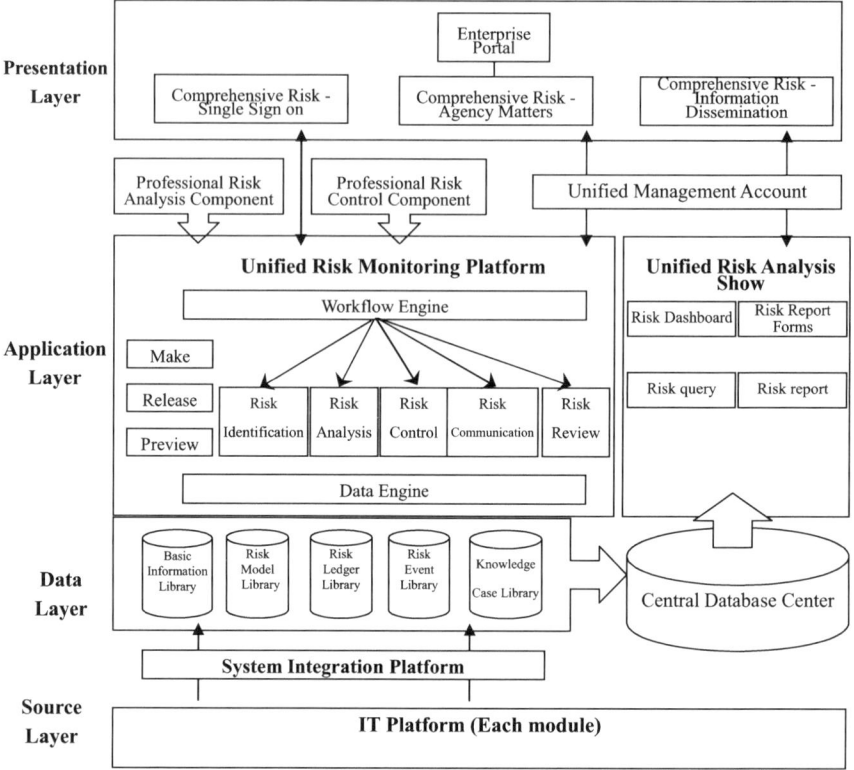

Fig. 95.4 The local state owned investment finance companies risk management information system general technology frame

4. Create Unique Risk Management Culture (Culture)

Local state-owned investment Finance Companies must build comprehensive risk management culture in order to achieve all-round, full process and full participation risk management. The construction of a comprehensive risk management culture should pay attention to the following aspects: (1) Propose Modern Enterprise Risk Management Idea and Behavior. The local state-owned investment Finance Companies should pay close attention to the risk. The reason is that involves and across multiple industries, many participants, professional requirements are complex. (2) Improve Risk Management Knowledge System. The local state-owned investment Finance Companies should strengthen and improve enterprise risk management techniques and methods. DR. ARORA (2013) [1] explains that Credit Risk Management (CRM) is one of the most demanding and important function in a commercial bank and enterprises. Enterprises should learn from the international risk control management theory and methods from the financial industry and other industries, takes its essence. Make it change with the actual enterprise combination. On the other hand, enterprises should strengthen the risk management information construction, to build the risk management information system including improve the risk management database, the risk analysis tool library, risk decision support system and communication system, which can be use to improve the risk management efficiency and ensure the consistency of management. Furthermore, enterprises should establish risk enterprises responsibility distribution system, includes improve the enterprise performance evaluation method and the risk management contents. (3) Risk Management System. Enterprises should establish local state-owned investment Finance Companies risk control and management system framework in order to come into being the system of unified standards, which include the internal control system and incentive mechanism. On the other hand, enterprises should mobilize and organize the whole departments and subordinate units to establish risk control standards and management measures. Enterprises also need to improve the work linking, such as the risk identification, evaluation, decision-making, control as well as to focus on long-term, scientific and reasonable incentives and business assessment methods.

95.7 Conclusion

Local State-owned Investment Finance Companies play irreplaceable roles in guiding the social investment, promoting the adjustment of economic structure, strengthening controlling force of state-owned economy and increasing the value of state-owned assets. The comprehensive risk management system for local state-owned investment and Finance Companies include organizational structure, business processes, information systems and management culture, which may provide references for the government to conduct a comprehensive risk management for local state-owned investment Finance Companies.

References

1. Arora A (2013) Credit risk management process and credit risk management framework in a commercial bank: an integrated view. Asia Pacific J Res Bus Manage 4(7):1
2. COSO II (2004) Enterprise risk management-integrated framework. Committee of Sponsoring Organizations of the Treadway Commission
3. Ganguly K (2013) A case study approach for understanding inbound supply risk assessment. Decision 40(1–2):85–97
4. Kochetova-Kozloski N, Kozloski TM, Messier WF (2013) Auditor business process analysis and linkages among auditor risk judgments. Auditing J Pract Theor 32:123–139
5. Lin WS (2010) The local government investment and financing platform: risk suggestions. Financ Econ 3:8–10
6. Raj B, Sindhu D (2013) Managing non-financial risks: business & growth. SCMS J Indian Manage 10(4):63
7. Wang P (2007) The State-owned Invest-holding company comprehensive risk management application research. Tianjing University, China
8. Wei X, Shao LQ, Ju MT (2010) Environmental risk information system based on comprehensive risk management construction research eris. Environ Pollut Control 32(2):102–107
9. Yang W, Huang TT (2010) Risk analysis local government investment and financing platform. China Financ 6:80–81

10. Yun HL (2007) Measure and strategy of comprehensive risk management in modern enterprises. Mod Financ Econ 27(4):33–35
11. Zhang Q, Chen LQ (2009) Enterprise risk management (ERM) to construct the theory and frame. Mod Financ Econ 31(7):25–32
12. Zhao SX, Ding RJ, Xin CH (2012) Risk management framework based on social responsibility construction of enterprise. Bus Res 32(2):102–107

Chapter 96
An Evaluation Method of the Development Level of Low-carbon Economy

Dan Zhang, Shuying Deng and Li Wang

Abstract The paper presented an evaluation method of the development level of low-carbon economy. First, we applied GMDH model for selecting the key indicators to evaluate the development level of low-carbon economy, and then we implement clustering analysis for the development level of the low-carbon economy towards two different years' evaluation objects by using the AC clustering method and evaluation indicators. The result shows that the development level of low-carbon economy of Sichuan in 2009 is roughly in according with Shanghai in 2003, and lags behind two stages compared to Shanghai in 2009. Thus, in the process of developing the low-carbon economy, Sichuan can refer to Shanghai in low-carbon industries, energy conservation and other policy measures to accelerate economic transformation.

Keywords Low-carbon economy · GMDH · AC clustering · Elastic decoupling indicators

96.1 Introduction

In 2003, the British government [2] published the white paper entitled "Our Energy Future: Creating a low-carbon economy", proposing the "low-carbon economy" that caused widespread concern in the international community firstly. Subsequently, terms such as "low-carbon life", "carbon footprint", "low-carbon consumption", "low-carbon energy", "low-carbon technologies", "low-carbon industries" are brought out more frequently in our social life. The international community is setting off an upsurge of transformation to a low-carbon economic development model, which is the fourth wave after agriculture, industrialization, informatization, called the wave of low-carbonization. Low-carbon economy is an economic

D. Zhang · S. Deng (✉) · L. Wang
Business School, Sichuan University, Chengdu 610064, People's Republic of China
e-mail: 313592349@qq.com

J. Xu et al. (eds.), *Proceedings of the Eighth International Conference on Management Science and Engineering Management*, Advances in Intelligent Systems and Computing 281, DOI: 10.1007/978-3-642-55122-2_96, © Springer-Verlag Berlin Heidelberg 2014

development model with strong sustainability based on low-energy, low-emission and low-pollution, which can improve the self-regulating ability of earth ecological systems. This will become the inevitable choice for the world economy development.

Currently, an increasing number of scholars have joined the research group on the low-carbon economy. These studies can be broadly divided into two directions: qualitative research and quantitative research. The qualitative research conducts descriptive studies on the connotation of low-carbon economy, the history of low-carbon economic development, macro factors affecting the development of low-carbon economy and policy measures may be taken and so on. Cao [1] concluded the main contents, methods, mechanisms and paths of low-carbon economy. Wang [9] put forward policy innovations and suggestions on the domestic development of low-carbon economy. The quantitative research models and analyzes existing economic data, energy data using techniques, evaluates the development level of low-carbon economic scientifically and objectively, and explores key factors affecting the development of low-carbon economy. Zhuang [13] analyzed decoupling characteristics from 20 greenhouse gases emitters including China in different periods by using Tapio elastic decoupling indicators, and compared their development level of low-carbon economy. Li et al. [5] comprehensive evaluated development situation of low-carbon economy of four municipalities in China by using fuzzy AHP method and principal components analysis. The results of those methods varied significantly with different weights because of strong subjectivity.

This paper proposed a method combined GMDH (Group Method of Data Handling) and AC (Analog Complexion) clustering method to evaluate the development level of low-carbon economy, in turn compared the level of Sichuan and Shanghai as example.

96.2 Methods

In this section, we introduced the main research method of this paper, and proposed a calculating method of carbon emission according to references.

96.2.1 GMDH Method

The self-organization theory, also known as Group Method of Data Handling (GMDH), was generated and developed based on the rapid development of neutral network and computer science. Similar to a biological neural network, self-organizing modeling method combines the black-box, the biological neuron method, the induction and the probability theory organically. It implements the unification between automatic control and pattern recognition theory. It also avoids people involving the process of understanding greatly and thus more objectivity and impartiality. Self-organizing modeling thoughts was first presented by the Ukrainian cybernetics scientist A. G. Ivakhnenko in 1967, and developed in the efforts of Adolf Mueller and other scientists. Now it has become an effective and practical data mining tool. Self-organizing modeling process is essentially a process of seeking and determining the optimal complications model of a system. The objects it processes are certain of input variables. One or more output variables constitute a closed system. And in this system, the correlations among variables are to be determined. Input variables generate numerous candidate models with each other. Then a number of optimal models chosen by external criteria are bound to get the next generation. All the above steps are repeated until the new model will be no more superior to the last one. Therefore the optimal model of penultimate generation is the optimal complications model.

The basic algorithms of GMDH are as follows [4, 7]:

(1) Pre-process the data sample and divide it into a training set and a testing set.
(2) Establish a correlation between independent variables and dependent variables as a "reference function" which is discrete form of Kolmogorov-Gabor polynomials generally.
(3) Select one (or several) criterion as the objective function also called external criteria from selection criteria with external supplement properties.
(4) Use the internal criteria to parameter estimate on the training set to get the intermediate candidate model set.
(5) Use the external criteria on the testing set to select several intermediate candidate models to be input variables of the next layer.
(6) Repeat 4 and 5 steps until the value of external criteria cannot be improved, and ultimately get the explicit optimal complications model.

96.2.2 AC Clustering Method

Similar synthesis algorithm AC developed by Laurence was first applied to meteorology forecast successfully. AC method can be regarded as a sequence pattern recognition method for prediction, clustering and classification of complex objects. It is based on the assumption that the typical situation of time process would be repeated in some form, which means that there would be one or more periods in development history similar to the current period for a given multidimensional time process. Thus, the similarity of two periods can be calculated to cluster. The differences between two modes of AC method are usually expressed in Euclidean distance or Hamming distance. In order to measure the similarity, the transform from candidate modes to reference modes must be found to describe these differences because every period may have different mean value and standard deviation. In this case usually take a linear transformation.

AC clustering method regards the data set consist of all samples as state space. The clustering for each sample is considered as clustering state space. State space takes variables x_1, x_2, $\cdots$, x_m as axis. Object O_i is the points in the space to be classified. Each object O_i is more or less different from others. This difference (distance) can be calculated by $S_{kh}^1, k = 1, , 2, \cdots, N, h_i = 1, 2, \cdots, N$. Hence the basis of clustering is symmetric similar matric $S_{NN}^1 = \{S_{kh}^1\}$. Clustering is to divide state space into n similar classes.

Usually AC method includes four steps [12]:

1. Produce candidate modes.
2. Transform candidate modes.
3. Select similar modes.
4. Cluster similar modes.

96.2.3 Elastic Decoupling Indicators and Calculating Method of Carbon Emission

To evaluate carbon emission level, we used Tapio elastic decoupling indicators and an indirect calculation method.

(1) Tapio elastic decoupling indicators Tapio [8] introduced elastic conception to construct decoupling indicators when he did decoupling research in European transportation energy and carbon dioxide in 1970–2001. Within a specific time, the percentage change in traffic volume when GDP changes one percentage point is shown as (96.1):

$$r_{v,\text{GDP}} = \frac{\%\Delta V/V}{\%\Delta \text{GDP}/\text{GDP}}, \tag{96.1}$$

r is elastic value of traffic volume. V is the traffic volume. This elasticity shows an increase traffic volume caused by economic growth. The decoupling elasticity between traffic volume and carbon dioxide emissions generated by traffic is shown as (96.2):

$$m_{\text{CO}_2,\text{GDP}} = \frac{\%\Delta \text{CO}_2/\text{CO}_2}{\%\Delta V/V}. \tag{96.2}$$

Multiply (96.1) and (96.2) to get the elastic decoupling indicator shown as (96.3):

$$t_{\text{CO}_2,\text{GDP}} = \frac{\%\Delta \text{CO}_2/\text{CO}_2}{\%\Delta \text{GDP}/\text{GDP}}. \tag{96.3}$$

(2) Calculating method of carbon emission There is still no direct monitoring data of CO_2 emissions in China at present, so most of the researches to estimate CO_2 emissions are based on amount of energy consumption. This paper proposed an Eq. (96.4) based on the total energy consumption data of China, which referred to carbon emissions decomposition model algorithm from Xu et al. [10].

$$\text{CO}_2 = \sum_i S_i \times F_i \times E, \tag{96.4}$$

Table 96.1 Carbon emission factors of energy

Category	Coal	Oil	Gas	Hydropower and nuclear power
F_i (A ton of carbon/Million tons of coal)	0.7476	0.5825	0.4435	0.0000

Information source Energy Research Institute National Development and Reform Commission. Scenario Analysis of Energy and Carbon Emissions for Sustainable Development in China 2003 [3]

E is the total consumption of primary energy in China. F_i is the carbon emission intensity for the energy class i. S_i is the proportion of energy class i of total energy. The values of F_i are shown in Table 96.1.

96.3 Empirical Analysis

In this section, we use GMDH model with the selected data of Sichuan's and Shanghai's different period to compare their development level of low-carbon economy. And we conclude that Sichuan could partially take example by Shanghai to reform to the low-carbon economy mode.

1. Selecting and Calculating the Input Index

Considering four aspects of economic development, social development, scientific and technological development and environmental development,18 indicators have been formulated as the initial evaluation index system of low-carbon economy referred to reference [5, 6, 11]: the first industry GDP (X_1), the second industry GDP (X_2), the third industry GDP (X_3), GDP per capita (X_4), the employment of third industry (X_5), the comprehensive utilization rate of industrial solid waste (X_6), the Engel coefficient of urban residents (X_7), the annual per capita disposable income of urban households (X_8), the per capita net income of rural households (X_9), the passenger transport of highways (X_{10}), the freight transport of highways (X_{11}), the civilian car ownership (X_{12}), the per-capita living space of urban residents (X_{13}), the coal consumption (X_{14}), the proportion of hydropower, nuclear power and wind power of total energy consumption (X_{15}), decontamination rate of urban refuse (X_{16}), the forest coverage rate (X_{17}) and the coverage rate of afforestation in developed area (X_{18}). All those indicators served as the input index of GMDH model.

We selected a total of 14 groups of low-carbon economic development statistics in Sichuan and Shanghai from 2003 to 2009 (The data is from Sichuan Province Statistical Yearbook 2003–2009, Shanghai Province Statistical Yearbook 2003–2009, "Collection of Sixty-Year Statistical Data of New China"), and got 14 groups with 18 indicators: x_{ij} ($i = 1, \cdots, 14; j = 1, \cdots, 18$). These dimensions of these indicators are not the same, which need to be normalized:

Table 96.2 The CO_2 Tapio elastic decoupling indicators of Sichuan from 2003 to 2009

Year	Total energy consumption[a]	CO_2 emissions (T)	ΔCO_2 (T)	GDP (billion)	ΔGDP (billion)	Indicator[b]
2003	9203.00	4787.27	1045.89	533.309	60.808	2.17218
2004	10699.00	5281.04	493.77	637.963	104.654	0.52561
2005	11300.00	5339.70	58.67	738.510	100.547	0.07049
2006	12538.00	5941.19	601.48	869.024	130.514	0.63739
2007	13685.00	6519.53	578.35	1056.239	187.215	0.45186
2008	14558.00	6863.06	343.52	1260.123	203.884	0.27297
2009	16322.00	8036.47	1173.42	1415.128	155.005	1.38996

[a] 10 thousand tons of standard coal. [b] CO_2 Tapio elastic decoupling

Table 96.3 The CO_2 Tapio elastic decoupling indicators of Shanghai from 2003 to 2009

Year	Total energy consumption[a]	CO_2 emissions (T)	ΔCO_2 (T)	GDP (billion)	ΔGDP (billion)	Indicator[b]
2003	6796.30	4505.12	309.91	669.423	95.320	0.44492
2004	7405.60	4743.07	237.95	807.283	137.860	0.25647
2005	8312.10	5113.48	370.41	924.766	117.483	0.53663
2006	8967.40	5344.66	231.18	1057.224	132.458	0.31564
2007	9767.70	5665.58	320.92	1249.401	192.177	0.33033
2008	10314.20	6039.32	373.74	1406.987	157.586	0.52301
2009	10367.00	6168.84	129.51	1504.645	97.658	0.30897

[a] 10 thousand tons of standard coal. [b] CO_2 Tapio elastic decoupling

$$X_{ij} = \frac{X_{j\max} - X_{ij}}{X_{j\max} - X_{j\min}} \quad (i = 1, \cdots, 14; \ j = 1, \cdots, 18). \tag{96.5}$$

Since the data is too large, the data sheets are saved in this paper.

2. Selecting and Calculating the Output Index

In recent years, more and more scholars used the elastic decoupling indicators in the field of greenhouse gas emissions reduction. The elastic decoupling indicator mainly reflects the uncertain correlation between economic growth and ecological environment protection. It also reflects another uncertain correlation between economic growth and material investment or consumption. The indicator can also measure the pressure relation between these two aspects. So it is suitable for low-carbon economic development assessment. Therefore we used CO_2 Tapio elastic decoupling indicators as GMDH self-organizing data mining output indicators (Y), and analyzed the main factors affecting CO_2 Tapio elastic decoupling indicators through GMDH models. According to the formula of Tapio elastic decoupling indicators (96.3) and the carbon emissions calculation formula (96.4), we calculated the elastic Tapio decoupling indicators of CO_2 Femissions for each year from 2003 to 2009 in Sichuan and Shanghai based on the constitution and total of energy consumption. The results are shown in Tables 96.2 and 96.3.

3. Building and Analyzing the GMDH Model of Low-carbon Economy Development

We used 14 groups of low-carbon economic development comprehensive evaluation index values in Sichuan and Shanghai as input variables X, took 14 CO_2 Tapio elastic decoupling indicators computing by Tapio elastic decoupling models as output variables Y, and selected the optimal complications models by software Knowledge Miner as (96.6).

$$Y = 5.28 - 0.56X_1 - 4.57X_6 - 4.33X_{15} + X_16 - 0.24X_{18}. \quad (96.6)$$

The prediction error sum of square (PESS) of (96.6) is only 0.1276, and the sample coefficient of determination (R^2) reaches 0.8468. It indicates that the models fits well, and reflects the correlation between CO_2 Tapio elastic decoupling indicators and indicators objectively and fairly. We can conclude that the 5 selected indexes are main factors affecting the development of low-carbon economy. The relatively large absolute value of coefficients of X_6 and X_{15} shows these two indexes associated with Tapio elastic decoupling indicators most closely. There are some period conclusions shown below.

(1) The comprehensive utilization rate of industrial solid waste (X_6) not only reflects the level of development of science and technology in a country or region, but also shows the capacity of govern environmental pollution. The purpose of the development of low-carbon economy is to ease the pressure on the environment caused by economic development. To achieve this goal, we are supposed to improve the environment, reduce damage to the environment and develop low-carbon economic model through the development of science and technology.

(2) The proportion of hydropower, nuclear power and wind power of total energy consumption (X_{15}) reflects the clean energy consumption structure of a country or region. This kind of structure is the main factor affecting the development of low-carbon economy. The structure is directly related to how much of a region's carbon emissions, and has an important impact on the development of low-carbon economy. To develop a low-carbon economy and circular economy needs to change the structure of energy consumption, reduce the consumption of fossil fuels such as coal, and vigorously develop renewable energy sources such as hydropower, wind power, nuclear power and bio-energy, etc.

4. Clustering Analysis for the Development Level of a Low-carbon Economy

According to the result of (96.6), we extracted 14 groups corresponding to each year of X_1, X_6, X_{15}, X_{16} and X_{18} these five indicators from 2003 to 2009 in Sichuan and Shanghai as a sample data for AC clustering. Then we used the self-organizing data mining software for clustering analysis, and clustered into three categories accordance with 90 % similar standard:

The first category: Sichuan 2003, Sichuan 2004, Sichuan 2005, Sichuan 2006, Sichuan 2007, Sichuan 2008, Sichuan 2009, Shanghai 2003.

The second category: Shanghai 2006, Shanghai 2005.

The third category: Shanghai 2006, Shanghai 2007, Shanghai 2008, Shanghai 2009. From the clustering results we can see:

- The period of Sichuan 2003–2009 and Shanghai 2003 is in the first category indicating that the level of low-carbon economic development of Sichuan in 2009 is roughly the same as Shanghai in 2003.
- The period of Shanghai 2004 and Shanghai 2005 is in the second category. These are the transition period for energy conservation and low-carbon economy development of Shanghai. Before and after 2004, Shanghai began to pay attention to low-carbon economy. Energy saving content as an individual project is more frequent shown in all types of files of Shanghai. Especially since the concept of low-carbon economy put forward in 2005, Shanghai has done a lot of work and reform measures. For example, eliminate, close, stop or change some high energy consumption and high pollution industries.
- Located in the third category is the developing period of low-carbon economy in Shanghai, 2006–2009. Shanghai completed a total of 517 industrial restructuring project in 2007, saving energy about 100 million tons of standard coal. It is noteworthy that Shanghai is building the world's first sustainable development eco-city, Dongtan of Chongming Island in Shanghai. The purpose of the design is to approach "carbon neutral" within the limits of feasible economic conditions. As we can see, Shanghai is in a positive trend towards a low carbon economy.

The development level of low-carbon economic in Shanghai in 2009 has already far exceeded the level in Sichuan with two stages ahead. Sichuan can partially refer to Shanghai in the development model of low-carbon economy, substitute human capital with natural capital, and reduce the use of fossil fuels relying on energy conservation and new energy technologies. Sichuan also should gradually promote the transformational to a low-carbon economy development model by adjusting the industrial structure, promoting the development of low-carbon industries, limiting high-carbon industries, encouraging new energy industries and reconstructing the national economy.

96.4 Conclusions

First, we combined research results of the development level comprehensive assessment system of low-carbon economy in recent years, and constructed a low-carbon economy development level evaluation comprehensive assessment system with 18 indicators. Second, we established the self-organizing model of a low-carbon economy using GMDH method to select 5 main indexes reflecting the development level of low-carbon economy. Finally, we used these 5 main indexes to do AC clustering analysis. The results show that: the level of low-carbon economic development of Sichuan in 2009 is roughly the same as Shanghai in 2003. The model of developing low-carbon economy and the policy measures in Shanghai can provide a reference for the development of low-carbon economy in Sichuan.

As a cutting-edge concept of low-carbon economy, the theoretical system has a vast research space to improve. In the future we can learn from foreign research results, to explore the application of low-carbon economy policy tools.

Acknowledgments The authors acknowledge the financial support Humanities and Social Science Planning Fund Project of the Ministry of Education of China, project No. 11YJA630029. The authors are grateful to the anonymous referee for a careful checking of the details and for helpful comments that improved this paper.

References

1. Cao H, Zhang F (2010) Advances in low-carbon economy at home and abroad. Prod Res 3:1–6 (In Chinese)
2. Dti U (2003) Energy white paper: our energy future–creating a low carbon economy. DTI, London (In Chinese)
3. Energy Research Institute National Development and Reform Commission (2003) Scenario analysis of energy and carbon emissions for sustainable development in China (In Chinese)
4. He C (2005) Self-organizing data mining and economic forecasting. Science Press, Beijing (In Chinese)
5. Li X, Deng L (2010) Exploring the comprehensive evaluation of urban low-carbon economy. Mod Econ Res 2:82–85 (In Chinese)
6. Liu Z, Xu Z, Li Y (2010) Low-carbon economy evaluation system and empirical research. Econ Tribune 5:37–41
7. Mueller J, Lemke F (2003) Self-organising data mining. Syst Anal Model Simul 43(2):231–240
8. Tapio P (2005) Towards a theory of decoupling: degrees of decoupling in the EU and the case of road traffic in Finland between 1970 and 2001. J Transp Policy 12(2):137–151 (In Chinese)
9. Wang Y (2010) An innovative analysis of the low-carbon economic development in China. Economist 11:15–20
10. Xu G, Liu Z, Jiang Z (2006) The factor decomposition model and empirical research of carbon emissions for China: 1995–2004. China Popul Resour Environ 6:158–161 (In Chinese)
11. Xu D, Ou Y (2010) Constructing a low-carbon economy statistical evaluation system. Stat Decis 22:21–24 (In Chinese)
12. Zhang Z, He C (2005) A comparative study of ac clustering method and hierarchical clustering method. Sci-Tech Inf Dev Econ 19:168–169 (In Chinese)
13. Zhuang G (2007) Low-carbon economy: the Chinese road of development under the background of climate change. China Meteorological Press (In Chinese)

Chapter 97
Healthcare Knowledge Management: Integrating Knowledge with Evidence-based Practice

Maria do Rosário Cabrita, Ana Miriam Cabrita and Virgílio António Cruz-Machado

Abstract Healthcare is experiencing a significant growth in the scientific understanding and practical approach of diseases, care pathways, treatments and clinical decisions. However, the literature reveals that this exponential growth of knowledge is not consistent with the users' ability to effectively disseminate, transfer and apply healthcare knowledge in clinical practice. Healthcare is intensive in knowledge and its efficient use can profoundly impact the quality of patient care decisions and health outcomes. Over the past decade Knowledge Management (KM), as a concept and a set of practices, has penetrated increasingly into the fabric of managerial processes in organizations all over the world. KM refers to strategies and processes for identifying, capturing, structuring, sharing, storing and applying an organization's knowledge to extract sustainable competitive advantages. KM in healthcare may be seen as a set of methodologies and techniques to facilitate the creation, acquisition, development, dissemination and utilization of healthcare knowledge assets. The goal of Healthcare Knowledge Management (HKM) is to structure, provide and promote timely and effectively healthcare knowledge to healthcare professionals, patients, individuals and policy makers when and where they need it in order to help them to take high quality, and cost-effective care decisions. The Evidence-Based Practice (EBP) approach focuses on the need for clinicians to keep up to date and improve not only their own skills in seeking the evidence, but also to build on their own knowledge base of what effective practice is. KM can only improve healthcare when knowledge has been successfully integrated with EBP. KM in the context of evidence-based

M. R. Cabrita (✉) · V. A. Cruz-Machado
UNIDEMI, Department of Mechanical and Industrial Engineering, Faculty of Science and Technology, FCT, Universidade Nova de Lisboa, 2829-516 Caparica, Portugal
e-mail: m.cabrita@fct.unl.pt

A. M. Cabrita
Faculty of Electrical Engineering, Mathematics and Computer Science, Telemedicine group, Roessingh Research and Development, Enschede, The Netherlands

A. M. Cabrita
Telemedicine group, University of Twente, Enschede, The Netherlands

J. Xu et al. (eds.), *Proceedings of the Eighth International Conference on Management Science and Engineering Management*, Advances in Intelligent Systems and Computing 281, DOI: 10.1007/978-3-642-55122-2_97, © Springer-Verlag Berlin Heidelberg 2014

healthcare creates a learning environment and ensures that best practice is captured and disseminated. This work aims to explore how KM practices can leverage different types of healthcare knowledge in the context of EBP. This research is theoretical in nature and seeks to contribute to understand the numerous challenges that exist to fully realize the HKM portfolio, namely knowledge processes that can improve the quality of patient care.

Keywords Clinical decisions · Evidence-based practice · Healthcare knowledge · Healthcare knowledge management · Practical approach of diseases

97.1 Introduction

Healthcare organizations are source of knowledge creation; yet, much of healthcare knowledge is under-utilized at the point-of-care and point-of-need. In such organizations, clinical staff is one of the key sources of knowledge creation however the inability of physicians to access and apply current and relevant knowledge leads, sometimes, to the delivery of suboptimal care to patients. Studies in the field [15, 17, 24] reveal that the large amount of healthcare knowledge is dispersed across different tools which makes difficult for healthcare professionals to apply timely the relevant knowledge to make the best patient care decisions. In 1999, the Institute of Medicine published the "To Err Is Human" report, which estimating that up to 98,000 patients a year die as a consequence of preventable errors [14]. More recently, a study reports that the number must be much higher–between 210,000 and 440,000 patients each year who go to the hospital for care suffer some type of preventable harm that contributes to their death [13]. The literature provides evidence that the under-utilization of healthcare knowledge contributes to medical errors, incorrect clinical decisions, high healthcare delivery costs and sub-optimal utilization of resources. Healthcare knowledge is central to clinical decision making processand then it is critical for organizations to make the most of their internal knowledge in order to offer the best possible health care. While there is no theory associated with the Evidence-Based Practice (EBP) [22], it is described as clinical practice consistent with the current best evidence. Studies in the field suggest that it is possible to conceptualize EBP as an evolving heuristic structure that helps improve patient outcomes, accounting for concepts such as knowledge acquisition, knowledge development and knowledge use. These concepts are part of the Knowledge Management (KM) cycle offering a structured process for the generation, development, sharing, distribution and utilization of knowledge, in order to generate value from it. This includes both tacit knowledge (personal experience) and explicit knowledge (evidence). In this sense, KM can provide an effective and efficient way of organizing what is known, reinforcing the EBP of healthcare professionals. In addition, clinical practitioners need to acquire proficiency in understanding and interpreting clinical information so as to attain knowledge and wisdom when dealing with large amounts of clinical data. Integrating KM paradigm with the healthcare system, in a manner where technology, people and processes are in harmony, can provide a holistic picture of HKM. We believe that a critical understanding of the concept of KM can provide an important perspective into how EBP can be more effectively implemented within a healthcare organization. On the basis of the KM and healthcare literature, this paper aims to develop a framework that integrates KM practices with EBP to provide an explanation of the application and impact of KM practices in healthcare delivery.

97.2 Theoretical Development

In a technically and intellectually based economy the rules of economics have been transforming the way we live and the way we work. The knowledge economy is seen as an external KM promoter, which influences every organization within this economy. Knowledge is described as a "capacity to act" [23] which suggests that the link between knowledge and outputs/outcomes in organizations becomes a critical issue to be addressed in the business and societal fields. In the healthcare field, knowledge is defined as "capacity to act competently" [25] signaling the importance to managing healthcare knowledge through systematic mechanisms. The literature highlights the potential of the concept "KM" to the healthcare domain. It is recognized that sound research requires a conceptual framework of the empirical reality being analyzed. Therefore, a key theme in which progress must be made is associated with the modeling of the processes of healthcare knowledge creation and diffusion.

An overview of the health and business/management literature on KM in healthcare will be undertaken. Theoretical assumptions for this work start by characterizing healthcare knowledge. Having gained an understanding of healthcare knowledge, we follow by discussing some KM and EBP concepts and processes and their application in HKM.

1. Healthcare Knowledge

Healthcare is knowledge rich, being generated at rapid pace. The information explosion in healthcare in the last decades has adversely affected the ability of healthcare professionals, particularly physicians, in providing accurate and timely medical decisions. Healthcare industry is in a state of flux, likely to be further accentuated by advances in biomedical knowledge and genetic engineering. The large volume of healthcare knowledge is often dispersed across different mediums which make it extremely difficult for healthcare professionals to be aware of the relevant knowledge to make the best patient care decisions [2]. One of the most recurring approaches of the literature on KM in the healthcare is the discussion around the distinctive nature of knowing in healthcare. The first issue is the highly fragmented and distributed nature of healthcare knowledge [10]. The healthcare practice is knowledge intensive and the large amount of knowledge is still tacit. Additional difficulties arising from the distributed nature of knowing in the healthcare sector have been discussed in the literature [7, 11], such as: (1) the presence of strong professional boundaries which retard the spread of innovations and makes knowledge sharing very difficult to happen in practice, and; (2) the presence of different groups with specific rules, job representations, behaviors, and values, which makes it difficult to see the whole knowledge process because of the distinctly different way their organizations and their work practices are structured. The second consideration is the reference to the proliferation of knowledge within the sector. It is observed that the digital era is revolutionizing the healthcare industry, providing an over abundance of complex medical knowledge launching medicine at a crisis point. Some authors [12, 19] claim that doctors can no longer memorize or effectively apply the vast amounts of scientific knowledge that are relevant to their clinical practice. A third theme relates to the importance of local knowledge in the medical decisions processes. As healthcare decisions come from various different sources and types of knowledge it is common a preference for local knowledge and tacit knowledge [6]. At this respect, the literature on evidence-based medicine [9] emphasizes that the integration of individual clinical expertise with the best available external clinical evidence (based on all valid and relevant information) that comes from a systematic research, moves medical practices toward evidence faster, more consistently, and more efficiently than evidence-based individual decision making alone.

Additionally, some works observed that health literacy skills are increasingly important for both health and healthcare. Patients with inadequate health literacy who have chronic diseases, e.g. diabetes, hypertension or asthma have less understanding of their disease than patients with adequate literacy. Healthcare knowledge concerns are at both the point-of-care and the point-of-need. Studies on social marketing for healthcare purposes have proven that centric-customer programs are well succeeded to promote healthy behaviors, preventing diseases, with impact on many aspects of a person's welfare, housing services, unemployment and lifestyle [4].

The literature identifies an assortment of knowledge types that contribute to clinical decision-making and care planning. Abidi has defined eight different types of knowledge within healthcare, as described in Table 97.1.

2. Knowledge Management

Within the context of the new world order, the challenges of prosperity and sustainability are essentially determined by our ability to wisely use knowledge, a global resource that is the embodiment of human intellectual capital and technology. The literature discusses several approaches for integrating KM with business processes and strategies in organizations. From the healthcare management perspective, the focus of research has been to examine different healthcare management concepts such as evidence-based medicine (EBM) and KM, both of which could potentially alleviate the problem of health-care information overload.

Knowledge is managed, structured and categorized information accessible by the right people at the right time [3]. Knowledge combines data and information, in addition to past experiences of experts' knowledge to support decision-making. Being described as the capacity to act, it is suggested that organizational knowledge is uniquely linked to action. Knowledge in organizations exists in multiple experiences and perspectives, being classified as: (1) tacit knowledge, and; (2) explicit knowledge. Explicit knowledge can be embodied in a code or language and as a consequence it can be verbalized and communicated, processed, transmitted and stored relatively easily. It is public and can be shared in the form of data, scientific formula, man-

Table 97.1 Types of healthcare knowledge

Type of knowledge	Description
Medical knowledge	It is knowledge domain that describes the theories abouthealth and healthcare, healthcare delivery models and processes
Patient knowledge	Refers to a clear description of the health status of the patient. It comprises medical observations of the patient and the inferences drawn by physicians, which are coded in the medical record, to provide a complete picture of the patient
Practitioner knowledge	Entails practice-related tacit knowledge that is exercised by the practitioner whilst discharging patient care. Practitioner knowledge is acquired through active learning, internship, observations and experiences
Organizational knowledge	This domain comprises knowledge flows within the organization such as a variety of knowledge from medical diagnostic systems, text-based materials, and other medical professionals with medical specialties
Process knowledge	Concerns institution-specific care pathways (or work flows) that determine the stipulated discourse of care for specific medical conditions within a healthcare setting
Resource knowledge	Refers to care delivery resources and infrastructure available within a healthcare setting, such as medical diagnostic devices and tools, drugs, support staff, nurses, hospital beds or surgical facilities
Relationship knowledge	Reflects the social capital withheld within an organization. It entails the communication mechanisms and contacts between multiple departments and institutions for the purposes of patient information sharing
Measurement knowledge	Describes the metrics, criterion and standards to measure success of a healthcare delivery process/system and the associated health outcomes

Source Adapted from Abidi [2]

uals, books, journals, and mass media such as newspapers, television, internet, etc. In a business context, patents may be considered an ideal example of explicit knowledge. In the healthcare domain, an example of explicit knowledge is the EBM literature, reviews, case studies, clinical practice guidelines, and so on. In contrast, tacit knowledge is embedded within an individual's experiences, beliefs, perspectives, values and instincts and is mostly inexpressible. Because it is rooted in action, commitment, values and emotions, it is hard to formalize. Tacit knowledge is well communicated by face-to-face encounters and it is acquired by sharing experiences, by observation and imitation. In healthcare, tacit knowledge of practitioners is manifested in terms of their problem-solving skills, judgment and intuition. Tacit and explicit knowledge are complementary, which means both types of knowledge are essential to knowledge creation. Explicit knowledge without tacit insight quickly loses its meaning. Knowledge is created through interactions between tacit and explicit knowledge and not from either tacit or explicit knowledge alone [18].

There are a number of management studies that stress the influence of KM in business performance [5], innovation [8] and sustainability [16]. Everybody discusses KM, but how can it be used and how can we successfully apply it? This question has its root in a practical problem experienced by many organizations (pri-

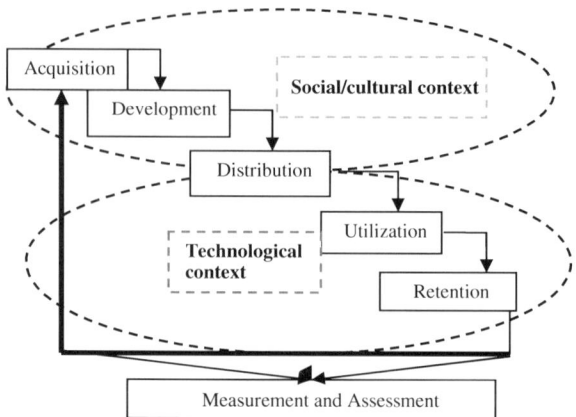

Fig. 97.1 Knowledge management key activities

vate or public, profit or non-profit) that are seeking to understand and deploy KM for their business. Several definitions of KM exist in the literature. KM is a complex and multidisciplinary concept that encompasses everything an organization does to make knowledge available to the business, such as embedding key information in systems, processes and products, applying incentives to motivate employees, interpreting and absorbing customer's wishes and forging alliances to combine the business with new knowledge. The objective of an organization applying KM is simply to make the right knowledge available at the right time at the right place. Therefore, KM relates to the processes and practices through which organizations create knowledge-based value.

Fundamental Approaches to Knowledge Management

The literatures stresses two fundamental approaches to KM: the process approach and the practice approach. The process approach attempts to codify organizational knowledge through formalized controls, processes, and technologies. In contrast, the practice approach to KM assumes that a great deal of organizational knowledge is tacit in nature, and that formal controls, processes, and technologies are not suitable for transmitting this type of understanding. The flow of knowledge depends on people and the social environment they operate in Fig. 97.1 illustrates this conceptual linkage between socio-cultural context (practical approach) and technological context (process approach).

Key KM activities are: knowledge acquisition (from customer, supplier, competitor and partner relations), knowledge development (directed toward creation of new skills and products, better ideas and improved processes), knowledge distribution (exchange and dissemination of knowledge from an individual to a group or the organization), knowledge utilization (productive use for the benefit of the organization), knowledge retention (selection, storage and updating of information, documents and experience) and measurement and assessment of knowledge. Therefore, KM represents a systematic approach towards searching and using the knowledge

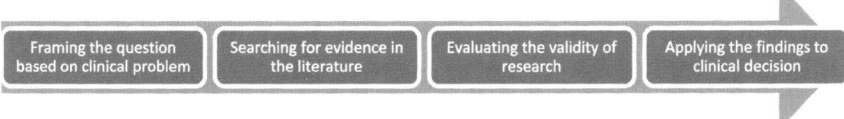

Fig. 97.2 Basic steps in evidence-based practice

on behalf of creating value. The goal of KM initiative in a healthcare setting is to provide the professional healthcare with appropriate tools, technologies, strategies and processes in order to make healthcare delivery more effective and efficient, and thereby maximize the full potential of all healthcare knowledge assets.

3. Evidence-Based Practice

Evidence-based practice (EBP) is the conscientious, explicit and judicious use of current best evidence in making decisions about the care of individual clients [21]. Evidence-based practice is meant to integrate individual clinical expertise and the best external evidence found in research. Hence, medical knowledge should be made available to practitioners. While evidence based medicine is a key aspect of today's medical practice, the abundance of information can keep a health professional from finding the right information. The need is to deliver the right information, at the right time, to the right person, and in the right format. Failing to do so is an impediment to the implementation of evidence based medicine. In this context, KM can play an important role by organizing knowledge and making it accessible.

The evidence-based method aims to turn clinical problems into questions and then systematically locate up-to-date research findings to produce qualitative appraisals or quantitative summary statistics as the basis for recommendations for clinical practice [20].

Implementation of EBP mainly involves four sequential steps, as depicted in Fig. 97.2: first, framing a clear question based on a clinical problem; second, searching for relevant evidence in the literature; third, critically evaluating the validity of contemporary research, and; fourth, applying the findings to clinical decisionmaking.

Despite the huge amount of information held in the healthcare knowledge database, it is unable to successfully apply the information across the entire spectrum of health-care delivery. Information provided in order to support evidence-based decision making in healthcare is a complex and non-structured component of KM. Using the best available evidence means identifying and integrating the most current research and practice results for effective care in order to support clinical decision making of the healthcare practitioners. EBP approach aims to understand how health resources can be used most effectively to improve health outcomes and the quality of patient care.

4. Healthcare Knowledge Management

There is much debate in healthcare over the use of the term "knowledge management" (KM), particularly when applied to healthcare operations. The reason behind this is that KM is a topic associated to business and industry, may be unfamiliar to many healthcare workers, but largely used by non-healthcare industries to achieve improved performance both for the individual and the organization. In addition, it is argued that previous healthcare management paradigms were unable to offer solutions to the information management crisis in healthcare. The information explosion in the last decade has adversely affected the ability of healthcare professionals, particularly physicians, in providing accurate and timely medical diagnosis and treatment.

Knowledge management research in healthcare over the years has focused on three topics: (1) the nature of knowledge in healthcare sector; (2) the type of KM tools and initiatives that are suitable for the healthcare domain, and; (3) the barriers and enablers to adoption of KM practices. More recently, researchers have begun to examine the theories and practices of KM applied to healthcare. Healthcare is a knowledge intensive business and KM initiatives hold the promise of improved efficiency in this sector. Given the universal pressures on healthcare resources worldwide, there is a clear need to examine whether an approach to KM could bring benefits to health services. However, we need an approach that recognizes the whole picture and embraces holism, rather than reductionism, in order to understand the complexity of human cognition; in other words a systems understanding of Healthcare Knowledge Management (HKM).

At present, KM in health care has largely concentrated on the generation of evidence from research (explicit knowledge) and the provision of evidence at the point of clinical decision making. According to Abidi [1] HKM can be defined as the systematic creation, modeling, sharing, operationalisation and translation of healthcare knowledge to improve the quality of patient care. This definition excludes a number of processes which may support and facilitate the knowledge flow. Knowledge flows comprise the set of processes, events and activities through which data, information, knowledge and meta-knowledge are transformed from one state to another (tacit to explicit; explicit to tacit). The literature describes various KM processes, varying according to the organization's context. To simplify the analysis of knowledge flows, the framework described in this paper is based primarily on a general knowledge model. The model organizes knowledge flows into four primary activity areas: knowledge acquisition, retention, distribution and utilization (Fig. 97.3).

Knowledge acquisition: This comprises activities associated with the entry of new knowledge into the system, and includes knowledge development, discovery and capture.

Knowledge retention: This includes all activities that preserve knowledge and allow it to remain in the system once introduced. It also includes those activities that maintain the viability of knowledge within the system.

Knowledge distribution: This refers to activities associated with the flow of knowledge on sharing acquired knowledge. This includes communication, translation, conversion, filtering and rendering.

Knowledge utilization: This includes the activities and events connected with the application of knowledge to business processes. Utilization is regarded as the capacity of the organization in applying knowledge generated in new situations. Within each activity phase exists other, smaller knowledge flows and cycles. These layers span a wide range of macro and micro behaviors, ranging from broad organizational

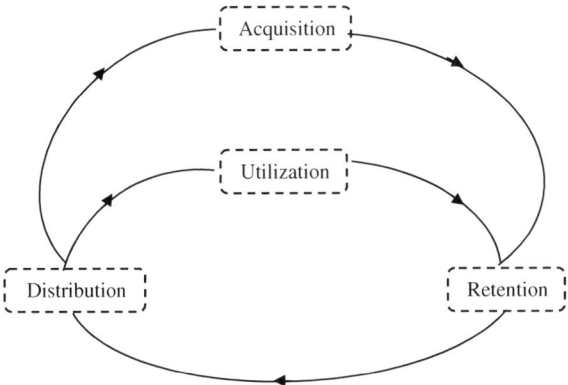

Fig. 97.3 The general knowledge model

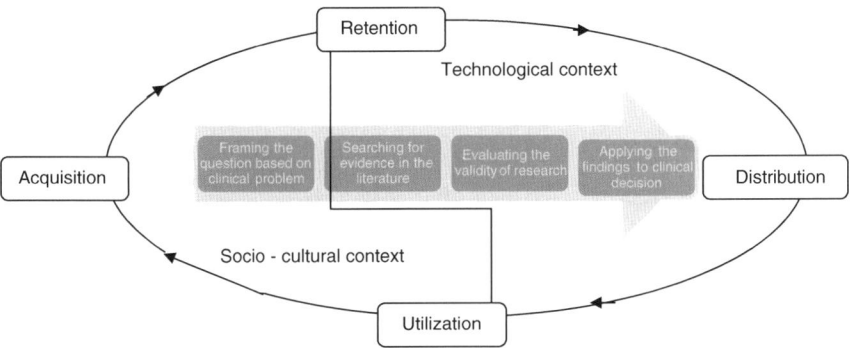

Fig. 97.4 Integrating knowledge and evidence-based practice

processes to discrete actions and decisions, and include all the various intervening layers: activities, tasks, work flows, systems, interfaces and transformations.

The purpose of the HKM is to promote and provide optimum health knowledge, timely, effective and pragmatic for health professionals (and even to patients and individuals) where and when they need to help them create high quality, well-informed patient care decisions and cost savings.

On the basis of the literature review, we suggest the framework depicted in Fig. 97.4, which combines the primary phases of general knowledge model with the EBP sequential phases.

One of the main features of EBP is the reliance on the partnership among hard scientific evidence, clinical expertise and individual patient needs and choices. As a process, EBP is about finding, appraising, retrieving, and applying scientific evidence to the treatment and management of healthcare. Ultimately EBP is the formalization of the care process that the best clinicians have practiced for generations. Its aim is to support practitioners in their decision making to eliminate/mitigate the use of ineffective, inappropriate, too expensive and potentially dangerous practices.

Theoretically, evidence-based medical practice is premised upon both explicit and tacit knowledge use. Steps of promoting adoption of EBP can be aligned with KM practices. Knowledge acquisition can conduct the research and then retaining and packaging relevant research findings into products that can be put into action—such as specific recommendations—thereby increasing the likelihood that research evidence will find its way into practice. Librarians play an important role in the spread of EBP because of the importance of identifying and retrieving appropriate literature from various sources for use in making health care decisions.

The findings from this study have implications for the provision of an evidence based practice as part of systematic KM as a way of improving decision making. We argue that KM in the context of evidence-based healthcare creates a learning environment and ensures that "best practice" is captured and disseminated.

97.3 Conclusions

There is an increasing consensus that healthcare decisions should be based on the best possible evidence, ensuring that healthcare is both effective and efficient. The literature reports the production and dissemination of evidence-based guidelines as a frequently used approach. At the same time, an increasing interest has been emerging in KM as an approach to increase the effectiveness of organizations. Experience of KM initiatives in non-health care organizations can offer useful insights and strategies to implement evidence-based practice in healthcare. KM can play a vital role in organizing, structuring and supporting evidence-based health decision making. KM is able to assist in medical errors reduction, and consequently their cost, by providing a decision support for practitioners.

There are many reasons for adopting the knowledge management practices in EBP, like patient safety, supporting care and reducing treatment cost are the factors for knowledge management adoption.

Acknowledgments We gratefully acknowledge the support given by UNIDEMI, R&D unit in Mechanical and Industrial Engineering in the Faculty of Science and Technology, FCT, New University of Lisbon, Portugal.

References

1. Abidi S (2001) Knowledge management in healthcare: towards knowledge driven decision support services. Int J Med Inf 63(1–2):5–18
2. Abidi S (2008) Healthcare knowledge management: the art of the possible. Springer, Heidelberg
3. Bose R (2002) Knowledge management capabilities and infrastructure for e-commerce. J Comput Inf Syst 42(5):40–49
4. Cabrita M (2014) Applying social marketing to healthcare: Opportunities and challenges. In: Kapoor A, Kulshrestha C (eds) Dynamics of competitive advantage and consumer perception in social marketing, IGI Global, USA, Chap 4, pp 78–97

5. Cabrita M, Cruz-Machado V, Grilo A (2010) Leveraging knowledge management with the balanced scorecard. In: Proceedings of international conference on industrial engineering and engineering management, Macau
6. Clarke C, Wilcockson J (2002) Seeing need and developing care: exploring knowledge for and from practice. Int J Nurs Studies 39(4):397–406
7. Currie G, Suhomlinova O (2006) The impact of institutional forces upon knowledge sharing in the UK NHS: the triumph of professional power and the inconsistency of policy. Public Adm 84(1):1–30
8. Du Plessis M (2007) The role of knowledge management in innovation. J Knowl Manag 11(4):20–29
9. Eddy D (2005) Evidence-based medicine: a unified approach. Health Aff 24(1):9–17
10. Edwards J, Hall M, Shaw D (2005) Proposing a systems vision of knowledge management in emergency care. J Oper Res Soc 56(2):180–192
11. Ferlie E, Fitzgerald L et al (2005) The (non) spread of innovations: the mediating role of professionals. Acad Manag J 48(1):117–134
12. Heathfield H, Louw G (1999) New challenges for clinical informatics: knowledge management tools. Health Inf J 5(2):67–73
13. James J (2013) A new evidence-based estimate of patient harms associated with hospital care. J Patient Safety 9(3):122–128
14. Kohn L, Corrigan J, Donaldson M (1999) To err is human: building a safer health system. National Academy Press, Washington DC
15. Lansisalmi H, Kivimaki M et al (2006) Innovation in healthcare: a systematic review of recent research. Nurs Sci Quart 19:66–72
16. Mohamed M, Stankosky M (2009) An empirical assessment of knowledge management criticality for sustainable development. J Knowl Manag 13(5):271–286
17. Nicolini D, Powell J et al (2008) Managing knowledge in thehealthcare sector: a review. Int J Manag Rev 10(3):245–263
18. Nonaka I (1991) The knowledge creating company. Harv Bus Rev 69(6):96–104
19. Pope C, Smith A et al (2003) Passing on tacit knowledge in anesthesia: a qualitative study. Med Educ 37(7):650–655
20. Rosenberg W, Donald A (1995) Evidence-based medicine: an approach to clinical problem solving. Brit Med J 310:1122–1125
21. Sackett D, Rosenberg W et al (1996) Evidence-based medicine: what is it and what isn't it? Brit Med J 312:71–72
22. Straus S, Richardson W et al (2005) Evidence-based medicine—how to practice and teach EBM, 3rd edn. Elsevier, Edinburgh
23. Sveiby K (2001) A knowledge-based theory of the firm to guide strategy formulation. J Intellect Capital 2(4)
24. Tucker A, Edmondson A (2003) Why hospitals don't learn from failures: organizational and psychological dynamics that inhibit system change. Calif Manag Rev 45(2):55–72
25. Wickramasinghe N, Gupta J, Sharma S (2005) Creating knowledge-based healthcare organizations. Idea Group Pub, Hershey

Chapter 98
The Compensation Plan on Doctors Considering the Contextual Performance

Yuguo Yuan and Sheng Zhong

Abstract The doctor's performance can be divided into two kinds of task performance and contextual performance, and current articles on contextual performance are less. This paper is studied by Team Principal-agent theory. The benchmark mode is established for contrast, in which agents don't make efforts on contextual performance. And then, according to whether accounting cost respectively, we set up two models that only one agent has contextual performance. The conclusion shows contextual performance is obvious in medical institutions, and it benefits to the whole team. Therefore, we advocate using team performance compensation plan, and then explain the necessity of government subsidies for medical institutions.

Keywords Contextual performance · Team output · Principal-agent theory

98.1 Introduction

Medical resource allocation efficiency in China is relatively low. Many hospitals at the level of Grade III is long-term operating at full capacity, and even overload, while the number of patients in basic-level hospitals was just the opposite. To solve this problem, some scholars have proposed the "hospital alliance" [10]. The specific mode of operation is that through the core and basic-level hospital alliance, core hospitals provide technical support for basic hospitals, and basic hospitals pay management fees to core hospitals. Core hospitals break the constraints of time, space, and expand the service radius; basic hospitals improve the quality of service, and increase the quantity of patients [15]. In many cases, patients may select a hospital because of a doctor's reputation, and obviously the influence of these doctors can bring additional value to the hospital. These phenomena are due to the influence of the famous

Y. Yuan · S. Zhong (✉)
Business School, Sichuan University, Chengdu 610065, People's Republic of China
e-mail: zhongshengscu@163.com

J. Xu et al. (eds.), *Proceedings of the Eighth International Conference on Management Science and Engineering Management*, Advances in Intelligent Systems and Computing 281, DOI: 10.1007/978-3-642-55122-2_98, © Springer-Verlag Berlin Heidelberg 2014

doctor, we call it contextual performance. Unlike task performance, contextual performance is difficult to quantify directly. As early as 60 years ago, scholars realized that the behavior of a person in the team not only helped him to complete organizational tasks, but also indirectly affect the development of the organization, and they call this behavior contextual performance or surrounding performance [14]. Specifically, contextual performance includes behaviors such as helping and cooperating with others, volunteering to do additional job, persisting to accomplish tasks and so on. Both types of performance contribute to overall organization. Task performance contributes directly through the production of goods and services, while contextual performance makes contributions indirectly [2]. Nowadays, we found domestic articles on contextual performance of medical institutions are very few. Regardless of the hospital alliance or hospital having famous doctors, is a team, so Team Principal-agent theory is used to suggest that contextual performance in medical institutions is obvious, and can be beneficial to the team.

Since the 1980s, a group of economists which regarded Holmstrom [8] as pioneer used game theory and information economics analysis method to discuss how to solve the difficult problems in team production, so formed the basic framework of team incentive theory. Performance pay schemes are thought to generate additional productivity and earnings in return for additional effort [13]. Heywood and Tsertsvadze [7] considered in the team, if an agent's salary depends on another agent output, then there is the problem of profit sharing. The principal motivating agents to helping others caused this dependence. Actually, profit sharing is a kind of incentive to increase cooperation among team members. However, on the other hand, unreasonable profit sharing will also increase the pressure of the team members, lower job satisfaction, and distorted actions [12]. FitzRoy and Kraft [6] discussed the valuable cooperative behavior and believed that profit sharing could motivate cooperation to increase productivity when work organization facilitates interaction. Drago and Turnbull [4] proposed profit sharing provides incentives for helping on the job in the teamwork, since each worker's income depends, in part, on the output of colleagues. Thus, both the firm and the employees can gain through profit sharing. FitzRoy and Kraft [5] provides a model with a worker utility function that differs from that of Drago and Turnbull, and got the same conclusion that profit sharing can promote common efforts, increase the output and the staff's working satisfaction. Economists Itoh [9] researched whether principal should encourage agents to help peers in his work team, after finish his own task. Itoh proved that, if the agents of their own work efforts and help fellow efforts on the cost function are independent, but are complementary to the work, principal uses incentives to lure the teamwork is optimal. Chao and Siqueira [3] illustrated the optimistic impact of contracts on team output in a partnership when production depends on the efforts of agents in their own tasks as well as their efforts in helping other teammates. Kato [11] conducted a case study to show the different incentive plan led to different productivity but changing the plan was costly. But Baker et al. [1] had been aware that team performance salary system is powerful incentive to human behavior, but profit sharing strategy has the problem of free riding. Then Nalbantian and Schotter [16] pointed out in the mode of team operation, the performance of the team members performance is difficulty to measure and assess. The asymmetric information between managers and team members is easy to cause the free riding and the loss of productivity.

In this paper, based on the model of Itoh, through a series of hypothetical inference, then shows motivating agents to help others is always beneficial in medical institutions in our country. The paper is organized as follows. This section describes the framework of the paper; In Sect. 98.1, we establish the benchmark mode for contrast. In the model of Sect. 98.2, one of the two doctor agents has contextual performance, and the costs of agents' task performance and contextual performance are dependent. Section 98.3 considers the opposite situation. The costs of doctor agents' task performance and contextual performance is independent. Section 98.4 is the comprehensive analysis of the three models, then given the conclusions. Section 98.5 summarized the full text.

98.2 The Benchmark Model

The benchmark mode is regarded as mode (98.1). In this section, two doctor agents don't make efforts on contextual performance, and both are risk averse. $\eta_i\,(\eta_i > 0)$ represents the agent's absolute risk aversion coefficient. Suppose that a_i represents task performance of agent i $(i = 1, 2)$. The doctor agent's task performance can be measured by some observable aspects, such the number of patients, medical records, academic research and so on, and thus more realistic reflection of task performance output. Thus we can get a more accurate task performance output. So, we believe that task performance output is equal to the task performance, namely, $q_i = a_i$. The cost function of agent i is: $\varphi_i = \frac{1}{2}c_i a_i^2$. Among this, c_i represents the marginal performance cost of coefficient of agent i, reflecting the ability of agent i to complete the task. The principal provides linear incentive contract to the agents: $w_i = t_i + m_i q_i\,(i = 1, 2)$, t_i represents the fixed salary of agent i, determined by the wage system. m_i is the task performance incentives of agent i, determined by the principal. So $w_1 = t_1 + m_1 a_1$, $w_2 = t_2 + m_2 a_2$.

The model we established as follows:

$$\begin{cases} \max\limits_{m_i,\eta_i} E[q_1 + q_2 - w_1 - w_2] \\ \text{s.t.} \begin{cases} E(-e^{-\eta_i[w_i-\varphi_i(a_i)]}) \geq \mu(\varpi_i, a_i) \\ a_i \in \arg_{a_i} \max E(-e^{-\eta_i[w_i-\varphi_i(a_i)]}), \end{cases} \end{cases} \qquad (98.1)$$

ϖ_i is the minimum wage of the contract which agent i can accept, and $u(\varpi_i, a_i)$ is the reservation utility of agent i.

- Given the principal to agents' optimal incentive: $m_1^1 = 1, m_2^1 = 1$.
- The doctor agent's optimal effort: $a_1^1 = \frac{1}{c_1}, a_2^1 = \frac{1}{c_2}$.
- The doctor agent's optimal compensation: $w_1^1 = t_1 + m_1 a_1 = t_1 + \frac{1}{c_1}, w_2^1 = t_2 + m_2 a_2 = t_2 + \frac{1}{c_2}$.

- The principal's optimal income: $E^1[q_1 + q_2 - w_1 - w_2] = -t_1 - t_2$.
- The total output of the department: $Q^1 = q_1 + q_2 = a_1 + a_2 = \frac{1}{c_1} + \frac{1}{c_2}$.

98.3 Cost Dependent Model

98.3.1 Assumption

Only one doctor agent makes efforts on contextual performance in the model of this section. We suppose it is agent 1. λa_i is the contextual performance which agent i contributes to agents j, $(i \neq j)$. Among this, λ indicates the correlation of contextual performance and task performance. And λ can be also recognized the impact factor of contextual performance, which represents the ability of agents helping others. We just consider this situation contextual performance makes positive contribution to the team, and one's contextual performance can't more than his task performance, namely $\lambda \subseteq [0, 1]$. In order to simplify model, we regard the relationship between task performance and contextual performance is linear correlation. Actually, the reality is more complex. In addition, because of inertia, one person can't change his habits soon. So, λ is constant in a salary payment period.

The doctor's contextual performance is invisible influence. Unlike the task performance, contextual performance isn't so easy to measure, but also influenced by outside interference. Therefore, we think contextual performance's output equals the efforts making on contextual performance and random noise. So, the output of agent 1 is $\lambda a_1 + \varepsilon$. Random noise part ε obeys a normal distribution whose mean is 0 and variance is σ^2. σ^2 means the metrizability of contextual performance. And the smaller σ^2 is, the easier it is measured. The output of agent i is: $q_1 = a_1, q_2 = a_2 + \lambda a_1 + \varepsilon$. The cost functions of agents don't change. The principal provides linear incentive contract to the doctor agents: So, $w_1 = t_1 + m_1 a_1 + n_1(\lambda a_1 + \varepsilon), w_2 = t_2 + m_2(a_2 + \lambda a_1 + \varepsilon)$. To doctor agent 1 who makes efforts on contextual performance, the principal give him n_1 as incentives. Although the principal doesn't motivate doctor agents 2 to make efforts on contextual performance, the salary of agent 2 increased by the influence of contextual performance of agent 1.

98.3.2 Modeling and Solving

As a team, our target is the team to maximize the return. So, the model we established as follows:

$$\begin{cases} \max_{m_i, n_i} E[q_1 + q_2 - w_1 - w_2] \\ \text{s.t.} \begin{cases} E(-e^{-\eta_i[w_i - \varphi_i(a_i, \lambda)]}) \geq \mu(\varpi_i, a_i) \\ (a_i, \lambda a_i) \in \arg_{a_i, \lambda a_i} \max E(-e^{-\eta_i[w_i - \varphi_i(a_i, \lambda)]}). \end{cases} \end{cases} \quad (98.2)$$

- Given the principal to agents' optimal incentive: $m_1^2 = 1 + \lambda, n_1^2 = 0, m_2^2 = \frac{1}{1+\eta_2 c_2 \sigma^2}$.
- The doctor agent's optimal effort:

$$a_1^2 = \frac{1+\lambda}{c_1}, a_2^2 = \frac{1}{c_2(1 + \eta_2 c_2 \sigma^2)}.$$

- The doctor agent's optimal compensation:

$$w_1^2 = t_1 + m_1 a_1 + n_1 \lambda a_1 = t_1 + \frac{(1+\lambda)^2}{c_1},$$

$$w_2^2 = t_1 + m_2(a_2 + \lambda a_1) = t_2 + \frac{1}{c_2(1 + \eta_2 c_2 \sigma^2)^2} + \frac{\lambda(1 + \lambda)}{c_1(1 + \eta_2 c_2 \sigma^2)}.$$

- The principal's optimal income:

$$E^2[q_1 + q_2 - w_1 - w_2] = \frac{\eta_2 c_2 \sigma^2}{c_2(1 + \eta_2 c_2 \sigma^2)^2} - \frac{\lambda(1 + \lambda)}{c_1(1 + \eta_2 c_2 \sigma^2)} - t_1 - t_2.$$

- Finally, the output of the department:

$$Q^2 = q_1 + q_2 = \frac{(1+\lambda)^2}{c_1} + \frac{1}{c_2(1 + \eta_2 c_2 \sigma^2)}.$$

Specific solution procedure is in the appendix.

98.4 Cost Independent Model

Now we consider the cost of task performance and the performance are independent, and still only doctor agent 1 makes efforts on contextual performance. Other assumptions are unchanged, the same model. After the change the assumptions are as follows: Doctor agent 1: Task performance: a_{11} Contextual performance: a_{12}. Output: $q_1 = a_{11}$. Cost function: $\varphi_1 = \frac{1}{2}c_1 a_{11}^2 + \frac{1}{2}d_1 a_{12}^2$. Incentive contract: $w_1 = t_1 + m_1 a_{11} + n_1 a_{12}$. Doctor agent 2: Task performance: a_2. Output: $q_2 = a_2 + a_{12}\lambda + \varepsilon$. Cost function: $\varphi_2 = \frac{1}{2}c_2 a_2^2$. Incentive contract: $w_2 = t_2 + m_2(a_2 + a_{12}\lambda + \varepsilon)$.

Solving method is similar to Sect. 98.2 and omitted from the solution process.

- Given the principal to agents' optimal incentive: $m_1^3 = 1, n_1^3 = \lambda, m_2^3 = \frac{1}{1+\eta_2 c_2 \sigma^2}$.
- The agent's optimal effort:

$$a_{11}^3 = \frac{m_1}{c_1} = \frac{1}{c_1}, a_{12}^3 = \frac{n_1}{d_1} = \frac{\lambda}{d_1}, a_2^3 = \frac{m_2}{c_2} = \frac{1}{c_2(1 + \eta_2 c_2 \sigma^2)}.$$

- The agent's optimal compensation:

$$w_1^3 = t_1 + \frac{1}{c_1} + \frac{\lambda^2}{d_1}, \quad w_2^3 = t_2 + \frac{1}{c_2(1 + \eta_2 c_2 \sigma^2)^2} + \frac{\lambda^2}{d_1(1 + \eta_2 c_2 \sigma^2)}.$$

- The principal's optimal income:

$$E^3[q_1 + q_2 - w_1 - w_2] = \frac{\eta_2 c_2 \sigma^2 d_1 - \lambda^2 c_2(1 + \eta_2 c_2 \sigma^2)}{c_2 d_1(1 + \eta_2 c_2 \sigma^2)^2} - t_1 - t_2.$$

- The total output of the department:

$$Q^3 = q_1 + q_2 = a_{11} + a_{12}\lambda + a_2 = \frac{1}{c_1} + \frac{\lambda^2}{d_1} + \frac{1}{c_2(1 + \eta_2 c_2 \sigma^2)}.$$

98.5 Comprehensive Analysis

Based on model (98.1), we analyze separately from five aspects respectively, which are incentives, agent's efforts, agent's compensation, team output as well as the principal gains.

Conclusion 1: The principal is willing to pay for the contextual performance, which depends on the method of cost accounting.

In the model (98.2), the costs of agents' performance are dependent. So the task performance is $n_1^2 = 0$, which indicates the principal don't compensate the contextual performance of agents alone. But $m_1^2 = 1 + \lambda$, the task performance incentives increased λ based on the benchmark model (98.1). To model (98.3), where the costs of agents' performance are independent, the task performance incentives of agent 1 isn't changed, and contextual performance incentives is $n_1^3 = \lambda$, which is just right the extra task performance incentives of agent 1 in the model (98.2).

This also explains why it is necessary that basic hospital should pay for management fees to core hospital in the hospital alliance, which is composed of fixed and variable. Thus, doctors can also provide advisory services to get a certain amount of consulting fees. When $0 \prec c_1 \prec \frac{\lambda}{\lambda+2} \cdot d_1$, separate payment on contextual performance become more. In the basic hospitals, it was easy to do, because the contextual performance cost of doctors there is obviously much less than task performance cost. However, the core hospital is not the case. Due to the limited medical resources, especially it is not lack of patient in the core hospital, increased patient will increase the cost of the hospital, so of course, it will not give doctors consulting fees.

So, whether medical institutions pay the doctor's contextual performance alone or not in realty? We visited 42 departments in 19 Grand III hospitals in Chengdu, a city in China which has 31 Grand III hospitals. The result shows that 73.8% cannot pay the doctor's contextual performance alone, especially consulting fees. Also we visited on the 26 basic-level hospitals, only 8 cannot pay separately, but it will be included in the registration fee or inquiry fees. This is consistent with theoretical analysis above.

Conclusion 2: The efforts of doctor agents who have contextual performance has positive correlation with contextual performance impact factor λ, and negative correlation with the cost of efforts.

In the model (98.2), $\frac{\partial a_1^2}{\partial \lambda} = \frac{1}{c_1} > 0$, the efforts of doctor agent 1 will increase with the increase of contextual performance impact factor λ, which shows that if principal award, the agents would love to make more efforts in return. As to model (98.3), the task performance efforts of agent 1 is the same to the benchmark model (98.1), and contextual performance efforts is $a_{12}^3 = \frac{\lambda}{d_1}$. When $d_1 = c_1$, the efforts of doctor agent 1 are the same in the two models. However, the efforts of doctor agent 1 will reduces with the increase of the marginal cost of contextual performance d_1. This condition is consistent with the reality. If the cost of increase peer's output is too expensive, employees would tend to his task.

Conclude 3: The compensation of all doctor agents in team will increase as long as there are the agents who making efforts on contextual performance. And the increase part has positively related to the contextual performance impact factor λ.

In the model (98.2), the salary of doctor agent 1 increases $\frac{\lambda^2 + 2\lambda}{c_1}$ compared to benchmark model (98.1), in addition, $\frac{\partial w_1^2}{\partial \lambda} = \frac{2(\lambda+1)}{c_1}$, which has positively related to the performance impact factor λ. When the measurement error is very small and can be ignored, the salary of doctor agent 2 increases

$$\frac{\lambda(1 + \lambda)}{c_1(1 + \eta_2 c_2 \sigma^2)},$$

which is due to the contextual performance of doctor agent 1,

$$\left(\frac{\partial w_2^2}{\partial \lambda} = \frac{2\lambda + 1}{c_1(1 + \eta_2 c_2 \sigma^2)} \right).$$

The similar conclusion can be obtained in the model (98.3). Both models (98.2) and the model (98.3), the compensation of doctor agent 1 isn't and related to the performance of doctor agent 2, while the compensation of doctor agent 2 has relationship with the performance of doctor agent 1. In such situation, doctor agent 2 relies on doctor agent 1, leading to free-riding problem.

Conclude 4: From the angle of the whole team, the doctor agents who make efforts on contextual performance increase the total output.

When the measurement error is very small and can be ignored, total output increase $\frac{\lambda^2 + 2\lambda}{c_1}$ in the model (98.2), while increase $\frac{\lambda^2}{d_1}$ in the model (98.3). The increase of total output is because of contextual performance in both two models. Moreover, the increase part is equal to the increased salary of doctor agent 1. All the medical market increased so much service, can serve more patients, which alleviates the pressure of current medical treatment. As a result, we should appeal on contextual performance in public hospital. It benefits patients, doctors, hospital as well as the government.

Conclude 5: The income of principal is negative, indicating that it is necessary of the government paying for financial subsidies to public hospital. In both models (98.2) and (98.3), the increased part of total output is equal to the increased compensation of doctor agent 1, and the compensation of doctor agent 2 is also increased. So the principal's income is reduced, which part is the compensation of agent 2. The income of principal is still negative. In that way, whether can we say contextual performance have negative benefits? Actually otherwise, contextual performance increases the amount of social services, which is beneficial to the whole market. In order to make the entire market operate well, government financial subsidies to hospitals is necessary. To ensure that all patients can go to see a doctor timely, the government makes the price on medical service and promises the price is afford by most of citizen. In this situation, the charges for most services in public hospitals can not meet the costs. Consequently, the result is more adverse to the masses. There has been appeared induction consumer behavior, such as more drugs, more check, etc. So, we appeal on contextual performance in hospital, at the meantime, the government has to give a certain degree of financial allowance. Then, the hospitals will provide better service to the masses.

98.6 Conclusion

Actually, Hospital Alliance takes advantages of doctors' contextual performance in the core hospitals. Similarly, contextual performance of well-known doctors is also useful to hospitals. Through the proving, we concluded that contextual performance can increase the output of the team. So we advocate team performance pay plan. In order to improve the quality of medical service, more than ten countries, such as the United States, Germany, Canada and so on, develop performance compensation plan to pay for the doctors [18]. Of course, it is defective, such as undermining productivity, condoning on-the-job leisure, fostering a bureaucratic mentality, passing the buck, and so on [17].

Appendix

As a team, our target is the team to maximize the return. So, the model we established as follows:

$$\begin{cases} \max\limits_{m_i,n_i} E[q_1 + q_2 - w_1 - w_2] \\ \text{s.t.} \begin{cases} E(-e^{-\eta_i[w_i-\varphi_i(a_i,\lambda)]}) \geq \mu(\varpi_i, a_i) \\ (a_i, \lambda a_i) \in \arg_{a_i,\lambda a_i} \max E(-e^{-\eta_i[w_i-\varphi_i(a_i,\lambda)]}). \end{cases} \end{cases} \quad (98.3)$$

Among them, ϖ_i is the minimum wage of the contract which agent i can accept, and $\mu(\overline{\varpi}_i, a_i)$ is the reservation utility of agent i. To simplify the constraints: When $i = 1$: $E(-e^{-\eta_1[w_1-\varphi_1(a_1,\lambda)]}) = E(-e^{\eta_1[t_1+m_1a_1+n_1(\lambda a_1+\varepsilon)-\frac{1}{2}c_1a_1^2]})$
$= (-e^{-\eta_1[t_1+m_1a_1+n_1\lambda a_1-\frac{1}{2}c_1a_1^2]}) \cdot E(e^{-\eta_1n_1\varepsilon})$.

Random variable obeys normal distribution which mean is 0 and variance is σ^2, for any β: $E(e^{\beta\varepsilon}) = e^{\frac{\beta^2\sigma^2}{2}}$. Therefore $E(e^{-\eta_1n_1\varepsilon_1}) = e^{\frac{1}{2}\eta_1^2n_1^2\sigma^2}$. So,

$$E(-e^{-\eta_1[w_1-\varphi_1(a_1,\lambda)]}) = -e^{-\eta_1[t_1+m_1a_1+n_1\lambda a_1-\frac{1}{2}c_1a_1^2-\frac{1}{2}\eta_1n_1^2\sigma^2]}.$$

To maximize principal utility, so that

$$-e^{\eta_1[t_1+m_1a_1+n_1\lambda a_1-\frac{1}{2}c_1a_1^2-\frac{1}{2}\eta_1n_1^2\sigma^2]} = E(-e^{-\eta_1\varpi_1}),$$

when $i = 1$,

$$\varpi_1 = t_1 + m_1a_1 + n_1\lambda a_1 - \frac{1}{2}c_1a_1^2 - \frac{1}{2}\eta_1n_1^2\sigma^2. \tag{98.4}$$

Similarly, when $i = 2$:

$$\varpi_2 = t_2 + m_2(a_2 + \lambda a_1) - \frac{1}{2}a_2^2c_2 - \frac{1}{2}\eta_2m_2^2\sigma^2. \tag{98.5}$$

In other words, the agent is willing to accept the lowest salary equal to the remuneration paid by the principal to remove his cost and the risk of a variety of reasons in reality Agent i will choose the task performance level of effort a_i to maximize salary. Make the derivations of (98.4), (98.5) equal to zero:

$$\frac{\partial\varpi_1}{\partial a_1} = m_1 + n_1\lambda - a_1c_1 = 0, \quad \frac{\partial\varpi_2}{\partial a_2} = m_2 - a_2c_2 = 0. \tag{98.6}$$

By Eq. (98.6), we have:

$$a_1 = \frac{m_1 + n_1\lambda}{c_1}, a_2 = \frac{m_2}{c_2}. \tag{98.7}$$

In addition, we also can get the behavior of an agent how to influence another agent's salary:

$$\frac{\partial\varpi_1}{a_2} = 0, \quad \frac{\partial\varpi_2}{a_1} = m_2\lambda. \tag{98.8}$$

By Eq. (98.8), we have that the performance of the agent 1 will affect salary of agent 2 in the only two agent models. But in turn does not affect.

If $a_i = 0(i = 1, 2)$, m_i, n_i are also 0, there is no significance to study. Therefore, in the following discussion, we consider only $a_i > 0$.

The principal's expected utility is:

$$E[q_1 + q_2 - w_1 - w_2] = a_1 + a_2 + \lambda_1 a_1 - [t_1 + m_1 a_1 + n_1 \lambda a_1] - [t_2 + m_2(a_2 + \lambda a_1)].$$

Principal chose m_i, n_i to maximize his utility. Based on the above solution, the problem becomes:

$$\begin{cases} \max\limits_{m_i, n_i} E[q_1 + q_2 - w_1 - w_2] \\ \text{s.t.} \begin{cases} a_1 = \frac{m_1 + n_1 \lambda}{c_1} \\ a_2 = \frac{m_2}{c_2}, \end{cases} \end{cases} \tag{98.9}$$

$$t_1 + m_1 a_1 + n_1 \lambda a_1 - \frac{1}{2} c_1 a_1^2 - \frac{1}{2} \eta_1 n_1^2 \sigma^2 \geq \varpi_1, \tag{98.10}$$

$$t_2 + m_2(a_2 + \lambda a_1) - \frac{1}{2} a_2^2 c_2 - \frac{1}{2} \eta_2 m_2^2 \sigma^2 \geq \varpi_2. \tag{98.11}$$

In the most advantage, should take tight restrictions, so we can replace $a_i, t_1 \, (i = 1, 2)$, which translate constraint problem into unconstrained optimization problem:

$$\max\limits_{m_i, n_i} E[q_1 + q_2 - w_1 - w_2]$$

$$= \max\limits_{m_i, n_i} a_1 + (a_2 + \lambda a_1) - [t_1 + m_1 a_1 + n_1 \lambda a_1] - [t_2 + m_2(a_2 + \lambda a_1)]$$

$$= \max\limits_{m_i, n_i} (1 + \lambda) a_1 + a_2 - \left[\varpi_1 + \frac{1}{2} c_1 a_1^2 + \frac{1}{2} \eta_1 n_1^2 \sigma^2 \right] - \left[\varpi_2 + \frac{1}{2} c_2 a_2^2 + \frac{1}{2} \eta_2 m_2^2 \sigma^2 \right].$$

We can see distinctly, the higher minimum salary ϖ agents expect, the smaller the principal's utility expected. And the minimum salary of an agent acceptable is fixed in a certain period. So (98.7) into:

$$\max\limits_{m_i, n_i} \frac{m_1 + n_1 \lambda}{c_1} (1 + \lambda) + \frac{m_2}{c_2} - \left[\varpi_1 + \frac{1}{2} c_1 \left(\frac{m_1 + n_1 \lambda}{c_1} \right)^2 + \frac{1}{2} \eta_1 n_1^2 \sigma^2 \right]$$

$$- \left[\varpi_2 + \frac{1}{2} c_2 \left(\frac{m_2}{c_2} \right)^2 + \frac{1}{2} \eta_2 m_2^2 \sigma^2 \right].$$

We get the first order conditions: when $i = 1$,

$$\frac{\partial E}{\partial m_1} = \frac{1 + \lambda}{c_1} - \frac{m_1 + n_1 \lambda}{c_1} = 0, \quad \frac{\partial E}{\partial n_1} = \frac{(1 + \lambda)\lambda}{c_1} - \frac{(m_1 + n_1 \lambda)\lambda}{c_1} - \eta_1 \sigma^2 n_1 = 0.$$

Solution: $m_1^* = 1 + \lambda, n_1^* = 0$. Similarly, when $i = 2$: $m_2^* = \frac{1}{1 + \eta_2 c_2 \sigma^2}$. So we obtain the incentive the principal gives agents when the principal's return is maximized. At the same time, also get the agents' optimal tasks performance, namely: $a_1^* = \frac{1+\lambda}{c_1}, a_2^* = \frac{1}{c_2(1+\eta_2 c_2 \sigma^2)}$. And, agents' optimal compensation are:

$$w_1^* = t_1 + m_1 a_1 + n_1 \lambda a_1 = t_1 + \frac{(1+\lambda)^2}{c_1},$$

$$w_2^* = t_2 + m_2(a_2 + \lambda a_1) = t_2 + \frac{1}{c_2(1+\eta_2 c_2 \sigma^2)^2} + \frac{\lambda(1+\lambda)}{c_1(1+\eta_2 c_2 \sigma^2)}.$$

The optimal gain of the principal is:

$$E[q_1 + q_2 - w_1 - w_2] = \frac{\eta_2 c_2 \sigma^2}{c_2(1+\eta_2 c_2 \sigma^2)^2} - \frac{\lambda(1+\lambda)}{c_1(1+\eta_2 c_2 \sigma^2)} - t_1 - t_2.$$

Finally, the department's total output as follows:

$$q_1 + q_2 = \frac{(1+\lambda)^2}{c_1} + \frac{1}{c_2(1+\eta_2 c_2 \sigma^2)}.$$

Acknowledgments This work was supported by the National Natural Science Foundation of China (No. 71131006).

References

1. Baker GP, Jensen MC, Murphy KJ (1988) Compensation and incentives: practice vs theory. J Finance 43(3):593–616 NULL
2. Borman WC, Motowidlo SJ (1997) Task performance and contextual performance: the meaning for personnel selection research. Human Perform 10(2):99–109
3. Chao H, Siqueira K (2013) Mixed incentive contracts in partnerships. Int J Econ Theory 9(2):147–159
4. Drago R, Turnbull GK (1988) Individual versus group piece rates under team technologies. J Jpn Int Econ 2(1):1–10
5. FitzRoy FR, Kraft K (1986) Profitability and profit-sharing. J Ind Econ 113–130
6. FitzRoy FR, Kraft K (1987) Cooperation, productivity, and profit sharing. Quart J Econ 102(1):23–35
7. Heywood JS, Jirjahn U, Tsertsvadze G (2005) Getting along with colleagues—does profit sharing help or hurt? Kyklos 58(4):557–573
8. Holmstrom B (1982) Moral hazard in teams. Bell J Econ 324–340
9. Itoh H (1991) Incentives to help in multi-agent situations. Econometrica J Econ Soc 611–636
10. Jiao JH, Zhao JY (2004) Strategic union: a visual angle for basic modernization of hospital. Chin Hosp 8(12):4–7
11. Kato T, Kauhanen A, Kujansuu E (2013) The performance effects of individual and group incentives
12. Kauhanen A, Napari S (2012) Performance measurement and incentive plans. Ind Relat J Econ Soc 51(3):645–669
13. Lazear EP (1996) Performance pay and productivity. Technical report, National Bureau of Economic Research

14. LePine JA, Van Dyne L (2001) Voice and cooperative behavior as contrasting forms of contextual performance: evidence of differential relationships with big five personality characteristics and cognitive ability. J Appl Psychol 86(2):326
15. Liang JY, Li Y (2007) The key factors affecting knowledge transfer success in hospital alliance. Forecasting 26(1):27–32
16. Nalbantian HR, Schotter A (1997) Productivity under group incentives: an experimental study. Am Econ Rev 314–341
17. Robinson JC (2001) Theory and practice in the design of physician payment incentives. Milbank Quart 79(2):149–177
18. Smith PC, Stepan A, Valdmanis V (1997) Principal-agent problems in health care systems: an international perspective. Health Policy 41(1):37–60

Chapter 99
Research of Contractor's Incentive Problem Under the Ternary Structure

Lei Zhao and Hongmei Guo

Abstract In order to solve the problem of insufficient incentives for the contractor due to the construction of information asymmetry, the use of agency theory, under the ternary structure of owners, contractors and supervisor, based on the information available verifiable and information partial verify, the conditions were established under the supervision of the owners of the side effects of moral hazard model contractor, conducted a quantitative research project management. The results show that: in the BOQ model is based on the condition part of verifiable information will increase the risk of the contractor's risk aversion, and it will make the owners to obtain a lower expected utility, size and risk of the contractor and the contractor supervisor's ability to work is negatively correlated with the cost of the contractor's work is positively correlated.

Keywords Ternary structure · Contractors · Supervisor · Moral hazard · Incentives

99.1 Introduction

Construction industry is one of the pillar industries for China's national economy. With the rapid development of urbanization in China, the more broad market for construction industry to develop is coming. As a result, the sound development of construction industry is extremely important for the national economy. However, there are many problems in this industry which threaten the health development of construction industry, such as the chaos of management system, outstanding

L. Zhao
College of Environment and Civil Engineering, Chengdu University of Technology, Chengdu 610059, People's Republic of China

H. Guo (✉)
School of Management, Sichuan University, Chengdu 610064, People's Republic of China
e-mail: guohongmei8888@163.com

J. Xu et al. (eds.), *Proceedings of the Eighth International Conference on Management Science and Engineering Management*, Advances in Intelligent Systems and Computing 281, DOI: 10.1007/978-3-642-55122-2_99, © Springer-Verlag Berlin Heidelberg 2014

payments, low profit margins for contracting companies, weak enterprise competitiveness, quality defects and so on.

To prevent such vicious incident, China has absorbed the advanced experience of western countries and formulated the corresponding institutional measures to guarantee the health development of construction industry. Since 1988, the supervision system for engineering was implemented and continuously developed for perfect in aspects of laws and regulations. At present, engineering management system of China is mainly formed the following four basic systems: responsibility system of engineering legal person, tendering and bidding system, engineering supervision system and contract management system. Correspondingly, the former dual structure of construction market has changed into ternary structure of the owner, contractor and supervisor. Practice has proved that these systems played significant effect to ensure the quality of the engineering construction, completion on schedule and improvement of level of engineering construction.

But in the process of project management in the actual operation, there have been many problems, construction of information asymmetry still exists, especially between the owner and the contractor, the contractor's own efforts to conceal information is widespread, so they formed a typical moral hazard problem. For this equation of problem, there are many corresponding research. Among them, from an empirical perspective [1, 12] and the risk of [2, 14] for the study. Use methods, mostly in the framework of a principal and an agent of the study, the main research questions using game theory [6, 8] and incentive theory [15, 16] by one game under complete information and asymmetric information in the case under the framework of a comparative study.

In reality, the contractor can not only conceal their efforts to information, and because the non-professional landlords, for the quality of construction can also be viewed as private information of the contractor (which in the quality of the project can be broadly understood as the construction of final results, including the quality of the project, the construction period, the project cost, engineering safety, etc.). Owners in order to effectively obtain information about the quality of the project, need to ask the supervisor to participate in the construction process, due to the characteristics of professional supervision, it can be more effective access to appropriate information. The other hand, the uncertainty makes construction contractor may not even be hard to obtain high quality, even if the effort is not necessarily the same owners will get the right quality of the project information, so that makes the quality of the project has become a part of verifiable information. In this case, the owners of incentive for the contractor what kind of change?

Verifiable information for some of the concepts, Grossman [5] first raised this argument, he does not think the presence of a certain probability test out its true quality signal when the products are tested, such as product quality can be verified with a certain probability information is known as part of its verifiable information. Green and Laffont [4] believes the traditional commission adverse selection model agent, and the agent is always designed according to the client to choose the contract, then the agent would not choose to lie, but agents may lie, the article gives the agent when possible lie within a certain range, the principal game with a certain probability model of supervision and inspection, the article will be verified certain probability principal agent of the information referred to partial verifiable information. Later scholars [9, 10, 13] mostly follows the Green and Laffont verifiable information on the part of the definition, but there are some studies [3, 11] discussed the moral hazard model partial verifiable information.

In this paper, the tender contract model [7] as the background, assuming that the amount of the contract is determined after bidding, consider commissioning an effort by the owners and have chosen contractor composed-agent model, which is the principal owners, contractors are its agents people. First, the establishment of moral hazard model verifiable information conditions as a benchmark to establish the moral hazard model last half verifiable conditions under more realistic conditions by calculating and comparing the two situations, the best analysis would pay under different circumstances what kind of change, how different conditions affect the optimal payments.

Game timing tripartite contract as follows: (1) Owners contract; (2) Contractor to reject or accept the contract; (3) Contractor choose whether efforts; (4) Contractor for construction; (5) The direction of the Commissioner of the owners send the report; (6) Execute the contract.

99.2 Assumptions

Assumption 1 The owners as principal is risk neutral. Contractor as the agent is risk averse, their valuable effort level can take two possible values, we were normalized to a zero level of effort and positive effort level. Contractor's utility function between effort and money is separable, $U = u(t) - \varphi(e)$. While the utility function $u(t)$. Is increasing, concave, $u' > 0$, $u'' < 0$. Definitions $\overline{u} = u(\overline{t})$ and $\underline{u} = u(\underline{t})$, equivalently there $\overline{t} = h(\overline{u})$ and $\underline{t} = h(\underline{u})$.

Assumption 2 The randomness of the process of building contractors, effort level will affect the quality of the project, random quality level can only take two results: high quality and low quality. Contractor will be hard enough to achieve a certain probability of high quality, i.e., $pr(q = \overline{q}|e_1 = 1) \geq \frac{1}{2}$, this also reflects the probability of actually working ability of the contractor. Assume contractors do not work hard, then it will be of low quality, $pr(q = \overline{q}|e_1 = 0) = 0$, $pr(q = \underline{q}|e_1 = 0) = 1$.

Assumption 3 Supervisor supervision process is random, the effort level will affect the accuracy of the report, the report only two values, high and low quality, low quality of the project and the supervisor efforts, it will observe a certain probability of correct to make the low quality of the project report, which $pr(s = \underline{s}|e_2 = 1, q = \underline{q}) \geq \frac{1}{2}$, this probability can also be seen as a reflection of the commissioner of the ability to work. To facilitate the study, simplifying variable, high-quality engineering is assumed, the supervisor of hard work and the low quality of the project can not find the problem, so will send the high quality of the project report, which $pr(s = \bar{s}|e_2 = 1, q = \bar{q}) = 1$. The same time, when supervisor no effort, no matter the level of quality of the project, the supervisor will choose to send disclaimer high quality of the project report, which $pr(s = \bar{s}|e_2 = 0, q = \bar{q}) = pr(s = \bar{s}|e_2 = 0, q = \underline{q}) = 1$.

Assumption 4 To determine the owners paid the contractor under the contract and report the results of supervision. Specifically, if the commissioner's report was sent to the owners of good quality, then get a high transfer payments to the contractor under the contract, if the commissioner's report was sent to the owners of poor quality, then the contractor will be subject to appropriate penalties, get low transfer payment under the contract. Contractor can choose the complaint after the parties informed the commissioner of reporting results, then assume the correctness of the owners neutral body through external intervention to fully determine the supervisor's report, if the report is wrong supervisor, then their fines to this party is prevented from poor supervision and maliciously sent rewarded quality of the project report.

99.3 Model

In both cases the problem of discussion, the first verifiable information analysis conditions, which is a standard model of moral hazard; next partial verifiable information analyzing conditions, a comparison of the former model, taking into account where the supervisor factors that affect the ability of owners and contractors and utility revenue.

1. Model of Engineering Quality Verifiable Conditions Due to motivate the contractor makes efforts must ultimately choose to implement, so the income of owners of two cases discussed, probability of the two cases were π_1 and $1 - \pi_1$. So, the owners ultimately benefits are: $V = \pi_1(\bar{Q} - h(\bar{u}) + (1 - \pi_1)(Q - h(\underline{u})))$. Similarly, the income of the contractor is also divided into these two cases. So you can arrive at a final effort to gain the contractor in case of: $U = \pi_1\bar{u} + (1 - \pi_1)\underline{u} - \varphi_1$. The plan to establish the following:

$$\max_{\{(\bar{u},u)\}} \pi_1(\bar{Q} - h(\bar{u})) + (1 - \pi_1)(Q - h(\underline{u})),$$

$$\pi_1\bar{u} + (1 - \pi_1)\underline{u} - \varphi_1 \geq 0, \tag{99.1}$$

$$\pi_1\bar{u} + (1 - \pi_1)\underline{u} - \varphi_1 \geq \underline{u}. \tag{99.2}$$

which Eq. (99.1) for the participation constraint, the expected return after the contractor efforts is greater than 0, in order to make the contractor may accept the contract owners. Equation (99.2) for the incentive, the contractor select effort there will be greater utility, select the contractor to guide efforts.

Order λ_1 and λ_2 respectively, Eqs. (99.1) and (99.2) of the Lagrange multiplier. Select $\bar{u}$ and optimization $\underline{u}$, the following first-order conditions:

$$-\pi_1 h'(\bar{u}^1) + \lambda_1 \pi_1 + \lambda_2 \pi_1 = 0, \tag{99.3}$$

$$-(1-\pi_1)h'(u^1) + \lambda_1(1-\pi_1) - \lambda_2\pi_1 = 0. \tag{99.4}$$

Superscript symbols which represent the best results under a plan, the same below. From Eqs. (99.3) and (99.4) obtained by adding:

$$\lambda_1 = \pi_1 h'(\bar{u}^1) + (1-\pi_1)h'(u^1) = \frac{\pi_1}{u'(\bar{t}^1)} + \frac{1-\pi_1}{u'(t^1)} > 0. \tag{99.5}$$

The Eqs. (99.5) into (99.3) we have:

$$\lambda_2 = (1-\pi_1)\left(h'(\bar{u}^1) - h'(\underline{u}^1)\right) = (1-\pi_1)\left(\frac{1}{u'(\bar{t}^1)} - \frac{1}{u'(\underline{t}^1)}\right). \tag{99.6}$$

Because there $u' > 0$, $u'' < 0$, we can see Eq. (99.6) the right side is greater than 0, $\lambda_2 > 0$. So the corresponding constraint Eqs. (99.1) and (99.2) are tight, through their compactness, can be calculated $u(\bar{t}^1)$ and $u(\underline{t}^1)$ values: $u(\bar{t}^1) = \varphi_1 + \frac{1-\pi_1}{\pi_1}\varphi_1$, $u(\underline{t}^1) = 0$. Or direct payment is optimal:

$$\bar{t}^1 = h\left(\varphi_1 + \frac{1-\pi_1}{\pi_1}\varphi_1\right), \tag{99.7}$$

$$\underline{t}^1 = h(0). \tag{99.8}$$

In this case, the owners of expected utility can be obtained as follows:

$$V^1 = \pi_1 \bar{Q} + (1-\pi_1)\underline{Q} - \pi_1 h\left(\frac{\varphi_1}{\pi_1}\right) - (1-\pi_1)h(0). \tag{99.9}$$

2. Model of Engineering Quality Partial Verifiable Conditions

Under this condition, the owners can not be fully observed by the contractor supervisor to engineering quality information, the commissioner may only be observed with a certain probability quality information.

At this time the contractor's effort can not be observed directly by the owners and supervision side, but can not force contractors to implement effort. Even so, the owners can make through incentives contractor chose to implement effort. Due to motivate the contractor makes an inevitable choice for the implementation effort, so the income of owners of three cases discussed, the probability of the three cases were π_1, $(1-\pi_1)\pi_2$ and $(1-\pi_1)(1-\pi_2)$. So, the owners ultimately benefits are: $V = \pi_1(\bar{Q}-h(\bar{u}))+(1-\pi_1)\pi_2(\underline{Q}-h(\underline{u}))+(1-\pi_1)(1-\pi_2)(\underline{Q}-h(\bar{u}))$. Similarly, the income of the contractor is also divided into three cases. So eventually gain can be drawn when the contractor efforts to: $U_1 = \pi_1\bar{u} + (1-\pi_1)\pi_2\underline{u} + (1-\pi_1)(1-\pi_2)\bar{u} - \varphi_1$.

Establish the following plan:

$$\max_{\{(\bar{u},\underline{u})\}} \pi_1(\bar{Q} - h(\bar{u})) + (1 - \pi_1)\pi_2(Q - h(\underline{u})) + (1 - \pi_1)(1 - \pi_2)(\underline{Q} - h(\bar{u})),$$

$$\pi_1\bar{u} + (1 - \pi_1)\pi_2\underline{u} + (1 - \pi_1)(1 - \pi_2)\bar{u} - \varphi_1 \geq (1 - \pi_2)\bar{u} + \pi_2\underline{u}, \tag{99.10}$$

$$\pi_1\bar{u} + (1 - \pi_1)\pi_2\underline{u} + (1 - \pi_1)(1 - \pi_2)\bar{u} - \varphi_1 \geq 0. \tag{99.11}$$

which Eq. (99.11) for the participation constraint, the expected return after the contractor efforts is greater than 0, in order to make the contractor may accept the contract owners; Eq. (99.10) for the incentive, the contractor select effort there will be greater utility, select the contractor to guide efforts.

Order u_1 and u_2 respectively, Eqs. (99.10) and (99.11) lagrange multiplier. Were selected and optimization, the following first-order conditions:

$$-[1 - (1 - \pi_1)\pi_2]h'(\bar{u}^2) + \pi_1\pi_2\mu_1 + [1 - (1 - \pi_1)\pi_2]\mu_2 = 0, \tag{99.12}$$

$$-(1 - \pi_1)\pi_2 h'(\underline{u}^2) - \pi_1\pi_2\mu_1 + (1 - \pi_1)\pi_2\mu_2 = 0. \tag{99.13}$$

Equations (99.12) and (99.13) obtained by adding:

$$\mu_2 = [1 - (1 - \pi_1)\pi_2]h'(\bar{u}^2) + (1 - \pi_1)\pi_2 h'(\underline{u}^2) = \frac{1 - (1 - \pi_1)\pi_2}{u'(\bar{t}^2)} + \frac{(1 - \pi_1)\pi_2}{u'(\underline{t}^2)} > 0. \tag{99.14}$$

Thus, participation constraint Eq. (99.11) must be tight. The Eqs. (99.14) into (99.13), you can get:

$$\mu_1 = \frac{(1 - \pi_1)[1 - (1 - \pi_1)\pi_2]}{\pi_1}(h'(\bar{u}^2) - h'(\underline{u}^2)) = \frac{(1 - \pi_1)[1 - (1 - \pi_1)\pi_2]}{\pi_1}\left(\frac{1}{u'(\bar{t}^2)} - \frac{1}{u'(\underline{t}^2)}\right). \tag{99.15}$$

Because $\bar{u} \geq \underline{u}$ and $u'' < 0$, we can see Eq. (99.15) on the right is strictly positive, so u_1 corresponding incentive Eq. (99.10) is also tight. Using Eqs. (99.10) and (99.11) compactness, can be calculated $\bar{u}^2$ and $\underline{u}^2$ values:

$$\bar{u}^2 = \frac{\varphi_1}{\pi_1}, \quad \underline{u}^2 = -\frac{1 - \pi_2}{\pi_1\pi_2}\varphi_1. \tag{99.16}$$

Or written as:

$$\bar{t}^2 = h\left(\varphi_1 + \frac{1 - \pi_1}{\pi_1}\varphi_1\right), \tag{99.17}$$

$$\underline{t}^2 = h\left(\varphi_1 - \frac{1 - (1 - \pi_1)\pi_2}{\pi_1\pi_2}\varphi_1\right). \tag{99.18}$$

Owners utility is:

$$V^2 = \pi_1 \bar{Q} + (1-\pi_1)Q - [1-(1-\pi_1)\pi_2]h\left(\frac{1}{\pi_1}\varphi_1\right) - (1-\pi_1)\pi_2 h\left(\frac{1-\pi_2}{\pi_1\pi_2}\varphi_1\right). \quad (99.19)$$

99.4 Model to Analyze and Compare

First, we can see from the model under the condition of verifiable information, the owners will be part of the transfer of risk to the contractor, and when compared to the information is complete, the contractor no longer get full insurance.

Proposition 99.1 *Under conditions of verifiable information, the contractor will not get full coverage, but will face some risks. Supervisor's decision to report the payment, specifically:* $\bar{t}^1 = h(\varphi_1 + \frac{1-\pi_1}{\pi_1}\varphi_1), \underline{t}^1 = h(0).$
Under the condition of the model and compare verifiable portion verifiable conditions model Eqs. (99.7) and (99.17) shows that, in both cases, a high contracting party transfers is the same.

Proposition 99.2 *Either partially or verifiable conditions under verifiable conditions, when the Commissioner sends high-quality reports, transfer payments are the same contractor, are* $h(\varphi_1 + \frac{1-\pi_1}{\pi_1}\varphi_1)$, *supervisor own factors will not affect this result.*

Compare verifiable conditions under model conditions and some verifiable models, Eqs. (99.8) and (99.18), we can see the lower part of verifiable conditions, low transfers contractor becomes lower. This means that the engineering contractor with greater risk.

Comparing Eqs. (99.9) and (99.19) shows that:

$$V^1 - V^2 = (1-\pi_1)(1-\pi_2)h\left(\frac{1}{\pi_1}\varphi_1\right) + (1-\pi_1)\pi_2 h\left(\frac{1-\pi_2}{\pi_1\pi_2}\varphi_1\right) - (1-\pi_1)h(0)$$

$$\geq (1-\pi_1)(1-\pi_2)h\left(\frac{1}{\pi_1}\varphi_1\right) + h\left(\frac{(1-\pi_1)(1-\pi_2)}{\pi_1}\varphi_1\right) \geq 0. \quad (99.20)$$

Also due to the risk aversion of the contractor, the owner needs to increase the cost to motivate the contractor still accepted contract. Expected revenue owners and thus also reduced.

Proposition 99.3 *In the contractor under the conditions of risk aversion, some verifiable condition information will increase the risk of contracting parties, which would make the owners to obtain a lower expected utility.*

Observation Eq. (99.16), it is obvious, π_1 *larger, higher transfer payments to the contractor smaller, the greater the lower transfer payments, which means that the risk of contracting the smaller pair; contrary to* φ_1, *the size of this risk with the contractor, he was a positive correlation.*

Equation (99.16) for derivative can be obtained: $\frac{\partial u_2}{\partial \pi_2} = \frac{\varphi_1}{\pi_1\pi_2^2} > 0.$

As can be seen from the above equation, π_2 larger, lower transfer payments to the contractor greater, corresponding to the risk of contracting the smaller aspects. In fact, when $\pi_2 = 1$, part of the conditions of the model verifiable information becomes verifiable information conditions.

Proposition 99.4 *Risk size and ability to work with the contractor or the contractor's supervision negative correlation between the parties, the contractor's cost and effort were positively correlated.*

99.5 Conclusion

This paper discusses a landlord to consider the impact of Supervisor-Contractor's principal-agent relationship, through the establishment of appropriate moral hazard models were analyzed under conditions verifiable information and verifiable information partial, the owner of the contracting parties to the most excellent incentive policy choice and to optimal results under these two conditions were compared. Supervisor occurs discovery may make the quality of the information verifiable part of the project. Part of the risk of such verifiable conditions, risk aversion itself will expand the contractor faced, for the owners, this means more incentives need to pay the cost, lowering his expected return. Their ability to work simultaneously supervisor will also affect the risk of contracting aspects. In particular, the ability of the size of the supervisor, will reverse the effects of the size of this risk. The results of the study contractor incentive problems owners have theoretical significance.

References

1. Arditi D, Yasamis F (1998) Incentive/disincentive contracts: perceptions of owners and contractors. J Constr Eng Manage 124(5):361–373
2. Berends TC (2000) Cost plus incentive fee contracting-experiences and structuring. Int J Project Manage 18(3):165–171
3. Ghosh S (2001) Financial stability and public policy: an overview. Library University of Munich, Germany
4. Green JR, Laffont JJ (1986) Partially verifiable information and mechanism design. Rev Econ Stud 53(3):447–456
5. Grossman SJ (1981) The informational role of warranties and private disclosure about product quality. J Law Econ 24(3):461–483
6. Hancher DE, Lambert SE (2002) Quality-based prequalification of contractors. Transp Res Record: J Transp Res Board 1813(1):260–274
7. Hong S, Qian Z, Zhang D (2007) Analysis of supply chain principal-agent incentive contract. Int J ManageSci Eng Manage 2(2):155–160
8. Hughes RK, Ahmed SA (1991) Highway construction quality management in oklahoma. Transp Res Record 13(10):20–26
9. Korn E (2004) Voluntary disclosure of partially verifiable information. Schmalenbach Bus Rev 56(2):139–163

10. Lipman BL, Seppi DJ (1995) Robust inference in communication games with partial provability. J EconTheory 66(2):370–405
11. Martimort D, Pouyet J (2008) To build or not to build: normative and positive theories of public-private partnerships. Int J Ind Organ 26(2):393–411
12. Tang W, Qiang M, Duffield CF (2008) Incentives in the chinese construction industry. J Constr Eng Manage 134(7):457–467
13. Wagner PA (2011) Unmediated communication with partially verifiable types. J Math Econ 47(1):99–107
14. Ward SC, Chapman CB (1995) Evaluating fixed price incentive contracts. Omega 23(1):49–62
15. Yang Q (2009) Principal-agent relationship of owner and construction agent of substitute expressway. China J Highw Transp 5:105–110
16. Zhu B, Li QM (2005) Analysis of the moral risk model during construction process under asymmetric information. J Chongqing Jianzhu Univ 27(4):102–105

Chapter 100
Improvement of Decisions about Project Risk Management Based on Bayesian Network

Jinpeng He and Hongchang Mei

Abstract The project risk management is a hot topic in the field of current management. As to the problems of lacking data to construct the decision model and lacking quantitative ways to inspect model,this paper sets about from the Bayesian network and puts forward a method of constructing Bayesian belief network with chain inference method, diagnosis inference method and the guidance of experts under the condition of lacking data. The network are not only used to display project risk directly in the form of graph but also used to calculate local risk and overall risk. In addition, the paper has discussed the way of using Bayesian belief construction algorithm to check out the similarity of the network structure and sample in a quantitative way combined with B-Course tool. In order to save a large amount of time of testing, the paper also optimizes the traditional mathematical Bayesian algorithm with the method of compression. Thus the decision maker can improve the utilized efficiency of time. Finally, this paper has summarized the advantages and disadvantages of this new method and provides a different way of thinking on decision-making of risk management.

Keywords Project risk · Bayesian network · Quantitative test · Compressive algorithm

100.1 Background of Researching

The founding work of Bayesian school is based on Bayes's paper "comments on the solution of several probability problem" [12]. This thesis had not been published during his lifetime because he felt that his academic research had some drawbacks.

J. He (✉) · H. Mei
Management School of Chongqing, Technology and Business University,
Chongqing 400067, People's Republic of China
e-mail: 2598911286@qq.com

J. Xu et al. (eds.), *Proceedings of the Eighth International Conference on Management Science and Engineering Management*, Advances in Intelligent Systems and Computing 281, DOI: 10.1007/978-3-642-55122-2_100, © Springer-Verlag Berlin Heidelberg 2014

But after his death, it is published by his friend. The famous mathematician Laplace had derived an important rule called "succession rule" in 1812 using Bayesian method.So Bayesian method and the theory were gradually understood and valued at the 20th century. The Italian Firat and the British Geoffroy had made important contribution to the theory of Bayesian School [14]. After the second world war, Wardle had proposed statistical decision theory and Bayes theorem occupies an important position in this theory [4]. In the 1950s, represented by Robbins, some people had put forward the combination of empirical Bayes method and the classic methods. The combination had aroused wide attention from all walks of life. It showed its advantages soon and became an active way in some directions [12]; In the late 80s, Bayesian networks is successfully applied to the expression of expert system about the knowledge of uncertainty system and its technology of reasoning is also developed successfully; Into the 90s, with the emergence of technology of data mining, Bayesian networks were started to use in the research of data mining [5].

There have been many researches on using Bayesian network to solve practical problems in recent years in China. They Mainly focus on the foundation of risk management thought combined with probability, statistics, computer technology, linear programming, mathematical differential, etc. They also have some shortcomings, for example, measuring risk just from the perspective of a single variable, being not timely and accurate to deal with the uncertainty information, getting accurate results only in the case of having sufficient historical data, lacking quantitative ways to inspect model and so on.

This paper sets about from the Bayesian network to solve two major problems in the field of theory. One is lacking historical data to model in project risk management and another is failing to use the quantitative ways to verify the key factors that affect risks.

100.2 Application of Bayesian Network

Bayesian network is also called Bayesian belief network [10]. It is often applied in project risk management, and in other words it is a reaction to the causality network diagram of an event. It mainly includes two aspects. The first part is the network structure in the form of graph.It is a directed acyclic graph. The nodes of each network in the graph represent risk events and the arcs between nodes represent the causal relationship between the risk events. The second part is a set of conditional probability. Every risk event has a related set of conditional probability. As shown in Fig. 100.1

Risk events in figure are signified by X_i. $Pa(X_i)$ signifies the sets of X_i's all father nodes. $P(X_i|Pa(X_i))$ signifies that father nodes are the direct causes of child nodes risk incidents and the probability of risk events which occur following in the happening of their father nodes.

If we want to Use the Bayesian networks in the project risk management, we must determine the model of network structure and the set of conditional probability of

Fig. 100.1 The structure of
Bayesian Network

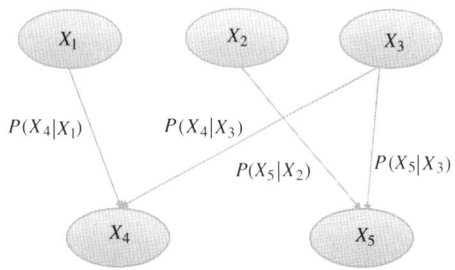

each node firstly. Then we can determine the risk probability of the project according to the following formula: $P(X_1, X_2 \cdots X_n) = \prod_{i=1}^{n} P(X_i | Pa(X_i))$, where $P(X_1, X_2 \cdots X_n)$ is the incidence of the state of $(X_1, X_2 \cdots X_n)$; $\prod_{i=1}^{n}$ is the multiplication from 1 to n; $Pa(X_i)$ is the state of the father nodes of X_i.

100.3 Construction of Bayesian Network

After consulting a large amount of information on the Bayesian network structure, now this paper puts forward a relative perfect, clear and concise way of building combined with their advantages of network building mode. The construction of a Bayesian network is a problem of combined explosion. That is to say if the Bayesian network has N nodes, it may be the network with $N!$ kind of structures. Therefore, to build a reasonable network structure, we must do it according to certain principles and methods.

There will be no adequate data to let us learn the building of network because the occurrence of risk is uncertain In project risk management [7]. So the establishment and application of Bayesian network should be carried out in view of the current actual situation.

100.3.1 Determining the Content of the Bayesian Network Nodes

Bayesian network is composed of various nodes. In project risk management, each node represents a different risk event, such as in the project, they will be construction scheme, construction progress and construction quality, material purchasing supply, etc. [11].

When we are analyzing the risk events, we should do deep discussion and analyze the influential factors and the properties of the corresponding factors from all levels if the events belong to one of the important areas of knowledge. Otherwise, if the events belong to the area that their impacts on the project are not big, we just need analyses of shallow levels [8]. The so-called level can also be called a hierarchy.

It is made up of all risk events that contribute to the success of project in one area of knowledge [2].

100.3.2 Determining the Relationship Between the Bayesian Network Nodes

It mainly relies on experts to determine the dependencies among the nodes when we are determining the relationships of them which on behalf of all risk events [1]. Then we can get the network structure of the project risk events for the inference of Bayes. We must pay attention to prevent the ring when determining the network structure.

We often use chain inference method and diagnosis inference method in the inference of Bayesian network. Chain inference method—Also called causal inference method. It is a way of deducing results from reasons. Firstly, find out the factors that are related to the predicted object or predicted targeton on the basis of theoretical analysis and practical experience, especially the main factors that have direct relationships with the forecast object. And then infer them according to the inner causal relationship of related events. Diagnosis inference method—It is a way of Infering the reason from the results. That's to say, when risk occurs we find out the causes of the risk. The purpose is to find the source of risk to control it.

100.3.3 Determining the Probability Distribution of Bayesian Network Nodes

That determining the probability distribution of Bayesian network nodes needs experts with plenty of knowledge of risk management, and it is usually operated by the project management experts and risk management experts. So it can ensure that the system already contains the knowledge of authoritative experts at the same time of construction of Bayesian network.

100.4 Instance Analysis

After the Bayesian network is completed, the operator can make use of the network diagram to assess the major causes of risk and the probability of comprehensive risk directly according to the conditional probability table. We don't have to look up a lot of historical data after the risk has happened.

Now a Bayesian network diagram is assumed as Fig. 100.2.

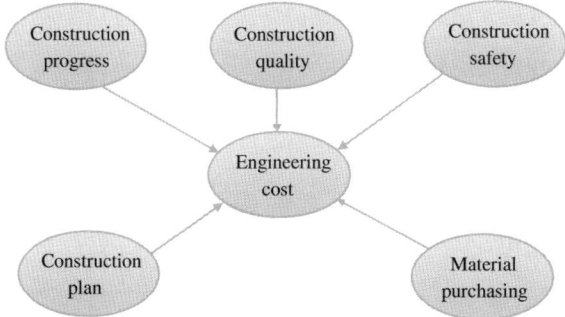

Fig. 100.2 The assumed Bayesian Network

In this network structure, the father nodes of engineering cost are construction progress, construction quality, construction safety, construction plan and material purchasing. The Conditional probability table is assumed as follows:

- The probability of risk of construction progress: $P(S_2) = 0.2$.
- The probability of risk of construction quality: $P(K) = 0.3$.
- The probability of risk of construction safety: $P(S_3) = 0.3$.
- The probability of risk of construction plan: $P(S_4) = 0.1$.
- The probability of risk of material purchasing: $P(X) = 0.5$.
- The Conditional probabilities of engineering cost: $P(J|S_2) = 0.4$, $P(J|K) = 0.4$, $P(J|S_3) = 0.2$, $P(J|S_4) = 0.1$, $P(J|X) = 0.3..$

100.4.1 The Basic Assessment

If the event of cost risk has happened, the possibility of being caused by construction progress is as follows:

$$P_1 = P(S_2|J) = \frac{P(S_2 \cap J)}{P(J)}$$

$$= \frac{P(J|S_2)P(S_2)}{P(S_2)P(J|S_2)+P(K)P(J|K)+P(S_3)P(J|S_3)+P(S_4)P(J|S_4)+P(X)P(J|X)}$$

$$= \frac{0.3 \times 0.5}{0.4 \times 0.2 + 0.4 \times 0.3 + 0.2 \times 0.3 + 0.1 \times 0.1 + 0.3 \times 0.5} = 0.36.$$

Similarly, the possibility of being caused by construction quality is as follows:

$$P_2 = P(X|J) = \frac{P(X \cap J)}{P(J)}$$

$$= \frac{0.4 \times 0.3}{0.4 \times 0.2 + 0.4 \times 0.3 + 0.2 \times 0.3 + 0.1 \times 0.1 + 0.3 \times 0.5} = 0.29,$$

the possibility of being caused by construction safety is as follows:

$$P_3 = P(S_3 \mid J) = \frac{P(S_3 \cap J)}{P(J)}$$

$$= \frac{0.2 \times 0.3}{0.4 \times 0.2 + 0.4 \times 0.3 + 0.2 \times 0.3 + 0.1 \times 0.1 + 0.3 \times 0.5} = 0.14,$$

the possibility of being caused by construction plan is as follows:

$$P_4 = P(S_4 \mid J) = \frac{P(S_4 \cap J)}{P(J)}$$

$$= \frac{0.1 \times 0.1}{0.4 \times 0.2 + 0.4 \times 0.3 + 0.2 \times 0.3 + 0.1 \times 0.1 + 0.3 \times 0.5} = 0.02,$$

the possibility of being caused by material purchasing is as follows:

$$P_5 = P(S_5 \mid J) = \frac{P(S_5 \cap J)}{P(J)}$$

$$= \frac{0.3 \times 0.5}{0.4 \times 0.2 + 0.4 \times 0.3 + 0.2 \times 0.3 + 0.1 \times 0.1 + 0.3 \times 0.5} = 0.36.$$

100.4.2 The Overall Assessment

According to the above analysis, the evaluated conclusion of risk level of the current cost is as follows:

$$P(J) = \sum P(J \mid P_a(J))$$
$$= P(S_2)P(J \mid S_2) + P(K)P(J \mid K) + P(S_3)P(J \mid S_3) + P(S_4)P(J \mid S_4) + P(X)P(J \mid X)$$
$$= 0.4 \times 0.2 + 0.4 \times 0.3 + 0.2 \times 0.3 + 0.1 \times 0.1 + 0.3 \times 0.5 = 0.42.$$

Here, this paper has presented the way of using the Bayesian network to evaluate the single risk. If we need to evaluate the compound risk, we may calculate the single risk firstly and then make use of the formula $P(X_1, X_2 \cdots X_n) = \prod_{i=1}^{n} P(X_i \mid Pa(X_i))$ to accomplish it.

In project risk management, that using Bayesian network structure model not only solves the problem of lacking research data effectively, but also found the key causes of the risk through certain qualitative analyses. In addition, the rest of this paper also discusses the method of how to use quantitative analyses with the real data to verify the effectiveness of the Bayesian network.

100.5 Testing the Effectiveness of Bayesian Networks

We can find out the key factors in the project risk by using the theory of Bayesian network structure. But that how to use real data to verify whether the key factors are right with quantitative analysis is a topic which has been explored all the time in the area of risk management. The rest of this thesis will discuss the data mining algorithm of Bayes and how to use the compression optimization, combined with B-COURSE tools, to find out the similarity between Bayesian network structure and samples of data. So according to the result of arithmetic, we can clearly deduce whether the key factors that cause risk are right.

100.5.1 Testing Tool

Here because of the reason that we will analyze the model of multivariate probabilistic dependencies, the paper chooses B-COURSE tool to model online considering its advantages.

B-Course is a web Bayesian network tool for educators and researchers to use freely [13]. It is a powerful analysis tool by exploiting several theoretically elaborate results developed recently in the fields of Bayesian and causal modeling. B-Course can be used with most web-browsers, and the facilities include features such as matic missing date handling and discretization, a flexible graphical interface for probabilistic inference on the constructed Bayesian network models, automatic pretty-printed layout for the networks, exportation of the models, and analysis of the importance of the derived dependencies. If you enter your data, you can use the tool to model online, and to reason simply according to the established model.

100.5.2 Bayesian Belief Construction Algorithm

Bayesian theory and method has many applications in the field of data mining algorithm [14]. Here the paper will use Bayesian belief construction algorithm to check effectiveness of network structure, and optimize the problem of spending too much time on the algorithm. That's to say using compression ways to reduce the time of inspection on key risk factors.

In the condition of giving a risk event group $D = \{X_1, X_2 \cdots X_n\}$, and the conditional probability distribution table of each event, Bayesian belief network expresses a joint probability distribution about D, where represents the risk events in the project risk management. The definition of Bayesian belief network is:

$$B = \langle G, \theta \rangle. \tag{100.1}$$

Among them, G is a directed acyclic graph (as shown in Fig. 100.1). Its vertex corresponds to risk events $X_1, X_2, \cdots X_n$ in the D. Its arcs represent kinds of relationships of dependence. If there is an arc through event X_1 to X_2, X_1 is the parent or direct precursor of X_2, and X_2 is the successor of X_1.

Once given its parents, each risk event node in the acyclic graph is independent of the other nodes which are not its successors. The X_i's all parent events set are denoted by the set $Pa(X_i)$ in G.

In Eq. (100.1), θ represents a set of parameters that are used to quantify network. $\theta_{X_i|Pa(X_i)} = P(X_i|Pa(X_i))$. It points out the conditional probability of X_i's occurrence when $Pa(X_i)$ has happened.

So, in fact, a Bayesian belief network has given the joint probability distribution of the risk events in the set D: $P(X_1, X_2 \cdots X_n) = \prod_{i=1}^{n} P(X_i|Pa(X_i))$.

Bayesian belief construction algorithm can be expressed like this: given the group of risk event $D = \{x_1, x_2, \cdots x_n\}$, x_i is the instance of X_i. we will Look for a Bayesian belief network that match the sample mostly.

100.5.3 Bayesian Algorithm of Compression

Here we do not use the traditional algorithm, because traditional Bayesian belief construction algorithm will search X_i's father nodes from $n-1$ candidate nodes one by one. It does not take into account the connection between elements. It is unreasonable and costs a lot of time to test the candidate nodes. Such as the following implications: $X \to Y \to Z$.

We can find that there exists dependencies among X, Y and Z. But when we consider the X and Y is the father nodes of Z, we can find that once we view Y as the father node of Z, X is not of any help to the occurrence of Z. Based on the above ideas, This paper puts forward the method of compressing candidat.

We measure the dependence of two variables through a metric function $I(X, Y)$ [6]. If the value of $I(X, Y)$ is big, the association between variable X and Y is strong, and X and Y are more likely to have a father-son relationship. On the contrary, the possibility of the father-son relationship is small. Formula is as follows:

$$I(X, Y) = D_{KL}(P(X, Y)|P(X)P(Y)), \tag{100.2}$$

$$D_{KL}(P(X)|P(Y)) = \sum_{X} p(X) \log \frac{P(X)}{P(Y)}. \tag{100.3}$$

Therefore, we can choose the centralized scanning on those which are most likely to be the father nodes of X_i when we want to find out the set of the father nodes of risk events through calculating the dependencies between risk events. For example: $Y_{i1}, Y_{i2}, \cdots Y_{ik}, k \ll n$.

Based on the above ideas, combined with B-Course tool, algorithm is put forward as follows: Input:

Risk events set: $D = \{x_1, x_2, \cdots x_n\}$.

Initializing the network: B_0.

Evaluation function: $S(B|D) = \sum_i S(X_i|Pa(X_i), D)$.

Parameter: K.

In the above, $S(B|D)$ is an evaluation function of Bayesian algorithm (Commonly used evaluation functions are Bayesian weight matrix and the minimum description length function). We use this function to evaluate the similarity between every certain network structure and samples.

Output: The optimal network.

From 1, 2, to n.

- Compression

  Use candidate compression according to D and B_{n-1}. Select a set $C_i^n (|C_i^n| \leq k)$ of candidate parents for the event X_i from the set of events "$x_1, x_2 \cdots x_n$". Here defines a directed graph: $H_n = (\chi, E)$, $E = X_j \rightarrow X_i | \forall i, j, X_j \in C_i^n$.

- Maximizing

  Looking for a Bayesian network $B_n = \langle G_n, \theta_n \rangle$ that can maximize evaluation function $S(B_n|D)$. Among them, $G_n \subset H_n, \forall X_i, Pa^{G_n}(X_i) \subseteq C_i^n$.

The crucial step of this algorithm is that using the compression method to screen possible father set $Pa(X_i)$ and Choosing k nodes which are most likely to be the father nodes of X_i. This paper uses the dependence function $I(X, Y)$ to calculate the contact tightness between risk nodes. Combined with B-COURSE tool, that using the candidate compression algorithm of the above defined dependence function can be described as follows:

Input:

Risk events set: $D = \{x_1, x_2, \ldots x_n\}$.

Initializing the network: B_n.

Evaluation function: $S(B|D)$.

Parameter: K.

Output:

For each variable X_i, it can return a set of candidate father C_i.

- For each X_i, calculating $I(X_i, X_j)$, $X_i \neq X_j$ and $X_j \notin Pa(X_i)$.
- Selecting $k - l$ elements with the highest weight, $l = |Pa(X_i)|$.

The candidate sets $C_i = Pa(X_i) \cup \{x_1, \ldots x_{k-l}\}$.

Returning $\{C_i\}$.

According to the result of compression algorithm, we can compare with the previous Bayesian network structure, and reason out whether the key factors that cause risk are right.

100.6 Conclusion and Implication

In this paper, the Bayesian theory is introduced into the project risk management. We can effectively find out the key factors that cause risk through the establishment of the Bayesian network structure in the case of lacking a large amount of historical data, and use the Bayesian belief construction algorithm, the quantitative way to calculate the similarity between network structures and risk events. Thus we can effectively point out whether the judgment to the key risk factors is reasonable.

In addition,there is a striking feature of the Bayesian approach that we can understand the assumptions through observing the results. That is to say, when we know very little about the prior knowledge, or have no idea about the case, the Bayesian method has the incomparable advantages than other methods.

At the same time, because of the reason of lacking real data, this thesis only bring Bayesian belief construction algorithm to the project risk management from the area of theory. In actual work we should combine with B-Course tool to operate and let Bayesian theory become a truly effective project risk management tool.

References

1. Avent T (2013) Foundational issues in risk assessment and risk management. Risk Anal 32(10):1647–1656 NULL
2. Hu ZG (2005) Project management. Publishing House of Mechanical Industry, Beijing, pp 126–153 (in Chinese)
3. Huang X, Qian YT (2012) Risk analysis and countermeasures in construction enterprise. J Bus (Study Finance Econ) 2:102–113 (in Chinese)
4. Li QZ, Wang Q (2011) Method of software project risk management based on Bayesian network. Appl Comput Syst 2:87–96 (in Chinese)
5. Li XM, Cai X (2011) Extreme risk measure model based on the bayes statistical inference. Stat Decis Mak 12:68–89 (In Chinese)
6. Mao SS (2013) Bayesian statistics. Publishing House of China statistics, Beijing, pp 38–49 (in Chinese)
7. Project Management Institute (2013) A guide to the project management body of knowledge: guide, 5th edn. Project Management Institute, Oregon
8. Shen JM (2009) Project risk management. Publishing house of mechanical industry, Beijing, pp 224–263 (in Chinese)
9. Wang ZX (2012) Researching of project investment risk decision-making method in the water conservancy construction. J Water Conservancy Shanxi Prov 4:68–72 (In Chinese)
10. Wang S (2011) The evaluation of electric enterprise management based on Bayesian decision method. J Wuhan Univ (Eng Sci) 4:124–132
11. Wang X, Xu YQ, Gao Y (2011) Risk assessment of large construction project based on Bayesian network. J Eng Manage 5:113–127
12. Weber P, Medina OG et al (2013) Overview on Bayesian networks applications for dependability, risk analysis and maintenance areas. Eng Appl Artif Intel 25(4):671–682
13. Xiong RQ, Li XS (2011) Risk analysis of the real estate project based on Bayesian network inference. J Zhengzhou Univ (Philos Soc Sci Ed) 4:92–108 (in Chinese)
14. Zhao Y, Zhang Y (2012) Risk analysis of rules of it project based on the theory of rough set and bayes. J Shenyang Inst Eng (Nat Sci Ed) 1:81–97

Chapter 101
Risk-Oriented Assessment Model for Project Bidding Selection in Construction Industry of Pakistan Based on Fuzzy AHP and TOPSIS Methods

Muhammad Nazam, Jamil Ahmad, Muhammad Kashif Javed, Muhammad Hashim and Liming Yao

Abstract Risk management for the selection of complex multiple projects during the bidding process is one of the most significant problem in construction industries all over the world. This article develops an evaluation model based on fuzzy set theory, analytical hierarchy process (AHP) and the technique for order performance by similarity to ideal solution (TOPSIS) methods. The criteria weight is achieved by adopting Fuzzy set theory and Analytical Hierarchy Process (AHP). Then, with the minimized risk as the objective, the technique for order performance by similarity to ideal solution (TOPSIS) is applied to determine the final ranking level of the bidding projects according to their closeness coefficient. Finally, a real world application of National Construction Limited (NCL), a largest government oriented company of Pakistan, is conducted to demonstrate the utilization of the proposed model. The results indicate that the proposed model is feasible for risk assessment of project bidding selection in construction industry.

Keywords Risk management · Multi-projects selection · Criteria weights · Fuzzy set · Fuzzy TOPSIS

101.1 Introduction

Construction project is plagued by risk and often suffers poor performance as a result. Risk evaluation of the bidding project is of great significance for construction enterprizes. Tamosaitien and Zavadskas proposed that Risk assessment of the construction project is the major challenge task for the public sectors corporations throughout the world, particularly in developing countries [10]. Ebrahimnejad et

M. Nazam (✉) · J. Ahmad · M. K. Javed · M. Hashim · L. Yao
Uncertainty Decision-Making Laboratory, Sichuan University, Chengdu 610064, People's Republic of China
e-mail: nazam_ehsas@yahoo.com

J. Xu et al. (eds.), *Proceedings of the Eighth International Conference on Management Science and Engineering Management*, Advances in Intelligent Systems and Computing 281, DOI: 10.1007/978-3-642-55122-2_101, © Springer-Verlag Berlin Heidelberg 2014

al. put forward a novel two phase group decision making approach for a fuzzy environment and applied it to the risk control in construction project selection [4]. Chauhan proposed the multiple criteria decision making approach as the underlying concept of risk management is to manage risks effectively, risk management is a critical part of project management [1]. Reza and Mousavi supposed that risk assessment is the critical procedure of risk management in construction industries. Many risk factors affect the selection process of the suitable risk assessment model [8]. El-Sayegh defined that One of the major steps in project risk management is to identify and assess the potential risks [5]. Xu and Tiong supposed quantities as random variables and proposed a general method of risk evaluation of bidding strategies using risk management procedures and stochastic programming methods [13]. Tah put forward a Fuzzy logic technique of risk evaluation and applied it to the risk control of Large-scale projects [11]. Some scholars apply case-based reasoning approach to analyze the bidding risks [2, 7]. In addition, the projects risk analysis methods include grey theory, hybrid rough sets, artificial neural network approach, case based reasoning methods, etc [3, 6, 9, 12]. For the models and methods mentioned above, some of them have so strong assumption conditions that it is difficult to obtain the data needed to apply them; some models are too subjective and the evaluation results are affected largely by personal experiences and subjective bias. In order to overcome the two aspects of faults of the existing models, in this study, the Fuzzy set theory and TOPSIS methods are combined and form a new risk assessment model.

101.2 Key Problem Description

With the development of economic globalization, project selection in construction corporations and in some other domains plays a basic and important role. A good project selection not only make a lot of economic benefits, but also creates a reliable system which can provides a powerful support for one project. This paper presents the multiple criteria decision making problem which shows that how to select the optimal bidding project in a construction industry under fuzzy environment where the vagueness and subjectivity are handled with linguistic values parameterized by triangular fuzzy numbers. A proper project selection is a very important issue for construction industries due to the fact that improper project selection can negatively affect the overall performance and productivity of a construction system. Selecting new project is a time-consuming and difficult process, requiring advanced knowledge and deep experience. For a proper and effective evaluation, the decision-maker may need a large amount of data to be analyzed and many factors to be considered. The main objective of this study is to propose a systematic evaluation model to help the actors in construction industries for the selection of an optimal project among a set of available alternatives.

Risk assessment model involves two other key modeling aspects: First, construction is described as a collaborative teamwork process where parties with different interests, functions, and objectives, share a common goal, which is successful completion of a project. A second important consideration of the risk assessment project model selection is that much knowledge in the real world is imprecise than precise, thus the preference information provided to model selection may be imprecise or incomplete. As a result, multiple factors, which are either quantitative or qualitative and may be in conflict with each other, impact the project bidding risk assessment model selection problem and problem arises in group setting with incomplete, vague and uncertain information. In line with the multidimensional characteristics of the risk assessment model selection, the problem is a kind of multi-criteria decision-making (MCDM) problem, which requires MCDM methods for an effective problem solving. It can rank different methods when they are compared in terms of their overall performance. In order to overcome this problem Fuzzy TOPSIS method is extended in this paper for selecting a proper risk assessment mode. Therefore, this study utilizes a MCDM method (AHP) to determine the importance weights of evaluation criteria, and fuzzy TOPSIS to obtain the performance ratings of the feasible alternatives in linguistic values parameterized with triangular fuzzy numbers. This paper attempts to address this limitation and the gap in the current literature and provide a framework for determining optimal risk assessment project.

The rest of this paper is structured as follows. In Sect. 101.3, proposed risk assessment model for optimal bidding project selection is presented and the different phases of the proposed approach are explained in detail. In Sect. 101.4, briefly introduced the proposed methodologies and fuzzy set theory. A real case study of a Pakistani construction company is presented to demonstrate the computational procedure of the proposed framework in Sect. 101.5. Some results analysis and conclusions are made finally.

101.3 The Proposed Risk Assessment Model

The proposed risk assessment model for the project bidding selection problem is composed of AHP, fuzzy set theory and fuzzy TOPSIS methods. It consists of three basic phases, in first phase we identify the criteria to be used in the model. After that using Matlab software AHP computations and in the final stage project ranks are determined by fuzzy TOPSIS method.

In the first stage, alternative projects and the criteria which will be used in their assessment are determined and the decision hierarchy structure is formulated. AHP model is constructed such that the objective is in the first phase, criteria are in the second phase and projects are on the third phase. In the last step of the first phase, the decision hierarchy structure is approved by decision-making team. After the approval of decision hierarchy, criteria used in project selection are assigned weights using AHP in the second stage. In this phase, pairwise comparison matrices are formed to determine the criteria weights. Using expert's scoring method and combining it with the data from National Construction Limited (NCL), to determine the values of the elements of pairwise comparison matrices. Calculating the geometric mean of the values obtained from individual evaluations, a final pairwise comparison matrix

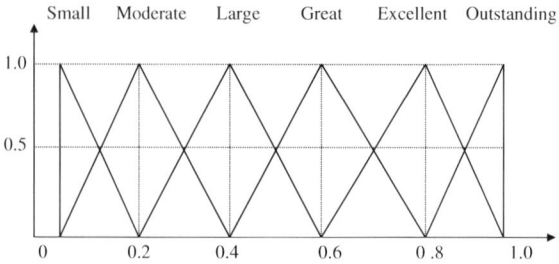

Fig. 101.1 Membership function of linguistic variables for criteria rating

Table 101.1 Linguistic values and fuzzy numbers

Linguistic values	Smaller	Moderate	Large	Great	Excellent	Outstanding
Fuzzy numbers	(0,0,0.2)	(0,0.2,0.4)	(0.2,0.4,0.6)	(0.4,0.6,0.8)	(0.6,0.8,1)	(0.8,1,1)

determined. The weights of the criteria are calculated based on this final comparison matrix. In the last step of this phase, calculated weights of the criteria are approved by decision making team. Project ranks are determined by using fuzzy TOPSIS method in the third phase. Linguistic values are used for evaluation of alternative projects in this step. The membership functions of these linguistic values are shown at Fig. 101.1, and the triangular fuzzy numbers related with these variables are shown at (Table 101.1). The project having the maximum CC_j^+ value is determined as the optimal project according to the calculations by Fuzzy TOPSIS. Ranking of the other projects is determined according to CC_j in descending order. Structural diagram of the proposed model for project selection is provided in Fig. 101.2.

101.4 Solution Approach

1. Fuzzy Set Theory

In order to deal with the vagueness of human thought, Zadeh [14] first introduced the fuzzy set theory. Here the main outline of this theory is briefly summarized in order to solve the desired problem.

(1) Fuzzy Set

A fuzzy set is a class of objects with a continuum of grades of membership. Such a set is characterized by a membership function which assigns to each object a grade of membership ranging [0, 1]. A tilde~will be placed above a symbol if the symbol represents a fuzzy set. Therefore, $\tilde{a}, \tilde{b}$ are fuzzy sets. The membership function for these fuzzy sets will be denoted by $\mu_{\tilde{a}(x)}, \mu_{\tilde{b}(x)}$ respectively in this article.

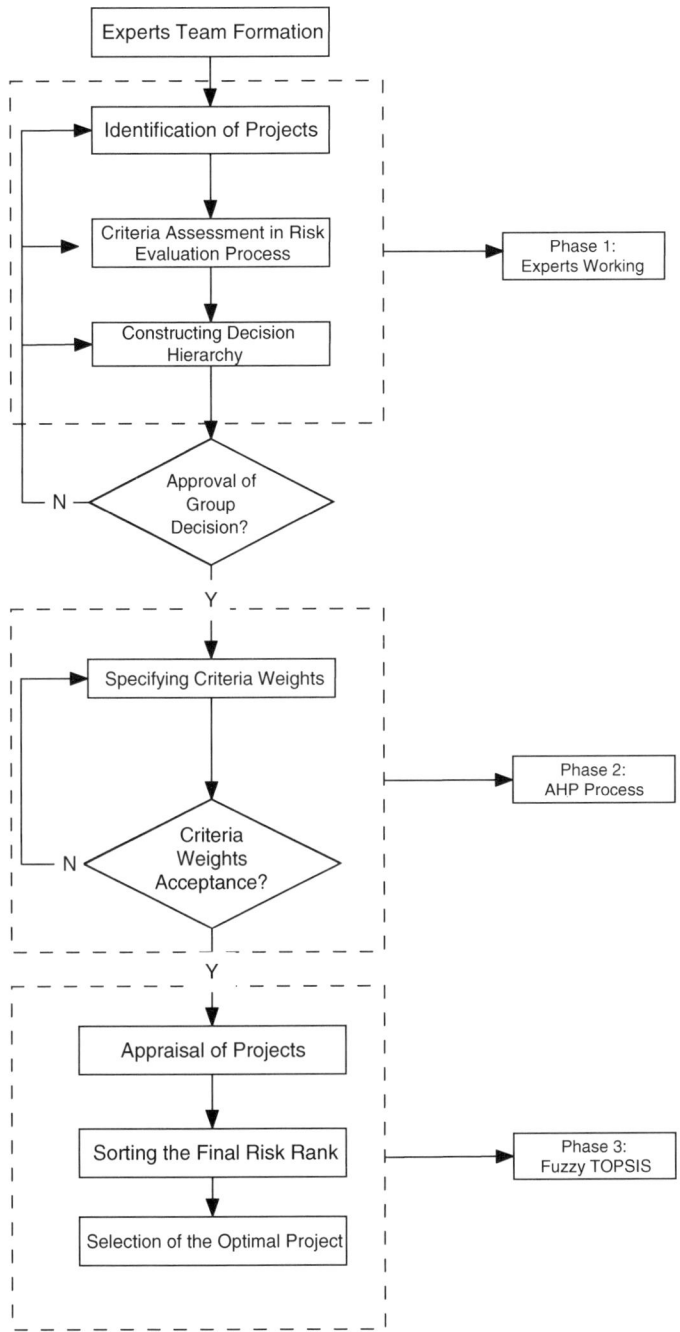

Fig. 101.2 Structural diagram of the proposed model

Fig. 101.3 A triangular fuzzy number

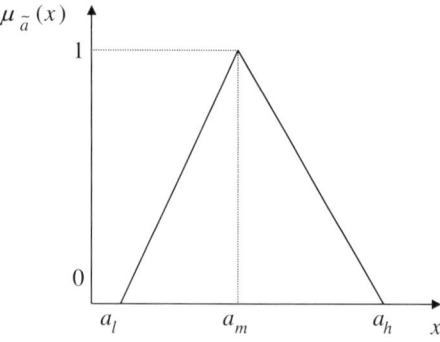

(2) Linguistic Variables

A linguistic variable is a variable whose values are words or sentences in a natural or artificial language.

(3) Fuzzy Numbers

Triangular fuzzy numbers are special class of fuzzy numbers and they are useful in promoting representation and information processing in a fuzzy environment. A triangular fuzzy number $\tilde{a}$ can be defined by triplet (a_l, a_m, a_h) shown in Fig. 101.3. The membership function $\mu_{\tilde{a}(x)}$ is defined as:

$$\mu_{\tilde{a}(x)} = \begin{cases} 0, & x < a_l \\ \frac{x-a_l}{a_m-a_l}, & a_l \leq x \leq a_m \\ \frac{a_h-x}{a_h-a_m}, & a_m \leq x \leq a_h \\ 0, & x > a_h, \end{cases} \tag{101.1}$$

where $a_l \leq a_m \leq a_h$ and these values denote the smallest possible value, the most promising and largest possible value of a fuzzy event, respectively. Let assume two positive triangular fuzzy numbers $\tilde{a} = (a_l, a_m, a_h)$, $\tilde{b} = (b_l, b_m, b_h)$ and a positive real number r, the algebraic operation s of the triangular fuzzy numbers can be expressed as follows:

$$\tilde{a} \oplus \tilde{b} = [a_l + b_l, a_m + b_m, a_h + b_h], \tilde{a} \ominus \tilde{b} = [a_l - b_h, a_m - b_m, a_h - b_l],$$
$$\tag{101.2}$$

$$\tilde{a} \otimes \tilde{b} = [a_l b_l, a_m b_m, a_h b_h], r \otimes \tilde{a} = [r a_l, r a_m, r a_h]. \tag{101.3}$$

The distance between two triangular fuzzy numbers can be calculated by using the vertex method as: $d(\tilde{a}, \tilde{b}) = \sqrt{\frac{1}{3}[(a_l - b_l)^2 + (a_m - b_m)^2 + (a_h - b_h)^2]}$.

2. Fuzzy TOPSIS Based on Fuzzy Theory and MCDM Method

A systematic approach to extend the TOPSIS is proposed to solve the risk assessment model selection problem under a fuzzy environment in this section.

The difference between TOPSIS and fuzzy TOPSIS chiefly lies in rating approaches. The merit of fuzzy TOPSIS is using fuzzy numbers instead of precise numbers [2]. Fuzzy TOPSIS method is extended in this paper for selecting a proper risk assessment model. The main steps of the Fuzzy Set-TOPSIS bidding project risk assessment model are as follows.

Step 1. Build up a normalized decision matrix for the performance ranking of each alternative with respect to the criteria. In this phase the normalized value is calculated by this equation:

$$R = (r_{kj}) = \frac{f_{kj}}{\sqrt{\sum_{j=1}^{m} f_{kj}^2}} \text{for } j = 1, 2, \cdots, J, k = 1, 2, \cdots, m. \quad (101.4)$$

Step 2. Construct the weighted normalized decision matrix. Multiply each column of the normalized decision matrix by its associated weights. The weighted normalized value can be calculated as:

$$\tilde{V} = [\tilde{v}_{kj}]_{n \times J}, k = 1, 2, \cdots, m, j = 1, 2, \cdots, J, \quad (101.5)$$

where $V_{kj} = w_k \times \tilde{x}_{kj}$.

A set of importance weights of each criterion $w_k = (k = 1, 2, \cdots, n)$. A set of performance ratings of $P_j (j = 1, 2, \cdots, J)$ with respect to criteria $C_k (k = 1, 2, \cdots, n)$ called $\tilde{X} = \{\tilde{x}_{kj}, k = 1, 2, \cdots, n, j = 1, 2, \cdots, J\}$. Here, w_k shows the weight of the criterion.

Step 3. In this step identify the fuzzy positive ideal solution (FPIS, P^+) and fuzzy negative ideal solution (FNIS, P^-) by using these equations:

$$P^+ = \{\tilde{v}_1^+, \tilde{v}_2^+, \cdots, \tilde{v}_i^+\} = \{(\max \tilde{v}_{kj} | k \in K') \times (\min \tilde{v}_{kj} | k \in K'')\}, \quad (101.6)$$

$$P^- = \{\tilde{v}_1^-, \tilde{v}_2^-, \cdots, \tilde{v}_k^-\} = \{(\min \tilde{v}_{kj} | k \in K') \times (\max \tilde{v}_{kj} | k \in K'')\}, \quad (101.7)$$

where $k = 1, 2, \cdots, n, j = 1, 2, \cdots J$. In this formula, P^+ represents the most ideal, P^- represents the most undesirable, K' represents the benefit criteria, the bigger the index value, the better it is and K'' represents the cost criteria, the smaller the index value, the better it is.

Step 4. Calculate the separation measures, using the n-dimensional Euclidean distance. The separation of each alternative from the positive ideal solution (L_j^+) is given as: $L_j^+ = \sqrt{\sum_{k=1}^{n} (v_k^+ - v_{kj})^2}, j = 1, 2, \cdots, J.$

Similarly, the separation of each alternative from the negative ideal solution (L_j^-) is as follows: $L_j^- = \sqrt{\sum_{k=1}^{n} (v_k^- - v_{kj})^2}, j = 1, 2, \cdots, J.$ Compute the distance of each alternative from P^+ and P^- using the below mentioned equations:

$$L_j^+ = \sum_{j=1}^{n} d(\tilde{v}_{kj}, v_k^+), \ j=1, 2, \cdots, J, \quad L_j^- = \sum_{j=1}^{n} d(\tilde{v}_{kj}, v_k^-), \ j=1, 2, \cdots, J.$$

$$(101.8)$$

Step 5. Calculating the relative closeness degree of each alternative project from the positive ideal solution using the following equation:

$$CC_j = \frac{L_j^-}{L_j^+ + L_j^-}, j = 1, 2, \cdots, J, \qquad (101.9)$$

where $CC_j \in [0, 1], \forall j = 1, \cdots, n$,

Step 6. Finally the preferred orders can be obtained according to the similarities to the value of CC_j in descending order to choose the best project. If any project has the highest CC_j value, then it is the most important project.

101.5 Case Study

In this section, a practical example in Pakistan is considered to show the whole process of the proposed modeling as follow.

1. Background Review

National Construction Limited, which is a multi-discipline company, has in its fold the experience and background of sophisticated construction technology. Recently (NCL), has taken a huge project in road construction. For the application, an expert team was formed from three junior managers of the Pakistani construction industry and the authors of this paper. The economical projects selection make significant improvements in the construction capabilities of the construction industries. Therefore, selecting the most proper project is of great importance for the construction companies. But it is hard to choose the most suitable one among the set of different alternatives which dominate each other in different characteristics. The application performed is based on the steps provided in previous section and explained step by step together with the results.

2. Identification of the Risk Assessment Criteria and Weights in Bidding Projects

Criteria to be considered in the risk assessment model of projects are determined by the expert team. There are six construction projects to choose from. The construction enterprize analyzes the technical and economic situation in detail of each project and regards the grade of the risk and benefit level as the main basis for selection. We apply Fuzzy TOPSIS model to evaluate the risks of six construction projects. Six criteria are defined from a number of risk factors, forming a risk assessment index system, denoted by $C = C_1, C_2, \cdots, C_6$, where C_1 is project scale, expressed with the amount of project cost (10 Million Rupees); C_2 is the start-up capital (10 Million Rupees); C_3 is the usable construction period in a year; C_4 is the terms and conditions of the contract either its feasible or not; C_5 is the competitiveness of

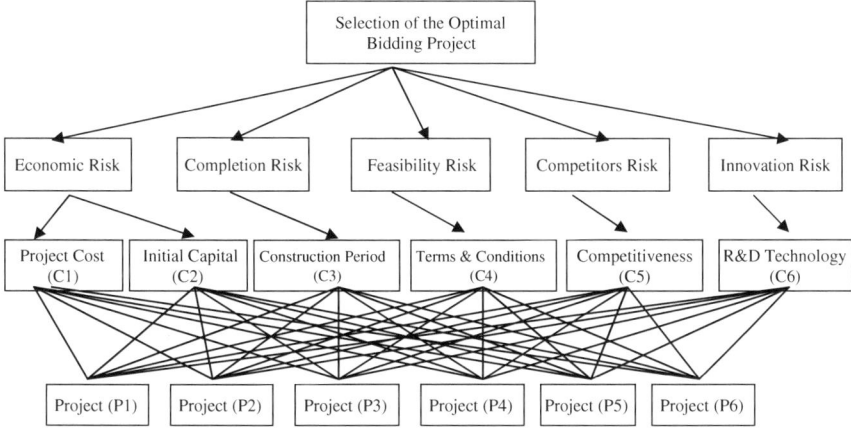

Fig. 101.4 The decision problem structure of project selection

bidding project, which can be measured by the number of competitors participating in the bidding project of the same project; C_6 is degree of maturity of the construction technology, which can be measured by whether the construction enterprize has the building experience of the similar project or not. Decision hierarchy structured with the determined alternative projects and criteria is provided in Fig. 101.4.

With the formation of the decision hierarchy for the problem, the weights of the criteria to be used in evaluation process are calculated by using AHP method. Using expert's scoring method and combining it with the data from National Construction Limited (NCL), to determine the values of the elements of pairwise comparison matrices. Geometric means of these values are found to obtain the pairwise comparison matrix on which there is a consensus among the experts and analysts. The C_4, C_2 and C_5 are determined as the three most important criteria in the project selection process by AHP. Consistency ratio of the pairwise comparison matrix is calculated as $0.037 < 0.1$. So the weights are shown to be consistent and they are used in the selection process.

3. Establish the Fuzzy Evaluation Matrix of Bidding Projects

Fuzzy evaluation matrix established by the linguistic variables as in Table 101.1, is presented in Table 101.2. Linguistic variables are in the upper section and the lower section is composed of fuzzy triangular numbers which are equivalent to linguistic variables.

4. Determine the Fuzzy Positive and Negative Ideal Solution

We can explain the fuzzy positive and negative ideal solution as $\tilde{v}_1^+ = (1, 1, 1)$ and $\tilde{v}_1^- = (0, 0, 0)$ for benefit criterion, and $\tilde{v}_1^+ = (0, 0, 0)$ and $\tilde{v}_1^- = (1, 1, 1)$ for cost criterion. C_1 and C_2 are cost criterion in this problem whereas the other criteria are benefit criteria. The distance of each project can be calculated by using the Eq. (101.8).

Table 101.2 Fuzzy linguistic and numeral evaluation matrix for the bidding projects

	C_1	C_2	C_3	C_4	C_5	C_6
P1	Large	Moderate	Outstanding	Excellent	Large	Great
P2	Great	Great	Large	Excellent	Excellent	Large
P3	Outstanding	Excellent	Great	Large	Great	Moderate
P4	Smaller	Great	Large	Outstanding	Moderate	Great
P5	Smaller	Excellent	Moderate	Large	Outstanding	Excellent
P6	Moderate	Large	Outstanding	Great	Large	Excellent
P1	(0.2,0.4,0.6)	(0,0.2,0.4)	(0.8,1.0,1.0)	(0.6,0.8,1.0)	(0.2, 0.4,0.6)	(0.4,0.6,0.8)
P2	(0.4,0.6,0.8)	(0.4,0.6,0.8)	(0.2,0.4,0.6)	(0.6,0.8,1.0)	(0.6,0.8,1)	(0.2,0.4,0.6)
P3	(0.8,1.0,1.0)	(0.6,0.8,1.0)	(0.4,0.6,0.8)	(0.2,0.4,0.6)	(0.4,0.6,0.8)	(0,0.2,0.4)
P4	(0,0,0.2)	(0.4,0.6,0.8)	(0.2,0.4,0.6)	(0.8,1.0,1.0)	(0,0.2,0.4)	(0.4,0.6,0.8)
P5	(0,0,0.2)	(0.6,0.8,1)	(0,0.2,0.4)	(0.2,0.4,0.6)	(0.8,1.0,1.0)	(0.6,0.8,1.0)
P6	(0,0.2,0.4)	(0.2,0.4,0.6)	(0.8,1,1)	(0.4,0.6,0.8)	(0.2,0.4,0.6)	(0.6,0.8,1.0)
Weight	0.065	0.270	0.086	0.273	0.159	0.141

Table 101.3 The pairwise comparison and weights of each criterion by AHP using MATLAB

	C_1	C_2	C_3	C_4	C_5	C_6	Weights (w)	Judgment procedure
C_1	1.00	0.35	0.50	0.30	0.43	0.40	0.065	$\lambda_{max} = 6.235$
C_2	2.80	1.00	1.80	1.70	2.40	2.30	0.270	$RI = 1.24$
C_3	2.00	0.50	1.00	0.40	0.30	0.40	0.086	$CI = 0.046$
C_4	3.30	1.42	2.50	1.00	2.70	1.60	0.273	$CR = 0.037 < 0.1$
C_5	2.30	0.41	3.33	0.37	1.00	1.70	0.159	
C_6	2.50	0.43	2.50	0.62	0.58	1.00	0.141	

5. Weighted Evaluation for the Bidding Projects

After the fuzzy evaluation matrix was determined, the second step is to obtain a fuzzy weighted decision table. Using the criteria weights calculated by AHP (Table 101.3) in this step, the Weighted Evaluation Matrix is established with Eq. (101.5). The resulting fuzzy weighted decision matrix is shown in Tables 101.4 and 101.5. According to (Tables 101.4 and 101.5), it is seen that the elements v_{kj} are normalized positive triangular fuzzy numbers and their ranges belong to the closed interval [0, 1].

6. Compute the Similarities Between Projects and Ideal Solution

To explain the calculation procedure of the proposed method, we used the steps 3, 4 and 5 to get the optimal results in risk bidding projects. The calculations of the Project 1 to get weighted and un-weighted CC_1 value is done as an example to identify the risk rank. Same calculations can be done for the remaining projects. The results of other projects are clearly mentioned in Tables 101.6 and 101.7.

By using Eq. (101.9) we get the weighted and un-weighted coefficient of closeness for projects 1: $CC_j = \dfrac{L_j^-}{L_j^+ + L_j^-} = \dfrac{2.37}{3.63 + 2.37} = 0.395.$

Table 101.4 Fuzzy weighted evaluation matrix for the projects

	C_1	C_2	C_3
P1	(0.013,0.026,0.039)	(0.000,0.054,0.108)	(0.068,0.086,0.086)
P2	(0.026,0.039,0.052)	(0.108,0.162,0.216)	(0.017,0.034,0.051)
P3	(0.052,0.065,0.065)	(0.162,0.216,0.270)	(0.034,0.051,0.068)
P4	(0.000,0.000,0.013)	(0.108,0.162,0.216)	(0.017,0.034,0.051)
P5	(0.000,0.000,0.013)	(0.162,0.216,0.270)	(0.000,0.017,0.034)
P6	(0.000,0.013,0.026)	(0.054,0.108,0.162)	(0.068,0.086,0.086)
FPIS$^+$, P^+	$\tilde{v}_1^+ = (0,0,0)$	$\tilde{v}_2^+ = (0,0,0)$	$\tilde{v}_3^+ = (1,1,1)$
FNIS$^-$, P^-	$\tilde{v}_1^- = (1,1,1)$	$\tilde{v}_2^- = (1,1,1)$	$\tilde{v}_3^- = (0,0,0)$

Table 101.5 Fuzzy weighted evaluation matrix for the projects

	C_4	C_5	C_6
P1	(0.163,0.218,0.273)	(0.031,0.063,0.095)	(0.056,0.0846,0.112)
P2	(0.163,0.218,0.273)	(0.095,0.127,0.159)	(0.028,0.056,0.0846)
P3	(0.054,0.109,0.163)	(0.063,0.095,0.127)	(0.000,0.028,0.0560)
P4	(0.218,0.273,0.273)	(0.000,0.031,0.063)	(0.056,0.0846,0.112)
P5	((0.054,0.109,0.163)	(0.127,0.159,0.159)	(0.084,0.112,0.141)
P6	(0.109,0.163,0.218)	(0.031,0.063,0.095)	(0.084,0.112,0.141)
FPIS$^+$, P^+	$\tilde{v}_4^+ = (1,1,1)$	$\tilde{v}_5^+ = (1,1,1)$	$\tilde{v}_6^+ = (1,1,1)$
FNIS$^-$, P^-	$\tilde{v}_4^- = (0,0,0)$	$\tilde{v}_5^- = (0,0,0)$	$\tilde{v}_6^- = (0,0,0)$

Table 101.6 Results obtained with fuzzy TOPSIS and weighted risk assessment level of projects

Projects	L_j^+	L_j^-	CC_J	Risk ranks	Weighted CC_J	Weighted risk ranking
P1	3.635	2.379	0.395	1	0.451	P6
P2	3.784	2.249	0.373	2	0.395	P1
P3	3.998	2.029	0.338	3	0.374	P4
P4	3.772	2.255	0.374	4	0.373	P2
P5	3.843	2.184	0.362	5	0.362	P5
P6	3.716	3.049	0.451	6	0.338	P3

7. Result Analysis

Based on CC_j values, the ranking of the projects in descending order are $P6 > P1 > P4 > P2 > P5 > P3$. Proposed model results indicate that P6 is the best project with CC_j value of 0.451. Project six has the highest coefficient of closeness with lower risk. From the calculating results we can see that National Construction Ltd should select the sixth project firstly, and then select the project first P1 to bid. This will enable the enterprize to achieve the better economic results with lower risk. The case in which criteria weights are not considered and the un-weighted CC_j values obtained in this condition are presented in Table 101.7 with their comparisons with previous values. Based on un-weighted CC_j values, the ranking of the projects

Table 101.7 Results obtained with fuzzy TOPSIS and un-weighted risk assessment level of projects

Projects	L_j^+	L_j^-	CC_J	Risk ranks	Un-weighted CC_J	Un-weighted risk ranking
P1	2.117	4.246	0.667	1	0.668	P1
P2	3.003	3.361	0.528	2	0.667	P6
P3	4.057	2.307	0.362	3	0.571	P4
P4	2.723	3.620	0.571	4	0.570	P5
P5	2.743	3.641	0.570	5	0.528	P2
P6	2.117	4.246	0.667	6	0.362	P3

in descending order are $P1 > P6 > P4 > P5 > P2 > P3$. The best project has changed according to the un-weighted ranking results. The change in the optimal project when criteria weights are considered into account has shown that criteria weights found consistently constitute an important phase in decision-making process. The result indicates that Fuzzy set and TOPSIS method can reflect the randomicity and fuzziness of risk information in the selection of bidding projects and make the risk assessment more scientific and reasonable. According to the results and calculation analysis of case study, it is proved that the criterion weights are most important to assess the risk ranking of projects.

101.6 Concluding Remarks

Many construction project risk assessment techniques are currently used in the construction industry but insufficient attention has been paid by researchers to a select suitable risk assessment bidding project. To address this decision problem, in this paper a group based fuzzy TOPSIS approach is developed with an effective algorithm to improve the quality and effectiveness of decision making. In this article, presenting a scientific framework to evaluate project bidding in construction, uses triangular fuzzy numbers to express the linguistic values that consider the subjective judgments of evaluators and then adopts fuzzy multiple criteria decision making approach to synthesize the group decision. The proposed model differs from the present project bidding selection literature. AHP is used to assign weights to the criteria to be used in project selection, while fuzzy TOPSIS method is applied to the risk assessment and ranking order for the bidding project and is beneficial to construction enterprises to make the decisions according to their own situation on the project. Finally, in the case study results, it is shown that calculation of the criteria weights is most important in TOPSIS method and they could change order of the ranking. It indicates that model is applicable to multiple projects risk analysis and comparisons. Therefore development of a group decision making system for construction industry is very useful and economical.

This model with a little modification can be employed in other decision making problems in construction industries of other countries. With the proposed model some optimization and mathematical models can be combined to get some better results. Further research can apply this method to assess the risk in construction industry, like material selection, performance selection, supplier selection etc.

Acknowledgments The authors wish to thank the anonymous referees for their helpful and constructive comments and suggestions. The work is supported by the National Natural Science Foundation of China (Grant No. 71301109), the Western and Frontier Region Project of Humanity and Social Sciences Research, Ministry of Education of China (Grant No. 13XJC630018), the Philosophy and Social Sciences Planning Project of Sichuan province (Grant No. SC12BJ05), and the Initial Funding for Young Teachers of Sichuan University (Grant No. 2013SCU11014).

References

1. Chauhan A, Vanish R (2012) Magnetic material selection using multiple attribute decision making approach. Mater Des 36:1–5
2. Chen TY, Tsao CY (2008) The interval-valued fuzzy TOPSIS method and experimental analysis. Fuzzy Sets Syst 159:1410–1428
3. Deng XD, Wang CY (2007) The Grey Fuzzy theory appraises the risk of the project. J China Three Gorges Univ (Natural Sciences) 29:12
4. Ebrahimnejad S, Mousavi SM, Moghaddam T (2012) A novel two phase gruoup decision making approach for construction project selection in a fuzzy environment. Appl Math Model 36:4197–4217
5. El-Sayegh SM (2008) Risk assessment and allocation in the UAE construction industry. Int J Project Manage 26(4):31–38
6. Hassan Y, Eiichiro T (2002) Decision making using hybrid rough sets and neural networks. Int J Neural Syst 12:35–46
7. Ngai EWT, Wat FKT (2005) Fuzzy decision support system for risk analysis in e-commerce development. Decis Support Syst 40:235–255
8. Reza KA, Mousavi N (2011) Risk assessment model selection in construction industry. Expert Syst Appl 38:105–111
9. Shi KB, Zhang LX, Shi GQ (2005) Study on risk index system and comprehensive evaluation in construction project. J Xinjiang Agriculture Univ 28:76–80
10. Tamosaitien J, Zavadskas EK (2013) Multi-criteria risk assessment of a construction project. Procedia Comput Sci 17:129–133
11. Tah JHM, Carr V (2000) A proposal for construction project risk assessment using fuzzy logic. J Constr Manage Econ 18:491–500
12. Wang KC (2007) Modelling risk allocation decision in construction contracts. Int J Project Manage 25(4):85–93
13. Xu TJ, Tiong LK (2001) Risk assessment on contractors pricing stratigies. J Constr Manage Econ 19:77–84
14. Zadeh LA (1965) Fuzzy sets. Inf Control 8:338–353

Chapter 102
Risk Management in Maritime Structures

Maria Teresa Leal Gonsalves Veloso dos Reis, Pedro Gonçalo Guerra Poseiro, Conceição Juana Espinosa Morais Fortes, José Manuel Paixão Conde, Eric Lionel Didier, André Miguel Guedelha Sabino and Maria Armanda Simenta Rodrigues Grueau

Abstract Maritime structures (breakwaters, groynes, ocean wave/wind energy converters, etc.) are exposed to wave attack, which generates common emergency situations with serious environmental and economic consequences. This paper presents a methodology developed at LNEC for forecasting and early warning of wave overtopping in ports/coastal areas to prevent emergency situations and support their management and the long-term planning of interventions in the study area. It is implemented in a risk management tool. The methodology uses numerical models to propagate waves from offshore to port/coastal areas to obtain the required input for overtopping assessment methods, such as empirical formulas and artificial neural networks. The calculated overtopping values are compared with pre-established admissible values in order to define the warning levels. These admissible values are derived from general international recommendations and information from local authorities. They depend on the characteristics of the overtopped structure and of the protected area, and on the activities developed there. The warning methodology is applied to Praia da Vitória Port (Azores-Portugal) as an illustrative example. Future developments include the use of complex numerical models (e.g. Navier-Stokes equations solvers; particle methods) to calculate wave overtopping and to extend the methodology to warn of risks associated with wave energy production failure.

M. T. L. G. V. dos Reis · P. G. G. Poseiro · C. J. E. M. Fortes
Hydraulics and Environment Department (DHA), National Laboratory for Civil Engineering (LNEC), Av. do Brasil, 101, 1700-066 Lisbon, Portugal

J. M. P. Conde (✉) · E. L. Didier
UNIDEMI, Faculdade de Ciências e Tecnologia (FCT), Universidade Nova de Lisboa (UNL), 2829-516 Caparica, Portugal
e-mail: jpc@fct.unl.pt

A. M. G. Sabino · M. A. S. R. Grueau
CITI/DI, Faculdade de Ciências e Tecnologia (FCT), Universidade Nova de Lisboa (UNL), 2829-516 Caparica, Portugal

J. Xu et al. (eds.), *Proceedings of the Eighth International Conference on Management Science and Engineering Management*, Advances in Intelligent Systems and Computing 281, DOI: 10.1007/978-3-642-55122-2_102, © Springer-Verlag Berlin Heidelberg 2014

Keywords Risk management · Maritime structures · Ports · Wave overtopping · Forecast and warning

102.1 Introduction

Emergency situations caused by adverse sea conditions incident on maritime structures (breakwaters, groynes, ocean wave/wind energy converts and others) are frequent and endanger the safety of people and goods, lead to structural damage and wave energy production failure, with negative impacts for the society, the economy and the environment. Therefore, risk management in maritime structures is an important issue and the development of a methodology to warn of emergency situations in port/coastal areas is essential for a proper planning and management of these areas.

In particular, in Portugal, it is extremely relevant to study wave induced risks, and especially wave overtopping, due to its long coastline, the importance of the socio-economic activities in port/coastal areas and the severity of the sea conditions. In this context, the National Laboratory for Civil Engineering (LNEC), Portugal, has been developing the HIDRALERTA system [1–4], which is a set of integrated decision-support tools for port and coastal management, whose focus is to prevent and support the management of emergency situations and the long-term planning of interventions in the study area.

This system has coupled several wave propagation models, which transfer offshore wave conditions to inshore, and then calculates wave overtopping and flooding through artificial neuronal networks and empirical formulas. Moreover, it compares estimated values of wave overtopping with pre-established admissible values in order to define different warning levels. Although, the main goal of this paper is to present the warning module of the system, HIDRALERTA is also able to perform risk assessment, based on evaluation of historical wave data. Therefore, it is intended as a tool for both long term planning, and forecasting and warning. As a long-term planning tool, the system uses long-term (years) time series of sea-wave characteristics and/or predefined scenarios, and evaluates the sea-wave risks for the protected areas, allowing the construction of GIS based risk maps. These maps aim to support decision-making of the responsible entities regarding long-term management. As a forecast and early warning tool, the system uses numerical forecasts of sea-wave characteristics, that allow the identification, in advance, of the occurrence of emergency situations and enables the adoption of measures by those entities to avoid loss of lives and to minimize damage. It is also worth noticing that such a system does contribute to the fulfilment of the stipulated in directive 2007/60/EC of the European Parliament and of the Council of 23/10/2007, which recommends the development of risk maps by 2013 and flood risk management plans, including the establishment of systems of forecasting and early-warning, by 2015.

The warning system module is already running at LNEC on a daily basis for Praia da Vitória Port, in Terceira Island (Azores, Portugal) and it is being prepared to run for S. João da Caparica beach, in Costa da Caparica, Portugal.

This paper analyses the case of Praia da Vitória Port, as an illustrative application of the warning methodology. The system downloads sea-wave characteristics predicted offshore, up to 180 h, every day. These data correspond to results obtained through WAVEWATCH III [5], which is a regional model. To transfer to the port entrance and then into the port, SWAN [6] and DREAMS [7] are applied, which are a spectral wave model and a mild slope wave model, respectively. DREAMS model provides wave characteristics in front of each structure, which are then used as input to the neural network tool NN_OVERTOPPING2 [8], together with cross-section characteristics of the maritime structures. NN_OVERTOPPING2 gives an estimate of mean overtopping discharges per unit length of the structure crest [9]. In this case, different maritime structures were considered: the south breakwater, the north breakwater and the seawall that protects Praia da Vitória Bay. Thus, taking into account the limits of the mean overtopping discharge described in [10] and the recommendations from the local authorities, different thresholds were adopted specifically for each structure, bearing in mind the characteristics of the overtopped structure and of the protected area, and the activities developed there. The issue of warnings is activated whenever thresholds are exceeded.

After this introduction, the paper describes the steps of the methodology implemented in the HIDRALERTA system to warn of wave overtopping. An example of application of this methodology is next presented for Praia da Vitória Port. Finally, future developments for the system are discussed, including the use of complex numerical models to calculate overtopping and to extend the methodology to warn of risks associated with wave energy production failure.

102.2 Methodology

The methodology implemented in the HIDRALERTA system [1, 3] to assess wave overtopping risk and to warn of inadmissible overtopping events follows four steps (Fig. 102.1): I: Sea-wave characterization; II: Wave overtopping determination; III: Risk assessment; IV: Warning system. The next sections describe, in more detail, steps I, II and IV.

1. Sea-wave Characterization

In the HIDRALERTA system, the sea-wave regime within a port or at the coast is obtained from numerical models for sea-wave propagation. The use of one or more numerical models for the propagation depends on the study region characteristics and on the phenomena involved in the propagation. In the case of open coastal areas, the offshore wave characteristics are either obtained from buoy measurements or predicted by WAVEWATCH III [5], which is a numerical model for sea-wave prediction at regional level. The offshore wave characteristics are then propagated to the coast with the SWAN model [6], which is a spectral wave model. In case of sheltered areas, such as port areas, to perform the wave propagation into the port the DREAMS model [7], which is a mild slope wave model, is applied after SWAN.

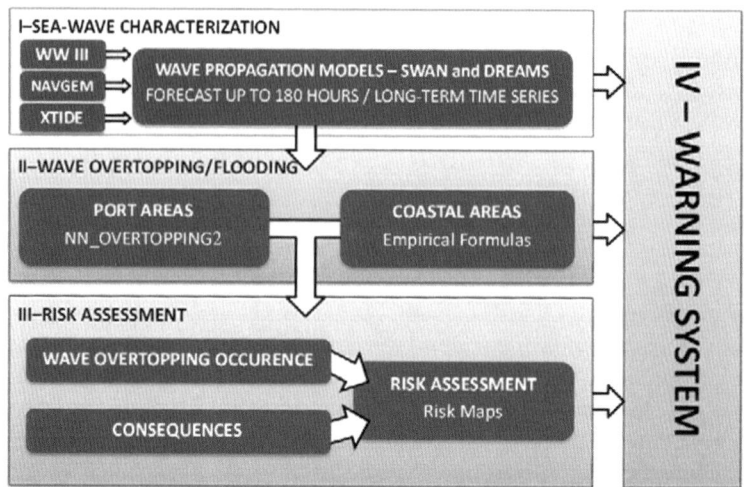

Fig. 102.1 HIDRALERTA scheme

Besides WAVEWATCH III input, numerical models used in HIDRALERTA also need to consider wind fields at a regional level and astronomical tide level, which are obtained from NAVGEM [11] and XTide [12], respectively. The offshore wave conditions from WAVEWATCH III and the wind field from NAVGEM are both provided through The Fleet Numerical Meteorology and Oceanography Center (FNMOC). FNMOC delivers forecast data for WAVEWATCH III up to 180 h and historic data since September 2003, with 1° resolution. It also delivers data for NAVGEM up to 180 h and historic data since January 2004, with 0.5° resolution.

So far, the storm surge has been considered in an approximate manner (considering a constant value), unless there are water level measurements available from tide gauges.

2. Wave Overtopping Determination

In the HIDRALERTA system, wave overtopping determination follows two different approaches, in case of port or coastal areas. For port areas, a tool based on neural network modeling is employed, NN_OVERTOPPING2 [8]. This tool was developed as part of the CLASH European project [10] to predict Froude-scaled mean wave overtopping discharges, q, and the associated confidence intervals for a wide range of coastal structure types (such as dikes, rubble-mound breakwaters and caisson structures). To run NN_OVERTOPPING2, the input needed includes the wave/water level conditions in front of each structure and its geometrical characterization. For coastal areas (whether simple beaches or beaches with coastal defence structures), empirical formulas are applied to evaluate wave run-up/overtopping.

3. Warning System

The warning system integrates all the information from the other modules and issues warning messages. It comprises two components: the data evaluation component, which integrates and processes the data from the other modules, and the user interface component. Whenever the pre-set wave overtopping thresholds are exceeded in a specific area, the system issues warning messages to the responsible authorities, by e-mail and/or by sms.

Thresholds are pre-defined for each structure based on the limits described in [10] and on the information provided by the local authorities to LNEC. Table 102.1 shows the recommendations presented in [10], which depend on the characteristics of the overtopped structure and of the protected area, and on the activities developed there.

Up to now, only the limits concerning the mean overtopping discharge have been used for the threshold definition. In the near future, LNEC aims to apply the limits provided on the maximum volume as well, since individual maximums are more adequate to characterize local overtopping hazard, due to the randomness of the overtopping spectrum [10].

102.3 Case Study: Praia Da Vitória (Azores)

1. Study Area

The port and bay of Praia da Vitória (Fig. 102.2) are located at the Terceira Island, the second largest of the Azores archipelago.

The so-called north breakwater was built to protect the port facilities that support the Lajes airbase. It is a rubble-mound breakwater, 560 m long, with a north-south alignment, rooted in the Ponta do Espírito Santo (Fig. 102.3a). The second rubble-mound breakwater (south breakwater) is rooted on the south side of the bay, near the Santa Catarina fort (Fig. 102.3b). The breakwater is approximately 1300 m long, with a straight alignment (north-south) that bends close to its shore connection. It protects the facilities (commercial sector and fishing port) of the Praia da Vitória Port.

The bay shoreline has a coastal road protected by a seawall which is 1 km long (Fig. 102.3c). In front of the port entrance and rooted to the seawall there is a field of five groynes. These groynes do not have the same length but they have approximately the same alignment (WSW-ENE).

In the port area, there are now several sea-wave measuring devices that can characterize the sea wave regime within the port. In fact, within the scope of the CLIMAAT project [13], a directional wave-buoy was deployed 4 km northeast from the port, in a region about 100 m deep, whose data were used to validate the methodology for wave propagation applied in this study.

2. Illustration of the Warning Methodology

Currently, the warning system is running permanently for Praia da Vitória. The first module (I: Sea wave characterization) runs every day to predict 180 h of wave propagation at the port entrance and into the port, together with wind field and tide level predictions. For each 3 h, the system creates the following layout for each model

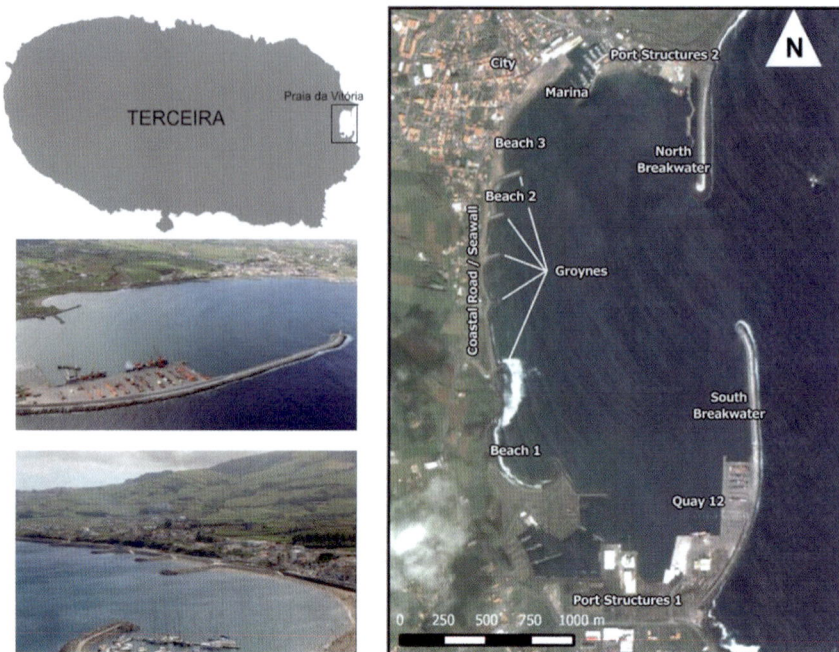

Fig. 102.2 Praia da vitória, Azores

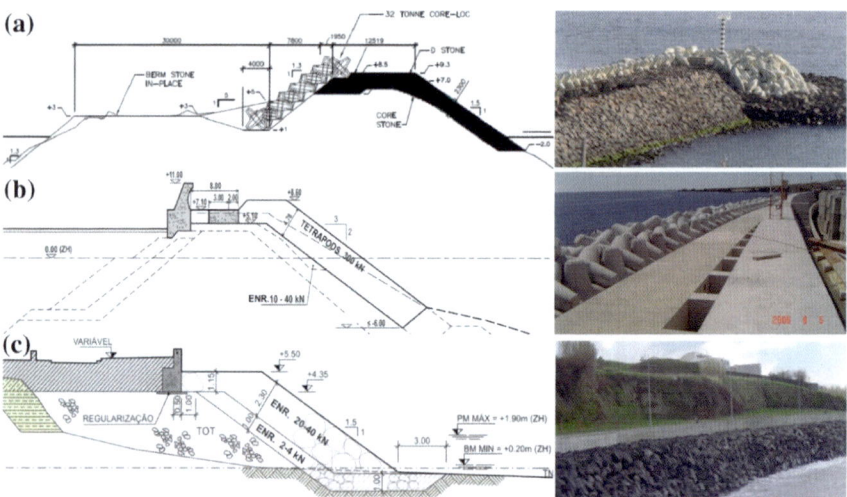

Fig. 102.3 Typical cross-sections of main structures of Praia da Vitória. **a** North breakwater, **b** South breakwater, and **c** Seawall at the bay

Table 102.1 Suggested limits for mean overtopping discharges and maximum volumes (adapted from [10])

Hazard type and reason	Mean dis-charge (l/s/m)	Maximum Volume (l/m)
Pedestrians		
Trained staff, well shod and protected, expecting to get wet, overtopping flows at lower levels only, no falling jet, low danger of fall from walkway	1−10	500 at low level
Aware pedestrian, clear view of the sea, not easily upset or frightened, able to tolerate getting wet, wider walkway	0.1	20−50 at high level/velocity
Unaware pedestrian, no clear view of the sea, easily upset or frightened, not dressed to get wet, narrow walkway or close proximity to edge	0.03	2−5 at high level/velocity
Vehicles		
Driving at low speed, overtopping by pulsating flows at low flow depths, no falling jets, vehicle not immersed	10−50	100−1000
Driving at moderate or high speed, impulsive overtopping giving falling or high velocity jets	0.01−0.05	5−50 at high level/velocity
Property		
Significant damage or sinking of larger yachts	50	5000−50000
Sinking small boats set 5−10 m from the wall Damage to larger yachts	10	1000−10000
Building structure elements	1	−
Damage to equipment set back 5−10 m	0.4	−
Maritime structure		
Embankment seawalls		
No damage if crest and rear slope are well protected	50−200	−
No damage to crest and rear face of grass covered embankment of clay	1−10	−
No damage to crest and rear face of embankment if not protected	0.1	−
Promenade or revetment seawalls		
Damage to paved or armoured promenade behind seawall	200	−
Damage to grassed or lightly protected promenade or reclamation cover	50	−

(Fig. 102.4) with Significant Height (Hs) and Wave Direction (θ). It is also possible to create a layout with Peak Period (Tp).

Once wave characteristics in the port are available, for every 6 h, the second module is applied (II: Wave overtopping/flooding). For each set of wave/water level

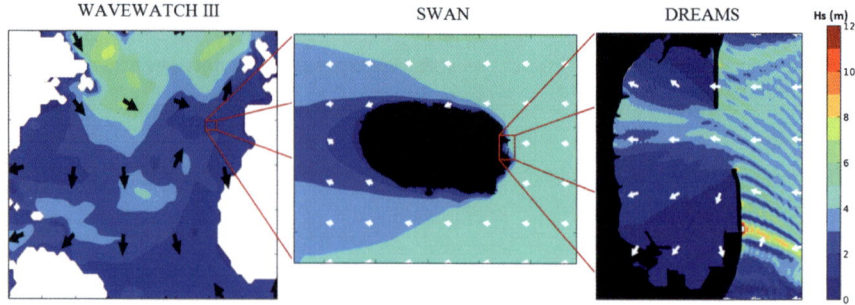

Fig. 102.4 Example of a layout created by the HIDRALERTA system for each wave propagation models

Fig. 102.5 Example of a layout created by the HIDRALERTA system for NN_OVERTOPPING2 results at 16/10/2010, 12 am

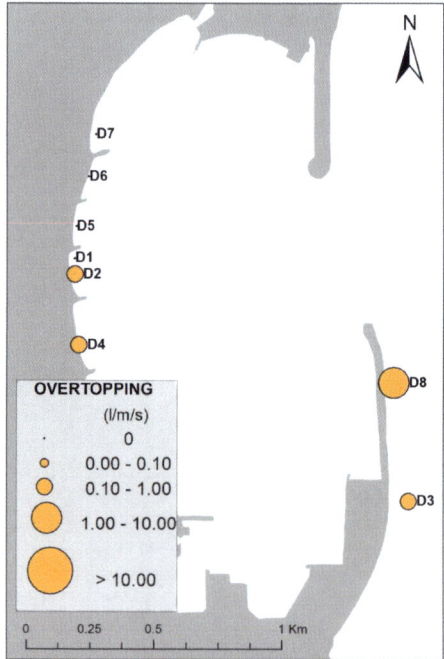

characteristics, NN_OVERTOPPING2 provides information on mean wave overtopping discharges, q, for each of the studied cross-sections of the structures. An example of the layout created by the HIDRALERTA system for NN_OVERTOPPING2 results is presented in Fig. 102.5.

Once the mean overtopping discharges are evaluated, the forth module (IV: Warning system) is applied for threshold definition for q for each of the studied cross-sections of the structures.

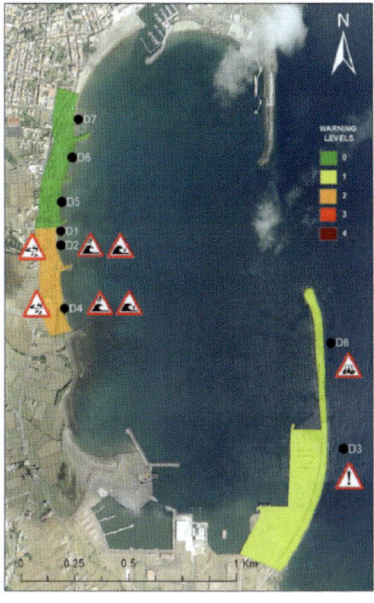

Sign	Pedestrians - Hazard type and reason	Mean discharge (l/s/m)
	Trained staff, well shod and protected, expecting to get wet, overtopping flows at lower levels only, no falling jet, low danger of fall from walkway	1-10
	Aware pedestrian, clear view of the sea, not easily upset or frightened, able to tolerate getting wet, wider walkway	0.1
	Unaware pedestrian, no clear view of the sea, easily upset or frightened, not dressed to get wet, narrow walkway or close proximity to edge	0.03
	Vehicles - Hazard type and reason	
	Driving at moderate or high speed, impulsive overtopping giving falling or high velocity jets	0.01-0.05
	Property - Hazard type and reason	
	Damage to equipment set back 5-10 m	0.4

Fig. 102.6 Example of a warning layout created by the HIDRALERTA system when thresholds are exceeded

Thus, considering as an example the cross-section of the south breakwater protecting quay 12 (D3 in Fig. 102.5), taking into account the limits of the mean overtopping discharges described in [10] and the information from local authorities, the following thresholds were adopted:

1. Users of quay 12 are not easily disturbed or frightened by overtopping events and they move in a large area—0.1 l/s/m;
2. Containers on quay 12 are 5−10 m away from the structure—0.4 l/s/m;
3. All vehicles travel at low speed—10 l/s/m;
4. For the analyzed stretch of the south breakwater the limit for a promenade/revetment seawall has been used—200 l/s/m.

Since quay 12 is approximately 130 m wide, ships moored there are too far away from the overtopped structure to be disturbed by overtopping events.

In module 4, after threshold definition, the calculated mean discharges obtained from the second module are compared with the threshold values outlined for the different cross-sections of the structures. A warning message to the responsible authorities is sent by e-mail and/or by sms whenever these thresholds are exceeded (Fig. 102.6).

102.4 Future Developments

The HIDRALERTA system is currently based on the coupling of several wave propagation models, which transfer offshore wave conditions to inshore, and then calculates wave overtopping and flooding through artificial neural networks and empirical formulas. These two kinds of overtopping models have the great advantage of being very fast, but results are very dependent on good determination of all model parameters and the breakwater characteristics, since these models are developed for some classical breakwaters. Due to the strong increase of computation power that occurred during the last years, it is possible, nowadays, to use complex 3D numerical models based on RANS equations for modeling free surface flows interacting with complex port/coastal structures. These models include specific procedures for free surface flow and are based on a different numerical method: Eulerian formulation, including Volume of Fluid approach for free surface flow, such as OpenFoam [14] and FLUENT [15]; Lagrangian formulation based on a Smoothed Particle Hydrodynamics (SPH) approach, such as SPHysics [16] and SPHyCE [17]; ALE (Arbitrary Lagrangian Eulerian) formulation, such as FLUINCO [18]; and Particle Finite Element Method approach, such as PFEM [19].

These models allow considering all type of port/coastal structures, such as porous/impermeable breakwaters and Oscillating Water Column (OWC) Ocean Wave Energy Converters (OWEC). They output several important coupled parameters simultaneously, as: free surface level near and above the port/coastal structure; wave overtopping; velocity fields inside and outside the structure; pressures and forces on structures; energy production. As these values are instantaneous, it is possible to obtain, mean and maximum values of these parameters.

The OWC is considered to be one of the most technically known OWEC, due to the large research effort that has been subject to in recent years. This device is also one of the first to have reached the status of full-sized prototype deployed in the real sea. An OWC-OWEC consists of a partially submerged structure, open below the water free surface. Within this structure, an air pocket above the free surface is trapped. The oscillating movement of the free surface inside the pneumatic chamber, produced by the incident waves, forces the air to flow through the turbine, which is directly coupled to an electrical generator [20]. Figure 102.7 presents the Pico Island (Azores-Portugal) OWC-OWEC (left), a computational mesh of this OWC-OWEC (middle) and the resulting free surface flow RANS simulation (right). It is not a common practice to use a single numerical code to simulate all the fluid dynamics effects present in this type of device. This code should accurately simulate the 3D wave propagation and its transformation when subject to the OWC-OWEC influence, the water inflow and outflow in the device, the air flow in the pneumatic chamber and the damping caused by the pressure loss at the turbine. A correct simulation of these flows is essential to evaluate the design of the pneumatic chamber and to determine the operating conditions of the turbine [21–23].

In the near future, to perform risk management in OWC-OWECs, in particular to extend the HIDRALERTA warning system to this device, one should first identify the

 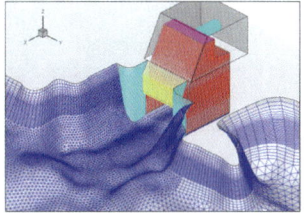 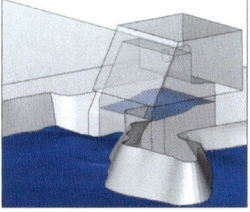

Fig. 102.7 Pico OWC device (Azores, Portugal): device, 3D mesh, numerical modeling

main failure modes of the OWC-OWEC, such as flooding of the installation, turbine failure, etc., and characterise their consequences in the electricity production. The aim is to produce a tool for business decision, as well as providing an effective technology development strategy.

Acknowledgments FCT support through projects PTDC/ECM-HID/1719/2012, PTDC/AAC-AMB/120702/2010, PTDC/ECM/114109/2009 and PTDC/CTE-GIX/111230/2009 is acknowledged. The authors are grateful for the information on Praia da Vitória (port and bay) provided by Portos dos Açores, S.A., Anabela Simões and Eduardo Azevedo from Universidade dos Açores, and Conceição Rodrigues from Azorina—Sociedade de Gestão Ambiental e Conservação da Natureza, S.A.

References

1. Fortes CJEM, Reis MT, Poseiro P et al (2013) The HIDRALERTA projet: Flood forecast and alert system in coastal and port areas. In: Proceedings of the 8as jornadas portuguesas de engenharia costeira e portuária (In Portuguese)
2. Poseiro P, Santos JA, Fortes CJEM et al (2013) Application of the analytic hierarchy process (AHP) to evaluate consequences of wave overtopping occurrence. Study case of Praia da Vitória bay. In: Proceedings of the 8as jornadas portuguesas de engenharia costeira e portuária (In Portuguese)
3. Raposeiro PD, Fortes CJEM, Capitão R et al (2013), Preliminary phases of the HIDRALERTA system: Assessment of the flood levels at S. João da Caparica beach, Portugal. J Coast Res SI65:808–813
4. Zózimo AC, Fortes CJEM, Neves DRCB (2008) Guiomar: geographical user interface for coastal and marine modeling. Recent developments and assessment of potential geographical errors. In: Proceedings med days of coastal and port engineering
5. Tolman HL (1999) User manual and system documentation of WAVEWATCH III version 1.18. Technical Note 166, NOAA/NWS/NCEP/OMB
6. Booij N, Ris RC, Holthuijsen LH (1999) A third-generation wave model for coastal regions: 1. model description and validation. J Geophys Res 104(C4):7649–7666
7. Fortes CJEM (2002) Nonlinear wave transformations in harbors. A finite element analysis. Ph.D. thesis, Mechanical Engineering, IST, Lisbon, Portugal (In Portuguese)
8. Coeveld EM, Van Gent MRA, Pozueta B (2005) Neural network manual NN_OVER-TOPPING2. CLASH WP8, WL-Delft Hydraulics, Delft, The Netherlands
9. Neves DR, Santos JA, Reis MT et al (2012) Risk assessment methodology for the overtopping of maritime structures. Application to the port and bay of Praia da Vitória, Azores, Portugal. J Int Coast Zone Manage 12(3):291–312 (In Portuguese)

10. Pullen T, Allsop NWH, Bruce T et al (2007) EurOtop: wave overtopping of sea defences and related structures: assessment manual. Environment Agency, UK, Expertise NetwerkWaterkeren, NL, and Kuratorium fur Forschungim Kusteningenieurwesen, DE.

11. Whitcomb T (2012) Navy global forecast system, NAVGEM: distribution and user support. In: Proceedings of the 2nd scientific workshop on ONR DRI: unified parameterization for extended range prediction

12. Flater D (1998) XTide manual: Harmonic tide clock and tide predictor. EUA

13. Azevedo EB, Mendes P, Gonçalo V (2008) Projects CLIMAAT and CLIMARCOST climate and meteorology of the Atlantic archipelagos, maritime and coastal climate. In: Proceedings of the i workshop internacional sobre clima e recursos naturais nos países de língua portuguesa-WSCRA08 (In Portuguese)

14. OpenFOAM (2013) OpenFOAM user guide. The OpenFOAM Foundation. http://www.openfoam.org/docs/user/

15. ANSYS (2006) ANSYSFLUENT 6.3 user's guide

16. Gomez-Gesteira M, Rogers BD, Crespo AJC et al (2012) Sphysics-development of a free-surface fluid solver-part 1: theory and formulations. Comput Geosci 48:289–299

17. Didier E, Neves MG (2012) A semi-infinite numerical wave flume using smoothed particle hydrodynamics. Int J Offshore Polar Eng 22(3):193–199

18. Teixeira PRF, Davyt DP, Didier E et al (2013) Numerical simulation of an oscillating water column device using a code based on navier—stokes equations. Energy 61:513–530

19. Oñate E, Idelsohn SR, Del Pin F et al (2004) The particle finite element method. an overview. Int J Comput Methods 1(2):267–307

20. Falnes J (2002) Ocean waves and oscillating systems: linear interactions including wave-energy extraction. Cambridge Univ Press, Cambridge

21. Paixão Conde JM, Gato LMC (2008) Numerical study of the air-flow in an oscillating water column wave energy converter. Renew Energy 33(12):2637–2644

22. Paixão Conde JM, Teixeira PRF, Didier E (2011) Numerical simulation of an oscillating water column wave energy converter: Comparison of two numerical codes. In: Proceedings of the twenty-first international offshore and polar engineering conference, pp 668–674

23. Didier E, Paixão Conde JM, Teixeira PRF (2011) Numerical simulation of anoscillating water columnwave energy converter with and without damping. In: Proceedings of MARINE 2011—computational methods in marine engineering IV, pp 216–217

Chapter 103
Research Performance Evaluation of Scientists: A Multi-Attribute Approach

Lili Liu

Abstract In this paper, we highlight the fact that we cannot find a perfect index to evaluate output completely fairly and reasonably, and the research evaluation is a multi-attribute problem. This paper studies the method of multi-attribute comprehensive evaluation of scientists. Firstly, this paper chooses appropriate bibliometric indicators to evaluate research output. Following this, TOPSIS method is used to make a comprehensive research evaluation. Numerical examples are made regarding the purpose of testing the feasibility of the evaluation indicators and the evaluation method. Compared with traditional evaluation approaches on research performance, multi-attribute evaluation is more comprehensive and persuasive. It can overcome one-sidedness and reduce the bias of single indicator effectively.

Keywords Multi-attribute evaluation · Research performance · TOPSIS · Bibliometric indicators

103.1 Introduction

The objective research includes two categories of approaches, namely single-indicator evaluation and multiple attribute evaluation (MAE). Typical single indicators used in evaluations include the total number of papers published (N_n), total number of citations garnered (N_c), the journals where the papers were published, their impact parameter, the mean number of citations per paper etc., especially, the proposal of $h$index for individuals [8] has taken the world of research assessment by storm. On the basis of the index, scientists have proposed several 'h-type' indicators with the intention of either replacing or complementing the original h index. For

L. Liu (✉)
Business School, Sichuan University, Chengdu 610064, People's Republic of China
e-mail: liulili@scu.edu.cn

J. Xu et al. (eds.), *Proceedings of the Eighth International Conference on Management Science and Engineering Management*, Advances in Intelligent Systems and Computing 281, DOI: 10.1007/978-3-642-55122-2_103, © Springer-Verlag Berlin Heidelberg 2014

Table 103.1 Definition of some representative indicators

Indicators		Definition
Normal-index	N_n	The total number of papers published
	p	The number of papers published by the scientist that have been cited at least once
	N_c	Total number of citations garnered
	n_c	$n_c = \frac{N_c}{N_n}$, the mean number of citations per paper
	y	Number of "significant papers", defined as the number of papers with y citations
	q	Number of citations to each of the q most-cited papers
	Max	The number of citations of the most cited paper published by the scientist
h-type index	h	A scientist has index h if h of his or her N_n papers have at least citations each and the other $(N_n - h)$ papers have $\leq h$ citations each [8]
	ch	h index corrected for self-citations [12]
	m	$\frac{h}{y}$, where h is h index, y is number of years since publishing the first paper
	g	The g index g is the largest rank (where papers are arranged in decreasing order of the number of citations they received) such that the first g papers have (together) at least g^2 citations [3]
	A	$\frac{1}{h} \sum_{j=1}^{h} cit_j$, where h is h index, cit is citation counts [10]
	R	$R = \sqrt{\sum_{j=1}^{h} cit_j}$, where h is h index, cit is citation counts, and clearly, $R = \sqrt{AR}$ [7]
	AR	$\sqrt{\sum(\frac{cit}{a_j})}$, where h is h index, cit is citation counts, a is number of years since publishing [11]
	$h(2)$	A scientist's $h(2)$ index is defined as the highest natural number such that his $h(2)$ most-cited apers received each at least $[h(2)^2]$ citations [12]
	$ch(2)$	$h(2)$ index corrected for self citations [12]
	h_w	$\sqrt{\sum_{j=1}^{r} cit_j}$, where r_0 is the largest row index j such that $r_w \leq cit_j$. The h index weighted by citation impact [4]

instance, Egghe'g index [3], Jin et al. index [11], and Komulski's $h(2)$-index [12] etc. Some representative indicators are summarized in Table 103.1.

MAE is a mathematical aggregation of a set of individual indicators that measure multi-dimensional concepts but usually have no common units of measurement [14]. Despite the ceaseless debate on its use, MAE has been increasingly used for performance monitoring, benchmarking, policy analysis and public communication in wide ranging fields including the economy, environment and society by many national and international organizations. MAE is essentially concerned with the problem of how to evaluate and rank a finite set of alternatives in terms of a number of decision criteria. Most popular MAE approaches currently used are: Weighted Sum Model, Weighted Product Model, and Technique for Order Preference by Similarity to Ideal Solutions (TOPSIS), Data Envelopment Analysis, and Elimination Et Choice Translating Reality [17]. Although, it is easy to think about using multiple complementary indicators to evaluate research performance, there are little studies about the comprehensive utilization of these different indicators, let alone the establishment of evaluation index system. This situation may be mainly due to the reason that all empirical studies that have tested the various indicators for scientists have reported high correlation coefficients. According to the exclusiveness principle, it seems a redundancy using various indicators [1]. In fact, it is indeed redundant to use several indicators to measure only one attribute of research performance, while, it is necessary to choose appropriate various indicators to make a multiple-attribute evaluation of research performance. On the other hand, although all indicators are of great significance to evaluate scientists' research performance quantitatively and promote healthy development of science, every index does have problems of this kind or that kind. Moreover, the evaluation of IRO is a multi-attribute problem. A lot of scholars of bibliometric have been aware of this situation:

- Obviously a single number can never give more than a rough approximation to an individual's multifaceted profile, and many other factors should be considered in combination in evaluating an individual [8].
- It is not wise to force the assessment of researchers or of research groups into just one specific measure. It is even dangerous, because it reinforces the opinion of administrators and politicians that scientific performance can be expressed simply by one note. That is why we always stress that a consistent set of several indicators is necessary, in order to illuminate different aspects of performance [16].
- Yet, all these indicators suffer from the same aspect of roughness: they are all single indicators, hereby reducing the evaluation of a researcher's scientific career, to a single (i.e. one-dimensional) observation [5].
- The publication set of a scientist, journal, research group or scientific facility should always be described using many indicators such as the number of publications with zero citations, the number of highly cited papers and the number of papers for which the scientist is first or last author [1].
- Scientific performance should not be measured by a one-dimensional metric such as publication, since it is a multi-dimensional phenomenon [15].

103.2 The Choice of Indicators

As it is generally accepted, this paper will evaluate the research performance from the following four aspects: the number of paper published (C_1), Journal influence (C_2), Timeliness (C_3), scientific prestige (C_4) and Impact of paper (C_5).

In order to make a calculable evaluation, we have to use bibliometric indicators which have proved to be a useful yardstick for the measurement of scientific outputs. More and more indicators have been presented (see Table 103.1) for the purpose of evaluating research output scientifically and making the assessment method accord with facts. To make a practical and convenient evaluation, we will consider the advantages and disadvantages of different indicators synthetically, especially their easy availability in the process of selecting bibliometric indicators.

Table 103.2 Advantages and disadvantages of the normal-index

Indicators	Advantages	Disadvantages
N_p	(1) Easy to get; (2) Measure productivity	N_n gives every author full credit in the multiple authors' paper
p	Measure of productivity	p does not measure productivity perfectly and pays little attention to impact
N_c	The simplest measure that takes into account both the productivity and the impact	(1) Hard to find; (2) May be inflated by a small number of 'big hits' [8]. (3) N_c gives undue weight to highly cited review articles versus original paper
n_c	Allows comparison of scientists of different ages	(1) Hard to find; (2) Rewards low productivity, penalizes high productivity
y	Gives an idea of broad and sustained impact	(1) y is arbitrary and will randomly favor or disfavor individuals; (2) y needs to be adjusted for different levels of seniority
q	(1) Measure of impact; (2) Easy to find	(1) Difficult to obtain and compare; (2) q is arbitrary and will randomly favor and disfavor individuals
Max	Measure of impact	May be inflated by a small number of 'big hits'

1. The number of paper published (C_1) indicator

Normal-index is used most widely in measuring research quantity. It is very easy to point out the advantages and disadvantages of normal-index, because of their direct meanings.

From Table 103.2, we can see that, the total number of papers published (N_n) is the best bibliometric index in measuring productivity. N_n is very easy to get and it does not measure importance nor impact of papers.

2. Journal influence (C_2) indicator

Most research concerning journal evaluation is about measuring the influence of journal quantitatively. Although there are many indicators to measure the journal influence, such as Impact factor, Total cites, 5-Year Impact Factor, Cites per Doc. (ny), Eigenfactor Score, Article Influence Score, h index etc., the most used and accepted indicator is Impact factor (IF). If is calculated as the ratio between the number of citations in a given year to any item published in that journal in the previous 2 years and the number of research items published in the same journal in the same 2 years. In this article, (C_2) is calculated as the average IF of all the journals author has published on: $C_2 = \overline{IF} = \frac{1}{N_p} \sum_{i=1}^{N_p} IF_i$.

3. Timeliness (C_3) indicator

Besides, timeliness is also an important evaluation element. Papers in some journals become "out of date" in a short time, and papers in some journals receive no attention for a long time until an occasional opportunity arises. These are not good situations. In this paper, we will use Cited half-life and Immediacy Index to measure the timeliness of a journal. The Cited half-life is a measure of citation survival measuring the number of years, going back from the current year, that covers 50 % of the citations in the current year of the journal [6]. An immediacy index is a measure of how topical and urgent work published in a scientific journal is. The average of these two indicators of author are expressed as $\overline{J}_i^C$ and $\overline{J}_i^I$ respectively.

$$C_3 = \lambda_1 \times \overline{J}_i^C + \lambda_2 \times \overline{J}_i^I, \lambda_1 + \lambda_2 = 1, \lambda_1, \lambda_2 \in [0, 1]. \qquad (103.1)$$

λ_1, λ_2 in Eq. (103.1) maybe different among different fields. For example, if it is accepted that the Cited half-life can well characterize the Timeliness in this field, which the researchers being evaluated belong to, therefore $\lambda_1 = 1, \lambda_2 = 0$, otherwise, $\lambda_1 = 0, \lambda_2 = 1$. Parameter λ_i can be determined by valuators.

4. Scientific prestige (C_4) indicators

As we know, the impact of a journal is affected by a lot of elements, and the journal influence is not the same as journal prestige. There is a journal prestige indicator—SCImago Journal Rank (SJR). The SJR indicator is a size-independent metric aimed at measuring the current "average prestige per paper" of journals for use in research evaluation processes. It has already been studied as a tool for evaluating the journals in the Scopus database [7]. Journal influence indeed has something to do with the journal prestige, but journal influence can be very different from journal prestige. When evaluate IRO, we have to consider both the influence and the prestige of journals where the papers are published. In this article, (C_4) is calculated as the average SJR of all the journals author has published.

5. Impact of paper (C_5) indicators

The h index of any researcher, a purported single-number measurement of the impact of an individual scientific career, is now easily available from the Web of Science. It performs better than other single-number criteria commonly used to evaluate the scientific output of a researcher. It is now a well-established standard tool for the evaluation of the scientific performance of researchers. The advantages and disadvantages have already been pointed out by Hirsch himself and many other scholars [2]. Not only the h index but also its variants, namely h-type indicators have proved to be good measurement of the impact of research papers.

When choosing an index, we should take specific field into consideration. Different indicators may have different applicable fields. Kosmulski [12] pointed out that the original h index is probably appropriate in the fields, where the typical number of citations per article is relatively low, e.g., in mathematics or astronomy. The $h(2)$ index is favored in chemistry and physics. In this article, when there is no any given field, we choose the original h index.

The IF, Cited half-life, immediacy index and h index can be found in the Journal Citation Reports. SJR can be found in Scopus database.

103.3 Multi-Attribute Evaluation of Research Performance

This article used TOPSIS method to make a comprehensive evaluation of IRO. Hwang and Yoon [9] presented the technique for order preference by similarity to TOPSIS. TOPSIS takes advantage of the positive-ideal solution (PIS) and negative-ideal solution (NIS) of multi-attribute problems to rank the plans sets. In the N-dimensional space, in the plans set, compare the distances of the alternatives between PIS and NIS, if the plan both nears the PIS and far from the NIS is the best plan.

In this section, we will discuss the indifference curve and marginal rate of substitution (MRS). The indifference curve and MRS have special meanings for IRO multi-attribute evaluation. Mathematically, if we have an indifference curve: $f(v_1, v_2) = c$, where, f is a value function, c is a constant, v_1 and v_2 are two attributes' value, so that we have MRS at (v_1, v_2):

$$\lambda = -\frac{dv_1}{dv_2}\bigg|_{v_1,v_2} = \frac{\partial f/\partial v_2}{\partial f/\partial v_1}\bigg|_{v_1,v_2}. \tag{103.2}$$

TOPSIS method with Euclid distance measurement has a value function:

$$f(v_1, v_2) = \frac{S_{i-}}{S_{i+} + S_{i-}} = \frac{\sqrt{(v_1 - v_1^-)^2 + (v_2 - v_2^-)^2}}{\sqrt{(v_1 - v_1^-)^2 + (v_2 - v_2^-)^2} + \sqrt{(v_1 - v_1^+)^2 + (v_2 - v_2^+)^2}} = c. \tag{103.3}$$

So that the MRS can be calculated as:

$$\lambda = \frac{S_{i-}^2 (v_2^+ - v_2) + S_{i+}^2 (v_2 - v_2^-)}{S_{i-}^2 (v_1^+ - v_1) + S_{i+}^2 (v_1 - v_1^-)}. \tag{103.4}$$

When $S_{t+} = S_{t-}$,

$$\lambda = \frac{v_2^+ - v_2}{v_1^+ - v_1}. \tag{103.5}$$

Because of Eqs. (103.3) and (103.4), the value function can be written as $cS_{t+}^2 - (1-c)S_{t-}^2 = 0$, where $0 < c < 1$. It reflects a hyperbolic-type situation. Some typical indifference curves are shown in Fig. 103.1. The attributes are simplified as the number of paper published (C_1), Journal influence (C_2). All curves with $c_+ \geq 0.5$ are convex to the preference origin, while all curves with $c_+ \leq 0.5$ are concave to the reference origin. In this case, if the valuator has to choose an author as an academic leader, and he thinks all the authors are not very good both in their research quality and research quantity, he may apt to the author with a better performance in the worst attribute.

Figure 103.2 shows the MRS of C_1 and C_2. MRS usually relies on the grade of C_1 and C_2, namely (b_1, b_2). If the MRS is λ_b at (b_1, b_2) as shown in Fig. 103.2. C_1 remains unchanged, we may find MRS increases while decreasing C_2, and MRS decreases while increasing C_2. It shows the situation, when a researcher's research quality is very high, he disdains to publish superficial papers to increase his research quantity. On the other hand, when evaluating researchers with very high research quality, evaluators won't care too much about their research quantity.

Figures 103.3 and 103.4 show a group of indifference curves under a couple of independent attributes respectively. In Fig. 103.3, A_1 has more papers published than A_2, while A_2 has more prestige papers published than A_1. The valuator thinks the research output of A_1 and A_2 is undifferentiated. Similarly, the research output of A_4 and A_5 is undifferentiated.

Fig. 103.1 The indifference
curve

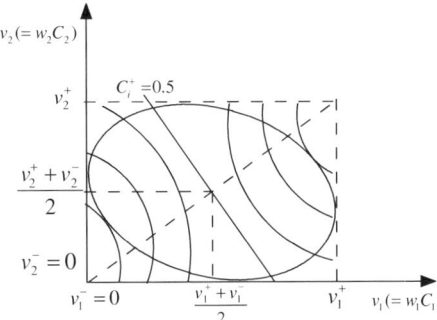

Fig. 103.2 The marginal rate
of substitution of C_1 and C_2

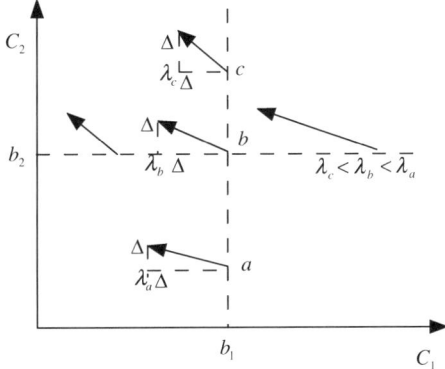

Fig. 103.3 A group of
indifference curve under
attribute C_1 and C_2

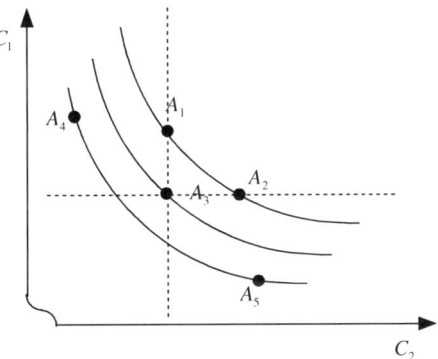

The journal prestige of A_1 is the same as A_3, however, A_1 has more papers published;
therefore, the evaluator think the research output of A_1 is better than A_3. The productivity of
A_2 and A_3 is indifferent, however A_2 has superiority on the attribute C_2, so that evaluator
think the research output of A_2 is better than A_3. In Fig. 103.4, if the valuator has to choose
a researcher as an academic leader, and there are two basic requirements: (1) the number of
papers published has to be more than a, and (2) the h index has to be greater than b. So that,

Fig. 103.4 A group of indifference curve under attribute C_1 and C_2

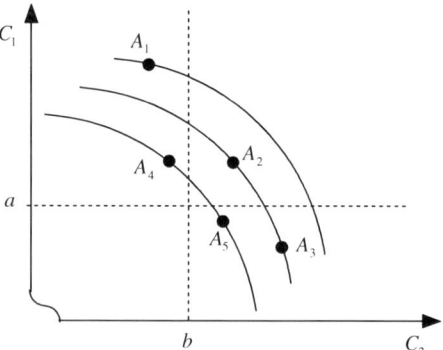

although A_2 is inferior to A_1 in general, and he is undifferentiated with A_3, the valuator will choose A_2.

103.4 Numerical Examples

To better understand this multi-attribute comprehensive evaluation method, let us consider following examples.

Querying the data on JCR and we can get journals' IF: $(IF_{O_1}, \cdots, IF_{O_9}) = (0.895, 0.83, 2.844, 0.89, 0.564, 0.562, 1.114, 2.328, 2.056)$. Scientist has published one paper on ACI STRUCT J, one paper on OPER RES LETT, two papers on IEEE T INTELL TRANSP, and one paper on IEEE J OCEANIC ENG, we can obtain:

$$C_{2,A} = \overline{IF} = \frac{1}{N_p} \sum_{i=1}^{N_p} IF_i = 8.303/5 = 1.6606,$$

$$C_{3,A} = \lambda_6 \bar{J}_A^6 + \lambda_7 \bar{J}_A^7 = 0.5 \times 7.14 + 0.5 \times 0.1278 = 3.6339,$$

$$C_{4,A} = \bar{J}_A^5 = 0.064.$$

Similarly, $C_{2,B} = 0.7008, C_{3,B} = 3.896, C_{4,B} = 0.0423, C_{2,C} = 0.7938, C_{3,C} = 3.616, C_{4,C} = 0.0593$. According to the definition of h index, $h_A = 5, h_B = 6, h_C = 6$. So that, according to the TOPSIS, we can get the weight vector: $w_j = (0.0345, 0.7795, 0.0058, 0.1456, 0.0346)$. Therefore, the PIS and the NIS are:

$$A^+ = \{\max v_{ij} | j \in \{1, 2, \cdots, 5\}\} = \{v_1^+, v_2^+, v_3^+, v_4^+, v_5^+\} = \{0.022005,$$
$$0.657211, 0.003496, 0.096075, 0.02113\},$$

$$A^- = \{0, 0, 0, 0, 0\}.$$

Then we can calculate the Euclid distance of these authors between the ideal point and the worst point:

Table 103.3 Numerical examples

Scientists	N_p	Journal	Citations	Scientists	N_p	Journal	Citations
A	5	O_1	45	C	10	O_7	21
		O_2	35			O_6	18
		O_3	25			O_6	23
		O_3	17			O_6	6
		O_4	13			O_6	5
B	8	O_5	5			O_8	5
		O_5	8			O_6	10
		O_6	12			O_6	12
		O_6	7			O_6	5
		O_5	10			O_6	3
		O_7	9				
		O_7	7				
		O_6	5				

Notes O_1 ACI STRUCT J, O_2 OPER RES LETT, O_3 IEEE T INTELL TRA NSP, O_4 IEEE J OCEANIC ENG, O_5 J CONSTR ENG M ASCE, O_6 J MANAGE ENG, O_7 J COMPUT CIVIL ENG, O_8 SCIENTOMETRICS

$$S_{A+} = \sqrt{\sum_{j=1}^{5}(v_{Aj} - v_j^+)^2} = 0.005389,$$

$$S_{A-} = \sqrt{\sum_{j=1}^{5}(v_{Aj} - v_j^-)^2} = \sqrt{\sum_{j=1}^{5}(v_{Aj})^2} = 0.66468, \quad C_{A+} = \frac{S_{A-}}{S_{A-}+S_{A+}} = 0.991957.$$

Similarly, $S_{B+} = 0.381259$, $S_{B-} = 0.286011$, $C_{B+} = 0.428629$, $S_{C+} = 0.343124$, $S_{C-} = 0.327967$, $C_{C+} = 0.488707$. Since $C_{A+} > C_{C+} > C_{B+}$, so that scientist A has the best research output, next is scientist C, and scientist B has the worst research output.

In the numerical example shown in Table 103.3, if consider the research quantity, scientist C is the best, and $N_{p,C} = 2N_{p,A}$, however, we cannot simply regard that the research output of C is two times better than A. When considering the journal where papers are published, the journal influence of C is the best, the timeliness of B is the best, while the scientific prestige of A is the best. We cannot compare which one is better. C and B have the same h index, we cannot compare C and B. By contrast, the result of a comprehensive evaluation is more persuasive.

103.5 Conclusions

As pointed out by Hirsch [8], a single number can never give more than a rough approximation of an individual's multifaceted profile, and many other factors should be considered in evaluating an individual scientist. Single indicator is unavoidable to be biased. Leeuwen [13] has tested that even h index is strongly biased across disciplines and a bias still can occur within one field. The contribution of this research is that we are trying to use multi-attribute comprehensive evaluation method to reduce bias and overcome one-sidedness of traditional single-indicator evaluation of research performance.

References

1. Bornmann L, Daniel HD (2009) The state of h index research. EMBO Rep 10:2–6
2. Costas R, Bordons M (2007) The h index: advantages, limitations and its relation with other bibliometric indicators at the micro level. J Informetrics 1:193–203
3. Egghe L (2006) Theory and practise of the g index. Scientometrics 69:131–152
4. Egghe L, Rousseau R (2008) An h-index weighted by citation impact. Inf Process Manage 44:770–780
5. Egghe L (2009) Characteristic scores and scales based on h-type indices. J Informetrics 4:14–22
6. Garfield E (2001) Interview with eugene garfield. Cortex 37:575–577
7. Guz AN, Rushchitsky JJ (2009) Scopus: a system for the evaluation of scientific journals. Int Appl Mech 45:351–362
8. Hirsch JE (2005) An index to quantify an individual's scientific research output. Nature 444:1003–1004
9. Hwang CL, Yoon K (1981) Multiple attribute decision making: methods and applications. Springer, New York NULL
10. Jin BH (2006) h-index: an evaluation indicator proposed by scientist. Sci Focus 1:8–9 (In Chinese)
11. Jin BH, Liang LM et al (2007) The R-and AR-indices: complementing the h-index. Chin Sci Bull 52:855–863
12. Kosmulski M (2006) A new hirsch-type index saves time and works equally well as the original h index. ISSI Newsl 2:4–6
13. Leeuwen TV (2008) Testing the validity of the hirsch-index for research assessment purposes. Res Eval 17:157–160
14. Park JH, Cho HJ et al (2013) Extension of the vikor method to dynamic intuitionistic fuzzy multiple attribute decision making. Comput Math Appl 65:731–744
15. Schmoch U, Schubert T et al (2010) How to use indicators to measure scientific performance: a balanced approach. Res Eval 19:2–18
16. Van Raan AFJ (2006) Comparison of the hirsch-index with standard bibliometric indicators and with peer judgment for 147 chemistry research groups. Scientometrics 67:491–502
17. Yu L, Chen Y et al (2005) Research on the evaluation of academic journals based on structural equation modeling. J Informetrics 3:304–311

Chapter 104
Multi-objective Optimization Model for Supplier Selection Problem in Fuzzy Environment

Muhammad Hashim, Liming Yao, Abid Hussain Nadeem, Muhammad Nazim and Muhammad Nazam

Abstract Supplier selection decisions are typically multi-objectives in nature and it is an important component of production and logistics management for many firms. The present study mainly investigate a multi-objective supplier selection planing problem in fuzzy environment and the uncertain model is converted into deterministic form by the expected value measure (EVM). This paper aims at multi-objective optimization for minimizing cost and maximizing product quality level. For solving the multi-objective problem a weighted sum base genetic algorithm is applied and the best solution is provided using fuzzy simulation. Finally, a numerical example is used to illustrate the effectiveness of the proposed model and solution approach.

Keywords Multiobjective model · Supplier selection · Supply chain · Fuzzy simulation · Genetic algorithm

104.1 Introduction

Supplier selection is a critical strategic issue for developing an effective supply chain. Supplier selection is a key process of supply chain management and also the right suppliers play a significant role for improving the overall performance. In designing a supply chain, a decision maker must consider decisions regarding the selection of the right suppliers and their order allocations (quota). Supply chain management (SCM) can be defined as a process with the flow and transportation and services from the production point to consumption point [1]. In fact, the suppliers play an important role in achieving the objectives of the supply chain and the success of a supply chain is highly dependent on selection of good suppliers. The right supplier

M. Hashim (✉) · L. Yao · A. H. Nadeem · M. Nazim · M. Nazam
Uncertainty Decision-Making Laboratory, Sichuan University, Chengdu 610064, People's Republic of China
e-mail: hashimscu@gmail.com

J. Xu et al. (eds.), *Proceedings of the Eighth International Conference on Management Science and Engineering Management*, Advances in Intelligent Systems and Computing 281, DOI: 10.1007/978-3-642-55122-2_104, © Springer-Verlag Berlin Heidelberg 2014

enhance the customer satisfaction and improve the supply chain performance in many ways including cost reduction, improving quality to achieve zero defects, improving flexibility to meet the needs of the end-consumers, reducing delivery time at different stages of the supply chain.

The supplier selection problem has received considerable attention in academic research [1–5]. Gaballa [6] is the first author who proposed mixed integer programming model for minimizing the total discounted price of allocated items to the suppliers for supplier selection problem in a real case. He developed a single-objective, mixed-integer programming for minimizing the sum of purchasing, transportation and inventory costs by considering the vendors delivery, quality and capacity. Weber and Current [7] presented a multi-objective approach to systematically analyze the trade-offs between conflicting criteria in supplier selection problems. Karpak et al. [8] proposed a goal programming model for minimizing costs and maximizing the delivery reliability and quality in supplier selection when assigning the order quanties to each supplier. Degraeve and Roodhooft [9] investigated a total cost approach with mathematical programming to treat supplier selection using activity-based cost information. Ghodsypour and Obrien [10] proposed a mixed-integer non-linear programming model for minimizing total cost of logistics, including net price, storage, ordering costs and transportation in supplier selection.

However, in a real situation for supplier selection problem many input information related to parameters are not known precisely and the decision makers usually confront with a high degree of uncertainties. These challenges increased the importance of stochastic and fuzzy programming for solving the real world problems where data are not known precisely. Deterministic models can not be suitable in this situation to obtain an effective solution. In these cases, fuzzy set theory is one of the best tool for handling the uncertainty. For example, it is very difficult for decision maker to determine the demand, cost and quality in advance because the values of these parameters changed with the passage of time and not remain the same. In this situation the fuzzy set theory can be used due to the presence of vagueness and imprecision of information in the supplier selection problem. Many scholars have been studied this uncertainty and imprecision by using fuzzy theory [11, 12]. Such as Jafar et al. [13] investigated fuzzy group decision making linear programming framework for supplier selection and order allocation, Xu and Yan [14] presented a comprehensive multi-objective supplier selection model under stochastic demand conditions considering demand quantity and timing uncertainties. Feng et al. [15] proposed an approach using fuzzy decision making and AHP method. Kilic [16] investigated an integrated approach for supplier selection in multi-item/multi-supplier environment. He developed a mixed integer liner programming model and fuzzy topsis method to select the best supplier in a multi-supplier environment. Liou et al. [17] presented a novel fuzzy integer-based model that addresses the interdependence among the various criteria and employs the non-additive gap-weighed analysis. Above all, the research models about supplier selection problem played an important role for solving the real problems. This paper contributes to current research as follows: first, a multi-objectives model is proposed which considers two objective functions which solve supplier selection planing problem. In addition, fuzzy variables are used to describe the demand, costs and quality which assists decision makers to make more effective and precise decisions.

The remainder of the paper is organized as follows: Sect. 104.2 explains the research problem about supplier selection under uncertain environment, Sect. 104.3 describes mathematical programming model under fuzzy environment and expected value operator to deal with fuzzy parameters for converting the multi-objective programming model into the expected value model, Sect. 104.4 states the solution approach for solving the model, Sect. 104.5 presents a numerical example and results, and finally, the concluding remarks are given in Sect. 104.6.

104.2 Problem Statement

Generally, the supplier selection problem (SSP) deals with issues related to the selection of right suppliers and their quota allocations. A good supplier selection can not only enhance the supply chain performance but also create a reliable system which can provide a powerful support to buyer for achieving his goals. The SSP is a complex problem due to several reasons. By nature, the SSP is a multi-criterion decision making problem. So the managers may be adopted different criteria for different material. How to choose some right (proper) suppliers always confuse the managers in case of many kind of candidates and materials. However, as to the supplier selection problem, it is hard to describes the cost, quality and demand as known values because there is not sufficient data available to analyze, which usually cause uncertainty, imprecise or vague situation [14].

For solving the above problem this paper considers that a buyer (producer) wants to buy a product from i number of suppliers with following objects: (1) Minimum cost; (2) Maximum quality. Different suppliers have different characteristics, for example, a supplier who can supply an item for the least per unit price may not have the better quality than the other competing suppliers. It is clear that there can be differences in the quality of different product cases provided by a different supplier. That's, some product of a supplier can have better quality but can be more expensive when compared to similar products of the other suppliers. Therefore, it is very difficult for a decision maker to balance the cost and quality when he is selecting the potential number of suppliers and their order allocations with goals of minimizing cost and maximizing the quality.

In real situation, decision making takes place in an environment where the objectives, constraints or parameters are not known precisely. It is difficult to describe the problem parameters as known due to the complexity of social and economic environment as well as some unpredictable factors such as bad weather and vehicle breakdowns. These challenges increased the importance of stochastic and fuzzy programming for solving the real world problems where data are not known precisely. Many scholars have been presented stochastic supplier selection models that are closer to real situations. Most of them have modeled the uncertainty (e.g., demand) by using probability distribution. Although probability theory has been proved to be a useful tool for dealing with uncertainty, some times it may not be suitable when the historical data is not available. So many parameters like demand and cost are usually uncertain rather than deterministic and it is very difficult to determine exact figure of these parameters due to the fluctuation in the values. In fact, the decision makers cannot collect the perfect information for each parameter. As different people have different feeling about the uncertain demand and cost and since there is no clear definition of this change. Thus in such kind of problem, it can be characterized by uncertainty of fuzziness. Stochastic models may not be suitable. Some scholars have observed this uncertainty and imprecision and dealt with them using fuzzy theory [16, 18]. However, supplier selection problem often faced with uncertain environment where fuzziness exists in a decision making process. Therefore, fuzzy variables that can take into account fuzziness are favored by decision-makers to describe the uncertainty.

In this research, demand, cost and quality are uncertain parameters and represented by fuzzy variables that can be further characterized by triangular fuzzy numbers. The decision makers can not get the exact information for each uncertain variable. So in this situation, the decision maker can describe the parameters into triangular fuzzy numbers that are more suitable to explain the uncertainty such as demand is about ζ_m but definitely not less than ζ_l and grater than ζ_u.

104.3 Model Formulation

In this paper the problem is formulated as a multi-objectives programming problem with fuzzy coefficients, in which the decision makers has two objectives first minimize the cost and second maximize quality.

104.3.1 Assumptions

The supplier selection problem in fuzzy environment can be stated as follows: the manager can choose the supplier from a i candidates and some differentiations between them are price and quality. The manager wants to know which suppliers should be chosen and how much of the quantity X_{ij} should be ordered considering the objectives of minimize cost and maximize quality. To model the supplier selection problem for material supply in a fuzzy environment in this paper, the following assumptions are as follows:

1. Quantity discount is not considers.
2. Purchasing is limited to three kind of goods.
3. Cost, demand, and quality are characterized by fuzzy variables.

104.3.2 Notations

Suppose that there is one buyer, i suppliers. The task is to allocate the orders to suppliers for minimizing cost and maximizing the material quality.

Indices

Ω	: set of suppliers, i is an index, $i \in \Omega = \{1, 2, 3, \cdots, I\}$,
Υ	: set of products, j is an index, $j \in \Upsilon = \{1, 2, 3, \cdots, J\}$.

Parameters

$\widetilde{\zeta}_j$	: aggregate demand of the item over the period,
$\widetilde{P}_{ij}$	: unit price of the order quantity X_{ij} from the supplier i,
$\widetilde{Q}_{ij}$	: product quality of supplier i for product j,
B	: budget allocated to suppliers i,
C_{ij}	: capacity of supplier i for product j.

Decision variable

X_{ij}	: this variable denotes the order quantity for the supplier i.

For the proposed problem, there is a need to allocate the order to suppliers from the potential set $\{1, 2, 3, \cdots, I\}$.

104.3.3 Modelling Formulation

The mathematical formulations of objectives are as follows. In this model, the first objective function is to minimize the total cost

$$\min Z_1 = \sum_{i=1}^{I} \sum_{j=1}^{J} \widetilde{P}_i X_{ij}. \qquad (104.1)$$

Second objective function maximizes the supplers item quality.

$$\max Z_2 = \sum_{i=1}^{I} \sum_{j=1}^{J} \widetilde{Q}_{ij} X_{ij}. \qquad (104.2)$$

Generally speaking, some mandatory conditions must be satisfied when the decision maker makes the decision and these are listed below.

The first constraint states that the total supply should be greater than or equal to expected demand.

$$\sum_{i=1}^{I} \sum_{j=1}^{J} X_{ij} \geq \widetilde{\zeta}_j. \qquad (104.3)$$

Second constraint states that the total cost must be less than or equal to the expected budget amount which will be allocated to suppliers.

$$\sum_{i=1}^{I}\sum_{j=1}^{J}\widetilde{P}_{ij}X_{ij} \leq B_j. \tag{104.4}$$

Third constraint states that order quantity should be less than the suppler capacity.

$$X_{ij} \leq C_{ij}. \tag{104.5}$$

Fourth constraint states the non-negativity constraints on decision variable.

$$X_{ij} \geq 0 \text{ and integer} . \tag{104.6}$$

From the above discussion, by integration of Eqs. (104.1)–(104.6), a fuzzy multi-objective expected value model for supplier selection problem can be formulated as follows:

$$\min Z_1 = \sum_{i=1}^{I}\sum_{j=1}^{J} E[\widetilde{P}_{ij}]X_{ij}$$

$$\max Z_2 = \sum_{i=1}^{I}\sum_{j=1}^{J} E[\widetilde{Q}_{ij}]X_{ij}$$

$$\text{s.t.} \begin{cases} \sum_{i=1}^{I}\sum_{j=1}^{J} X_{ij} = E[\widetilde{\zeta}] \\ \sum_{i=1}^{I}\sum_{j=1}^{J} E[\widetilde{P}_{ij}]X_{ij} \leq B \\ X_{ij} \leq C_{ij} \\ X_{ij} \geq 0 \text{ and integer.} \end{cases}$$

104.3.4 Dealing with Fuzzy Variables

It is generally a very difficult task to solve the optimization problems with multi-objective under uncertainty (fuzzy environment). For solving the fuzzy model it is necessary to convert it into a deterministic one. In this paper, expected value model based on *Me* is adopted to solve the proposed problem. For calculating the expected values of the triangular variables, a new fuzzy measure with an optimistic-pessimistic adjusting index is applied to characterize fuzzy parameters. The definition of this fuzzy measure *Me* can be found in [19]. It is a convex combination of possibility (*Pos*) and necessity (*Nec*) and the basic knowledge for measure (*Pos*) and necessity (*Nec*) can be found in [20]. Let $\widetilde{\zeta} = (\gamma_1, \gamma_2, \gamma_3)$ denotes a triangular fuzzy variable. Based on the definition and properties of expected value operator of fuzzy variable using measure Me [19], if the fuzzy variable $\tilde{\zeta} = (\gamma_1, \gamma_2, \gamma_3)$, where $\gamma_1, \gamma_2, \gamma_3 > 0$, then the expected value of $\tilde{\zeta}$ should be: $E^{Me}[\tilde{\zeta}] = \frac{1-\lambda}{2}(\gamma_1 + \gamma_2) + \frac{\lambda}{2}(\gamma_2 + \gamma_3)$.

104.3.5 Weighted Sum Approach

However, it is not easy to find an optimal solution to the above model since it is a multi-objective optimization. In this paper, the weighted sum method is adopted to deal with the multi-objective model. The above model can be transformed into a single-objective one by using the following process. To solve a multi-objective optimization problem there is a need to assign a weight to each normalized objective function. The problem will be converted into a single objective by using the following process.

Suppose that w_i is a weighted coefficient for $f_i(x)$ objectives functions and the sum of w_i will be equal to 1. It expresses the importance of objective functions in the proposed model. Model can be converted into single objectives by using the following process: $f(x) = \sum_{i=1}^{m} w_i f_i(x)$. The weighted problem can be written as: $\min f(x) = \min_{x \in X} w_1 f_1(x) - \max_{x \in X} w_2 f_2(x) = \sum_{i=1}^{m} w_i f_i(x)$.

104.4 Solution Approach

There are many kinds of evolutionary computation methods available to solve the hard optimization problems. Among them, genetic algorithm (GA) is a well known class of evolutionary algorithms. It does not require the information expressed in terms of gradient of the optimization objective functions and can provide a number of potential solutions for a given problem, leaving the user to make the final decision. It has been employed considerable success in providing good solutions to many complex optimization problems containing fuzzy parameters [21, 22]. It is a stochastic search method of optimization problems based on the mechanics of natural selection and natural genetics-survival of the fitness function value. The advantage of this approach enable us to obtain the global optimal solution fairly. In addition, it is not required the specific mathematical analysis of optimization problems, which makes it easily coded by users who is not necessarily good at mathematical and algorithms. It has been applied to solve a wide variety of problems, such as transportation problems, facility layout problems, scheduling, network optimization and so on. Wang and Fang [23] is the first researcher who used GA for solving linear and non-linear transportation/distribution problems. In his method, he represented each chromosome of the problem by using $m \times n$ matrix. So, GA is a best solution approach for getting the good result of the proposed model. The following steps (procedure) of GA are used for solving the proposed model:

Step 1. Initialization: Randomly generate an initial population called chromosomes. These chromosomes represent the solution of the proposed model. In this research the chromosomes are denoted by vector x, $x = (x_1, x_2, x_3, \cdots, x_{N_{pop}})$.

Step 2. Constraint checking: Applying the fuzzy simulation for checking the chromosome generated by genetic operator for ensuring that the chromosomes are in the feasible region. Generate random vector x in the feasible region until a feasible one is accepted as a chromosome. Repeat the above process for $N_{pop-size}$ times for getting the initial feasible chromosomes.

Step 3. Evolution: For the evolution of each chromosomes the following function is applied for calculating the fitness function values for each chromosomes,

$$eval(x) = \sum_{k=1}^{m} \frac{E[f_k(x, \xi)] - z_k^{max}}{z_k^{max} - z_k^{min}}.$$

Step 4. Selection: The selection process is based on spinning the roulette wheel $N_{pop-size}$ times. Each time a single chromosome for a new population is selected in the following way: Calculate the cumulative probability q_i for each chromosome, $q_0 = 0, q_i = \sum_{j=1}^{i} eval(x^j), i = 1, 2, \cdots, N_{pop-size}$. Generate a random number γ in $[0, 1]$ and select the ith chromosome x^i. Repeat the above process $N_{pop-size}$ times for getting the $N_{pop-size}$ pairs of chromosomes.

Step 5. Crossover and Mutation: Update the chromosomes by applying the crossover and mutation. In crossover, the operation is started from generating a random number r from the open interval $(0, 1)$ and then chromosomes x^i are selected as a parent if $\gamma < P_c$ in which P_c is defined as the crossover probability parameter. Denote the selected parents by $x^1, x^2, \cdots$. First, generate a random number from the open interval $(0, 1)$ then the crossover operator on x^1 and x^2 will produce two children D^1 and D^2 as follows: $D^1 = cx^1 + (1 - c)x^2, D^2 = cx^2 + (1 - c)x^1$. If both children are feasible, then we will replace the parents with them. Otherwise, keep the feasible one if it exists and redo the crossover operation with another random number c until two feasible children are obtained or a given number of cycles is finished.

In the mutation operation, repeat the same process (steps) like crossover process from $i = 1$ to pop-size, the chromosome x^i is selected as a parent to undergo the mutation operation provided that random number $\gamma < P_m$ in which P_m is defined as the mutation probability parameter $P_m \cdot N_{pop-size}$ chromosomes are expected to be selected after repeating the process $N_{pop-size}$ times. Change each selected parent, denoted by $S = (x^1, x^2, \cdots, x^i)$, in the following process. Randomly choose a mutation direction $d \epsilon R^n$. Replace S with $S + M.d$ if $S + M.d$ is feasible, otherwise set M as a random number between 0 and M until it is feasible or a given number of cycles is finished. Here, M is an appropriately large positive number.

Step 6. For each chromosome $x = (x_1, x_2, x_3, \cdots, x_{N_{pop}})$, calculate the two objectives values.

Step 7. Repeat steps from 3 to 5.

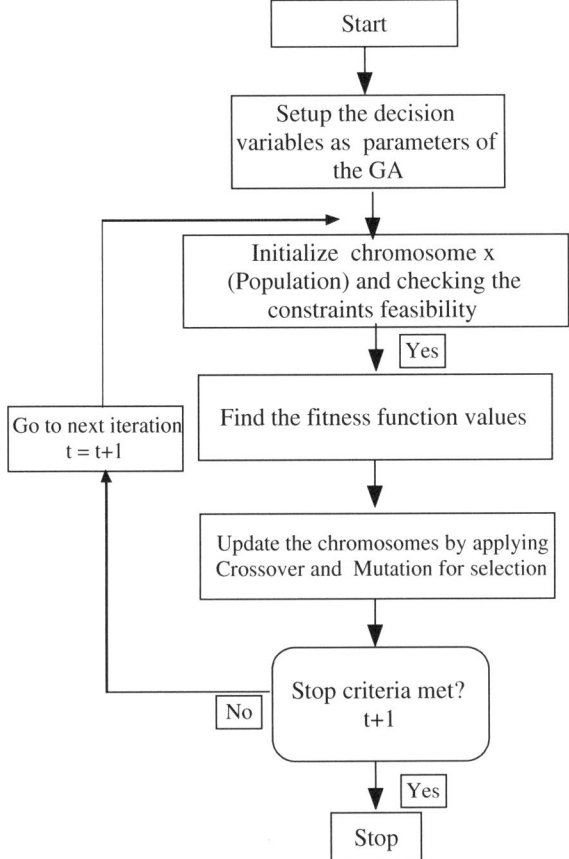

Fig. 104.1   GA procedure for solving the problem

Step 8. Terminate the whole procedure when a given number of cycles is completed and account the best chromosomes as the best solution.The steps for solving the proposed model can be seen in Fig. 104.1.

104.5 Numerical Example

In this section, a numerical example is presented to illustrate the application of the multi-objectives programming model and its solution approach for the optimal supplier selection. For the application of the proposed method, we consider five suppliers and three kinds of products. The available budget (B) is 44,000. The other data is shown in the Tables 104.1 and 104.2.

Table 104.1 The aggregate demand ($\widetilde{\zeta}$)

Product	1	2	3
Demand	(2000,2200,2400)	(2000,2200,2400)	(2000,2200,2400)

Table 104.2 Suppliers information

Suppliers	Product	Price	Quality	Capacity
1	1	(4.1, 4.2, 4.3)	(0.81, 0.82, 0.83)	(500)
	2	(5.1, 5.2, 5.4)	(0.81, 0.82, 0.83)	(1,000)
	3	(6.1, 6.2, 6.4)	(0.81, 0.82, 0.83)	(1,500)
2	1	(4.2, 4.3, 4.4)	(0.90, 0.91, 0.92)	(500)
	2	(5.1, 5.3, 5.4)	(0.90, 0.91, 0.92)	(1,000)
	3	(6.2, 6.1, 6.4)	(0.90, 0.91, 0.92)	(1,500)
3	1	(4.3, 4.4, 4.5)	(0.92, 0.93, 0.94)	(500)
	2	(5.3, 5.4, 5.5)	(0.92, 0.93, 0.94)	(1,000)
	3	(6.1, 6.4, 6.5)	(0.92, 0.93, 0.94)	(1,500)
4	1	(4.4, 4.5, 4.6)	(0.92, 0.93, 0.94)	(500)
	2	(5.4, 5.5, 5.6)	(0.92, 0.93, 0.94)	(1,000)
	3	(6.4, 6.5, 6.6)	(0.92, 0.93, 0.94)	(1,500)
5	1	(4.5, 4.6, 4.7)	(0.92, 0.93, 0.94)	(500)
	2	(5.5, 5.6, 5.7)	(0.92, 0.93, 0.94)	(1,000)
	3	(6.5, 6.6, 6.7)	(0.92, 0.93, 0.94)	(1,500)

104.5.1 The Result of the Numerical Example

Taking the above data into the proposed model, we adopted GA using MATLAB 7.9 on Pentium 4.2 GHz clock pulse with 2048 MB memory, and tested the performance of the method with the proposed example. we get the satisfactory solution after running the programme, and the best results are shown in Tables 104.3 and 104.4. The optimal solution is obtained on based *Me* fuzzy measure and set optimistic-pessimistic index $\lambda = 0.5$, pop-size is 20, crossover probability is 0.3, mutation probability is 0.2, the genetic algorithm based on the fuzzy simulation at the 100 generations for getting the optimal solution. The optimal solution in Table 104.2 is showed that 5 supplier get the most of the order quantity and appropriate price while first and third suppliers get the some of order quantity as compared to other suppliers du to its acceptable quality and low price. Suppliers first and fourth get a little quantity order relatively than other suppliers du to low quality. The GA search process for the optimal solution can be seen in Fig. 104.2. The optimal solution dependent on the value of λ and the different weights assigned to both objects. The decision maker can get the different objective values by changing these parameters.

Table 104.3 The optimal objective values the case problem

w_1	w_2	Z_1^*	Z_2^*
0.5	0.5	34812	6507

Table 104.4 The optimal solution the case problem

Vendor	Item		
	1	2	3
1	$X_{11} = 414$	$X_{12} = 254$	$X_{13} = 183$
2	$X_{21} = 227$	$X_{22} = 535$	$X_{23} = 289$
3	$X_{31} = 103$	$X_{32} = 127$	$X_{33} = 103$
4	$X_{41} = 614$	$X_{42} = 185$	$X_{43} = 665$
5	$X_{51} = 842$	$X_{52} = 1099$	$X_{53} = 960$

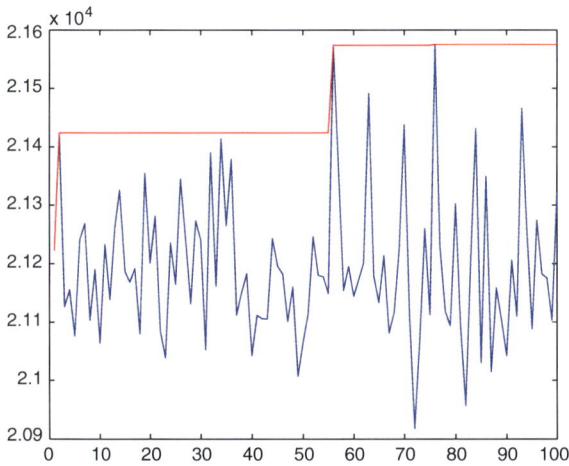

Fig. 104.2 Optimal solution layout (Evolutionary process of the proposed GA approach)

104.6 Conclusion and Remarks

Supplier selection is one of the most important decision taking by purchasing depart-ment in supply chain. The key role suppliers perform in terms of managing the quality, costs and services at desire level, which affect the outcome in the buyer's company. This research mainly investigated a multi-objectives expected value programming model with considering fuzzy coefficients for solving the supplier selection problem. Genetic algorithm method is proposed to get the optimal order allocation to suppli-ers for minimizing cost and maximizing quality. For checking the effectiveness of the proposed model and solution approach is tested by a numerical example, which proved to be effective and feasible to solve the presented numerical example. It can help the decision maker to find out the appropriate ordering from each supplier, and allows purchasing manager to make a good decision for managing supply chain performance on cost and quality.

Acknowledgments The authors wish to thank the anonymous referees for their helpful and constructive comments and suggestions. The work is supported by the National Natural Science Foundation of China (No. 71301109), the Western and Frontier Region Project of Humanity and Social Sciences Research, Ministry of Education of China (No. 13XJC630018), the Philosophy and Social Sciences Planning Project of Sichuan province (NO. SC12BJ05), and the Initial Funding for Young Teachers of Sichuan University (No. 2013SCU11014).

References

1. Buyukozkan G, Cifci G (2011) A novel fuzzy multi-criteria decision framework for sustainable supplier selection with incomplete information. Comput Ind 62(2):164–174
2. Kaslnqam R, Glee CP (2006) Selection of vendors-a mixed-integer programming approach. Comput Ind Eng 31(1):347–350
3. Lia Z, Rittscher J (2007) A multi-objective supplier selection model under stochastic demand condition. Int J Prod Econ 105:150–159
4. Weber CA, Currint JR (2003) A multi-objective approach to vendor selection. Eur J Oper Res 68(2):173–184
5. Wu D (2008) Supply chain risk, simulation and vender selection. Int J Prod Econ 114:646–655
6. Gaballa AA (1974) Minimum cost allocation of tenders. Oper Res Q 25(3):389–398
7. Weber CA, Current JR (1993) A multiobjective approach to vendor selection. Eur J Oper Res 68:173–184
8. Karpak B, Kumcu E, Kasuganti R (1999) An application of visual interactive goal programming: a case in vendor selection decisions. J Multi-Criteria Decis Anal 8:93–105
9. Degraeve Z, Roodhooft F (2000) A mathematical programming approach for procurement using activity based costing. J Bus Financ Acc 27(1–2):69–98
10. Ghodsypour SH, Obrien C (2001) The total cost of logistic in supplier selection, under conditions of multiple sourcing, multiple criteria and capacity constraint. Int J Prod Econ 73:15–27
11. Amid A, Ghodsypour SH, Brien OC (2006) Fuzzy multiobjective linear model for supplier selection in a supply chain. Int J Prod Econ 104:394–407
12. Kumar M, Vrat P, Shankar R (2006) A fuzzy programming approach for vener selection problem in a supply chain. Int J Prod Econ 101:273–285
13. Jafar R, Songhori M, Mohammad K (2009) Fuzzy group decision making/fuzzy linear programming (FGDMLP) framework for supplier selection and order allocation. Int J Adv Manuf Technol 105:150–159
14. Xu JP, Yan F (2011) A multi-objective decision making model for the vender selection problem in a bifuzzy environment. Expert Syst Appl 38:9684–9695
15. Feng D, Chen L, Jiang M (2005) Vendor selection in supply chain system: an approach using fuzzy decision and AHP. In: International conference on services systems and services Management Beijing, China, pp 721–725
16. Kilic HS (2013) An integrated approach for supplier selection in multi-item/multi-supplier environment. Appl Math Model 37:7752–7763
17. Liou JJH, Chuang CY, Tzeng HG (2013) A fuzzy integrad-based model for supplier evaluation and improvement. Inf Sci. doi:10.1016/j.ins.2013.09.025
18. Su RH, Yang DY, Pearn WL (2011) Decision-making in a single-period inventory environment with fuzzy demand. Expert Syst Appl 38:1909–1916
19. Xu JP, Zhou XY (2011) Fuzzy like multi objective decision making. Springer, Berlin

20. Dubois D, Prade H (1994) Possibility theory: an approach to computerized processing of uncertainty. Plenum Press, New York
21. Owen SH, Daskin MS (1998) Strategic facility location: a review. Eur J Oper Res 111:423–447
22. Pedrycz W (1997) Fuzzy evolutionary computation. Kluwer Academic Publishers, Boston, pp 318–327
23. Wang D, Fang SC (1997) A genetics-based approach or aggregate production planning in fuzzy environment. IEEE Trans Syst Man Cybern 27:636–645

Part VII
Industrial Engineering

Chapter 105
Process Developments in Friction Surfacing for Coatings Applications

Rosa M. Miranda, João Gandra and Pedro Vilaça

Abstract The present work focused on the production of coatings of AISI1024, AISI1045 and AISIH13 over mild steel and AA6082-T6 over AA2024-T3 by friction surfacing. Performance criteria were established to quantify material deposition rate and specific energy consumption aiming to compare friction surfacing with most direct concurrent technologies such as laser and arc welding based cladding. Simple analytical models were developed to estimate the rod consumption and the power and energy consumption rates. Friction surfacing is a competitive technology considering energy efficiency and power consumption in comparison with mainstream processes. However, coating processes based on wire or powder feed are more competitive, since they allow continuous feeding of deposited material. The effect of process parameters was also assessed and it was seen they have different impacts on process efficiency and deposition rate for each material combination.

Keywords Friction surfacing · Coating · Steel · Aluminium · Performance criteria

R. M. Miranda (✉)
UNIDEMI, Faculdade de Ciências e Tecnologia, Universidade Nova de Lisboa,
2829-516 Caparica, Portugal
e-mail: rmiranda@fct.unl.pt

J. Gandra
Instituto Superior Técnico, Universidade Técnica de Lisboa, Av. Rovisco Pais 1,
1049-001 Lisbon, Portugal

P. Vilaça
Department of Engineering Design and Production, School of Engineering,
Aalto University, P.O. Box 1420, 00076 Aalto, Finland

J. Xu et al. (eds.), *Proceedings of the Eighth International Conference on Management Science and Engineering Management*, Advances in Intelligent Systems and Computing 281, DOI: 10.1007/978-3-642-55122-2_105, © Springer-Verlag Berlin Heidelberg 2014

105.1 Introduction

The constant drive for more cost effective and energy efficient manufacturing processes has recently pushed research to develop friction based technologies for several joining applications. Friction based processes have grown very popular as they allow to join without external heat sources, while processing the base materials into an enhanced metallurgical condition. The parts are subjected to relative motion under pressure and heat is brought up by friction at the interface of the faying surfaces. A significant energy is also dissipated from the bulk viscoplastic deformation of the adjacent regions. Friction is increasingly being considered as a cheap energy source to join and process materials, while retaining the advantages inherent to solid-state materials processing.

Several recent investigations propose that friction-based processes allow reducing the coating manufacturing costs associated with energy consumption and consumables. These processes are also considered more environmental friendly. This is especially attractive when considering the current increase of eco-friendly regulations, as well as, the popularity and marketing advantages of endorsing such policies.

Friction surfacing (FS) is a solid state technology with increasing applications in the context of localized surface engineering [1, 4, 5]. FS has been investigated mainly for producing fine grained coatings, which exhibit superior wear and corrosion properties. Since no bulk melting takes place, this process allows the dissimilar joining of materials that would be otherwise incompatible or difficult to deposit by fusion based methods. Several studies also emphasize its energy efficiency and low environmental impact as key advantages when compared with other alternative technologies. Main applications include the repair of worn or damaged surfaces through building up or crack sealing. It has also been applied to enhance surface properties at specific areas in the manufacturing of parts and tools. A wide range of materials combinations has been deposited by FS, mainly alloy and stainless steels. Aluminium, magnesium and titanium alloys have also been investigated, including the production of metal matrix composites [3, 9, 12].

Friction Surfacing currently struggles with several technical and productivity issues which contribute to a limited range of engineering applications. The main drawbacks of this technology are related the unconstrained material flow that leads to the formation of a revolving flash and poor joining at the coating edges [2, 13, 14]. Additionally, this technique offers limited control over the coating cross section geometry, as post-processing by milling or gridding is required to achieve the desired geometry. Although FS is frequently suggested as a leaner alternative, few investigations support this assumption with a quantitative study. A realistic comparison between coating processes must consider the quantification of deposition rates and energy efficiency, which have not been determined previously for friction surfacing.

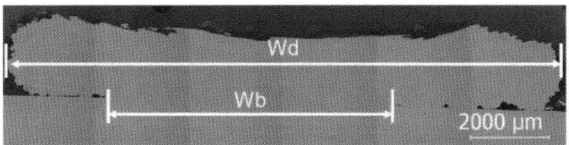

Fig. 105.1 Unbounded regions and joining efficiency nomenclature. Joined width (*Wb*) and the maximum coating width (*Wd*)

105.2 Material Transfer

Volumetric rod consumption rate (CR_{vol}) is determined by multiplying the rod plunging speed (V_z) by its cross section area (A_r), where r is the rod radius, using the following equation: $CR_{vol}[\mathrm{m}^3/\mathrm{s}] = A_r V_z = \pi r^2 V_z$. Likewise, the product between the travel speed (v) and the deposited cross section area (A_d) expresses the volumetric deposition rate (DR_{vol}) throughout the FS process, as given by equation: $CR_{vol}[\mathrm{m}^3/\mathrm{s}] = A_d v$.

Considering the consumable rod material density (ρ), CR and DR can be rewritten in order to express the mass flow, as depicted by Eq. (105.1). A steel density of $0.00785\,\mathrm{g/mm}^3$ was considered in order to estimate the mass flow from the volumetric deposition rate. For aluminium alloys the reference value was of $0.0027\,\mathrm{g/mm}^3$.

$$CR[\mathrm{kg/s}] = CR_{vol} \cdot \rho, \quad DR[\mathrm{kg/s}] = DR_{vol} \cdot \rho. \tag{105.1}$$

In order to determine the fraction of consumed material deposited and that is transferred to flash, the deposition efficiency ($\eta_{\text{deposition}}$) can be defined as the ratio between DR and CR, as given by Eq. (105.2):

$$\eta_{\text{deposition}} = \frac{DR}{CR}. \tag{105.2}$$

However, due to the formation of side unbonded regions, just a part of the deposited material is effectively joined, as shown schematically in Fig. 105.1. As such, the joining efficiency (η_{joining}) is given by the ratio between the joined width (W_j) and the maximum coating width (W_c), as expressed by Eq. (105.3):

$$\eta_{\text{joining}} = \frac{W_j}{W_c}. \tag{105.3}$$

Thus, the effective coating efficiency (η_{coating}) reflects the fraction of consumed rod that actually becomes joined to the substrate and is estimated by multiplying Eqs. (105.2) and (105.3), thereby obtaining Eq. (105.4).

$$\eta_{\text{coating}} = \eta_{\text{deposition}} \cdot \eta_{\text{joining}} = \frac{A_d v}{\pi r^2 V_z} \cdot \frac{W_b}{W_d}. \tag{105.4}$$

Fig. 105.2 Set up for steel depositions

105.3 Experimental Procedure

Friction surfacing was performed using an *ESAB LEGIO^{TM}* 3UL numeric control friction stir welding machine. The steel consumable rods were 65 mm long and 10 mm diameter and AA6082 consumable rods had 20 mm diameter and 120 mm long. The mild steel and AA2024 substrates were 6 and 4 mm thick, respectively. Figures 105.2 and 105.3 depict the set-up for performing FS depositions for the steel and aluminium combinations, respectively. Consumable rods and substrates were degreased prior to deposition. Depositions of 50 and 100 mm where performed in steel and aluminium, respectively.

Tests were performed varying the axial forging force, the rotation speed and the traverse speed. Table 105.1 summarizes the range of processing parameter for the material combinations under study that were further analyzed under optical microscopy to measure the joined width and the clad thickness.

105.4 Results and Discussion

1. Deposition and Consumption Rates Table 105.2

Summarizes data collected from the measurements performed on the samples produced and values computed of joining efficiency, deposition rate and energy consumption.

(1) Deposition of AISI 1024, AISI 1045 and AISI H13 over mild steel
Figures 105.4, 105.5, 105.6 present deposition (DR) and consumption rates (CR) for the tested conditions, regarding the deposition of AISI1020, AISI1045 and AISI H13 over mild steel. Deposition rates were seen to vary from 0.5–1.6 g/s to 0.4–2.6 g/s,

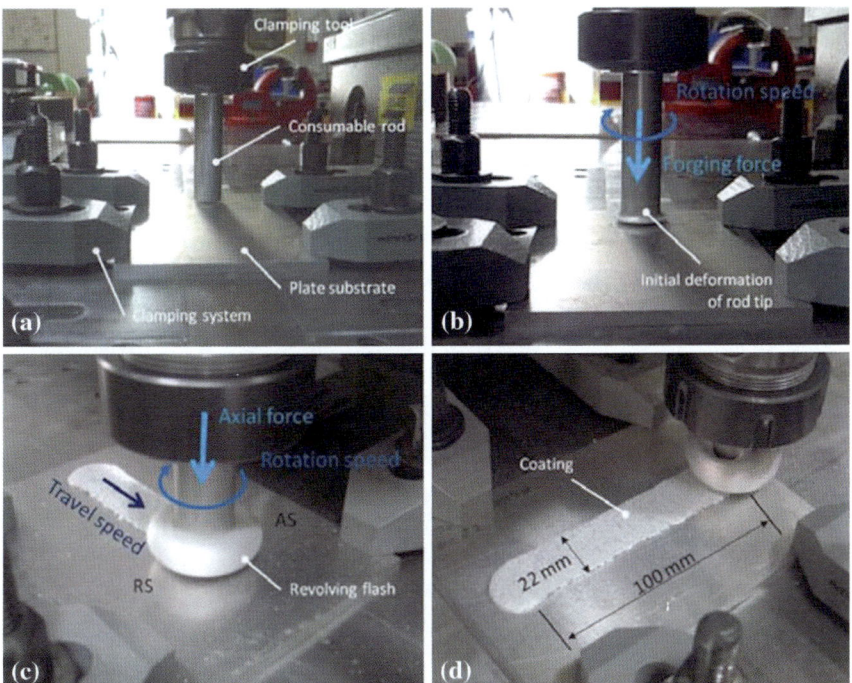

Fig. 105.3 Deposition of AA 6082-T651 aluminium alloy by friction surfacing; **a** experimental setup; **b** initial deformation stage; **c** deposition; **d** sample produced

Table 105.1 Process parameter window for each material combination

Coating/substrate	Force (kN)	Rotation speed (rpm)	Travel speed (mm/s)
IN St52/MS	5–6	2,000–2,500	4.2–10.8
DIN Ck45/MS	6–7	2,500–3,000	4.2–10.8
AISI H13/MS	9–10	2,000–2,500	4.2–10.8
AA6082-T6/AA2024-T3/MS	5–7	2,000–2,500	4.2–10.8

regarding the FS of 1020 and 1045, respectively. FS of AISI H13 generally presents lower values of deposition rate, compared to other steel combinations (from 0.14 to 1.13 g/s). Similar findings were reported by Shinoda et al. [7], who obtained a deposition rate of 0.28 g/s using martensitic stainless steel AISI 440 rods and structural steel plates. Thomas et al. [10] reported deposition rates of 1.38 and 1.94 g/s, when depositing austenitic stainless steel and mild steel, respectively.

A similar effect of process parameters was observed in all the steel combinations. Deposition rate increased for higher axial forces, following the resulting increase of coating width. Higher rotation speeds resulted in a reduction of coating thickness and width, thereby decreasing the deposition rate. The highest values of deposition rate were observed for the lowest rotation speeds. However, higher travel speeds led

Table 105.2 Process parameter window for each material combination

Coating/substrate	Force (kN)	Rotation speed (rpm)	Travel speed (mm/s)	Thickness (mm)	Joined. width (mm)	Joining η (%)	DR (mm³/s)	EC (J/mm³)
DIN St52/MS	5–6	2,000–2,500	4.2–10.8	1–2.5	8–10	60–75	64–200	24–35
DIN Ck45/MS	6–7	2,500–3,000	4.2–10.8	0.5–1.2	8–10	80–90	50–330	31–55
AISI H13/MS	9–10	2,000–2,500	4.2–10.8	0.4–1	8–11	80–95	12–150	63–94
AA6082-T6/ AA2024-T3/ MS	5–7	2,000–2,500	4.2–10.8	1–3	15–20	60–80	150–740	11–22

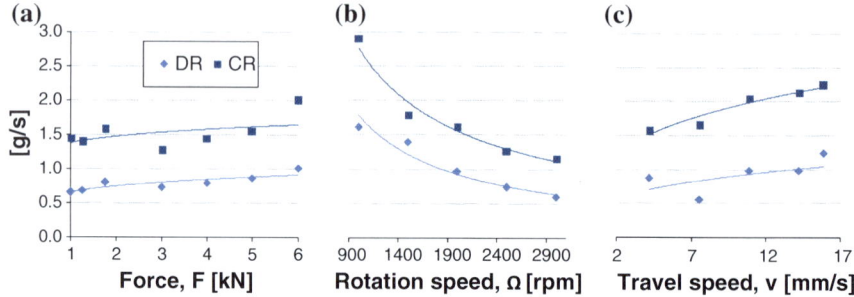

Fig. 105.4 Effect of process parameters on deposition rate (*DR*) and consumption rate (*CR*), regarding the FS of DIN St52 over mild steel. Process parameters: **a** Ω = 2,500 rpm, v = 4.2 mm/s, **b** F = 3 kN, v = 4.2 mm/s, **c** F = 5 kN axial force, Ω = 2,500 rpm

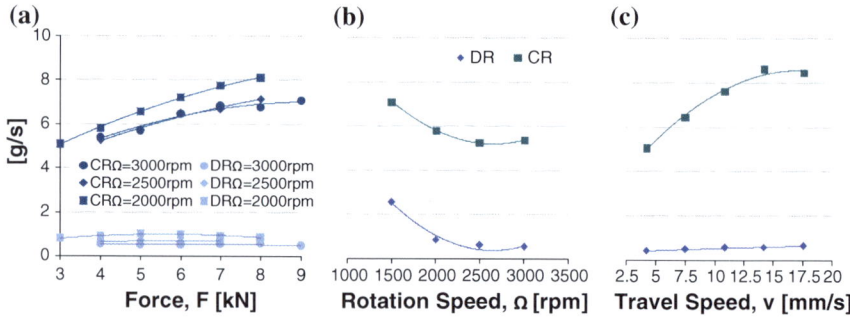

Fig. 105.5 Effect of process parameters on deposition rate (*DR*) and consumption rate (*CR*), regarding the FS of DIN Ck45 over mild steel. Process parameters: **a** v = 7.5 mm/s, **b** F = 4 kN, v = 7.5 mm/s, **c** F = 6 kN, Ω = 3,000 rpm

to an increase of deposition rate, despite the decrease in coating width and thickness. Travel speed influences the rate at which the material is deposited (in a direct relation) and, apparently, this effect is more significant than the resulting reduction in the coating cross section area. From figures below, it can be seen that the consumption rate is always higher than the deposition rate and closely follows its variation. The

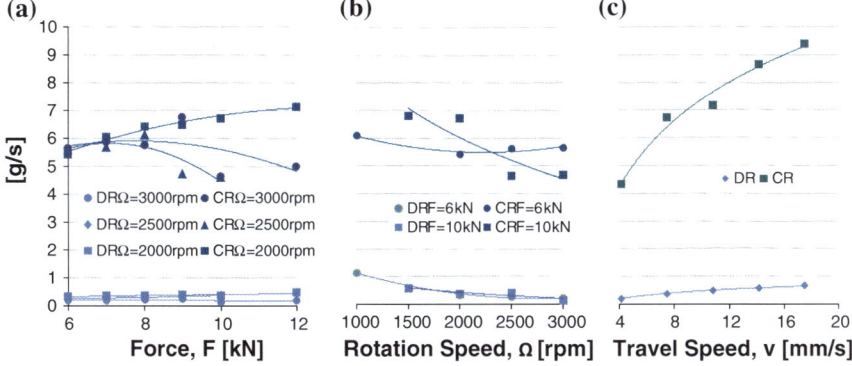

Fig. 105.6 Effect of process parameters on deposition rate (*DR*) and consumption rate (*CR*), regarding the FS of AISI H13 over mild steel. **a** $v = 7.5$ mm/s travel speed, **b** $v = 7.5$ mm/s travel speed, **c** $F = 10$ kN, $\Omega = 2,000$ rpm

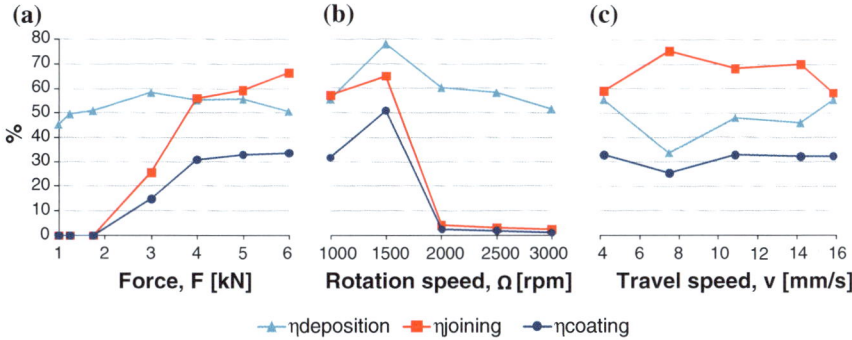

Fig. 105.7 Effect of process parameters on deposition, joining and coating efficiency, regarding the FS of DIN St52 over mild steel. Process parameters: **a** $\Omega = 2,500$ rpm, $v = 4.2$ mm/s; **b** $F = 3$ kN, $v = 4.2$ mm/s; **c** $F = 5$ kN, $\Omega = 2,500$ rpm

consumption rates in the deposition of AISI1045 and AISI H13 are significantly higher than those observed in mild steel. Since the deposition rates remain similar to those of AISI1020, this indicates that the flash formed is more significant. This finding suggests that the FS of materials with higher mechanical strength requires enhanced pressure boundary conditions.

Figures 105.7, 105.8, 105.9 present the influence of process parameters on the deposition, joining and coating efficiency. The effect of axial force, rotation and travel speeds has a different impact for each material combination. There some common effects, but a direct comparison between coating materials is not truthful as each material has its own process widow.

The increase of force contributes to improve the coating efficiency, whenever it results in the significant improvement of joined width and joining efficiency. This trend is clear in case of the FS of DIN St52 and AISI H13. However, excessive forces

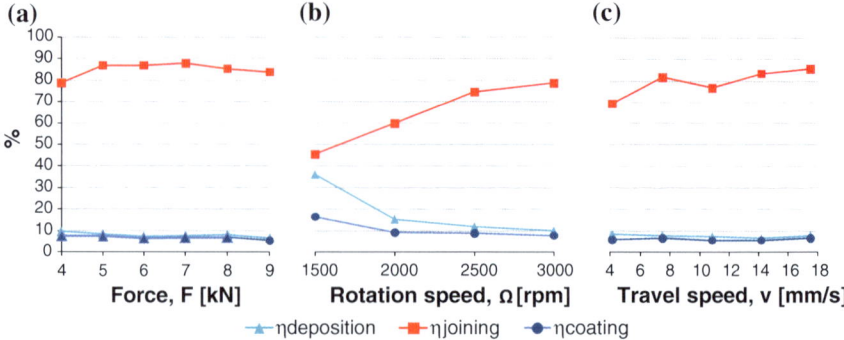

Fig. 105.8 Effect of process parameters on deposition, joining and coating efficiency, regarding the FS of DIN Ck45 over mild steel. Process parameters: **a** $\Omega = 3{,}000$ rpm, $v = 7.5$ mm/s, **b** $F = 4$ kN, $v = 7.5$ mm/s, **c** $F = 6$ kN, $\Omega = 3{,}000$ rpm

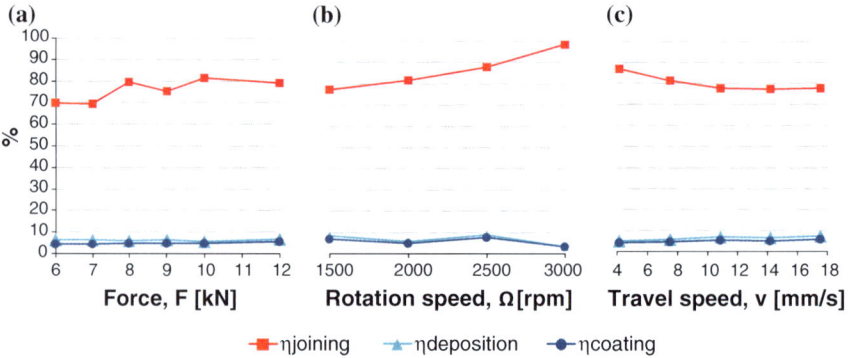

Fig. 105.9 Effect of process parameters on deposition, joining and coating efficiency, regarding the FS of AISI H13 over mild steel. Process parameters: **a** $\Omega = 2{,}000$ rpm, $v = 7.5$ mm/s, **b** $F = 10$ kN, $v = 7.5$ mm/s, **c** $F = 10$ kN, $\Omega = 2{,}000$ rpm

are also known to result in excessive flash formation, as evidenced by a significant increase on the consumption rate relatively to the deposition rate, in AISI 1045. In this case, a decrease on deposition efficiency (and consequently on the coating efficiency) can also be expected for excessive forces. The ratio between the coating width and the effectively joined width seems to improve with increasing force (thereby improving the joining efficiency), as seen in the deposition of DIN St52 and AISI H13. The deposition of AISI1045 also evidences this effect for axial forces between 4 and 7 kN. Excessive forces can result in the extension of coating width beyond the consumable rod diameter, thereby contributing to the deterioration of joining quality at the coating edges. It is believed that this phenomenon causes the decrease of joining efficiency in the deposition of this steel for axial forces from 8 to 9 kN.

The increase of rotation speed led to the decrease of deposition rate in all steel combinations. This is evidence of a more significant loss of material in the flash. The

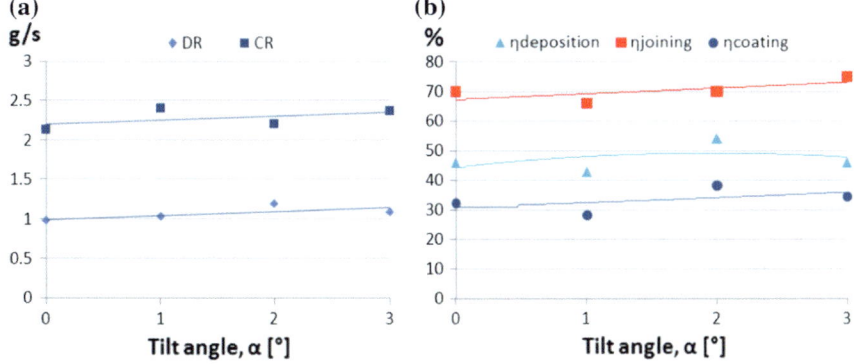

Fig. 105.10 Effect of tilt angle on the FS of DIN St52 over mild steel. **a** Deposition rate (*DR*) and consumption rate (*CR*); **b** Process efficiency. Process parameters: $F = 5\,kN$, $v = 14.2\,mm/s$ and $\Omega = 2{,}500\,rpm$

increase of rotation speed resulted in an increase of joining efficiency in the deposition of DIN Ck45 and AISI H13. Although the rotation speed resulted in the combined decrease of coating width and the joined width, the increase of joining efficiency indicates that the proportion between the two is improving. In the deposition of DIN St52, the increase of rotation speed resulted in a more significant decrease of joined width, thereby leading to a reduction of joining efficiency.

The coating and deposition efficiencies do not evidence a clear trend under the variation of travel speed for any material combination. Joining efficiency was improved depending on whether there was a positive effect on the ratio between the coating and joined width.

Flash represents 40–60 % of the material consumption in the friction surfacing of DIN St52. This percentage increases significantly in the FS of the two other steels, varying between 80 and 95 %. Nevertheless, the FS of AISI1045 and AISI H13 enabled higher joining efficiencies, reaching a maximum of 90 and 98 %, respectively. Process parameter optimization was capable of improving the DIN St52 joining efficiency to 75 %.

The overall coating efficiencies of DIN Ck45 and AISI H13 are lower than those observed for DIN St52. This is expectable considering the higher consumption rates inherent to the FS of these materials, as discussed previously. Coating efficiency generally varies from 25 to 35 % in the deposition of DIN St52, achieving a maximum value of 50 % for lower rotation speeds. Coating efficiency rarely exceeds 10 %, regarding the deposition of DIN Ck45 and AISI H13.

The increase of tilt angle from 1 to 3° led to an increase of deposition rate of DIN St52 in about 12 %. As seen from Fig. 105.10, the joining efficiency was also slightly improved, resulting in a coating efficiency improvement of about 8 %. The tilting of the consumable rod relatively to the substrate surface, in opposite direction to the travel movement, provides an enhanced forcing pressure for the deposited material.

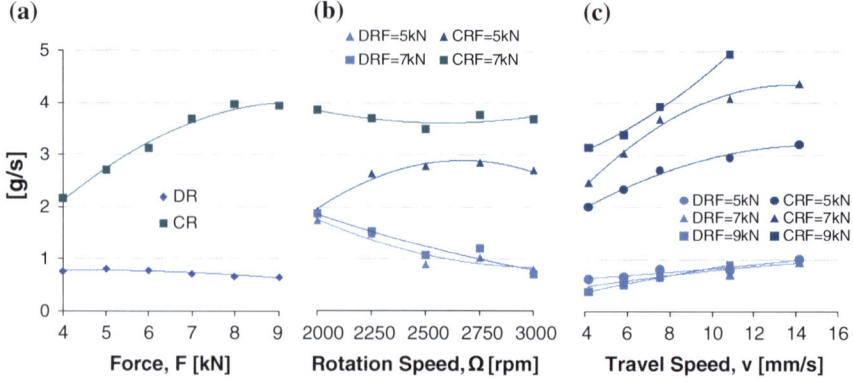

Fig. 105.11 Effect of process parameters on deposition rate (*DR*) and consumption rate (*CR*), regarding the FS of AA6082-T6 over AA2024-T3. Process parameters: **a** $\Omega = 3{,}000$ rpm, $v = 7.5$ mm/s, **b** $v = 7.5$ mm/s, **c** $\Omega = 3{,}000$ rpm

(2) Deposition of AA6082-T6 over AA2024-T3

Figure 105.11 depicts the effect of force, rotation and travel speeds on the deposition and consumption rates in the deposition of AA6082-T6 over AA2024-T3. Deposition rates vary from 0.38 to 1.88 g/s (140–697 mm³/s). It decreased for higher loads and rotation speeds due to the reduction of coating thickness. Similar to the steel combinations, the travel speed led the improvement of deposition rate, despite the decrease in coating thickness and width. Consumption rate increases significantly for higher axial forces and travel speeds, evidencing a more pronounced flash development.

As depicted by Fig. 105.12, overall coating efficiency reaches 10 and 20 %, while flash formation generally varies from 65 to 85 %. Joining efficiency improves for higher axial forces (7–9 kN), lower rotation speeds (2,000–2,500 rpm) and intermediate travel speeds (7–11 mm/s), reaching a maximum value of 80 %. Deposition and coating efficiency decreases with increasing axial forces and rotation speeds, due to a more significant flash formation. The coating and deposition efficiencies do not evidence a clear trend under the variation of travel speed. Suhuddin et al. [8], Tokisue et al. [11] and Sakihama et al. [6] reported similar deposition efficiencies from 20 to 60 % in the FS of similar and dissimilar combinations of aluminium alloys.

2. Energy Consumption

The mechanical power supplied by the equipment (W_e) can be divided into three main contributions regarding rod rotation (W_r), axial plunging (W_z) and travel (W_x), as determined by equation: $W_e[\text{J/s}] = W_r + W_z + W_x = \frac{2\pi\Omega}{60}T + F_z V_z + F_x v$. In the present investigation, the torque (T) and rotation speed (Ω) refer to values applied by the machine spindle motor of the equipment. The rotation speed of the spindle motor (Ω) is transmitted to the consumable rod by a belt-pulley system with 0.822 ratio. The axial plunging power (W_z) was determined based on the average value for the axial displacement speed (V_z) and applied axial force (F_z) for a steady state deposition. The travel movement power (W_x) was determined based on an estimated

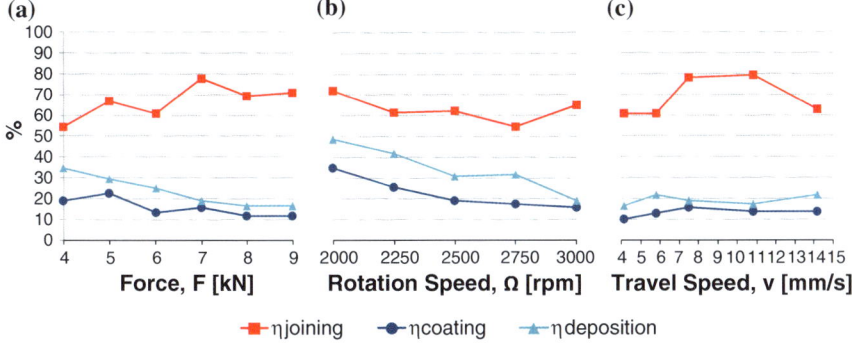

Fig. 105.12 Effect of process parameters on deposition, joining and coating efficiency, regarding the FS of AA6082-T6 over AA2024-T3. Process parameters: **a** $\Omega = 3{,}000$ rpm, $v = 7.5$ mm/s, **b** $F = 7$ kN, $v = 7.5$ mm/s, **c** $F = 7$ kN, $\Omega = 3{,}000$ rpm

value of the force applied to impel the consumable rod along that direction (F_x), multiplied by the travel speed (v). Energy consumption per deposited unit of mass or specific energy consumption (EC) is determined by Eq. (105.5):

$$EC[\text{J/kg}] = W_e/DR. \tag{105.5}$$

(1) Deposition of DIN St52, DIN Ck45 and AISI H13 over mild steel

It can be observed that the power consumption regarding the travel and the axial plunging movements are negligible compared to the spindle power, for all material combinations. Spindle power reaches about 3 and 4.5 kW. In contrast, the power required to impel the travel and plunging linear movements varies from 0.01 to 0.05 kW. This means that the power consumption for impelling the linear movements is at least 98 % lower than the spindle power. The force to impel the consumable along the travel direction (F_x) is estimated to reach from 1 to 1.2 kN, reaching its maximum values for the fastest travel speeds. The estimated travel movement power seems to be lowest of the three contributions.

Figure 105.13 depicts the variation of power for the steel combinations. The power spent by the welding equipment increases for higher axial forces and travel speeds. The increase of rotation speed tends to reduce the power consumption. This phenomenon results from the increased coupling between the consumable and the substrate observed for lower rotations. The equipment will have to exert additional mechanical power to impel the consumable. Power consumption is strongly affected by the axial force and the rotation speed, varying generally between 3 and 4.5 kW for all steel combinations.

As shown in Fig. 105.14 the specific energy consumption is higher in the deposition of AISI H13, DIN Ck45 and DIN St52. This is expectable considering the lower deposition efficiencies reported for AISI H13 and DIN Ck45. Specific energy consumption varies from 3 to 4.5 kJ/g in the FS of DIN St52 over mild steel. The FS of

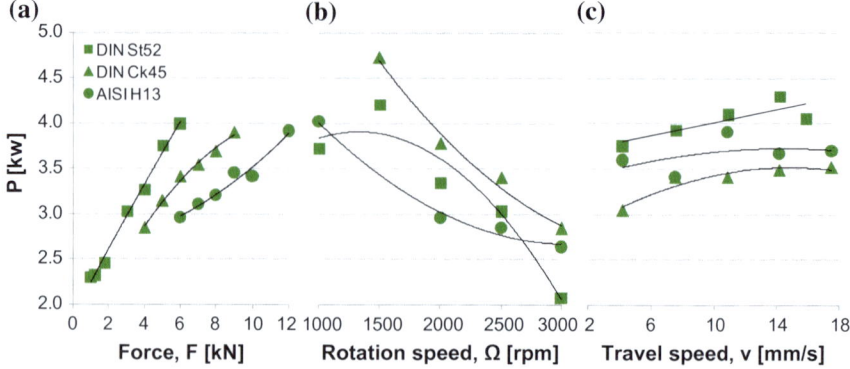

Fig. 105.13 Effect of process parameters on power consumption, regarding the deposition of steels. Process parameters: **FS of DIN St52** over mild steel, **a** $\Omega = 2,500$ rpm, $v = 4.2$ mm/s; **b** $F = 3$ kN, $v = 4.2$ mm/s; **c** $F = 5$ kN, $\Omega = 2,500$ rpm. **FS of DIN Ck45** over mild steel, **a** $\Omega = 3,000$ rpm, $v = 7.5$ mm/s; **b** $F = 4$ kN, $v = 7.5$ mm/s; **c** $F = 6$ kN, $\Omega = 3,000$ rpm. **FS of AISI H13** over mild steel, **a** $\Omega = 2,000$ rpm, $v = 7.5$ mm/s; **b** $F = 6$ kN, $v = 7.5$ mm/s; **c** $F = 10$ kN, $\Omega = 2,000$ rpm

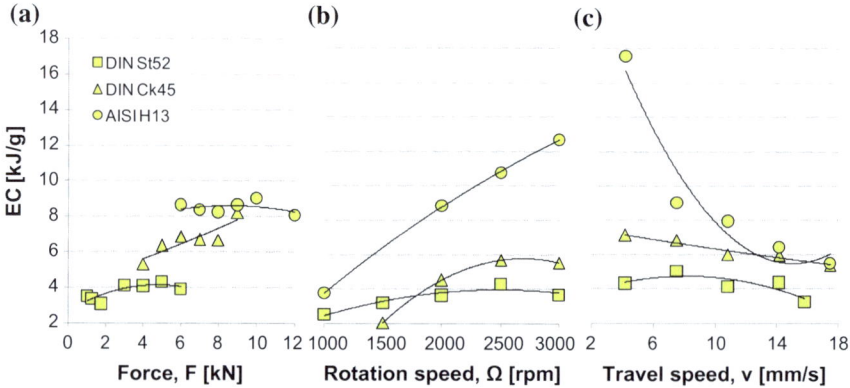

Fig. 105.14 Effect of process parameters on the specific energy consumption regarding the FS of steels. Process parameters: FS of DIN St52 over mild steel, **a** $\Omega = 2,500$ rpm, $v = 4.2$ mm/s; **b** $F = 3$ kN, $v = 4.2$ mm/s; **c** $F = 5$ kN, $\Omega = 2,500$ rpm. FS of DIN Ck45 over mild steel, **a** $\Omega = 3,000$ rpm, $v = 7.5$ mm/s; **b** $F = 4$ kN, $v = 7.5$ mm/s; **c** $F = 6$ kN, $\Omega = 3,000$ rpm. FS of AISI H13 over mild steel, **a** $\Omega = 2,000$ rpm, $v = 7.5$ mm/s; **b** $F = 6$ kN, $v = 7.5$ mm/s; **c** $F = 10$ kN, $\Omega = 2,000$ rpm

DIN Ck45 over mild steel required energy consumptions from 4 to 7 kJ/g, while the deposition of AISI H13 required typically 8 to 12 kJ/g. Specific energy consumption increases for higher axial forces, as shown in Fig. 105.14. For high rotation speeds, although both the required power and the deposition rate drop, specific energy consumption rises (Fig. 105.14). According to Eq. (105.5), this means that the decrease in deposition rate is more significant than the decrease in power consumption.

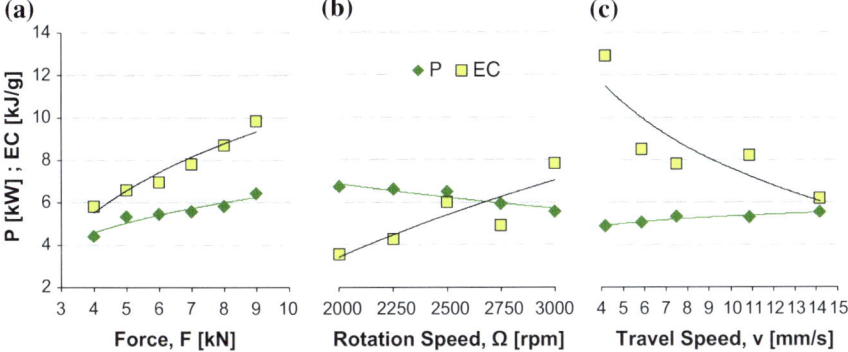

Fig. 105.15 Effect of process parameters on power (P) and specific energy consumption (EC) for FS deposition of AA6082-T6 over AA2024-T3. Process parameters: **a** $\Omega = 3{,}000$ rpm, $v = 7.5$ mm/s, **b** $F = 7$ kN, $v = 7.5$ mm/s, **c** $F = 7$ kN, $\Omega = 3{,}000$ rpm

Despite the increase in power, specific energy consumption decreases with increasing travel speed (Fig. 105.14c), due a more significant improvement of deposition rate. Hence, faster travel speeds allow to improve the deposition rates while decreasing specific energy consumption. However, faster travel speeds can result in aexcessive reduction of thickness and joined width. By adjusting the tilt angle, the specific energy consumption and the power decrease, increasing the deposition rate and joining efficiency.

(2) Deposition of AA6082-T6 over AA2024-T3

The FS of AA6082-T6 over AA2024-T3 presents similar effects to those reported for the steel combinations. Spindle power reached from 4 to 7 kW, while the power required by the travel and axial plunging of the rod (combined) varies from 0.02 to 0.065 kW. As reported for the steel combinations, this makes the power consumption of the linear motions quite negligible for the overall balance. The force to impel the travel movement (F_x) is estimated to reach from 1 to 1.1 kN, depending on the travel speed.

As shown in Fig. 105.15, the power consumption generally varies from 4 to 6.5 kW, increasing for higher forces and travel speeds. Power decreases with increasing rotation speed. The specific energy consumption generally varies from 4 to 8 kJ/g. It increases for higher axial forces and rotation speeds, while decreasing for higher travel speeds.

105.5 Conclusions

From the work performed the following can be concluded that:

- In FS of steels, higher mechanical strength seems to produce thinner coatings with improved joining efficiencies and joined width. However, a significant amount of the consumable rod material is lost in form of flash. Flash formation presents more than half of the material consumed.

- The highest volumetric deposition rates were achieved for the FS of aluminium, due to the use of a consumable rod with a larger diameter. By comparing the energy consumption per unit of deposited volume, it can also be observed that the FS of aluminium alloys requires less energy input than the steel combinations. This is expectable considering that the AA6082 aluminium alloy is a softer material with a lower melting temperature than the steel alloys under study. In all material combinations, the power consumption required to impel the linear motions can be considered negligible in comparison with the spindle power.
- The effect of process parameters has different impacts on process efficiency and deposition rate for each material combination.
- FS presents intermediate deposition rates and lower energy consumptions, compared to laser- and arc-welding based technologies.

Acknowledgments The authors would like to acknowledge FCT/MCTES funding for the project FRISURF (PTDC/EME-TME/103543/2008). JG acknowledges FCT/MCTES for funding its PhD. grant SFRH/BD/78539/2011. RM acknowledge Pest OE/EME/UI0667/2011.

References

1. Dunkerton SB, Thomas WM (1984) Repair by friction welding. In: Proceedings of the conference on repair and reclamation, London, 24–25 Sept
2. Gandra J, Miranda R et al (2011) Functionally graded materials produced by friction stir processing. J Mater Process Technol 211(11):1659–1668
3. Hsu C, Kao P, Ho N (2007) Intermetallic-reinforced aluminum matrix composites produced in situ by friction stir processing. Mater Lett 61(6):1315–1318
4. Nicholas E (2003) Friction processing technologies. Weld World 47(11–12):2–9
5. Olson DL (1993) ASM handbook: welding, brazing, and soldering, vol 6. ASM International (In Chinese)
6. Sakihama H, Tokisue H, Katoh K (2003) Mechanical properties of friction surfaced 5052 aluminum alloy. Mater Trans 44(12):2688–2694
7. Shinoda T, Okamoto S et al (1996) Deposition of hard surfacing layer by friction surfacing. Weld Int 10(4):288–294
8. Suhuddin U, Mironov S et al (2012) Microstructural evolution during friction surfacing of dissimilar aluminum alloys. Metall Mater Trans A 43(13):5224–5231
9. Thomas W, Nicholas E et al (2002) Friction based welding technology for aluminium. Mater Sci Forum (Trans Tech Publications) 396:1543–1548
10. Thomas WN (1987) Friction surfacing. In: Proceedings of the 2nd international conference on flash-butt and friction welding, September, pp 124–140 (In Chinese)
11. Tokisue H, Katoh K et al (2006) Mechanical properties of 5052/2017 dissimilar aluminum alloys deposit by friction surfacing. Mater Trans 47(3):874
12. Vilaça P, Gandra J et al (2012) Aluminium alloys—new trends in fabrication and applications. Intech, pp 159–197 (In Chinese)
13. Vitanov V, Javaid N, Stephenson DJ (2010) Application of response surface methodology for the optimisation of micro friction surfacing process. Surf Coat Technol 204(21):3501–3508
14. Voutchkov I, Jaworski B et al (2001) An integrated approach to friction surfacing process optimisation. Surf Coat Technol 141(1):26–33

Chapter 106
The Comprehensive Ensemble Pattern of Macroeconomic Early-Warning Information Management Cockpit

Yue He and Lingxi Song

Abstract This paper combines management cockpit thoughts and macroeconomic early-warning theory, and designs the system structure of macroeconomic early-warning information management cockpit, including metope display system and flight deck. We organically integrate economic early-warning, economic forecasting, economic simulation and economic monitoring model, and propose the comprehensive ensemble pattern of macroeconomic early-warning information management cockpit. This paper can draw a line of thought for scientific macroeconomic early-warning.

Keywords Management cockpit · Economy early-warning · Macroeconomic · System design · Comprehensive ensemble

106.1 Introduction

The concept of management cockpit is first put forward by German company SAP in its products in 1990's of 20 century, which is an innovative solution developed by neural scholars and computer scientists. At present Chinese scholars mainly describe and analyze the generation and origin of management cockpit theory as well as its role in the enterprise's strategic decision, and the application is limited in the aspect of enterprise management.

Huang [9], Sun [16] and Chen [2] summarized the generation, the function and the component of the management cockpit, and analyzed the managerial value of the "management cockpit" in the implementation of enterprise strategic decision. Qiu et al. [4, 10, 12, 15, 19] proposed data mining as a new method for the macroeconomic forecasting and simulation, and constructed a framework of the macro-control

Y. He (✉) · L. Song
Business School, Sichuan University, Chengdu 610064, People's Republic of China
e-mail: yuehe321@126.com

J. Xu et al. (eds.), *Proceedings of the Eighth International Conference on Management Science and Engineering Management*, Advances in Intelligent Systems and Computing 281, DOI: 10.1007/978-3-642-55122-2_106, © Springer-Verlag Berlin Heidelberg 2014

decision support system based on data warehouse. Liu et al. [3, 7, 8, 11, 13, 14, 17, 20, 21] apply the neural network algorithm, fuzzy C clustering algorithm, Rough collection, maximum entropy method and moving weighted average algorithm to the research of macroeconomic early-warning. There are little foreign research literature about management cockpit. Phornchanok et al. [1, 5, 6, 18] proposed the useful and leading indicators in economic crises, and constructed an early-warning system for predicting economic and financial crises.

Literature research above study economic forecasting, simulation and early-warning, but there is no research that Integrates macroeconomic early-warning, forecasting and simulation into management cockpit.

There still exists some urgent problems in economic early-warning. These problems are mainly in the following two aspects:

Firstly, it can't reflect the effects of macroeconomic policy. Resent economic early-warning is mostly the multi-index comprehensive early-warning, not considering the influence of the change of economic policy on future economic development. This economic early-warning method is not applicable for the future economic development. That's because once people found that the current economic situation out of the track of normal development, they tend to adopt policies and measures to stimulate economic development. The effect would appear with several time period lagged. However economic forecasting is based on historical data. It does not take the effect of economic stimulus measures into account, which may change the trajectory of the original economic development. So this method do not fully consider the impact of the economic policies on economic development.

Secondly, the lack of basis for qualitative analysis. People usually adopt the opinions of experts to predict the future economic development. However, when facing the unprecedented world economic crisis, who have enough experience? So this method is not comprehensive since its lack of quantitative analysis of the macro-economy. It can not explain how much effect that economic policies take on stimulate economy.

As the world is confronted with unprecedented crisis, the traditional methods can not make scientific early-warning. It's necessary to establish new early-warning methods and thinkings. Therefore, we believe that in the guidance of Integrated theory, we can Integrate economic early-warning theory method with management cockpit, and establish macroeconomic early-warning information management cockpit, which can provide new tools and methods for scientific early-warning.

This paper uses integrated theory and method, aiming at the problems in the economic early-warning, to integrate economic early-warning, economic forecasting and economic simulation with management cockpit, and constructs the macroeconomic early-warning information management cockpit.

Table 106.1 The key indicators of different walls

Walls	Key indicators
The black wall	GDP, industrial added value, sales income of industrial enterprises, fixed asset investment
The red wall	Per capita disposable income of residents, financial income, enterprise deposits balance, the narrower money supply M1
The blue wall	Retail sales of consumer goods, price index of industrial product, import and export
The white wall	Total retail sales of social consumer goods, per capita consumption of residents

106.2 Design of Cockpit Basic System

106.2.1 System Structure

Macroeconomic early warning information management cockpit consists of metope display system and flight bridge.

(1) Metope Display System

Metope display system generally consists of 24 display, each of which reflects to each index of a decision problem respectively. There are six display arranged as two rows and three columns on four metope respectively, and each display shows six different views, so the macroeconomic early-warning information management cockpit can display 144 different indicators. The six views are also called logical views. The combination of these views represent the circumstances of economic management indicators. The content of these views are logically related, so the circumstances of economic management indicators are more detailed reflected.

The array of indicators are not disordered, but displayed in different partitions according to its different content. The frontal wall of macroeconomic early-warning information management cockpit is the black wall, which shows key indicators of development status of production, expressing the current situation and development trend. The left is the red wall, which shows the development status of distribution. The right is the blue wall, providing the developing status of circulation. The white wall behind displays the development status of consumption. Table 106.1 shows the key indicators of different walls.

The analysis steps of the above key indicators of each wall:

- Describe the status of total amount and growth rate of the key indicators.
- Analyze the stability of key indicators in the time series.
- Analysis the structure of each indicator to find out the influence of local growth rate to the overall growth rate.
- Make static and dynamic comparisons of growth rate with the national average level, the average level of eastern and western major cities.

- Analyze the sort of the growth rates of key indicators in administrative regions at the same level.

The display of index views are not fixed. They are usually designed for the users' actual customization of macroeconomic early-warning information management cockpit, so as to match as much as possible with the condition of indicators which managers need to understand. So managers can set displays according to the key issues of current project. The number of indicators may be less than 144 or more than 144. In a word, through let managers can get to know what they want to know from the displays of the 4 walls.

(2) Flight bridge

The flight bridge of macroeconomic early-warning information management cockpit is the core of economic information management. We can collect and treat economic information with the flight bridge. Decision makers can query real-time information of economic development through the flight bridge. And managers make decisions on this basis. The support of macroeconomic early-warning information management cockpit to decision making is also reflected in the decision simulation, i.e. when managers make a decision, they can make a forecast of data after decision through the simulation model of economy, in order to understand the impact on other aspects after some economic indicators change. When there are multiple choices, managers can make a decision which is relatively good for economic development through the analysis of prediction. Flight bridge is the key to understand and manipulate the whole economy, and it is the important man-machine interaction window of decision, as well as the power equipment of macroeconomic early-warning information management cockpit.

The flight bridge system of macroeconomic early-warning information management cockpit, is a computer system with the operation of the simulation function based on decision support system. Its basic structure are shown in Fig. 106.1.

1. Database system. The database is the basic data source of macroeconomic early-warning information management cockpit, it consists of the enterprise data, industrial park data, industry data and other economic data. From the time dimension, the data is divided into monthly data, quarterly data and annual data; from the administrative point of view, the data can be divided into classes according to different administrative area. The database also stores the survey data, for the qualitative and quantitative analysis of economic early-warning. In general, the database system in the management cockpit has the following functions:

   a. Supporting the memory. Database can be used to save intermediate results, indexes between data links and data flip flop for decision makers to perform operation.
   b. Supporting data merge, including agglomeration of records and fields in database. It needs a large amount of data merge and abstraction to make a decision.

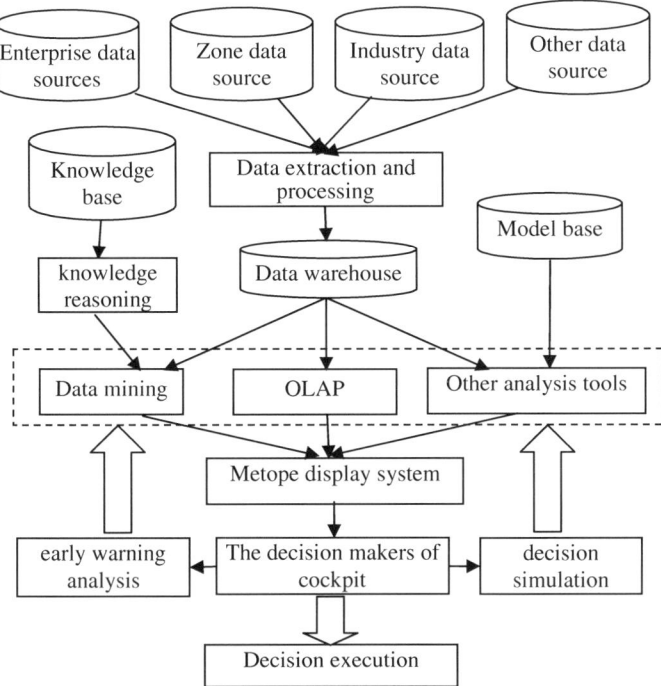

Fig. 106.1 The flight bridge system of macroeconomic early-warning information management cockpit

c. With multiple sources of data, the data for decision making may be from external or internal data sources. Even the internal data may come from different departments.

d. A wide range of time, the data in database is not only the data now, but also the past data. There are enough time series data to ensure the need of decision.

e. With good system interface. As the database system is a part of the management cockpit, it is necessary for it to combine with other parts to solve decision problems.

2. Model base system. Model base is the basic analysis tool of macroeconomic early-warning information management cockpit, mainly including the comprehensive early-warning models and boom early-warning models, multiple forecasting models, and economic simulation models. In general, the model base system in management cockpit has the following functions:

a. With the ability to create new model quickly and easily.

b. With the ability to access and integrate model blocks.

c. With the ability to support different users to classify and analyze problems.

    d. With the ability to connect models according to rules of model connection methods.

    e. With the ability to modify, add and delete, and operate the models.

3. Knowledge base system. The knowledge base is a collection of declarative knowledge and process knowledge in specific domain. The difference with the traditional database is that it contains not only a large number of simple facts, but also rules and process knowledge. The knowledge base is the expert knowledge source of the management cockpit. Some international and domestic political events have large impact on the economic trend, and the effect on the economy is difficult to quantify. So integrating the experts knowledge into knowledge base can achieve the qualitative and quantitative combined economic early-warning.

4. Data Warehouse. Data warehouse is the set of data that is subject-oriented, and it's a collection of data that extracted and arranged from the basic database for economic early-warning according to the actual needs. It contains the data collection of time sequence required in the comprehensive early-warning, economic forecasts and economic simulation.

5. On-Line Analysis Processing (OLAP). Decision analysts need to analyze the data from different angles. Multidimensional data analysis is the main content of decision. In this management cockpit, we can use OLAP technology to query multidimensional data conveniently. It is specifically designed to support the complex analysis, focusing on the support to analysts and senior managers. It can quickly and flexibly complete a complex inquiry process of a large amount of data at the request of analysts. And the query results are provided for decision makers in an understandable form, so that they can accurately grasp the economic status, and make correct decisions.

6. Data mining technology. Data mining is a new technology to discover and extract information hidden in the interior from the large database or data warehouse. The purpose is to help the decision makers to find potential relationship among data, and to find the elements that managers ignored, which are useful for forecasting and decision making.

Data warehouse in the economic early-warning information cockpit has saved a large amount of data. How to find the organic link among the data? The application of data mining technology is required to discover the rules from large amounts of data.

106.2.2 Display Content

Metope display system is the direct display of macroeconomic early-warning information management cockpit. When macroeconomic early-warning information management cockpit scheme is implemented, the figures displayed on the walls are different. Metope display system changes the displayed content according to the following factors:

Index data and information in the management cockpit are displayed in five forms.

1. Major administrative region. The various provinces and cities are selected to analyzed emphatically. For example, we can analyze the stability and volatility of the GDP development speed, the growth rate of industrial added value, profit and tax situation and etc of various regions.
2. Enterprises groups. Enterprises groups are the leading to promote economic development. The management cockpit monitors and predicts the development of corporate champions. The development trend of enterprises are displayed dynamically in the management cockpit.
3. Competitive industries and pillar industries. Competitive industries and pillar industries should be encouraged to develop in one area. Macroeconomic early-warning information management cockpit need to focus on the development of competitive industries and pillar industries in this area according to the characteristics of different region. The important task is to analyze which are competitive industries, and take dynamic monitoring and assess the competitive industries.
4. Seven types of enterprise registration type. We need to analyze the development status and the difference of all kinds of enterprises according to the type of the enterprise registration (enterprises funded in foreign countries or in Hong Kong, Macao, Taiwan, private enterprises, joint-stock cooperative enterprises, limited liability company, incorporated company, state-owned enterprises, collective enterprises). It is necessary to take measures to encourage and support the development of some categories of enterprises according to the national policy.
5. Industrial Park. The industrial park means a modern industrial division production area that a country or a regional government highlights the characteristics of industry, and optimizes the functional layout through the integration in a certain space range, so as to adapt it to the market competition and industry upgrade. The management cockpit will assess the development status of Industrial Park, monitor the operation status of Industrial Park.

106.3 The Integrated Model of Cockpit

The macroeconomic early-warning information management cockpit comprehensive ensemble model is formed by combining the management cockpit technology, economic early-warning methods and integrated theory organically.

The integrated model is a man-machine symbiosis. The management cockpit needs decision makers to control. It needs to extract data including statistical data, prosperity data and related data from other systems into the data warehouse for decision. Decision makers will control the flight bridge, use various models of flight bridge, analyze the economic status, predict the future status, start early-warning mechanism, simulate decision making process, and display various decision information through the metope display system.

106.3.1 System Function

Macroeconomic early-warning information management cockpit needs to realize macroeconomic early-warning and provide support to aid decision making. Therefore, the overall function should contain six systems as economic prosperity early-warning system, economic forecasting system, economic simulation system, economic operation monitoring system, expert resources system, online analysis tools.

(1) Economic early-warning system

Macroeconomic early-warning means that analysts compile indicators using several indicators to describe economic cycle fluctuation phenomena, and reflect the development direction and amplitude of fluctuation of economic phenomenon in advance, so the system would send out warning signals before a large economic change to provide the basis for the relevant departments to carry out macro-control.

This system will conduct real-time monitoring of the changes of prosperity indicators (leading indicator, coincident indicators, early-warning indicators) selected from regional GDP, the three industrial added value, industrial growth rate, the amount of fixed asset investment, real estate investment, total retail sales of social consumer goods, total import and export volume, deposit balance of financial institutions, loan balance of financial institutions, consumer price index and other economic indicators. Then the system will plot warning signal figures (using red light, yellow light, green light, light blue light and blue light to represent overheating interval, thermotaxis interval, appropriate interval, cooling interval and partial overcooling interval respectively) and index trend charts according to the monitoring results, which will vividly depicts the overall macroeconomic situation.

(2) Economic Forecasting System

Macroeconomic forecasting is defined that analysts establish several regional macroeconomic (monthly, quarterly, annual) prediction models according to the regional economic development planning and actual situation, and forecast some indicators such as above-scale industrial added value, GDP, total retail sales of social consumer goods, import and export volume, consumer price index. Then we should analyze the fitness of the prediction results to the actual situation, and choose the best fitting model as forecast results.

(3) Simulation system

The economic simulation system and prediction system are aimed at studying the impact of macroeconomic policy changes on the economy to realize the policy simulation and the economic prediction function in short term, medium term and long term. In macroscopic view, we can use the models to exert regional advantages most effectively in the condition of their own natural resources, regional superiority, technological level, economic structure and etc. In microscopic view, we can use the models to make a quantitative analysis of the interaction role of economic variables (including economic policies and important production factors) in the development of economy.

From the application point of view, the model can have a simulation and evaluation of the effect of macroeconomic policy on the regional economic, and also can study the relationship between the regional economy and the national economy. For the simulation of each case, a user can adjust the amplitude of variation of exogenous variables (policy variables), and compare and analyze the simulation results. For the policy simulation, in addition to the existing exogenous variables (policy variables), a user can also change some endogenous policy variables into exogenous variables as the policy evaluation required, and study its effects on economic results. Specifically, we can analyze the impact of the change of indicators such as fixed assets investment, government consumption, resident income, the national price level, total retail sales of social consumer goods, above-scale industrial added value and other indicators on macroeconomic operation.

(4) Economic monitoring system

Economic monitoring system based on the macroeconomic statistical data, shows the development trend of the macroeconomic operation through graphics, data, text and geographical information positioning. Economic operation monitoring system uses data mining to provide a long-term, stable and dynamic observation platform for analysts.

Economic operation monitoring system supports flexible expansion of topics, and provides a variety of application that charts can be dynamically generated, exported, printed and collected. Economic operation monitoring system has graphic display and text description to provide users with more vivid system interface.

(5) Online analysis system

Online-data analysis tools provide a series of efficient, practical data mining analysis methods, and provide the function to analyze the original data form multiple angles.

According to the basic requirements of macroeconomic monitoring, early-warning and forecasting, online analysis system tools contain three kinds of analysis methods as "index analysis", "statistical tools", "measuring tools".

Index analysis tools can realize some simple operation such as index preconditioning and index calculation, specifically including: missing value processing, data standardization, data transformation, growth rate, ratio, structure, smooth, elasticity and other tools; measuring tools use regression analysis, trend extrapolation forecasting, time series, combination forecasting and etc to achieve online analysis system function; statistical tools enable users to call database online, and statistically analyze the called data, such as correlation analysis, principal component analysis, factor analysis and cluster analysis.

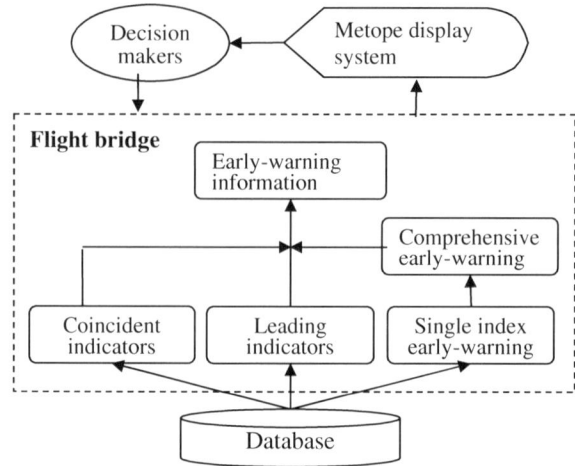

Fig. 106.2 The economic early-warning ensemble pattern

106.3.2 Economic Early-Warning Ensemble Pattern

Macroeconomic early-warning requires the application of integrated theory and methods to combine the indicators which reflect macroeconomic development trend into the management cockpit, and make scientific early-warning about the macro-economic trend. Economic early-warning comprehensive ensemble pattern is shown in Fig. 106.2.

Economic early-warning ensemble pattern consists of the coincident indicators, leading indicators, single index early-warning and comprehensive early-warning, and uses the various size of integrated data in data warehouse to generate economic early-warning coincident indicators and leading indicators through the models of diffusion indicators and composite indicators. Similarly, it extracts data from data warehouse, and give the coincident indicators a certain weight to obtain comprehensive early-warning model, and then integrate the information of the coincident indicators, leading indicators and comprehensive early-warning which result in early-warning integrated information. All the information is shown on the metope display system for reference.

106.3.3 The Process of Cockpit Integration Pattern

The core of macroeconomic early-warning information management cockpit is to achieve economic early-warning. Management cockpit needs to call the economic forecasting system and the economic simulation system to assist the realization of economic early-warning. The process of cockpit early-warning system is shown in Fig. 106.3.

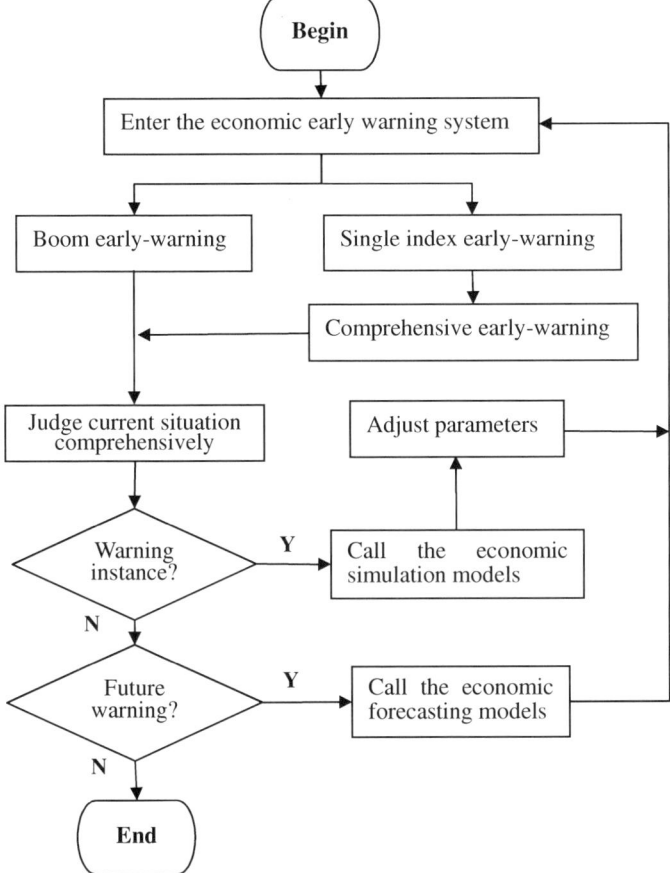

Fig. 106.3 Cockpit warning business processes

After entering the economic early-warning system, the system will call boom early-warning model and comprehensive early-warning model respectively, and then combine the two models together to realize early warning. If there is any warning instance, we can call the macroeconomic simulation model, and adjust main macroeconomic indicators. Then we need to simulate the situation after the indicators adjustment, and make a comprehensive early warning, until the warning instances are eliminated. Thus the impact of macroeconomic policy changes on the warning instance has been simulated. If it is necessary to conduct early warning for the future economic development, we can call the economic forecasting model, bringing the predictive indicators into the comprehensive early-warning model for early warning analysis.

106.4 Conclusions

This paper uses integrated theory and methods, and integrate economic early-warning, economic forecasting, economic simulation and management cockpit together to construct the macroeconomic early-warning information management cockpit. We have designed the system structure of the cockpit, dividing the cockpit into two parts: metope display system and the flight bridge, and we have separately determined the structure and function of each part. Finally, we have designed the comprehensive ensemble pattern of the cockpit, which integrate the economic early-warning model, economic forecasting model and economic simulation model with data warehouse technology, providing a new idea for the macroeconomic early warning.

This paper simply analyzed and designed the macroeconomic early-warning information management cockpit, but it has not fully realized. Such a cockpit system should be developed in future. In addition, in the study of economic early warning, the early warning of boom indicators and comprehensive early warning should be combined, in order to make more practical early warning.

Acknowledgments The authors acknowledge the financial support Humanities and Social Science Planning Fund Project of the Ministry of Education of China, project No. 11YJA630029. The author is grateful to the anonymous referee for a careful checking of the details and for helpful comments that improved this paper.

References

1. Babecký J, Havránek T et al (2013) Leading indicators of crisis incidence: evidence from developed countries. J Int Money Financ 35:1–19
2. Chen X (2003) Management cockpit–the new tool of strategic decision of enterprises. Manage Pract 1:52–54
3. Chen Y (2010) Economic monitoring and empirical research based on fuzzy mathematics theory. Stat Decis 1:43–45
4. Cheng J, Yang X (2006) The construction of decision support system for macroeconomic control. China Soft Sci 2:68–73
5. Cumperayot P, Kouwenberg R (2013) Early-warning systems for currency crises: a multivariate extreme value approach. J Int Money Financ 36:151–171
6. El-Shagi M, Knedlik T, von Schweinitz G (2013) Predicting financial crises: the (statistical) significance of the signals approach. J Int Money Financ 35:76–103
7. Feng R, Han D, Gu B (2009) Study of index selection of macroeconomic early-warning based on competition neural network. Mod Manage Sci 1:76–78
8. He X (2009) Construction of macroeconomic intelligent early-warning system of China. Sci Technol Manage Res 11:195–198
9. Huang X (1999) Enterprises management cockpit. IT Manage World 14:44–47
10. Li Z (2011) Prediction model of macroeconomic research based on grey system theory and neural network. J Qinghai Normal Univ (Nat Sci Edn) 1:19–21
11. Liu A (2007) The construction of economic early-warning model of our country. Econ Probl 9:28–29
12. Liu S, Zhao Y (2012) A summary of economic simulation research based on the intelligent modeling. Mod Econ Inf 19:197–199

13. Luo E (2009) Study on industrial economic early warning. Stat Decis 3:162–164
14. Pang X, Feng Y, Pang Z (2007) Research on economic early warning based on maximum entropy. Comput Eng Appl 5:215–218
15. Qiu B (2008) The research and application of data mining technology in the macroeconomic intelligent decision support system. Nanchang University
16. Sun H (2003) Management cockpit based on data warehouse. J Shenyang Electr Power Inst 1:12–14
17. Sun J, Li H (2009) Financial distress early warning based on group decision making. Comput Oper Res 36:885–906
18. Yoon W, Park K (2013) A study on the market instability index and risk warning levels in early-warning system for economic crisis. Digit Signal Proc 29:35–44
19. Yu W (2012) Construction of economic model and research on prediction method. South China University of Technology
20. Zhang Z (2009) The construction of economic early-warning model based on moving weighted average comprehensive index. Commercial Times 27:62–63
21. Zhu Y, Wu T (2007) Economic early-warning model based on rough set and the constructive learning neural network. J Hefei Univ Technol 30(7):836–839, 843

Chapter 107
Strategic Resilience Development: A Study Using Delphi

Raphaela Vidal, Helena Carvalho and Virgílio António Cruz-Machado

Abstract Resilience in strategic perspective is related to a continuous organizational development, anticipating and adjusting to environment changes, fostering organizations with more capacity to bounce back from disturbance and thrive. This paper attempts to fill a gap in literature identifying critical factors that promote Strategic Resilience. First, the Strategic Resilience concept is reviewed using Contingency Theory and Organizational Development as theoretical lens. From this theoretical study was derived a set of 11 promoting factors Strategic Resilience. In a second stage, academics and experts in management field were consulted about the factors relevancy. The Delphi technique was used to guarantee a consensus of ideas from those specialists, allowing the topic discussion, but assuring anonymity. The results suggests the existence of 10 promoting Strategic Resilience factors, among them "Leadership" and "Capacity of change" achieved total unanimity among the experts, but "Robust strategy" and "Organizational structure" achieved the smallest consensus. This paper contributes to theory building by proposing factors that help organizations in strategic level to get preparedness, allowing the alignment of strategies, responding a crises situation and succeed.

Keywords Strategy · Resilience · Delphi · Contingency

107.1 Introduction

The environment that organizations are operating in is getting more dynamic and unstable day by day. Any actions in politic, economic and social sphere may affect organizations in this global field, even if they happen in another part of the planet.

R. Vidal (✉) · H. Carvalho · V. A. Cruz-Machado
UNIDEMI, Department of Mechanical and Industrial Engineering, Faculdade de Ciências e Tecnologia da Universidade Nova de Lisboa, Campus Universitáio, 2829-516 Caparica, Portugal
e-mail: r.vidal@campus.fct.unl.pt

There are countless threats that may affect organizations, from extreme natural disasters and terrorist attacks that cause havoc for organizations and the community in general until a simple supplier fail. Stephenson et al. [32] corroborate that the world is more complex being necessary to deal with a broad spectrum of crises arising from both natural and man-made causes. Dalziel and McManaus [9] observe there are enormous implications to organizations that are not prepared for high impact and unpredictable events. Therefore, organizations are concerned about the risk of vulnerabilities and interruptions that may affect its operations and try to figure out strategies for being ready to recover from adversity and to make them more competitive. In this sense, emerges Organizational Resilience as a powerful management tool, mostly what concern strategic perspective that is the paper's focus.

Some authors stress the set of critical factors in organizational Resilience mostly related a psychological and operational perspective. Thus, the main contribution of this paper is to fill the gap in Organizational Resilience field under strategic perspective, identifying critical factors that promote strategic Resilience. Additionally, it is essential to stress that Resilience is about a preemptive strategy, definitively not a strategy reactively [37]. Then, understand some factors that influence strategic Resilience gain a considerable importance.

The paper is organized as follows. The next section introduces literature review, followed by a section with a discussion about the research methods used. The following section is about research results. Lastly, conclusions of the research and future study.

107.2 Literature Review

The term Resilience comes from latin "resiliens" and means "bounce back" or "get back to a natural state". Its use began in the area of physics to refer to some materials' properties in returning to their natural state after suffering pressure. It is studied in some fields besides physics, such as: psychology, sociology, ecology, economics and management [28]. It has been studied and highlighted over the past decade, becoming a very popular manager research area [2, 10, 22] with growing interest within operations management [4], and relatively interest of disciplines as risk management and supply chain [28].

Resilience is the ability to bounce back from adversity [31, 33] and move forward stronger than ever [33]. It allows organizations maintain operations following a disruption [30] with minimal stress [21]. Still being more successful than before disruption (thriving) [13]. "Organizational Resilience is the ability and capacity to withstand systemic discontinuities and adapt to new risk environments" [35]. Moreover it is the capacity to overcome problems, responding effectively to unexpected disturbances [5].

A resilient organization has the capacity to continuously change [15]. Resilience is the capability to anticipate risk [26] survive, adapt, and grow in the face of turbulent

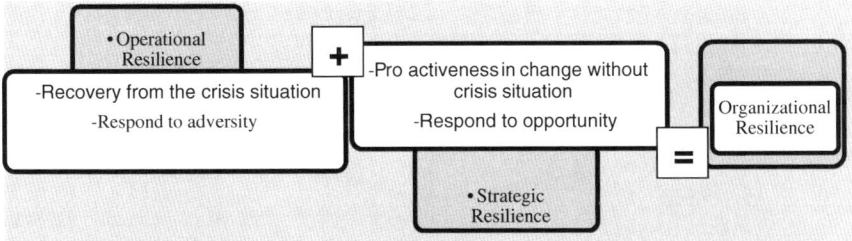

Fig. 107.1 Dimensions of organizational resilience

change [12, 26]. For Mallak [21] Resilience aims to change behaviors and attitudes to respond to internal and external demands.

In summary, McCann et al. [23] define Resilience as the capacity of resisting, absorbing and responding to a crisis, even to renewal itself it is required, to respond a unavoidable quick change.

Organizational Resilience is considered in 2 dimensions: operational and strategic (Fig. 107.1). Valikangas and Romme [38] define these two dimensions as following: "Operational Resilience is the ability to bounce back after a crisis or, more broadly, to respond to adversity. Strategic Resilience is the ability to turn threats into opportunities before it is too late - that is, to effectively respond to opportunity. Preparing for a response to adversity implies enhancing the organizational defenses, while the response to an opportunity involves engaging in exploration and experimentation that serve to build a portfolio of options for the future". Hamel and Valikangas [15] add "strategic Resilience is not about responding to a onetime crisis. It's not about rebounding from a setback. It's about continuously anticipating and adjusting to deep, secular trends that can permanently impair the earning power of a core business. It's about having the capacity to change before the case for change becomes desperately obvious". Valikangas e Merlyn [37] corroborate that Strategic Resilience is the capacity to renew itself before one crisis pushes organization to change.

107.2.1 Theoretical Lens

During the time, countless researchers focused to understand how organizations work. At first, their studies were developed considering only aspects inside organizations, after started consider organizations work inside environment and exchange resources and information with this, affect and are affected by environment. Considering these perspectives, Organizational Development and Contingency Theory fit well to have a deep understanding about Resilience.

Organizational Development (OD) focuses on development planned of organizations, encompassing change process, capacity of adaptation to the change and Organizational culture. For Chiavenato [6], Organizational Development models address

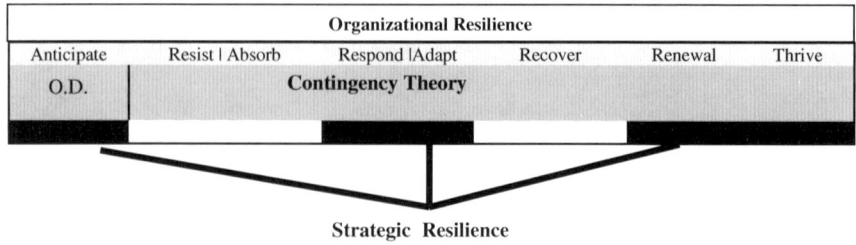

Fig. 107.2 Resilience and theories

to the follow aspects: turbulence inside environment; organization that suffer impact from environment; social group (leadership, communication and personal relations) and people (motivations and attitudes).

Contingency Theory considers organizations are inside environment where contingencies occur to be faced. Contingencies as unforeseen events that can affect deeply organizations. Uncertainties and risk increase over time, changing demand, changing consumer behavior, problems with suppliers and others changes in technological, social, economical, climate spheres require high attention to the environment conditions and continuous adaptation and adjust by organizations. Therefore, for theorists, there is no one single way to manage organizations. Chiavenato [6] stresses leaders must be flexible to follow changes that occur quickly in the environment. And adds the contingency theory aims to explain there is a functional relation between environment conditions and business practices.

In summary, Organizational Development refers a continuous state of planning change and Contingency Theory refers to recognize, diagnose and adapt to the changes and form together a fundamental pillar for Organizational Resilience. Observing what theories emphasize (considering changes in the environment, necessity to change, anticipating to change, leadership, communications, people, continuous state to change), it is possible understand how improve Strategic Resilience. The Fig. 107.2 illustrates how the principles of theories match with the concepts of organizational resilience proposed by authors in previous section. Furthermore, it highlights how theories match with the concept of Strategic Resilience that is related to anticipate opportunities, respond and adapt to a crisis, renewal itself and thrive.

107.3 Methodology

This research comprises 2 stages. In the first, a theoretical study was conducted encompassing Contingency Theory and Organizational Development Theory, emphasizing a strategic perspective. Based on literature, principles and assumptions about those theories were categorized and related to Strategic Resilience background to find promoting factors.

From this theoretical study was derived a set of 11 promoting factors Strategic Resilience. In a second stage, an exploratory research attempts to identify promoting factors Strategic Resilience based on personal judgments of panel (academics and experts) chosen according to their background in research area. For data gathering, an online questionnaires was used and Delphi technique for consolidating experts opinions.

Delphi is a technique that aims to get a consensual opinion among raters [25]. Linstone and Turoff [19] state to structure the study using Delphi is necessary define: Experts; number of rounds and how structured was each round.

This kind of technique (Delphi) generally involves 3–15 respondents [1] and requires from a group of experts: (1) availableness to respond rounds that are necessary; (2) commitment to results; (3) having a sound of knowledge about Resilience. Based on these criteria, 18 experts were chosen. An invitation was sent to 18 experts explaining research goals; the approximate duration of the research; tool for data gathering (online questionnaire); Emphasizing how important is sharing opinions among academics and experts; and assuring strongly the anonymity. In a total 18 experts, only 15 experts were considered to participate in this study. Two experts did not meet criteria (1) and (2), another one did not meet criteria (3). At least, 7 academics and 6 practitioners who work in Europe and South America were nominated to contribute to this research. It encompasses 31 % with doctored degree, 46 % with master degree and 23 % have specialization or MBA.

Adnan and Morledge [1] consider the number of rounds between two and seven to get a consensus of opinions. Thus this study was structured in two rounds of questions. The first round was structured in order to identify the profile of panelists; evaluate the importance of factors to promote Strategic Resilience and provide space for debating (panelist should justify the reason of evaluating any factor as not important, add another factor that consider relevant and make comments about the subject). The second round, a similar questionnaire (round#1) was sent to the group and panelists are given the opportunity to alter their previous opinion or just keep them.

107.4 Results

This part of paper demonstrates the results of research: Factors that promote Strategic Resilience. Each factor found is discussed relating them to Strategic Resilience. 1.

Phase I: The Factors That Promote Strategic Resilience

A theoretical study was made encompassing Contingency Theory and Organizational Development Theory. The principles and assumptions about these theories are highlighted. From this point it is possibleto relate to what authors state about Strategic Resilience. From this, a list of factors proposed is identified as follows: Leadership, Capacity of Change, Organizational Culture, Organizational Learning, Human Factor, Creativity, Risk Management, Innovation, Information System, Competitive Environment and Robust Strategy.

- Leadership—The results shows leadership achieved a total unanimity. Experts consider organizations need capable leaders to lead groups to achieve the same organizational goals. According to Everly [11] resilient leader will lead the organization in the direction of Resilience. For Vieira [40] the main challenges to achieving Resilience are related to managers. He asserts organizations will achieve Resilience only through professionals that are capable of developing it inside the organization as well. The leader is concerned with planning and managing for change that is a natural process for survive. For this, it is necessary capacity of change factor.

- Capacity of change—Organization survival is dependent on the way that each organization deals with its environment and how organizes itself to react to external changes. Adaptation is a basic condition for staying inside the environment. Organizations must be flexible to reformulate their strategies according to their environment, opportunities and threats. Additionally, organizational change must follow environmental changes. Organizations that change in slow motion and with gradual processes while the environment is changing quickly are always in crisis, and for longer. Hamel [16] emphasizes that successful organizations work at the same speed than the change, because they are resilient. Porter [27] adds that organizations must be flexible and respond promptly to market changes, seeking better practices than ever. Capacity of change encompasses flexibility to adapt and speed of change and help to achieve Strategic Resilience.

- Organizational culture is related to the history of the organization, its experiences, victories and failures. It is influenced by its working culture and by all organizations that it relates. It is appropriate to discuss organizational change considering the concept of organizational culture, because the only way to change an organization is to change its culture. Corporate culture is a decisive factor that makes a difference between organizations that are able to recover more quickly from a problem than other [7]. Everly Jr. [11] defends the idea that organizations can develop a culture focused on Resilience. Thus, a culture of Resilience is developed by resilient leaders.

  An organizational culture focused on Resilience is one that creates incentives to innovation, encourages initiative to try something new and does not focus on aspects of punishment when mistakes occur what is essential for strategic perspective.

- Organizational Learning—In times of constant changes, Organizational Learning is an essential discipline. Organizational Learning is the capacity to learn for change. Organizations only learn through the experiences of individuals. And it is related to an organizational culture that is supportive and stimulates such learning. Sutcliffe and Vogus [36] stress that one of fields that is most dedicated to understanding Organizational Resilience is the literature that encompasses Organizational Learning.

  Organizational Learning is a feature of organization that adapt to internal or external changes. The organizational change emerges easily in the organizations when organization learns from the past situations and provides feedback to its system.

Therefore, understanding the way organization learns and adapts will contribute for Strategic Resilience.

- Human Factor—According to Gittell et al. [13] relations among workers during crisis help the commitment and productivity. In other words, positive social relations at work are prerequisites to Organizational Resilience. Masten and Reed [20] corroborate positive relations help to develop Resilience in face to work stress. Lengnick-Hall, Beck and Lengnick-Hall [18] believe human resources management to achieve Resilience in organizations.

  One the other hand, human resources have the important role in recognize an opportunity in the market and take advantage of this. Thus, Haddadzadeh and Paghaleh [14] assure that a organizational plan should be devised in order to enhance contribution from employees to resolve crisis.

  For pro-activeness that is required in Strategic Resilience according to Valikangas and Romme [38], count with resilient people that has certain characteristics as such "a staunch acceptance of reality; a deep belief, often buttressed by strongly held values that life is meaningful; and an uncanny ability to improvise" [8] becomes fundamental.

- Creativity-is the ability to imagine, generate and see new possibilities with attitudes of open-mindedness and curiosity [41]. Moreover, creativity is the ability to construct diverse solutions and the flexibility to alter them if necessary [3]. In fact, in face to take advantage of an opportunity, creativity is a critical element for Strategic Resilience and it is fundamental to count upon a certain creativity to overcome challenges and changes in the environment.

- Risk Management—The business world is more complex. Uncertainties increase over time and affect organizations. It can be problem with supplier or a high dimension problem as a terrorist attack that affect organizations and community in general. Thus, risk management emerges to aid organizations to become resilient ones.

  Risk Management focuses on the probability of accidents happening that disturb organization operations and which are the consequence therefore [9]. An accident refers to any disorder of a stable system [17]. In Risk Management, an organization makes a list of possible risks that it may face during its productive activity, including market risks. For Seville [34] Risk Management considers a pro-active thinking way to manage situations, what fits well with Hamel and Valikangas asserted about Strategic Resilience has continuous anticipating state.

- Innovation—There is an extensive literature regarding innovation management [39]. Most academics and practitioners agree about achieving superior performance through innovation [29]. As Resilience is considered "the capability to self-renew over time through innovation" [29], without crisis, some organizations seek innovation to make something different in terms of products or processes and gaining market share, being more competitive.

  Besides, resilient organizations are innovative in times of adversity [11]. For withstanding crisis and get over, organizations combine creativity, having good ideas, with innovation, implementing creative ideas, being succeed and thriving.

- Information System—Mallak [21] considers the use of information as a critical factor for achieving Resilience. Resilient people know which information to access in chaos situations and know what implications for possible solutions. For this, it is necessary a technological support and get information from diverse sources for make sure of reliability. Furthermore, from reliable Information System, it is possible to plan change, to make decision, take the chance in the environment and turn threats in opportunities quickly.
- Competitive environment—Many rapid transformations are happening in the environment that organizations operate in, such as technological, geographic, political and social ones, which demonstrate a challenging scenario for organizations to succeed in and become competitive. Competition pushes organization to change, to learn how to be resilient. Turbulent market stimulates to organizations to be ready to change.
- Robust Strategy—In situations where it is difficult to forecast a potential disruption, it is acceptable for organizations to invest in robust strategies that will enable the organizational system work during the disorder. Establish a set of robust strategies help organizations to adapt to the rupture, promoting Strategic Resilience.

The next phase will provide the bounds of this study, how the data was collected and the Delphi technique used.

2. Phase 2: Delphi Study

Before the Delphi study, a presentation about Strategic Resilience, its key concept and literature that supports the choice of promoting factors Strategic Resilience, were sent to the group of experts by email in order to everyone has the same understanding about how each factors were treated in this study. Firstly, experts were asked to read this presentation and afterwards to access the link of the round 1 available in online questionnaire.

Round 1

In the round, experts were asked to: (a) give opinions about promoting factors Strategic Resilience from the list with 11 factors that was provided ; (b) justify the reason why they did not chose any factor; (c) add another factors that consider relevant to promote Strategic Resilience. Following completed the first round of questions, the data collected was analyzed, summarized of responses in a matrix and comments consolidated and posted in a website created for the purpose of giving feedback to the panelists. The feedback is condition for using Delphi. Thus, experts are able to analyze and keep or alter their previous opinion.

Round 2

In the second round, a similar questionnaire was sent to experts, adding 3 factors suggested by experts: Organizational Structure, Empowerment and Training in Risk Management. It was pointed Organizational Structure affect directly to organizational strategies. Ozsomer, Calantone and Di Benedetto [24] corroborate organizational structure affects the strategic positioning of organizations and contribute to make them more innovative. The second factor pointed was Empowerment that refers to re(design) jobs, including new attributions. Human resources gain more responsibilities about their tasks, increasing motivation and commitment.

Table 107.1 Results

Factors	Consensus (%)
Leardership	100
Capacity of change	100
Organizational culture	93
Organizational learning	93
Human factor	93
Creativity	87
Risk management	87
Empowerment	87
Innovation	80
Information system	73
Competitive environment	67
Training in risk management	67
Robust strategy	60
Organizational structure	60

Therefore, empowered jobs allow people in all organizational levels respond to new opportunities quickly what is imperative for Strategic Resilience. The third one is training in Risk Management as way of being pro-active and anticipating to change.

For filling this questionnaire, it is supposed the experts reflect about the responses from last round, available in the website, keep or change their opinion about the relevancy of the 14 factors identified in the second round of questions.

After the second round, the data indicates factors and respective consensus found (Table 107.1). The consensus of opinion set in a percentage equal and over 70%. Thus, it was just considered 10 factors that met 70% of consensus. This limit was suggested due to necessity of keep a high reliability.

107.5 Conclusion and Future Study

Resilience is one of topics more discussed in business world [8], configuring as a sustainable competitive advantage [26]. For this, the interest of organizations in Resilience is increasing. More than ever academics, leaders, managers and stake holders are interested to understand about Resilience and how it can make organization recover from a rupture, getting more competitive in the environment. In this sense, this paper provided some factors that promote Strategic Resilience considering expert opinions through Delphi technique. Therefore, based on research results, it is possible to assert that organizations have to focus specifically on these factors in order to build strategies stronger, improving themselves. In fact, this paper gives rise an opportunity to organizations understand factors such as work and from that knowledge to foster organizations any capacity of continuously anticipate and adjust to change and thrive.

Another contribution for this paper is related to theoretical approach. Theoretical basis is often missing in strategic perspective in the Resilience field. Here this deficit starts to decrease.

Despite of important contribution of this paper, limitation of this work should be mentioned. The research only comprised 2 rounds of Delphi technique, if more rounds were used, more debates among experts could be promoted. Moreover, it could provide more relevant data and the results of research could have higher reliability. Building from this study, a future study should provide weight to each factor identified. Obviously one factor can promote more Strategic Resilience over than another, being necessary further research.

References

1. Adnan H, Morledge R (2003) Application of Delphi method on critical success factors in joint venture projects in the Malaysian construction industry. In: Proceedings of CITCII conference, pp 10–12
2. Aleksić A, Arsovski S (2011) Resilience in supply chains. Center for Quality
3. Barrett F (2004) Coaching for resilience. Organ Dev J 22(1):93–96
4. Burnard K, Bhamra R (2011) Organisational resilience: development of a conceptual framework for organisational responses. Int J Prod Res 49(18):5581–5599
5. Carvalho H, Cruz_Machado V (2009) Lean, agile, resilient and green supply chain: a review. In: Proceedings of the 3rd international conference on management science and engineering management, November. World Academic Press, World Academic Union, Bangkok, Thailand, pp 2–4
6. Chiavenato I (1994). Administração de empresas: uma abordagem contigencial. Makron Books
7. Christopher M, Peck H (2004) Building the resilient supply chain. Int J Logist Manag 15(2): 1–14
8. Coutu DL (2002) How resilience works. Harvard Bus Rev 80(5):46–56
9. Dalziell EP, McManus ST (2004) Resilience, vulnerability, and adaptive capacity: implications for system performance. University of Canterbury, Civil and Natural Resources Engineering
10. Denhardt J, Denhardt R (2010) Building organizational resilience and adaptive management. In: Reich JW, Zautra AJ, Hall JS (eds) Handbook of Adult Resilience. Guilford Press, New York, pp 333–349
11. Everly G Jr (2011) Building a resilient organizational culture 2011. http://blogs.hbr.org/cs/2011/06/building_a_resilient_organizat.html
12. Fiksel J (2006) Sustainability and resilience: toward a systems approach. Sustain Sci Pract Policy 2(2):14–21
13. Gittell JH, Cameron KS, Lim S et al (2006) Relationships, layoffs, and organizational resilience. J Appl Behav Sci 42(3):300–329
14. Haddadzadeh A, Paghaleh MS (2012) he role of human resource system on crisis resolve. World Acad Sci Eng Technol 6:1764–1770
15. Hamel G, Valikangas L (2003) The quest for resilience. Harvard Bus Rev 81(9):52–65
16. Hamel G (2009) Todos podemos ser resilientes. HSM Management 50
17. Hollnagel E (2006) Resilience: the challenge of the unstable. Ashgate Publishing, Aldershot
18. Lengnick-Hall CA, Beck TE, Lengnick-Hall ML (2011) Developing a capacity for organizational resilience through strategic human resource management. Human Resour Manag Rev 21(3):243–255
19. Linstone HA, Turoff M (1979) The Delphi Method: tecniques and applications. Addison-Wesley, Massachusetts, p 29

20. Masten AS, Reed MGJ (2002) Resilience in development. In: Snyder CR, Lopez SJ (eds) Handbook of Positive, Psychology. Oxford University Press, New York, pp 74–88
21. Mallak L (1998) Putting organizational resilience to work. Ind Manag 40(6):8–13
22. Mamula-Seadon L (2009) CDEM, integrated planning and resilience: what is the connection? Tephra, Ministry of Civil Defence and Emergency Management
23. McCann J, Selsky J, Lee J (2009) Building agility, resilience and performance in turbulent environments. People Strategy 32(3):44–51
24. Özsomer A, Calantone RJ, Di Bonetto A (1997) What makes firms more innovative? A look at organizational and environmental factors. J Bus Ind Mark 12(6):400–416
25. Passig D (1997) Imen-Delphi: a Delphi variant procedure for emergence. Human Organ 56(1):53–63
26. Plodinec J (2009) Definitions of resilience: an analysis. Community and Regional Resilience Institute (CARRI), Oak Ridge
27. Porter ME (1996) What is strategy? Harvard Bus Rev 74(6):1–21 (November)
28. Ponomarov SY, Holcomb MC (2009) Understanding the concept of supply chain resilience. Int J Logist Manag 20(1):124–143
29. Reinmoeller P, Van Baardwijk N (2005) The link between diversity and resilience. MIT Sloan Manag Rev 46(4):61
30. Rice JB, Caniato F (2003) Building a secure resilient supply network. Supply Chain Manag Rev 7(5):22–30
31. Sheffi Y, Rice J (2005) A supply chain view of the resilient enterprise. MIT Sloan Manag Rev 47(1):41–48
32. Stephenson A, Seville E, et al (2010) Benchmark resilience: a study of the resilience of organisations in the Auckland region. Resilient organisations research report
33. Stolker RJM, Karydas DM, Rouvroye JL (2008) A comprehensive approach to assess operational resilience. In: Proceedings of the third resilience engineering symposium, pp 28–30
34. Seville E (2008) Resilience: great concept but what does it mean? University of Canterbury, Civil and Natural Resources Engineering
35. Starr R, Newfrock J, Delurey M (2003) Enterprise resilience: managing risk in the networked economy. Strategy Bus 30:70–79
36. Sutcliffe KM, Vogus TJ (2003) Organizing for resilience. Positive organizational scholarship: foundations of a new discipline. Berrett-Koehler, San Francisco, pp 94–110
37. Välikangas L, Merlyn P (2005) Strategic resilience: staying ahead of a crisis. Handb Bus Strategy 6(1):55–58
38. Välikangas L, Romme AGL (2012) Building resilience capabilities at "Big Brown Box Inc". Strategy Leadersh 40(4):43–45
39. Välikangas L, Hoegl M, Gibbert M (2009) Why learning from failure isn't easy (and what to do about it): Innovation trauma at Sun Microsystems. Eur Manag J 27(4):225–233
40. Vieira L (2006) A nova ordem da resiliência. HSM Management, Update 38
41. Wilson SM, Ferch SR (2005) Enhancing resilience in the workplace through the practice of caring relationships. Organ Dev J 23(4):45

Chapter 108
The Industrial Development of Beibu Gulf Economic Zone of Guangxi: Issues and Measures

Xiaoguo Xiong and Yongjun Tang

Abstract China is the ASEAN's largest trading partner and ASEAN is China's third largest trading partner. The leading role of BGEZ makes the coastal areas of Guangxi an important strategic district for China's western development and international regional economic cooperation. However, due to geopolitical factors and the desire for benefits of offshore energy resources, certain problems still exist between China and ASEAN countries. There are four main issues regarding the BGEZ's development: overemphasis on industry policies rather than market mechanism, overemphasis on introducing major projects rather than intensive utilizing resources and environmental protection, overemphasis on expansion of industrial scale rather than optimizing industrial structure and layout, and overemphasis on preferential policies rather than factor endowment promotion. The following measures should be taken in order to further improve the BGEZ: (1) to develop industrial clusters which established comparative advantages, (2) to develop leading industries that oriented by domestic demand, and (3) to development environmentally friendly industries as well as services in the production process.

Keywords Beibu gulf economic zone · CAFTA · Industrial policy · Industrial structure

X. Xiong (✉)
School of Business Administration, South China University of Technology,
Guangzhou 510006, People's Republic of China
e-mail: evia@vip.sina.com

X. Xiong
Business School, Guangxi University, Nanning 530004, People's Republic of China

Y. Tang
Baise University, Baise 533000, People's Republic of China

J. Xu et al. (eds.), *Proceedings of the Eighth International Conference on Management Science and Engineering Management*, Advances in Intelligent Systems and Computing 281, DOI: 10.1007/978-3-642-55122-2_108, © Springer-Verlag Berlin Heidelberg 2014

108.1 Introduction

In January 2008, the Beibu Gulf Economic Zone of Guangxi Development (BGEZ) Plan was ratified by the State Council, marking the upgrading of BGEZ's development and opening up to a national strategy. As proposed by the State Council in "Guidelines Regarding Development of the Beibu Gulf Economic Zone in Guangxi" in December 2009, BGEZ plays an important strategic role in China's western development and international regional economic cooperation. By virtue of the "11th Five-year Plan", the next 5–10 years is a critical period for BGEZ's economy to transform its development model and optimize industrial structure.

According to the structuralism theory of economic growth, economic growth not only comes from the increase of total input of production factors such as capital and labor, but also comes from the resources reconfiguration of production factors in different industry sectors, i.e. structural change [2], which is more important to the potential and significance of economic growth for developing countries [10]. Although some scholars investigated on the relationship of industrial structure and economic performance in that industry [4], there is little literature in using structuralism theory to analyze the regional industrial clusters in BGEZ, which plays an important role between China and ASEAN countries.

This article aims to identify current trends in international and domestic environmental and to carry out in-depth analysis of the problems in its industrial development. Based on both internal and external analysis, we then come out a sustainable path of industrial progression in alignment with local comparative advantages.

108.2 Analysis of the Internal and External Environment of BGEZ

1. Consensus of Accelerating Progress of China-ASEAN Market Integration

The annual bilateral trade volume between China and ASEAN countries increased over 20 % since 1990. The ASEAN countries' share in China's foreign trade rose from 6.18 % in 1991 to 9.8 % in 2010; meanwhile China also increased its share in ASEAN countries' foreign trade from 2.1 % in 1994 to 11.6 % in 2009. Currently, China is the ASEAN's largest trading partner and ASEAN is China's third largest trading partner [1]. With the establishment of China-ASEAN Free Trade Area (CAFTA), bilateral trade maintains a trend of continued growth and diversified trade categories. We will gradually stride towards regional market integration, featuring the free-flow of all factors of production.

However, due to geopolitical factors and the desire for benefits of offshore energy resources, certain problems still exist between China and ASEAN countries, such as border disputes, islands ownership disputes and territorial seas disputes [5]. BGEZ is an outstanding strategic position for both sea channels and neighboring land areas between China and ASEAN countries. The geopolitical factors are opportunities as well as challenges for the development of BGEZ. On the whole, cross-border sub-regional economic cooperation is irreversible trend for the common interests. It is also an effective way to resolve potential conflict and alleviate regional tensions.

2. More Decision-making Power of Local Government's Institutional Innovation

During the "11th Five-year Plan" period, the central government successively approved a series of programs for new areas, experimental areas and economic zones, including "Guangxi Beibu Gulf Economic Zone Development Scheme" approved in 2008. As a national level approved regional program, more support will be given to infrastructure construction, industrial layout and large engineering projects by the central government. Different from the previous "top-down" pattern of the Special Economic Zones, the new area programs will give local governments more independent rights for system innovation and tests as well as bring about greater effects through proper application [6].

3. Increased Pressure of Industrial Transformation and Intense Regional Competitions

With the development of production and technology as well as the catalysis of the global financial crisis, China has been entering a period of economic transformation. The main direction of future development is to realize the endogenous economic growth model: industries will become increasingly high-end and low-carbon, and will depend on technological innovation and simulating domestic demands instead of an export-oriented model. Therefore, on one hand, BGEZ should proactively facilitate the industrial shift from eastern coastal areas and seize all opportunities to develop; on the other hand, in order to deal with the rapid changing environment both at home and abroad, we need to constantly optimize the local industrial structure and promote market competitiveness of local enterprises. Industries like petrochemicals or steel have excessive capacity. In addition, a certain number of regions attract investment through preferential policies concerning land and tax reduction or exemption. This has caused a growing number of rural residents to lose their lands and the aggravation of social conflict and conflict.

4. Developments of Low-carbon Economy and Environmental Restrictions

Energy shortage and climate change has become the focus of international concerns. Under this background, on one hand, the developed countries vigorously push forward the carbon-trade market; on the other hand, they increase more financial support for new energy, new materials and other emerging industries. In order to deal with the international competition in future, Chinese manufacturing enterprises on the global industrial chain and the local enterprises with hope of participating in global competition should strengthen the risk assessment of carbon management and corporate environment [7]. However, the energy-intensive industry is still at a stage of high-speed development in most areas of China and the transition of economic growth pattern is extremely difficult. Moreover, resource and environment capacity has been on the decline in recent years. Destruction of the environment and climate as well as pollution of water, land and air in certain districts is at a difficult level to treat.

Energy security and environmental contamination are serious problems for China to face and solve in the future. If local governments do not change the current economic growth pattern, which features high-consumption, high-emission and high-pollution, through autonomous transition, greater energy resources limitation and environmental risk will be fostered in the future. Once the Carbon Standard, formulated by developed countries becomes mainstream, the current industrial development pattern in China will be confronted with its significant impact.

108.3 Issues of BGEZ's Industrial Development

108.3.1 Overemphasis on Industry Policies Rather than Market Mechanism

The local government serves as the policy maker for regional leading industries. Through industrial policies, the local government intervenes in the distribution progress of resources among various industries, thus compensating for deficiencies in market development. However, as the independent maneuvering space of local governments expands and local revenue display encouraging growth, the selection of the leading industry seems to be increasingly dependent on the government and ignoring trends caused by market forces. The former mainly manifests itself in the disproportionately high choices of the leading industry and excessive homogeneity. Take the "Twelfth Five-Year Plan" of various cities as an example; both Nanning and Qinzhou selected eight leading Industries, while Fangcheng Port also picked as many as six leading industries. At least three Guangxi cities choose petrochemical, paper-making, wood processing, electronic information, construction materials, and mechanical equipment manufacturing as their leading industries; while ship producing and maintenance, chemical engineering, agricultural goods processing, cereals and oils etc. were also selected by at least two cities as their leading industries. The result is: on one hand, when the government takes the place of market mechanism in electing leading industries, enterprises' enthusiasm concerning in technology innovation might be discouraged. On the other hand, an excessive number of leading industries and the high level of similarities in the selection between different areas cannot bring about benefits of economies of scale of the lowest cost. Thus the distribution of limited resources would have low efficiency. Neglect of market forces is mainly expressed through distortion of the factor market and insufficient protection of market subjects. For example, China's household registration system has caused the segmentation of the labor market, while administrative monopoly has brought about price distortion of resources and a high-level of administrative control in the land market. Furthermore, local governments often favor large-scale companies such as central-established and state-owned enterprises when providing financial credit and support, thus overlooking small and middle-sized enterprises and private companies. As the main body of the market, consumers have also long been at a weak position. This manifests itself most clearly in the turmoil of the food and medicine market as well as clogged channels of consumer rights protection [8].

108.3.2 Overemphasis on Introducing Major Projects Rather than Intensive Utilizing Resources and Environmental Protection

Introducing major projects is beneficial for creating intensive effects of scale in a short period, boosting the advancement of underdeveloped areas by leaps and bounds. In order to attract such projects, local governments often lower resource and environmental standards. Qinzhou can be taken as an example: Through Petro China's ten-million ton oil refining project, Qinzhou's added value in the petroleum processing industry increased 283.4 % in 2010, taking up 27 % of total added value for industrial enterprises above designated size. However, the petroleum processing industry causes high energy-consumption, high water consumption and high levels of pollution. Introducing this project poses a threat to local ecological environment, marine resources and tourism resources. Moreover, it is estimated that by 2013, China will have over 200 million ton oil refining bases; by 2015, China's oil refining capacity will reach 750 million tons, which may lead to excess production capacity. Another example is Fangcheng Port, which initiated 537 construction projects in the past 3 years, with its 2010 fixed asset investment nearly three-fold that of the total completed during the "11th Five-Year Plan". Most investors target Fangcheng's inexpensive land and low environmental costs, and the major projects introduced are mainly those with a high level of resource consumption and environmental damage, such as iron and steel, non-ferrous metal smelting, energy projects and so on. When the underlying reasons are probed, one finds that on one hand, low price policies have long been applied to resource and production factors, while corresponding compensation measures are lacking when environmental damage occurs. Thus the relation between market supply and demand, degree of resource scarcity and cost of environmental damage is not represented, which encourages extensive expansion. On the other hand, the local government of all levels is the main body responsible for developing the local economy. Fiscal predicaments, taxation, and government officials' assessment system all lead to increased local competition, causing issues such as blindly investing in projects, unhealthy one-sided pursuit of GDP growth, excessive consumption of resources and damage of environment, and extensive economic growth measures.

108.3.3 Overemphasis on Expansion of Industrial Scale Rather than Optimizing Industrial Structure and Layout

The economic foundation of underdeveloped areas is relatively weak, and needs large amounts of infrastructure construction and expansion of industrial scale to support its economic development. However, placing excessive importance on the expansion of industrial scale and neglecting to optimize industrial quality and spatial layout is unbeneficial for the sustainable development of regional economy. Thus future costs will be caused when adjusting the structure. According to statistics, from 2006 to 2010, the regional gross production of the BGEZ grew at an annual rate of 16.3 %. In 2010, the gross production reached 302173 CNY, increasing at a rate of 15.6 % over the past year. This was 1.4 % higher than the Guangxi Autonomous Region average and contributed 32.9 % to the GDP growth of Guangxi [3].

Even though such achievements were made, certain issues can be seen when examining the industrial structure. Firstly, the primary industry takes up an exceedingly large proportion. In 2009, the proportion was as high as 20.8 %, whereas in the same period, the share of primary industry in the Pearl River Delta and Yangtze River Delta Zones were under 5 %. Secondly, the secondary and tertiary industries are in a "misleadingly high" situation. Looking at the proportion of secondary and tertiary industries, one can see that in 2009, the secondary industry of BGEZ was at 33.3 %, with 44.6 % for the tertiary industry. The tertiary industry displays a typical "misleadingly high" situation to the relatively low level of the secondary industry. This results from a low degree of value chain, relatively low added value, high energy-consumption, high material consumption, high pollutant emission and a lack of technological content and competitiveness. The tertiary industry is mainly composed of wholesale, catering, retail and other such life related services. The development of production related services such as the financial sector and logistics lags behind. Thirdly, it is harder for the second and tertiary industries to absorb rural surplus labor. In 2009, the non-agricultural population took up 23.7 % among the four municipalities of Beibu Gulf. According to the standard of the labor force taking up 72 % of the total population, the non-agricultural labor force of BGEZ's four municipalities should be 4,236,000. At the same time, the employed population of the four cities and surrounding counties totaled a mere 961,000 in 2009. The huge gap in the non–agricultural population and the urban employed population shows that there is great instability in the transfer of rural residents to urban areas and the transformation of non–agricultural industries. The residents who have transferred out of rural areas are largely without official employment in the economic zone, and their identities are not either agricultural or urban [9].

From the perspective of industrial layout, there are several issues. Firstly, there is a high level of homogeneity between the industrial layouts of the four Beibu Gulf municipalities. Industries with a high degree of similarity include coastal industries, port shipping and others with high taxes and profits. The converging of similar layouts is closely related to the government's industrial development strategy and industry policy formulation. Secondly, port function positioning for the three cities of

Beihai, Qinzhou and Fangcheng Port is vague and indistinct. The establishment of the Guangxi Beibu Gulf International Port Group was an advantageous step to optimizing port resource. But viewed from an overall perspective, competition still outweighs cooperation, and it is difficult for the ports to form a scale effect. In addition, the supporting ability of industries is weak. Many major projects introduced to BGEZ in recent years has seen low levels of integration with local markets and enterprises, with local enterprises often not being able to meet the demands of quality of materials and components, delivery time limits, and cost.

108.3.4 Overemphasis on Preferential Policies Rather than Factor Endowment Promotion

In the process of the rapid development of industrialization, local governments often compete through offering very favorable land and capital prices as well as tax cuts to attract outside capital. In 2010 the "Policies Concerning Promotion of the Development of Guangxi Beibu Gulf Economic Zone" was issued, establishing incentive measures on industrial project investment, tax reductions and land use. Since land resources in developed areas are scarce and the cost of land has grown continuously, these preferential policies can significantly attract investment and secure a large amount of capital flow in the short term. But viewed in the long run, it can be seen that on one hand, the gap in preferential policies, salaries, and land prices has a weakening influence in attracting investment. Solely depending upon the "race-to-the-bottom" competition of preferential policies will only lead to increasingly similar policies between different areas. On the other hand, tax and land price reduction policies will cause price distortion of production forces, so that the true cost of enterprises cannot be correctly reflected. A false and seemingly low cost will cause enterprises to shift their focus from pursuing efficiency to finding cheaper resources. This distortion in prices will lead to abusing of resources and environmental deterioration. The issue is essentially achieving economic growth at the cost of damage to the environment, waste of resources and the sacrifice of residents' welfare.

If implementing tax breaks for attracting capital is a "race-to-the-bottom" strategy, then regional competition to attract human capital can improve the residents' welfare level; thus it is a kind of benign competition. However, the BGEZ is quite weak concerning human capital reserve; this is mainly displayed in the following aspects: First, inadequate population inventory. According to data from China's Sixth National Census, the total current population of BGEZ is 12.1475 million, accounting for 26.39 % of the total population in Guangxi; in contrast, the population of the Pearl River delta of Guangdong province is 56.1184 million, accounting for 53.80 % of the total population. Second, there is a lack of highly educated talented persons, with such talents mainly concentrated in the city of Nanning. The number of people that received higher education is 1.0516 million for BGEZ, accounting for 8.66 % of the total current population. Of these people, 778,700 are in Nanning, accounting

Table 108.1 Population distribution and education level of BGEZ (by the end of 2010)

City	University/ college	High school/secondary vocational school	Secondary school education	Primary school education	Population
Qinzhou	112,000	278,200	1,106,100	1,127,900	3,079,700
Beihai	111,100	241,800	625,300	379,500	1,539,300
Fangcheng	49,800	106,400	352,800	243,100	866,900
Nanning	778,700	966,700	2,603,500	1,657,200	6,661,600
Total	1,051,600	1,593,100	4,687,700	3,407,700	12,147,500

Source of data Based on Guangxi Statistical Year Book 2010 [9]

for 74 % of those in BGEZ. In contrast, the Pearl River Delta Economic Zone has 6.6883 million personnel that received higher education, which accounts for 11.92 % of the total population. Third, labor outflow to other areas is relatively heavy. According to incomplete statistics, among the secondary vocational schools in BGEZ, nearly two thirds of the graduates work in the Pearl River Delta, nearly half of the graduates from higher vocational colleges work in other provinces, while nearly a third of universities and college graduates work outside of Guangxi.

In addition, the utilization efficiency and investment structure of capital elements in the BGEZ also remain to be improved. According to statistics, the share of capital formation in Guangxi's GDP rose from 33.7 % in 2001 to 74.7 % in 2009, which shows that in recent years Guangxi's economic growth displayed capital-driven characteristics. Other data imply that total capital stock of Guangxi rose from 80.1 billion RMB in 2001 to 303.1 billion RMB in 2009, with an average annual growth rate of 18 %. The growth rate of capital formation is faster than the growth rate of GDP, reflecting the low efficiency of capital output. We predict the same trend for BGEZ. Taking Qinzhou as an example, in 2010, investment in the transportation industry, manufacturing and real estate were the top three industries, and respectively accounted for 25.96, 22.97, and 21.24 %. By comparison, investment level in education, medical care, social security and other social public services was very low. The investment structure is geared towards areas with obvious short-term benefits, while inadequate attention is paid towards long-term investment in public service where endogenous growth is marshaled.

108.4 Key Choices of the Industrial Development in BGEZ

1. Development of Industrial Clusters That Established Comparative Advantages

The high level of similarities with surrounding areas and excessive strategic planning for emerging industries has led to an inefficient use of resources, as well as low-end homogenized competition between different areas. By contrast, the traditional industry was often formed on the basis of comparative advantage, such as agricultural byproduct processing, wood processing and wood product manufacturing, pharmaceutical manufacturing, etc. These traditional industries of BGEZ have formed certain industrial foundations in previous years of development, and their added value has increased throughout the years. An increase of invested resources and enhanced policy support such as reducing taxes, providing a fair environment for competition, etc., will be greatly benefit these industries and help them improve and expand. Concerning policy orientation, the local government should on one hand guide the innovation and upgrading of the value chain for industrial clusters which are based on the low end of the chain with low cost advantages. Thus the value of the entire chain will see a rise. For example, the processing industry of agricultural byproducts can optimize and combine various factors of production through strengthening supply chain management, consequently transforming from manufacturing primary products to deep processing. Through building a base for approval of standardized systems, the market positioning and domestic and international competitiveness of products can be enhanced. On the other hand, the local government should also cultivate proprietary production factors such as professional personnel and labor, while establishing effective mechanisms for international cooperation, so as to maintain the vitality for innovation and upward development impetus for these industrial clusters.

2. Development of Leading Industries That Oriented by Domestic Demand

Fostering leading industries which are geared towards domestic demand is in line with the market situation, and beneficial for opening up advanced markets while resisting risks of insufficient external demand. The "11th Five-Year Plan" also put forward to continue releasing consumer potential of urban and rural residents in the next 5 years, thus gradually promoting China's domestic market scale to rank among the top globally. The BGEZ also should conform to changes in market conditions, and focus on the development of domestic leading industries. The focal point should be tapping the domestic demand in adjacent areas of southwest China that harbor huge potential for future development. According to statistics, during the "11th Five-Year Plan" period, the four provinces and one city of southwest China has maintained an average urban and rural resident income growth of above 10 %, almost all of which is higher than the national average. In addition, social gross retail sales of consumer goods for these areas accounted for 10 % of China, while the vast majority of provinces in southwest China enjoyed higher levels of growth in 2008–2009 than the national average. The consumption structure exhibits a tendency to develop in a diversified manner, while the level of consumption is transforming to development and enjoyment and upgrading from low-grade to high-grade. Home facilities and services, transportation and communication expenses, entertainment, education and culture services have become the new growth point for consumer spending. Relying on its advantage in location, the BGEZ should focus on developing industries that meet new characteristics in domestic demand as well as those with sizable market potential. Thus the domestic market of greater southwest China will be opened up and the effects can radiate to the entire Chinese market as well as the ASEAN market.

3. Emphasis on Development of Environmentally Friendly Industries

Environmentally friendly industries not only include tertiary industries with low pollution and high added value such as tourism, commerce and trade, and exhibitions, but also consist of modern manufacturing industries with circular economy and clean production as the main characteristics. Both Guangxi and BGEZ are at an intermediate stage of industrialization, and it is difficult to fundamentally change the current situation in the short term: Heavy industry is currently the main body of the industrial structure, and raw coal is the main body of energy consumption structure. Certain high energy consumption, high pollution and high emission industries that are growing rapidly such as equipment manufacturing, petrochemical, metallurgy, paper-making etc. have also increased the risk of energy, environmental, and ecological issues in BGEZ. BGEZ should put energy into developing the environmentally friendly industry. This is to facilitate a natural ecological environment for Beibu Gulf while reducing the restriction on industrial development that energy scarcity and introducing environmental standards brings. On the one hand, when choosing industries, energy consumption and environmental impact should be important criteria. Introduction of major products, strict inspection and approval, raising admittance thresholds should be done to reduce resource consumption and environmental cost as much as possible. On the other hand, concerning construction of industrial parks, measures should be taken to promote clean production, introduce quantitative evaluation of full life cycle to assess the environmental impact of industries and enterprises, arrange layouts according to industrial metabolism and co-existence, and use water, raw materials, energy etc. as efficiently as possible. Thus, minimizing pollution and maximizing resource utilization efficiency can be achieved.

4. Emphasis on Development of Services in the Production Process

The rapid development of modern manufacturing requires suitable supportive services during the production process. Compared with manufacturing and consumer services, services in the production process enjoy higher profitability rates and production rates. In recent years, its proportion in the total service output has grown steadily. The BGEZ can select the key fields for development in services in the production process according to the following factors: advantages and disadvantages, industrial foundation, geological positioning, infrastructure, urban environment and so on. Using modern logistics as an example, the three cities of Beihai, Qinzhou and Fangcheng Port can give play to their advantages of ports and vigorously progress port and shipping industries, as well as international bonding logistics. They can also advance logistics of agricultural goods and cold-chain which stems from processing of agricultural byproducts. The development of services in the production process is on one hand, beneficial for upgrading the manufacturing sector of BGEZ and increasing its competitiveness; on the other hand, it also enhances the position of BGEZ among international and domestic industrial distribution, making it more dynamically competitive, so more relative profit is earned.

108.5 Conclusion

China is the ASEAN's largest trading partner and ASEAN is China's third largest trading partner. Although there are certain problems still exist between China and ASEAN countries, it is undeniable that the whole region of BGEZ and enterprises related to it have great opportunities, especially on geographic characteristics and institutional context. We have identified four main issues regarding the BGEZ's development, i.e. overemphasis on industry policies rather than market mechanism, overemphasis on introducing major projects rather than intensive utilizing resources and environmental protection, overemphasis on expansion of industrial scale rather than optimizing industrial structure and layouts, and overemphasis on preferential policies rather than factor endowment promotion. In order to address these issues, we propose (1) to develop industrial clusters that established comparative advantages, (2) to develop leading industries that oriented by domestic demand, and (3) to development environmentally friendly industries as well as services in the production process.

References

1. China-ASEAN Trade Deal Begins Today. Jakarta Globe. Bloomberg (2010). http://www.thejakartaglobe.com/archive/china-asean-trade-deal-begins-today/350274/
2. Demurger S (2001) Infrastructure development and economic growth: an explanation for regional disparities in China. J Comp Econ 29:95–117
3. Jiang H (2008) On constructing harmonious culture-circle in Pan-Beibu Gulf Area. J Nanning Polytech 4:30–33 (In Chinese)
4. Joshua D, Edward F (2012) Regional industrial structure and agglomeration economies: an analysis of productivity in three manufacturing industries. Reg Sci Urban Econ 42:1–14
5. Xiong X (2013) Strategic development of Beibu Gulf economic zone of Guangxi: the perspective of low carbon economy. In: Proceedings of the 7th International Conference on Management Science and Engineering, pp 1251–1260 (In Chinese)
6. Tang X (2008) A study of the expectations, methodology and policies for China-ASEAN education collaboration. High Educ Forum 1:4–12 (In Chinese)
7. Tang Y, Xiong X (2009) A study on operational strategy and innovation of implementation model for environment protection in business. In: Proceedings of the 3rd International Conference on Management Science and Engineering Management, pp 288–292
8. Wang J (2011) The economic cooperation mechanism and the concept of economic integration—also on the cross-strait economic cooperation mechanism. J Beijing Union Univ (Humanit Soc Sci) 1:64–67 (In Chinese)
9. Wang X, Tan Y (2008) The new development and promoting strategies of China-ASEAN mutual investment under the CAFTA framework. Int Econ Trade Res 6:21–24 (In Chinese)
10. Zhao Q, Niu M (2013) Influence analysis of FDI on China's industrial structure optimization. Procedia Comput Sci 17:1015–1022

Chapter 109
Impact of Human Capital Investment on Firm Performance: An Empirical Study of Chinese Industrial Firms

Qian Li, Xiaoye Qian, Shiyang Gong and Zhimiao Tao

Abstract As the technological innovation becomes the vitality of enterprise development, firms are increasingly realizing that, employees' knowledge and skills are the unique resources to gain the competitive advantages. Human capital is becoming the driving force to promote the enterprise development. Using data from Chinese Industrial Enterprises Database, this paper examines how human capital investment (including education and training investment) affects the firm performance. The results show that, employees' educational level has a significantly positive impact on firm performance; an inverted "U" shaped relationship exists between training investment and firm performance. Also, there is a negative interaction between employees' education and training investment. The lower employees' education is, the higher impact training would have on the firm performance. In addition, we also investigate the impact of industry types and find that, in the lower labor-intensive industries, employees' educational level will have a greater impact on performance, while in the high labor-intensive industries, the effect of training investment will be greater.

Keywords Human capital investment · Education · Training · Firm performance · Labor intensity

109.1 Introduction

With the development of knowledge economy, talents and knowledge embedded in talents are becoming the crucial resources for firms' innovation and development. Firms are increasingly realizing that, employees' knowledge and skills are the unique sources to gain the competitive advantages. Human capital is becoming the driving

Q. Li · S. Gong
School of Economics and Management, Tsinghua University, Beijing 100000,
People's Republic of China

X. Qian (✉) · Z. Tao
Business School, Sichuan University, Chengdu 610064, People's Republic of China
e-mail: xyqian@scu.edu.cn

J. Xu et al. (eds.), *Proceedings of the Eighth International Conference on Management Science and Engineering Management*, Advances in Intelligent Systems and Computing 281, DOI: 10.1007/978-3-642-55122-2_109, © Springer-Verlag Berlin Heidelberg 2014

force to promote the enterprise development. A research conducted by World Bank provides the evidence that, firms' return rate of human capital investment is up to 8.6 %, which is almost as high as the return rate of fixed capital investment. As the technological innovation becomes the vitality of enterprise development in China, the contribution of human capital investment to firm performance is predicted to be greater in future [1].

The return on human capital investment is always being a hot topic since the human capital concept was brought up [5, 24]. Abundant studies have provided evidence regarding to the positive contribution of human capital investment on individuals' income growth, career success [15, 19], as well as the economic growth and productivity enhancement in a region or country [8, 13]. In last decades, the issues of human capital investment start to raise the interests of scholars in management field. Grant [14] suggested that a firm's competitive advantage came from the knowledge embedded in their employees. Huselid [16] also proposed that talent selection and staff training were two important components for firms to establish the high performance human resource management system. Nevertheless, in comparison to the abundant studies on micro and macro level, there is relatively less empirical evidence of human capital investment on firm level. Firstly, most of the current studies in China focus on the particular types of firms or firms in specific industries. Such as the accounting firm; listing companies in manufacturing, information technology and real estate industries [21], the generalization of the research findings may be restricted; Secondly, most of the studies only examine the correlation between the education and firm performance, neglecting the important role of specific human capital. Thirdly, existing researches have ignored the differences of investment returns among varied industries. Lastly, the current literature mainly employs the small-scale surveys. A related study indicates that this method has certain errors [3].

Adopting the 2004–2005 year data from China Industry Business Performance Database, this study provides further empirical evidence on how human capital investment (including education and training investment) affects the firm performance. Moreover, considering the research context of Chinese industry sector, we also discuss how these correlations differ among industries with different labor intensities.

109.2 Theory and Hypotheses

109.2.1 Human Capital Investment and Firm Performance

Current literature draws the conclusion that formal education has a significant impact on individual's productivity and income [15]. There are also studies examining the relationship between formal education and firm performance. Black and Lynch [6] found that, 1 year increase of employees' average schooling is associated with 8.5 % increase of productivity in manufacturing industries, and 12.7 % increase in non-manufacturing industries. Liu and Zheng [20] also found the significant contribution of employees' educational level to the productivity of State-owned enterprises in China. This positive correlation between employees' education and firm performance can be explained by the general ability increase and cognitive skill enhancement caused by receiving higher education. Therefore, we propose that:

Hypothesis 1. Employees' educational level has a positive impact on firm performance.

With regard to the correlation between on-the-job training and firm performance, most studies find out that, on-the-job training can significantly improve the firm performance. Bartel [4] used the data from a firm which conducted a new employee training program in 1983 and found that, the introduction of new training program have significantly positive impact on firm performance in the next three years. Barrett and O'Connell [2] conducted a survey in 215 companies in Ireland and provided the evidence that employees' on-the-job training provided by firms can significantly increase the productivity. Tharenou et al. [25] did a META analysis on the 67 studies related to the training and performance, asserting that training does have a significant contribution to the accumulation of firm's human capital, hence increases the firm performance.

However, some researchers pointed out that, the correlation between on-the-job training and firm performance is not always linear. Pennings et al. [23] studied the operation of accounting firms in German from 1880 to 1990, and found that the specific human capital has a U-shaped effect on the organization dissolution. We propose that such nonlinear effect may also exist in the relation between training investment and firm performance. On one hand, people's absorb capacity of knowledge and learning ability have limitations [10]. Trainings beyond the individual's ability may not be successfully transferred to the accumulation of human capital. On the contrary, they will bring individuals with much pressure and decrease their passion of learning as well as the productivity [22]. On the other hand, holding the other factors in production constant, the marginal effect of training input contributing to the firms' productivity may subject by Law of Diminishing Marginal Returns, demonstrating the decreasing trend along with the increase of training input. Therefore, we propose that:

Hypothesis 2. An inverted "U" shape exists between the firm's training investment and performance.

Although education and on-the-job training are the two major ways for firms to accumulate human capital, there is limited evidence regarding to the correlation between these two ways. There might be two different interaction mechanisms between education and on-the-job training. One mechanism is the complementary effect, which means the higher employees' education is, the lager impact training would have on the firm performance. This is because, well-educated employees are usually endowed with higher learning and cognitive capability, hence can better absorb the knowledge and skills acquired during trainings in the workplace. Under this circumstance, training investment is going to have a larger impact on firm performance.

Another possible mechanism is the substitution effect. The lower the educational level is, the higher impact training investment would have. If the employees have lower education in overall, the formal training provided by the firm may compensate for the lack of knowledge they should acquire during formal education. Blundell [7] pointed out that, keeping the training level constant, employees with lower-level education can gain more returns from formal training, especially for those who received less education due to education budget constraint or have limit access to higher education resources. It is the substitution effect other than complementary effect more fit to the China's context. In China, a large amount of manufacturing workers come from the rural areas, who have less opportunity to get higher education. Once the firm plays the role as the education resource provider, employees with lower education will greatly benefit from the training program and diminish their knowledge gap with higher-educated employees. Hence, we propose that:

Hypothesis 3. Educational level and training investment has a negative interaction effect on firm performance. Training investment has a larger effect on firm performance when employees have lower-level education.

109.2.2 The Impact of Industrial Labor Intensity

Although current studies rarely discuss the different returns of human capital investment among industries, this paper will analyze the impact of industrial labor intensity. Datta et al. [11] did a survey analysis targeting to the capital-intensive manufacturing industries in western countries and found that the lower the capital intensity is, the greater impact high performance human resource management system will have on labor productivity. They argued that, high performance human resource management system put more emphasis on training, which equipped employees with higher skills, therefore significantly contributed to the lower capital-intensive industries. Moreover, many studies found that physical capital is complementary to human capital [18]. The higher investment in physical capital leads to the larger demand of higher-educated workers. Based on the predication of commentary effect between physical capital and human capital, in the less labor-intensive industries, employees need higher educational level to be qualified for their job tasks, or even to do some innovations to improve productivity. In this case, the impact of educational level to firm performance is greater than that in higher labor-intensive industries. While in the industries of higher labor intensity, such as textile industry, due to the work content is usually not high-tech, the demand of knowledge level is not high. Therefore, as long as employees can master the skills acquired through job training, their work efficiency will be improved. Hence, we propose that:

Hypothesis 4. In industries with lower labor intensity, employees' educational level has a larger impact on firm performance. In industries with higher labor intensity, firm's training investment has a larger impact on firm performance.

Table 109.1 Summary of the sample

Feature of the firms		Observations	
		Number	Ratio (%)
Firm's age	Less than 5 years	6360	28.9420
	5–9 years	8463	38.5119
	10–19 years	5492	24.9920
	Over 20 years	1660	7.5540
Firm ownership	State-owned enterprises (SOEs)	1702	7.7452
	Non-SOEs	20273	92.2548
Industry	Mining	335	1.5245
	Manufacturing	20978	95.4630
	Power, thermal, gas and water supply	662	3.0125
	Northeastern area	1188	5.4061
	Eastern area	18252	83.0580
	Middle area	962	4.3777
	Western area	1573	7.158

109.3 Data and Variable

109.3.1 Data

The data listed in Table 109.1 we used in this study comes from the China Industry Business Performance Database collected by China's National Bureau of Statistics. As the largest longitudinal database of China's manufacturing firms, it provides the information of more than 300 thousand firms. We use the data from year 2004 to 2005. This is because the information of human capital investment was only collected in 2004, hence we used the 1-year-lagged data of firm performance to relieve the endogenous problem in this study. After merging the 2 year's data, dropping the missing values as well as the firms which are not currently in operation or have too fewer employees (less than 10), we finally obtain a sample of 21,975 manufacturing firms. Following the previous studies [1, 9, 12, 20, 23], we used the variables listed in Table 109.2.

109.3.2 Descriptive Analysis

In Table 109.3, we report the results of descriptive analysis. Here are the findings: Firstly, the rate of return on assets is 6.8%, indicating that the sample firms are profitable in 2005. Secondly, there are variations of employees' educational level across firms. The mean ratio of college-educated employees is 12.5%, indicating that in every 100 employees, 12 of them have college and above degrees. The training investment in average is 132 Yuan per capita. Thirdly, the sample firms have an

Table 109.2 Definitions of variables

Variable	Unit	Definitions	Description
ROA	Percentage	Dependent variable, used to measure the return on assets	ROA = net profit/average total assets
Edu	Percentage	Ratio of employees with college, undergraduate, graduate and above degrees	Edu = (number of employees with college, undergraduate, graduate and above degrees)/total number of employees
Training	Thousand Yuan	Training investment per capita	Training = total education expenses/total number of employees
Size	Trillion Yuan	The average of total assets was measured	Size = (average of total assets in 2003 + average of total assets in 2004)/2
Age	Year	Length of established time	Age = 2004 − established time
Leverage	Percentage	Asset-liability ratio	Leverage = (long-term liabilities + current liabilities)/total assets
People	Thousand	Total number of employees	
Salary	Thousand Yuan	The average wage of employees	Salary = total payable wages/total number of employees
Welfare		Including labor and unemployment insurance (x_1), housing fund and subsidies fee (x_2), pension and medical insurance (x_3), payable welfare (x_4) per capita. Due to space limitations, the regression results are not listed separately in the model	x_1 = labor and unemployment insurance/total number of employees x_2 = housing fund and subsidies fee/total number of employees x_3 = pension and medical insurance/total number of employees x_4 = payable welfare/total number of employees
Union		Whether a union was established in the firm. Dummy variable	Firms with union are assigned to be 1, without union are assigned to be 0
Ownership		Registration type. Dummy variables	State-owned firms are assigned to be 1, non-state-owned firms are assigned to be 0
Industry		Industry type. Dummy variables	39 industries generate 38 industry dummies
Province		Province in which firm locates. Dummy variables	32 provinces generate 31 province dummies

average asset of 110 million Yuan and 333 employees. Last but not least, we find that the financial leverage is 54.8 %, and the yearly salary is CNY 15694. Table 109.3 also demonstrates the correlation matrix of the major variables. The ROA (return on assets) and edu (ratio of college-educated employees) are positively correlated.

Table 109.3 Descriptive and correlation analysis

Variables	Mean	S.D	1	2	3	4	5	6	7	8
ROA	6.7897	15.4834	1.0000							
Edu	12.4676	16.1502	**0.0160**	1.0000						
Training	0.1324	0.5233	**0.0172**	**0.1567**	1.0000					
Size	0.1106	1.1422	−0.0016	**0.0957**	**0.0460**	1.0000				
Age	9.4053	9.9694	**−0.0782**	**0.0866**	**0.0483**	**0.1061**	1.0000			
Leverage	54.7874	23.5998	**−0.1475**	**−0.0878**	**−0.0261**	−0.0105	**−0.0289**	1.0000		
People	0.3318	1.3056	−0.0096	0.0095	0.0069	**0.4968**	**0.1659**	−0.0028	1.0000	
Salary	15.6947	14.4887	**0.0398**	**0.3825**	**0.3176**	**0.1261**	**0.0853**	**−0.1099**	0.0341	1.0000

Note Sample $N = 21975$; bold figures represent significant level of 10 % and below

Meanwhile, the ROA is also significantly positively correlated with per capita training investment.

109.4 Regression Analysis

109.4.1 Influence of Human Capital Investment on Firm Performance

In Table 109.4, we estimate the impact of human capital investment on firm performance. The dependent variable is the logarithm of ROA. The specification (1) is only with control variables, including firm's size, age, financial leverage, size of employment, yearly salary and welfare, whether having a union or not, the ownership status, and the industries and provinces firms belong to. In specification (2), the edu (ratio of college-educated employees) is introduced into the model to examine the influence of general human capital to firm performance. In specification (3), training is included to examine whether the contribution of education to firm performance is still consistent after controlling for the specific human capital. In specification (4), we further include the quadratic form of edu to examine the non-linear correlation between ratio of college-educated employees and firm performance.

Specification (5) are estimated to examine the hypothesis 2. The quadratic form of training is added into the model to examine the non-linear effect of training on firm performance. Furthermore, we introduce the interaction term of edu and training into the specification (6) to provide evidence to support hypothesis 3. We also ran the VIF test for all the models, and the results suggest that multicollinearity is not an issue we need to worry in these models.

Demonstrated in Table 109.4, estimates of control variables in column (1) are consistent with our expectation. The firm size has no significant predict power on firm performance, while the age and financial leverage is negatively correlated with ROA, indicating that the longer a firm was set up, the heavier a firm has debts, the less

Table 109.4 The influence of human capital on firm performance

	Dependent variableln (ROA)					
	(1)	(2)	(3)	(4)	(5)	(6)
Edu		0.0018^c	0.0018^c	−0.0006	0.0013^b	0.0018^c
Edu^2				0.00004^a		
Training			0.0433^b	0.0439^b	0.1429^c	0.0717^c
$Training^2$					−0.0053^c	
Edu*training						−0.0012^a
Size	0.0055	0.0042	0.0043	0.0043	0.0031	0.0042
Age	−0.0072^c	−0.0071^c	−0.0071^c	−0.0070^c	−0.0071^c	−0.0071^c
Leverage	−0.0105^c	−0.0105^c	−0.0105^c	−0.0105^c	−0.0104^c	−0.0105^c
People	0.0322^c	0.0332^c	0.0332^c	0.0334^c	0.0335^c	0.0332^c
Salary	0.0064^c	0.0058^c	0.0053^c	0.0053^c	0.0057^c	0.0055^c
Welfare	Yes	Yes	Yes	Yes	Yes	Yes
Union dummy	Yes	Yes	Yes	Yes	Yes	Yes
Ownership	Yes	Yes	Yes	Yes	Yes	Yes
Industry	Yes	Yes	Yes	Yes	Yes	Yes
Province	Yes	Yes	Yes	Yes	Yes	Yes
Constant	1.7413^b	1.7006^b	1.7050^b	1.7396^b	1.6892^b	1.6993^b
R^2	0.1125	0.1129	0.1131	0.1133	0.1141	0.1133
Adjusted R^2	0.1093	0.1096	0.1098	0.1099	0.1108	0.1099
N	21975	21975	21975	21975	21975	21975

Notes (1) a, b, c represent 10, 5 and 1 % level of significance respectively;
(2) Figures in brackets are standard error of the estimated coefficients;
(3) Due to space limitations, estimated coefficients of the per capita welfare, union, ownership, industry and province are not listed

profitable it will be. We also find the size of employment and the level of salary are positively correlated with ROA. Without any explanatory variables, regression model including control variables can explain 11 % of the variation of ROA. Specification (2)–(4) examine the hypothesis1.

The coefficient estimate of edu in column (2) is significantly positive, indicating that the firms with higher educational level of employees have better financial performance. The estimated coefficient is 0.0018, suggesting that, with every 1 % increase of employees with college and above degrees, the firm's ROA will increase by 0.18 %. After controlling for the training investment in specification (3), the level and the significance of edu's estimated coefficient haven't changed. This results also suggests that the contribution of training to firm's profitability is significant but small. This finding is also evident by the fact that the R square of specification (3) is merely slightly increase by 0.0002 from specification (4). After controlling the quadratic form of edu in column (4), we found that the estimated coefficient of edu is no longer significant. The coefficient estimates of quadratic term (edu^2) is only marginal significant, therefore we did not find the non-linear effect of education on firm performance.

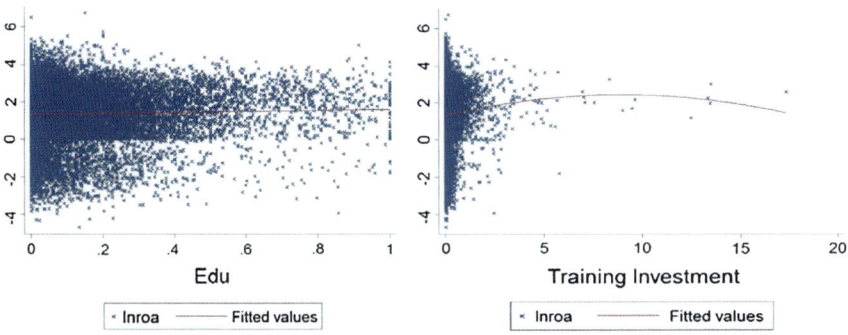

Fig. 109.1 The correlation between employee's educational level and firm performance as well as the correlation between training investment per capita and firm performance

In column (5), we examine the non-linear effect between training and ROA. The estimated coefficient of training is significantly positive ($\varepsilon_1 = 0.1429$, $p < 0.01$), while the quadratic form of training is significantly negative ($\varepsilon_2 = -0.0053$, $p < 0.01$). Those estimates indicate that the correlation between training investment and firm performance is an inversed U shape, which means the return of investment on specific human capital firstly increase with the training input growth, beyond a certain point of input, the contribution of training to ROA will start to decrease.

We also draw the scatter figures to demonstrate the correlations between employee's educational level and ROA, and between training investment and ROA as well to provide with a more intuitive understanding. As the left side figure in Fig. 109.1 shows, the correlation between edu and lnROA is linear. The right side figure in Fig. 109.1 demonstrates that the fitted curve of scatter plot reflecting the correlation between training investment and ROA is an inversed U shape, as we predicated in hypothesis 2.

The estimation in specification (6) evident the hypothesis 3. We introduce the interaction term of educational level and training investment (both centralized), and found that the estimated coefficient of interaction term is significantly negative ($\sigma_1 = -0.0012$, $p < 0.10$), suggesting that the lower education level employees have, the larger contribution training investment will have on ROA. Training program's return is especially high for firms with less-educated employees.

109.4.2 The Moderating Effect of Industrial Labor Intensity

We examine the mode rating effect of industrial labor intensity on the correlation between human capital investment and firm performance. Following Ji and Wei's [17] method, we firstly adopt the data from input-output statement of China's 42 sectors to calculate the labor intensity indicator. The calculating rule is that the labor intensity of a certain sector equals to the ratio of total labor's compensation to the

Table 109.5 The estimation of regressions including industrial labor intensity

	Dependent variable: ln(ROA)		
	(7)	(8)	(9)
Edu	0.0019^c	0.0015^b	0.0014^b
Training	0.0512^c	0.0507^c	0.0732^c
Labor	0.2752	0.1743	0.1913
Edu*labor		−0.0707^b	−0.0816^c
Training*labor			2.9776^b
Size	0.0039	0.0032	0.0033
Age	−0.0064^c	−0.0064^c	−0.0064^c
Leverage	−0.0104^c	−0.0104^c	−0.0104^c
People	0.0348^c	0.0347^c	0.0347^c
Salary	0.0058^c	0.0058^c	0.0060^c
Welfare dummies	Yes	Yes	Yes
Union dummy	Yes	Yes	Yes
Ownership dummies	Yes	Yes	Yes
Industry dummies	No	No	No
Province dummies	Yes	Yes	Yes
Constant	1.6680^c	1.6691^c	1.6629^c
R^2	0.0984	0.0986	0.0989
Adjusted R^2	0.0965	0.0967	0.0969
N	21975	21975	21975

Notes The same with Table 109.4

total input. In the regression model, we also centralize the variables of education, training investment and labor intensity to deal with the multicollinearity issue.

The estimation results are reported in Table 109.5. Specification (7) introduces the labor intensity into the regression. The estimation suggests that, industries with different labor intensities have no significant difference-son ROA. Estimations in column (8) and (9) demonstrate that the interaction term of labor intensity and educational level is significantly negative ($\gamma_1 = -0.0707$, $p < 0.05$). For that of labor intensity and training investment, the estimated coefficient is significantly positive ($\theta_1 = 2.9776$, $p < 0.05$). These results suggest that, education has a lager impact on firms in the lower labor-intensive industries, but for those firms in higher labor-intensive industries, the contribution of training to firm performance is greater. The hypothesis 4 has been supported.

109.5 Conclusions and Policy Suggestions

Using longitudinal data of 21975 manufacturing firms from China Industrial Enterprise Database, this study provides the empirical evidence on the returns of human capital investment in China's manufacturing firms. We obtain three major findings: Firstly, the educational level of employees has a significantly positive impact on firm performance, the higher the employees' educational level is, the higher firm's return on assets will be. Moreover, the correlation between firm's per capita training investment and firm performance shows an inversed "U" shape. The training investment is not always follow the rule of "the more the better". When a firm's training investment locates in the left side of the curve, investing in the training program can greatly increase the firm's return on assets. However, when the training investment exceeds a certain point, it will fall to the right side of the U curve. In that circumstance, the contribution of training to firm performance will be decreasing.

Secondly, we find educational level and training investment has a negative interaction effect on firm performance. When a firm's staff has a lower level of education, the training investment will have a greater contribution to the next year's ROA. One explanation to the finding is that, a great amount of manufacturing workers in China are subject to the budget constraint or have limit access to higher education resource. Under this circumstance, once the firm plays the role to offer them with opportunity to learn knowledge and skills, employees with lower educational level can quickly accumulate their human capital through training program, and finally achieve better performance.

Thirdly, we find that, the labor intensity of different industries plays a moderating role in the relationship between human capital investment and firm performance. When firms are in the industries with higher labor intensity, the contribution of employees' educational level to firm performance will be greater, which is largely because the non-labor-intensive job has a higher requirement of employee's knowledge, ability and other capacity developed in schooling, suggested by the capital-skill complementarity theory [18]. On the other side, when firms are in industries with higher labor intensity, the training investment to employees will be more effective, indicating that in high labor-intensive industries, such as textile and paper making, the skills and experience employees learn from the workplace is more important to the productivity.

Based on the above findings, we propose the following policy advice:

Firstly, higher-educated employee is a valuable asset for firms, especially those with college and above degrees. Secondly, firm's training investment is not "the more the better". Firms need to choose the best level of training adjusting to their own conditions. Thirdly, for firms in different industries, the human resource management operation should be different. For those in higher labor-intensive industries, firms should put more effort in providing formal training program while in the lower labor-intensive industries, firms should put more effort in attracting higher-educated employees.

References

1. Almeida R, Carneiro P (2009) The return to firm investments in human capital. Labour Econ 16(1):97–106
2. Barrett A, O'Connell PJ (2001) Does training generally work? the returns to in-company training. Ind Labor Relat Rev 54(3):647–662

3. Barron JM, Berger MC, Black DA (1997) How well do we measure training? J Labor Econ 15(3):507–528
4. Bartel AP (1994) Productivity gains from the implementation of employee training programs. Ind Relat A J Econ Soc 33(4):411–425
5. Becker B, Gerhart B (1996) The impact of human resource management on organizational performance: progress and prospects. Acad Manag J 39(4):779–801
6. Black SE, Lynch LM (1996) Human-capital investments and productivity. Am Econ Rev 86(2):263–267
7. Blundell R, Dearden L, Meghir C (1999) Human capital investment: the returns from education and training to the individual, the firm and the economy. Fiscal Stud 20(1):1–23
8. Chi W (2008) The role of human capital in china's economic development: review and new evidence. China Econ Rev 19(3):421–436
9. Chuliang L, Shi L (2007) The human capital, the characteristics and the inequality in income of industries. Manage World 1(10):005 (In Chinese)
10. Cohen WM, Levinthal DA (1990) Absorptive capacity: a new perspective on learning and innovation. Adm Sci Q 128–152
11. Datta DK, Guthrie JP, Wright PM (2005) Human resource management and labor productivity: does industry matter? Acad Manag J 48(1):135–145
12. Delaney JT, Huselid MA (1996) The impact of human resource management practices on perceptions of organizational performance. Acad Manag J 39(4):949–969
13. Ding S, Knight J (2009) Can the augmented solow model explain china's remarkable economic growth? a cross-country panel data analysis. J Comp Econ 37(3):432–452
14. Grant RM (1996) Toward a knowledge-based theory of the firm. Strateg Manag J 17:109–122
15. Harmon C, Walker I (1995) Estimates of the economic return to schooling for the united kingdom. Am Econ Rev 85(5):1278–1286
16. Huselid MA (1995) The impact of human resource management practices on turnover, productivity, and corporate financial performance. Acad Manag J 38(3):635–672
17. Ji L, Wei SJ (2013) Learning from an apparent surprise: when can strong labor protection improve productivity? NBER working papers
18. Krusell P, Ohanian LE et al (2000) Capital-skill complementarity and inequality: a macroeconomic analysis. Econometrica 68(5):1029–1053
19. Li X, Qian X, Chi W (2012) Can the identification for the professional qualification increase employee's salary? a case study on effect of the identification for the professional qualification on income. Manag World 9:100–110 (In Chinese)
20. Liu X, Zheng J (1998) The determinants of state-owned enterprises performance: evidence from 1985–1994. Econ Res 1:37–47 (In Chinese)
21. Lu X, Huang S (2009) Research on the effectiveness of intellectual capital in driving business performance- and empirical study based on manufacturing, it and real estate industries. Acc Res 2:66–75 (In Chinese)
22. O'Reilly CA (1980) Individuals and information overload in organizations: is more necessarily better? Acad Manag J 23(4):684–696
23. Pennings JM, Lee K, Van Witteloostuijn A (1998) Human capital, social capital, and firm dissolution. Acad Manag J 41(4):425–440
24. Schultz TW (1961) Investment in human capital. Am Econ Rev 51(1):1–17
25. Tharenou P, Saks AM, Moore C (2007) A review and critique of research on training and organizational-level outcomes. Hum Resour Manag Rev 17(3):251–273

Chapter 110
Analysis on Joint-Purchasing of Small and Medium-Sized Enterprises in Western China

Liming Zhang and Yating Wang

Abstract In an increasingly informational and global era, small and medium-sized enterprises should explore and practice new joint-purchasing model to prepare for the coming opportunities and challenges. This article builds the joint-purchasing model by analyzing inner and outer environment, sponsor, supplier of the purchasing organization. It provides one of the available opinions of purchasing model innovation for the small and medium-sized enterprises in western China.

Keywords Western of China · Small and medium-sized enterprises · Joint-purchasing

110.1 Introduction

Up to now, there is still no explicit and complete definition of consortium purchasing. But in some related document exist clues of Consortium purchasing theory. Aljian hold that purchasing process should make sure the company gain all the necessary goods, service, competence and knowledge, which is essential for the most favorable condition of the company's operation, maintenance and management. Harrison deems that supply chain is a functional network chain which implements the procurement of raw materials, converts them into intermediate and finished products, and sell finished products to users [1]. The above definitions reflect that purchasing is a critical part in supply chain management and that consortium purchasing is operated among homogeneous companies. Centralized procurement emphasize centralized procurement management within the Group or corporate while consortium purchasing is the purchase behavior of a number of alliances among enterprises. Therefore,

L. Zhang (✉) · Y. Wang
Business School, Sichuan University, Chengdu 610064, People's Republic of China
e-mail: zhangliming@vip.163.com

naturally, consortium purchasing is the further extension of centralized procurement [2]. The author holds that consortium purchasing refers to the purchasing behavior of multiple homogeneous companies that form alliances.

110.2 Macro Factors Analysis

Accompany with the Western Development Strategy execution, economy of the west achieved a great evolution but still fell far behind the East. From both international and domestic, the "25-year" provides great opportunities to the west to join international division and speed up the change of its development mode.

1. Economic Factors

   Enterprises, especially the small and medium-sized ones, as the micro carriers of economic development, their growth would be greatly influenced when the economic environment changes. With Western Development Strategy and entering into WTO since 1999 and 2001, China is opening to the world which helps enterprises take part in international competition.

   Limited by their development level and management power, small and medium-sized enterprises should find a way to establish a union among them. Financial Crisis in American in 2008 had brought this world a deep influence to its economy system. Meanwhile, it brought huge impact to import-export business of China. Competition between two enterprises is gradually taken place by that between supply chain and company union with the deepening of economy globalization. It's wise to find a more effective cooperation mode to improve competitiveness through assessing the situation, keeping flexible and responsive, following the market trend.

2. Politic Factors

   Stable politic environment is important guaranty for economic development. The former chainman Jiang mentioned twice in 1999: the condition has possessed and time is ripe for speeding up mid-west development. He first proposed "Western Development Strategy". The strategy was put into implementation in 2000, provinces, autonomous regions and municipalities in the west responded positively to it. The strategy proceeded in three phases: laying foundation, accelerating development, promoting modernization comprehensively. The key missions of laying foundation (2001–2010) are adjusting structure, improving infrastructure, perfecting market mechanism, cultivating industry growth engine. The key missions of accelerating development (2010–2030) are consolidating the foundation, fostering characteristic industry, promoting marketization, optimizing industrial structure. The key missions of promoting modernization comprehensively (2031–2050) are accelerating development of mountainous and pastoral regions, improving living standards of the west, narrowing the gap of the rich and the poor [3].

   In such period, government helped strengthen enterprises' union especially between small and medium-sized ones. Joint-purchasing among small and medium-sized enterprises plays full role in competition and also made a significant contribution to the development of western economy. To this, the government should advocate to build and improve enterprise credit system, by which some unfair competition such as "transshipment" would be forbidden and normal friction or conflict would be actively coordinated or solved.

3. Social and Cultural Factors

China is a big family composed of 56 ethnic groups, most of ethnic minorities live in the western region of the country. In Xinjiang, the Uighur accounted for 99.8 % of the national Uighur compatriots and uzbekistan, kazakhstan and other ethnic minorities could be find there. There are 45 ethnic minorities in Gansu and large number of Tibetan live there. Tibetan, Mongolian, hui, soil, salar ethnic minorities accounted for 42.1 % of the population in Qinghai province. Large number of the hui and almost 80 % of hereby. So in general, minorities in the western region account for over 70 % and minority species accounted for more than 80 % of the total number of nationwide [4].

Many minorities live together in the west, communication there will be affected by language, customs and habits. Purchasing, production, shipment and sales of enterprises especially small and medium-sized ones will be hindered due to the lack of understanding local culture, habits and customs and beliefs. Strengthening the exchange of culture, resources within the western regions, promoting common development are important foundation to improve the development of western economy and are also the direction of national policy, the inevitable developing trend of the international situation.

4. Scientific and Technical Factors

An Shuwei (2009), from Northwestern university, Research Center of the Economic Development of Western China hold that during the implementation of the Western Development Strategy, the scientific and technological achievements in the western region increased year by year, but compared with the developed eastern region, still less. According to national authoritative department statistics, in 2007, the number of patent grants and the numbers of Invention Patent Grants in the west were respectively 28,611 and 3,177, accounting for 10.08 and 10.08 % of the whole nation. Fewer scientific and technological achievements in Western regions is caused by unsound market system in the Western region, unable to protect the legitimate rights and interests of traders; intellectual property protection system is imperfect; lack of effective communication between enterprises and scientific research institutions, universities, and between regions, especially the less scientific and technological exchanges between the Eastern and Western areas [5].

During the "25-year", enterprises the western region should seize the opportunity to cooperate actively with scientific research institutes and Universities, speed up to build and perfect the technology innovation system which need to be enterprise-led, market-based and production, education and research-combined. At the same time, responding to the national policy, interacting with the eastern regions for learning, cooperation and promoting the competitiveness as soon as possible build a foundation for long-term development of enterprises. Small and medium-sized enterprises in the western regions play an important role in the economic development. The degree of technology innovation, speed of the informationization directly affects the market share and its competitiveness which indirectly affect the overall level of economic development in western China.

110.3 Internal Factors Analysis

Joint-purchasing is not only a purchasing strategy, but also the development strategy. The key to successfully implement the strategy are establishing joint-purchasing department, building and perfecting the platform, sharing purchasing resource and finally evaluating and optimizing the joint-purchasing model.

110.3.1 Partner Selection

Qian, White, Hao, Chen, Ling argued that small and medium-sized enterprises should buy the same specifications and standards of raw materials or the same series of products together from suppliers to have a high discount [6–9]. Xu put forward partners from joint-purchasing group could be either within the same industry or could be the similar. They set a particular purchase order together to gain discount [10]. Chen studying on the joint-purchasing mode of Jian De Jia Fang textile base in Zhejiang Province found there were two ways to select joint-purchasing partners. One was contacting with counterparts by purchasing department managers according to their personal working experience to see if it has a similar raw material purchasing list could be combined. The second was checking contact information of early cooperators to look for cooperation opportunities [11]. Integrated the above point of view, this article argues that partners of joint-purchasing should be neither within the same industry or be the similar. Small and medium-sized enterprises obtained information of partners (with production capacity, financial strength, social reputation, information technology, etc) who may have the same purchasing needs through industry association, government, university alumni association, reunion of the President (CEO).

110.3.2 Establishment of Joint-Purchasing Organization

The establishment of the joint-purchasing group is not the simply making their order together but the combination the recognized organization structure, responsibility division, purchasing principle, professional procurement staff, etc.

Joint-purchasing group could be established by industry association or by enterprises through reasonable and effective negotiation and responsibility division, or even by government or a third party. No matter in what form, stable organization structure and clear responsibility is a must. Joint-purchasing should always follow the principles of transparent, fair competition, honest and trustworthy. For managers and workers, should not only be familiar with professional knowledge, loyalty, integrity, dedication, patience, objective and fair, but also have the ability of analysis, judgment, forecast, communication and coordination and the sense of team work.

The structure is determined by industries, character of the goods and materials, specifications. There is not the only type of structure. Therefore, specific forms of organizational structure should be decided by actual situation of the cooperative enterprise which premised on low price and fine quality [12].

110.3.3 The Perfection of Joint-Purchasing Platform

In the early stage of Joint-purchasing, division of responsibilities and the purchasing process management were more perfect, but in actual operation, unexpected events will still appear unavoidable and without prediction. On the other side, it is the existence of these events, will enable the joint procurement platform continue to mature and improve. In today's supply chain management and information throughout the global era, part of small and medium-sized enterprises established their own ERP or MRP system, to ensure that enterprises in the procurement, production, transportation, sales process efficient operation, the informatization level increases. Therefore, enterprises in the transfer of the material purchase list to the joint procurement organizations are uniform use of joint procurement platform, in order to implement joint procurement. Good joint procurement organizations should fully consider the characteristics of small and medium-sized enterprises, the rapid reaction mechanism. For example, different requirements for the same raw material of different small and medium enterprises delivery time, this requires the joint procurement organizations in the procurement and supplier agreement is good, reasonable arrangement of the delivery time, in order to ensure the normal operation plan of small and medium sized enterprises. In addition, because of the small and medium-sized enterprise flexibility is relatively large, because the other event happens suddenly, may temporarily change the purchase orders, emergency measures which requires joint trading platform has the necessary, and maximize the reduction of small and medium enterprises loss [13].

110.3.4 Evaluation and Optimization of Joint-Purchasing

Joint-purchasing organization of small and medium-sized enterprises in the western region with suppliers as agent, has a dual identity, it is the small and medium-sized enterprise supplier and product or service supplier buyer. Therefore, small and medium-sized enterprises in the western region to evaluate the joint procurement organizations in, is the joint procurement organization and suppliers as a whole evaluation, if the final evaluation results than the small and medium-sized enterprise alone when making a purchase, they will continue to participate in, support joint procurement; if the final evaluation results inferior to the individual purchase of small and medium-sized enterprises. The evaluation results, will give up to participate in joint purchasing organization, eventually led to the collapse of joint purchasing organization.

The western region of small and medium-sized enterprises in the evaluation of the procurement, focus is the quality, the price of purchased products, delivery on time, patency of information transfer, joint procurement organization service quality etc. Therefore, the western region of small and medium-sized enterprises in the joint purchase, must maintain the performance evaluation. This paper suggest "5S" evaluation principle as the center, the appropriate price, timely, appropriate and suitable, essence, quantitative index for evaluating scale.

Price index of performance has important practical significance for enterprises to reduce costs, mainly by the ratio of actual price and standard purchase price difference and the purchase price and the base of purchase price ratio with the current price index with base price index comparison.

Time performance indicators, to measure the efficiency of procurement, supplier alliance deal with the delivery date on time.The delay will affect the normal production planning of small and medium-sized enterprises, causing shortage phenomenon; advance the time of delivery will cause unnecessary inventory and related expenses. Among them, the delay in delivery index = delay in delivery times/total purchase frequency; advance the time of delivery times = early delivery number/total purchase frequency.

Quality performance indicators, generally through the purchase of qualified acceptance rate index and in-process acceptance judging index, the former mainly refers to the acceptance of the products in the supplier delivery, the latter mainly refers to the percentage of unqualified products found in the production process. Among them, the purchase of qualified acceptance rate = qualified number/total number of purchases; product qualified acceptance rate = actual available amount/purchase quantity.

Quantitative performance indicators for evaluation, purchase amount is appropriate, avoid excessive inventory, has led, waste status report, assessment mainly through cost index and report the loss index of storage and waste disposal. Among them, storage cost index = stock interest + existing inventory carrying cost − normal inventory standard interest − normal inventory carrying cost; to stay and scrap loss index = processing ailiao and waste income − to stay and scrap purchasing and storage cost.

Delivery performance indicators, mainly refers to the actual delivery and reasonable requirements of delivery distance multiplied by the local, the transportation costs (labor costs, transportation costs) assessment.

Evaluation of joint procurement organization of small and medium-sized enterprises are not limited to the above five indicators, and the evaluation index of other organizations such as the ability of technology, management and coordination ability, training services etc. Joint procurement organization received the small and medium-sized enterprise evaluation information, timely adjustment and improvement according to the corresponding index feedback information of the original work or deficiencies, and to improve the quality and efficiency of services for SMEs.

110.4 Main Organization Body of Joint Procurement

Joint procurement organization initiator has many kinds, Chen [11, 14] in the foreign large enterprises joint procurement on the basis of the model, put forward by the industry associations, leading the enterprises set up the third party alliance and operation joint procurement form [11, 15]. Xu points out that the main sponsor of joint purchasing in addition to the above three, can the government initiated the formation [10]. Xin think, joint procurement organizations can be members of other enterprises entrust one has special advantages of enterprises to purchase, also can be the member enterprise cooperation alliance to purchase, also can be a member enterprises jointly invest to establish a separate purchasing organization to purchase [16]. The scholars at home and abroad and the practice of joint procurement, this paper think the mainstay of the medium and small-sized enterprises in western region joint procurement organization can be summarized into three categories, namely the spontaneous organization formed among small and medium enterprises joint procurement organization, industry association leading coalition purchasing organization and the third party organization set up joint purchasing organization.

110.4.1 Spontaneous Formation

The spontaneous organization of small and medium-sized enterprises in the West form way is to entrust, agreement or joint venture sourcing company through ways such as between small and medium-sized enterprises refers to the western region, joint procurement organization form and form for the common goal. Members of the enterprise joint procurement organization largely will be added to the procurement activities, together with information collection, the vendor selection and procurement activities of the control.

But the disadvantages of the joint procurement mode of small and medium-sized enterprises (SMEs) are the lack of effective supervision and rewards and punishment mechanism in the process of joint procurement as well as deficiency of a better checks and balances method. This will bring huge uncertainty in the actual purchasing process, which is likely to lead to the collapse of the joint procurement organization.

As a result, the joint procurement mode is suitable for those with specific, definite purchase target or SMEs which have a certain appeal in the industry with peers or SMEs at the same stage in the western region [11].

110.4.2 Industrial Association

Industrial association has strong regional characteristics and is a non-profit and very professional organization. It consists of most large, medium and small enterprises in the industry within this area. Meanwhile it has the extremely high professionalism in the industry, master the latest trends of the industry, and comprehensive market information and be able to timely grasp the direction of government's policy and conduct effective cooperation with domestic and foreign multinationals or regional group, and then can provide members with the latest and the most valuable information, as well as offer more development opportunities and platform, playing a role of leading the escort for the regional industry development, so the Industrial association has lots of advantages which individual SMEs don't have. The joint procurement organization of SMEs led by industrial association in the western region has a good reputation and stable strength, can maintain a long-term cooperation relationship with material suppliers, and have a stronger bargaining advantage, which can make the enterprise purchase goods or service with lower cost. In addition the joint procurement organization can better regulate and manage the friction and conflict between enterprises in real purchasing process through industrial association which would definitely enhance the effectiveness and timeliness of management.

Therefore, the joint procurement organization mode is suitable for newly established SMEs in western region. Because at this stage, western SMEs have little power, and are lack of comprehensive understanding for other enterprises, industry and market, at the same time, other companies or suppliers will have doubts upon their financial strength and integrity. These western SMEs have even no bargaining power when purchasing, so price is very high; when they encounter capital shortages, there is lack of effective guarantor, so suppliers will not ensure the sufficient supply of goods for them. In order to ensure they can get stable supplies at a lower price, the joint procurement organization led by industrial association is the ideal choice [15].

110.4.3 Third-Party Association

The body of joint procurement organization set up by the third-party association can be dominated by the government, and also can be a specialized profit organization. The third party organization set up joint procurement organization,all the members are not directly involved in procurement activities, undertaken by the third party [17]. Joint procurement organization of SMEs led by the government is strongly related with policy. If SMEs involved in could not fully display their initiative, the procurement organization will eventually become a mere formality and can't play a real role. If the body of joint procurement organizations the procurement service providers (profit organization), through the collection of many small SMEs orders, the scale advantages of procurement will greatly reduce purchase and circulation costs by directly trading with product manufacturer or service provider. At the same time, in order to maintain its reputation, win the trust of SMEs in the society and attract more SMEs in the western region to join and maintain a good lasting relationship with supplier, the procurement service providers (profit organization) will definitely take the initiative to guarantee for SMEs and suppliers. The main way is to ensure the quality, price of the products, and delivery on time, stable supplies, ensure payment in a timely manner and convenience. In addition, as a professional procurement third-party organization can provide management and decision-making reference for SMEs to ensure their scientific decisions.

Therefore, this joint procurement organization is more suitable for those SMEs in western China who already have a certain scale, and certain market competitiveness and want to further enhance their core competitiveness, and be developed into a large enterprise. At this stage, they will pay more attention to professional procurement of goods and service. If the companies don't have the professional procurement ability, the joint procurement organization organized a third-party will be the best choice [11].

110.5 Building Joint-Purchasing Model

Based on the above analysis, this paper set politic, economy, culture, technology as given viriables. Public information is shared by enterprises who want to have a place in the market. Therefore, joint-purchasing system in this article is composed of small and medium-sized enterprises (purchasing entity), purchasing organization (spontaneous group/leading industry association or third party) and suppliers. The three parts of the system are connected and combined by cash flow, logistics and information, as shown in Fig. 110.1 [14].

As shown in Fig. 110.1, enterprises send their purchasing list to the organization through the information system. Workers from the organization collect the list to classify and summary and then post message to outer suppliers. After comprehensive comparison, the most suitable suppliers will be chosen. Finally, detail information such as quality, quantity, delivery time, place will be transferred to supplier and at the same time buyers will be informed of price to pay in advance or on time after receipt of the goods (payment methods depend on the type of goods or on the negotiation). It's pointed that the model of joint-purchasing system should operate on a perfect information system which can not only collect but also sort and analyze information to provide basis for enterprise of decision-making. Joint-purchasing model enables enterprises to buy goods or service at a lower cost through information sharing which also help reduce management risk and improve competitiveness.

It can be drawn from the Fig. 110.1, elements of joint procurement organization mode of small and medium enterprises in the western region has three: Western Small and medium-sized enterprise, joint procurement, supplier.

(1) Western Small and Medium-sized Enterprise

Procurement is the foundation and necessary precondition for company's production and operation. The extensively applied purchasing model today cannot adapt to the needs of the development of the small and medium companies in the west part

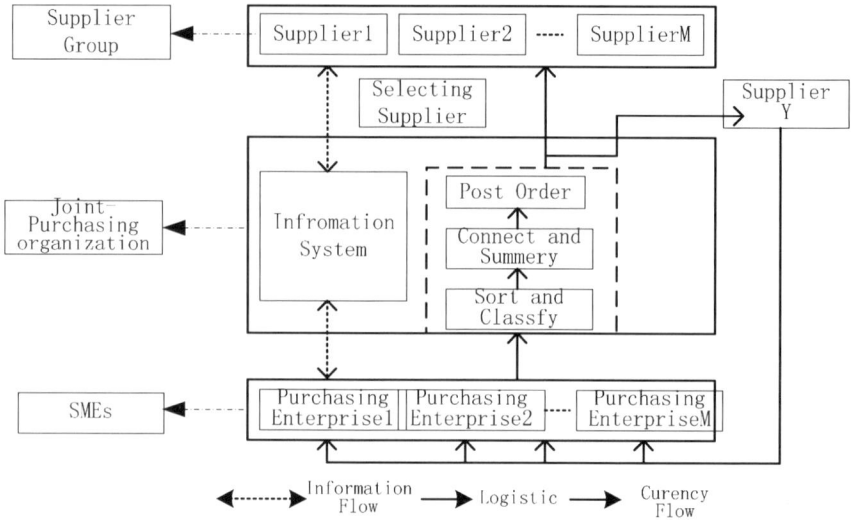

Fig. 110.1 Joint-purchasing model of SMEs

of China. As a new and effective model, consortium purchasing is just favorable to their development. The small and medium companies in the west part of China is the party in request, i.e., the buyer. Their production type and business scopes are different, as well as their scales. Besides their ability to bargain is weak. Consortium purchasing can bring down the purchasing cost. They can spare more capital, materials and human resources and invest them to improving the innovation on technology and sale to further strengthen the core competitiveness.

(2) Joint Procurement

Joint procurement is the crucial element of consortium purchasing. It is the core link, which integrates the purchase needs of homogeneous small and medium companies in the west part of China. Through this way, it sends an integrated purchasing message to the supplier and manages to get a lower price by scale purchase. In the meantime, it offers more purchasing services for the small and medium companies in the west part of China. Chart 2 demonstrates that the information advantage of consortium purchasing organizations. Consortium purchasing organizations are established in three ways. They can be organized spontaneously, organized by the industry association, or organized by a third party. But the operation of the three types are different.

(3) Supplier

Supplier mentioned here is that who directly offer products and service to SMEs or the joint procurement of SMEs and we do not discuss suppliers for a second round of joint procurement. The principle to choose suppliers for joint procurement organization is to only select suppliers who could provide optimal integrated conditions and does not distinguish between supplier's geographic location and property and the size of the ownership of enterprises. Practice has proved that as long as to ensure that the purchase orders continue to increase, it will continue to win the supplier's attention. So, SMEs in the western region through joint procurement organization could attract more attention from suppliers, reduce procurement cost, at the same time the long-term strategic cooperative relationship between suppliers and SMEs joint procurement organization could be established (joint making purchasing and supply plan, actively providing information about raw materials, or even achieving joint participation in the technical development of raw materials), which can guarantee the stability of supplier's funding cycle.

110.6 Conclusion

The present paper discusses 4 aspects of consortium purchasing, the macro-economic elements, micro-economic elements, the subject organizations and the suppliers. Besides, the present paper establish the mode of consortium purchasing in small and medium companies in the west part of China. In this mode, consortium purchasing organizations are the critical part. These companies should reasonably select the subject organizations of consortium purchasing according to their own condition (extra or intra conditions). The present paper, to some extent, helps companies clarify whether they are suitable to consortium purchasing and offers them a new purchase option. This new mode in the present paper is applicable widely. However, it lacks detailed discussion for the corresponding systems of different consortium purchasing organizations. Suggestions are proposed for further research and improvement.

Acknowledgments This research is supported by the Central College Topic: The condition of company brand and strategic choice under uncertainty relations research (Grand No. SKqy201209).

References

1. Harland C (1997) Supply chain operational performance roles. Integr Manuf Syst 8(2):70–75
2. Yang NM, Cao XH (2001) Joint enterprise procurement mechanism research. Aviat Maintenance Eng 2:33–35 (in Chinese)
3. The three-phase of the western development strategy. The Tenth Anniversary of the Western Development, [EB/OL]. http://www.xinhuanet.com/politics/xbdkf10zn/ (in Chinese)
4. Jie XW, Cao XH (2006) Joint enterprise procurement mechanism research. Southwestern University of Finance and Economics Press (in Chinese)
5. Yao HQ, Ren ZZ (2009) Report of economy development of western China. Social Science Press, Beijing, pp 15–16 (in Chinese)
6. Chen JM, LV L, (2008) Research on model and operation of jonit-purchasing union of SMEs. Chin For Econ 1:19–20 (in Chinese)
7. Hao YX, Cao XH (2003) Modern logistic purchasing management. Zhongshan University Press, Guangzhou
8. Ling HKR (2010) Research on purchasing union operation of smes. Technol Dev Enterprises 11:80–81 (in Chinese)

9. White S, Lui SY (2000) Interaction costs: the cooperative side of an internal tension in alliances. In: The Asia academy of management conference, Singapore
10. Xu ZJ (2010) Discuss on joint-purchasing of small and medium sized enterprises. SMEs Manage Technol 11:31 (in Chinese)
11. Chen JM (2006) Research on purchasing system of SMEs-discuss on purchasing mode of jiande textile base in Zhejiang province. Nanjing Forestry University, pp 29–33 (in Chinese)
12. Han GJ (2007) Purchasing management. Capital University of Economics and Business Press, Beijing (in Chinese)
13. China Enterprises Management Association (2000) Enterprise management. Economy and Science Press, Beijing (in Chinese)
14. Chen YQ (2008) Research on purchasing strategy of SMEs based on SCM. Zhejiang Industrial University, pp 26–28 (in Chinese)
15. Chen YQ (2008) Research on purchasing strategy of SMEs based on SCM. Zhejiang Industrial University, pp 26–28 (in Chinese)
16. Xin LG (2008) Research on multiple strategies of joint-purchasing. Central South University, pp 35–36 (in Chinese)
17. Nollet J, Beaulieu M (2005) Should an organisation join a purchasing group? Supply Chain Manage Int J 10(1):11–17

Chapter 111
The Case Study on the Cultural Identity in Virtual Community

Jing Yang and Han Li

Abstract The rapid growth of the internet has been influencing people's daily lives deeply and widely. As one of the basic human activity organizations, the social community also has made a profound change. Virtual community, which has been created by it, is a new model of social community. Twenty-first century is an era of culture management. With the development of globalization and informatization, people pay more attention to the cultural identity in virtual community. Starting from the cognition and emotional dimension of cultural identity, this study is based on the network interpersonal interaction. Through the research of the specific virtual community "Exempt Exam Ban", it finds out that the participants of the virtual community establish their virtual identity and cultural identity through interactive symbol displaying and interaction. Meanwhile, after a long period of information sharing and interpersonal communication, the participants have formed the emotional attachment and identification.

Keywords Virtual community · Cultural identity · Interpersonal interaction

111.1 Introduction

A new organization of human society and survival mode caused by internet comes into our lives quietly. It has built a huge living space which is beyond time and space. Different from traditional communities, a new human living community has been formed which is called virtual community. This particular internet interactive mode and social relationships are different from reality. Besides, it has special form and function. The new community has dissociated space-time conception in reality

J. Yang (✉) · H. Li
School of Political Science and Public Administration, University of Electronic Science
and Technology of China, Chengdu 610054, People's Republic of China
e-mail: yangjin3720@sina.com

J. Xu et al. (eds.), *Proceedings of the Eighth International Conference on Management Science and Engineering Management*, Advances in Intelligent Systems and Computing 281, DOI: 10.1007/978-3-642-55122-2_111, © Springer-Verlag Berlin Heidelberg 2014

through a virtual space created by information technology, accompanied with participants who have common interests and actively interact [1]. It has formed many groups and each has certain identity and relative stability. Formally, it is a community form that participants participate in it freely with interests. Internally, it is a cultural community which has psychological contracts and organizational norms.

This study is trying to make an empirical research on the specific virtual community, then it inspecting essential relationships of interpersonal interaction in virtual community and characteristics of cultural identity in social networks which has been formed based on it. Meanwhile, the paper analyzes a unique scene which has been taken by interpersonal relationships in a virtual community (The community is characterized by body absent). Finally, it explores the influence on human society which is caused by the internal cultural identity in the new community mode.

111.2 Theoretical Bases and Research Hypothesis

Interpersonal interaction is a significant way for our living and developing. Virtual community is a platform for people from all concerns of world to get together. It can supply not only the discussing about information and opinions, but also interpersonal interaction. Based on the interpersonal interaction theory, this paper hypothecate hypothesis from two aspects of cultural identity: cognition and emotion.

1. Theoretical Basis

This research suggests that interpersonal interaction theory can interpret the cultural identity in virtual community deeply and accurately. The theory holds the position that obtaining information is the primary motivation for interpersonal interaction. Secondly, establishing social cooperation is an important motivation for it. If a person wants to make an efficient social cooperation, the basic premise is the person should understand both him and others, at the same time, he also makes himself understood. Therefore, the theory states that the third basic motivation is self-cognition and mutual cognition, and identification in virtual community which has been formed based on these. It does not mean that interaction must bring about community relationships. Sallette, an interpersonal interaction scholar, asserts that an interaction will bring about a community when it under certain situations: It should help participants develop themselves and pursue individual identities, and it should preserve community and establish common values and norms of behavior among participants [2].

2. Research Hypothesis

According to the interpersonal interaction theory, we hypothecate hypothesis from two aspects of cultural identity: cognition and emotion.
Hypothesis 1. In virtual community, the interactions among participants could promote self-identity, help participants establish individual identities and improve cultural identity.

Hypothesis 2. In virtual community, faithfulness and contribution could improve cultural identity through strengthening participants' emotional recognition.

111.3 Case Analyses

Starting from the cognition and emotional dimension of cultural identity, this study is based on the network interpersonal interaction. Through the research of the specific virtual community "Exempt Exam Ban", it inspects essential relationships of interpersonal interaction in virtual community and characteristics of cultural identity in social networks which has been formed based on it, and explores the influence on human society which is caused by the internal cultural identity in the new community mode.

1. Case Introduction and Variables

This paper quotes from a case of "Exempt Exam Ban" (http://www.Eeban.com/forum.php). It is a platform which is built according to key words, exempt exam (postgraduate recommendation). There are 191,535 registered users and 774,245 posts up to June 6th, 2013. Opened on February 15th, 2009, EEBan focused on imparting experience and sharing information for postgraduate recommendation. It provides information, such as how to contract a tutor, ranks of postgraduate recommendation, should I go to other school or stay own, how to prepare materials of recommendation, how to achieve a good result in second test, departments in other schools and institutes, and etc. By these means, it offers professional information for Chinese students to be recommended successfully. In EEBan, EEers (used to address participants in EEBan) from talking about recommendation and politics to communicating and confiding emotion, they not only help each other warmly, but also discuss rapidly and heatedly. Through characters, expressional symbols, videos, pictures, and etc, EEers display a vivid spectacle and sketches an intricate relationship picture in virtual community.

According to the hypothesis, we choose vitality to measure the self-identity of EEbers, and use loyalty and contribution to measure the emotional identity to organization.

2. Analysis of Cultural Identity in EEBan

In EEBan, cultural identity involves various aspects. This paper makes a concrete analysis from self-identity and emotional identity.

(1) Self-identity in Interaction

In a community, having its common traits is important for positioning identity. Each individual could not obtain a sense of belonging unless they have community traits in the socialization process. Community identity is an identification for common images, it solves a problem "who are we" and helps individual answer a belonging question "where am I" [3].

To individuals, virtual community provides a virtual platform that they could show their values; at the same time, it is an ideal place for them to establish self-identity. These are the values of virtual community. In EEBan, participants show themselves and understand others by special function. According to these, they could establish self-identity and groups' cultural identity.

Interviewee A: In EEBan, people pay more attention to "domain", they visit it mutually. I am famous in it. There are two kinds of posts: digest posts and water posts. Replies are usually numbered. Someone would reply and comment when I posted here. This is a strong interaction. My posts have written by myself, also my friends'. Sometimes, although a new post will be posted after half a month, we would comment in detail. By this way, we could have more communication of the mind.

Interviewee B: When I have posted here, I would often see the replies. If there is a reply, I would be happy. There are many restrictions in EEBan, such as how many posts should be posted, how long does it take to level up, how long does it take to possess a privilege. Therefore, when I have upgraded in a higher-level, it warmed the cockles of my heart.

According to the interviewees' answers, we can conclude that all initiators hope to get more replies and read them repeatedly. Most initiators may have a feeling of satisfaction which is based on others' identification. Other participants' reactions will influence initiators' participation more or less, maybe they are more enthusiastic, or less interests. It is very rare that initiators are completely indifferent to others' reactions and comments. It reflects that people need to search and establish self-subjectivity from the eyes of others. To EEers, especially actives, they need to an approved satisfaction from the eyes of others. According to this, they could intensify and confirm their images. The participants who often express their opinions and views are the leaders of opinions in community. Due to have more channels of information, they get news first; at the same time, they have an ability of independent thinking and could give their views and promote the related discussion about all subjects in EEBan. They also make more participants discover things they have never heard.

Interviewee C: In EEBan, I have not only communicated postgraduate recommendation information with other EEers in public boards, but also participated in the branch of Wuhan and Wuhan University groups. The reason is simple: I observed the rules and participated in groups based on areas and schools. I want to participate in the community by this way so as to pursuit self-worth.

EEBan consists of many boards, the functions of them have embodied people's richness and variety through participating in particular boards (Fig. 111.1). To be on a board is equivalent to pronouncing that your identity is same as others', it is also equivalent to having a label on themselves. This action is a part of showing personality. Conversely, boards play an important role in judging identity. Mediated by interests, the functions of boards relay on same interests and similar background of knowledge structure, gather participants who have common interests but could not meet at ordinary times and build a highly cultural identity in community on the basis of self-identity.

Interviewee D: When I participated in EEBan, my bonus was 0. According to the rules, I was a pupil. I communicated with other EEers actively for getting highly reading permissions and privileges. There are many effective posts, they are long but excellent. Commentators' ideas are quite original and words are almost elegant. The atmosphere is nice. Participants would post when they have had a careful thinking. There is a little water post. Replies are earnest, even some of them are longer than original posts. They are all EEers' thinking and feeling. I have been an academician now. For this title, I would communicate with others more actively. Because of the title, others would like to consult me. It gives me a great satisfaction.

The participants, including pupils, middle school students, university students, masters, doctors, lecturers, professors, doctoral supervisors and academicians, identify the community culture. Different EEers enjoy different rights as their identities. In a serious of communications and interactions, EEers identify their virtual roles: little talent and less learning, or a mind enriched by books. Participants will be gathered together by different virtual roles. When EEers are constructing and identifying by themselves, they establish senses of belonging and identification for culture in virtual community.

1#EE lecturer: logonku—[experience and data] [postgraduate recommendation composition in 2011] Experience and suggestions of being recommended to the master program at PKU exempt for examinations. I hope it will be helpful to you! (Original)

Postgraduate recommendation, a confused experience came to an end last Saturday. I have to be admitted into Ph.D. program in Institute of Physical Electronics, school of electronics engineering and computer science, Peking University. So far I have often mooned. I want to look back my experiences and give some suggestions for future students. Thank to EEBan, I want to express my grateful to all people who have helped me.

[Summary]

First, I will introduce basic information of mine. I am graduated in a 211, 985 university. Majoring in micro-electronics, it is relatively weak in this school. My credits ranking is 6/30, it is not forward. Integrative ranking is 4/30. I have got the dual-degree, finance. The marks over 80 a little, it is not good enough. I have passed the CET4 and CET6, the results were good. I have not any experience of research, only a little course design. I have not published papers and won in contests. I am an ordinary student. Believing in yourselves, not only the great students could be admitted into Peking University and Tsinghua University. We should try our best. I have not gotten a quota to other schools, only to be admitted into PhD program in Institute of Physical Electronics, school of electronics engineering and computer science, Peking University. The long process could be divided into several stages: frustrated in Summer Camp, met various problems in contacting with tutors during holidays, strove for a quota confusedly and succeeded in the interview.

2#EE doctor: Ning Xi—Thanking the master's sharing. When I looked up the documents of Ministry of Education, I found that dual-degree could not be admitted in CHSI. Is that right? I have thought about it recently.

3#EE master: xjj188—The master is a good man, I should learn from you. Thank you. If I succeed in postgraduate recommendation next year, I will share my experience like you.

    ...

37#EE university students: CNTs—Congratulations! It is a good example proves that human effort is the decisive factor!

It is an original post in EEBan, EE lecturer logonku has shared and innovated information with EEers by interaction. Participants fulfilled themselves in it. Although the EEBan has not been built based on completing a specific task, there is a central area in community. Participants could communicate, discuss and share experience according to a common theme. In a discussion, professional participants join us easily. They provide cases, response or different perspectives and make it possible for more participants to understand them. In EEBan, the virtual community becomes a platform for sharing information. It is not restricted by time and gathers various information easily. In a virtual community, different participants have different experience, so everyone has unique view to share. Participants could develop the field of vision and have deep consideration based on others' views; at the same time, they integrate others' views in their thoughts.

According to the study on EEBan, we found that choice of cultural medium has embodied the cultural ability and cultural identity of individual. EEers identify the cultural significance which is embodied by EEBan, and they could get along well with it. Compared with Renren and Pengyou, EEBan possesses more token capital. It conveys information that virtual community is a small group which is unique and reflects its individuality. They all reflect the cultural identity in virtual community.

(2) Emotional Identity in Interaction

In EEBan, analysis of emotional identity in interaction includes three aspects, participants' leave rate, participants' original rate and theme activities.

a. Analysis of Participants' Leave Rate in EEBan

In "Individual SHOW", 100 IDs were taken at random and numbered from 1 to 100. Then, we observed them for three months (from February 1st, 2012 to May 1st, 2012). During three months, if an ID has not any activity, we acquiesced he has left. We could study the leave rate by this way [4]. We have been observing them for three months, and here is what we got. The sampling survey shows, IDs who have 50–100 interaction activities were most in "individual SHOW", which is about 43 per. During three months, ID who has not any interaction activity only about 3 per and who removed by himself about 1 per, or leave rate is 4 per. We consider that the lower leave rate, the higher emotional identity in community.

b. Analysis of Participants' Original Rate in EEBan

Different from reprinting, most of posts in EEBan have been created by EEers. By May of 2012, there were more than 800,000 posts in EEBan. Over 1,000 posts have been posted every day, and more than 80 per were original. The numbers of replies are as much as 10,000. Such abundant originals originated from internet elites who are very creative. These users settled in EEBan permanently, they have high loyalty and identity to it. It is the core of advantages that other Web sites have not. At present, homogeneity is a trend on the internet. EEBan attracts new and old users by its abundant and high quality originals, they create great values through integrating and editing online and offline.

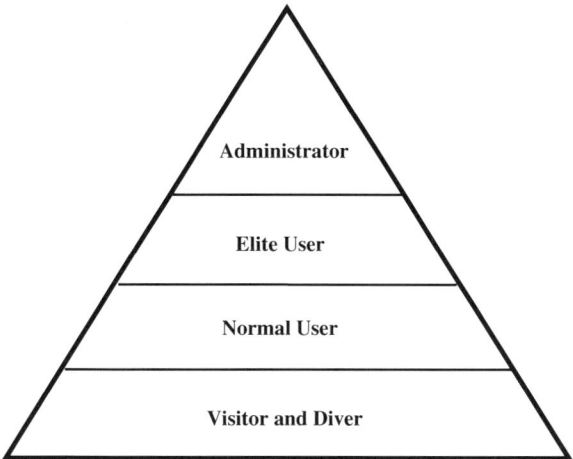

Fig. 111.1 Structure of internet BBS user

c. Analysis of Participants' Original Rate in EEBan

Interviewee E: I make a lot of friends in EEBan, we are friendly. I miss the days that we talked about postgraduate recommendation. When it was all over, I often miss them. Because I would like to know what they are doing, I often go to EEBan to see what happened and read my posts or replies.

Interviewee F: We have known each other for years. Beginning with postgraduate recommendation, we have talked about many topics, from parents to teachers and studying to working. We would like to talk together if we have problems. If there are happy things or bad, we would like to share with them at first time. In the end, we became friends according to recommendation. We begin to having parties if we are in the same city.

Interviewee G: Sister Maizi, I received the offer of Fudan University yesterday, I was very happy and sent an E-mail to you. Such a long time, we first to meet each other, then to know each other, then cling on to each other. As you said, we did not participate in it until you have communicated with us. We have special feeling to you, and we could not forget it. Even you left, we would not forget you!

These statements express that EEBan is used to not only exchanging information and sharing knowledge, but also helping participants communicate feelings and establishing internet relationships. Besides exchanging information and discussing together, participants also know each other and speak informally in virtual community. Therefore, social interaction has been established among participants (Fig. 111.2).

According to interaction, participants establish a sense of trust and identity of value in community. Based on it, virtual community is filled with humanization. Meanwhile, participants could meet in reality, not virtual. It promotes confidence and emotion between them, thus strengthens cultural identity.

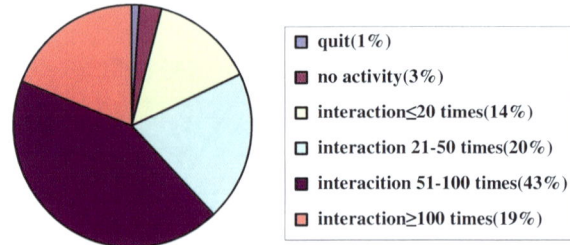

Fig. 111.2 Interaction statistics of individual show group in EEban

Table 111.1 Research vitalities and indexes of cultural identity

Concept	Variable	Index
Cognition (Self-identity)	Vitality	Log in frequency
Emotion (Emotional identity)	Loyalty	Leave rate
		Monoid consciousness
	Contribution	Original net paste
		Theme activities
		Post/reply frequency

1# master of board safely—EEBan Anniversary T-shirt Reservation Announcement, 2012

Did you remember the happiness when you received the T-shirt last year? Did you remember the excitement when you showed the T-shirt? Did you remember the T-shirts which were imitated valentine's shirts? Did you remember what we witnessed with it?

This year!! 2012!!

For boards party, graduation and opening ceremony, EE T-shirts lasting happiness and moving!

In senior year, we will show our T-shirts as them!

Remembering the meeting with you, remembering the moving because of you...

EE T-shirts, it is to be a witness of our common experience!

Let life remember our unique Every Effort!

For commemorating EEBan's growth and development in 2012, safely, the master of board launched an activity of reserving T-shirts again. Reserving T-shirts is a manifestation of participants' emotional identity. Only identifying the community culture could actively participate in the activity and have had the T-shirts which witness the emotions of EEers. It reveals the group identification which is established by participants, and it also shows the common views and emotions which are built based on cognition and appraisal by participants in community. That is to say, according to common believes, emotions, demands and benefits, participants have the same cognitions and values, and they should be consistent by themselves in community.

Table 111.2 Net paste statistics of eight elite members for 1 month

ID	Post number	Reply number	ID	Post number	Reply number
Warm blood ice-man	15	1721	Autumn peach	21	2055
Scu619794890	7	577	Wind bean	5	163
Flysky754	3	213	Dream in cloud	5	380
Slofhust	9	867	Sishisiren	12	1478

Table 111.3 Attraction rank of thread in EEBan

Thread type	Attractive title	Many replying thread	High click rate	High quality	Newly updated	Introduced by master	Stick post	Active writer	High-level writer	Other
Rate %	52.6	50.2	46	45.6	37.8	37.7	36.6	26.6	21.3	0.2

111.4 Conclusions and Discussion

Based on self-identity and emotional identity, this paper makes a concrete analysis and concludes that interaction of participants in virtual community could improve its self-satisfaction and cultural identity (Tables 111.1, 111.2 and 111.3).

111.4.1 Conclusions

According to the analysis of user architecture and self-identity in EEBan, this paper has demonstrated the research hypothesis. It has confirmed that the interaction of participants in virtual community could improve its self-satisfaction and help them establish individual identity in the community. It will then raise more approval to the community culture. Through the analysis of leave rate, original theme and emotional identification in EEBan, it has further demonstrated the loyalty and contribution by participants, which could strengthen emotional identity of participants to the community, and then, raise cultural identity.

111.4.2 Discussion

Through the analysis of actual case, this issue could be explored further from two aspects, self-identity and emotional identity in internet interaction.

(1) Discussion about Self-Identity in Internet Interaction

In EEBan, participants' internet interaction activities, including posting, replying and commenting, have shown that internet interaction has at least 3 significant aspects to establish self-identity. First of all, participants' internet interactions influence the formation of virtual self-identity. Virtual community provides opportunities for participants to express themselves on internet. Participants have established virtual identities through posting and replying, signing on posts, humorous quoting behind posts, and etc. Others know and understand master's views and personality through these posts. This basic interaction is a central form in virtual community. The policy optimization and interaction platform of virtual community will enable interaction more easily, and attract more participants in order to raise connections with each other [5]. They win identity and position through interaction. Community participants finally enhance a sense of duty and prefer to fulfill upon their roles in these activities.

Secondly, participants developed themselves in virtual community interaction. The anonymity of internet provides an equal, free and opening environment for netizens who are restrained by real identity. Under the IP network and online nickname, netizens have a psychology to evade reality. According to a variety of self-expression, they rebuild themselves different from reality in order to possess another identities and personalities, and then get self-satisfaction from them. In virtual community, self-construction could be achieved through communicating and emotionally connecting with others. Participants realize self-construction through the theme of virtual community, and the self-construction includes expressing themselves, sharing experience and connecting emotion [6].

Thirdly, individuals have established culture identity in virtual community interaction. The essence of virtual community is social relationships. Besides the functional requirements as obtaining information and contents, participants need to social supports through participating in a community, and then establish communication with others and groups' awareness, it is the most important. On the one hand, virtual community culture is in the nature of group homogeneity. The formation of virtual community is based on common experience. For a virtual community, it is built based on similar conception of community theme [7]. On the other hand, virtual community culture has exclusiveness. Though the boundary of virtual community is vague and dynamic, it also could rule out "alien" effectively. The exclusiveness will better determine the extension of internal cultural identity in community.

(2) Discussion about Emotional Identity in Internet Interaction

Under EEBan's atmosphere, we could conclude that emotional relationship is important in virtual community. Scholar Rheingold has raised the conception of "virtual community" [8], he asserts if enough people have put enough feeling in virtual space and last long enough, then the virtual community is coming.

That is to say, as a "society", virtual community needs to interpersonal emotion for social cohesion. Internet creates a new virtual environment by its own way and rich contents. Going into internet space will be like going into an immense sea with information, knowledge and interesting numbers. It stimulates people's heads, attracts the human mind and makes people have feelings for it. People's emotion to internet provides an emotional foundation of the virtual community [9].

Emotional basis which has been formed in internet could continue in virtual community on principle. In modern society, because of the pace of life, difference in careers and values shifts, the connections between people become less and less [10]. Life without true emotion, it is lack of vitality. Because of the demands were not met in reality, a strong internal motive power has been emerged, which should encourage individuals to seek satisfaction in virtual community. Netizens get together by common interests and hobbies in virtual community. Through a series of activities, on the one hand, they satisfy their needs, on the other hand, they establish a virtual community with long-term relationships in order to make it becomes more emotional, and then they attach the community in emotion and deepen the cultural identity in virtual community.

Acknowledgments This research is financed by Funding Project of Education Ministry for Liberal Arts and Social Science (Project No. 10YJC810053) and Research Funds for the Central Universities (Project No. ZYGX2011X022).

References

1. Rheingold H (1993) The virtual community: homesteading on the electronic frontier. Basic Books, New York
2. Romm C, Pliskin N, Clarke R (1997) Virtual communities and society: toward an integrative three phase model. Int J Inf Manage 17(4):261–270
3. Hagel J (1999) Net gain: expanding markets through virtual communities. J Interact Mark 13(1):55–65
4. Duff WM, Cherry JM, Singh N (2006) Perceptions of the information professions: a study of students in the master of information studies program at a canadian university. Arch Sci 6(2):171–192
5. Ghaziani A, Fine GA (2008) Infighting and ideology: how conflict informs the local culture of the Chicago Dyke March. Int J Polit Cult Soc 20(1–4):51–67
6. Ngai EWT, Ryan S, et al. (2012) Sentiment analysis-based decision support system for Chinese media monitoring. In: Proceedings of conference on web based business management, pp 15–18.
7. Al-Gahtani SS (2003) Computer technology adoption in Saudi Arabia: correlates of perceived innovation attributes. Inf Technol Dev 10(1):57–69
8. Howard R (2000) The virtual community: homesteading on the electronic frontier issue 28. MIT Press, Cambridge
9. Tajfel H (1974) Social identity and intergroup behaviour: social science information. Sage Publications, London
10. Johnson ML, Wall TL et al (2002) The psychometric properties of the orthogonal cultural identification scale in Asian Americans. J Multicultural Couns Dev 30(3):181–191

Chapter 112
Research on the Integrated Management and Technology Advancement of the Reinforced Widening Embankment with No Extra Land Acquisition

Xiaoping Li, Zhineng Zheng, Tianqing Ling and Bin Zhou

Abstract Resource-conservative traffic is the formidable requirement of the modern society towards traffic development. This paper, targeted at the expressway expansion project, came up with the project management and technical scheme of the reinforced widening embankment with no extra land acquisition by keeping in line with the framework of Conservative Traffic and introducing the integrated management approach. With the example of Fokai expressway, this paper, conducted an analysis of the widening embankment with no extra land acquisition project by adopting methods such as geotechnical centrifugal model test, numerical analysis and on-spot monitoring. Furthermore, this paper tested that the high feasibility of the expressway reinforced widening embankment project, but also that this project managed to save money.

Keywords Resource-conservative traffic · Integrated management · Reinforced widening embankment · Project management · Fokai expressway

112.1 Introduction

In recent years, with the development of economy and society, China's total passenger transport logistics went through a rapid increase, rendering an unprecedented

X. Li (✉)
Business School, Sichuan University, Chengdu 610064, People's Republic of China
e-mail: lixiaoping@scu.edu.cn

Z. Zheng · T. Ling
Department of Civil Engineering and Architecture, Chongqing Jiaotong University,
Chongqing 400074, People's Republic of China

B. Zhou
Guangdong Provincial Expressway Development Co., LTD, Guangzhou 510100, People's
Republic of China

J. Xu et al. (eds.), *Proceedings of the Eighth International Conference on Management Science and Engineering Management*, Advances in Intelligent Systems and Computing 281, DOI: 10.1007/978-3-642-55122-2_112, © Springer-Verlag Berlin Heidelberg 2014

pressure on the infrastructure construction of transportation. As an important part of China's transportation system, road traffic (namely, railroad and highroad) is characteristic of having a low dependency on the environment, a short construction period, and an easy construction. In recent years, the idea of a conservation-oriented transport enjoyed a widespread application in road construction and the carrying capacity of many roads have been enhanced as a result of road reconstruction and extension engineering. The essence of a conservation-oriented transport is to decrease the consumption of transport inner resources and outer resources such as land for transport under the premise to meet the demands of social and economic development, especially the development of a resource-saving society and an environment-friendly society. The implementation of road expansion with no extra land acquisition in those favorable geographical conditions can substantially lower the costs such as environmental-impact assessment, geological exploration and design. Compared with new road with the same carrying capacity, its cost of investment is 30 % cheaper [6]. However, road expansion and reconstruction needs not only to overcome technical problems (especially difficult ones under unfavorable environmental conditions such as overseas highway, twisting mountain road, etc.,) but also to deal properly with relevant engineering and management problems such as land expropriation for expansion, transfer of ancillary facilities and vehicle flow arrangements, etc.

The original number four lane of Fokai expressway cannot meet the demands of the fast grow of economy. According to the predictions of the transport agency and the requirements towards the project, the two lanes of the original expressway was widened, changing the original bidirectional four roadways into eight. In recent years, many scholars analyzed the common diseases [4, 13] in the construction of widening subgrade and came up with the technological measures [8] based on an in-depth study of the deterioration model and design index [3], the differential settlement of the widening subgrade and its control standards [1, 5]. In those technical measures, reinforced materials including geogrid were treated as uneven settlement in the junction of the digging and filling [2, 7], used to strengthen the soft soil foundation [10] or reinforced slopes [11, 12]. However, the usage of those reinforced material had always been onefold.

In order to meet the demands of the a strengthened expressway carrying without extra land acquisition, the research suggested combing the idea of intergraded management by Xuesen Qian and the technical scheme of reinforced widening roadbed. As a result, the land resources have been saved and the road carrying capacity of expressway strengthened.

112.2 Theory and Methods

The reinforced widening roadbed of Fokai expressway (from Fozhou city to Kaiping city) is characteristic of the following features: (1) the problem with land acquisition is huge. As the area along Fokai expressway is highly economically developed with a high level of urbanization and expensive land, the cost of demolition is inanely expensive and how to save land is an increasingly prominent problem.

(2) The roadbed had high filled soil. The area along the Fokai expressway is a region with densely covered rivers and more than 50 % of the area is covered with bridges and culverts. The short length of the roadbed between different bridges has resulted in the high filling soil. (3) The soft soil along the expressway is not favorable to road construction. The soft soil in fact are super soft and weak soft clay with the features of being "three-high, two-low and one-strong". In this paper, the value of the integrated management approach by Xuesen Qian to the engineering construction of zero extra land acquisition and the design ideas of the reinforced widening roadbed in technical application with the example of the road expansion and reconstruction of Fokai expressway will be highlighted.

1. The Integrated Management Theories

Xuesen Qian is the forerunner in creating the integrated system theory in the management of engineering projects. His idea of integrated management has provided an effective weapon towards solving complicated system problems in project construction and management. The existing research accomplishments in system science indicated that the inner structure and external environment, together with their ways of connection determined the wholeness and functions of the system [9]. Theoretically speaking, the basic task of system theory is to research on how the inner structure and external environment of a system can determine the wholeness and functions of a system and to disclose the general rule of the existence, evolution, synergy, and development of a system. The functions are determined by the structures and influenced by environment. Consequently, the inner structure and external environment of a system can be changed to achieve the ideal objective. Since the environment is not easily subject to alteration, the only way is left to change the structure to achieve functional optimization, especially by changing the relationship within the combination of a system and the relationship between different levels.

The integrated management approach inspired us to come up with the idea of zero extra land acquisition, which is not only a technical problem, but more involved with management issues. Undoubtingly, the zero extra land acquisition and reinforced widening embankment is a huge engineering system. The attempt to double the road area and double the carrying capacity of expressway with no extra land acquisition is in itself a bold conception and its successful application relies on the successful combined of the integrated management approach by Xuesen Qian and relevant engineering techniques.

2. The Scheme of Reinforced Widening Embankment with No Extra Land Acquisition

Based on the features of the Fokai expressway widening project, this paper proposed a technical scheme of reinforced widening embankment with the uneven sedimentation of the widened foundation, reinforced soft foundation and slopes and no extra land acquisition. As a result, this scheme managed to deal with expressway widening through proper technical plan under the framework of the integrated management idea with no or little extra land acquisition.

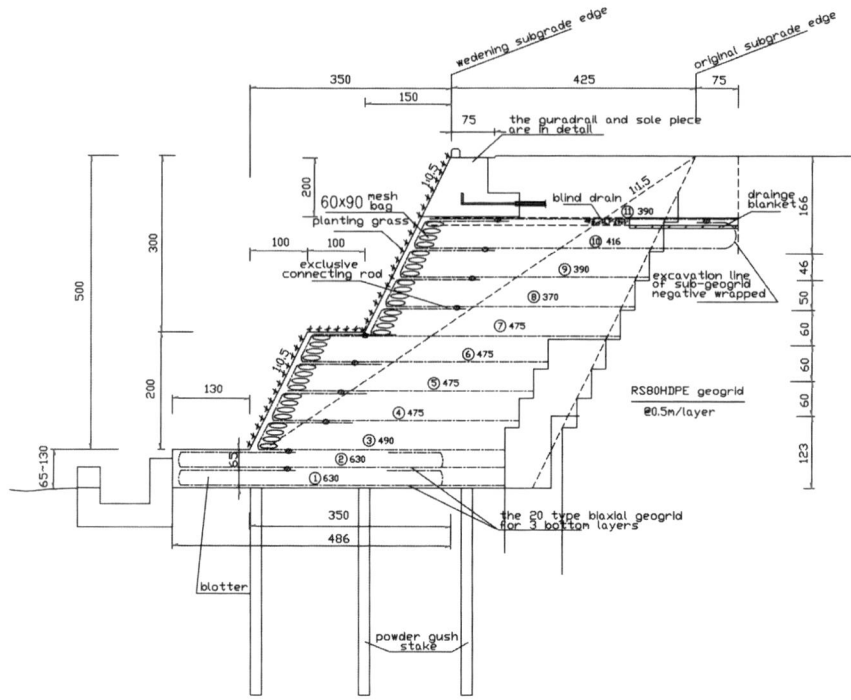

Fig. 112.1 The example scheme of the reinforced widening embankment with no extra land acquisition applied in Fokai expressway. *Notes* (1) The unit of measurement in this figure is centimeter. (2) The reinforced widening embankment requests the subgrade capacity must reach 120 kpa and the internal friction angle of the filling inside the wall is larger than 30°. (3) Construction should be strictly conducted in accordance with the design requirements and implementation of relevant norms. For details of method and craft, you can refer to the design notes. (4) The treatment of the soft ground below subcrust is not marked. (5) The inner side of the sub-geogrid should be negatively wrapped and connected to the top-geogrid by rods. (6) The excavation and stability of foundation pit directly affect the safety of the existing traffic and road construction, so great attention should be paid to the stability of the original embankment, and practical and reliable support measures should be taken to ensure stability and security of the temporary pit wall

The scheme of the reinforced widening with no extra land acquisition goes like this: firstly, the original embankment slope will be dug to serve as a step and then the soft land foundation will be dealt with by powder gush stake and then the de-geo-grid reinforced slope will be built, its representative structure showed in Fig. 112.1. The area of the bank slope was 1:0.5. To expand vertical stress, resist shearing slip and increase the integrity and the water drainage capacity of the walls, 65 cm thick gravel cushion layer and three tiers of geo-grid will be inserted in the bottom of the slope. Within the slope, from the bottom to up, a tier of geo-grid will be inserted every 50 cm, and the top tier of the geo-grid should not intrude the pavement base. The de-geo-grid will be connected and reinforced by pitman. The packings behind the wall should be sandy soil and limit loam under densification at each layer and should meet the requirements of Highway Subgrade Engineering Design Specifications (JTG D30-2004). A stone-laying drain will be installed at the foot of the reinforced retaining wall and should be complished before the retaining wall was built to the height of 1.5 m. Considering the implications of the new widening part on the sedimentation of the old road, the soft soil foundation should be dealt with by power gush stake.

112.3 Analysis and Results

Through geotechnical centrifuge model test, numerical analysis and experimental engineering monitoring, the technical problem resulted by the widening of expressway on soft foundation has been solved.

1. Geotechnical Model Test
According to the demands of the engineer, a geotechnical centrifuge model was conducted on the expressway widening and slope reinforcement to testify the stability and deformation law of the scheme. A model casing was used in the experiment and its inside size was length × width × height = 60 cm × 35 cm × 55 cm. Considering the height of the original embankment and the size of the model casing, the ratio of similitude of the experimental model casing was set at 20. And the stuffing adopted sandy soil used in the project, the geo-grid was replaced by screen window according to the ratio of similitude. Altogether 4 geotechnical centrifuge models were created. After experiment with the 3 models above, the following conclusions can be drawn:

(1) The stability and the deformation performance of the reinforced widening embankment with no extra land acquisition can meet the demands of the widening subgrade.
(2) A comparison of the results of Model 1 and Model 2 shows that the maximum lateral displacement of the reinforced slope decreased to 34 mm from the 50 mm before reinforcement, down 32 %. The maximum vertical displacement of the reinforced top embankment decreased to 44 mm from the 100 mm before reinforcement, down 56 %. After reinforcement, not only does the stability of the bank slope strengthened, but also the top sedimentation and the lateral displacement underwent a remarkable decrease.
(3) A comparison of the results of Model 1 and Model 3 shows that the changes of water content of the stuffing has a clear influence on the deformation of the cross-range of the embankment. In terms of the edge of the road shoulder, Model 3 is 23 mm wider than Model 1. So in actual project, the design and construction of the drainage engineering should be paid special attention (Fig. 112.2).

2. Numerical Analysis
In this part, the finite difference FLAC software platform will be adopted to analysis and calculate the land acquisition of the reinforced widening embankment. In the process of calculation, various data and illustrations were withdrawn such as the vertical displacement value of the embankment top, lateral displacement of the bank slope, the lateral displacement and change of the reinforced material. Furthermore,

Fig. 112.2 The comparison after the project plan was carried out 120 mins at the weight of 20 g

Table 112.1 Geotechnical centrifuge model

Pattern number	The ratio of similitude	Simulation of the actual bar spacing	Simulation of the actual height of the embankment (m)	Water content of the widened part of the embankment
Model1	20	50 cm	5	11.5
Model2	20	no	5	11.5
Model3	20	50 cm	5	19.5

the influence of water content and degree of compaction on the project of reinforced widening embankment with no extra land acquisition were discussed. In numerical analysis, the soil adopted the Mohr-Coulomb elastoplastic constitutive model and the geo-grid adopted the model of two-dimensional belt anchor model (Table 112.1). From analysis, the following conclusion can be drawn:

From the results of the numerical analysis, we can see that the technology and scheme of the reinforced widening embankment with no extra land acquisition is safe.

(1) The maximum vertical displacement of the top embankment was within 1–2 m of the outer edge of the embankment; when the the water content was 11.5 %, the maximum vertical displacement of the embankment top was within 22–23 mm; when the water content was 19.5 %, the maximum vertical displacement of the embankment top was within 38–40 mm. The maximum vertical displacement of the embankment top increases with the increase of the water content of the stuffing. Control the water content of the embankment properly and prevent the underground water from rising will enhance the stability of the reinforced soil project and decrease the deformation of the embankment (Fig. 112.3).

(2) The maximum lateral displacement of the bank slope was all within 0.4–0.5 m under the top of the embankment. The lateral displacement of the slope along the embankment height was subject to an increase change first and then a decrease change. From the position of the step downwards, the lateral displacement value of the slope was subject to an increase change first and then a decrease change (Fig. 112.4).

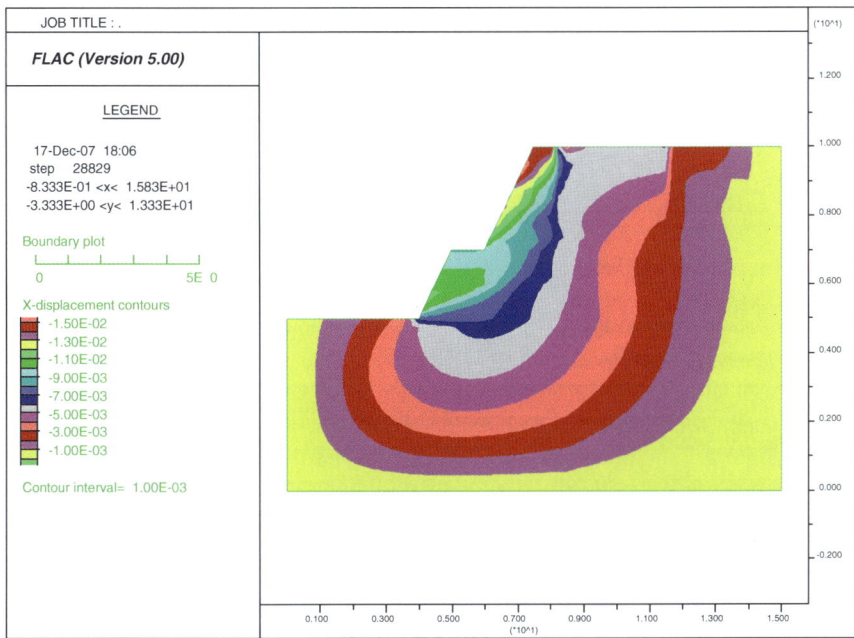

Fig. 112.3 The vertical displacement of the engineering plan

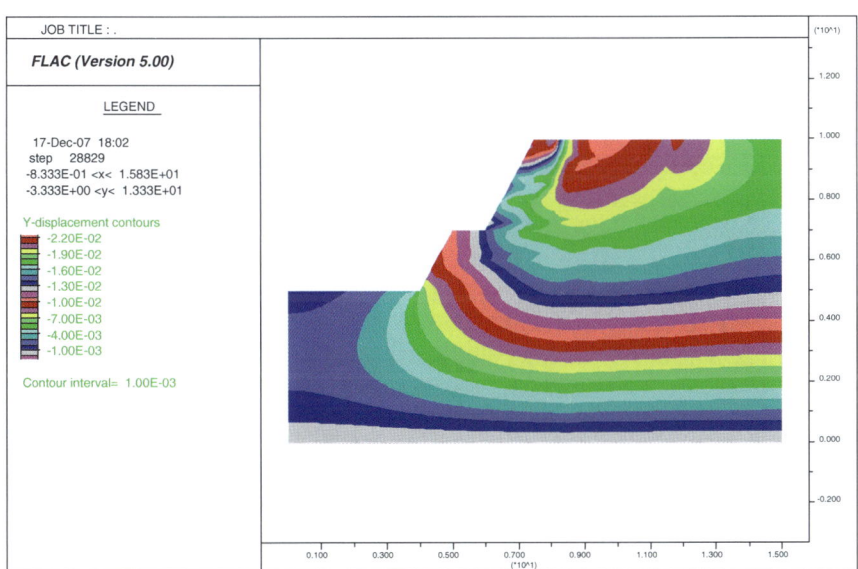

Fig. 112.4 The lateral displacement of the engineering plan

When the water content was 11.5 %, the maximum lateral displacement above the step platform was within 15–16 mm while the maximum lateral displacement below the step platform was within 10–11 mm; when the water content was 19.5 %, the maximum lateral displacement above the step platform was within 36–38 mm while the maximum lateral displacement below the step platform was within 20–22 mm.

(3) When the degree of compaction was 93 %, the maximum vertical displacement of the embankment top was within 20–22 mm; when the degree of compaction was 90 %, the maximum vertical displacement of the embankment top was within 22–24 mm; when the degree of compaction was 85 %, the maximum vertical displacement of the embankment top was within 32–34 mm. The maximum vertical displacement of the embankment top decreases with the increase of the degree of compaction.

(4) When the degree of compaction was 93 %, the maximum lateral displacement above the step platform was within 14–15 mm and the maximum lateral displacement below the step platform was within 10–11 mm; when the degree of compaction was 85 %, the maximum lateral displacement above the step platform was within 30–32 mm and the maximum lateral displacement below the step platform was within 16–18 mm. The maximum lateral displacement of the slope decreases with the increase of the degree of compaction and when the degree of compaction was below 90 %, the increase of the lateral displacement of the slope will be marked.

3. Monitoring Analysis

The two cross sections of K12+905 and k13+025 of the Fokai expressway have been chosen as monitoring cross sections. The monitoring of each cross section includes sedimentation, horizontal displacement, pore water pressure, soil pressure and other 4 items. The following conclusions can be drawn from the integrated analysis of each monitoring data.

According to each monitoring data, in general all monitoring data are tending towards stability, subjecting to no changes and the status of the widened embankment has been stabilized.

(1) The variations of each monitoring spot and each monitoring value happened mainly during the period of embankment building and within the three months after the embankment has been built. After the accomplishment of the construction, the monitoring value only subject to little changes. For example, the accumulated maxim lateral displacement was about 20 mm, but within the accumulated lateral displacement, about 15 mm happened during the building period of the embankment. The scheme plan and the building plan can guarantee the deformation and the stability of the widening project.

(2) Test results from pore pressure and soil pressure showed that using power gush stake to attend to soft soil foundation, with the power gush stake being rigid pile, the stake and the soil between the stakes was subject to no coordinated

deformation. In fact, the power gush stake served as the main force in bearing the upper load.

(3) A comparison of various monitoring index showed that during rainy seasons, the sedimentation of the cross section was subject to marked changes while other indexes (lateral horizontal displacement, soil pressure, and pore water pressure, etc.) basically had no change.

4. Results

From the geotechnical model test, the numerical analysis and the monitoring analysis, the following results can be concluded:

(1) From the results of the numerical analysis, the conclusion that the technology and project of the reinforced widening embankment with no extra land acquisition is safe can be drawn.

(2) In order to maintain the long-term deformation performance of the widening project, the dealings of the soft soil roadbed and its sedimentation should be strictly controlled.

112.4 Conclusions

Based on the research analysis of the roadbed widening technology and reinforced soil technology both home and abroad, this paper came up with the expressway reinforced widening technology with no extra land acquisition targeted to the Fokai expressway widening embankment project characterizing the features of difficult in land acquisition and low carrying capacity of soft foundation; the geotechnical centrifuge model, numerical analysis and on-spot monitoring showed that for soft soil foundation, the reinforced widening embankment technology is of high degree of safety and reliability and can meet the demand of reinforced widening embankment project; the technology of expressway with no extra land acquisition is feasible and the research results have provided a technical reference for similar engineering projects.

The integrated management and technology of reinforced widening embankment with no extra land acquisition started with technology and lowered the consumption of external resources, especially the land in road expansion and is exemplary of the application of technology management innovation and the integrated management.

References

1. Fu ZH, Wang XC, Chen XG (2007) The sedimentation features and influential factors of the widening road foundation. J Transp Eng 1:54–57 (In Chinese)
2. Hao XZ, Shen L, Wang XC (2007) Research on the uneven sedimentation technology of the digging and filling road foundation in geogrid. J Road both Home Abroad 27(5):56–58 (In Chinese)

3. Huang QL, Ling JM (2004) The mechanism and disease characteristics of the widening of old road. J Tongji Univ (Science Edition) 2:197–201 (In Chinese)
4. Lu YY (2012) The treatment of common disease of the highway widening embankment. J Heilongjiang Traffic Sci Technol 7:44 (In Chinese)
5. Nie P, Qu XJ, Liu FQ (2005) Research on the uneven sedimentation index of the expanded expressway of shenda. J Energy-efficient Traffic Environ Prot 11:18–20 (In Chinese)
6. Shi J (2006) Road construction and the resource-conservative traffic. J Energy-efficient Traffic Environ Prot 1:11–13 (In Chinese)
7. Sun QY (2013) The application of geogrid in highway engineering. J Highw Eng 6:32–34 (In Chinese)
8. Tian B (2013) The analysis of construction technique of highway widening embankment. J Traffic Build Sci 5:168–171 (In Chinese)
9. Von Bertalanffy L (1968) General system theory: foundations, development, applications. George Braziller, New York
10. Wei HB (2013) The research of anchoring reinforcement technology in widening embankment of highway extension. J Road Eng 2:38–40 (In Chinese)
11. Yang XW, Ouyang ZC (2000) Research on the centrifugal model of the reinforced embankment and slope. J China Civil Eng 5:88–91 (In Chinese)
12. Yang XW, Xu JJ, Wang DY (2000) The application of reinforced slope and embankment in yuchang road. J Chongqing Jiaotong Univ 4:55–58 (In Chinese)
13. Zhou Y, Xiao YM (2006) The analysis of common disease characteristics of the widening embankment project. J Chongqing Jiaotong Univ 1:71–74 (In Chinese)

Chapter 113
Quantifying Mutual Correlations Between Supply Chain Practices and Customer Values: A Case Study in the Food Industry

Meysam Maleki and Virgílio António Cruz-Machado

Abstract Supply chains set their ultimate aim as to satisfy their end customers. In other words, they implement a set of practices along their chain in order to generate value for customers. Therefore, aligning supply chain practices with customer values is a core objective of supply chains. The current research employs Bayesian Network (BN) and Analytic Network Process (ANP) to quantify mutual correlations between supply chain practices and customer values. In the first phase, it collects and analyzes data about six specific customer values; in the second phase, the importance of practices is evaluated by expert in the respective industry using ANP; and in the final phase, the output of the two analysis meet in a BN to generate a model which is capable of quantifying their mutual correlations given the input data. In addition, the proposed approach can handle scenarios to identify in case a specific customer value is preferred by the end customer what practice should be implemented; or vice versa, in case a specific practice is implemented how does it contribute to customer values. This approach is applied to a case study in the food industry to present its application in practice.

Keywords Supply chain management · Practices · Customer value · Food industry

113.1 Introduction

During the 1990s, many manufacturing and service firms collaborated with their strategic suppliers to upgrade traditional supply and materials management functions and integrate them as part of corporate strategy. Correspondingly, wholesalers and retailers also integrated their logistics activities with other functional areas to elab-

M. Maleki (✉) · V. A. Cruz-Machado
UNIDEMI, Department of Mechanical and Industrial Engineering, Faculty of Science and Technology, Universidade Nova de Lisboa, Lisbon, Portugal
e-mail: maleki@fct.unl.pt

J. Xu et al. (eds.), *Proceedings of the Eighth International Conference on Management Science and Engineering Management*, Advances in Intelligent Systems and Computing 281, DOI: 10.1007/978-3-642-55122-2_113, © Springer-Verlag Berlin Heidelberg 2014

orate on their competitive advantage. Eventually, these two traditional supporting functions of corporate strategy evolved and merged into a holistic and strategic approach to materials and logistics management, nowadays known as Supply Chain Management (SCM) [1]. The dominant belief is that the most successful companies are those that are operating closely within their supply chain and have carefully linked their internal processes to external suppliers and customers [2]. Evolving from the economic theory of vertical integration and the operational theory of product life cycle, SCM has been a major source of competitive advantage in the global economy. SCM is the management of a network of interconnected businesses involved in the provision of product and service packages required by the end customers in a chain of firms [3]. This definition requires SCM's to integrate internal activities with expectations of the end customers. Development of the integrated supply chain is the most significant contribution to the delivery of goods and services [4].

Supply chains are facing variety of challenges such as customer service, cost control, partner relation management, fragmented chain, lack of visibility, and coordination difficulties. Looking into such challenges, stresses the existing problem of enterprises in effectively integrating their activities with their supply chain partners as well as aligning their products with the end customer expectations. Integration, as a key factor in achieving improvements, has been one of the main themes in the SCM literature, therefore it is frequently examined by researchers [5].

The need for a comprehensive integration model is stressed in the SCM literature. Although there have been some works on this fields, there are gaps due to its diversity. The purpose of this research is to fill in some gaps through developing an integration model which quantitatively addresses relations between SCM practices and customer values. In order to do so, a combination of Analytic Network Process (ANP) and Bayesian Network (BN) is proposed in the model development procedure. The proposed model imports comparative data about preferences of end customers and priorities of SCM practices through interview with expert; then it generates quantitative output about their relations.

The current research is the extension of the two previous research work of authors. We have deeply investigated SCI literature and identified its gaps in a paper entitled "An Empirical Review On Supply Chain Integration" [5]. Thereafter, the SCI model is presented in the next paper entitled "Development of Supply Chain Integration model through application of analytic network process and Bayesian network" [6]. In the current research, the proposed model is implemented in a supply chain in the food industry.

113.2 Supply Chain Integration Model

Two modeling approaches are used in this research: BN and ANP. BN is used to data mine customer value data and identify casual relations and patterns among them. ANP is used on the other side to quantify experts' tacit knowledge about SCM practices. Thereafter, the SCI model is built through taking the inputs of these models and developing another BN (Fig. 113.1).

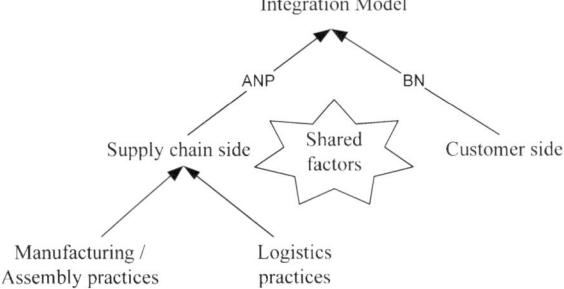

Fig. 113.1 Modeling approaches

Development of the SCI model takes place in three phases. Phase 1 focuses on data collection and data mining of customer values. This phase benefits from BN in data mining. The second phase concentrates on interview with experts and prioritization of SCM practices according to interviews with respect to customer values. Customer values have the role of bond between phase one and two. ANP is employed in this phase in order to achieve a quantitative prioritization. Both phases use pairwise structure in data collection therefore if the number of elements to be compared is I, the total number of possible pairwise comparisons (number of questions) is $I \times (I-1)/2$. Phase three receives inputs from the preceding phases in order to build up the model. In this phase customer values and practices are connected through a BN model.

1. Phase I: Customer values

Customer value data is collected through an innovatively designed questionnaire in which pairwise comparisons among customer values are investigated. Five different states are given to the respondent to select according to his/her preferences. As the respondent picks one state two digits will be stored. For example, in case if quality is much more important than cost to the respondent then quality receives a score of 4 and cost receives score of 0 that are stored in the database. In the Fig. 113.2 the closest importance level to each side of comparison is "significantly more important", after that there is "more important", then there is "the same importance" in the middle. Therefore, this figure should be read by starting from the customer value which is closer to the bullet.

2. Phase II: Interview with expert

Interviews are conducted with SCM experts who satisfy two criteria: interviewee should have practical knowledge about SCM practices; and should be in touch with marketing departments to have sufficient knowledge about customer expectations. Interviews were conducted through video conference meetings and data was exchanged through e-mails. However, in order to make sure geographical barriers don't harm the research, experts were kept posted about the progress of the research. Interviewee received different phases of the case study to ensure that the output presents real case scenarios.

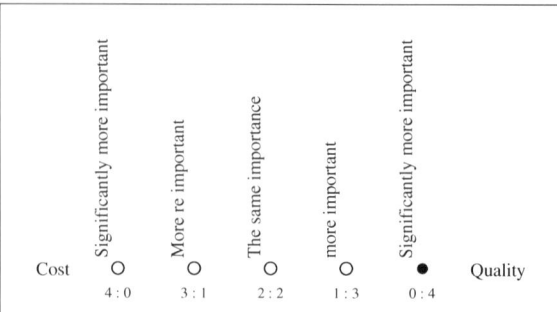

Fig. 113.2 The customer value questionnaire design

3. Phase III: Development of the conceptual model

The third phase takes inputs from previous phases into a BN model to identify relations between SCM practiced and customer values. Figure 113.3 illustrates the framework through which the conceptual model is developed. This model provides quantitative output which can lead to visual output through BN platforms. In addition, the potential of doing sensitivity analysis and planning scenarios in the BN model, makes it a strong decision making tool. It works in both directions from SCM practices to customer values and vice versa. In other words, the model gives quantitative outputs to questions such as: if we implement one specific practice, how does it contribute to the customer values? Or, if the aim to contribute to one specific customer value, which SCM practices should be implemented? Consequently, the output of the model is limited to the introduced SCM practices and customer values to it.

The conceptual model starts with selecting the corresponding industry which will be the context of the integration model. Then, the customer values (CVn) of this industry will be identified and comparative data about them will be collected from end customers. Data analysis of customer values will be done using BN to quantify correlations among them. In parallel, interview with experts take place to find out relative importance of manufacturing practices (PMi) as well as logistics practices (PLj) in the selected industry sector. ANP is used to calculative priorities and synergies among practices. Comparison among practices goes through pairwise analysis with respect to customer values. Thus, customer values are considered as shared values which put together SCM practices and end customer preferences. SCM practices (from ANP model) will be represented as nodes on the network in BN. Each SCM practices gets two states as "recommended" and "not recommended". The value of each state depends whether or not that node (SCM practice) changes in that node leads to changes in the related CVn or not. In other words, in case a customer value is positively sensitive to application of a SCM practice, that node gets the state of "recommended". The conceptual model of SCI includes both SCM practices and

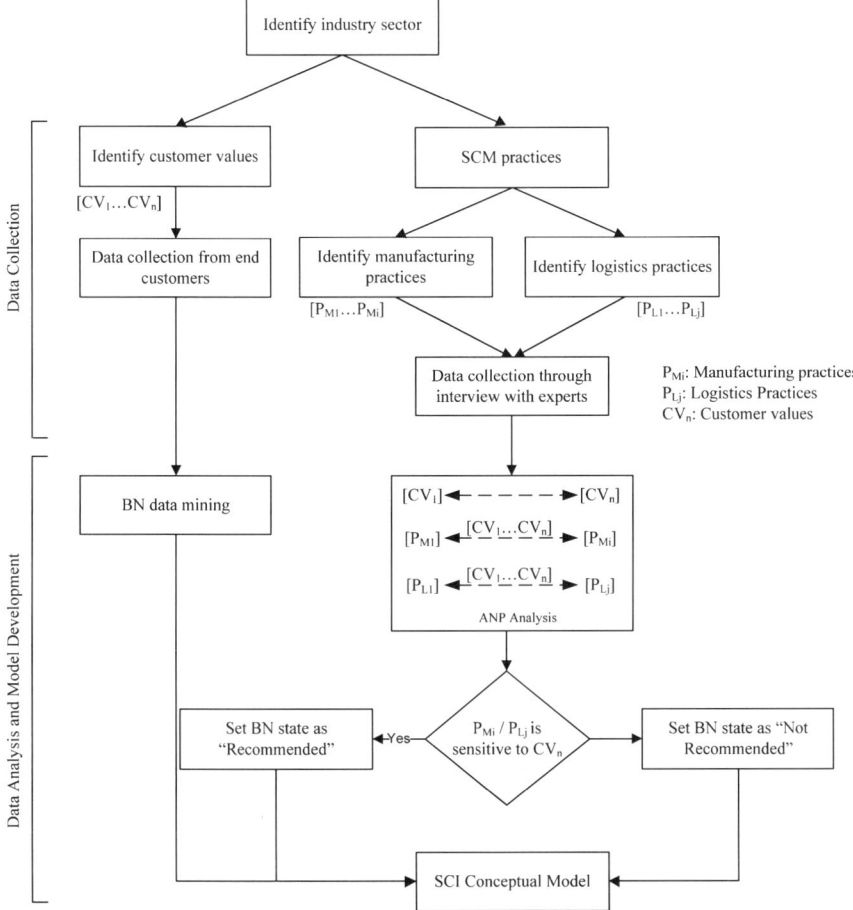

Fig. 113.3 The customer value questionnaire design

customer values presented as nodes of the network. This model which is constructed with BN illustrates relations between SCM practices and customer values. The SCI conceptual model can be used to conduct sensitivity analysis and scenario planning through grounding one (or more) nodes and monitoring the influence on the rest of the network.

113.3 Case Study in the Food Industry

1. Customer values data collection

The case company is situated in New Zealand. Customer value data benefits from 131 respondents collected in a pairwise approach where each pair of customer values is compared. This research categorizes customer values into six factors as: time, quality,

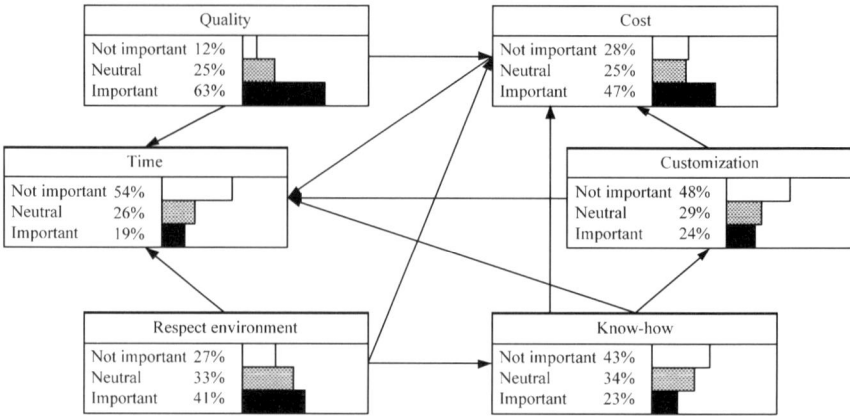

Fig. 113.4 Bayesian network of the end customer values in the food industry-Generated by GeNIe 2.0

cost, customization, know-how, respect environment. Since the objective is to explore trade-offs among factors, the Friedman test is used to identify the importance order of values from end customer perspectives. Results show that the Chi-squared value was significant at 0.01 level ($p < 0.01$). This indicates that the mean ranks significantly differ among time, quality, cost, customization, know-how and respect-environment factors and the highest rank is devoted to quality. The next important factors are respect-environment, cost, know-how, customization and time, respectively (Fig. 113.4).

2. The ANP model

The ANP inputs are prepared through interview with director of the case company. Since food industry is mostly influenced by local neighborhoods, a small scale case company is selected which is located in New Zealand. The Interviewee is the director of the case company, he is also consultant of some other companies in the food industry. Due to his expertise in this specific industry sector, the interview is conducted with him. The expert is asked to make pairwise comparison of practices with respect to customer values. There are five production practices and four logistics practices which go through comparison procedure (Fig. 113.5). There are three types of comparison: firstly, customer values are compared to one another, secondly production practices are compared with respect to customer values, and in the third step logistics practices are compared with those values (Appendix A).

3. The SCI model

In the last phase—in order to put together experts knowledge and customer preferences-synthesized values from ANP model are used to identify prior probabilities of practices and construct the BN model including both practices and customer values. In the SCI model presented in the Fig. 113.6 customer values have three different states as: not important, neutral, and important whereas SCM practices has two states

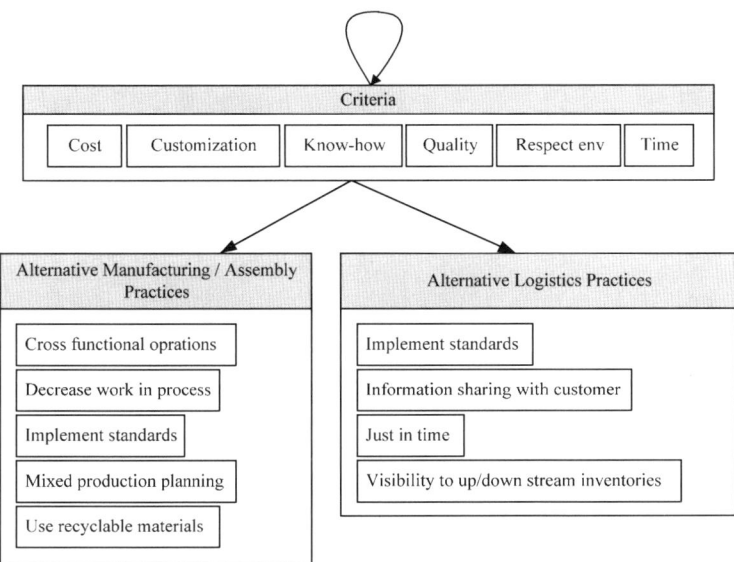

Fig. 113.5 Criteria and alternatives of the analytic network process model: Food industry

as: not recommended and recommended. This model operates in both directions from child nodes to parent nodes and vice versa. Therefore, in case of grounding one of the practices, the status of other nodes will be dynamically changed according to it. In addition, if a root node is grounded in one of its states, the rest of the network will be changes accordingly. It worth noticing that the status of customer value nodes are calculated based on the customer data which was collected and analyzed in the first case study. Since the market conditions are dynamic and variety of factors may influence customer preferences, therefore practitioners are suggested to keep this data updated. On the other side, SCM practices nodes are calculated based on expert's comparative analysis with respect to customer values. Since these are technical practices, this analysis can be used for longer time and doesn't require frequent update however practitioners are recommended to double check them and adjust the network in case amendments are required.

The importance order of customer values when none of nodes are grounded is respectively quality, cost, respect environment, know-how, customization, and time. According to Fig. 113.6 the most recommended manufacturing practice to increase quality is implementation of standards. The most recommended logistic practice for this customer value is information sharing with customer. Details of the importance level of customer values and SCM practices in food industry are presented in the Appendix.

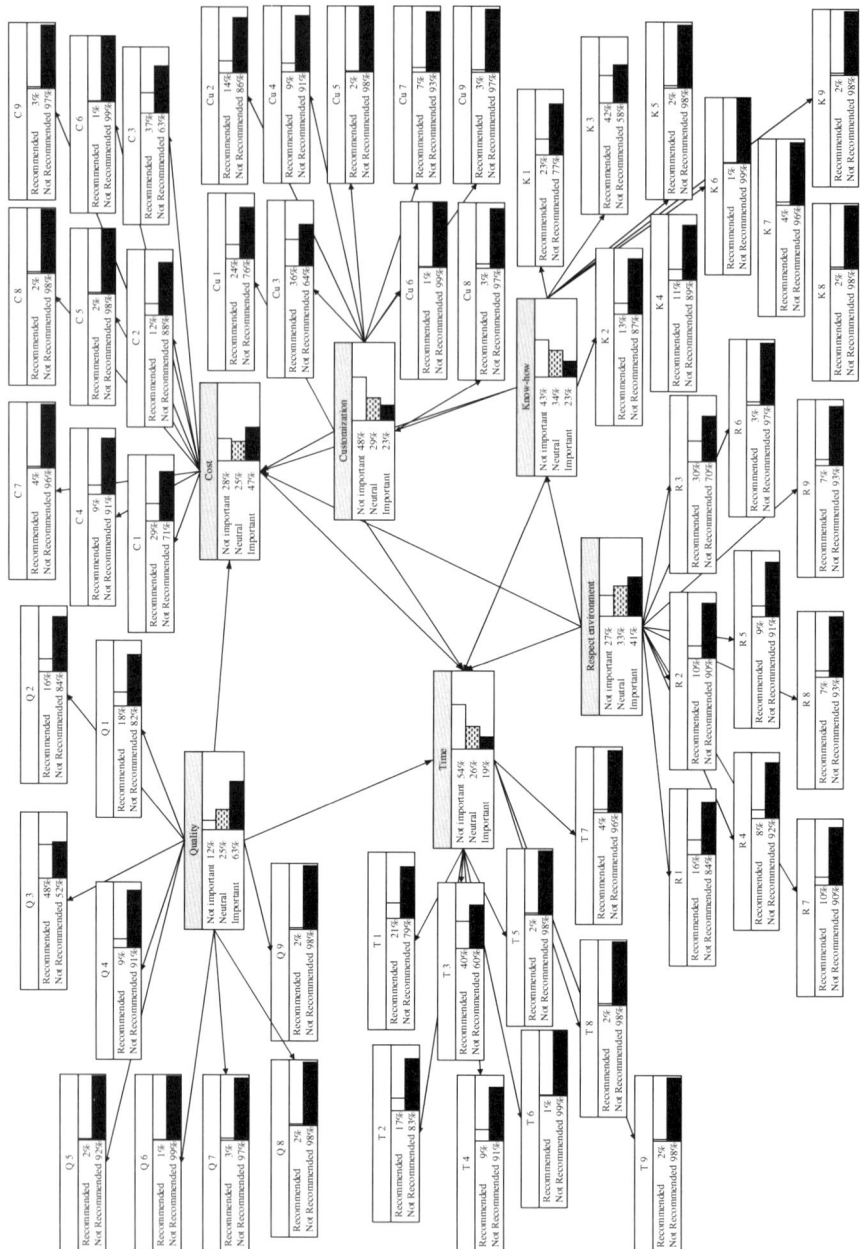

Fig. 113.6 Integration of supply chain practices with customer values in the food industry - Generated by GeNIe 2.0 (use Table 113.1 to read this figure)

Table 113.1 Manual of Fig. 113.6

Node names (supply chain practices and customer values)		Initial
Manufacturing practices	Cross functional operations	1
	Decrease work in process	2
	Implement standards	3
	Mixed production planning	4
	Use recyclable materials	5
Logistics practices	Implementing logistics standards	6
	Information sharing with customer	7
	Just in time	8
	Visibility to upstream/downstream inventories	9
Customer values	Cost	C
	Customization	Cu
	Know-how	K
	Quality	Q
	Respect environment	R
	Time	T

113.4 Conclusion

The current research presented a quantitative method to explore mutual connection between practices and customer values in the SCM context. The proposed approach is applied in a case study in the food industry in three phases. The proposed SCI model can be looked from theoretical, managerial, and practical perspectives. Theoretical perspectives was concerned with theory building to address gaps in SCI which can be considered as one step forward in solving obstacles in this context. Managerial perspective is concerned with applicability of the proposed model in decision making procedures. Managers can benefit from these findings in the design, analysis, and improvement of their supply chain in order to have an integrated supply chain which contributes to end customer values. Through applying the proposed approach, they can avoid mismatching between their applied practices and expected outcomes. Although it is theoretically possible, a practical recommendation is not to combine technical and strategic entities in one network. It is due to the fact that it increases the complexity of the network resulting in an unrealistic inference. In both cases the proposed approach and the presented case study provide foundation to develop strategic and technical networks.

Acknowledgments Authors would like to acknowledge the support by Fundação para a Ciência e a Tecnologia, Project MIT-Pt/EDAM-IASC/0033/2008. We also appreciate the time and availability of director of the case company.

Table 113.2 Pairwise comparison of customer values

	Customization	Know-how	Quality	Respect env.	Time
Cost	←6	← 9	↑ 7	← 9	← 4
Customization		↑ 7	↑ 9	← 9	← 2
Know-how			↑ 8	← 9	← 2
Quality				← 9	←4
Respect env.					↑ 9

Table 113.3 Comparisons with respect to "Cost"

	Decrease work in process	Implement standards	Mixed production planning	Use recyclable materials
Cross functional operations	← 8	← 4	← 4	← 9
Decrease work in process		↑ 6	← 1	← 4
Implement standards			← 6	← 9
Mixed production planning				← 9

Table 113.4 Comparisons with respect to "Customization"

	Decrease work in process	Implement standards	Mixed production planning	Use recyclable materials
Cross functional operations	← 5	← 8	← 9	← 9
Decrease work in process		← 7	← 3	← 9
Implement standards			↑ 3	← 7
Mixed production planning				← 9

Appendix A: ANP Data Collected from Interview

1. Pairwise comparison of customer values (Table 113.2)
2. Pairwise comparison of manufacturing practices with respect to factors (Tables 113.3, 113.4, 113.5, 113.6, 113.7 and 113.8)
3. Pairwise comparison of logistics practices with respect to factors (Tables 113.9, 113.10, 113.11, 113.12, 113.13, 113.14 and 113.15).

Table 113.5 Comparisons with respect to "Know-how"

	Decrease work in process	Implement standards	Mixed production planning	Use recyclable materials
Cross functional operations	← 8	↑ 8	← 4	← 9
Decrease work in process		↑ 6	↑ 6	← 9
Implement standards			↑ 2	← 9
Mixed production planning				← 9

Table 113.6 Comparisons with respect to "Quality"

	Decrease work in process	Implement standards	Mixed production planning	Use recyclable materials
Cross functional operations	↑ 6	↑ 7	← 5	← 5
Decrease work in process		↑ 7	← 1	← 8
Implement standards			← 9	← 9
Mixed production planning				← 5

Table 113.7 Comparisons with respect to "Respect environment"

	Decrease work in process	Implement standards	Mixed production planning	Use recyclable materials
Cross functional operations	← 8	↑ 3	↑ 4	↑ 9
Decrease work in process		↑ 4	↑ 2	↑ 8
Implement standards			← 2	↑ 9
Mixed production planning				↑ 9

Table 113.8 Comparisons with respect to "Time"

	Decrease work in process	Implement standards	Mixed production planning	Use recyclable materials
Cross functional operations	↑ 8	↑ 5	← 1	← 9
Decrease work in process		← 1	← 5	← 9
Implement standards			← 3	← 9

Table 113.9 Comparisons with respect to "Cost"

	Information sharing with customer	Just in time	Visibility to up/down stream inventories
Implementation of standards	↑ 3	↑ 3	↑ 4
Information sharing with customer		← 3	← 1
Just in time			← 2

Table 113.10 Comparisons with respect to "Customization"

	Information sharing with customer	Just in time	Visibility to up/down stream inventories
Implementation of standards	↑ 7	↑ 8	← 1
Information sharing with customer		← 5	← 7
Just in time			↑ 3

Table 113.11 Comparisons with respect to "Know-how"

	Information sharing with customer	Just in time	Visibility to up/down stream inventories
Implementation of standards	↑ 8	← 7	← 5
Information sharing with customer		← 5	← 5
Just in time			← 1

Table 113.12 Comparisons with respect to "Quality"

	Information sharing with customer	Just in time	Visibility to up/down stream inventories
Implementation of standards	↑ 3	↑ 3	↑ 4
Information sharing with customer		← 3	← 1
Just in time			← 2

Table 113.13 Comparisons with respect to "Respect environment"

	Information sharing with customer	Just in time	Visibility to up/down stream inventories
Implementation of standards	↑ 3	↑ 3	↑ 4
Information sharing with customer		← 3	← 1
Just in time			← 2

Table 113.14 Comparisons with respect to "Time"

	Information sharing with customer	Just in time	Visibility to up/down stream inventories
Implementation of standards	↑ 3	↑ 3	↑ 4
Information sharing with customer		← 3	← 1
Just in time			← 2

Table 113.15 Unweighted super matrix in the food industry—Calculated by Super Decisions 2.2.6

		Cost	Custom-ization	Know-how	Quality	Respect env.	Time
Log Practices	Implement standards	0.01	0.02	0	0.01	0.08	0.01
	Information sharing with customer	0.04	0.24	0.01	0.01	0.22	0.06
	Just in time	0.03	0.07	0	0.01	0.16	0.04
	Visibility of up/down stream inventories	0.03	0.07	0	0.01	0.16	0.04
Man Practices	Cross functional operations	0.45	0.35	0.25	0.11	0.03	0.08
	Decrease work in process	0.06	0.14	0.05	0.19	0.01	0.36
	Implementation of standards	0.26	0.04	0.44	0.57	0.05	0.28
	Mixed Production planning	0.09	0.07	0.23	0.07	0.04	0.1
	Use recyclable materials	0.02	0.02	0.02	0.02	0.23	0.02

References

1. Tan KC (2002) Supply chain management: practices, concerns, and performance issues. J Supply Chain Manage 38:42–53
2. Mitra S, Singhal V (2008) Supply chain integration and shareholder value: evidence from consortium based industry exchanges. J Oper Manage 26:96–114
3. Harland CM (1996) Supply chain management, purchasing and supply management, logistics, vertical integration, materials management and supply chain dynamics. In: Blackwell encyclopedic dictionary of operations management, Blackwell
4. Stonebraker PW, Liao J (2006) Supply chain integration: exploring product and environmental contingencies. Supply Chain Manage: An Int J 11:34–43
5. Maleki M, Cruz-Machado V (2013) An empirical review on supply chain integration. Manage Prod Eng Rev 4:85–96
6. Maleki M, Shevtshenko E, Cruz-Machado V (2013) Development of supply chain integration model through application of analytic network process and bayesian network. Int J Integr Supply Manage 8:67–89

Chapter 114
Establishing the System of How to Select the Contractor in a Large-Scale Construction Project

Wuming Xu, Xiaowen Tang, Hongyan Dan and Huan Liu

Abstract Construction is very important to the success of large-scale construction projects. How to choose the construction contractor is of great significance in engineering project management. The system of how to choose the construction contractor is established in the paper. At first, according to engineering practice, the input selective properties of the contractor, which is categorized into technical attribute set and business attribute set, are proposed and optimized. Moreover, artificial neural network is applied to the actual construction's future output prediction, and a appropriate neural network structure is established to predict the future output of the contractors. Finally, the construction contractor is selected based on more accurate output of forecasts of the contractors. The writer brings up the algorithm's basic idea and its steps, uses Matlab as experimental tool and selects some cases of model prediction and choice. The experiment shows that the model has self-learning ability and high prediction accuracy and can be used to select contractors. It is effectively helpful for selecting contractor in large construction projects, and will be quite promising in the future.

Keywords Project management · Construction contractor · Selection · Prediction · Attribute · Neural network · Matlab

W. Xu (✉) · H. Liu
School of Management, Xihua University, Chengdu 610039, People's Republic of China
e-mail: 147610753@qq.com

X. Tang
School of Business, Sichuan University, Chengdu 610064, People's Republic of China

H. Dan
School of Economics and Trade, Xihua University, Chengdu 610039, People's Republic of China

J. Xu et al. (eds.), *Proceedings of the Eighth International Conference on Management Science and Engineering Management*, Advances in Intelligent Systems and Computing 281, DOI: 10.1007/978-3-642-55122-2_114, © Springer-Verlag Berlin Heidelberg 2014

114.1 Introduction

Construction is very important to the success of large-scale construction projects. The key to the major projects is the selection of the construction contractors.

The methods that choose the construction contractor are the price factor method and the comprehensive evaluation method in the actual [1, 9, 10, 14, 19]. The price factor method is that the contractor is to be selected according to the lowest price or the closest to the base bid price and other price factors under meeting the whole requirements of the owner to the contractor qualification, performance. The comprehensive evaluation method is to meet the substantive requirements under the premise of various factors on the qualitative and quantitative analysis. Through a comprehensive selection to determine. The project owner and the parties involved tend to use quantitative mufti-index selection method currently [10]. There are many practical uses simple weighted average method, the study has adopted AHP [19] a useful multivariate regression [6], there is the use of fuzzy theory [3, 8, 15], the use of rough set theory [7], and so on.

In the past, the research and practice of select engineering contractor, there are three inadequacies. The first is over-based pricing, such as price factors in the choice of law based primarily on quotes, obviously this is easy to make mistakes. The second is the selection of contractors is not based on the contractor's actual output. Did not effectively predict and methods of measurement for the future of real output, and thus it is not based on actual production to select contractors. The owners and the parties really care about is the future of real output. The third is the contractor selected research and practical choice in the tender, subjectivity is too large, between the various elements and evaluation of the results of the value that people assume that a certain function, but it is failed to prove a specific function and confirmed.

The selection of large construction contractors. In essence, first, handle mufti-attribute decision making problems [16]. Second, the problem of how to predict the actual output. Therefore, to find more useful and more scientific indicators (is property) is suitable choice for large-scale construction contractors, but also the right to find the corresponding indicator function of various input and output between the actual future. An alternative method with some smart of construction contractors for the project, it is especially significant for large-scale project management and success.

114.2 Solution Ideas

Neural networks are nonlinear dynamical systems refers to methods using mathematical simulation of biological neural networks in a massively parallel [4, 18]. BP (Back Propagation) neural network can right itself between the layers of neural network which can be self-corrected weight. And it has been proven in the N-dimensional closed interval to any one-dimensional continuous function M can be used contain a hidden layer neural network to approximate that on a 3-layer neural network arbitrary precision can be achieved by the N-dimensional mapping of the M-dimensional [5, 18]. Specific neural networks without the need for specific rules in advance a clear grasp of the corresponding neural network capable of self-learning self-correction features, find the corresponding functions for the correct relationship between the various properties and construction of real output between large construction contractor provides the possibility provides a scientific method to predict. The neural network science to predict the actual output of construction contractors can effectively solve the above-mentioned inadequacies. It is a large construction contractor selection to help find more effective ways.

The basic idea of the algorithm is: first of all, the first train the neural network data sample set until certain conditions are met so far, it has the ability to predict. Use it to predict the actual output of the construction contractor. Then select the index of real output into construction contractor selection.

Step of the algorithm can be described as follows:

1. Create an input attribute set based on the actual construction of major projects.
2. Construct neural network, and then determine the structure and type of neural network as well as the number of hidden layers and nodes of each layer, range of the weights, training algorithm and the number and meaning of the output layer nodes.
3. Determine the original input data, and then preprocessing the data to select the training set, test sample set and sample set which is to be selected. Training set is used to train the neural network; Test sample set is used to Test the neural network which has been trained. The sample set which is to be selected is used to test actual situation and choose construction contractor.
4. The number of nodes within the range, the weight value, generates a random network structure, from the input layer to the output layer, the training algorithm for training purposes. Meet certain conditions to terminate.
5. With a test sample set simulation test to predict the effect of the trained neural network.
6. The actual output of the construction contractor to be selected sample set used to predict and achieve construction contractor selection on this basis.

114.3 Building Attribute Set

The elements of large-scale construction project contractor selection, that is, attributes (or indexes), can be classified as technical attributes set, business attribute set, so as to in accordance with the approach in practice. The former is based on the technical documents supplied by construction contractor, and the latter is based on quoted price, contractor situations and service promise, and so on. Refer to relevant regulations and information [11–13], according to consult practice, the paper builds the technical attributes set as Table 114.1, business attribute set shown as Table 114.2. The attribute value can be classified C-continuous, and S-discrete.

114.4 Neural Network Structure

The paper chooses MATLAB 7.1 as the development tool [17].

The neural network includes 3 layer BP network, which are input layer, hidden units processing layer and output layer [5, 18]. There are 36, 73 and 8 nodes in input layer, hidden layer and output layer respectively.

The nodes in the input layer are $I = (I_1, I_2, \cdots, I_{36})$, as there are total 33 attributes, including technical and business attributes, which maps 33 nodes $(I_1, I_2, \cdots, I_{33})$. 2 nodes (I_{34}, I_{35}) denote project type, as shown in Table 114.3. The last node is threshold value.

The number of nodes in hidden layer is calculated by $2n + 1$ [4, 16, 18], which is 73, where, 72 nodes take in the results from input layer, and the other one is threshold value of hidden units.

The output layer is mainly corresponding to the construction output, which is measured by the target output method and stakeholder satisfaction [2]. The nodes are $O = (O_1, O_2, O_3, O_4, O_5, O_6, O_7, O_8)$, where, O_1 = project settlement score, O_2 = construction time score, O_3 = construction quality score, O_4 = owner satisfaction, O_5 = supervisor satisfaction, O_6 = government satisfaction, O_7 = operator satisfaction, O_8 = community satisfaction.

Hidden layer adopts logsig function, output layer adopts purelin function. Initial weights of the threshold points sets in weight range $(1, 1)$, the other points in weight range $[0, 1]$. The initial weights are randomly generated within the range.

114.5 Case Study

Choose the bid-winning enterprises and engineering (completed) as examples from the bidding data set in 2002–2006 of the S provincial tendering companies. Select 12 training samples and 6 testing samples. Choose 3 bidding enterprise of a project from 2008 data set, as 3 alternative samples.

114.5.1 Data Preprocessing

The input layer refers to Tables 114.1 and 114.2, 1–33 nodes are corresponding to 1–33 attributes. Most of the data are assigned according to certain rules, some data are assigned in bid evaluation process, and other through scoring by 5 experts. For

Table 114.1 The technical attribute set of contractor

Attribute subset	Attribute	No.	Type	Range	Brief description
Response and promise	Technical standards and requirements	1	S	$\{0, 1\}$	Response as 1, otherwise 0
	Project duration	2	S	$\{0, 1\}$	Response as 1, otherwise 0
	Project quality	3	S	$\{0, 1\}$	Response as 1, otherwise 0
Human resources and equipments, etc	Aptitude of project leader	4	S	$\{0, 0.5, 1\}$	Class 1 construction engineer as 1, Class 2 construction engineer as 0.5, or else 0
	If the leader presided over similar projects	5	S	$\{0, 0.5, 1\}$	More than once as 1, once as 0.5, or else 0
	Title of technical leader	6	S	$\{0, \cdots, 3\}$	Professor senior engineer as 3, senior engineer as 2, intermediate engineer as 1, or else 0
	If technical leader presided technically similar projects	7	S	$\{0, 0.5, 1\}$	More than once as 1, once as 0.5, or else 0
	The allocation of other main personnel	8	S	$\{0, 0.8, 1\}$	Complete as 1, basically complete as 0.8, or else 0
	Labor force	9	S	$\{0, 0.8, 1\}$	Enough as 1, basically enough as 0.8, or else 0
	Construction equipments	10	S	$\{0, 0.8, 1\}$	Enough as 1, basically enough as 0.8, or else 0
	Testing, detecting instruments	11	S	$\{0, 0.8, 1\}$	Enough as 1, basically enough as 0.8, or else 0
	Resource planning	12	C	$[0, 1]$	Score by experts
Construction and management	Construction scheme and technical measures	13	C	$[0, 1]$	Score by experts
	Construction plane layout	14	C	$[0, 1]$	Score by experts
	Quality management system and measures	15	C	$[0, 1]$	Score by experts
	Occupational health and safety management system and measures	16	C	$[0, 1]$	Score by experts
	Environmental protection management system and measures	17	C	$[0, 1]$	Score by experts
	The progress plan and measures	18	C	$[0, 1]$	Score by experts
Documentation	Integrity of technical documents and the level	19	S	$\{0, 0.8, 1\}$	Complete as 1, basically complete as 0.8, or else 0

Table 114.2 The business attribute set of contractor

Attribute subset	Attribute	No.	Type	Range	Brief description
Quoted price	Total quoted price	20	C	{0, 0.8, 1}	Million yuan
	Engineering quantity list priced	21	S		Complete as 1, basically complete as 0.8, or else 0
Situations of enterprises	Construction qualification	22	S	{0, · · ·, 5}	Overall contract: special grade as 5, first grade as 4, second grade 3, third grade 2. Specialty contract: first grade as 3, second grade 2, third grade 1, other 0
	Age limit of qualification	23	S	Natural number	
	Reputation, litigation and arbitration in recent years	24	C	[0, 1]	Score by experts
	Registered capital fund	25	C		Ten thousand yuan
	Assets and liabilities at the end of the previous year	26	C		Net asset, ten thousand yuan
	Profit in the previous year	27	C		Ten thousand yuan
	Registered staff size	28	S	Natural number	
	If enterprise adopted project management method	29	S	{0, 1}	Adopted as 1, otherwise 0
	If enterprise passed through the ISO quality certification, and operated for a while	30	S	{0, 0.5, 1}	Not passed as 0, passed as 0.5, operated for a while as 1
	If enterprise completed similar projects in recent years	31	S	{0, 0.8, 1}	More than once as 1, once as 0.8 or else 0
	If similar projects have completed inspection and acceptance	32	S	{0, 1}	Yes as 1, otherwise 0
Service promise	Promise of biding documents, follow-up service, energy conservation, and so on	33	S	{0, 1}	Assurance service as 1, otherwise 0

Table 114.3 Project type nodes

I_{34}	Industrial project	Yes as 1, other as 0
I_{35}	Civil project	Yes as 1, other as 0

Table 114.4 The magnitude of satisfaction

Satisfaction	Very sat.	Quite sat.	Sat.	Generally sat.	Intermediate state	Generally dissat.	Dissat.	Quite dissat.	Very dissat.
Magnitude	0.9	0.8	0.7	0.6	0.5	0.4	0.3	0.2	0.1

Sat. refers to satisfied
Dissat. refers to dissatisfied

the training samples and testing samples, the output layer of each training sample is dealt with by the following processing.

O_1 = quoted price/settlement price, project settlement price is according to the project price by actual contract completion (unit: million yuan), quoted price is I_{20}.

Consider that if the construction time is in accordance with the provisions, if yes, $O_2 = 1$, otherwise $O_2 = 0$.

$O_3 = 1$ is construction quality score. The construction quality is scored by 5 experts according to construction material and evaluation previously. The scores are within the range (0, 1), take the arithmetic mean.

Satisfactions of stakeholders of O_4, O_5, O_6, O_7, O_8, are their satisfactions about construction and construction contractor. Through distributing and recycling questionnaires, the satisfactions take the arithmetic mean of effective grading. The magnitude of satisfaction is shown in Table 114.4.

114.5.2 Training

To improve the training efficiency, we choose batching training, select the trained function in Matlab (i.e., the steepest descent method) as the training algorithm, and select the MSE (mean square error) as the performance function.

Use the preprocessing 12 training samples for batching training, the results converges steadily after 40 cycles, shown as in Fig. 114.1.

To train 100 times. So far, the predictor of construction output of contractors is built up.

114.5.3 Testing

On the basis of perfect forecast, take the predicting actual output as section indexes, $O_1, O_2, O_3, O_4, O_5, O_6, O_7, O_8, I_{13}, I_{20}$ as selection index set, to select the con-

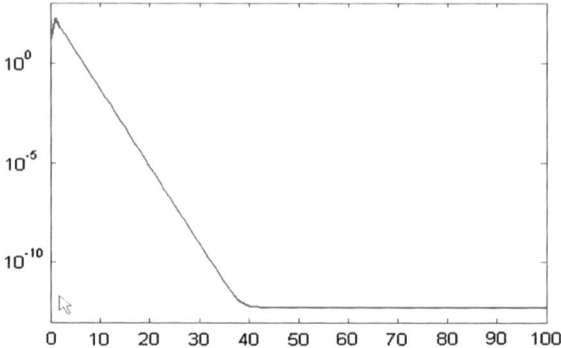

Fig. 114.1 Training sketch of the neural network

struction contractor. Take arithmetic mean method, and adjust O_1, I_{20} as follows:

$$O_1 \longrightarrow O_1 - |O_1 - 1|; I_{20} \longrightarrow \frac{\max\{I_{20}\} - I_{20}}{\max\{I_{20}\} - \min\{I_{20}\}}, \tag{114.1}$$

where $\max\{I_{20}\}$ is the highest quoted price of 3 construction contractors, $\min\{I_{20}\}$ is the highest quoted price.

Applying the above method to predict the construction contractor based on the 3 alternative samples, 10 indexes of each sample are obtained, the weighted average values are 0.927, 0.905, 0.882, so choose the first W company. In practice, W company is also the construction contractor of the project.

114.6 Conclusions

For large-scale project, the owners really concern the construction output of contractor, that is, that really influences construction projects is construction output, such as construction quality, engineering cost (budget), etc. Construction qualification and last year's profits don't have a lot to do with engineering, which are just indexes that used for predicting construction output. So it is necessary to establish a scientific and accurate prediction method, also very valuable.

This paper creatively utilized the neural network, which has the characteristics that has no need to master concrete rules in advance, and can learn by itself, to construct the predictor of actual output, it can predict actual construction output more accurately. On the basis of perfect forecast, choose actual output as index to select construction contractor, which is more reasonable. The optimization of input attributes also has been carried and realized by using Matlab in this paper. The method in the paper, therefore, overcame what was unreasonable, unscientific in logic and operation of previous contractor selection, and made it more scientific and more rigorous.

The further work, one is to subdivide large-scale project constructions in accordance with industry engineering types, such as water conservancy and hydropower, railway, electric power, building, etc, to make indexes and neural network structure more targeted; the other one is to collect more data for network training, to improve the predictive accuracy of neural network, consequently maker it better be used into large-scale construction projects, help to select the construction contractor, optimize engineering construction, and improve the economic and social benefits of large-scale construction projects.

Acknowledgments General Project of the Education Section of Sichuan Province (13SB0060): Ecological Risk Analysis and Countermeasures Research in the Construction of Major Projects in Sichuan. Key Project of Xihua University (zw1221504): Study on Ecological Environmental Risk of Large-scale Project.

References

1. Chen Y, Zhang N, Yang Q (2010) Literature review on project delivery systems. Constr Manage Modernization 24(5):473–478 (In Chinese)
2. Daft RL (2009) Organization theory and design. Cengage Learning, Mason
3. Guan B (2009) Decision making of types of general contractors in alternative projects. Constr Manage Modernization 23(3):217–220 (In Chinese)
4. Han L (2006) Artificial neural network tutorial. Posts and Telecommunications University Press, Beijing (In Chinese)
5. Hecht-Nielsen R (1989) Theory of the back propagation neural network. In: IEEE international joint conference on neural networks IJCNN, 1989, pp 593–605
6. Holt GD (1995) A methodology for predicting the performance of construction contractors. PhD Thesis, University of Wolverhampton, Wolverhampton
7. Hou J, Chi H, Li Y (2008) The evaluation on bidding choice of construction project contractor based on rough set theory. Commercial Res 1:31–33 (In Chinese)
8. Li F, Shi D (2008) The fuzzy comprehensive decision in the evaluation of bid on architectural design schemes. J Xi'an Pet Univ (Social Science Edition) 17(3):45–49 (In Chinese)
9. Liu Y (2002) Construction project bidding and contract management. Northern Jiaotong University Press, Beijing (In Chinese)
10. Mahdi IM, Riley MJ et al (2009) A multi-criteria approach to contractor selection. Eng Constr Architectural Manage 9(1):29–37
11. National Development Planning Commission and Other Seven Ministries (2003) Engineering construction project bidding approach. National Development Planning Commission and Other Seven Ministries 30 orders (In Chinese)
12. People set the standard documentation standard construction tender pre-qualification documents (2007) edition. China Planning Press, Beijing (In Chinese)
13. People set the standard documentation standard construction tender documents (2007) edition. China Planning Press, Beijing (In Chinese)
14. Russell JS (1992) Decision models for analysis and evaluation of construction contractors. Constr Manage Econ 10(3):185–202
15. Shi J, Tian S (2006) Intermediary organization evaluation system. Techno-economic 1:91–93 (In Chinese)

16. Xu J, Wu W (2006) Theory and method of multi-attribute decision making. Tsinghua University Press, Beijing (In Chinese)
17. Yang G, Yuan B (2006) Proficient MATLAB7.0 mixed programming. Electronic Industry Press, Beijing (In Chinese)
18. Yin C, Yin H (2002) Artificial intelligence and expert systems. China Water Conservancy and Hydropower Press, Beijing (In Chinese)
19. Zhu Z, Liu Z, Wang F (2001) Ahp application in the selection of the project contractor. J Zhejiang Univ (Enginering Science) 35(5):567–571 (In Chinese)

Chapter 115
Research on GEM Companies' Investment Behaviors and Governance Factors' Incentive and Constraint Effects

Sheng Ma and Rui Wang

Abstract Corporate governance mechanism has a positive effect on improving enterprise investment decisions. The establishment of growth enterprise market (GEM) provides empirical evidence for the research on investment efficiency of SMEs and the improvement of relevant governance mechanism. The free cash flow is closely related to corporate investment behavior. A benign internal governance structure can significantly restrain over-investment; however, it has limited effect on relieving under-investment. The external governance environment can effectively improve corporate investment decision-making.

Keywords Over-investment · Under-investment · GEM company · Governance factors

115.1 Introduction

Starting from functions of incentive and constraint of corporate governance mechanism, this thesis plans to explore its effect on investment behavior of GEM companies, and analyze and examine the following three questions. Firstly, on one hand the information asymmetry [3] makes a company's situation is not known to outside world, and due to financing constraint, a company lacks sufficient cash

S. Ma (✉) · R. Wang
Business School, Sichuan University,
Chengdu 610064, People's Republic of China
e-mail: mahongyi76@163.com

S. Ma
Accounting Department, College of Chengdu,
Chengdu 610106, People's Republic of China

J. Xu et al. (eds.), *Proceedings of the Eighth International Conference on Management Science and Engineering Management*, Advances in Intelligent Systems and Computing 281, DOI: 10.1007/978-3-642-55122-2_115, © Springer-Verlag Berlin Heidelberg 2014

flow to invest in projects with positive NPV, which results in under-investment; on the other hand, agency problems [2] make managers to meet their own interests, and apply surplus cash flow to invest in projects with negative NPV, resulting in over-investment. Then, is there under-investment due to shortage of cash flow or over- investment due to surplus cash flow in GEM companies? Secondly, benign corporate governance mechanism contributes to improvement of corporate investment decisions, while functions of incentive and constraint of governance complement each other [7]. Can the internal governance structure play the role of incentive and constraint, so as to curb the inefficient investment? Finally, the external governance environment is as important as the internal governance structure. Then does a companys external governance environment have the same effect on inefficient investment behavior? This thesis attempts to comprehensively examine the effect of functions of incentive and constraint of corporate governance mechanism on investment decisions of SMEs, taking GEM companies as samples and using for reference the model of Richardson [4] to measure inefficient investment degree. Therefore, this thesis provides useful reference for increasing investment efficiency of GEM companies and improving corporate governance mechanism, and deepens and expands the research on the field of corporate investment.

115.2 Research Hypotheses

1. Does the Inefficient Investment Behavior Exist?

Hypothesis 1a. The more shortage of free cash flow a GEM company faces, the more serious the under-investment is.
Hypothesis 1b. The more surplus free cash flow a GEM company has, the more serious the over-investment is.

2. The Effect of Incentive And Constraint of Internal Governance Structure on Investment Behavior

Hypothesis 2. "High salary to nourish honesty" and managerial ownership have incentive effect, and curb inefficient investment behavior, if other conditions remain unchanged.
Hypothesis 3. If other conditions remain unchanged, the relationship between the share ratio of the largest shareholder and corporate inefficient investment is an inverted "U"-shaped; the greater the counterbalance of other major shareholders is, the stronger the constraint effect on inefficient investment behavior is.

3. The Effects of Incentive and Constraint of External Governance Environment on Investment Behavior

Hypothesis 4. If other conditions remain unchanged, better regional innovation capability has the incentive effect on relieving inefficient investment.
Hypothesis 5. If other conditions remain unchanged, a favorable legal environment has a constraint effect on inhibiting inefficient investments.

115.3 Research Design

1. Samples Selection

Since GEM was just established four years ago, and most variables to measure inefficiency model in this thesis are data from financial statement of the previous year, therefore, the initial samples selected by this thesis are companies listed on GEM in 2010–2011, excluding samples with IPO that year and samples missing data. We also processed extreme value samples by Winsorize, and ultimately selected 154 companies listed on GEM as the study samples. Financial data used in this thesis are from CSMAR database; regional innovation capability index and regional legal environment index in the external governance environment are from China's Regional Innovation Capability Report compiled by China's Science and Technology Development Strategy Research Team [5] and Report on Business Environment Index of China's Provinces compiled by Wang Xiaolu et al. Statistical software is STATA 12.0.

2. Model Building and Variable Definition

Richardson [4] model can estimate the expected newly increased business investment, subtract which from the actual newly increased investment, we can get the balance, the unexpected investment, i.e, the residue of model. The positive residual indicates over-investment, and negative residual represents under-investment. The greater the absolute value of residual is, the more serious the over-investment and under-investment. This econometric model is widely used in studies by domestic scholars, including Wei and Liu [6], Huang and Huang [1]. This thesis also adopts this model, and firstly estimates the extent of over-investment and under-investment (model 115.1), then calculates the surplus of free cash flow and the shortage of internal cash flow, which are respectively denoted by variables PFCF and NFCF. On this basis, the econometric model is built to examine the relations between over-investment and free cash flow surplus, and the relations between under-investment and internal cash flow shortage; and cross multiply items $PFCF_t * CG_t$ and $NFCF_t * CG_t$ constructed by internal and external incentive and bounded variable CG_t (Corporate Governance) of corporate governance is introduced to examine the effects of incentive and constraint factors of corporate governance on business investment behavior (models 115.2 and 115.3). If regression coefficients of $PFCF_t$ and $NFCF_t$ in models 115.2 and 115.3 are significantly positive, it shows that a company's free cash flow surplus results in over-investment and internal cash flow causes under-investment. If regression coefficients of $PFCF_t * CG_t$ and $NFCF_t * CG_t$ in models 115.2 and 115.3 are significantly negative, it indicates that corporate governance factors have incentive and constraint effects, otherwise, they are ineffective. The meanings of relevant variables in the model are shown in Table 115.1.

$$Invest_t = \beta_0 + \beta_1 Growth_{t-1} + \beta_2 Lev_{t-1} + \beta_3 Cash_{t-1} + \beta_4 Age_{t-1} + \beta_5 Size_{t-1}$$
$$+ \beta_6 EPS_{t-1} + \beta_7 Invest_{t-1} + Year + Industry + \varepsilon, \quad (115.1)$$

$$Overinvest_t = \beta_0 + \beta_1 PFCF_t + \beta_2 CG_t + \beta_3 PFCF_t \times CG_t + Year + Industry + \varepsilon, \quad (115.2)$$

Table 115.1 Relevant variables in models and their meanings

Variable name			Variable declaration and calculating method
Dependent variable	$Invest_t$		Newly increased investment (cash paid by purchasing fixed assets, intangible assets and other long-term assets—cash received by selling fixed assets, intangible assets and other long-term assets at this period)/total assets at the beginning of the year
	$Overinvest_t$		Overinvestment, residual greater than zero in model 115.1
	$Underinvest_t$		Underinvestment, residual less than zero, and absolute value is adopted
Independent variable	$Growth_{t-1}$		Investment opportunity, Tobin'Q replacement, Tobin'Q = (market price of tradable shares at the beginning of the year+non-tradable shares at the beginning of the year and book value of liabilities)/total assets at the beginning of the year
	Lev_{t-1}		Assets-liabilities ratio, total liabilities at the beginning of the year/total assets at the beginning of the year
	$Cash_{t-1}$		Capital source, Net cash flow from operating at the beginning of the year/total assets at the beginning of the year
	Age_{t-1}		Enterprise age, listed year as of the end of previous year
	$Size_{t-1}$		Enterprise scale, natural logarithm of total assets at the beginning of the year
	EPS_{t-1}		Earnings per share, indicates a company's profitability
	$PFCF_t$		Cash flow surplus, positive value of (Net cash flow from operating/total assets at the beginning of the year-newly increased investment)
	$NFCF_t$		Cash flow shortage, negative value of (Net cash flow from operating/total assets at the beginning of the year-newly increased investment), and absolute value is adopted
	CG_t	$In(TPay)_t$	Monetary remuneration, natural logarithm of total remuneration of top three senior executives
		$Mhold_t$	Management shareholding, total shareholding ratio of senior executives published in annual report
		$Top1_t$	Shareholding ratio of major shareholders
		$Top1_t^2$	Squared value of shareholding Ratio of major shareholders
		$Top2-10_t$	Sum of ratios of second largest to tenth largest shareholders, representing counterbalance of other major shareholders
		$Innovate_t$	Innovation capability index , higher value representing better regional innovation Capability
		$Legal_t$	Legal environment index, higher value representing better regional legal environment. Since there is no correspondence index of year 2011, this thesis estimates index of year 2011 by average method
	Year		Yearly dummy variable
	Industry		Industry dummy variable

$$\text{Underinvest}_t = \beta_0 + \beta_1 \text{NFCF}_t + \beta_2 CG_t + \beta_3 \text{NFCF}_t \times CG_t + \text{Year} + \text{Industry} + \varepsilon.$$
$$(115.3)$$

115.4 Empirical Results and Analysis

1. Inefficient Model Regression Result

The regression result in model 115.1 of Table 115.2 shows that enterprise financial leverage has a constraint effect on newly increased investment, and they have significantly negative correlation. The relationship between enterprise age and newly increased investment is negative correlation, and is notable at the level of 1 %. Effects of other variables on newly increased investment are consistent with expectation. There is the significantly positive correlation between the investment expenditure of the prior period and newly increased investment of the current period. Tobin Q value coefficient is positive, but not significantly. Meanwhile $Adj - R^2$ reaches 33.71 %, indicating a good model fitting.

2. Descriptive Statistic Analysis

Descriptive statistic result of variables is shown in Table 115.3. From the results we can see that 40 % of GEM companies have the problem of over-investment, with average over-investment volume accounting for 5.31 % of total assets; 60 % of GEM companies have the problem of under-investment, with average investment shortage accounting for 3.5 % of total assets. 67 % of GEM companies have positive cash flow, with the mean value accounting for 5.44 % of total assets; 37 % of GEM companies have cash flow shortages, with the mean value accounting for 5.22 % of total assets. The maximum of the sum of monetary remuneration of senior managers is 8,044,024 yuan, and minimum is 124,012 yuan, which shows the big difference in senior mangers' remuneration probably due to differences in company size, profitability and growth. Generally, senior managers have high ratios of shares, with the average sum of shareholding ratios being 38.21 %, maximum being 89.70 %, minimum being 0, which indicates that control of senior managers over GEM companies are different. The average shareholding ratio of largest shareholders is 33.53 %, which shows ownership concentration of GEM companies; the maximum of shareholding ratio is 65.21 %, which shows the presence of "sole majority shareholder". The mean value of shareholding ratios of top ten shareholders is 69.85 %, which shows that top ten shareholders can basically control over GEM companies.

3. Effects Analysis of Incentive and Constraint Factors of Corporate Governance on Corporate Investment Behavior

Firstly, to examine effect of governance factors on over-investment behavior of company with free cash flow surplus. It is shown in Table 115.4 that whether governance factor variable is taken into account or not, there is significant positive correlation between over-investment and free cash flow. The more free cash flow a company has, the more serious the over-investment is, which is in line with empirical results of Huang and Huang [1], Hypothesis 1a. has been verified.

Table 115.2 Model 115.1 inefficient investment model regression result

	Intercept	Growth$_{t-1}$	Lev$_{t-1}$	Cash$_{t-1}$	Age$_{t-1}$	Size$_{t-1}$	EPS$_{t-1}$	Invest$_{t-1}$
Expectation symbol	+	–	+	–	+	+	+	
Coefficient	0.3472	0.0154	−0.0143* (−3.0109)	0.0690** (2.0901)	−0.0491* (−2.6957)	0.0200***	0.0021	0.8170* (7.8945)
	(1.4511)	(0.9723)				(1.9506)	(1.1423)	

Notes *, **, *** respectively represents they are significant at the level of 1 %, 5 %, 10 %.
In brackets are *T* values of coefficients of correspondence. Coefficient of Adj − R^2 is 0.3371

Table 115.3 Descriptive statistic of variables

Variable	Mean value	Median	Standard deviation	Minimum	Maximum
Overinvest	0.0531	0.0391	0.0500	0.0011	0.2426
Underinvest	0.0353	0.0323	0.0261	0.0011	0.1172
PFCF	0.0544	0.0475	0.0363	0.0014	0.1683
NFCF	0.0522	0.0396	0.0454	0.0017	0.2427
TPay	110.1567	89.8321	86.2362	12.4012	804.4024
(10,000 yuan)					
Mhold	0.3821	0.4082	0.2182	0.0000	0.897
Top1	0.3353	0.3187	0.1281	0.0872	0.6521
Top2-10	0.3632	0.3660	0.1112	0.0272	0.6181

Notes The above data are from CSMAR database

According to column 3 of Table 115.4, there is a significant negative correlation between over-investment and monetary remuneration and shareholding of senior executives, which indicates that the internal incentive mechanism makes interests of executives are in line with that of shareholders and over-investment has been significantly alleviated. Column 3 examines the effect of the ownership structure on over-investment. Coefficients of Top1 and Top1 $*$ PFCF at the 1 % level are significantly positive, and the coefficient of Top12 is significantly negative, indicating there is an inverted "U"-shaped relationship between shareholding ratio of the largest shareholder and over-investment of the GEM company. With the increase of shareholding ratio of the largest shareholder, the separation of control power and power of cash flow becomes smaller, and goals of shareholders and company become consistent. Major shareholders will stop conspiring with executives, and supervise executives, so as to curb over-investment and protect their own interests. Coefficient signs of Top2-10 and Top2-10 $*$ PFCF are consistent with expectation, but significant of results are not high, which shows that there is not enough counterbalance of other shareholders to the largest shareholder. Column 4-5 examine effects of incentive and constraint of the external governance environment on over-investment behavior. There is a significant correlation between over-investment Innovate $*$ PFCF and Legal $*$ PFCF, and the significance level reaches 1 %, indicating a region with good innovation capability and favorable legal environment can effectively promote development of GEM companies, and make value maximization as the goal of enterprise's investment behavior. Therefore, most corporate governance factors in the research hypothesis have good incentive and constraint effects on over-investment behavior of GEM companies.

Secondly, Table 115.5 shows the effect of governance factors on under-investment of a company due to free cash flow shortage. It can be observed in Table 115.5 that there is a significantly negative correlation at the level of 1 % between under-investment and cross items Innovate $*$ PFCF and Legal $*$ PFCF which present the external governance environment, indicating that a region with better innovation capability and legal environment can provide favorable business environment and

Table 115.4 Model 115.2 Governance factors and over-investment due to cash flow surplus regression results

Independent variable	Expectation symbol	PFCF	CG^a	CG^b	CG^c	CG^d
Intercept		0.0521 (0.6435)	0.0501* (3.7100)	0.0861 (0.7234)	0.0501* (3.7322)	0.0475* (3.6510)
PFCF	+	0.1445* (3.9105)	0.1112*** 1.9641	0.4248* (2.7961)	0.7190* (2.7154)	0.1783 (1.5608)
ln(TPay) * PFCF	−		−2.020* (−2.8201)			
Mhold * PFCF	−		−0.3722* (−3.3494)			
Top1	+			0.2711* (4.3648)		
Top1^2	−			−0.6011*** (−1.9372)		
Top2-10	+			−0.1429*** (−1.8972)		
Top1 * PFCF	+			0.5306* (5.4171)		
Top2-10 * PFCF	−			−0.2715 (−0.6200)		
Innovate * PFCF	−				−0.0138* (−4.8170)	
Legal * PFCF	−					−1.6208* (−3.4321)
Year		Control	Control	Control	Control	Control
Industry		Control	Control	Control	Control	Control
Adj-R^2		0.1194	0.1360	0.1431	0.1212	0.1187

Notes *, **, *** respectively represents they are significant at the level of 1 %, 5 %, 10 %. In brackets are T values of coefficients of correspondence. ^{a} Remuneration Incentive; ^{b} Shareholding Constraint; ^{c} Innovation capability Incentive; ^{d} Legal Environment Constraint

Table 115.5 Model 115.3 Regression result of governance factors and under-investment due to cash flow shortage

Independent variable	Expectation symbol	NFCF	CG[a]	CG[b]	CG[c]	CG[d]
Intercept		0.0347* (5.2891)	0.0388* (5.0900)	0.0391 (0.4977)	0.3312* (4.8801)	0.0321* (4.5915)
NFCF	+	0.0441*** (1.8437)	0.7613*** (1.8539)	0.0428 (0.7332)	0.1823*** (1.9537)	0.2206 (1.5195)
ln(TPay) * NFCF	−		−0.0573 (−1.3809)			
Mhold * NFCF	−		−0.2347 (−1.5021)			
Top1	+			0.1134*** (1.8890)		
Top1²	−			−0.2103 (−0.7982)		
Top2-10	−			−0.0167 (−0.5329)		
Top1 * NFCF	+			0.0482 (1.1278)		
Top2-10 * NFCF	−			−0.0253 (−0.4981)		
Innovate * NFCF	−				−0.0031* (−3.5671)	
Legal * NFCF	−					−0.3710* (−2.6891)
Year		Control	Control	Control	Control	Control
Industry		Control	Control	Control	Control	Control
Adj − R²		0.0782	0.1213	0.1591	0.1160	0.1257

Notes *, **, *** respectively represents they are significant at the level of 1 %, 5 %, 10 %. In brackets are T values of coefficients of correspondence. [a] Remuneration Incentive; [b] Shareholding Constraint; [c] Innovation capability Incentive; [d] Legal Environment Constraint

financial and technological support, guide and encourage GEM companies to make efficient investment decisions, so as to alleviate under-investment.

However, we can also see that the variable cross item In(TPay) $*$ NFCF which is related to internal incentive mechanism and Mhold $*$ NFCF coefficient sign are consistent with expectations, but results are not significant, which indicates that monetary remuneration and executives shareholding mechanism cannot effectively alleviate under-investment. This result is in line with the research of Xin Qingquan et al. Moreover, results of shareholding variables of Top1, Top12 and Top2-10 representing internal constraint mechanism are not significant. In summary, the external governance environment can effectively alleviate under-investment of enterprises, but there is only weak evidence proves the hypothesis that the internal governance factors can alleviate under-investment.

115.5 Research Conclusion and Suggestions

Like the main board market of A-share, the information asymmetry and agency problems will also result in under-investment or over-investment of GEM companies with private background and high-growth. However, due to these features of GEM companies, we found that by designing effective remuneration incentive, establishing good ownership structures, enhancing regional innovation capabilities and improving relevant legal system, more efficient enterprise investment decisions can be made. Taking GEM companies listed in 2010–2011 as samples, this thesis starts with effects of incentive and constraint resulting from corporate governance, examines the relationship between enterprise investment behavior and internal and external governance factors, and finds that: firstly, GEM companies are faced with the problems of over-investment due to cash flow surplus and under-investment due to cash flow shortage; secondly, as for the internal governance factors, favorable remuneration design and shareholding structure have effects of incentive and constraint, so as to curb over-investment of GEM companies, but have limited impact on enterprise's under-investment; at last, in regard of the external governance environment, better regional innovation capabilities and legal environment can guide GEM companies to make efficient investment decisions.

In view of the above research results, this thesis proposes the following suggestions.

1. To Improve the Construction of Multi-level Capital Market, and Develop an Open and Transparent Financing Platform of GEM

Currently, domestic economic growth slows down gradually. In order to maintain the sustained and healthy economic development, China has carried out industrial restructuring and introduced a number of policies to support emerging industries. Although direct investment and subsidies of state fund plays a role in guiding the market, it only accounts for a small part of total social investment. Meanwhile, innovation activities are uncertain, the characteristic of "small size and high risk" of

SMEs make they cannot secure funding in the more active publicly trading market, which requires the construction of the "ladder" financing platform to improve the downward extension of capital market. The advantages of Angel funds and venture capital can help start-up enterprises to solve difficulties in human resources, technology and funding, etc., therefore, the government should issue policies to support the development of these market powers, and construct high-tech industrial park with good software and hardware facilities for these enterprises. Through massive trial and error of start-ups enterprises by social funds, high-tech enterprises which are in line with national economic transition and have good growth abilities can be found to be cultivated objects, and ultimately go public in GEM, so as to reduce systematic risk of economic transition. On the other hand, enterprises listed in GEM have higher risks compared to those listed in the main board market, and have some problems in business compliance and information disclosure due to their private-owned background, which need regulators including exchanges and CSRC to further improve development of GEM system, and strengthen supervision on enterprise, improve the quality of information disclosure, so as to make GEM an open and transparent financing platform.

2. To Improve the Protection System for Investors, and Enhance Judicial Protective Efficiency

Most GEM companies are private-owned enterprises, with the largest shareholder being an individual or a family and shareholding concentration. On the premise of existence of capital majority principle, large shareholders provide major support for a company's development, and their interests are in line with a company's operation prospect, therefore, large shareholders naturally have more decision-making power of a company. However, when large shareholders are not honest and make decision illegally to seek personal gains, the interests of minority shareholders will be damaged. In the course of judicial litigation, minority shareholders have high litigation cost, while violation cost of large shareholders is low, which severely hampers the efficiency of judicial protection. Therefore, it is necessary to increase violation cost of large shareholders, develop convenient judicial proceedings of investors, expand investor protection channel, and create specific voting mechanisms including voting-avoidance of relevant shareholders and vote-accumulating to make necessary restriction on large shareholders' rights, so as to grant equal juridical status for large shareholders and minority shareholders, and let them comply with their own legal obligation, thereby to improve efficiency of judicial aid.

References

1. Huang J, Huang N (2012) Over-investment, debt structure and governance effect-empirical evidence from China's listed real estate companies. Acc Study 9:67–72
2. Jensen MC (1986) Agency costs of free cash flow, corporate finance and takeovers. Am Econ Rev 76:323–329

3. Myers SC, Majluf NS (1984) Corporate financing and investment decisions when firms have information that investors do not have. J Fin Econ 13(2):187–221
4. Richardson S (2006) Over-investment of free cash flow. Rev Acc Stud 11(2–3):159–189
5. China's Technological Development Strategy Research Team (2011) China's regional innovation capabilities report 2010: research on innovation system of Pearl River Delta. Science Press, China
6. Wei M, Liu J (2007) SOEs' dividends, governance factors and over-investment. Manag World 4:88–95
7. Xie Z (2008) Some problems involved in corporate governance. Acc Study 12:62–68

Chapter 116
Earthquake Emergency Supplies Scheduling Order Based on Grey Correlation–Fuzzy Evaluation

Fumin Deng, Ting Wang, Guizhi Zeng and Xuedong Liang

Abstract In recent years, the earthquake disasters occurred frequently in our country, and caused huge losses in aspects of property and personnel. After the earthquake, it is very important to dispatch the emergency supplies timely and efficiently. Various and kinds of emergency supplies are required in the process of rescue. It needs scientific and efficient transport scheduling, in order to reduce earthquake damage as much as possible. In this paper, the scheduling order was obtained by grey relation method and fuzzy comprehensive evaluation method on the base of analyzing the factors which influence the urgency of the emergency supplies. Case study shows that using this method can not only obtain a scientific and reasonable emergency supplies scheduling order, but also provide a reference to the earthquake emergency resource scheduling.

Keywords Earthquake disaster · Emergency supplies · Scheduling order · Grey relation method · Fuzzy comprehensive evaluation method

116.1 Introduction

In recent years, the earthquake disasters occurred frequently in our country, and caused huge losses in aspects of property and personnel. For example, Ya'an earthquake happened in 2013 took the lives of nearly 200 people; East Japan earthquake in 2011 caused more than 25,000 people to be dead [2]; more than 220,000 people were killed in the earthquake in Haiti in 2010 [1]; More than 87,000 people died and missed in Wenchuan earthquake [10]. Because the rich experience in the years of confrontation against nature, people gradually realized the importance of dispatching the emergency supplies timely and efficiently in the earthquake relief. However, the

F. Deng (✉) · T. Wang · G. Zeng · X. Liang
Business School, Sichuan University, Chengdu 610064, People's Republic of China
e-mail: dengfum@sina.com

J. Xu et al. (eds.), *Proceedings of the Eighth International Conference on Management Science and Engineering Management*, Advances in Intelligent Systems and Computing 281, DOI: 10.1007/978-3-642-55122-2_116, © Springer-Verlag Berlin Heidelberg 2014

correct analysis and understanding of the urgency of the emergency supplies is the premise of the smooth scheduling.

Research on the problem of emergency supplies scheduling started from the 1950s, and scholars studied from various aspects. After decades of research and development, many new scheduling method and theory has been formed. Scholars who studied on the distribution of emergency supplies and scheduling decisions mainly used the methods of mathematical programming model, the network flow model, the game theory model, combinatorial mathematics and so on [6]. According to the characteristics of the disaster emergency rescue management and the number of emergency point, Zhou divided the whole process of emergency supplies into different stages by dynamic programming method, constructed the mathematical model and configured and dispatched for each stage [11]. Tian described the demand of emergency supplies by means of triangular fuzzy number in fuzzy mathematics, simulated the real dynamic network traffic using the speed time dependent function, and set up a multi-objective mathematical model of emergency supplies dynamic scheduling [8]. Sheu put forward a dynamic demand management model of emergency supplies when the information resources are not enough. This model mainly included three steps: demand forecasting based on data fusion, fuzzy clustering on the affected area and sorting the priority [7]. Lin and Batta put forward a multi-objective integer programming model for the operation of the key distribution problems in the disaster relief [5].

116.2 Analysis of Influencing Factors in Demand Urgency of Emergency Supplies

The demand urgency of emergency supplies is not only influenced by a certain factor. In this paper, the factors include the following three aspects through the study of relevant literature: the inadequate extent of emergency supplies, the irreplaceable extent of emergency supplies, the extent of harm when emergency supplies is not enough [9].

1. The inadequate extent of emergency supplies. It refers to the residual quantity that the total number of some goods minus the existing number in the process of rescue. It needs to consider from two aspects when evaluate this factor: the number of gap and the gap rate. It can't reflect the gap of supplies objectively when only consider the number of gap. In general, the more insufficient the emergency supplies are, the more urgent it is.
2. The irreplaceable extent of emergency supplies. It refers to the extent that goods can't be replaced by other supplies which have the same function or effect in the process of rescue. It is also need to consider from two aspects when evaluate this factor: the function of the supplies and the existing quantity of the supplies. In general, the more irreplaceable the emergency supplies are, the more urgent it is.

3. The extent of harm when emergency supplies are not enough. It refers to the extent that the safety of the injured is threatened by the absence of some goods. It will seriously affect the life and health when some emergency supplies are in shortage. If these supplies can't be supplied promptly, it will lead to death in a short time. For example, medical supplies, food, etc. Some supplies' shortage has less threat to the health. For example, the flashlight, communication equipment, etc. In general, the more harm when emergency supplies is not enough, the more urgent it is.

116.3 Determine Index Weight Using the Grey Correlation Method

The indexes of evaluating the emergency supplies' urgency are obtained from the above analysis. When the experts evaluate the weights of indexes, they usually judge by the past experience. It can not accurately reflect the relative importance of each target. The essence of this method is to determine the reference sequence which composed the maximum of each expert's experience judgment value, and then seek correlation among them. The greater the correlation is, the greater the weight of the index is [4]. The main calculation method and steps are as follows.

1. The expert determine index weight according to the experience

Assuming that through the analysis, the indicator system which influence the urgency of emergency supplies is $X = \{X_1, X_2, X_3\}$, which X_1 represents the inadequate extent of emergency supplies, X_2 represents the irreplaceable extent of emergency supplies, X_3 represents the extent of harm when emergency supplies is not enough. Invite m experts to judge the weights of three indexes at the same time, and obtain the experience judgment data column of each index weight. It is expressed as respectively: $X_1 = (x_1(1), x_1(2), \cdots, x_1(m))$, $X_2 = (x_2(1), x_2(2), \cdots, x_2(m))$, $X_3 = (x_3(1), x_3(2), \cdots, x_3(m))$.

2. Determine the reference sequence

Find the maximum of each expert's experience judgment value, and compose the reference sequence X_0: $X_0 = (x_0(1), x_0(2), \cdots, x_0(m))$.

3. Calculate the correlation coefficient and correlation

Calculate the correlation coefficient $\varepsilon_{0i}(k)$ and correlation γ_{0i} between the reference sequence X_0 and the weight value by experience judgment using the formula.

$$\varepsilon_{0i}(k) = \frac{\min_{i} \min_{k} |X_0(K) - X_i(K)| + \rho \max_{i} \max_{k} |X_0(K) - X_i(K)|}{|X_0(K) - X_i(K)| + \rho \max_{i} \max_{k} |X_0(K) - X_i(K)|},$$

$$\gamma_{0i} = \frac{1}{m} \sum_{k=1}^{m} \varepsilon_{0i}(k).$$

In these formulas, ρ is Resolution Coefficient and ρ take 0.5 normally.

The size of each sequence correlation directly reflects the importance of each evaluation index relative to the set sequence. The greater the correlation is, the greater the weight of the index.

4. Normalized the correlation

Each index weight will be obtained by normalizing the correlation: $\omega_i = \frac{\gamma_{0i}}{\sum_{i=1}^{n} \gamma_{0i}}$.

116.4 The Fuzzy Comprehensive Evaluation Model

Many different kinds of emergency supplies s are needed in the process of rescue. The urgency of the emergency supplies s is a result combined with multiple factors. It can't just be judged accurately by personal subjective feeling. Fuzzy comprehensive evaluation is a method of multifactor decision-making that can reach a comprehensive and effective evaluation for things which affected by many factors based on the fuzzy mathematics. Fuzzy comprehensive evaluation method has the characteristics of systemic and clear. It can effectively combine quantitative and qualitative evaluation, control the influence of artificial factors and provides a feasible way to solve the problem of emergency supplies' urgent demand [3]. Fuzzy comprehensive evaluation includes the following steps.

1. Determine the evaluation index set

According to previous research, the indicator system which influence the urgency of emergency supplies is $X = \{X_1, X_2, X_3\}$, which X_1 represents the inadequate extent of emergency supplies, X_2 represents the irreplaceable extent of emergency supplies, X_3 represents the extent of harm when emergency supplies is not enough.

In the process of evaluation, three factors will be taken into account in order to reduce the error as much as possible. However, the three factors have different effects on the urgency of emergency supplies. So, it is need to determine the weight of each factor further more.

2. Determine the weights of evaluation indexes

Determine the weights of evaluation indexes ω using the grey correlation method.

3. Build index evaluation level set

According to the evaluation purpose, establish the evaluation level L: $L = L_1, L_2, L_3$, in which, $L_1 =$ veryurgent, $L_2 =$ urgent, $L_3 =$ general.

4. Determine the subordinate relations matrix M

Subordinate relationship matrix is also called the single factor evaluation matrix. It is been determined according to evaluation index set and the index evaluation level set:

Table 116.1 Experience judgment of each evaluation index weight value

Index	Expert							
	1	2	3	4	5	6	7	8
X_1	0.2	0.2	0.4	0.3	0.2	0.2	0.1	0.5
X_2	0.1	0.2	0.3	0.2	0.3	0.2	0.2	0.1
X_3	0.7	0.6	0.3	0.5	0.5	0.6	0.7	0.4

Table 116.2 Absolute difference sequence

Index	Expert							
	1	2	3	4	5	6	7	8
X_1	0.5	0.4	0	0.2	0.3	0.4	0.6	0
X_2	0.6	0.4	0.1	0.3	0.2	0.4	0.5	0.4
X_3	0	0	0.1	0	0	0	0	0.1

$$M = \begin{bmatrix} m_{11} & m_{12} & m_{13} \\ m_{21} & m_{22} & m_{23} \\ m_{31} & m_{32} & m_{33} \end{bmatrix}, \quad (0 \le m_{ij} \le 1).$$

5. Comprehensive assessment

The weight vector ω and the subordinate relations matrix M will be fuzzy composed using synthetic operator. The following is specific computing formula: $e_j = \sum_{i=1}^{n}(\omega_i m_{ij})$. It will get evaluation vector E of each kind of emergency supplies.

6. Determine urgency of the order of emergency supplies

Give the corresponding score S to the three index evaluation level. $S = s_1, s_2, s_3 = 30, 20, 10$, then calculate using the following formula: $f_t = \sum_{j=1}^{3}(e_j \times s_j)$.

    Rank F_t from big to small. The urgency of the order about many kinds of emergency supplies was obtained.

116.5 Application Analysis

Assuming that in a sudden earthquake emergency rescue, eight kinds of emergency supplies are needed: bottled water, food, tents, medical supplies, hydraulic shears, clothes, excavator, shovel.

    First, determine the evaluation index set $X = \{X_1, X_2, X_3\}$. Then determine the weights of evaluation indexes. Invite eight experts to judge the weights of three indexes at the same time. The results are shown in Table 116.1.

    The column is the weight of each index given by each expert. Find out the largest value judgment and set the reference sequence $X_0 = (0.7, 0.6, 0.4, 0.5, 0.5, 0.6,$

Table 116.3 Correlation coefficient

Index	Expert							
	1	2	3	4	5	6	7	8
X_1	0.375	0.429	1	0.6	0.5	0.429	0.333	1
X_2	0.333	0.429	0.75	0.5	0.6	0.429	0.375	0.429
X_3	1	1	0.75	1	1	1	1	0.75

Table 116.4 The membership degree matrix for each supply

Bottled water		V_1	V_2	V_3	Food		V_1	V_2	V_3
	u_1	0.7	0.3	0		u_1	0.2	0.8	0
	u_2	0.9	0.1	0		u_2	0	0.9	0.1
	u_3	1	0	0		u_3	1	0	0
Tents					Medical supplies				
	u_1	0.2	0.4	0.4		u_1	1	0	0
	u_2	0.1	0.5	0.4		u_2	1	0	0
	u_3	0.2	0.6	0.2		u_3	1	0	0
Hydraulic shears					Clothes				
	u_1	0.9	0.1	0		u_1	0.3	0.6	0.1
	u_2	1	0	0		u_2	0.2	0.5	0.3
	u_3	1	0	0		u_3	0.2	0.7	0.1
Shovel					Excavator				
	u_1	0	0.1	0.9		u_1	0.8	0.1	0.1
	u_2	0.1	0.1	0.8		u_2	0.3	0.5	0.2
	u_3	0	0.2	0.8		u_3	0.3	0.5	0.2

0.7, 0.4). Calculate in Table 116.1 according to the formula $\Delta_{0i}(K) = |X_0(K) - X_i(K)|$. And obtain the absolute difference sequence.

Table 116.2 illustrates $\min\limits_{i} \min\limits_{k} |X_0(K) - X_i(K)| = 0$, $\max\limits_{i} \max\limits_{k} |X_0(K) - X_i(K)| = 0.6$.

According to the correlation coefficient formula:

$$\varepsilon_{0i}(k) = \frac{\min\limits_{i} \min\limits_{k} |X_0(K) - X_i(K)| + \rho \max\limits_{i} \max\limits_{k} |X_0(K) - X_i(K)|}{|X_0(K) - X_i(K)| + \rho \max\limits_{i} \max\limits_{k} |X_0(K) - X_i(K)|},$$

and Table 116.2, the correlation coefficient between X_i and the reference sequence X_0 can be got and shown in Table 116.3.

Calculate according to the correlation formula. $\gamma_{0i} = \frac{1}{m} \sum_{k=1}^{m} \varepsilon_{0i}(k)$, $\gamma_{01} = 0.583$, $\gamma_{02} = 0.481$, $\gamma_{03} = 0.938$. Normalize the correlation. The weight of the index system was obtained. $\omega = (0.291, 0.240, 0.469)$. Decision-maker determined the membership degree matrix for each kind of supplies according to the actual demand as shown in Table 116.4.

Then make the synthetic operation with the weight of factors, each evaluation vector of emergency supplies will be obtained:

$$e_1 = (0.291, 0.240, 0.469) \begin{bmatrix} 0.7 & 0.3 & 0 \\ 0.9 & 0.1 & 0 \\ 1 & 0 & 0 \end{bmatrix} = (0.8887, 0.1113, 0),$$

$$e_2 = (0.291, 0.240, 0.469) \begin{bmatrix} 0.2 & 0.8 & 0 \\ 0 & 0.9 & 0.1 \\ 1 & 0 & 0 \end{bmatrix} = (0.5272, 0.4488, 0.0240),$$

$$e_3 = (0.291, 0.240, 0.469) \begin{bmatrix} 0.2 & 0.4 & 0.4 \\ 0.1 & 0.5 & 0.4 \\ 0.2 & 0.6 & 0.2 \end{bmatrix} = (0.1760, 0.5178, 0.3062),$$

$$e_4 = (0.291, 0.240, 0.469) \begin{bmatrix} 1 & 0 & 0 \\ 1 & 0 & 0 \\ 1 & 0 & 0 \end{bmatrix} = (1, 0, 0),$$

$$e_5 = (0.291, 0.240, 0.469) \begin{bmatrix} 0.9 & 0.1 & 0 \\ 1 & 0 & 0 \\ 1 & 0 & 0 \end{bmatrix} = (0.9709, 0.0291, 0),$$

$$e_6 = (0.291, 0.240, 0.469) \begin{bmatrix} 0.3 & 0.6 & 0.1 \\ 0.2 & 0.5 & 0.3 \\ 0.2 & 0.7 & 0.1 \end{bmatrix} = (0.2291, 0.6229, 0.1480),$$

$$e_7 = (0.291, 0.240, 0.469) \begin{bmatrix} 0 & 0.1 & 0.9 \\ 0.1 & 0.1 & 0.8 \\ 0 & 0.2 & 0.8 \end{bmatrix} = (0.0240, 0.1469, 0.8291),$$

$$e_8 = (0.291, 0.240, 0.469) \begin{bmatrix} 0.8 & 0.1 & 0.1 \\ 0.3 & 0.5 & 0.2 \\ 0.3 & 0.5 & 0.2 \end{bmatrix} = (0.4455, 0.3836, 0.1709).$$

At last, calculate the urgency score of emergency supplies using the formula:

$$f_t = \sum_{j=1}^{3} (e_j \times s_j),$$

$f_1 = (0.8887, 0.1113, 0)(30, 20, 10)^T = 28.887,$
$f_2 = (0.5272, 0.4488, 0.0240)(30, 20, 10)^T = 25.032,$
$f_3 = (0.1760, 0.5178, 0.3062)(30, 20, 10)^T = 18.698,$
$f_4 = (1, 0, 0)(30, 20, 10)^T = 30,$
$f_5 = (0.9709, 0.0291, 0)(30, 20, 10)^T = 29.709,$
$f_6 = (0.2291, 0.6229, 0.1480)(30, 20, 10)^T = 20.811,$
$f_7 = (0.0240, 0.1469, 0.8291)(30, 20, 10)^T = 11.949,$
$f_8 = (0.4455, 0.3836, 0.1709)(30, 20, 10)^T = 22.746.$

The results shows that the scores of bottled water, food, tents, medical supplies, hydraulic shears, clothes, shovel and excavator are respectively 28.887, 25.032, 18.698, 30, 29.709, 20.811, 11.949, 22.746. Sort these score from higher to lower. It can be conclude that the urgency sequence of the eight kinds of emergency supplies is medical supplies, hydraulic shears, bottled water, food, excavator, clothing, tents, and shovel.

116.6 Conclusion

The dispatch of emergency supplies is the key in the relief process of earthquake disaster, and completing this task successfully depends on exact analysis and ascertain of the urgency of emergency supplies. This paper proposed a sorting method for the emergency supplies schedule based on the grey relation-fuzzy evaluation. Furthermore, the analysis of application indicates that this method is feasibility and effectiveness. So this paper offers a solution to the decision makers which can dispatch the emergency supplies properly and improve the rescue efficiency significantly. However, this paper will provide a beneficial reference for scientific and effective management of the emergency supplies.

Acknowledgments This research is funded by the National Nature Science Foundation of China (Grand No. 71131006), China Postdoctoral Science Foundation Funded Project (Grand No. 2012M521705), Postdoctoral Science Special Foundation of Sichuan Province and the Fundamental Research Funds for the Central Universities (Grand No. skzx2013-dz07).

References

1. Chen H, Wang Z, Li C (2011) Haiti's earthquake disaster and its experience. Int Earthq Dyn 9:9
2. Dong Z, Cheng J, Guo G (2012) Japan's magnitude 9 earthquake and its influence on china earthquake activity. Inland Earthq 26(1):1–9
3. Gao C, Chen X, Wei C (2006) Application of entropy weight and fuzzy synthetic evaluation in urban ecological security assessment. J Appl Ecol 17(10):1923–1927
4. Jiang L, Lv S (2008) Credit risk evaluation based on grey correlation model. Mod Econ 1:9
5. Lin Y, Batta R (2011) A logistics model for emergency supply of critical items in the aftermath of a disaster. Socio-Economic Plann Sci 45(4):132–145
6. Pang H, Liu N (2012) Game models for incomplete put-out distribution of emergency supplies for natural disasters. J Zhejiang Univ: Eng Sci 46(11):2068–2072
7. Shen J (2010) Dynamic relief-demand management for emergency logistics operations under large-scale disasters. Transp Res Part E 46:1–17
8. Tian J, Ma W et al (2011) Emergency supplies distributing and vehicle routes programming based on particle swarm optimization. Syst Eng Theor Pract 31(5):898–906
9. Wang J, Wang H (2012) Gradation for demand urgency of emergency supplies in emergency relief. Comput Eng Appl 12:5
10. Wang T (2008) Statistical data in the wenchuan earthquake. China's Stat 6:42–43
11. Zhou X, Jiang L, Zhang Y (2007) Quantitative model research of the emergency resources optimization under the sudden accident. J Saf Environ 7(6):113–115

Chapter 117
A Class of Expected Linear Bi-level Programming with Random Fuzzy Coefficients

Xiaoyang Zhou, Yan Tu and Kin Keung Lai

Abstract In this paper, we consider a class of linear bi-level programming with random fuzzy coefficients, which has no mathematical meaning because of the uncertain factors. So in order to make it solvable, we introduced the linear bi-level model with expected objectives and constraints. And some theorems are proposed to obtain the equivalent model. Then we employ the interactive programming technique to deal with the bi-level equivalent model. At last an illustrative example is present to show the efficiency.

Keywords Linear bi-level programming · Random fuzzy variable · Expected objective · Expected constraints · Interactive programming technique

117.1 Introduction

Bi-level programming problem is a hierarchical optimization problem where a subset of the variables is constrained to be a solution of a given optimization problem parameterized by the remaining variables. The BP problem is a multilevel programming

X. Zhou (✉)
International Business School, Shanxi Normal University, Xi'an 710062, People's Republic of China
e-mail: x.y.zhou@foxmail.com

Y. Tu
Uncertainty Decision-Making Laboratory, Sichuan University, Chengdu 610064, People's Republic of China

Y. Tu
Decision Sciences Department, LeBow College of Business, Drexel University, Philadelphia, PA 19104, USA

K. Lai
Department of Management Sciences, City University of Hong Kong, Tat Chee Avenue, Kowloon, Hong Kong

J. Xu et al. (eds.), *Proceedings of the Eighth International Conference on Management Science and Engineering Management*, Advances in Intelligent Systems and Computing 281, DOI: 10.1007/978-3-642-55122-2_117, © Springer-Verlag Berlin Heidelberg 2014

problem with two levels. The hierarchical optimization structure appears naturally in many applications when lower level actions depend on upper level decisions.

The original formulation for bi-level programming appeared in 1973, in a paper authored by Bracken and McGil [3], although it was Candler and Norton [4] that first used the designation bi-level and multilevel programming. However, it was not until the early eighties that these problems started receiving the attention they deserve. Motivated by the game theory of Stackelberg [18], several authors studied bi-level programming intensively and contributed to its proliferation in the mathematical programming community.

The applications of bi-level and multi level programming include transportation (taxation, network design) [2, 11, 14, 19], management (coordination of multi-divisional firms, facility location, credit allocation) [1, 5], planning (agricultural policies electric utility) [8, 9], and engineering design [10].

In many realistic situations, uncertainty of data comes form two sources: randomness and fuzziness. Randomness models the stochastic variability of all possible outcomes of an experiment and fuzziness describes the vagueness of the given or just realized outcome. So in modelling realistic situations, fuzziness is often tied to randomness since possible random outcomes have to be described by fuzzy sets, especially in the case of linguistically expressed outcomes. The advantage of considering random fuzzy variables is as follows. At the level of experimentation it is often possible to use the experience of the experimenter for a justified modelling of the fuzziness of the outcomes. Statistical inference with random fuzzy variables transfers the fuzziness, e.g., into parameter estimators. Now, at this level, the level of decision, it may be necessary to defuzzify the vague parameter estimate, but it can be done in a more responsible way as on the level of experimentation because consequences of wrong decisions can be taken into account, Hence, the vagueness of experimental outcomes is carried with a random fuzzy variable and the associated statistical procedures up to the level of decision. Roughly speaking, the philosophy of random fuzzy variables if: do not defuzzify at the level of experimentation, take care of a fair transfer of fuzziness and defuzzify at the level of decision.

In realistic world, people need to make decisions on the basis of some uncertain information. For dealing with such decision-making problems including uncertainty, many scholars have introduced some models including random or fuzzy or rough variables to formulate the uncertainty. In [7, 15], the authors discussed some basic approaches to modelling of stochastic optimization problems. In [6], Charnes and Cooper proposed the stochastic chance constrained programming. In [12], Liu presented chance constrained programming with fuzzy coefficients. In [20], Amelia Bilbao Terol designed flexible decision making models in the distance metric optimization framework for problems including parameters which are represented by fuzzy numbers. In [17], Slowinski applied the method of rough sets to solve the uncertain problem in the medical domain. However, in a decision-making process, we may face a hybrid uncertain environment. In such situations, the use of fuzzy set theory to represent unknown parameters provides an interesting help. Thus, these coefficients may be dealt with a random variable whose parameters are assumed to

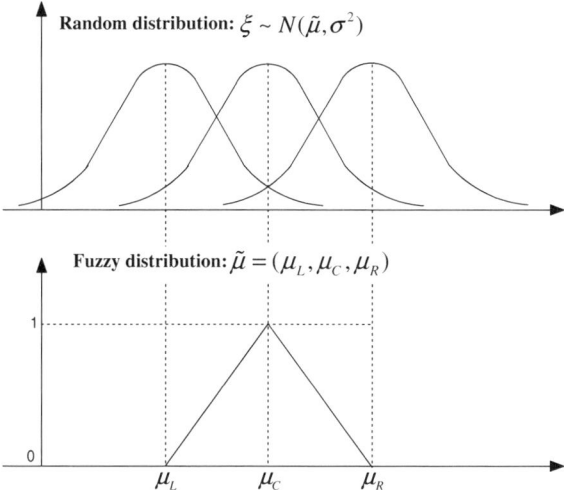

Fig. 117.1 Representation of a random fuzzy variable

be fuzzy numbers due to the decision maker's subjectivity. In other words, we may employ random fuzzy theory to deal with this combined uncertainty of randomness and fuzziness.

With respect to these mathematical programming problems including randomness and fuzziness, since they are not well-defined problems due to randomness and fuzziness, it is necessary that we consider a certain optimization criterion so as to transform these problems into well-defined problems. However, since they are usually transformed into nonlinear programming problems, it is difficult to find a global optimal solution directly. Therefore, in this paper, we construct an efficient solution method to find a global optimal solution of deterministic equivalent problem including more complicated constraints. This stimulates the author to employ the latest development of mathematics on uncertainty theory to study the bi-level programming problem in random fuzzy environment.

The rest of this paper is organized as follows. In Sect. 117.2 we recall some preliminaries of random fuzzy variables. In Sect. 117.3, we build the random fuzzy bi-level programming with expected objectives and constraints. The details of how to transform the random fuzzy bi-level model with expected objectives and constraints into its equivalent model are shown in Sect. 117.4. In Sect. 117.5, an illustrative example is presented to show the feasibility and effectiveness of the proposed models and methods. Finally, the conclusion has been made in Sect. 117.6.

117.2 Random Fuzzy Linear Expected Bi-level Programming

Roughly speaking, a random fuzzy variable is a random variable with a fuzzy parameter.

Example 117.1 Let $\xi \sim \mathcal{N}(\mu, \sigma^2)$ be a normally distributed random variable with
fuzzy μ and crisp σ, then ξ is a random fuzzy variable, and the situation is depicted
in Fig. 117.1.

As indicated in Fig. 117.1, the mean of the distribution of ξ is the triangular-shaped
fuzzy number (μ_L, μ_C, μ_R). As a consequence, μ may fall as far to the left as μ_L,
resulting in the left-hand distribution for ξ, and as far to the right as μ_R, resulting in
the right-hand distribution for ξ, although the highest possibility is associated with
μ_C.

The general linear bi-level programming with random fuzzy coefficients as follows:

$$
\max_x F(x, y) = \bar{\bar{a}}_1^T x + \bar{\bar{b}}_1^T y
$$

$$
\text{s.t.}
\begin{cases}
\bar{\bar{c}}_{1r_1}^T x + \bar{\bar{d}}_{1r_1}^T y \leq \bar{\bar{e}}_{1r_1}, \ r_1 = 1, 2, \cdots, p_1 \\
\text{where } y \text{ solves:} \\
\max_y f(x, y) = \bar{\bar{a}}_2^T x + \bar{\bar{b}}_2^T y \\
\quad \text{s.t.}
\begin{cases}
\bar{\bar{c}}_{2r_2}^T x + \bar{\bar{d}}_{2r_2}^T y \leq \bar{\bar{e}}_{2r_2}, \ r_2 = 1, 2, \cdots, p_2 \\
x, y \geq 0,
\end{cases}
\end{cases}
\tag{117.1}
$$

where $F(x, y)$ and $f(x, y)$ are called the objective functions of the upper level and
the lower level, respectively; $\bar{\bar{a}}_1, \bar{\bar{a}}_2, \bar{\bar{c}}_{1r_1}, \bar{\bar{c}}_{2r_2} \in R^{n_1}$, and $\bar{\bar{b}}_1, \bar{\bar{b}}_2, \bar{\bar{d}}_{1r_1}, \bar{\bar{d}}_{2r_2} \in R^{n_2}$
are random fuzzy vectors; $\bar{\bar{e}}_{1r_1}, \bar{\bar{e}}_{2r_2}$ are fuzzy random variables; $x \in R^{n_1}, y \in R^{n_2}$
are the decision variables of the upper level and lower level, respectively.

Note that there are random fuzzy variables in the objective functions and constraints of the upper level and lower level, Eq. (117.1) has ambiguous explanation.
That is, the model is not well-defined, and has not a mathematical meaning because
it has different interpretations. So in the following, we propose the bi-level programming with expected objectives and constraints based on the philosophy: selecting
the decision by optimizing the expected objective values subject to some expected
constraints.

Then we can propose the random fuzzy expected bi-level programming as follows:

$$
\max_x E[F(x, y)] = E[\bar{\bar{a}}_1^T x + \bar{\bar{b}}_1^T y]
$$

$$
\text{s.t.}
\begin{cases}
E[\bar{\bar{c}}_{1r_1}^T x + \bar{\bar{d}}_{1r_1}^T y] \leq E[\bar{\bar{e}}_{1r_1}], \ r_1 = 1, 2, \cdots, p_1 \\
\text{where } y \text{ solves:} \\
\max_y E[f(x, y)] = E[\bar{\bar{a}}_2^T x + \bar{\bar{b}}_2^T y] \\
\quad \text{s.t.}
\begin{cases}
E[\bar{\bar{c}}_{2r_2}^T x + \bar{\bar{d}}_{2r_2}^T y] \leq E[\bar{\bar{e}}_{2r_2}], \ r_2 = 1, 2, \cdots, p_2 \\
x, y \geq 0.
\end{cases}
\end{cases}
\tag{117.2}
$$

In this paper, we mainly consider the random fuzzy variables deduced from the continuous random variables. First, we introduce the basic definitions of expected value for random fuzzy variables as follows:

Definition 117.1 [13] Let ξ be a continuous random fuzzy variable on $(\Theta, \mathscr{P}(\Theta), Pos)$ with the density function $\tilde{p}(x)$, then its expected value can be defined as follows,

$$E[\xi] = \int_0^\infty Cr \left\{ \int_{x \in \Theta} x \tilde{p}(x) dx \geq r \right\} dr - \int_{-\infty}^0 Cr \left\{ \int_{x \in \Theta} x \tilde{p}(x) dx \leq r \right\} dr, \quad (117.3)$$

where $\tilde{p}(x)$ is a density function with fuzzy parameters defined on $(\Theta, \mathscr{P}(\Theta), Pos)$.

Then we introduce the following concepts of optimal solutions to the multi-objective linear bi-level programming with random fuzzy coefficients. For the sake of simplicity, we set

$$\tilde{X} \times \tilde{Y} = \left\{ (x, y) \left| \begin{array}{l} E[\tilde{\tilde{c}}_{1r_1}^T x + \tilde{\tilde{d}}_{1r_1}^T y] \leq E[\tilde{\tilde{e}}_{1r_1}], \ r_1 = 1, 2, \cdots, p_1; \\ E[\tilde{\tilde{c}}_{2r_2}^T x + \tilde{\tilde{d}}_{2r_2}^T y] \leq E[\tilde{\tilde{e}}_{2r_2}], \ r_2 = 1, 2, \cdots, p_2. \end{array} \right. \right\},$$

and assume that $\tilde{X} \times \tilde{Y}$ is compact.

Definition 117.2 A solution $(x, y) \in \tilde{X} \times \tilde{Y}$ is called an expected feasible solution.

Definition 117.3 A solution $(x^*, y^*) \in \tilde{X} \times \tilde{Y}$ is called an expected complete optimal solution if there does not exist $(x, y) \in \tilde{X} \times \tilde{Y}$ such that $E[F(x, y)] \geq E[F(x^*, y^*)]$ and $E[f(x, y)] \geq E[f(x^*, y^*)]$ hold.

117.3 Equivalent Equation

In this section, we focus on the special random fuzzy variables with normal distributions whose mean values are LR fuzzy random variables, and introduce the equivalent formulas as follows.

Theorem 117.1 *Let $\xi \sim \mathcal{N}(\tilde{u}, \sigma^2)$ be a normally distributed random fuzzy variable, where $\tilde{u}$ is a fuzzy variable on $(\Theta, \mathscr{P}(\Theta), Pos)$ with the following membership function:*

$$\mu_{\tilde{u}}(t) = \begin{cases} L\left(\frac{u-t}{\alpha}\right), & t \leq u, \alpha > 0 \\ R\left(\frac{t-u}{\beta}\right), & t \geq u, \beta > 0, \end{cases}$$

where μ is the "mean" value of $\tilde{u}$, α and β are the left and right spreads of $\tilde{u}$, respectively. Reference functions $L, R : [0, 1] \rightarrow [0, 1]$ with $L(1) = R(1) = 0$ and $L(0) = R(0) = 1$ are non-increasing, continuous functions. Then we have

$$E[\xi] = u - \frac{\alpha}{2}[\Upsilon(1) - \Upsilon(0)] + \frac{\beta}{2}[\chi(1) - \chi(0)]. \tag{117.4}$$

Proof By Definition 117.1, we know

$$E[\xi] = \int_0^\infty Pr\{\theta \in \Theta | E[\xi(\theta)] \ge t\} dt - \int_{-\infty}^0 Pr\{\theta \in \Theta | E[\xi(\theta)] \le t\} dt. \tag{117.5}$$

Since $\xi \sim N(\tilde{u}, \sigma^2)$, and obviously we know that $E[\xi(\theta)] = \tilde{u}$.
Since $\tilde{u}$ is a LR fuzzy variable with the following membership function,

$$\mu_{\tilde{u}}(t) = \begin{cases} L\left(\frac{u-t}{\alpha}\right), & t \le u, 0 < \alpha < 1 \\ R\left(\frac{t-u}{\beta}\right), & t \ge u, 0 < \beta < 1. \end{cases}$$

Then we have

$$Cr\{\theta | \xi(\theta) \ge t\} = \begin{cases} 1, & t \le u - \alpha \\ 1 - \frac{1}{2}L\left(\frac{u-t}{\alpha}\right), & u - \alpha \le t < u \\ \frac{1}{2}R\left(\frac{t-u}{\beta}\right), & u \le t \le u + \beta \\ 0, & t > u + \beta \end{cases}$$

and

$$Cr\{\theta | \xi(\theta) \le t\} = \begin{cases} 0, & t \le u - \alpha \\ \frac{1}{2}L\left(\frac{u-t}{\alpha}\right), & u - \alpha \le t < u \\ 1 - \frac{1}{2}L\left(\frac{t-u}{\beta}\right), & u \le t \le u + \beta \\ 1, & t > u + \beta. \end{cases}$$

It follows from Definition 117.1 that

$$\int_{-\infty}^0 Cr\{\theta | \xi(\theta) \le t\} dt = 0$$

and

$$E[\tilde{u}(\theta)] = \int_0^{+\infty} Cr\{\theta | \xi(\theta) \ge t\} dt - \int_{-\infty}^0 Cr\{\theta | \xi(\theta) \le t\} dt$$

$$= \int_0^{u-\alpha} 1 dt + \int_{u-\alpha}^u \left[1 - \frac{1}{2}L\left(\frac{u-t}{\alpha}\right)\right] dt + \int_u^{u+\beta} \frac{1}{2}R\left(\frac{t-u}{\beta}\right) dt$$

$$= u - \frac{\alpha}{2}[\Upsilon(1) - \Upsilon(0)] + \frac{\beta}{2}[\chi(1) - \chi(0)],$$

where $\Upsilon(x)$ is a continuous function on $[u - \alpha, u]$ and $\frac{\partial \Upsilon(x)}{\partial x} = L(x)$. $\chi(x)$ is a continuous function on $[u, u + \beta]$ and $\frac{\partial \chi(x)}{\partial x} = R(x)$.

So we have

$$E[\xi] = E[\xi(\theta)] = E[\tilde{u}] = u - \frac{\alpha}{2}[\Upsilon(1) - \Upsilon(0)] + \frac{\beta}{2}[\chi(1) - \chi(0)].$$

This completes the proof.

Remark 117.1 Let $\zeta \sim \mathcal{N}(\tilde{u}, \sigma^2)$ be a normally distributed random fuzzy variable, where $\tilde{u} = (u - \alpha, u, u + \beta)$ is a triangular LR fuzzy variable. So the reference functions of $\tilde{u}$ are $L(x) = R(x) = 1 - x(x \in [0, 1])$, so we can obtain the $\Upsilon(x)$ and $\chi(x)$ by the integration, i.e., $\Upsilon(x) = \chi(x) = -\frac{1}{2}x^2 + x + C$.

Then by using Theorem 117.1, we can get that

$$E[\zeta] = u - \frac{\alpha}{2}\left[-\frac{1}{2} + 1\right] + \frac{\beta}{2}\left[-\frac{1}{2} + 1\right] = u + \frac{\beta - \alpha}{4}.$$

Lemma 117.1 [13] *Assume that ξ and η are random fuzzy variables with finite expected values. Then for any real numbers a and b, we have*

$$E[a\xi + b\eta] = aE[\xi] + bE[\eta].$$

Suppose the coefficients of objective functions and constraints in Eq. (117.2) are as described as in Remark 117.1, then we can get the equivalent crisp model for Eq. (117.2) as follows using Theorem 117.1 and Lemma 117.1.

$$\max_x \left[\left(u^{a_1} - \frac{\alpha^{a_1}}{4} + \frac{\beta^{a_1}}{4}\right)^T x + \left(u^{b_1} - \frac{\alpha^{b_1}}{4} + \frac{\beta^{b_1}}{4}\right)^T y\right]$$

$$\text{s.t.} \begin{cases} \left(u^{c_{1r_1}} - \frac{\alpha^{c_{1r_1}}}{4} + \frac{\beta^{c_{1r_1}}}{4}\right)^T x + \left(u^{d_{1r_1}} - \frac{\alpha^{d_{1r_1}}}{4} + \frac{\beta^{d_{1r_1}}}{4}\right)^T y \\ \quad \leq u^{e_{1r_1}} - \frac{\alpha^{e_{1r_1}}}{4} + \frac{\beta^{e_{1r_1}}}{4}, \; r_1 = 1, 2, \cdots, p_1 \\ \text{where } y \text{ solves:} \\ \max_y \left[\left(u^{a_2} - \frac{\alpha^{a_2}}{4} + \frac{\beta^{a_2}}{4}\right)^T x + \left(u^{b_2} - \frac{\alpha^{b_2}}{4} + \frac{\beta^{b_2}}{4}\right)^T y\right] \\ \text{s.t.} \begin{cases} \left(u^{c_{2r_2}} - \frac{\alpha^{c_{2r_2}}}{4} + \frac{\beta^{c_{2r_2}}}{4}\right)^T x + \left(u^{d_{2r_2}} - \frac{\alpha^{d_{2r_2}}}{4} + \frac{\beta^{d_{2r_2}}}{4}\right)^T y \\ \quad \leq u^{e_{2r_2}} - \frac{\alpha^{e_{2r_2}}}{4} + \frac{\beta^{e_{2r_2}}}{4}, \; r_2 = 1, 2, \cdots, p_2 \\ x, y \geq 0. \end{cases} \end{cases}$$

(117.6)

117.4 Interactive Programming Technique

In this section, the interactive programming technique proposed in [16] is applied to solve the bi-level programming.

It's natural that we can take the uncertain objective function to evaluate the DMs imprecise consideration. For the objective function of every level in problem (117.6), decision maker has fuzzy goals such as "the goal should be more than or equal to a certain value". Let

$$
H_1(x, y) = \left[\left(u^{a_1} - \tfrac{\alpha^{a_1}}{4} + \tfrac{\beta^{a_1}}{4} \right)^T x + \left(u^{b_1} - \tfrac{\alpha^{b_1}}{4} + \tfrac{\beta^{b_1}}{4} \right)^T y \right],
$$
$$
H_2(x, y) = \left[\left(u^{a_2} - \tfrac{\alpha^{a_2}}{4} + \tfrac{\beta^{a_2}}{4} \right)^T x + \left(u^{b_2} - \tfrac{\alpha^{b_2}}{4} + \tfrac{\beta^{b_2}}{4} \right)^T y \right]
$$

and

$$
S = \left\{ (x, y) \geq 0 \;\middle|\;
\begin{aligned}
& \left(u^{c_{1r_1}} - \tfrac{\alpha^{c_{1r_1}}}{4} + \tfrac{\beta^{c_{1r_1}}}{4} \right)^T x + \left(u^{d_{1r_1}} - \tfrac{\alpha^{d_{1r_1}}}{4} + \tfrac{\beta^{d_{1r_1}}}{4} \right)^T y \\
& \leq u^{e_{1r_1}} - \tfrac{\alpha^{e_{1r_1}}}{4} + \tfrac{\beta^{e_{1r_1}}}{4}, \quad r_1 = 1, 2, \cdots, p_1; \\
& \left(u^{c_{2r_2}} - \tfrac{\alpha^{c_{2r_2}}}{4} + \tfrac{\beta^{c_{2r_2}}}{4} \right)^T x + \left(u^{d_{2r_2}} - \tfrac{\alpha^{d_{2r_2}}}{4} + \tfrac{\beta^{d_{2r_2}}}{4} \right)^T y \\
& \leq u^{e_{2r_2}} - \tfrac{\alpha^{e_{2r_2}}}{4} + \tfrac{\beta^{e_{2r_2}}}{4}, \quad r_2 = 1, 2, \cdots, p_2.
\end{aligned}
\right\}.
$$

We can denote the maximum and minimum values of each objective functions as follows:

$$
H_1^{\max} = \max_{(x,y)\in S} H_1(x, y), \quad H_1^{\min} = \min_{(x,y)\in S} H_1(x, y);
$$
$$
H_2^{\max} = \max_{(x,y)\in S} H_1(x, y), \quad H_2^{\min} = \min_{(x,y)\in S} H_1(x, y).
$$

The functions $\mu_i(H_i(x, y))$, $i = 1, 2$ vary strictly between $H_i^{\min}$ and $H_i^{\max}$. For the sake of simplicity, we can take the linear function to characterize the goal at each level. They can be defined as follows:

$$
\mu_i(H_i(x, y)) = \begin{cases} 1, & H_i(x, y) \geq H_i^{\max} \\ \frac{H_i(x,y) - H_i^{\min}}{H_i^{\max} - H_i^{\min}}, & H_i^{\min} \leq H_i(x, y) < H_i^{\max} \\ 0, & H_i(x, y) < H_i^{\min} \end{cases} \quad i = 1, 2. \quad (117.7)
$$

The objective of the upper level can be specified with a minimal satisfactory level $\varepsilon \in [0, 1]$ after introducing the membership functions. Then the lower level minimize the objective subjects to the additional condition $\mu_1(H_1(x, y)) \geq \varepsilon$, that is, the lower level should solve the following problem:

$$\max\ \mu_2(H_2(x, y))$$
$$\text{s.t.}\ \begin{cases} \mu_1(H_1(x, y)) \geq \varepsilon \\ (x, y) \in S. \end{cases} \tag{117.8}$$

In order to obtain the overall satisfactory optimal solution for both upper and lower levels, the upper level have to comprise with the lower level with consideration of his satisfactory level. Thus, a satisfactory degree for both upper and lower levels is defined as

$$\lambda = \min\{\mu_1(H_1(x, y)), \mu_2(H_2(x, y))\} \tag{117.9}$$

and Eq. (117.8) can be transformed into

$$\max\ \lambda$$
$$\text{s.t.}\ \begin{cases} \mu_1(H_1(x, y)) \geq \lambda \\ \mu_2(H_2(x, y)) \geq \lambda \\ (x, y) \in S. \end{cases} \tag{117.10}$$

By solving Eq. (117.10), we get the overall satisfactory solution for both upper and lower levels.

Theorem 117.2 *If Eq. (117.10) has optimal solutions, then solutions of Eq. (117.10) must be the solutions of Eq. (117.6).*

Proof We consider $x = (x_1, x_2, \cdots, x_{n_1})$, $y = (y_1, y_2, \cdots, y_{n_2})$ and λ as decision variables. Let $X^* = (x^*, y^*, \lambda^*)$ be an optimal solution of Eq. (117.10). Apparently, X^* satisfies all constraints. Assume that X^* is not the optimal solution of Eq. (117.6), so in the lower level, there exists a solution $y' = (y'_1, y'_2, \cdots, y'_{n_2})$ such that $H_2(x, y') < H_2(x, y^*)$. Then

$$\mu_2(H_2(x, y')) = \frac{H_2^{\max} - H_2(x, y')}{H_2^{\max} - H_2^{\min}} > \mu_2(H_2(x, y^*)) = \frac{H_2^{\max} - H_2(x, y^*)}{H_2^{\max} - H_2^{\min}}. \tag{117.11}$$

Similarly, in the upper level, x^* satisfies the constraints in the lower level. If x^* is not the optimal solution of Eq. (117.6), there exists x' such that $H_1(x', y') < H_1(x^*, y^*)$, then

$$\mu_1(H_1(x', y')) = \frac{H_1^{\max} - H_1(x', y')}{H_1^{\max} - H_1^{\min}} > \mu_1(H_1(x^*, y^*)) = \frac{H_1^{\max} - H_1(x^*, y^*)}{H_1^{\max} - H_1^{\min}}. \tag{117.12}$$

According to Eqs. (117.9), (117.11) and (117.12), we know that the optimal solution λ' for variables x', y' must be subjected to $\lambda' < \lambda^*$. It shows a conflict that λ^* is the optimal solution of Eq. (117.10). This completes the proof.

After proving the above theorem, we know that if Eq. (117.10) has an optimal solution, the bi-level programming must have optimal solutions. Next, we apply the genetic algorithm to find optimal solutions of Eq. (117.10).

117.5 Illustrative Example

The following illustrative example is given to show the application of the proposed models and algorithms.

$$
\max \ F(x, y) = \bar{\tilde{a}}_{11}x_1 + \bar{\tilde{a}}_{12}x_2 + \bar{\tilde{a}}_{13}x_3 + \bar{\tilde{b}}_{11}y_1 + \bar{\tilde{b}}_{12}y_2
$$

$$
\text{s.t.}
\begin{cases}
\bar{\tilde{c}}_{11}x_1 + \bar{\tilde{c}}_{12}x_2 + \bar{\tilde{c}}_{13}x_3 \leq y_1 \\
\bar{\tilde{c}}_{21}x_1 + \bar{\tilde{c}}_{22}x_2 + \bar{\tilde{c}}_{23}x_3 \leq y_2 \\
x_1, x_2, x_3 \geq 0 \\
\text{where } y \text{ solves:} \\
\quad \max \ f(x, y) = \bar{\tilde{a}}_{21}x_1 + \bar{\tilde{a}}_{22}x_2 + \bar{\tilde{a}}_{23}x_3 + \bar{\tilde{b}}_{21}y_1 + \bar{\tilde{b}}_{22}y_2 \\
\quad \text{s.t.}
\begin{cases}
\bar{\tilde{d}}_{11}x_1 + \bar{\tilde{d}}_{12}x_2 + \bar{\tilde{d}}_{13}x_3 + \bar{\tilde{d}}_{14}y_1 + \bar{\tilde{d}}_{15}y_2 \leq \bar{\tilde{e}}_1 \\
\bar{\tilde{d}}_{21}x_1 + \bar{\tilde{d}}_{22}x_2 + \bar{\tilde{d}}_{23}x_3 + \bar{\tilde{d}}_{24}y_1 + \bar{\tilde{d}}_{25}y_2 \leq \bar{\tilde{e}}_2 \\
y_1, y_2 \geq 0,
\end{cases}
\end{cases}
\tag{117.13}
$$

where

$\bar{\tilde{a}}_{11} \sim N(\tilde{a}_{11}, 1)$ with $\tilde{a}_{11} = (9, 11, 12)$, $\quad \bar{\tilde{a}}_{21} \sim N(\tilde{a}_{21}, 1)$ with $\tilde{a}_{21} = (2, 4, 5)$,
$\bar{\tilde{a}}_{12} \sim N(\tilde{a}_{12}, 4)$ with $\tilde{a}_{12} = (5, 6, 8)$, $\quad \bar{\tilde{a}}_{22} \sim N(\tilde{a}_{22}, 1)$ with $\tilde{a}_{22} = (1, 1, 3)$,
$\bar{\tilde{a}}_{13} \sim N(\tilde{a}_{13}, 2)$ with $\tilde{a}_{13} = (6, 7, 8)$, $\quad \bar{\tilde{a}}_{23} \sim N(\tilde{a}_{23}, 1)$ with $\tilde{a}_{23} = (3, 6, 9)$,
$\bar{\tilde{b}}_{11} \sim N(\tilde{b}_{11}, 1)$ with $\tilde{b}_{11} = (0, 2, 3)$, $\quad \bar{\tilde{b}}_{21} \sim N(\tilde{b}_{21}, 4)$ with $\tilde{b}_{21} = (1, 3, 4)$,
$\bar{\tilde{b}}_{12} \sim N(\tilde{b}_{12}, 1)$ with $\tilde{b}_{12} = (0, 1, 2)$, $\quad \bar{\tilde{b}}_{22} \sim N(\tilde{b}_{22}, 1)$ with $\tilde{b}_{22} = (1, 2, 3)$,
$\bar{\tilde{c}}_{11} \sim N(\tilde{c}_{11}, 3)$ with $\tilde{c}_{11} = (1, 2, 3)$, $\quad \bar{\tilde{c}}_{21} \sim N(\tilde{c}_{21}, 1)$ with $\tilde{c}_{21} = (3, 5, 7)$,
$\bar{\tilde{c}}_{12} \sim N(\tilde{c}_{12}, 1)$ with $\tilde{c}_{12} = (1, 2, 3)$, $\quad \bar{\tilde{c}}_{22} \sim N(\tilde{c}_{22}, 3)$ with $\tilde{c}_{22} = (1, 3, 5)$,
$\bar{\tilde{c}}_{13} \sim N(\tilde{c}_{13}, 1)$ with $\tilde{c}_{13} = (3, 4, 5)$, $\quad \bar{\tilde{c}}_{23} \sim N(\tilde{c}_{23}, 1)$ with $\tilde{c}_{23} = (3, 4, 5)$,
$\bar{\tilde{d}}_{11} \sim N(\tilde{d}_{11}, 1)$ with $\tilde{d}_{11} = (1, 2, 3)$, $\quad \bar{\tilde{d}}_{21} \sim N(\tilde{d}_{21}, 1)$ with $\tilde{d}_{21} = (1, 3, 5)$,
$\bar{\tilde{d}}_{12} \sim N(\tilde{d}_{12}, 1)$ with $\tilde{d}_{12} = (1.5, 2, 2.5)$, $\bar{\tilde{d}}_{22} \sim N(\tilde{d}_{22}, 1)$ with $\tilde{d}_{22} = (1, 2, 3)$,
$\bar{\tilde{d}}_{13} \sim N(\tilde{d}_{13}, 1)$ with $\tilde{d}_{13} = (1, 2, 3)$, $\quad \bar{\tilde{d}}_{23} \sim N(\tilde{d}_{23}, 1)$ with $\tilde{d}_{23} = (0, 1, 2)$,
$\bar{\tilde{d}}_{14} \sim N(\tilde{d}_{14}, 1)$ with $\tilde{d}_{14} = (1, 2, 3)$, $\quad \bar{\tilde{d}}_{24} \sim N(\tilde{d}_{24}, 1)$ with $\tilde{d}_{24} = (0.5, 1, 1.5)$,
$\bar{\tilde{d}}_{15} \sim N(\tilde{d}_{15}, 1)$ with $\tilde{d}_{15} = (0, 1, 2)$, $\quad \bar{\tilde{d}}_{25} \sim N(\tilde{d}_{25}, 1)$ with $\tilde{d}_{25} = (1, 2, 3)$,
$\bar{\tilde{e}}_1 \sim N(\tilde{e}_1, 9)$ with $\tilde{e}_1 = (450, 500, 550)$, $\bar{\tilde{e}}_2 \sim N(\tilde{e}_2, 9)$ with $\tilde{e}_2 = (700, 800, 900)$

are independent random fuzzy coefficients.

According to Eq. (117.2), we can obtain the following model, in which we use the expected operator to handle the objective functions and employ the chance constrained operator to deal with the constraints.

$$\max \ F_{(x, y)} = E[\bar{\bar{a}}_{11}x_1 + \bar{\bar{a}}_{12}x_2 + \bar{\bar{a}}_{13}x_3 + \bar{\bar{b}}_{11}y_1 + \bar{\bar{b}}_{12}y_2]$$

s.t. $\begin{cases} E[\bar{\bar{c}}_{11}x_1 + \bar{\bar{c}}_{12}x_2 + \bar{\bar{c}}_{13}x_3] \le y_1 \\ E[\bar{\bar{c}}_{21}x_1 + \bar{\bar{c}}_{22}x_2 + \bar{\bar{c}}_{23}x_3] \le y_2 \\ x_1, x_2, x_3 \ge 0 \\ \text{where } y \text{ solves:} \\ \max \ f(x, y) = E[\bar{\bar{a}}_{21}x_1 + \bar{\bar{a}}_{22}x_2 + \bar{\bar{a}}_{23}x_3 + \bar{\bar{b}}_{21}y_1 + \bar{\bar{b}}_{22}y_2] \\ \text{s.t.} \begin{cases} E[\bar{\bar{d}}_{11}x_1 + \bar{\bar{d}}_{12}x_2 + \bar{\bar{d}}_{13}x_3 + \bar{\bar{d}}_{14}y_1 + \bar{\bar{d}}_{15}y_2] \le E[\bar{\bar{e}}_1] \\ E[\bar{\bar{d}}_{21}x_1 + \bar{\bar{d}}_{22}x_2 + \bar{\bar{d}}_{23}x_3 + \bar{\bar{d}}_{24}y_1 + \bar{\bar{d}}_{25}y_2] \le E[\bar{\bar{e}}_2] \\ y_1, y_2 \ge 0. \end{cases} \end{cases}$ (117.14)

Based on the Theorem 117.1, we can get the crisp equivalent model for the above Eq. (117.14) as follows,

$$\max \ 7.75x_1 + 6.25x_2 + 7x_3 + 4.75y_1 + 3y_2$$

s.t. $\begin{cases} 2x_1 + 2x_2 + 4x_3 - y_1 \le 0 \\ 5x_1 + 3x_2 + 4x_3 - y_2 \le 0 \\ x_1, x_2, x_3 \ge 0 \\ \text{where } y \text{ solves:} \\ \max \ 3.75x_1 + 1.5x_2 + 3x_3 + 7.75y_1 + 6y_2 \\ \text{s.t.} \begin{cases} 2x_1 + 2x_2 + 2x_3 + 2y_1 + y_2 \le 500 \\ 3x_1 + 2x_2 + x_3 + y_1 + 2y_2 \le 800 \\ y_1, y_2 \ge 0. \end{cases} \end{cases}$ (117.15)

Then according to the interactive programming technique, we have

$$H_1(x, y) = 7.75x_1 + 6.25x_2 + 7x_3 + 4.75y_1 + 3y_2,$$
$$H_2(x, y) = 3.75x_1 + 1.5x_2 + 3x_3 + 7.75y_1 + 6y_2$$

and

$$S = \left\{ (x, y) \ge 0 \ \middle| \ \begin{array}{l} 2x_1 + 2x_2 + 4x_3 - y_1 \le 0 \\ 5x_1 + 3x_2 + 4x_3 - y_2 \le 0 \\ 2x_1 + 2x_2 + 2x_3 + 2y_1 + y_2 \le 500 \\ 3x_1 + 2x_2 + x_3 + y_1 + 2y_2 \le 800 \end{array} \right\}.$$

In order to get the value of $H_1^{\max}, H_1^{\min}, H_2^{\max}, H_2^{\min}$, we need to solve the following two models:

$$\begin{array}{cc} \max \ H_1(x, y) & \max \ H_2(x, y) \\ \text{s.t. } (x, y) \in S, & \text{and} \qquad \text{s.t. } (x, y) \in S. \end{array}$$

After solving these two models, we can obtain that

$$H_1^{\max} = 1478.57, \ H_1^{\min} = 1416.69,$$
$$(x_1, x_2, x_3) = (28.57, 0, 0), \ (y_1, y_2) = (57.14, 328.57)$$

and

$$H_2^{\max} = 2716.67, \quad H_2^{\min} = 2516.25,$$
$$(x_1, x_2, x_3) = (0, 0, 0), \quad (y_1, y_2) = (66.67, 366.67).$$

Following the interactive programming method, we get the following model,

$$\max \lambda$$
$$\text{s.t.} \begin{cases} (H_1(x, y) - 1416.69)/(1478.57 - 1416.69) \geq \lambda \\ (H_2(x, y) - 2516.25)/(2716.67 - 2516.25) \geq \lambda \\ (x, y) \in S. \end{cases} \tag{117.16}$$

After solving Eq. (117.16), we get the following results: $\lambda = 0.572$, $F(x, y) = 443.21$, $f(x, y) = 554.4$, $(x_1, x_2, x_3) = (4.857, 0, 11.067)$, $(y_1, y_2) = (53.981, 360.19)$.

117.6 Conclusion

In this paper, we introduced the linear bi-level programming with random fuzzy coefficients. In order to make the bi-level model solvable, we use the philosophy that optimize the expectation objectives subject to the chance constraints, and construct a bi-level model with expected objectives and constraints. Then we propose transform it into an equivalent model which is easy to solve. Interactive programming technique is employed to combine the bi-level model to a single level model. Finally, we apply the proposed models and solution methods to a numerical example. In the later research, we will apply the proposed approach to some realistic problems, such as cooperative purchase problem, transportation problem and so on.

References

1. Bard J (1983) Cordination of a multidivisional organization through two levels of management. Omega 1:457–468
2. Ben-Ayed O, Blair C et al (1992) Construction of a real-world bi-level inear programing model of the highway design problem. Ann Oper Res 34:219–254
3. Bracken J, McGil J (1973) Mathematical programs with optimization problems in the constraints. Oper Res 21:37–44
4. Candler W, Norton R (1977) Multilevel programing. Technical report, World Bank Development Research Center, Washington DC
5. Casidy R, Kirby M, Raike W (1971) Efficient distribution of resources through three levels of government. Manage Sci 17:462–473
6. Charnes A, Cooper W (1959) Chance-constrained programming. Manage Sci 6:73–79
7. Dupačová J (2002) Applications of stochastic programming: achievements and questions. Eur J Oper Res 140:281–290

8. Haurie A, Loulou R, Savard G (1990) A two-level systems analysis model of power cogeneration under asymetric pricing. In: Proceedings of IEEE Automatic Control Conference, SanDiego
9. Hobs B, Nelson S (1992) A nonlinear bi-level model for analysis of electric utility demand-side planing iisues. Ann Oper Res 34:255–274
10. Kocvara M, Outrata J (1992) A nondiffierentiable aproach to the solution of optimum design problems with variational inequalities. In: System Modeling and Optimization. Lecture Notes in Control and Information Sciences, vol 180, pp 364–373
11. Leblanc L, Boyce D (1986) A bi-level programing algorithm for exact solution of the net work design problem with user optimal flows. Transp Res 20:259–265
12. Liu B, Iwamura K (1998) Chance constrained programming with fuzzy parameters. Fuzzy Sets Syst 94:227–237
13. Liu Y, Liu B (2003) Expected value operator of random fuzzy varaible and random fuzzy expected value models. Int J Uncertainty Fuzziness Knowl Based Syst 13:195–215
14. Marcote P (1986) Network design problem with congestione effects: a case of bi-level programing. Math Program 34:142–162
15. Ruszczyński A, Shapiro A (2003) Stochastic programming models handbooks. Oper Res Manage Sci 10:1–64
16. Sakawa M, Nishizaki I, Uemura Y (1997) Interactive fuzzy programming for multilevel linear programming problems. Comput Math Appl 36(2):71–86
17. Slowinski K (1988) Rough sets approach to analysis of data from peritoneal lavage in acute pancreatitis. Med Inform 13:143–159
18. Stackelberg H (1952) The theory of the market economy. Oxford University Press, Oxford
19. Suh S, Kim T (1992) Solving nonlinear bi-level programing model sof the quilibrium network design problem: acomparativereview. Ann Oper Res 34:203–218
20. Terol A (2008) A new approach for multiobjective decision making based on fuzzy distance minimization. Math Comput Model 47:808–826

Chapter 118
Research on Factors Influencing the Sugarcane Production in Pakistan: A Case Study of Punjab

Jamil Ahmad, Muhammad Kashif Javed, Muhammad Nazam and Muhammad Nazim

Abstract Sugarcane is the one of the most important cash crop and industry of Pakistan and it is an important sugarcane producing countries in the world but, the Punjab province of Pakistan has highest share in sugarcane production. Like growers of other crops, sugarcane growers of the Punjab province are also facing economic, technical and social problems, but high rate of inputs are the primary problems of sugarcane growers. The efficacy checked by using Cobb-Douglas production function, Marginal value of product and allocated efficacy are calculated. The results show that the coefficient of multiple determination indicated that 93 % variation in the cost of inputs is explained by all variables and the regression model is well fitted. For this purpose, Data were collected during the period 2009–2010. Data were collected from 400 sugarcane growers from 5 districts of Punjab province. The purpose of this study is to explore and mechanisms for a progression to sustainable development of agriculture sector in Pakistan.

Keywords Influenced factors of sugarcane production · MVP value · Resource allocation efficiency · Cobb-Douglas function · Cost production analysis of Punjab Pakistan

118.1 Introduction

Pakistan is essentially an agricultural country. A major proportion of the population 65.9 % lives in the rural areas and directly or indirectly depend on agriculture. It provides employment to 44.8 % of labour force and contributes about 22.3 to GDP. Sugarcane is an important major cash crop. Pakistan is an important sugar-

J. Ahmad (✉) · M. K. Javed · M. Nazam · M. Nazim
Uncertainty Decision-Making Laboratory, Sichuan University, Chengdu 610064,
People's Republic of China
e-mail: jamil.ahmad040@gmail.com

J. Xu et al. (eds.), *Proceedings of the Eighth International Conference on Management Science and Engineering Management*, Advances in Intelligent Systems and Computing 281, DOI: 10.1007/978-3-642-55122-2_118, © Springer-Verlag Berlin Heidelberg 2014

Fig. 118.1 World top five
sugarcane producers (*Source*
FAPRI (USA), JCR-VIS
Research)

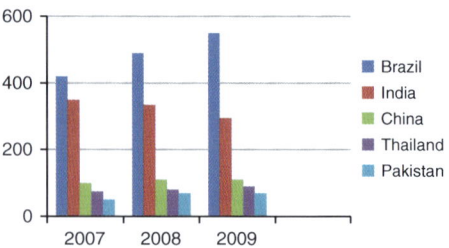

cane producing country and is ranked 5th in terms of area under sugar cultivation,
15th in sugar production and 60th in yield of the world. Sugarcane is cultivated on
over a million hectares and provides the raw material for Pakistan's 83 sugar mills
which comprise the country second largest agriculture industries after textiles. The
sugar sector constitutes 4.3 % of manufacturing. In size, the sugar sector matches the
cement sector. It accounts for 9.67 % in value added in agriculture It is an important
source of income and employment for the farming community. Exertive growth of
sugar industry has contributed to economic development of the country. Sugarcane
is a source of revenue to the government because this crop fetches billions of rupees
to the government in the form of duties and taxes. Sugar industry has an indirect
socio-economic impact in overall terms which is significantly larger than its direct
contribution to GDP because of it backward and forward linkages in the economy
(Fig. 118.1).

Pakistan is an important position in sugarcane producing countries of the world.
The top five producers of sugarcane, an average accounts for about 70 % of the total
world's production during the last 3 years. These include Brazil the record crop production in 2009, followed by India, China, Thailand and Pakistan. Top five sugarcane
producers and sugar beet account for about 55 % of world total sugar production in
2009. Brazil leads with 20.47 % followed by India and China by 14.43 % and 9.96 %
respectively.

Punjab province in Pakistan has highest share in sugarcane. Sugarcane acreage of
Punjab is 62 %, khyber Pakhtunkhwa Sindh is 16 % and Sindh is 26 %. The average
yield of sugarcane in the year 2005–2006 was 20105.35 kg/acre. The record yield
(20669.96 kg/acre) was reported in 1997–1998 [5]. Total production of Pakistan is
45229700 tons in which the share of Punjab is 59.43 %. Its yield is 16460.90 kg/acre
which is lower than that of national average i.e. 19341.56 kg/acre [5]. In the same
way the cost of production of sugarcane per acre is also high in Punjab than the other
provinces of Pakistan. Malik and Sardar [8] investigated into the cost of production
and contribution of various cost items in cultivation of sugarcane crop in Punjab and
Sindh during 1995–1996 and 1999–2000, and reported that overall cost of production
was higher in Punjab than Sindh.

Like growers of other crops, sugar cane growers of Punjab province are also facing
a host of social, technical and economic problems. Technical problems were related
to production techniques and technologies. Social constraints were related to theft of

Table 118.1 Sugarcane area and production by province

Province	Area/000 hectare			Production/000 MT		
	2007–2008	2008–2009	2009–2010	2007–2008	2008–2009	2009–2010
Punjab	827.00	675.00	689.00	40372.00	3200.00	33500.00
Sindh	310.00	264.00	280.00	18300.00	14760.00	15350.00
Kpk	105.00	105.00	110.00	4800.00	4700.00	4700.00
Baluc	0.50	0.80	1.00	28.00	40.00	50.00
Total	1242.50	1044.80	1080.00	63500.00	51500.00	53600.00

Sources Ministry of Food, Agriculture and Livestock; FAS/Islamabad

sugarcane, the economic problems were related to the financial difficulties and the lack of education of formers is also a primary problem of low yield of sugarcane. Raza Ramachandran [11] also indicated that farmer education improves the management skills. Nazir et al. [1] found that sugarcane production system has passed down from previous generation and is dominant among the growers. In Table 118.1 shows that the area and production of each province decreasing year by year. Pakistani Punjab farmers are able to get a yield of only 43 tonnes per acre, in the Indian Punjab the yield is over 63 tonnes per acre [13]. In 2007–2008 the total area of sugarcane production of Punjab was 827 thousand acreage but in 2008–2009 the area and production of sugarcane were decrease because of to these affecting factors. But the main factors for lesser production and area are maximum area under wheat crop during 2008–2009 restricted the sugarcane acreage, shortage of canal water, load shedding of electricity, pests disease and weeds, lower price of preceding season and high rate of inputs (cost of urea, cost of DAP, cost of land preparation, cost of FYM, cost of seeds cost of labor, cost of transportation and cost of weeding) discourage the farmers to grow more sugarcane crop. Timely provision of credit is essential as it has a positive impact on the productivity of sugarcane crop [2]. The area and production of sugarcane per hectare of the last 3 year are given in Table 118.1.

The purpose of this study is to identify the basic problems of sugarcane production facing by sugarcane growers and The data analyze by Cobb-Douglas production function. The Marginal Value of Product and its Ratio are also calculated. In the end, suggest the policies to the Pakistan Government and related organizations for solving these basic problems.

The purpose of this study is to identify the basic problems of sugarcane production facing by sugarcane growers and The data analyze by Cobb-Douglas production function. The Marginal Value of Product and its Ratio are also calculated. In the end, suggest the policies to the Pakistan Government and related organizations for solving these basic problems.

118.2 Problem Statement

Sugarcane production is a complex process and can be suspected as a function of many variables. Like growers of other crops, sugarcane growers also facing technical, social, educational and economic problems. Technical problems were related to production techniques and technologies, such as land preparation, pesticides and insecticides, natural calamities, seeds, lack of scientific knowledge and inadequate irrigation etc. Social constraints were related to theft of sugarcane, cutting tops, most of the farmers reported that the villagers cut the tops of sugarcane for using it as cattle feed. Sugarcane is tasty and an attractive crop, people especially children are generally attracted to it. Chewing of cane was third social problems reported by farmers in the study area. The economic problems were related to the financial difficulties, which were low price of output, lack of capital, and late payments etc and the lack of education of the formers is also a reason of low yield of sugarcane production, but almost sugarcane farmers reported that a high price of inputs are main problem in the way of practicing the production of sugarcane. Some important inputs of sugarcane are given below:

(1) Land Preparation. Sugarcane is a deep rooted crop so land preparation play an important role in the development of cane root system. If farmer prepared land in proper way then they can achieve optimal growth of the sugarcane. Land should be prepared by deep ploughing after every consequent year. The use of efficient equipment is important than simple cultivator. The use of more efficient equipment such as, goble plough, disc plough, leveler, cultivator and bullock etc, are better for land preparation. The use of these efficient equipment are very costly, especially for small and middle sugarcane growers and they also unable to archived required production of sugarcane.

(2) Farm Yard Manure. Farm yard manure is an other important input of sugarcane production. Farmers use farmyard manure to sugarcane crop in order to restore fertility for better yield as compared with other kharif crops. Well rotten farmyard manure should be applied for land preparation. The sugar industry is also an excellent source of organic matter and nutrients. In Khyber pakhtunkhwa sugarcane growers highly applied an average 4.2 tractor trolleys per acre of farmyard manure followed by Punjab 3.4 tractor trolleys per acre and in Sindh 0.9 trolleys per acre. The overall average usage of farm yard manure were recorded 2.83 tractor trolleys per acre. Most of the sugarcane growers are unable to use it in land for sugarcane production because of it's high price.

(3) Irrigation. Irrigation is an other important input of sugarcane production and sugarcane farmers are also facing this problem. Irrigation application and water distribution methods are the most neglected in the study area region. The farmers of some distributaries are so favorable. It is difficult for their haries to manage the surplus water. The cane fields get deluge and root zones remain submerged in water. It not only depresses tillering, growth and cane yield but also leads to water logging [8].

The recommended number of irrigation were 27–34 for autumn crop and 22–27 for spring crop. Data shows that 63.3 % used canal water, 35.5 from tube-well and only 8.3 % used tube well for irrigation purpose. Whereas the average availability of canal water per 7th turn was for 16 h, while 80.5 % annual, 12.6 % seasonal availability of canal water.

(4) Weeding. Weeding is also an other basic problem for sugarcane growers. Weeds restrict the nutrients, moisture and light to the sugarcane crop and also serve as alternative for many insect pests. These pests are affect the yield and t cane quality. The main key factor to control weeds is to manage the land properly. For proper weed control, Gesapax combe (80 WP) applied 1.9 kg/acre in medium soils and 2 kg/acre in heavy soils in 120–140 l of water. The weedicide devices used by the suggestion of technical experts. The use of weedicide is also expensive, so the, mostly sugarcane growers unable to use it.

(5) Seed and Application. Seed is also an other basic input of sugarcane. It has different varieties and application, seeds varieties also impact on sugarcane yield and production. Most of the farmers are disable to choose best variety of sugarcane seeds according to climate for high output because they have lack of education and also not familiar with best varieties. Some of good varieties are expensive so the most of farmers cannot use because of lack of finance.

(6) Fertilizer. Like other plant growers, sugarcane also need nutrients in the soil to grow healthy. Plants use up nutrients in the soil, growers generally replace these nutrients by adding fertilizers. One kg of fertilizer nutrients yield about 114 kg of stripped sugarcane [13]. With the passage of time, the scientific research has determined optimum fertilizer amounts for the healthy and productive sugarcane plants. Fertilizer application is important for achieving the optimum yield of sugarcane. But the use of chemical fertilizer is unbalanced and inadequate. Most of the farmers use only nitrogenous fertilizers and an unbalanced combination of N and P. The use of K is almost neglected in sugarcane crop. It is very important to use balanced doses of fertilizers to obtain the maximum yield of sugarcane. Department of Agriculture Sindh recommended the fertilizer doses of 220–320 kg, 105–130 kg P204 and 130–180 kg, K22 per hectares for various regions of the province [1]. It was inspected that, sugarcane growers applied urea overall average 4.11 bags per acre.

(7) Low Price of Output. Majority of the politician are the owners of sugar mills and almost every mill Board has a member in National Assembly. So in the price policy making process, these sugar barons affect the policies through their say in parliament. Almost 47 % sugar mill owners have more than one sugar mill and 60 % mill family members also have sugar mills. That's why the sugar mill owners have monopolistic power in sugar sector of Pakistan.

(8) Transportation. After harvesting, it is the duty of mill owners to caries sugarcane from growers farms to mills, but the process is totaly different. The farmers carry sugarcane in front of mill and wait for turn out the unloading sugarcane and also pay the transportation cost. Both large and small farmers face a same problem od transportation as the delivery of sugarcane transaction required a big carriers or trucks. This problem also discourage the growers from sugarcane

cultivation. Because of that, they are moving from the farming of sugarcane to other beneficial crops like, from low water consumption to early payment crops. As a result cultivation of sugarcane is decreasing each year.

(9) Labor. High price of labor for bowing and harvesting is also a problem for growers. Cutting and bowing cost represent the high portion of variable cost and decrease the revenue of farmers. The labor cost is about 13–14 % of the total cost, but it can be high on depending the required cutting and bowing days and number of persons.

118.3 Methodology

The research is established on the primary data from the target area (Punjab province) through a comprehensive questionnaire from 400 sugarcane growers. The growers were selected at randomly from 5 districts of Punjab province namely (Rajan pur, Rahimyar khan, Bahawalpur, Faisalabad and Muzaffargarh). A methodology has been used in this paper is based on the work of Fare et al. Coelli et al. Raheman et al. [3]. A wide range of problems can be investigated by using this approach.

The data source used in this paper consists of primary sources. The primary data was collected from the sugarcane growers by using a well structured pre-tested set of questionnaires. This research was conducted in five sugarcane producing districts of Punjab, during the crop year 2009–2010. From a total 400 growers, 70 sugarcane growers from each district were selected randomly as a sample size. This sampling is called stratified sampling [9]. Information was collected on farm size, cropping pattern, labor cost, input cost, credit source, transportation, processing cost. The data were collected the main regions of the country where sugarcane production was most important in terms of volume of production. Five major district of Punjab of Pakistan were selected for this study.

118.4 Data Analysis

After completing the survey, the worksheet prepared from the collecting data. Gives The variable names r to the numbers of each question in the questionnaire. To measure the profitability of sugarcane production, is based on the analysis of production cost. The term production function is mostly used as input and output relationship. More production function refers to the relationship between the output of product nd input factor services. The purpose of this analysis is to identify the sugarcane input and output relationship in mathematical form and to understand the influences of the various inputs on sugarcane output. A Cobb-Douglas production function applied for estimating the input and output relationship of sugarcane production.

The stochastic form of Cobb-Douglas production function may be expressed as:

$$Y = \alpha_1 z_{2i}^{\alpha_2} z_{3i}^{\alpha_3} e^{\omega_i}, \tag{118.1}$$

where e = Base of natural logarithm, ω = Stochastic disturbance term, z_3 = Capital input, z_2 = Labor input, Y = Output.

The equation show that the relation of input and output is non linear. Now we transfer log and then:

$$\ln y_i = \ln \alpha_1 + \alpha_2 \ln z_2 + \alpha_3 \ln z_3 + \omega_i = \alpha_0 + \alpha_2 \ln z_2 + \alpha_3 \ln z_3 + \omega_i, \tag{118.2}$$

where $\alpha_0 = \ln \alpha_1$.

It is a log linear model, the multiple regression counter part of the two variable log linear model [6].

118.4.1 Econometric Model

To calculate the variable input cost, Cobb Douglas function used. This form of regression model used for sugarcane production.

$$\ln Y = \alpha_0 + \alpha_1 \ln z_1 + \alpha_2 \ln z_2 + \alpha_3 \ln z_3 + \alpha_4 \ln z_4 + \alpha_5 \ln z_5$$
$$+ \alpha_6 \ln z_6 + \alpha_7 \ln z_7 + \alpha_8 \ln z_8 + \alpha_9 \ln z_9 + \omega_i, \tag{118.3}$$

where

$\alpha_1, \cdots, \alpha_7$: Coefficients of respective variable,	ω_i	: Stochastic disturbance term,
z_1 : Cost of irrigation/hectare,	z_2	: Cost of Farm Yard Manure,
z_3 : Cost of DAP/hectare,	z_4	: Cost of land preparation/hectare,
z_5 : Cost of weeding,	z_6	: Cost of Urea/hectare,
z_7 : Cost of seed and application/hectare,	z_8	: Cost of labor,
z_9 : Cost of transportation,	α_0	: Intercept,
Y : Net Return/hectare,	ln	: Natural logarithm.

The advantage of this function are that, it shows The diminishing marginal returns and it can also be used to estimate return to scale and it is easy to estimate. The possible disadvantages are that it cannot show both increasing and diminishing marginal returns in a single response curve, and that may lead to over-estimate of the economic optimum [10].

The Marginal Value Products is performed by deriving the following equation from the Cobb-Douglas function for assessing resource allocation efficiency.

$$MVP = T_i(Y/Z)P_y = Q_i P_i, \tag{118.4}$$

where

Q : The allocative efficiency parameters of the ith input,
P : Average price of ith input,
Z : Average cost of ith input,
Y : Mean of sugarcane net return/hectare,
T : Output elasticity of ith input,
MVP : Marginal value product of ith input, Z_i.

If $Q < 1$ it shows that the inputs are over used,and if the $Q > 1$ shows that the inputs used are under utilized, and $Q = 1$ means the inputs are efficiently used. The parameters of the regression equation are calculated using analytical software.

From the tables prove that, the cost of urea, labor, transportation, weeding and seed were over utilized and the cost of Land Preparation, FYM, irrigation and DAP were poorly utilized in the sugarcane production. The results also show that, there are opportunities to increase the out put of sugarcane production and sugarcane growers increase revenue by decreasing the cost of major inputs.

118.5 Results and Discussion

Sugarcane production is a complex process and can be conceived as a function of several variables. The production efficiency and production function analysis had been the cost of inputs and net returns relationships of the sugarcane producers in Pakistan. The Cobb-Douglas production function was used to estimate the production function from a data set from the sugarcane growers survey carried out during, 2009–2010. This approach was commonly used to assess input and output relationships [7, 10]. This method has easy to interpret results also provides a sufficient degree of freedom for statistical testing [7].

The least squares regression method is widely used to estimate input and output relationship. This method enables not only to find the line of best fit,but also to measure how good a fit it [10]. The factor were highly significant at 5 % level for the cost of sugarcane production. The cost of Urea, the cost of irrigation, cost of seeds, cost of DAP, seeds and its application and weeding were set in the analytical technique.

(1) Cost of Irrigation. Irrigation is an input of sugarcane production. In Table 118.3, result show that the coefficient of irrigation is 0.09595. It means the irrigation has positive impact on sugarcane production, but the ratio of MVP is greater then 1, It shows that the use of irrigation is poorly utilized by growers. Farmers facing this problem because some distributaries are favorable blessed that it is difficult for their haries to manage the surplus water.

(2) Cost of FYM and Application. In Table 118.3, result show that FYM and sugarcane production has portative relation because the coefficient of FYM is 0.08030. if sugarcane growers increase 1 % the use of FYM it would increase 0.08030

production of sugarcane at 5 % level of significance, holding the other factors constant. But the MVP ration of FYM in greater then 1, it means the use of FYM is poorly utilized by farmers.

(3) Cost of DAP. In Table 118.3, the relationship between the output of sugarcane production and the use of DAP is positive because the regression coefficient of the cost of DAP is positive (0.33610). If the sugarcane growers increase 1 % use of DAP, the output will increase by 0.7 %, if the other factor constant, but The value of MVP ratio is greater then one so the result show the use of DAP is poorly utilized. The result also show that the cost of DAP influenced the sugarcane revenue due to moderate use of DAP.

(4) Cost of Land Preparation. The result show the cost of land preparation should be reduced as it has positive impact on sugarcane revenue and returns. Because the regression coefficient of the variable of cost of land preparation is 0.97009 and significant at 5 % level of significance if the other variables are constant. The cost of land preparation has a big impact on the output of sugarcane production. In Table 118.2, result show that the value of MVP ratio is greater then 1 so the preparation of land is poorly prepared by growers.

(5) Cost of Weeding. In Table 118.3, result show that increasing the use of weedicide would help in decreasing weeds and shrubs from the sugarcane crop increasing the revenue and returns overall. Weeding has positive impact on sugarcane production, But in result the coefficient of weed is negative (-0.20475) at 5 % level of significance and ratio of MVP less then 1, it means the use of weeding is over used by growers. Here result shows negative impact because of over usage. SO if growers use the weeds in limit then it gives positive impact on sugarcane otherwise result will be going worse, other factors constant.

(6) Cost of Urea. In Table 118.3, result show that the regression coefficient of cost of urea is positive, it means the cost of urea and sugarcane revenue has positive impact. if the sugarcane growers increase 1 % use of urea would increase the output by 1.84 %, keeping the other factor constant. The sugarcane growers increase the revenue due to moderate use of urea. The cost of urea significantly influenced the sugarcane revenue of sugarcane growers, keeping other factors constant.

(7) Cost of Seed and Application. The regression coefficient of the variable cost of seed and application is negative (-0.09530), which is non-significant, which implied that 1 % increase in the use of seed and application would decrease the returns by -0.095 %, indicating that the cost of seed and application must be improved as it has positive impact on sugarcane production and revenue.

(8) Cost of Labor. The regression coefficient of cost of labor is negative (-0.17465), which is non significant. The result shows that, cost of labor and sugarcane output has negative impact and MVP ratio also show that the usage of labor is overused by farmers. The result also show that the cost of labor is decreasing the output of sugarcane and it is also a important factor for sugarcane bowing and harvesting. Cost of labor depends on number of persons and required time. So mostly in high season labor charge high price per mund. If the other factor keep constant then according to table result, it decrease the revenue of growers.

Table 118.2 Calculated value of coefficient and statistics of Cobb Douglas production function of sugarcane

Variables	Coefficient	MVP	Ratio of MVP
z_1	0.0955	121.49	12.24
z_2	0.08030	5.67	8.22
z_3	0.3361	48.99	7.11
z_4	0.97009	12.80	4.36
z_5	−0.20475	−63.48	−12.38
z_6	1.84616	6.21	0.83
z_7	−0.09539	−121.49	−24.69
z_8	−0.17465	−6.11	−0.72
z_9	−0.28525525	−6.05	−0.65

Table 118.3 Calculated value of coefficient and statistics of Cobb Douglas production function of sugarcane

Variables	Coefficient	Standard error	t-value	P-value
z_1	0.09595	0.02608	3.68	0.0000
z_2	0.08030	0.03058	2.63	0.0008
z_3	0.33610	0.03769	8.92	0.0000
z_4	0.97009	0.04988	19.45	0.0000
z_5	−0.20475*	0.04399	−4.65	0.0000
z_6	1.84616	0.3404	5.42	0.0000
z_7	−0.09530	0.08985	1.06	0.3967
z_8	−0.17465	0.03077	−5.68	0.0000
z_9	−0.28525	0.25305	−1.12725	0.0000
R^2	0.9350			
Adjusted R^2	0.9336			
R.MS	0.49563			
S.D	0.74749			

$* = 5\%$ level of significance

(9) Cost of Transportation. The transportation is high from sugarcane farms to mill. It requires bulk carriage or trucks, the owners of trucks charged high price in high season, because they have to wait for unloading their sugarcane at the mill. Because trucks unloaded on first come first serve policy. In the table the coefficient of transportation is negative (−0.28525) which is non significant, it shows that the cost of transportation and sugarcane revenue has negative impact. Table also show that the ratio is less then 1 it means the growers pay high cost for transportation. If the other factor constant then result show that the cost of transportation is also a main reason to decreasing the sugarcane farmers revenue.

118.6 Conclusion and Recommendation

There are many factors that effect the sugarcane production. The purpose of this study is to identify the some important factors which have huge impact on sugarcane production. There are many factors which contribute towards higher output of sugarcane. One of the important factor of low yield is lack of education in the farmers. Education plays a vital role to enhance the production. But unfortunately most of the farmers are illiterate and do not have enough knowledge and experience to tackle the problems and to enhance the production. In this study the cost input factors are the basic problem of low sugarcane production. For this purpose data were collected from 400 sugarcane growers from the 5 districts of Punjab province of Pakistan. The data were collected during the period 2007–2008. The collected data show that the main factor like, weeding, cost of irrigation, urea, DAP, FYM, land preparation, seed and its application are affect the sugarcane production. In this paper we used Cobb Douglas production function and Marginal Value Product. The result show that the cost of FYM, irrigation, cost of DAP, and cost of land preparation were poorly utilized and the cost of Urea, Weeding, seed, transportation, and labor were over utilized in sugarcane production. The results also show that the high cost of inputs, technical economic and social problems decrease the revenue of sugarcane growers.

Government and other related organizations work out on identified problems of the farmers and then growers can earn higher return from sugarcane production. Government and other related organization must be projected consumer friendly policy and productivity enhancement program must be constituted adjust and support prices. They should provided latest machinery to the farmers to increase per acre yield and Government also to erase the monopolistic powers of Mill owners and in police making process, avoids to get suggestion from those assembly members, they have own sugar mills or the member of mills. The results of this paper also show that there have opportunities to increase the output of sugarcane production by decreasing the cost of major inputs.

References

1. Adnan N, Ghulam AJ et al (2013) Factor affecting sugarcane production in pakistan. Pak J Commer Soc Sci 1:128–140
2. Bashir MK, Gill ZA et al (2007) Impact of credit disbursed by commercial banks on the productivity of sugarcane in faisalabad district. Pak J Agric Sci 44(2):361–363
3. Coelli T, Rao DS et al (1998) An introduction to efficiency and productivity analysis. Kluwer Academic Publishers, Boston
4. Fare R, Grosskopf S et al (1994) Productivity growth, technical progress, and efficiency change in industrialized countries. Am Econ Rev 84:66–83
5. Government of Economic Survey of Pakistan 2005–2006 (2006) Economic Advisor Wing, Finance Division, Islamabad, Pakistan
6. Gujarati ND (2003) Basic econometrics. Fourth international edition. McGraw-Hill Education, Asia

7. Heady EO, Dillon J (1961) Agricultural production function. Iowa State University Press, Ames Iowa
8. Malik KM, Gurmani MH (1999) Sugarcane production problems of lower sindh and measure to combat the problems. Dewan Sugar Mills Limited, Budho Talpur, District Thatta
9. McMillan JH (1999) Educational research: fundamental for the consumer, 3rd edn. Harper-Collins College Publisher, New York
10. Upton M (1996) The economics of tropical farming system. Cambridge University Press, Cambridge
11. Raza M, Ramachandran H (1990) Schooling and rural transformation. Vikas Publishing House PVT, New Delhi
12. Raheman A, Talat A et al (2008) Stimating total factor productivity and its components: evidence from major manufacturing industries of Pakistan. Pak Dev Rev 47(4):677–694
13. Syed MRZ, Ahmad S (2013) Cane growing capacity of the growers of Sindh and its consumption in sugar ibdustries a study of issues and solution. Interdisc J Contemp Res Bus 5(1). SSRN: http://ssrn.com/abstract=2326958
14. Syed MRZ, Ahmad S, et al (2013) Impact of low-sugar-cane-yield on sugar industry of Pakistan. Interdisc J Contemp Res Bus 4(12). SSRN: http://ssrn.com/abstract=2326957

Chapter 119
The Structure and Effects of the Collective Psychological Capital Research

Xia Luo and Weizheng Chen

Abstract It is a common concern to enterprise and employees that make employees have a healthy state of mind. Collective psychological capital as a group shared healthy state has not been studied, especially in domestic. This paper, through literature, interview method and open questionnaire investigation, found that the collective psychological capital is made up by collective hopes, collective resilience and collective efficacy, then developed a measurement scale with good reliability and validity. Finally, this article found that collective psychological capital can achieve win-win between organization and employee development.

Keywords Psychological capital · Collective psychological capital · Positive organizational behavior

119.1 Introduction

As to psychological capital theory, most of it focused on personal level, which means Psychological capital is the positive psychological state in the process of personal growth and development. But some researchers argue that it is the utmost flaw of psychological capital theory for collective level is absence [3, 9]. Therefore, this article tries to explore the psychological capital from the collective level and discuss its effect to the organization.

X. Luo (✉)
School of Management, Southwest University for Nationalities, Chengdu 610064, People's Republic of China
e-mail: luo_xia@126.com

W. Chen
Busniss School, Sichuan University, Chengdu 610064, People's Republic of China

J. Xu et al. (eds.), *Proceedings of the Eighth International Conference on Management Science and Engineering Management*, Advances in Intelligent Systems and Computing 281, DOI: 10.1007/978-3-642-55122-2_119, © Springer-Verlag Berlin Heidelberg 2014

119.2 Literature Review

The research about collective psychological capital is start, which is focused on the concept and measurement.

The collective psychological capital is defined as the result of the interaction and coordination activities [6]. The interaction mechanism makes the collective psychological capital is greater than the sum of the individual psychological capital. They think that the collective psychological capital is a group shared positive state of mind, with collective efficacy, collective optimism, collective hope and collective resilience. West [8] argues that the collective psychological capital can be regarded as the same form with individual psychological capital.

A survey of the collective psychological capital has two ways:

1. Summation method: measurement results will converge to the collective level through direct application of individual psychological capital scale [2];
2. Transfer the reference point consensus construct measurement method: the individual psychological capital scale adapted for collective psychological capital measurement scale.

For example, Walumbwa et al. [6] changed the PCQ-24 individual psychological capitals scale into collective psychological capital scale. The 24 items are reduced to eight items, namely each dimension two items. West et al. [8] used the same train of thought, basic reference to individual psychological capital scale measuring scale. In order to make the measurement for the collective psychological capital is not only the capital merging of individual psychology, every measurement item changed to questions from the collective level.

Summation method is simpler because don't have to develop new scale operation, but due to the collective psychological capital is not the sum of individuals in the collective psychological capital, so the measurement validity is poor. The second method compiled according to the concept of collective psychological capital scale in accordance with the characteristics of the collective psychological capital, also makes the measurement more accurate, but the existing literature does not give a better collective psychological capital scale, which is derived from the revision of the scale not rigorous.

To sum up, the collective psychological capital is the group Shared psychological state, which is the result of the interaction and coordination between members of activities. The interaction mechanism makes the collective psychological capital is greater than the sum of the individual psychological capital, but for the further research of the collective psychological capital is still very lack and the scale is imprecise.

In view of the difference of cultural background between China and the west, to construct localized collective psychological capital theory is very necessary, and to explore the connotation of the localization of the collective psychological capital and measuring tool is the primary work.

119.3 The Study of the Structure of the Collective Psychological Capital

To adopt the concepts of collective psychological capital of Walumbwa et al. [6], most scholars holds that the collective psychological capital is the group shared a positive state of mind.

119.3.1 Setting up the Initial Table

Considering both to inherit the existing research results from scholars, and to increase the content of localization, the collective psychological capital's initial measuring table consist of three aspects: firstly, through in-depth interviews and open question-naire survey to collect the collective psychological capital content of localization; secondly, keywords from the existing literature on the collective psychological cap-ital; thirdly, the individual psychological capital measurement subject was adapted for collective psychological capital measurement subject.

(1) In-depth interviews

In-depth interviews with the method of network chat, talk about an hour. Respondents had more than 3 years work experience, and also their work unit, region, position characteristics was considered. Each interviews conducted on a total of 5 people. The interview subject asked respondents to describe their organization's collective psychological capital characteristics. Prior to the interviewer, we explain the concept of collective psychological capital for respondents.

(2) Open questionnaire survey

Issuing open questionnaire to MBA students from the school of business, Sichuan university (work with more than half a year), to collect collective psychological cap-ital items. 101 questionnaires were recycled. The questionnaire uses the open key event description method, respondents were asked to describe their organization's collective psychological capital characteristics with 3–5 keywords. We explained the concept of collective psychological capital for respondents before they answered the questionnaire survey. Two doctoral student analyzed the data of in-depth inter-views and open questionnaire through open coding, spindle coding, finally got 7 terms, which according to the occurrence frequency from more to less were good vision, employees motivated, good prospects for development, employee engage-ment, toughness, confidence, employee vigorousness.

(3) Revising individual psychological capital scale

We revised the measurement-scale made by Luthans [4] according to the advice of Chan [1]: measurement index on the team should not just the sum of the individual, each item shall be converted into measurement of team. It also followed the method used by Walumbwa [6] and West [8]. As part of the subject was not suitable for adaptation to the collective level, so only 15 of 24 individual psychological capital measurement was adapted.

(4) Literature refinery

By reading the published literature related to the collective psychological capital after the positive organizational behavior was put forward, 7 keywords were defined that may express the content of collective psychological capital content, which Included collective hope, collective tenacity, collective optimism, collective efficacy, the collective subjective well-being, the collective emotional intelligence, collective virtue, and adapted for 11 questions according to the literature. Initial scale for collective psychological capital was eventually composed of 31 subjects.

119.3.2 The Exploratory Research on Scale

The questionnaire survey has been carried in Chengdu, Hainan, Guangdong, and so on. Respondents were all employees. We collected 336 of 350 questionnaires, and finally acquired 244 valid questionnaires after excluding 92 questionnaires filled by employees including from non-enterprise work, less than half a year and other waste volume. The recovery rate was 96%, and efficient rate 69.7%. Enterprises located in Sichuan accounted for 69%, 12% in Hainan, 12% in Guangdong and 5% in the rest area. Samples from state-owned or state-controlled companies accounted for 30%, private or private holding company (36.6%), foreign investment or foreign holding company (20.7%), public institution (12.7%). Sample distributed in multiple industries. Men and women accounted for 57.4% and 42.6%. Young employees from 20 to 40 years old was the most part, and 88% of employee have a college degree or above and averagely 5 years work experience. Employees from ordinary position accounted for 30.4%, management or technical personnel at the grass-roots level accounted for 22.8%, middle managers or middle-level technical staff accounted for 43.9%, senior managers or senior technical staff accounted for 3%.

The exploratory research on initial scale of collective psychological capital included that analysis the investigation results, exploratory factor analysis, and cancelled the topic according to the results of the analysis, tested questionnaire reliability, and finally got formal scale of the collective psychological capital. According to the principle of characteristic roots greater than 1 and oblique rotation method, we deleted the subject that its factor loading less than 0.4 and cross load more than 0.4. Through many times rotation, it showed that collective psychological capital consisted of three factors of 14 topics. Factor naming, composition and degree of variation explanation as shown in Table 119.1.

The Cronbach alpha coefficient of overall Collective psychological capital scale is 0.965. The Cronbach alpha coefficient of the collective hope, collective tenacity, collective efficacy was 0.895, 0.876, 0.843, respectively, and they were greater than 0.7, so the collective psychological capital scale has good reliability.

Table 119.1 Exploratory factor analysis results

Subjects	Factors		
	Collective hope	Collective resilience	Collective efficacy
My colleagues believe that our organization development prospect is very good	0.942	−0.158	0.019
My colleagues think our work is very meaningful	0.927	−0.024	−0.069
My colleagues Motivated	0.821	0.165	−0.102
My colleagues work hard in his own heart	0.741	0.115	−0.046
My colleagues think that our organization has a clear and positive perspective	0.711	−0.020	0.201
My colleagues understand organizational goals and believe will succeed	0.538	0.055	0.332
My colleagues are vigorous	0.476	0.321	0.099
My colleagues are able to deal with a lot of things at the same time	−0.096	0.908	−0.104
My colleagues can be relatively independent processing work	0.008	0.844	−0.047
My colleagues often try their best to solve the difficulties	0.187	0.749	−0.032
My colleagues always out of the difficulties in the work	0.063	0.704	0.144
My colleagues always very confidently set work goal	0.074	−0.129	0.855
My colleagues are always confident to communicate with people outside the company and discuss the problem	−0.206	0.226	0.851
My colleagues always very confidently discuss about the organization strategy	0.129	−0.251	0.850
My colleagues always very confidently introduce documents	−0.028	0.266	0.675
Characteristic value	8.196	1.258	1.029
Explained variance	56.641	8.389	6.860
The cumulative explain variance	56.641	63.030	69.889

119.3.3 Confirmatory Study

This study conducted a questionnaire survey in places such as Chengdu, Panzhi-hua, the investigation object is on-the-job employees from more than 10 companies. Respondents were defined according to nature of enterprise, industry, scale and other factors. A total of 419 of 460 questionnaires were collected (91 % recycling rate). Effective questionnaire 325 was acquired except 94 questionnaires from employee worked less than half a year, and other waste volume, effective rate was 77.6 %. Male respondents accounted for 66.7 %. 80 % respondents was younger than 40 year old,

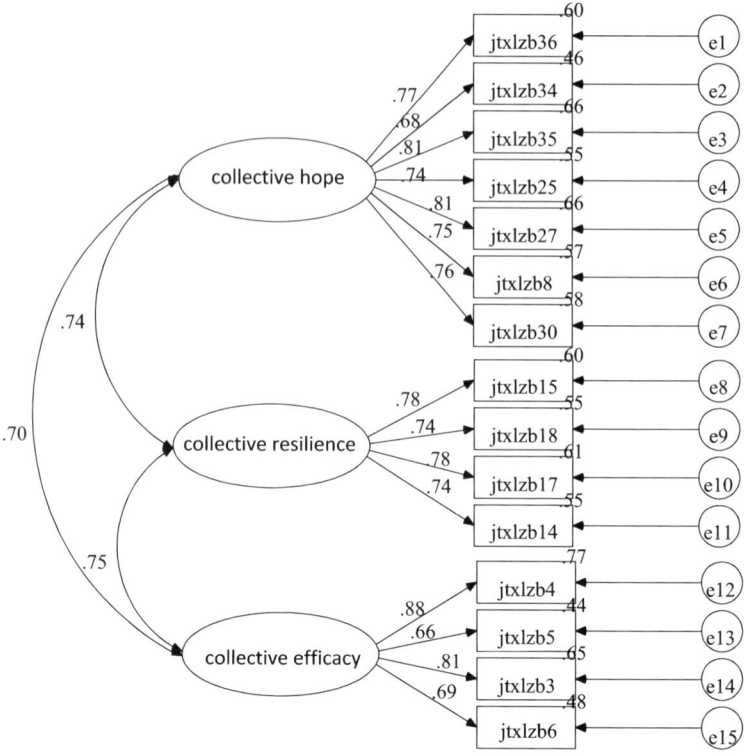

Fig. 119.1 Confirmatory factor analysis of collective psychological capital

95 % respondents have a college degree or above. Ordinary respondents accounted for 12.6 %, managers or technicians at the grass-roots level accounted for 32.1 %, senior managers or senior technical staff accounted for 53.3 %, senior managers or senior technical staff accounted for 2 %, the average working year was 8 years. The confirmatory study on initial scale of collective psychological capital included that violating estimated inspection, confirmatory factor analysis, model adaptation degree evaluation, competitive model comparison, test on reliability and validity. Confirmatory factor analysis results of the collective psychological capital scale of were shown in Fig. 119.1.

The analysis results showed that all item of standardized load coefficient of the collective psychological capital scale was greater than 0.66 and the t value was greater than 1.96, all these showed that the collective psychological capital scale has certain structure validity.

Three alternative models were compared:

- A one-dimensional model, assume that all 15 questions share a common latent variables-collective psychological capital;

Table 119.2 Compared results of factors competition model ($N = 325$)

Modle	x^2	df	x^2/df	RMSEA	NFI	GFI	IFI	CFI	NNFI
One-dimensional model	270.281	106	2.55	0.068	0.919	0.949	0.949	0.949	0.934
Two-dimensional model	252.992	105	2.409	0.064	0.908	0.924	0.954	0.954	0.940
3 d model	145.355	63	2.307	0.050	0.967	0.961	0.981	0.981	0.968

- Two-dimensional model, incorporating collective tenacity and collective efficacy for a dimension, collective hope for a dimension;
- 3 d model, collective hope, collective tenacity, collective efficacy consists of the 3 d model, respectively.

AMOS software was respectively used to model validation for the three alternative models. The verification results were shown in Table 119.2 that 3 d model had the best model index fitting effect.

Combination reliability CR of the latent variables of Collective psychological capital scale was between 0.846 and 0.906, all greater than 0.7, so the reliability test suggested the scale was good. In addition, mutational extraction (AVE) of each factor was between 0.580 and 0.587, were greater than 0.5, so the collective psychological capital scale has good convergent validity. Maximum correlation coefficients between latent variables was 0.75, the square value was 0.562, less than the AVE minimum value of 0.580, so the distinction validity of collective psychological capital scale was good.

According to above analysis, we concluded that collective psychological capital consist of the collective hope, collective tenacity, collective efficacy, and the developed collective psychological capital scale has good validity.

119.4 The Effect of the Collective Psychological Capital

Study assumes that the collective psychological capital can achieve harmony and win-win development between organization and employee, so choose the organizational performance and employee grow as prediction variables, as well as test criterion validity.

119.4.1 Research Hypothesis

Collective psychological capital's impact on organizational performance Collective psychological capital's impact on performance of group or organization has obtained some research support, such as Walumbwa [6] empirical study found that the collective psychological capital can effectively predict collective performance and organizational citizenship behavior. Smith [2] found that under the common action of integrity leadership 1 and subordinate Collective psychological capital, subordinates' trust in leadership and organizational financial performance improved significantly. Therefore, put forward the following hypothesis.

Hypothesis 1. The collective psychological capital has positive effect on organizational performance.

Collective psychological capital affects employee thriving if the collective psychological capital is only significant in improving organizational performance, then the collective psychological capital theory is not different to the traditional management theory. Collective psychological capital aim to improve organizational performance, and meantime the employees also obtain development. Employees' thriving is that employees feel energetic at work (vitality), and to get a kind of psychological experience of growing up [5], it reflects like instinct of person pursuit the self-realization, is people's most powerful positive force. Therefore, put forward the following hypothesis.

Hypothesis 2. The collective psychological capital has positive effect on employee thriving.

119.4.2 Methods

During the process of confirmatory factor analysis, data of organizational performance and employee thriving were also collected. Organizational performance measurement scale developed by Wang [7] was used. The scale is composed of one dimension, a total of seven topics, measuring organizational performance from two aspects of financial indicators and non-financial indicators. Respondents gave the score through compared their enterprise to other enterprise within the same industry. Questionnaire using 6 point scale method of likert, the higher the score, agreed with the more subjects described in the topic, the lower the score, agreed with the less subjects described in the topic. Scale of Cronbach's alpha value is 0.915, the reliability is very good.

Grow scale of employees was adopted the scale developed by Spreitzer [5]. By professional translators translated and back translated program, conducted exploratory factor analysis data from 244 samples. All 11 questions and two dimensions of original questionnaire was finally retained. Scale was made up of learning dimensions, dynamic dimension. Measurement learning was consisted of three questions, measuring dynamic was consists of eight topics. 6 point scale method of likert was used in questionnaire, the higher the score, agreed with the more subjects described in the topic, the lower the score, agreed with the less subjects described in the topic. Using confirmatory factor analysis to analyze employee thriving scale, the results show that two factors structure fitting is good. X^2/df a value was 2.87, RMSEA value was 0.064, CFI value and IFI were 0.987, the factor load standardized coefficient estimates of each item were significant at the 0.001 level.

Table 119.3 The correlation analysis between variables

Variables	Average value	Standard deviation	Collective psychological capital	Development of employee
Collective psychological capital	3.67	0.84		
Development of employee	4.26	0.96	0.542***	
Organizational performance	3.97	1.07	0.295***	0.342***

*** Means the coefficient in $P < 0.001$ significant level (two-tailed test)

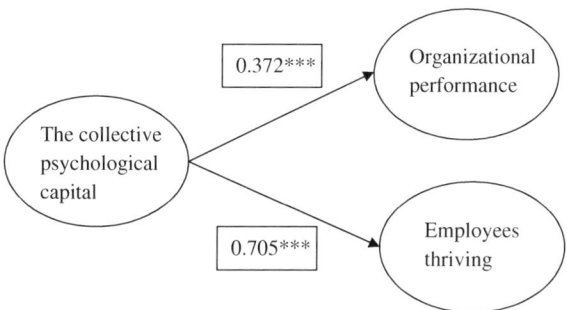

Fig. 119.2 Collective psychological capital affected on organizational performance, employees thriving

119.4.3 Results

Harman single factor test method was firstly used to test common method biases, all subjects was analyzed together. Not rotating factor analysis results show that there are 10 common factors' eigenvalues greater than 1, 66.4 % of the total variance was explained, one of the largest eigenvalue factor for 20.457 explained 33.54 % of the total variance, so did not appear the situation that a principal components explained most variation, so common method biases of this research data are not significant. Results of descriptive statistical analysis and correlation analysis of every variable were shown in Table 119.3. It is significant correlation between variables, which conform to the hypothesis.

Using structural equation model for further analysis on the relationship between the three variables, the results were shown in Fig. 119.2. The analysis results showed that the model fitting effect is good. The x^2/df was 1.896, RMSEA value was 0.044, CFI and IFI were 0.996. Collective psychological capital significantly positive affected on organizational performance ($b = 0.372$, $p < 0.001$), hypothesis 1 was verified; collective psychological capital significant positive affected on employee thriving ($b = 0.705$, $p < 0.001$), hypothesis 2 was verified.

119.5 Discussions

Collective psychological capital is a kind of group Shared positive psychological state. In accordance with the requirements of the grounded theory, this article get the collective psychological capital is divided into three dimensions, respectively is collective hope, collective resilience and collective efficacy. According to Table 119.1, collective hope explains the collective psychological capital 56.6 % of the total variance, is the largest eigenvalue. The research results show that in Chinese culture, the collective psychological capital is made up by the collective hopes, collective resilience and collective efficacy. Collective hope is one of the main factors. The result is different from western researchers'. West et al. [8] select collective efficacy, collective optimism and collective resilience three dimensions, and found that at the beginning of the development group, collective optimism is an important predictor variable to group satisfaction. Walumbwa et al. [6] select collective efficacy, collective resilience, collective hope and collective optimism four dimensions. Compared with the western research, in Chinese, collective hope is the most important part of the collective psychological capital and there is no collective optimism dimension. Possible reasons include under Chinese culture background, the contribution of the collective optimistic to the collective positive psychological capital is not enough compared with other three dimensions; optimistic is considered to be more of a individual characteristics and less by collective owned; collective optimistic are generally unimportance in the workplace; employees collective optimism is a lack of awareness.

The questions of the initial scale of local content, namely through in-depth interviews and open questionnaire, almost preserved and all reflected in the collective hope dimension. This finding reveal that employees feel hope for the future is very important, but the causes of hope are different from the west's. In the study of western, the factor of collective hope is the organization's main goal and success experience, while in China, bringing hope factors mainly rely on the development prospects and positive climate of the organization, as well as the work whether it makes sense.

Collective psychological capital is able to improve organizational performance and to promote employees thriving, which suggests that the collective psychological capital can achieve win-win harmony between organization and employee development.

The meaning of this article as follows:

1. Academic, to clear collective psychological capital structure and to develop collective psychological capital scale development lay the foundation for further study on the collective psychological capital theory;
2. Practice, to provide a reference for enterprises to improve the overall employees positive psychological capital, which is focused on improving employee collective hope, collective tenacity and collective efficacy.

Acknowledgments   This research is supported by NSFC (No.71272210), MOE Project of Humanities and Social Sciences (No. 10YJC630171) and National Higher-education Institution General Research and Development Funding (No. 11SZYQN54). The degree of construction project (No. 2011XWD-S1204).

References

1. Chan D (1998) Functional relations among constructs in the same content domain at different levels of analysis: a typology of composition models. J Appl Psychol 83(2):234–246
2. Clapp-Smith R, Vogelgesang GR, Avey JB (2009) Authentic leadership and positive psychological capital: the mediating role of trust at group leve of analysis. J Leadersh Organ Stud 15(3):227–240
3. Hackman JR (2009) The perils of positivity. J Organ Behav 30(2):309–319
4. Luthans F, Youssef XM (2004) Human, social, and now positive psychological capital management: investing in people for competitive advantage. Organ Dyn 33(2):143–160
5. Spreitzer G, Sutcliffe K et al (2005) A socially embedded model of thriving at work. Organ Sci 16(5):537–549
6. Walumbwa FO, Luthans F et al (2011) Authentically leading groups: the mediating role of collective psychological capital and trust. J Organ Behav 32(1):4–24
7. Wang D, Tsui AS et al (2003) Employment relationships and firm performance: evidence from an emerging ecomomy. J Organ Behav 24(5):511–535
8. West BJ, Patera JL, Carsten MK (2009) Team level positivity: investigating positive psychological capacities and team level outcomes. J Organ Behav 30(2):249–267
9. Zhao S (2011) Intellectual capital and the application research on theory and practice of psychological capita. Nanjing Soc Sci 2:11–17 (In Chinese)

Chapter 120
A Two-Stage Resource-Constrained Project Scheduling Model with Proactive and Reactive Strategies Under Uncertainty

Lu Chen and Zhe Zhang

Abstract The aim of this paper is to develop a two-stage model to obtain a proactive and reactive schedule in resource-constrained project scheduling problems (RCPSP) under uncertainty. In the proactive phase, the highest cumulative instability weight scheduling of resource buffering is selected to optimize the initial schedule, which is arrange the activities in decreasing order by the sum weights of the activity and its successors. For the reactive schedule, the tabu search is employed to ensure the scheduling process execution. Actually, in practice, the uncertain resource availabilities are inevitable in RCPSP. In this situation, the uncertain factors are considered as the fuzzy random variables in this paper, and some properties of fuzzy random variables are discussed. Subsequently, the Particle Swarm Optimization (PSO) algorithm that solve the two-stage model of RCPSP with proactive and reactive strategies under uncertainty is developed. Finally, by testing the example, the effectiveness of the proposed model and approach is validated by the computation results.

Keywords Proactive and reactive scheduling · Two-stage · RCPSP · Fuzzy random variable

120.1 Introduction

As one of the most representative problems in the project management field, resource-constrained project scheduling problems (RCPSP) is a classical NP-hard problem and most researchers assume that the generation of a baseline schedule is in a static and determined environment with complete crisp information, meanwhile the project can be executed smoothly and not disrupted if the initial schedule is determined. In

L. Chen · Z. Zhang (✉)
School Economics and Management, Nanjing University of Science and Technology,
Nanjing 210094, People's Republic of China
e-mail: zhangzhe@njust.edu.cn

J. Xu et al. (eds.), *Proceedings of the Eighth International Conference on Management Science and Engineering Management*, Advances in Intelligent Systems and Computing 281, DOI: 10.1007/978-3-642-55122-2_120, © Springer-Verlag Berlin Heidelberg 2014

fact, however, these assumptions will be hardly satisfied. The baseline schedule is formulated in a dynamic and uncertain environment such as machine breakdown, rush order, and so on. No matter how much the initial schedule will be protected from the effect of possible disruptions, the occurrence can not be completely eliminated. To avoid these situations, many researchers start to pay more attention on proactive schedule and reactive schedule, such as Nikulin [9], Lambrechts et al. [6], Honkomp et al. [4] , and so on. From these articles, some based knowledge can be concluded about proactive schedule and reactive schedule. Proactive schedule is an off-line scheduling process [11] and is based on the construction of a baseline schedule, which will guide schedule execution by determining each activity its planned starting time. The objective is to generate robust and stable schedule that takes potential disruptions into account and minimize the effect after the occurrence of disruption. Proactive schedule creates the baseline schedule based on given information and requirements and constrains prior and provided a possible way to maximize the baseline schedule's robustness by to include safety in the initial schedule in order to absorb the anticipated disruptions as well as possible. Contrary to proactive schedule, reactive schedule is an on-line scheduling process, and it gives on the construction of a initial schedule and merely depends on the use of scheduling tactics to maintain on-line, which activities are to be started in random time that occur serially through time [10]. The aim of it is to modify the baseline schedule during the schedule process to adapt to the environment changes. it involves revising or rescheduling when an occurrence of disruption. It consists of defining a process to react to disruptions which can not be absorbed by the baseline schedule. In the field of proactive schedule for RCPSP, some studies have been done. First, some approach and methodology are proposed by researchers. Jun-hyung et al. [13] use a parametric optimization approach to solve proactive scheduling problems under uncertainly. A stochastic methodology which aim to optimization a cost function that consists of the weighted expected activity starting time deviations and the penalties or bonuses associated with late or early project completion is proposed by Filip et al. [2]. In addition, because of there are eight optimization choices for proactive schedule, many studiers do some works about them. Rafat. et al. [12] develop a new time proactive buffer heuristics to generate a robust scheduling project. Vonder et al. [14] study a proactive heuristic to improve project scheduling robustness. In the field of reactive schedule, many researchers do a large number of research. For example, Filip [3] use a tabu search heuristic for repairing a disrupted schedule to evaluate some reactive schedule that assume no activity can be started before its baseline starting time. A branch-and-bound procedure with minimum and maximum time lags is used to cope with the multi-mode resource-constrained project scheduling problem by Heilmann [5]. Zhu et al. [18] discuss the disruptions for RCPSP and propose some management measures. But either proactive scheduling progress or reactive scheduling progress is considered individually, there may be some peoblem during the execution of scheduling project. If only proactive schedule is employed to generate a initial scheduling progress, during the schedule is executed, some disruptions is occur and could lead to significant cost increase. Similarly, just consider only reactive scheduling tactics will bring about exceed the deadline constraint and resource excessive or lacking.

So many researchers consider proactive schedule and reactive schedule simultaneously. Olivier et al. [10] generalize the methods to optimization proactive and reactive scheduling progress. A multi-agent-based proactive/reactive scheduling project for job shops is used to hedge against the uncertainties of dynamic manufacturing environment by Lou et al. [11]. Although proactive and reactive schedule for RCPSP is studied by more and more researcher, we note that most of them consider the schedule in the stationary resources or duration. Each activity's finish time is not explicit, thus a proactive-reactive schedule for RCPSP with duration time is set to fuzzy random variables is considered by us. some properties and formulas of fuzzy random variables are discussed to solve duration times of each activity. And a two-stages model is built to deal with the schedule problem. The two stages includes proactive stage and reactive stage. In the proactive stage, resource buffering is considered. the tabu search is regarded as a tactics to cope with reactive scheduling project. Until today, few scholars study proactive and reactive schedule, so in this paper, we not only study the schedule, but also consider it with a two-stage model and fuzzy random amount of resources.

This paper will propose a two-stage model to obtain a proactive and reactive schedule in RCPSP under fuzzy random environment. The remainder of the paper is organized as follows. Research problem and statement is presented in Sect. 120.2, including the explanation of motivation for employing the fuzzy random variables. Then a two-stage model will be developed in Sect. 120.3, and the details of dealing with the fuzzy random variables are also presented. In Sect. 120.4, a particle swarm optimization algorithm is designed for solving the two-stage model. The validity of the proposed model and algorithm is proven by an example in Sect. 120.5. Finally, concluding remarks and the discussion about further research are made in Sect. 120.6.

120.2 Research Problem and Statement

In this section, before develop a model for proactive and reactive schedule for RCPSP, we describe the problem and explain the motivation that select fuzzy random variables.

120.2.1 Problem Description

Assuming that an enterprise wants to manufacture a product, a project is arranged to achieve it. The project which the amount of resource is fuzzy random variables will be scheduled, the project manager should generate a scheduling process to finish the project. But considering the environment is complex and unpredictable, something maybe occur during the scheduling process is executed (such as equipment failure, treacherous weather, machine breakdown, and so on), the robustness and

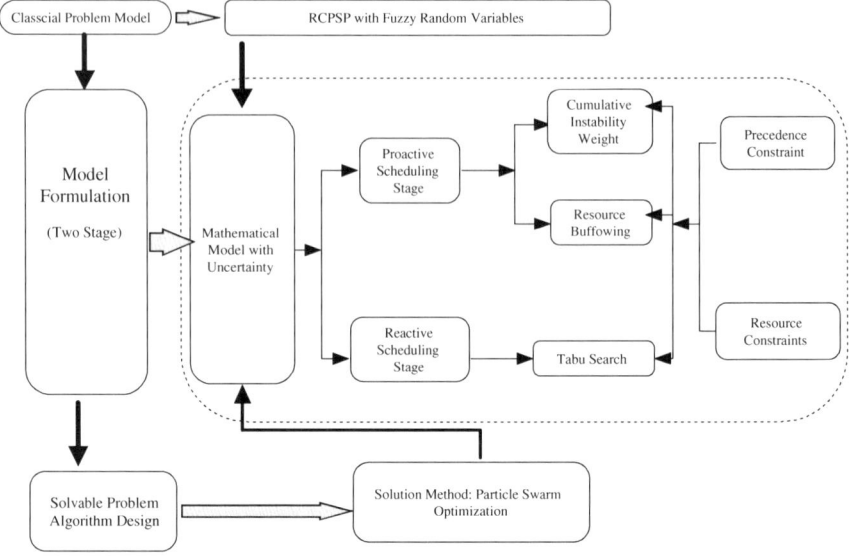

Fig. 120.1 Flow chart of the two-stage model

stability of the initial schedule is considered when the project manager generates it. The project manager can do it by a proactive scheduling stage. After the baseline schedule is ensured, the project is started to executed. At a time, a problem occurs, the scheduling process is disrupted. The manager must deal with it immediately. To minimize the deviation between the reality and the plan schedule, a reactive scheduling stage is considered. Generally, the proactive schedule is employed before starting project scheduling process, and is used to generate a baseline schedule for planning activities. The project manager can understand and predict the completion duration of the project and the efficiency of resource utilization according to the proactive schedule. Naturally, the realized situation and the predicted situation of the schedule are both of importance to the project manager. The initial schedule developed by the proactive scheduling tactics with deterministic information does not run well in dynamic project environments and can not guide the schedule effectively. To solve the scheduling problem in dynamic manufacture environments, a two-stage model for fuzzy random RCPSP is developed here, which includes the proactive scheduling stage and reactive scheduling stage, see Fig. 120.1.

120.2.2 Motivation for Employing Fuzzy Random Variables

The uncertain resource availabilities are inevitable in RCPSP. To cope with the hybrid uncertainty in the proactive and reactive scheduling process for RCPSP, we employ the fuzzy random variables in this study. Actually, the fuzzy random variable has been successful used in many areas, such as a large-scale water conservancy and hydropower construction project [17], queuing problems [15], inventory problems [1], portfolio problems [7], renewal reward processes [16], and so on. These studies show the efficiency of fuzzy random variables in handling a uncertain environment where includes fuzziness and randomness. So in this study, we also use fuzzy random variables to deal with the RCPSP, because it is hard to describe these problem parameters as crisp variables and the data provided by project managers are vague. For instance, the amount of resource is fuzziness and randomness co-exist. At present, as the price of the a resource is too high, the project manager want to reduce the use of this resource to reduce the cost. However, At the same time, the project manager hope that the reduction in the resource will not be too much that the qualities of the product will be compromised. In this case, the project manager give the range of an quantities in accordance with an expected value, such as viz $r = (a, \rho, b)$, with $\rho \sim N(\mu, \sigma^2)$. Therefore, a situation exists that may be fuzzy variables taking random parameters throughout the project indeed. So we employ fuzzy random variable to cope with those uncertain parameter of combining fuzziness and randomness and obtain more feasible schedule.

120.3 Two-Stage Modeling

In this section, a two-stage model will be developed, which includes a proactive scheduling stage and a reactive scheduling stage. In order to formulate the model, indices, certain parameters, fuzzy random coefficients, and the decision variable are introduced. Based on the assumptions and notations, we will propose the two-stage model for fuzzy random RCPSP.

1. Assumptions

The RCPSP can be stated as follows. A project consists of $i + 2$ activities where each activity has to be processed in order to accomplish the project. The $i+2$ activities are composed of a set of activities $i = 0, 1, 2, \cdots, I + 1$, Where the activity 0 and activity $I + 1$ are dummy, have no duration, and represent the initial and final activity (i.e., $d_0 = d_{I+1} = 0$) where d represents the activity's duration. There exists a set of resource $k = \{1, 2, \cdots, K\}$. No activity can be started before all its predecessors are finished. Each activity has a fixed duration and requires one or more styles of renewable or nonrenewable resources. In order to reflect each activity's importance of starting them at their planned starting time t_i, the weight ω_i ($i = 0, 1, \cdots, I + 1$) is assigned for activities and it denotes the marginal cost of deviating form the planned starting time during the initial schedule execution. In order to develop the proactive and reactive scheduling tactics for RCPSP under fuzzy random variables, some assumptions are follows

- A single project consists of a number of activities with several known execution modes;
- The start time of each activity is rely on the completion of its precedence activities (precedence constraints of activities). After completing a activity, the next activity must be started in the project;
- Resources are available in certain limited quantities, but the consumption of each resource is a fuzzy random variable.
- Activities are never started before their planned starting time.
- The objective function is to minimize the deviation from the initial baseline schedule.

2. Notations

The following notations are used to describe the two-stage model for RCPSP under fuzzy random environment.

i: activity index;

r: resource index;

t_i^S: starting time of activity i;

t_i^F: finish time of activity i;

t_i^D: the due time of activity i;

t_i: the processing time of activity i;

$\tilde{\bar{r}}_{ik}$: the activity i consumes resource units for resource k;

R_k: the maximum limited resource k;

ω_i: the importance of activity i;

here, $\tilde{\bar{r}}_{ik}$ is fuzzy random variable.

3. Mathematical Model

Based on the assumptions and notations above, we propose a proactive-reactive stage model of (RCPSP) under a fuzzy random environment. In the model, we minimize the deviation from the initial baseline schedule.

As to a RCPSP with resource-constraints under fuzzy random variables, the objective is to minimize the total project time. In this paper, we use the starting time of the last activity and its due time to represent the duration of the whole project. The objective function can be described as follows: $\min \sum_{i=1}^{I+1} \omega_i |t_i^s - t_i^D|$.

The start time of each activity is dependent on the completion of its precedence activities (precedence constraints of activities). After finishing a activity, the next activity must be started. It can be described as follows: $t_i^S + t_i \leq t_{i+1}^S$.

In addition, In order to strengthen the robustness of the schedule, resource buffering can be considered. it is important to limit the resource consumption of resources used by all activities in the project scheduling. In reality, each resource is available in limited quantities. Thus, it can be described as follows: $\sum_{i=1}^{I} \tilde{\bar{r}}_{ik} \leq R_k^*, \sum_{i=1}^{I} (t_i \tilde{\bar{r}}_{ik}) > R_k^* D$, where R_k^* can be calculated according to the following formulations:

$$R_k^* = \sum_{j=0}^{R_k} j p_r (r_k = j),$$ (120.1)

$$p_r (r_k = j) = C_{R_k}^j A_k^j (1 - A_k)^{R_K - j},$$ (120.2)

$$A_K = \frac{MTTF_k}{MTTF_k + MTTR_k}.$$ (120.3)

Here, $MTTF_k$ represents the mean time to failure and $MTTR_k$ represents the mean time to repair. D is the duration of totally project.

In the practical project scheduling, when the occurrence of a disruption, the activity i which the starting time would have a higher impact on instability should be scheduled than the activity j with a lower impact. we can rely on their cumulative instability weight (CIW_I) to arrange schedule.

$$CIW_i \geq CIW_{i+1}, CIW_i = \omega_i + \sum_{j \in succ_i} \omega_j,$$ (120.4)

here, $succ_i$ is the set of all successors of activity i, include direct and indirect.

In order to describe some non-negative variables for practical purposes, we use two equations as the time constraint and the resource constraint to indicate them,

$$t_i^S \geq 0,\ R_k \geq 0\ (i = 1, 2, \cdots, I, k = 1, 2, \cdots, K),\ \omega_i \geq 0,\ \sum_{i=0}^{I+1} \omega_i = 1.$$ (120.5)

So, by integration of above equations, the mathematical model can be stated as follows:

$$\min \sum_{i=1}^{I+1} \omega_i |t_i^S - t_i^D|$$

$$\text{s.t} \begin{cases} t_i^S + t_i \leq t_{i+1}^S \\ \sum_{i=1}^{I} \tilde{r}_{ik} \leq R_k^* \\ \sum_{i=1}^{I} (t_i \tilde{r}_{ik}) > R_k^* D \\ CIW_i \geq CIW_{i+1} \\ t_i^S \geq 0, R_k \geq 0 \\ \omega_i \geq 0 \\ i = 1, 2, \cdots, I, k = 1, 2, \cdots, K. \end{cases}$$

Since the model contains uncertain variables, the model can hardly be solved. In this situation, before the model is solved, we should deal with the uncertainty first.

In this paper, the element of the uncertainty is the fuzzy random coefficient $\tilde{\bar{r}}_{ik}$ and it should transformed to be a deterministic constraint. There we use the conclusion which is proved by Xu and Zhang [17]. These conclusion can be described as follows, if $\tilde{\bar{r}}_{ik} = (a, \rho, b)$, with $\rho \sim N(\mu, \sigma^2)$. Then,

$$\tilde{\bar{r}}_{ik} = \frac{a + \underline{\xi} + \bar{\xi} + b}{4}, \qquad (120.6)$$

where

$$\underline{\xi} = b - \beta \left[b - (\mu - \sqrt{-2\sigma^2 \ln(\alpha\sqrt{2\pi}\sigma))} \right] \qquad (120.7)$$

and

$$\bar{\xi} = a + \beta \left[(\mu + \sqrt{-2\sigma^2 \ln(\alpha\sqrt{2\pi}\sigma))} - a \right], \qquad (120.8)$$

where $\alpha \in [0, 1]$ be the given probability level of the random variable, and $\beta \in \left[\frac{b-a}{(\phi_\alpha^R - \phi_\alpha^L) + (b-a)}, 1 \right]$ be the possibility level of the fuzzy variable (where ϕ_α^R and ϕ_α^L are right and left border of α-level set for ρ).

120.4 Algorithm

In order to solve the problem, a particle swarm optimization (PSO) algorithm is designed. In fact, because the superior search performance and fast convergence, particle swarm optimization algorithm is considered as an effective tool for solving RCPSP under fuzzy random variables [8]. In PSO, the particles seek the best solution by the updating mechanism which it is:

$$v_{id}(t + 1) = \omega v_{id}(t) + c_1 \text{rand}_1 \left[p_{id}^{\text{best}}(t) - p_{id}(t) \right] + c_2 \text{rand}_2 \left[g_{id}^{\text{best}}(t) - g_{id}(t) \right],$$
$$x_{id}(t + 1) = x_{id}(t) + v_{id}(t + 1),$$

where $v_{id}(t)$ is the velocity of particle i at the dth dimension in the tth iteration, ω is an inertia weight, $p_{id}(t)$ is the position of particle i, $rand_1$ and $rand_2$ are random number in the range $[0, 1]$, c_1 is personal best position acceleration constant and c_2 is global best position acceleration constant, p_{id}^{best} is personal best position of particle i at the dth dimension and g_{id}^{best} is the global best position of particle i at the dth dimension.

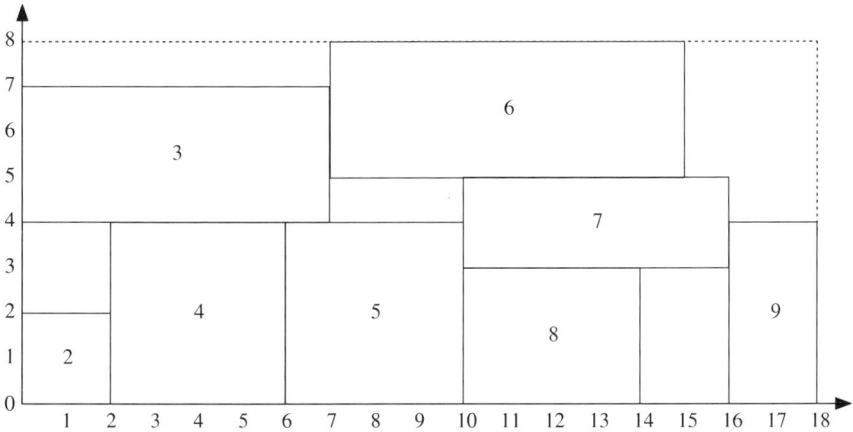

Fig. 120.2 Production scheduling order of Gantt Chart

120.5 Application

In this paper, the network diagram instance which is used by Olivier et al. [10] is used. But in order to meet the model of this paper, the resource consumption of each activity is set to be the fuzzy random variables. We assume those resource consumption is that:

$$\tilde{\bar{r}}_2 = [1.75, N(2, 1^2), 2.25]; \tilde{\bar{r}}_3 = [2.87, N(3, 2^2), 3.13]; \tilde{\bar{r}}_4 = [3.85, N(4, 2^2), 4.15],$$

$$\tilde{\bar{r}}_5 = [3.91, N(4, 3^2), 4.09]; \tilde{\bar{r}}_6 = [2.75, N(3, 1^2), 3.25]; \tilde{\bar{r}}_7 = [1.85, N(2, 2^2), 2.25],$$

$$\tilde{\bar{r}}_8 = [2.65, N(3, 4^2), 3.35]; \tilde{\bar{r}}_9 = [3.87, N(4, 2^2), 4.13]. \tag{120.9}$$

In addition, we take $\alpha = 0.4$, $\beta = 0.8$. After calculate by Eqs. (120.6) ~ (120.8), the value of resource is transform into determinately and their is that: $r_2 = 2$; $r_3 = 3$; $r_4 = 4$; $r_5 = 4$; $r_6 = 3$; $r_7 = 2$; $r_8 = 3$; $r_9 = 4$. And the $MTTF = 18$, $MTTR = 5$, then $A_k = \frac{18}{23}$, $R_k^* = \sum_{j=0}^{R_k} jp_r(r_k = j) = 8 \times \frac{18}{23} = 6.26 \approx 6$. But in order to meet formula Eq. (120.6), increasing R_k^* with one unit to 7. Other parameters are: $CIW_1 = 102$, $CIW_2 = 73$, $CIW_3 = 54$, $CIW_4 = 58$, $CIW_5 = 57$, $CIW_6 = 47$, $CIW_7 = 39$, $CIW_8 = 44$, $CIW_9 = 43$, $CIW_{10} = 38$. Above all, we substitute the data of that study [10] into the model of this paper. Then, the above algorithm is coded in MATLAB R2012a. Finally, we get the result: $t_2^S = 0$; $t_3^S = 1$; $t_4^S = 2$; $t_5^S = 0$; $t_6^S = 8$; $t_7^S = 4$; $t_8^S = 9$; $t_9^S = 15$; $t_{10}^S = 18$ and the min $\sum_{i=1}^{I+1} \omega_i |t_i^s - t_i^D| = 59$, the scheduling order is shown in Fig. 120.2.

120.6 Conclusions

In this paper, we consider the proactive-reactive scheduling tactics to deal with the resource-constrained project scheduling model under fuzzy random environment. In the propose model, we consider to minimize the duration of the whole project. The model employed fuzzy random variables to characterize the uncertain amount of resources where fuzzyness and randomness co-exist. For handling fuzzy random variables, we use the expected value model to cope with the constraints with fuzzy random coefficients. In the stage of proactive scheduling, we use the "highest CIW first" and resource buffering, and when a disruption occurs, tabu search tactic is used to deal with it in order to make the deviation as small as possible. And in the last of article, we tested a example and applied particle swarm optimization algorithms to resolve it.

One of the most important follow-up researches is the application of the proposed model and the two-stage tactics to the practical RCPSP. Besides, the proposed model under the fuzzy random variables can be also suitable for describing other uncertainties such as bi-random, fuzzy rough, rough random and so on. Other researchers could pay close attention on those uncertainty environment mentioned above.

References

1. Dutta P, Chakraborty D et al (2005) A single-period inventory model with fuzzy random variable demand. Math Comput Model 41:915–922
2. Filip D, Erik D et al (2011) Proactive policies for the stochastic resource-constrained project scheduling problem. Comput Oper Res 38:63–74
3. Filip D, Erik D et al (2010) Reactive scheduling in the multi-mode RCPSP. Comput Oper Res 38:63–74
4. Honkomp SJ, Mockus L et al (1997) Robust scheduling with processing time uncertainty. Comput Chem Eng 21:1055–1060
5. Heilmann R (2003) A branch-and-bound procedure for the multi-mode resource-constrained project scheduling problem with minimum and maximum time lags. Eur J Oper Res 127(2):348–365
6. Lambrechts O, Demeulemeester E, et al (2006) A tabu search procedure for generating robust project baseline schedules under stochastic resource availabilities. Research Report KBI 0604, FETEW, Katholieke Universiteit Leuven, Belgium
7. Li J, Xu J (2009) A novel portfolio selection model in a hybrid uncertain environment. Int J Manage Sci 37:439–449
8. Liu M, Cheng M et al (2008) Study on multi-mode resource-constrained project scheduling problem based on particle swarm optimization. J Agric Mach 39(2):134–138
9. Nikulin Y (2004) Robustness in combinatorial optimization and scheduling theory: an annotated bibliography. Research Report 583, Christian-Albrechts University in Kiel, Institute of Production and Logistics
10. Olivier L, Demeulemeester E et al (2008) Proactive and reactive strategies for resource-constrained projectscheduling with uncertain resource availabilities. J Sched 11:121–136
11. Ping L, Liu Q et al (2012) Multi-agent-based proactive Creactive scheduling for a job shop. Int J Adv Manuf Technol 59:311–324

12. Raafat E, Hidehiko Y (2012) New proactive time buffer heuristics for robust project scheduling. J Adv Mech Des Syst Manufact doi:10.1299/jamdsm.6.559
13. Ryu J, Dua V et al (2007) Proactive scheduling under uncertainty: a parametric optimization approach. Ind Eng Chem Res 46:8044–8049
14. Van de Vonder S, Demeulemeester E et al (2008) Proactive heuristic procedures for robust project scheduling: an experiment alanalysis. Eur J Oper Res 189(3):723–733
15. Wang S, Liu YK et al (2009) Fuzzy random renewal processwith queuing applications. Comput Math Appl 57:1232–1248
16. Wang S, Watada J (2009) Fuzzy random renewal reward process and its applications. Inf Sci 179:4057–4069
17. Xu J, Zhang Z (2012) A fuzzy random resource-constrained scheduling model with multiple projects and its application to a working procedure in a large-scale water conservancy and hydropower construction project. J Sched 15:253–272. doi:10.1007/s10951-010-0173-1
18. Zhu G, Bard JF et al (2005) Disruption management for resource-constrained project scheduling. J Oper Res Soc 56(4):365–381

Chapter 121
TRIZ Methodology Applied to Noise Comfort in Commercial Aircraft

João D. Molina, Helena V. G. Navas and Isabel L. Nunes

Abstract Acoustic comfort is currently one of the most important components in ensuring customer satisfaction promoting the well-being inside the aircraft. The cabin of the airplane is a space where the passenger remains closed during the trip, subject to noise from other passengers or from the aircraft itself. This noise can interfere with communication, disturb sleep, rest and relaxation. Currently, there are no means for select the sources of noise. There are only devices that target the total ear sound isolation. This paper presents the analysis of this problem (contradictions) and the design of a noise filtration system based on Theory of Inventive Problem Solving (TRIZ). TRIZ has the potential to aid in the creation of innovative systems. The main tool used in this methodology was the Substance-Field Analysis, which provided a detection of problems and "rough edges" in the original idea. The noise filtration system designed aims to eliminate the influence of unwanted noise in communication between people. It consists of sound headphones equipped with Passive Noise Control and a subsystem. The subsystem is also equipped with directional microphones "shotgun" capable of filtering out all the outside noise through Active Noise Control and band-pass filter.

Keywords TRIZ · Acoustic comfort · Headphones · Active noise control · Passive noise control · Substance-field analysis · Problem solving

J. D. Molina
Departamento de Engenharia Mecânica e Industrial, Faculdade de Ciências e Tecnologia, Universidade Nova de Lisboa, 2829-516 Caparica, Portugal

H. V. G. Navas (✉) · I. L. Nunes
UNIDEMI, Departamento de Engenharia Mecânica e Industrial, Faculdade de Ciências e Tecnologia, Universidade Nova de Lisboa, 2829-516 Caparica, Portugal
e-mail: hvgn@fct.unl.pt

J. Xu et al. (eds.), *Proceedings of the Eighth International Conference on Management Science and Engineering Management*, Advances in Intelligent Systems and Computing 281, DOI: 10.1007/978-3-642-55122-2_121, © Springer-Verlag Berlin Heidelberg 2014

121.1 Introduction

Commercial aviation is one of the most important means of transportation today, showing a high rate of growth. Also the number of airline companies has increased exponentially, originating a very strong competition among them. All this growth and competition are accompanied by the development of new aircrafts, which cannot disregard passengers comfort, thus making this question a constant challenge for aircraft and aircraft systems design. The main goal of aircraft manufacturers is to improve the efficiency and safety of air transportation systems; however, with the increased competition, customer satisfaction and passengers'comfort are also a main concern, therefore being one of the focus of manufacturers'attention.

From passengers and air crew comfort standpoint, noise is one of the most important issues. For air crews, cabin noise is an essential ergonomic concern, since the aircraft is their workplace. For passengers, this issue could be a determining factor in choosing the airline. This means that the strategies to improve aircraft cabin acoustic comfort should be constantly optimized [2]. Dealing with this issue calls for creativity in developing new ways to combat unwanted noise.

TRIZ is a human-oriented and knowledge-based systematic methodology, for inventive problem solving. Nowadays it has applications in different forms and in different areas, but the original idea behind this methodology is to help creation.

Therefore, since TRIZ applies in any matter involving creativity, this was the methodology selected to be used in this work, whose objective is the conceptualization and analysis of a unique and innovative idea, which aims to improve the acoustic comfort for passengers travelling in commercial aircrafts. The initial idea is to implement a system of special headsets, able to filter any unwanted noise, through mechanical and electronic systems, improving substantially the acoustic comfort inside the aircraft.

121.2 Acoustic Comfort

The main goal of Ergonomics is to ensure human well-being, health and safety while maximizing the performance of work systems or other aspects of our daily lives, like leisure or sport. One of its fields of intervention is related with noise.

Noise is an annoying and possibly harmful sound. According to Kryter [5] noise is "audible acoustic energy that adversely affects the physiological or psychological well-being".

In aircrafts noise can be generated by power engines and equipment or by passengers' conversation. Since noise is harmful it should be reduced and, if possible, eliminated. In a recent study [2] passengers have classified noise as one of the most important sources of aircraft discomfort. Other studies [14] show that the reduction of sound pressure level (background noise in the airplane) does not necessarily improve comfort satisfaction, since the reduction of this sound may enhance other audible sounds, for instance conversations of other passengers or babies crying. On other hand, discomfort associated with high frequency noise, such as the crying of children, is much more significant.

In the context of aviation, where safety is paramount, the elimination of noise will allow individuals to focus on safety-related information (for instance, safety announcements made by the air crew in the public address system). If the noise is too intrusive this concentration is affected. Therefore, the challenge for aircraft manufacturers and airlines is to create an environment inside the cabin where the noise is kept to a minimum.

At present improvements are already being made and the "low in-cab noise" concept is being promoted as a source of comfort in new aircraft models (e.g. A380 and Boeing 787) [7]. Since the average age of commercial aircraft is between 16 and 24 years [8], the options available for airlines to reduce noise in older aircrafts are limited. One of the most intuitive options, which was at the basis of the research presented in this paper, involves the use, at all phases of a flight, of headphones and earphones with noise cancellation and filtering. For some airlines this may require a radical change of procedures, namely on the way safety-related information is presented. However, such a change should not be dismissed because the receptivity of safety messages by passengers and crew would be substantially improved. Recent studies point out the importance of developing new alternatives in reducing noise in aircraft cabins, which can revolutionize the whole experience of traveling by air [7].

121.3 Theory of Inventive Problem Solving

The TRIZ, better known by its acronym (TRIZ), was developed by Genrich Altshuller in 1946 [1]. TRIZ is a theory that can assist any engineer in the inventing process.

The TRIZ methodology can be seen and used on several levels. At the highest level, the TRIZ can be seen as a science, as a philosophy or a way to be in life (a creative mode and a permanent search of continuous improvement). In more terms, the TRIZ can be seen as a set of analytical tools that assist both in the detection of contradictions on systems and in formulating and solving design problems through the elimination or mitigation of contradictions [10].

The TRIZ methodology is based on the following grounds:

- Technical systems
- Levels of innovation
- Law of ideality
- Contradictions.

Every system that performs a technical function is a technical system. Any technical system can contain one or more subsystems. The hierarchy of technical systems can be complex with many interactions. When a technical system produces harmful or inadequate effects, the system needs to be improved. Technical systems emerge,

Fig. 121.1 Steps of the
TRIZ's algorithm for problem
solving

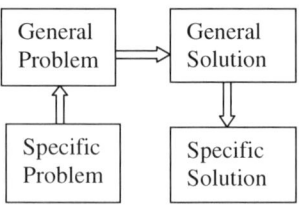

ripen to maturity, and die (they are replaced with new technical systems). TRIZ systematizes solutions that can be used for different technical fields and activities.

In TRIZ, the problems are divided into local and global problems [1]. The problem is considered to be local when it can be mitigated or eliminated by modifying a subsystem, keeping the remaining unchanged. The problem is classified as global when it can be solved only by the development of a new system based on a different principle of operation.

Over the past decades, TRIZ has developed into a set of different practical tools that can be used collectively or individually for technical problem solving and failure analysis.

Generally, the TRIZ's problem solving process is to define a specific problem, formalize it, identify the contradictions, find examples of how others have solved the contradiction or utilized the principles, and finally, apply those general solutions to the particular problem.

Figure 121.1 shows the steps of the TRIZ's problem solving.

Substance-Field Analysis is one of TRIZ analytical tools. It can be used in the solution of problems related to technical or design activities through functional models building [1].

Substance-Field Analysis is a useful tool for identifying problems in a technical system and finding innovative solutions to these identified problems. Recognized as one of the most valuable contributions of TRIZ, Substance-Field Analysis is able to model a system in a simple graphical approach, to identify problems and also to offer standard solutions for system improvement [6].

The process of functional models construction comprehends the following stages [15]:

• Survey of available information.
• Construction of Substance-Field diagram.
• Identification of problematic situation.
• Choice of a generic solution (standard solution).
• Development of a specific solution for the problem.

There are mainly five types of relationships among the substances: useful impact, harmful impact, excessive impact, insufficient impact and transformation [15]. Substance-Field Analysis has 76 standard solutions categorized into five classes [11]:

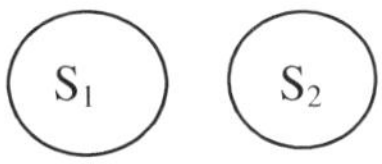

Fig. 121.2 Problematic
Situation 1-Incomplete model

Fig. 121.3 General solution
1 for problematic situation 1

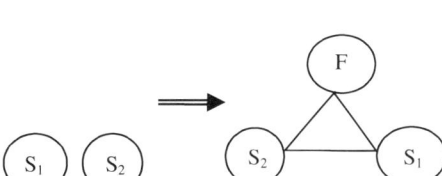

Class 1: Construct or destroy a substance-field (13 standard solutions)
Class 2: Develop a substance-field (23 standard solutions)
Class 3: Transition from a base system to a super-system or to a subsystem (6 standard solutions)
Class 4: Measure or detect anything within a technical system (17 standard solutions)
Class 5: Introduce substances or fields into a technical system (17 standard solutions)

These 76 solutions can be condensed and generalized into seven standard solutions. For example, Fig. 121.2 shows the Problematic Situation 1-Incomplete Model [15].

The Substance-Field Model is incomplete, a field is missing. The problem corresponds to Problematic Situation 1 and can be solved resorting to General Solution 1.

Fig. 121.3 shows the solution.

Then the model becomes complete.

121.4 Sound Filtering Model Development

A detailed analysis was performed in order to find possible solutions for the problem of noise inside the cabin of an airplane, which interferes directly with the comfort of the person traveling. A possible solution could be headphones equipped with passive noise control system that allows passenger to stop listening to everything that surrounds him, this way, all the unwanted noises (like the cry of a child, the noise of the plane, the boring conversation of another passenger, etc.) would be eliminated, but unfortunately up to eliminate unwanted noise, all noise are eliminated, disabling the passenger to listen to what he wants to hear such as communication with other passengers or the crew.

The solution can be materialized through a system (headphones fitted with passive noise control system), able to identify the noise, thus allowing the choice of those who want to listen and removing remaining. To create a really efficient system, it is necessary to distinguish the unwanted noise in order to know, specifically, what has to be eliminated.

Fig. 121.4 Controllable by the user, electronic system with directional microphones

Thus, several specific unwanted noise is divided into:

- background noise of the airplane itself;
- noise produced by other passengers (communication, food, snoring, etc.);
- cry of a child (this noise has very distinct sound characteristics).

A controllable by the user electronic system can be used. The system is able to recognize the external noise through directional microphones (Fig. 121.4), thus making the entire passenger hears, coming from abroad, had to first pass through the microphone.

The background noise of the plane is a noise whose sound wave is constant [4] and is easily clearable by active noise control. The system captures this constant sound wave and sends a counter wave to cancel the first wave. This system is not automatic, the passenger is free to cancel the noise or not, because there are customers who do not wish to annul such background noise.

Thus, the system consists of a pair of headphones fitted with passive noise control, which leaves no hear nothing, and an electronic device fitted microphone, which passes around the outside noise except the background noise of the plane. Moreover, the designed system has to cancel unwanted noise produced by other passengers. This noise is probably the most damaging and the most difficult to eliminate.

The system consist of directional microphones that allow selection of the area that the passenger wants to hear. These microphones have to belong to a very specific range that can be as directional as possible to fully capture only the sounds coming from the person to whom it is aimed. The microphone "shotgun" can be used (Fig. 121.5).

The positioning of the directional microphone "shotgun" must be calibrated because its focus is so sensitive so that if positioned exactly horizontal, it will capture the sounds from both the passenger who the user wants to listen and from the passengers back and forth.

One way to solve this problem is to riposte the microphone so that it does not capture the sounds from the back or front. If positioned in the middle of the chair

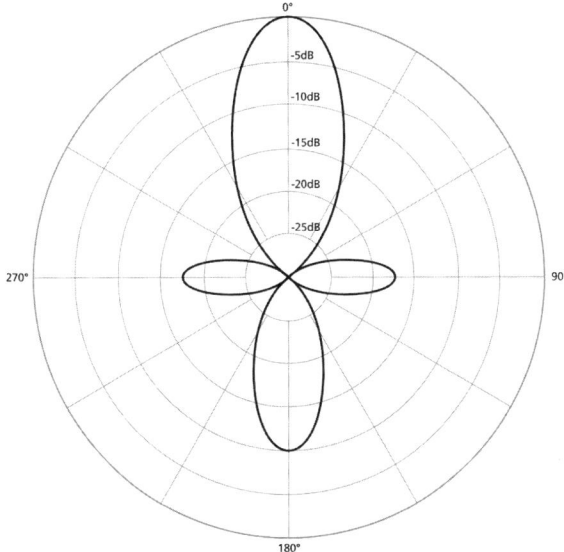

Fig. 121.5 "Shotgun" microphone directional diagram

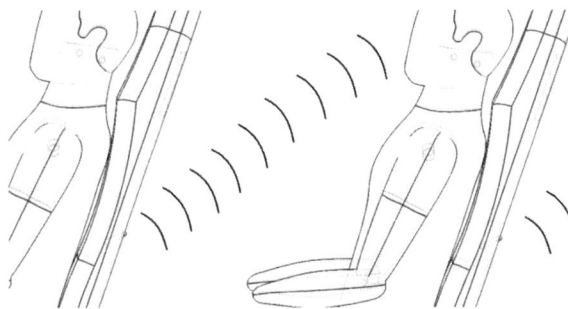

Fig. 121.6 A possible repositioning of the microphone "Shotgun"

in front, and pointed to his head, more or less, at an angle of 45°, can be reduced substantially, allowing it to attract unwanted noise from other passengers [3] it is pointed toward the top the aircraft is no sound source.

Figure 121.6 Illustrates a possible repositioning of the microphone "Shotgun".

Each passenger will have ahead a directional microphone capable of capturing only his own voice and his own noises. In Fig. 121.7, it is assumed that the user is sitting in the middle (shown without color) seat, the system consists of selecting which microphone you want to listen to it thus making it possible to hear what the fellow in green, the left side wants to tell, at the same time, have a snoring person (passenger red), ahead or to the side, without listening for only the selected microphone on the left.

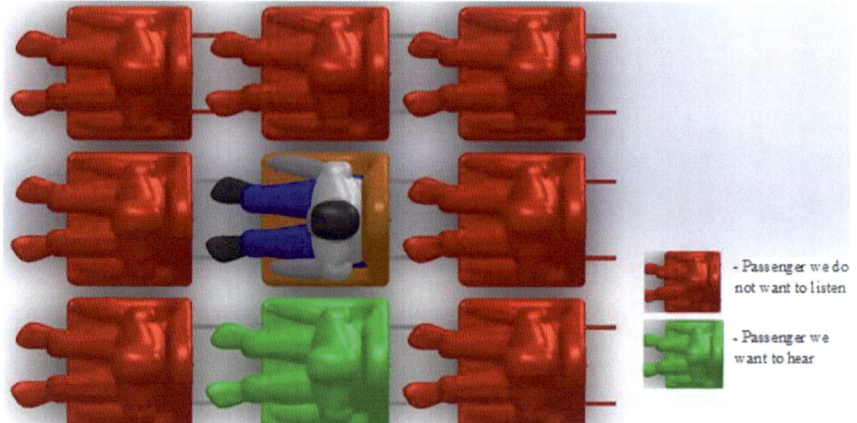

Fig. 121.7 Selection of passenger we want to hear

To have a dialogue, the passenger will have to enter the number of the microphone he wants to hear and the other player has to accept it through the system.

Once selected the microphone, a message appears on the screen asking another player if he intends to accept the invitation to an audio conversation. This fact eliminates the possibility of being listened to without the consent of both passengers, and be assured that, so no one listens to other people's conversations.

Another noise that need to be eliminated is the babies crying. This is a major problem that airlines face nowadays; its neutralization for a good acoustic comfort is required. Due to the intensity and frequency of baby crying, the above solutions do not apply because the microphones are more directional, would also capture any cry of a child that lies close.

The sound of the baby crying and the sound of human useful speech are at very different frequencies, which should create some ways to separate, being possible to neutralize the undesirable noises. Once the frequencies separated, a mechanism counteracts unwanted and leave "pass" the desirable. A sound technology that basically recognizes and neutralizes all frequencies that are not in the selected range was used.

The useful frequency range of the speech is between 100 and 120 Hz for males and 200–220 Hz for women [11], while the crying of babies is upwards of 350 Hz [9].

121.5 Results and Discussion

When analyzing with the TRIZ methodology, the problems referred in the study, it was concluded that the best way to neutralize any unwanted noises began to isolate the ear from any noise, so much desired as harmful. This sealing becomes possible when using headphones provided with passive noise control. The headphones play better this system are the "in-ear" type. The headphones are able to isolate fully up sounds 44 dB [12].

Once the ear totally isolated, it is necessary to capture the sounds otherwise. To this end, a microphone was used inserted in the front seat. The microphone that will adapt to the required needs of the "shotgun" that captures exclusively the person for which it is targeted and not the people who are sitting beside. The microphone will be positioned with an angle so as not to catch the people who are behind and ahead.

The background plane noise is a constant noise and therefore the application of an active noise control filter that basically emits a sound wave equal to the background noise, but out of phase, thus canceling out each other.

The sound of baby crying was mentioned here because it has a very high amplitude and directivity of the microphone ("shotgun") is not enough to neutralize it, i.e., if a child is crying nearby, the sound waves are not too strong to be captured, no matter what directional microphone. Thus, a band pass filter that neutralizes the desired frequencies being applied. More precisely and as the name indicates, it selects a frequency band that want to listen and void if the remaining. As the infant crying is at much higher frequencies than human speech, it becomes possible to neutralize them without neutralizing the speech of the person.

Being already all noises filtered, the missing issue is put the desired noise to the desired headphones. To this end, the system will feature an embedded single board computer on which it will be possible for passengers to select the intended recipient interface.

Then becomes possible a kind of innovative acoustic comfort in the cabins of aircraft. To be implemented this system, the passenger shall hear only what he wants. With the used methodology, new possibilities with potential solution were discovered. One of these new possibilities is the dialogue between passengers who are not near each other, i.e., they are sitting at distant places. Due to the directional microphone system, it becomes possible to talk to a person that is not next door, filtering, anyway, all harmful noises that surround both passengers.

This new option will be very useful, since the scenario of "overbooking" is increasingly used in airlines, causing many passengers who supposedly traveled together, if they have to sit apart, often in different wings of the plane. In long-distance travel, the possibility of meeting "onboard" is applicable often impossible because the actors are not able to sit together. Also apply to tour groups, such as scheduled trips to several people who have to sit mandatorily away from each other.

Another new possibility is that opens the dialogue with multiple stake holders. Imagine a group consisting of several people who claim to talk to each other and are not sitting close enough to communicate normally.

121.6 Conclusions

The significant increase in competition in the air transport passenger sector is forcing airlines operating to intensify the search for new forms and concepts of customer satisfaction. The solution of the problems of noise can significantly contribute to passenger comfort and, consequently, for their well-being during flight.

The passenger in the cab, is subject to different sound waves around you, some of them annoying or even harmful to health and may impact the quality of communication between the elements.

The study focused on solving problems related to acoustic comfort in the cabin of commercial aircraft. A system of acoustic filtering that allows the selection of sound sources was designed. The system consists of headphones, microphones, and filters. The headphones feature passive noise control that isolates the ear up to 44 db. The microphones used in the proposed solution are directional microphones "shotgun", which captures only the sounds of a selected area. The sound captured on the selected area is subject to Active Noise Control to cancel the background noise of the plane, and passes later, the band-pass filter in order to eliminate the acoustic noise of the crying child, which by their nature and characteristics continues to be captured in selected sound field.

The introduced innovation is aimed at the possibility of selecting the sound sources that passengers want to hear, thus allowing communication without unwanted interference. The currently existing systems only allow the isolation of acoustic noise.

Since this is an innovative project, some analytical tools of the TRIZ were used, and this is a particularly appropriate methodology for projects that require more creative solutions.

Substance-Field Analysis proved to be very important not only in the analysis of the problems and generating solutions, but also in the development of new capabilities not imagined before their implementation, such as the dialogue between two people who are not seated at one adjacent or dialogue between various stake holders.

The implementation of the system designed in this study can significantly contribute to the improvement of acoustic comfort, allowing airlines to ensure their customers are assured of a free flight of unwanted sound sources.

The reduction or elimination of noise will provide a less "stress" and less passenger's psychological fatigue, increased satisfaction and, hence improving the work site crew.

The baby crying is a problem sound discomfort so serious that at this point some airlines are organizing flights where children aged under 12 years are not allowed. This decision has generated controversy but it has a significant participation by people who frequently travel. With the implementation of the system developed in the study, the problem of baby crying is solved without resorting to restrictive measures on access to flights.

Acknowledgments   The authors acknowledge the support provided by the Portuguese Fundação para a Ciência e a Tecnologia (FCT) through the strategic project PEst-OE/EME/UI0667/2011, for the work carried out in the framework of the research centre UNIDEMI.

References

1. Altshuller GS (1984) Creativity as an exact science: the theory of the solution of inventive problems, vol. 5. CRC Press, New York
2. Bitencourt RFD (2012) Desempenho de métodos de avaliaço do conforto acústico no interior de aeronaves
3. Cinema Cd (2012) Microfones direcionais shotgunboom como usar e qual comprar. http://www.cursodecinema.com/microfones-direcionais-shotgunboom-como-usar-e-qual-comprar/
4. Ivošević J, Miljković D, Krajček K (2012) Comparative interior noise measurements in a large transport aircraft—T Urboprops versus Turbofan. In 5th Congress of the Alps Adria acoustics association
5. Kryter KD (1985) The effects of noise on man, 2nd edn. Academic Press, London
6. Mao X, Zhang X, AbouRizk S (2007) Generalized solutions for Su-Field analysis. TRIZ J
7. Molesworth BR, Burgess M (2013) Improving intelligibility at a safety critical point: in flight cabin safety. Saf Sci 51(1):11–16
8. Morrell P (2009) The potential for European aviation CO_2 emissions reduction through the use of larger jet aircraft. J Air Transp Manag 15(4):151–157
9. Daga RP, Panditrao AM (2011) Acoustical analysis of pain cries in neonates: fundamental frequency. IJCA Spec Issue Electron, Inf Commun, Eng ICEICE (3):18–21
10. Savransky SD (2002) Engineering of creativity: introduction to TRIZ methodology of inventive problem solving. CRC Press, UK
11. Simpson AP (2009) Phonetic differences between male and female speech. Lang Linguis Compass 3(2):621–640
12. Solutions E (2011) Comparison of sound isolating earphones and BOSE noise-cancelling headphones. http://www.earphonesolutions.com/coofsoiseaan.html
13. Terninko J, Domb E, Miller J (2000) The seventy-six standard solutions, with examples. TRIZ J. (online) http://www.triz-journal.com
14. van der Zwaag MD, Dijksterhuis C et al (2012) The influence of music on mood and performance while driving. Ergonomics 55(1):12–22
15. Zlotin B, Zusman A, et al. (1999) Tools of classical TRIZ. Ideation International Inc, p 266

Part VIII
Information Technology

Chapter 122
A Multi-attribute Large Group Decision-Making Method Based on Fuzzy Preference Dynamic Information Interacting

Xuanhua Xu and Huidi Wu

Abstract Given that the traditional fuzzy group decision-making model does not take the process of information interaction into consideration, this paper proposes a large group decision-making method which is based on fuzzy preference dynamic information interaction. This method adopts intuitionistic fuzzy sets (IFSs) to represent decision preference values, defines the similarity between two IFSs and calculates the similarity between the preference vectors of each expert and the ideal alternative. The average similarity of group and the decision deviations of experts are presented. If the decision deviation is greater than the given threshold, the experts are required to revise their decision preferences. Until the decision deviation of each expert is less than the given threshold, we come to the decision-making stage. The weighted similarity between each alternative and the ideal alternative is calculated, and then decision alternatives are ranked by sorting the weighted similarity. The optimal solution is selected according to the maximum principle. At last, an example is taken to simulate the implementation process and the result shows the feasibility and the validity of this method.

Keywords Preference information interacting · Multi-attribute · Large group · Decision-making method

122.1 Introduction

In the real world, the decision-making environment becomes increasingly complex and the society faces more and more uncertain elements. Consequently, more and more complex decision problems emerge, such as the decision-making problems caused by natural disasters and the decision-making problems of large-scale con-

X. Xu (✉) · H. Wu
School of Business, Central South University, Changsha 410083, People's Republic of China
e-mail: xuxh@csu.edu.cn

J. Xu et al. (eds.), *Proceedings of the Eighth International Conference on Management Science and Engineering Management*, Advances in Intelligent Systems and Computing 281, DOI: 10.1007/978-3-642-55122-2_122, © Springer-Verlag Berlin Heidelberg 2014

struction projects. The solutions of decision problems largely depend on the experts group in many different fields. However, with the increasing scale of experts, the preference differences and preference conflicts among decision-making group members tend to be more evident. Thus, the decision-making process requires coordination. Malone and Crowston [10] first proposed the "coordination theory" systematically. Thereafter, many domestic and foreign scholars carry out progressive research programs aimed at solving some specific decision problems, such as the coordination decision problems among special departments of a company, but these studies fail to present a standard rule on coordination decision and to establish a general coordination decision-making model. Shan and Guo [11] defined the fundamental concepts, and then proposed a general model based on the coordination decision of dynamic information interaction. For the numerical set in the group coordination evaluation, Xu [18] proposed a simple and practical method to reach a group opinion consensus; Dong et al. [5] further defined the indicators of measuring and judging matrice negotiation degree and proposed the row geometric mean prioritization method under the AHP group decision-making coordination model. Nevertheless, it is unlikely to apply these methods to solve complex large-scale group coordination decision problems, for many complex group decision-making problems undergo a long period with dynamic process of evolution in real life. What's more, they require many experts to form large-scale groups for achieving coordination. Therefore, the large group decision-making based on coordination becomes the focus of recent researches. The literature [15] synthesized the weight vector of each attribute and the preference matrix of large group, and then ranked the decision alternatives according to the comprehensive evaluation vector of each alternative. The literature [16] a large group decision-making method oriented towards utility value preference information. These methods fail to take into account the coordination among the members. Since individual members in the group have their own experiences, knowledge structure and value utility, etc., they often show different preferences at the initial stage of decision-making. During the process of group decision making, group members tend to reach a consistent preference with the information interaction among groups, which reflects the multi-stage information exchange and the feedback process in the group decision-making process.

It is hard to capture the explicit information to express preferences due to the complexity of human thought and the fuzziness and uncertainty of objective things, and thus fuzzy information has often been used to express the preference information, which can improve the accuracy of expression. As an important branch of modern decision theory, fuzzy multiple attribute decision making has won remarkable achievements. Since Zadeh [21] first proposed the theory of fuzzy sets in 1965, in the past 40 years, many domestic and foreign scholars has worked hard on this theory; Atanassov [1] defined the IFSs on the basis of Zadeh's fuzzy sets, which take into account both membership and non-membership information and can describe the uncertainty in a more accurate way compared with traditional fuzzy sets; Bellman and Zadeh [2] first built a general model of fuzzy decision-making problems based on fuzzy mathematics theory. After that, domestic and foreign scholars proposed several methods for fuzzy decision making problems, such as the method of weighted average operator [3], the method of geometric average operator [20], TOPSIS method [13], AHP method [12], gray theory analysis [14], the similarity method [7], etc. Although the group decision-making methods under fuzzy information have gained significant achievements, the methods of fuzzy group coordination decision are few. Chen and Li proposed fuzzy dynamic decision-making model in the literature [4], but this model can hardly solve the fuzzy coordination decision problems of large-scale groups.

For such problems mentioned above, this paper proposed a fuzzy dynamic preference information interaction model with the IFSs representing the preference information of decision makers. First, the paper defines the similarity of two IFSs, and then it calculates the similarity between each decision-maker's preference and ideal preference, followed by setting the threshold and defining decision deviations. If the deviation of a decision expert is greater than the given threshold, namely, this expert has great difference with other experts, he is requested to revise decision preference information after coordinating with other experts. Until each expert's decision deviation is less than the given threshold, and then we enter the decision-making stage. We calculate the weighted similarity of each decision alternative and ideal alternative, and the maximum weighted similarity corresponds to the optimal solution according to the maximum principle.

122.2 Similarity of Intuitionistic Fuzzy Sets

Atanassow, a Bulgarian scholar, extended the Zadeh's fuzzy set theory in the eighth literature [1], the traditional fuzzy sets that only consider the membership is extended to the IFSs which cover three aspects, not only membership, but also non-membership and hesitation, so the essentials of fuzzy things are described more accurately. IFS is defined as follows:

Definition 122.1 Let $X = \{x_1, x_2, \cdots, x_n\}$ be a nonempty set, and the IFS α in X is defined as [1]:

$$\alpha = \{\langle x_j, \mu_\alpha(x_j), \nu_\alpha(x_j)\rangle | x_j \in X\}. \tag{122.1}$$

In Eq. (122.1), μ_α and ν_α respectively denote membership degree and non-membership degree of which x_j belongs to the IFS α, and $\mu_\alpha(x_j), \nu_\alpha(x_j) \in [0, 1]$, with the condition $0 \leq \mu_\alpha(x_j) + \nu_\alpha(x_j) \leq 1$ for any x_j. For convenience, the IFS can be expressed briefly as $\alpha = \langle \mu_\alpha, \nu_\alpha \rangle$.

For each IFS, $\pi_\alpha(x_j) = 1 - \mu_\alpha(x_j) - \nu_\alpha(x_j)$, $x_j \in X$, $\pi_\alpha(x_j)$ is called hesitation degree of x_j to the IFS α. Especially, if $\pi_\alpha(x_j) = 0$ is tenable for each $x_j \in X$, the IFS α reduces to a fuzzy set.

Let $\alpha = \langle a, b \rangle, \beta = \langle c, d \rangle$ be two IFSs, and λ be a positive real number, then we have the following rule of operation: $\alpha + \beta = \langle a + c, b + d \rangle$, $\alpha\beta = \langle ac, bd \rangle$, $\lambda\alpha = \langle \lambda a, \lambda b \rangle$.

In terms of comparing uncertain information or fuzzy information, the similarity measure is a commonly used measuring method, which can be defined in two ways: the first one is to build similarity measure of two fuzzy sets on the basis of distance between two fuzzy sets, including the Hamming distance similarity measure [8], Hausdorff distance similarity measure [6], Euclidean distance similarity measure [17], and the similarity measure which is extension of the mathematical distance model; another definition regards fuzzy sets as a vector and constructs the similarity measure of two fuzzy sets on the basis of existing space vector model, such as Cosine similarity measure [19]. The latter method takes IFS as a vector, and constructs the similarity of two IFSs based on the existing J-similarity measure in the vector space.

Definition 122.2 Let $\alpha = \{\langle x_j, \mu_\alpha(x_j), \nu_\alpha(x_j)\rangle | x_j \in X\}$ and $\beta = \{\langle x_j, \mu_\beta(x_j), \nu_\beta(x_j)\rangle | x_j \in X\}$ be two IFSs on sets X, then the similarity of these two IFSs α and β can be defined as:

$$S(\alpha, \beta)$$
$$= \frac{1}{n} \sum_{j=1}^{n} \frac{\mu_\alpha(x_j) \cdot \mu_\beta(x_j) + \nu_\alpha(x_j) \cdot \nu_\beta(x_j)}{\mu_\alpha^2(x_j) + \nu_\alpha^2(x_j) + \mu_\alpha^2(x_j) + \nu_\beta^2(x_j) - \mu_\alpha(x_j) \cdot \mu_\beta(x_j) - \nu_\alpha(x_j) \cdot \nu_\beta(x_j)}.$$
$$(122.2)$$

Obviously, similarity $S(\alpha, \beta)$ meets the following properties, ($P1$) Boundedness: $0 \le S(\alpha, \beta) \le 1$; ($P2$) symmetry: $S(\alpha, \beta) = S(\beta, \alpha)$; $P3$ reflexivity: if $\alpha = \beta$, then $S(\alpha, \beta) = 1$.

Let $W = \{w_1, w_2, \cdots, w_n\}^T$ be the weight vector of $x_j (j = 1, 2, \cdots, n)$, in which $w_j \ge 0$, and $\sum_{j=1}^{n} w_j = 1$. Then the definition of the similarity of two IFSs considering weight can be expressed as:

$$WS(\alpha, \beta)$$
$$= \sum_{j=1}^{n} w_j \cdot \frac{\mu_\alpha(x_j)\mu_\beta(x_j) + \nu_\alpha(x_j)\nu_\beta(x_j)}{\mu_\alpha^2(x_j) + \nu_\alpha^2(x_j) + \mu_\alpha^2(x_j) + \nu_\beta^2(x_j) - \mu_\alpha(x_j)\mu_\beta(x_j) - \nu_\alpha(x_j)\nu_\beta(x_j)}.$$
$$(122.3)$$

Especially, when $w_j = \frac{1}{n}$, this is the similarity of Definition 122.2. Similarly, $WS(\alpha, \beta)$ also meets ($P1 - P3$).

122.3 Large Group Decision-Making Coordination Methods Based on Fuzzy Preference Dynamic Information Interaction

Suppose there are n attributes $C = \{C_1, C_2, \cdots, C_n\}$ and p alternatives constitute the alternative set $A = \{A_1, A_2, \cdots, A_p\}$ in the decision, and the decision group is composed of m experts $G = \{G_1, G_2, \cdots, G_m\}$ who make decisions on p alternatives above based on n attributes, the preference vector of the lth alternative evaluated by the ith decision-maker can be expressed by $V_l^i = (\langle C_1, \mu_{A_l}(C_1), \nu_{A_l}(C_1) \rangle, \langle C_2, \mu_{A_l}(C_2), \nu_{A_l}(C_2) \rangle, \cdots, \langle C_n, \mu_{A_l}(C_n), \nu_{A_l}(C_n) \rangle)$, in which component $\langle C_j, \mu_{A_l}(C_j), \nu_{A_l}(C_j) \rangle$ is called attribute value of the alternative A_l under the attribute C_j, $\mu_{A_l}(C_j)$ denotes the degree that the alternative A_l satisfies the attribute C_j, $\nu_{A_l}(C_j)$ denotes the degree that the alternative A_l does not satisfy the attribute C_j, in which $1 \le i \le m, 1 \le j \le n, 1 \le l \le p$.

The process of the large group decision-making method based on fuzzy dynamic information interaction coordination includes the following seven steps:

Step 1 M experts make decisions on p alternatives under the n attributes, and then the experts' preference vector sets are formulated.

The ideal alternative is not available in real life, but it is effective for making decision, which can provide effective reference for the decision makers to make judgment. Generally, the decision attribute can be divided into two types, benefit attribute and cost attribute. The greater the former is, the more effective the attribute value is, while the decision makers hope that the latter is as small as possible. Usually, when the ideal alternative for which the attribute value is IFS has been determined, the benefit attribute can be supposed as $v_{\text{benefit}} = \langle C_{\text{benefit}}, 0.99, 0.01 \rangle$, while the cost attribute as $v_{\text{cost}} = \langle C_{\text{cost}}, 0.01, 0.99 \rangle$.

For example, there are 3 decision attributes in a decision-making problem, where C_1, C_3 are benefit attributes, C_2 is a cost attribute, then the preference vector of its ideal alternative is $V_0 = (\langle C_1, 0.99, 0.01 \rangle, \langle C_2, 0.01, 0.99 \rangle, \langle C_3, 0.99, 0.01 \rangle)$.

Step 2 Classify the decision attribute. If it is a benefit attribute, then the attribute value is $(0.99, 0.01)$, and if it is a cost attribute, the attribute value is $(0.01, 0.99)$. Thus the preference vector of the ideal alternative is determined.

Step 3 As for alternative, utilize Eq. (122.2) to calculate the similarity S_l^i between each expert's preference vector and the ideal alternative's preference vector, then the average similarity of a group is calculated by $S_l = \frac{1}{m} \sum_{i=1}^{m} S_l^i$, so the decision deviation of an expert is shown in the following equation:

$$D_l^i = \left| S_l^i - S_l \right|. \tag{122.4}$$

Step 4 The threshold value δ is given to estimate the differences of experts' preferences, if the decision deviation of expert G_i making on alternative A_l is $D_l^i \ge \delta$, that is to say, the decision preference deviation of expert G_i is greater on alternative A_l compared with others', so coordination is needed.

Step 5 These experts whose decision deviation is greater than the threshold value need to coordinate with others. The method of the twenty-first literature [9] is used to revise a preference vector of alternative A_l, and we take steps 2–4 until the decision deviations of all the experts on all alternatives are less than the given threshold. Then we come to the decision-making stage.

Step 6 Let the weight vector of the decision attributes be $W = \{\omega_1, \omega_2, \cdots, \omega_n\}$, and each attribute G_j $(j = 1, 2, \cdots, n)$ corresponds to the element x_j $(1, 2, \cdots, n)$ of Eq. (122.3). The decision matrix is calculated by Eq. (122.3):

$$M = \begin{pmatrix} WS_1^1 & WS_2^1 & \cdots & WS_p^1 \\ WS_1^2 & WS_2^2 & \cdots & WS_p^2 \\ \vdots & \vdots & \ddots & \vdots \\ WS_1^m & WS_2^m & \cdots & WS_p^m \end{pmatrix}. \tag{122.5}$$

In the formula, WS_l^i is the weighted similarity between expert G_i's preference vector to alternative A_l and the preference vector of ideal alternative.

Step 7 Let the weight vector of m experts be $\omega = \{\omega_1, \omega_2, \cdots, \omega_m\}$, then

$$O = \omega M = (\omega_1, \omega_2, \cdots, \omega_m) \begin{pmatrix} WS_1^1 & WS_2^1 & \cdots & WS_p^1 \\ WS_1^2 & WS_2^2 & \cdots & WS_p^2 \\ \vdots & \vdots & \ddots & \vdots \\ WS_1^m & WS_2^m & \cdots & WS_p^m \end{pmatrix} = (o_1, o_2, \cdots, o_p). \tag{122.6}$$

Rank the alternatives by sorting each o_i value.

122.4 Example Analysis

An investment company has three investment projects: (1) real estate (A_1); (2) small automobile factory (A_2); (3) hotel (A_3). The board of directors hires 10 experts to make decisions on above options under the following four attributes: (1) the investment risk (C_1); (2) the economic and social benefits (C_2); (3) the degree of environmental pollution (C_3); (4) the development prospect (C_4). The decision-making process of this issue is as follows:

Step 1 Ten experts utilize the mark method [9] to grade above three projects, and the fuzzy preference vectors of each decision project is shown in Table 122.1.

Step 2 Decision makers classify the four decision attributes, among which attribute C_2 and C_4 are the profit attributes and attribute C_1 and C_3 are the cost attributes. Then the preference vectors of ideal alternative can be shown as follows: $A_p = (\langle C_1, 0.01, 0.99 \rangle, \langle C_2, 0.99, 0.01 \rangle, \langle C_3, 0.01, 0.99 \rangle, \langle C_4, 0.99, 0.01 \rangle)$.

Step 3 Utilize Eq. (122.2) to calculate the similarity of preference vectors between each expert and the ideal alternative, and then the average similarity of each alternative is calculated. Calculate the decision deviation of each expert on each alternative by Eq. (122.4), and the result is shown as follows:

Table 122.1 Fuzzy preference vectors of decision experts

Alternatives	Experts	Attribute C_1	Attribute C_2	Attribute C_3	Attribute C_4
A_1	1	(0.81, 0.12)	(0.63, 0.27)	(0.75, 0.15)	(0.02, 0.80)
	2	(0.40, 0.59)	(0.65, 0.05)	(0.91, 0.09)	(0.54, 0.39)
	3	(0.70, 0.24)	(0.14, 0.57)	(0.95, 0.03)	(0.49, 0.28)
	4	(0.71, 0.15)	(0.76, 0.20)	(0.09, 0.69)	(0.52, 0.43)
	5	(0.50, 0.34)	(0.73, 0.22)	(0.17, 0.38)	(0.04, 0.82)
	6	(0.31, 0.44)	(0.16, 0.79)	(0.18, 0.44)	(0.70, 0.27)
	7	(0.80, 0.17)	(0.64, 0.25)	(0.75, 0.16)	(0.72, 0.03)
	8	(0.42, 0.22)	(0.90, 0.10)	(0.09, 0.54)	(0.49, 0.06)
	9	(0.48, 0.19)	(0.90, 0.05)	(0.13, 0.57)	(0.14, 0.62)
	10	(0.16, 0.31)	(0.56, 0.26)	(0.68, 0.25)	(0.52, 0.20)
A_2	1	(0.65, 0.11)	(0.97, 0.02)	(0.75, 0.06)	(0.39, 0.48)
	2	(0.64, 0.27)	(0.67, 0.16)	(0.49, 0.34)	(0.38, 0.55)
	3	(0.69, 0.13)	(0.54, 0.14)	(0.84, 0.14)	(0.92, 0.06)
	4	(0.61, 0.35)	(0.58, 0.17)	(0.75, 0.13)	(0.25, 0.43)
	5	(0.24, 0.35)	(0.25, 0.73)	(0.30, 0.54)	(0.23, 0.75)
	6	(0.38, 0.07)	(0.53, 0.34)	(0.56, 0.19)	(0.62, 0.31)
	7	(0.57, 0.37)	(0.55, 0.27)	(0.48, 0.43)	(0.18, 0.70)
	8	(0.03, 0.75)	(0.17, 0.62)	(0.31, 0.38)	(0.48, 0.47)
	9	(0.84, 0.11)	(0.70, 0.07)	(0.44, 0.27)	(0.11, 0.85)
	10	(0.40, 0.39)	(0.03, 0.82)	(0.03, 0.79)	(0.23, 0.67)
A_3	1	(0.56, 0.26)	(0.68, 0.26)	(0.05, 0.77)	(0.29, 0.46)
	2	(0.33, 0.43)	(0.52, 0.20)	(0.41, 0.42)	(0.45, 0.22)
	3	(0.15, 0.53)	(0.78, 0.10)	(0.46, 0.17)	(0.44, 0.25)
	4	(0.43, 0.38)	(0.33, 0.58)	(0.61, 0.27)	(0.96, 0.04)
	5	(0.68, 0.30)	(0.80, 0.16)	(0.18, 0.55)	(0.86, 0.14)
	6	(0.53, 0.35)	(0.63, 0.13)	(0.57, 0.14)	(0.62, 0.12)
	7	(0.75, 0.09)	(0.14, 0.81)	(0.66, 0.17)	(0.91, 0.04)
	8	(0.25, 0.63)	(0.75, 0.23)	(0.25, 0.68)	(0.77, 0.14)
	9	(0.60, 0.32)	(0.84, 0.13)	(0.61, 0.33)	(0.75, 0.07)
	10	(0.69, 0.13)	(0.43, 0.35)	(0.47, 0.49	(0.75, 0.12)

Step 4 The threshold has been given by decision-maker as $\delta = 0.2$, as shown in Fig. 122.1. We can directly see that the evaluation results of experts on alternative A_1 have large differences, while the evaluation results of alternative A_2 are consistent. The decision deviations of expert 1, expert 3 and expert 8 on alternative A_1 as well as expert 7 and expert 8 on alternative A_3 are greater than the given threshold, so we reach the coordination stage.

Step 5 Expert 1, expert 3 and expert 8 coordinate with other experts, revise and present the decision preference vectors of alternative A_1 respectively:

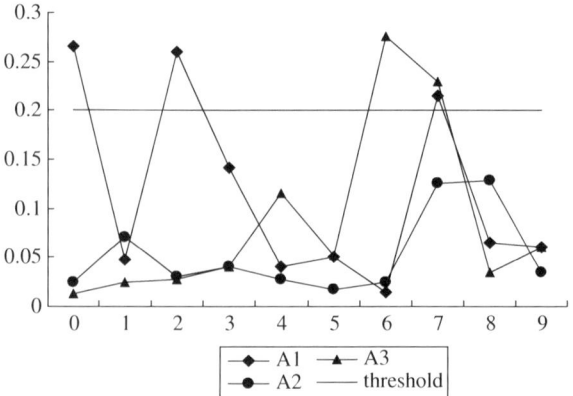

Fig. 122.1 The decision deviations of each expert on each alternative in the initial stage

Fig. 122.2 The decision deviations of each expert on each alternative after the first revision

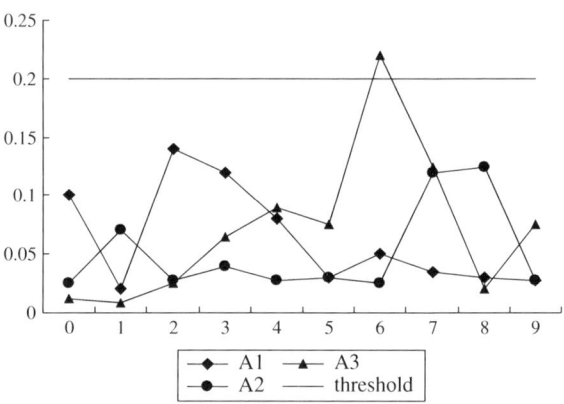

$$A_1^1 = (\langle C_1, 0.31, 0.44\rangle, \langle C_2, 0.76, 0.19\rangle, \langle C_3, 0.68, 0.24\rangle, \langle C_4, 0.57, 0.27\rangle),$$
$$A_1^3 = (\langle C_1, 0.20, 0.74\rangle, \langle C_2, 0.14, 0.57\rangle, \langle C_3, 0.95, 0.03\rangle, \langle C_4, 0.49, 0.28\rangle),$$
$$A_1^8 = (\langle C_1, 0.15, 0.53\rangle, \langle C_2, 0.28, 0.50\rangle, \langle C_3, 0.46, 0.17\rangle, \langle C_4, 0.44, 0.25\rangle).$$

Expert 3 and expert 8 coordinate with other experts similarly, revise and provide the decision preference vectors of alternative A_1 respectively:

$$A_3^7 = (\langle C_1, 0.15, 0.63\rangle, \langle C_2, 0.75, 0.23\rangle, \langle C_3, 0.25, 0.68\rangle, \langle C_4, 0.77, 0.14\rangle),$$
$$A_3^8 = (\langle C_1, 0.75, 0.09\rangle, \langle C_2, 0.64, 0.18\rangle, \langle C_3, 0.66, 0.17\rangle, \langle C_4, 0.91, 0.04\rangle).$$

The revised decision deviations of all experts on each alternative are shown in Fig. 122.2.

As shown in Fig. 122.2, we can directly see that the decision deviations of expert 7 on alternative A_3 are greater than the given threshold, and then we reach the

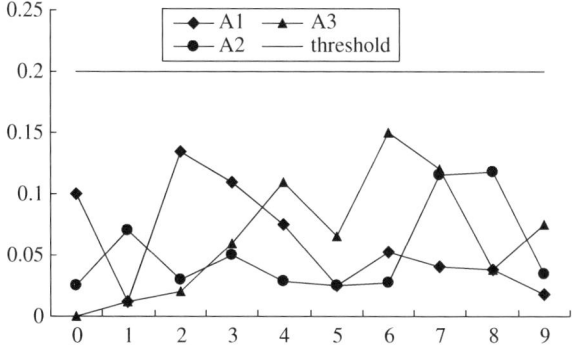

Fig. 122.3 The decision deviations of each expert on each alternative after the second revision

second revision stage. Expert 7 coordinates with other experts, revises the attribute value of C_1 which has a large difference and presents the decision preference vectors of alternative A_3 as $A_3^7 = (\langle C_1, 0.56, 0.43\rangle, \langle C_2, 0.75, 0.23\rangle, \langle C_3, 0.25, 0.68\rangle, \langle C_4, 0.77, 0.14\rangle)$.

According to step 3, the second revised decision deviations of all experts on each alternative are shown in Fig. 122.3.

In Fig. 122.3, the decision deviations of all experts are less than the given threshold, and the decision deviation curve of each expert is relatively flat, in other words, the deviation among experts is acceptable. Then we come to the decision-making stage. The threshold can be taken smaller if decision makers pursue better decision results, such as 0.15, 0.1, etc., so that the experts can make a better coordination to ensure that each expert decision deviation curve is close to a straight line which means the preferences of decision members are close.

Step 6 The weights of four decision attributes are given respectively by decision makers as: $0.3, 0.4, 0.2, 0.1$, and the decision matrix M is calculated by Eq. (122.3). For the convenience of expression, the transposed matrix is showed as follows:

$$M^T = \begin{pmatrix} 0.6265, 0.6067, 0.3778, 0.6243, 0.5491, 0.3997, 0.4620, 0.4063, 0.6118, 0.5011 \\ 0.4757, 0.5110, 0.4323, 0.4385, 0.3539, 0.3769, 0.4783, 0.4761, 0.4405, 0.3481 \\ 0.6131, 0.5694, 0.6710, 0.3918, 0.6850, 0.5323, 0.7288, 0.4697, 0.6209, 0.4248 \end{pmatrix}.$$

Step 7 The weights of 10 experts is $W = (0.10, 0.05, 0.10, 0.15, 0.05, 0.2, 0.05, 0.1, 0.10, 0.10)$. The decision result is calculated by Eq. (122.5) as $O = (0.5068, 0.4256, 0.5444)$, and the ranked order of decision alternatives is $A_3 \succ A_1 \succ A_2$, so the optimal decision is investing hotel.

122.5 Conclusion

Since the traditional group decision-making model can hardly solve the large-scale group coordination decision problems under the fuzzy information, this paper proposes a group coordination decision method based on fuzzy dynamic preference information interaction. The decision preferences of decision experts are expressed with IFSs, and through calculating the similarity of each expert's preference vector and the ideal preference vector, the average similarity of group and the decision deviations of decision experts are obtained. On the other hand, the threshold is used to discuss the preference differences of decision experts. If an expert's decision deviation is greater than the given threshold, the expert's decision result has a large difference from others', and thus he needs to coordinate with other experts, revises and presents each alternative preference. Furthermore, when the decision deviations of all experts on each alternative are less than the given threshold, we come to the decision stage, and the optimal solution is gained by calculating the weighted similarity of each alternative and the ideal alternative. Finally, an example is given to simulate the implementation process and illustrate the feasibility of the proposed method. The deviation of all experts on each alternative can be directly seen from the decision deviation graph, so the experts who need to coordinate with others and revise the preference vectors of the alternatives are easily found. The numerical results show that the method is feasible. The large-scale group decision-making method based on dynamic interaction coordination of fuzzy preference information proposed in this paper can be applied to the decision preference information with the expansion to interval intuitionistic fuzzy number, triangular fuzzy number, intuition triangular fuzzy number, interval triangular fuzzy number, trapezoidal fuzzy number, intuitionistic trapezoidal fuzzy number and other fuzzy sets as well as the linguistic representation of the expansion forms of such fuzzy sets. Therefore, the proposed method has strong expandability and practicality.

Acknowledgments The work was supported by a grant from Natural Science Foundation in China (71171202).

References

1. Atanassov K (1986) Intuitionistic fuzzy sets. Fuzzy Sets Syst 20:87–96
2. Bellman R, Zadeh L (1970) Decision making in a fuzzy environment. Manag Sci 17(4):141–164
3. Cao Q, Wu J (2011) The extended COWG operators and their application to multiple attributive group decision making problems with interval numbers. Appl Math Model 35:2075–2086
4. Chen Y, Li B (2011) Dynamic multi-attribute decision making model based on triangular intuitionistic fuzzy numbers. Scientia Iranica Trans B: Mech Eng 18(2):268–274
5. Dong Y, Zhang G et al (2010) Consensus models for AHP group decision making under row geometric mean prioritization method. Decis Support Syst 49(3):281–289
6. Hung W, Yang M (2004) Similarity measures of intuitionistic fuzzy sets based on Hausdorff distance. Pattern Recogn Lett 24:1603–1611

7. Kacprzyk J (2005) A new concept of a similarity measure for intuitionistic fuzzy sets and its use in group decision making. Model Decis Artif Intell 58:272–282
8. Lee S, Pedrycz W, Sohn G (2009) Design of similarity measures for fuzzy sets on the basis of distance measure. Int J Fuzzy Syst 11(2):67–72
9. Liu H, Wang G (2007) Multi-criteria decision-making methods based on intuitionistic fuzzy sets. Eur J Oper Res 179:220–233
10. Malone T, Crowston K (1994) The interdisciplinary study of coordination. ACM Comput Surv 26:87–119
11. Shan C, Guo Y (2003) A kind of decision-making method through coordination between functional departments based on dynamic information interaction. J Manag Sci China 6(3):8–12
12. Tang Y, Chang C (2012) Multicriteria decision-making based on goal programming and fuzzy analytic hierarchy process: an application to capital budgeting problem. Knowl-Based Syst 26:288–293
13. Wang Y, Elhag T (2006) Fuzzy topsis method based on alpha level sets with an application to bridge risk assessment. Expert Syst Appl 31:309–319
14. Wei G, Wei Y (2008) Model of grey relational analysis for interval multiple attribute decision making with preference information on alternatives. Chin J Manag Sci 16:158–162
15. Xu X, Chen X (2008) Research of a kind of method of multi-attributes and multi-schemes large group decision making. J Syst Eng 23(2):137–141
16. Xu X, Chen X, Wang H (2009) A kind of large group decision-making method oriented utility valued preference information. Control Decis 24(3):440–450
17. Xu Z (2007) Some similarity measures of intuitionistic fuzzy sets and their applications to multiple attribute decision making. Fuzzy Optim Decis Making 6:109–121
18. Xu Z (2009) An automatic approach to reaching consensus in multiple attribute group decision making. Comput Ind Eng 56(4):1369–1374
19. Ye J (2011) Cosine similarity measures for intuitionistic fuzzy sets and their applications. Math Comput Model 53:91–97
20. Yue Z (2011) An extended topsis for determining weights of decision makers with interval numbers. Knowl-Based Syst 24:146–153
21. Zadeh L (1965) Fuzzy sets. Inf Control 8:338–356

Chapter 123
How to Create Sustainable Competiveness Through Resilience and Innovation: A Practical Model Based on an Integrated Approach

Nuno Martins Cavaco and Virgílio António Cruz-Machado

Abstract The increasing sophistication of clients and the aggressively way how markets permanently act, introduce in organizations a constant need to change. The pressure to reduce decision time cycles and to be able to react and anticipate competitors is a requisite to survive and the key for success. Therefore, it is necessary to review existing models of competitiveness and to create new approaches that integrate new concepts and trends. This paper provides a model for the creation of competitive advantage that integrates the principles of sustainability (triple bottom line) and the concepts of resilience and innovation. It aims to contribute to the improvement of the strategic planning process of organizations, providing an alternative approach which also combines business evaluation models (such as EFQM and Shingo), with tools that reduce the strategic execution gap (for example Balanced Scorecard). Differentiated criteria for the evaluation of the current competitive positioning of organizations are established, parameters for implementing sustainable competitiveness factors are set and methods for monitoring and feedback proposed. In addition, this research establish a relation between resilience, innovation and sustainable competitiveness, and reveals that this approach is differentiating, able to add value (more focused and more efficient), and applicable to several business sectors and to different levels of technological sophistication, monitoring maturity, as well as adaptable to distinctive cultural environments.

Keywords Strategic planning · Sustainable competitiveness · Resilience · Innovation · Monitoring · Balanced scorecard · Shingo model · EFQM model

N. M. Cavaco (✉) · V. A. Cruz-Machado
Department of Mechanical and Industrial Engineering, Faculty of Science and Technology, UNIDEMI, FCT, Universidade Nova de Lisboa, 2829-516 Caparica, Portugal
e-mail: namc@fct.unl.pt

J. Xu et al. (eds.), *Proceedings of the Eighth International Conference on Management Science and Engineering Management*, Advances in Intelligent Systems and Computing 281, DOI: 10.1007/978-3-642-55122-2_123, © Springer-Verlag Berlin Heidelberg 2014

123.1 Introduction

The constant need to be ahead or prepared to react to competitors, aiming the achievement of positive results and generating stakeholders satisfaction, is the fundamental reason that drives companies to apply strategic planning processes. "Strategic planning concepts is the need for a framework to comprehensively understand industry structure and the behavior of competitors and to translate these into operational strategic recommendations" [11]. However, in spite of the different existing tools available to support management teams on their strategic planning activity, not always it is clear which tools are more suitable for each context. This is a source of inefficiency that can cause ineffectiveness. Another important fact is that "a universal and exact definition for competitiveness does not exist. As a result, competitiveness means different things to different organizations" [7]. This fact, besides constituting a factor of uncertainty it also makes competitiveness comparison and bench marking initiatives difficult. Additionally, the appearance of new concepts that could also be considered in strategic planning approaches is also a reason to develop a framework as more integrated and adapted to the new challenges of organizations. Assuming that competitiveness depends on "attributes, namely agility, flexibility, adaptability and connectivity, are frequently defined as supporting attributes of enterprise resilience" [6], and that "Competitiveness derives from the creation of the locally differentiated capabilities needed to sustain growth in an internationally competitive selection environment. Such capabilities are created through innovation, $\cdots$ " [2], concepts like resilience and innovation can base an alternative definition of competitiveness and establish principles to define a new framework for strategic planning. This reflection allows also to dispute about other used designations, such as competitive advantage and sustainable competitiveness. Considering that " $\cdots$ competitive advantage can arise from many sources, and shows how all advantage can be connected to specific activities and the way that activities relate to each other, to supplier activities, and to customer activities". Reference [12], it allows a reflection on new factors that contribute to create advantage. Additionally, if we take into account that " Empirically, sustained competitive advantage may, on average, last a long period of calendar time. However, it is not this period of calendar time that defines the existence of a sustained competitive advantage, but the inability of current and potential competitors to duplicate that strategy that makes a competitive advantage sustained". Reference [1], it make sense that competitiveness definition should not be based only in new concepts, but should also be able to define company's competitiveness positioning in an objective matter (preferentially quantitatively). This way it allows comparative analyses and loss of competitive advantage risk evaluation. So the development of such a model supposes the selection of competitiveness drivers that are representative of companies structural components and influence its competitiveness. Taking into account that this research aims an integrated framework, it is imperative to guarantee convergence between competitiveness drivers and measurement criteria used by the existing evaluation models, such as EFQM Excellence Model or Shingo Operational Excellence Model, and also to establish a relationship with monitoring systems like

Balanced Score Card. This approach wishes to be a practical tool to better understand competitiveness weaknesses and consequently to better support decision making on the definition of more efficient strategies.

123.2 Problem Statement

As mentioned, the appearance of new concepts, an inexistence definition for competitiveness and the accessibility to several strategic tools in a non-integrated approach, cause lack of focus, inefficiencies and are responsible for increase the cost of strategic planning processes. Therefore, it is presented a review of two fundamental concepts (sustainable competitiveness and competitive advantage) that illustrate how new concepts can be applied contributing to a better definition of competitiveness. For example: regarding this two concepts, it is reasonable to ask what is the difference between the two or if it is possible to define an relationship that integrates both? If we consider the following expression what conclusion can we obtain? What does it mean? Sustainable Competitiveness + Competitive Advantage = Sustainable Advantage.

1. Sustainable Competitiveness Concept Review

Instead of consider sustainability in terms of time, which is, the aptitude to be competitive in the future, this research develop an integrated concept based on the fact that time frame of competitiveness should be ensured trough the combination of the capability to be resilient (recover performance in time) and the ability to be innovative (increase performance in time), and following [9]. "The TBL adds social and environmental measures of performance to the economic measures typically used in most organization", sustainability should include the Triple Bottom Line principles (economic, social and environment). This hypothesis is a contribution to an evolution of sustainable competitiveness definition, integrating several concepts and establishing a direct relation between a modern definition of competitiveness with a overall definition of sustainability.

2. Competitive Advantage Concept Review

Companies aim to be more competitive, but that is not enough. The question is: to be more competitive than what? Does it mean: to be more competitive tomorrow than they are today? That doesn't mean they really are competitive, because competitiveness should be related with their capacity to be better than their competitors. So increasing competiveness should be defined as the capacity to increase their relative competitiveness positioning in comparison with the competitiveness positioning of their competitors. The achievement of this new positioning should establish the competitive advantage of the company, where the gap between the two, defines the intensity of this advantage (please see Fig. 123.3). It is important to be aware that this gap of advantage is continuously under pressure, existing a permanent risk of losing competitiveness.

3. Sustainable Advantage Concept

Considering the integration of the two previous concepts, a natural definition of sustainable advantage could be: "for how long is the company able to preserve its competitive advantage". Again this definition is based supposing that "Sustainable" as only a time frame principle. Since that "Advantage" already include this principle of time frame, as a result of considering resilience and innovation dimensions, it is possible to assume "Sustainable" based on sustainability principles. Thus, sustainable advantage is the capacity to preserve the advantage gap based on economic, social and environmental results.

123.3 Sustainable Competitiveness Model

This model establishes an alternative way to measure company's competitiveness and share guidance to increase competitiveness positioning and obtain sustainable advantage. It is based on a virtuous cycle of strategic competitiveness planning, translated into a practical framework that provides continuous action in a systematic approach. The model is based on an alternative definition for competitiveness that integrates resilience and innovation dimensions. It also establishes a matrix allowing companies to positioning themselves in one from four competitiveness quadrants, as a result of the measurement of their competitiveness metrics. At last it also provide the principles that assure results on a sustainability way.

123.3.1 A Continuous Action Framework

As mentioned the model promotes a continuous awareness about the company's competitiveness and allows taking actions concerning each evidence that is exposed. The framework fulfills strategic concerns and operational issues. It replies to management responsibilities in identifying competitiveness advantage and risk, and in defining the suitable strategies to maintain or increase company's positioning. Additionally, it deploys the strategy into operational actions that will be measured in terms of execution, impact gained. Achieving this integrated perspective the model assures an overall of the cause-effect between operational initiatives and competitiveness positioning (Fig. 123.1).

123.3.2 Resilience and Innovation as Competitiveness Dimensions

Considering that competitiveness can be measured through two parameters, namely, performance and time, and that or resilience or innovation can be expressed through

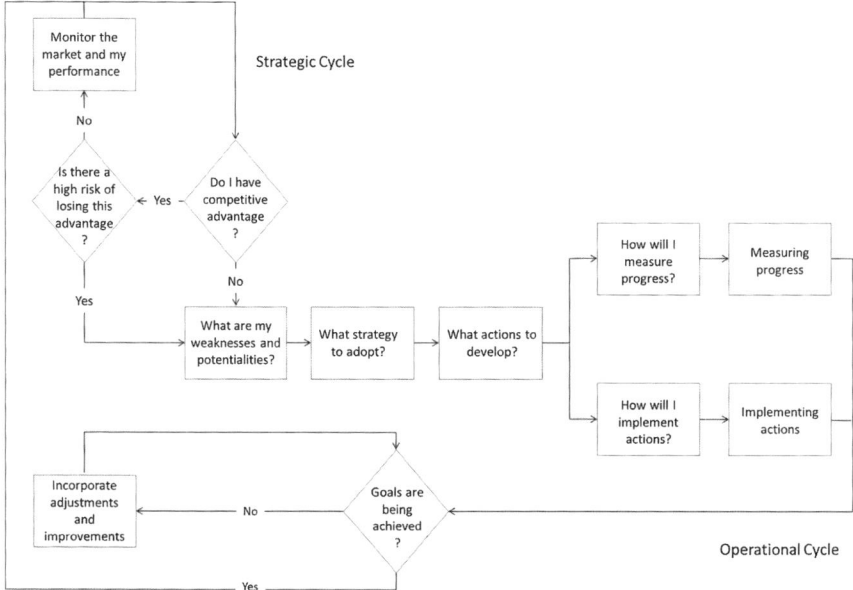

Fig. 123.1 Sustainable competitiveness framework

these two parameters, it is possible to establish a relation between them to support competitiveness definition. Regarding the "Resilience Triangle" and applying the same principle to the "Innovation S-Curve", competitiveness can be defined as the readiness to react to disturbances that affect company's performance and the willingness to leverage performance in a pro-active way. Considering Fig. 123.2, it means that competitiveness can be measured through the following expression: Competitiveness $= 1-$ (Resilience $+$ Innovation)*, $*$ normalizing the triangle area between 0 and 1, where: Resilience $= 1/2$ (Severity $*$ Recovery Time), Innovation $= 1/2$ (Intensity $*$ Advance Time).

So, to maximize competitiveness, taking into account that being more resilient means minimizing the triangle area, the same to innovation, companies should minimize de sum of being resilient and innovative. This can be achieved by normalizing the triangle area between 0 and 1 and reversing the sum expression of competitiveness.

123.3.3 Competitiveness and Competitive Advantage

Assuming that competitive advantage is based on resilience capability and innovation ability, it is possible to define four competitive positioning stages, as shown in Fig. 123.3.

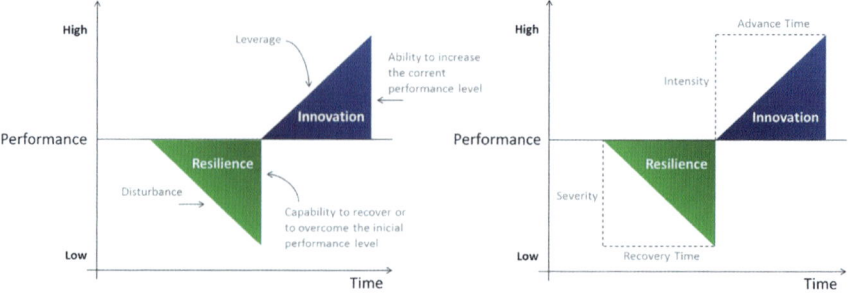

Fig. 123.2 Competitiveness definition based on resilience and innovation principles

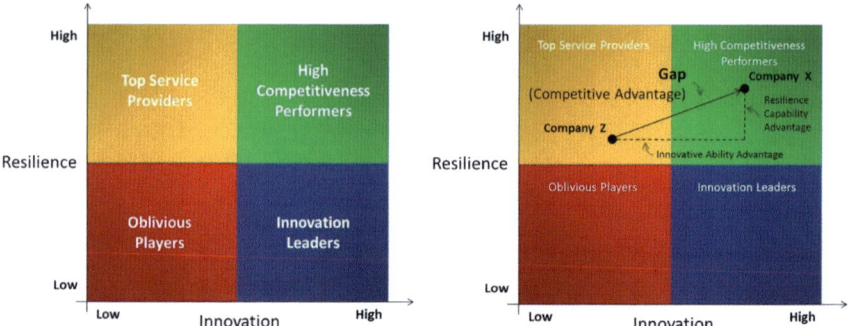

Fig. 123.3 Competitive positioning matrix

This matrix allows companies to identify their own positioning, justify its contribution (based on resilience or innovation variables) and distinguish their Competitive Gap (advantage) before competitors. Regarding the four kinds of competitive positioning defined by the matrix, it is relevant to describe each of them (Fig. 123.4).

Trough the matrix companies are able to have a more objective understanding about why they have competitive advantage and if there are risk of losing it. Risk of losing competitiveness means that there is a risk to be less innovative or less resilient. So it is understandable to assume that this risk can be expressed by the following equation: Risk = Prob (Less Innovative) * Impact (Less Innovative) + Prob (Less Resilient) * Impact (Less Resilient).

Combining the current level of competitiveness (Competitive Positioning) with the risk of losing it, we can define another matrix that illustrates the exposure of the company in terms of losing its competitive advantage (see Fig. 123.5). In this context, strategies can be defined to eliminate the risk factor inherent of losing resilience or innovation skills. Being so, resilience and innovation, as competitiveness dimensions, are risk minimizers or current competitiveness evaluators? They should be both, what involves a more careful definition of the metrics to be used.

		Competitive Positioning			
		Oblivious Players	Top Service Providers	Innovation Leaders	High Competitiveness Performers
Business Attitude	Market Perception	• My product/ service is unique, the best and timeless • My clients will always be loyal	• My market share depends on the quality of my service	• My market share depends on the differentiation of my product	• I'm never satisfied with my market share • My competitors are not sleeping
	Knowledge Management	No decision making based on real data (no risk assumption)	Decision making based on client reaction (low risk assumption)	Decision making based on market behavior (low risk assumption)	Decision making based on trend analysis (high risk assumption)
	Business Focus	Living from the success of the past (I'm already good)	Responsiveness (I can always be better) - Operation Driven -	Anticipation (I can always be better) - Product Driven -	Always visioning the future (I can always be different)
Competitiveness Dimensions	Resilience	• Accommodation to the usual service levels • Low concern to respond quickly to disturbances	• Solid procedures and routines to react to disturbances	• Low practice of dealing with disturbances	• Strong and deployed empowerment to provide quick responds to disturbances
	Innovation	• No motivation to develop new solutions	• Low practice of developing new solutions	• Motivation to accomplish new solutions (what clients want)	• High motivation to create disruptive solutions (what clients don't know they want)

Fig. 123.4 Competitive positioning levels description

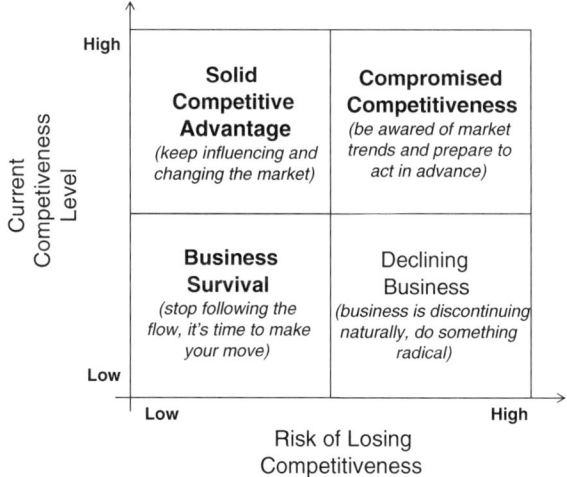

Fig. 123.5 Competitive advantage exposure matrix

Competitiveness Drivers	Features	Key findings	Competitiveness Dimensions (Variables) - EXAMPLES	
			Resilience (Failure Mode)	Innovation (Leverage Factor)
Corporate Behavior	• Ethics and solidarity • Leadership • Knowledge management (market share, clients satisfaction, complains, …) • Policies	• What are the main concerns of the board • How do they act • How do they deploy	• Corruption and personal scandals • Management changes (nominations and exonerations) • Strategic failures	• Visioning practices • Open innovation initiatives • Establishing strategic partnerships
Business Proposition	• Customers needs/ expectations • Product attractiveness • Service • Marketing (brand)	• What is the product/ service value added and its suitability to client expectations • What is the market recognition	• Sales decreasing • Crisis management (communication and brand) • Political instability of markets	• Research deployment • Product development
Financial Stability	• Return On Investment • Cash flow	• Is the business auto-sufficient	• Alternative business to distribute risk • Back-up practices to cash flow slippages	• New solutions to increase Return On Investment
Organization Wellbeing	• Culture and Leadership • Competencies and entrepreneurship • Motivation and empowerment	• What is the internal environment, employee satisfaction and labor capabilities to the future	• Change management routines • HR rotation and substitution plans • Social dynamics (e.g. strikes)	• Learning innovation • New social practices • New kind of acknowledge programs
Operational Leanness	• Logistics (planning, procurement, purchasing, storage, distribution) • Manufacturing/ service delivery • Maintenance	• How come are the operations efficient and effective	• Planning constraints • Capacity Shortage* • Material Shortage* • Quality assurance	• Implementing edge improvement methodologies • Adopting new partnerships over the business value chain
Technological Alignment	• Technological infra-structure • Communications • Technological applications	• How come technology satisfies the business needs	• Help desk capability • Disaster recovery • Business continuity planning	• Establishing collaborative initiatives with High Tech companies • Introduction of edge solutions (tech pioneering)
Facilities Suitability	• Installations • Equipment • Ergonomics	• How come facilities allows the proper on-going operations	• Catastrophes and disasters • Accidents and labor diseases	• Adapting newest facility solutions • Edging safer and efficient equipment

Fig. 123.6 Competitiveness drivers and examples of resilience and innovation variables

123.3.4 Measuring Competitiveness

The competitive positioning of a company should be obtained through the measurement of resilience and innovation metrics related to competitiveness drivers. The model proposes seven competitiveness drivers that support key elements of any kind of companies. Figure 123.6 describes each driver and also illustrates examples of possible relation with resilience and innovation variables [3].

Taking into account the competitiveness expression (mentioned in Sect. 123.3.2), to do the measurement properly it is supposed to calculate the triangle area. Therefore, it is necessary to unfold each variable and make use of severity and recovery time metrics regarding to resilience variables and intensity and advance time metrics concerning innovation variables.

123.3.5 Assuring Results on a Sustainability Way

After this description of the framework, we are able to make the relationship with sustainability concept. The basic principle relies on the expectation that competitiveness and having competitive advantage is achieved in a sustainable way. This means it generates positive impacts on Triple Bottom Line, which is being competitive trough low cost and creating value (economically), generating wellbeing (socially) and without compromising the environment (environmentally).

123.4 Application

This research is based on literature review and on professional experience. It as the intention to provide a framework that establishes the bases to the development of the global model. Further research will be focused on the definition of resilience and innovation metrics for each competitiveness drivers and on the design of overall calculation structure, considering principles from EFQM and Shingo models, as well as BSC system. In addition, it is also objective of this research to identify the relation between competitiveness drivers and the achievement of competitive positioning, e.g. in what way technological alignment contributes to be a high competitiveness performer? Or what are the resilience metrics from Corporate Behavior that define Top Service Providers?

This research generates opportunities to other studies, such as investigation about the suitability of the model to several business sectors or its aptness to measure competiveness of countries and regions. Another field of research could be finding out the value creation of applying such a model to distinctive cultural environments, in public sector or in companies with different levels of technological sophistication as well as monitoring maturity.

123.5 Conclusion

The current framework as shown that it is possible to establish an alternative definition for competitiveness and that is reasonable to assume that resilience and innovation can be key dimensions to measure it. It also become apparent that it make sense to develop an integrated model, based on these principles, with the ability to introduce major objectivity in the measurement of competitive advantage, improving the quality of strategic planning processes.

Acknowledgments In all of this, I would like to acknowledge the assistance and support I received from Professor Virgílio Cruz-Machado and Professor Helena Carvalho.

References

1. Barney J (1991) Firm resources and sustained competitive advantage. J Manag 17(1):99–120
2. Cantwell J (2003) Innovation and competitiveness. In: The Oxford handbook of innovation, revised edn, chap 21. Oxford University Press, Oxford, pp 543–567
3. Carvalho H, Cruz-Machado V (2013) Modeling resilience in supply chain. Technical report, Doctoral dissertation, UNL/FCT
4. Carvalho H, Azevedo S, Cruz-Machado V (2013) An innovative agile and resilient index for the automotive supply chain. Int J Agile Syst Manag 6(3):259–283
5. Christensen C (1992) Exploring the limits of the technology s-curve. Part I. Component technologies. Prod Oper Manag 1(4):334–357
6. Erol O, Sauser B, Mansouri M (2010) A framework for investigation into extended enterprise resilience. Enterp Inf Syst 4(2):111–136
7. Feurer R, Chaharbaghi K (1994) Defining competitivenes: a holistic approach. Manag Decis 32(2):49–58
8. Hart S, Milstein M (2003) Creating sustainable value. Acad Manag Perspect 17(2):56–67
9. Hubbard G (2009) Measuring organizational performance: beyond the triple bottom line. Bus Strategy Environ 18:177–191
10. Peck H (2005) Drivers of supply chain vulnerability: an integrated framework. Int J Phys Distrib 35(4):210–232
11. Porter M (1983) Industrial organization and the evolution of concepts for strategic planning: the new learning. Manag Decis Econ 4(3):172–180
12. Porter M (2008) Competitive advantage: creating and sustaining superior performance. Simon and Schuster, New York
13. Rosselet S (2011) Leveraging competitiveness to wage war against short-termism-building the house of sustainable competitiveness. Technical report, World Competitiveness Center
14. Schwab K (2012–2013) The global competitiveness report. Technical report, World Economic Forum
15. Shahin A, Dolatabadi H, Kouchekian M (2012) Proposing an integrated model of bsc and efqm and analyzing its influence on organizational strategies and performance—the case of Isfahan municipality complex. Int J Acad Res Econ Manag Sci 1(3):41–57
16. Sharma M, Kodali R (2008) Development of a framework for manufacturing excellence. Measuring Bus Excellence 12(4):50–66
17. Wang J, Gao F (2010) Measurement of resilience and its application to enterprise information systems. Enterp Inf Syst 4(2):215–223
18. Wikström P (2010) Sustainability and organizational activities-three approaches. Strateg Approaches Sustain Policy Manag 18(2):99–107
19. Zhang W, Lin Y (2010) On the principle of design of resilient systems—application to enterprise information systems. Enterp Inf Syst 4(2):99–110

Chapter 124
Optimization Design Strategies and Methods of Service Granularity

Pingping Wang and Zongmin Li

Abstract Service granularity design considers the complexity of the combination of services, the degree of service reuse, service portfolio to the needs of the business performance as well as the frequent change of adaptability. This paper researches into the relationship between elements of services, establishes the weighted directed graph to describe the logic of these relationships. Following this, this paper discusses the strategies of service granularity design with comprehensive considerations of coupling degree and cohesive degree. A mathematical model which maximize the coupling degree and minimize the cohesive degree is set up. Then this paper designs multi-objective particle swarm optimization to solve this problem. The approach is applied to a real realistic corporation to illustrate the effectiveness of the proposed model and algorithm.

Keywords Service granularity · Coupling degree · Cohesive degree · Multi-objective particle swarm optimization

124.1 Introduction

The so-called service granularity refers to the reasonable division of businesses of service based on the requirements of function and certain sizes of service. The service granularity represents the scale of service reuse and complexity of combination. Service granularity has profound influences to the organizational adaptability, software reuse degree and reuse efficiency.

The research on service research is still in the initial stage. Related researches mainly consider the software component and adopt clustering analysis methods and UML to improve the component granularities located within the degree of polymerization and reduce the coupling between components. These studies mainly focus on

P. Wang (✉) · Z. Li
Business School, Sichuan University, Chengdu 610064, People's Republic of China
e-mail: wangpp@scu.edu.cn

J. Xu et al. (eds.), *Proceedings of the Eighth International Conference on Management Science and Engineering Management*, Advances in Intelligent Systems and Computing 281, DOI: 10.1007/978-3-642-55122-2_124, © Springer-Verlag Berlin Heidelberg 2014

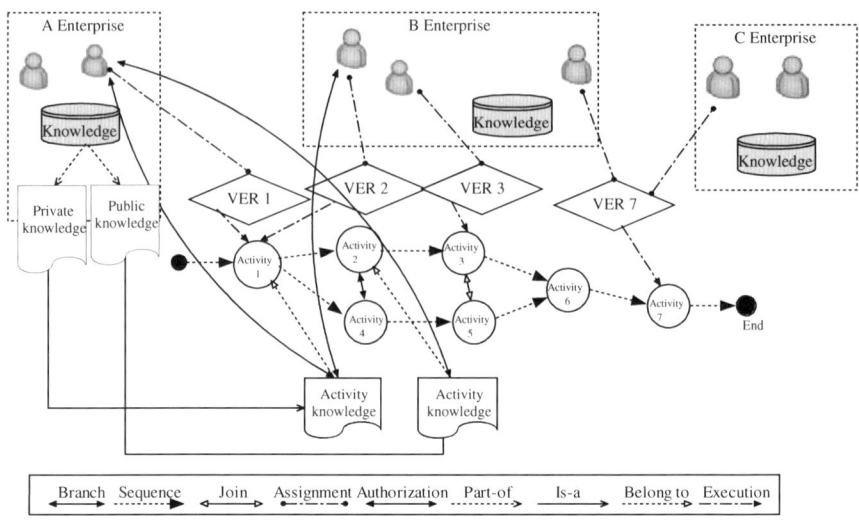

Fig. 124.1 The inter-organization e-commerce environment

the design and realization of software system and cannot improve business changes adaptability in inter-organization e-commerce environment. For example: IBM proposed the Component Business Model [7], Gordijn [10] proposed a service design method based modeling and the enterprise strategic objectives; Fang [9] proposed another service design method faced business process; and some other service design method which is from bottom-up method service. Chen [1] processed a quantitative calculation method from two dimensions, from which establish an analytical model of service granularity. However, these studies are lack of in-depth discussion of coupling degree and cohesive degree which are core of business process in the inter-organization e-commerce environment.

In inter-organization e-commerce environment [2], enterprise faces frequent changes of business partners and different business requirements, and enterprise must also adjusted its businesses according to these changes constantly, which is shown in Fig. 124.1.

This paper argues that the main factors affect service granularity are: the context relationships between business processes of service, the software design and implementation of service, as well as the business processes demands for service. Among those, the context relationships between business processes is the most important. Because it is has significant influences on the software design and implementation of service granularity, and also can make more efficient and flexible organizational to meet customers' needs. Therefore this paper will consider mainly on the context relationships between business processes of service, to establish service components extraction model, design intelligent algorithm, from the angle of quantitative research on services granularity.

The rest of the paper is organized as follows: Sect. 124.2 will introduce the basic relationships of business process; Sect. 124.3 will introduces the mathematical model of service granularity; The multi-objective particle swarm optimization for service granularity is presented in Sects. 124.4, 124.5 is a case study and Sect. 124.6 concludes with a summary.

124.2 The Basic Relationships of Business Process

Business processes of service can be divided into procedural business processes and entity type business processes. Procedural business processes correspond to the important business processes and its interrelated regulations, while the entity type business processes correspond to the important business entity and its interrelated regulations. According to the particularity of service area, this paper puts forward the basic relationships of business process in service area.

Include-of relationship. Include-of relationship is the relationship which means one business process contains another business process. i, j are the business process index.

Axiom 124.2.1

$$\forall i \in S, j \in S, include - of(i, j) \wedge include - of(i, k) \Rightarrow include - of(i, k),$$
$$\forall i \in S, j \in S, include - of(i, j) \Rightarrow !include - of(j, i).$$

Message-link relationship. Message-link relationship is the relationship which denotes the communication and information exchange between two processes.

Axiom 124.2.2

$$\forall i \in S, j \in S, message - link(i, j) \Rightarrow !message - link(j, i).$$

Create-link relationship. Create-link relationship is the relationship which denotes one business process creates another business process.

Read-link relationship. Read-link relationship is the relationship which denotes one business process can receive the message from another business process.

Axiom 124.2.3

$$\forall i \in S, j \in S, read - link(i, j) \Rightarrow !read - link(j, i).$$

Update-link. Update-link relationship is the relation which denotes one business process can update another business process.

Axiom 124.2.4

$$\forall i \in S, j \in S, update-link(i,j) \Rightarrow !read-link(j,i).$$

124.3 Mathematical Model of Service Granularity

The key question of service granularity is reasonable division of business process according to the function and the task of services. Service granularity represents the scale of service reuse and complexity. This paper establishes a mathematical model of service granularity with structure measurements, namely coupling degree and cohesive degree. There is a principle of low coupling for service, making the same size design service component of each composition business needs degree is high (high cohesion), and service interdependence among the components is lower (low coupling). According to the characteristics of service areas, this paper defines the coupling degree and cohesive degree.

124.3.1 Coupling Degree

Coupling degree of the service refers to the correlativity of different service components. The more correlativity between the service module, the more dependent on each other, the more complexity of the combination. This paper defines the coupling degree as follows.

Definition 124.1 The coupling degree of a service DO is defined as:

$$DO(G_i) = \begin{cases} 0, & R_{ij} = R_i \cup R_j \\ \sum_{r \in (R_{ij} - R_i \cup R_j)} \omega(r), & R_{ij} = R_i \cup R_j \subset R_{ij}, \end{cases}$$

where, R_{ij} is the edge collection of G_{ij}, namely $R_{ij} = \{(s_1, s_2, t)|s_1, s_2 \in S_i \cup S_j\}$. Then, as for the service granularity D, the coupling degree F_O can be defined as:

$$F_O(D) = \begin{cases} \sum_{i=1}^{m-1} DO(G_i)/m - 1, & m > 1 \\ 0, & m = 1. \end{cases}$$

124.3.2 Cohesive Degree

Cohesive degree refers to the connection strength of every internal business process. The higher the cohesive degree, the easier the service can be understood, and the easier the process can be modified and maintain. R denotes the relationship between business process, ω denotes the strength of association or the relationship.

Definition 124.2 The cohesive degree of a service DN can be defined as:

$$DN(G_i) = \begin{cases} 0, & R_i = \emptyset, l_i > 1 \\ 1, & l_i = 1 \\ \sum_{r \in R_i} \omega(r), & R_i \neq \emptyset, \end{cases}$$

where G_i (service module) is the subgraph of the function logic diagram, l_i is the number of the points (business processes) in the G_i.

Then, as for the service granularity D, the cohesive degree F_N can be defined as:

$$f_N = \begin{cases} DN(G_i), & l_i = 1 \\ \frac{DN(G_i)}{l_i \times (l_i-1)/2}, & l_i > 1 \end{cases} \text{ and } F_N(D) = \sum_{i=1}^{m} f_N(G_i)/m,$$

where, i is the number of G_i.

The mathematical model of service granularity has to balance the tradeoff between maximizing the coupling degree and minimizing the cohesive degree to find the best service granularity D. To solve this model, this paper proposes the following multi-objective particle swarm optimization in the next section.

124.4 Multi-objective Particle Swarm Optimization for Service Granularity

Particle swarm optimization (PSO) is a population-based self-adaptive search optimization technique that was proposed by Kennedy and Eberhart [15]. It simulates a social behavior such as birds flocking to a promising position for certain objectives in a multi-dimensional space [3, 8].

Like an evolutionary or meta-heuristic algorithm which evolves to find the global optimum of a real-valued function (called fitness function), PSO is based on a set of potential solutions defined in a given space (called search space). It conducts search using a fixed number population (called swarm) of individuals (called particles) that are updated from iteration to iteration. An n-dimensional position of a particle (called solution), initialized with a random position in a multidimensional search space, represents a solution of the problem, and it resembles the chromosome of a genetic algorithm [18]. The particles, which characterized by their positions and velocities [15], fly through the problem space by following the current optimum particles. Unlike other population-based algorithms, the velocity and position of each particle are dynamically adjusted according to the flying experiences or discoveries of its own and those of its companions. Kennedy and Eberhart [15] proposed as follows to update the position and velocity of each particle in 1995:

$$v_{ld}(\tau + 1) = w(\tau)v_{ld}(\tau) + c_p r_1 [p_{ld}^{best}(\tau) - p_{ld}(\tau)] + c_g r_2 [p_{gd}^{best(\tau)} - p_{ld}(\tau)],$$
$$p_{ld}(\tau + 1) = p_{ld}(\tau) + v_{ld}(\tau + 1),$$

where $v_{ld}(\tau+1)$ is the velocity of lth particle at the dth dimension in the τth iteration, w is an inertia weight, p_{ld}^{τ} is the position of lth particle at the dth dimension, r_1 and r_2 are random numbers in the range [0, 1], c_p and c_g are personal and global best position acceleration constant respectively, meanwhile, p_{ld}^{best} and p_{gd}^{best} are personal and global best position of lth particle at the dth dimension.

Since PSO can be implemented easily and effectively, it has been rapidly applied in solving real-world optimization problems in recent years, such as [14, 16, 19], etc. Researchers are also seeing PSO as a very strong competitor to other algorithms in solving multi-objective optimal problems [4] and it has been proved to be especially suitable for multi-objective optimization [5]. Therefore, a number of proposals have been suggested to extend PSO to handle multi-objective problems in the last few years. There are some studies reported in the literature that extend PSO to multi-objective problems, such as [11–13, 21]. The MOPSO approach uses the concept of Pareto dominance to determine the flight direction of a particle and it maintains previously found non-dominated vectors in a global repository that is later used by other particles to guide their own flight [4].

This paper designs MOPSO to solve the model of service granularity and the following notations are used:

τ	: iteration index, $\tau = 1, \cdots, T,$
i	: particle index, $i = 1, \cdots, L,$
k	: dimension index, $k = 1, \cdots, D,$
p_{ik}^{max}	: maximum position value of particle i at the kth dimension,
p_{ik}^{min}	: minimum position value of particle i at the kth dimension,
r_1, r_2	: uniform distributed random number within [0, 1],
$w(\tau)$	: inertia weight in the τth iteration,
$v_{ik}(\tau)$	: velocity of the ith particle at the kth dimension in the τth iteration,
$P_{ik}(\tau)$	: position of the ith particle at the kth dimension in the τth iteration,
$p_{ik}^{pbest}(\tau)$	: personal best position of the ith particle at the kth dimension in the τth iteration,
$G_{ik}^{gbest}(\tau)$	: global best position of the ith particle at the kth dimension in the τth iteration,
c_p	: personal best position acceleration constant,
c_g	: global best position acceleration constant,
$Fitness(P_{ik}(\tau))$	: the fitness value of the ith particle at the kth dimension in the τth iteration.

124.4.1 Solution Representation and Decoding Method

The solution representation and decoding method is shown in Fig. 124.2. In this paper, the permutation-based representation is used for the service granularity problem.

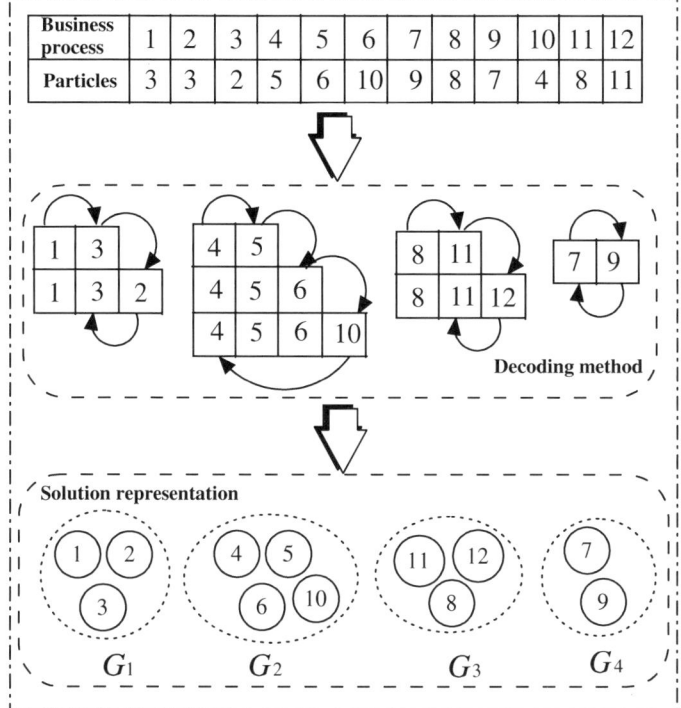

Fig. 124.2 Solution representation and decoding method

124.4.2 Fitness Value

The goal of service granularity is to find optimal service components division which the has the high value of $F_N(D)$ and low value of $F_O(D)$. The taxonomy that Reyes-Sierra and Coello [17] proposed to classify the current MOPSOs is aggregating approaches, lexicographic ordering, sub-population approaches, Pareto-based approaches, Combined approaches and Other approaches. This paper uses aggregating approach to handle the two objectives of service granularity model. w_1 and w_2 are weight the importance of the two objectives, $w_1 + w_2 = 1$. So the fitness value can be expressed as: $Fitness(D) = w_2 F_O(D) - w_1 F_N(D)$.

124.4.3 Framework of Algorithm

In order to guarantee the c onvergence of MOPSO, the parameters are selected on the basis of empirical results that are carried out to observe the behavior of the algorithm in different parameter settings. Empirical results have shown that the constant acceleration coefficients with $c_p = 1, c_g = 1.5$ and the adaptive inertia weights provide good convergent behavior in this study, which is in accordance with the results provided by Eberhart and Shi [6]. The adaptive inertia weights for the upper level (i.e., $e = 1$) and lower-level (i.e., $e = 2$) are set to be varying with iteration as follows:

$$w(\tau) = w(T) - \frac{\tau - T}{1 - T}[w(1) - w(T)], \tag{124.1}$$

where the iteration numbers $T = 500$, $w(1) = 0.9$ and $w(T) = 0.1$. Since the probability of becoming trapped in the stagnant state can be reduced dramatically by using a large number of particles [20], the population sizes are set to be 400. The framework of algorithm is shown in Fig. 124.3.

124.5 Application

The approach is applied to a real realistic corporation to illustrate the effectiveness of the proposed model and algorithm. There are 20 business processes in this service granularity problem. The strength of relationships between them are shown below.

$$W = \begin{pmatrix}
0 & 2 & 2 & 0 & 0 & 0 & 2 & 0 & 0 & 0 & 0 & 0 & 0 & 0 & 0 & 0 & 0 & 0 & 0 & 0 \\
0 & 0 & 0 & 0 & 0 & 0 & 0 & 0 & 0 & 2 & 0 & 0 & 0 & 0 & 0 & 0 & 0 & 0 & 0 & 0 \\
0 & 0 & 0 & 2 & 0 & 0 & 0 & 0 & 2 & 0 & 2 & 0 & 0 & 0 & 0 & 0 & 0 & 0 & 0 & 0 \\
0 & 0 & 2 & 0 & 0 & 0 & 0 & 2 & 0 & 0 & 0 & 0 & 0 & 0 & 0 & 0 & 0 & 0 & 0 & 0 \\
0 & 0 & 2 & 0 & 0 & 0 & 0 & 0 & 2 & 0 & 0 & 0 & 0 & 0 & 0 & 0 & 0 & 0 & 0 & 0 \\
0 & 2 & 0 & 0 & 0 & 0 & 0 & 0 & 0 & 2 & 0 & 0 & 0 & 0 & 0 & 0 & 0 & 0 & 0 & 0 \\
2 & 0 & 0 & 0 & 0 & 0 & 0 & 3 & 0 & 0 & 0 & 0 & 0 & 3 & 0 & 0 & 0 & 0 & 0 & 0 \\
0 & 0 & 0 & 0 & 0 & 0 & 0 & 0 & 0 & 0 & 0 & 0 & 0 & 0 & 0 & 0 & 0 & 0 & 0 & 0 \\
0 & 0 & 2 & 0 & 2 & 0 & 0 & 0 & 0 & 0 & 0 & 0 & 0 & 0 & 0 & 0 & 0 & 0 & 0 & 0 \\
0 & 2 & 0 & 0 & 0 & 2 & 0 & 0 & 0 & 0 & 0 & 0 & 3 & 0 & 0 & 0 & 0 & 0 & 0 & 0 \\
0 & 0 & 0 & 0 & 0 & 0 & 0 & 0 & 0 & 0 & 0 & 0 & 0 & 0 & 0 & 2 & 0 & 2 & 0 & 0 \\
0 & 0 & 0 & 0 & 0 & 0 & 0 & 0 & 0 & 0 & 0 & 0 & 0 & 0 & 0 & 0 & 0 & 2 & 0 & 0 \\
0 & 0 & 0 & 0 & 0 & 0 & 0 & 0 & 0 & 0 & 0 & 0 & 0 & 4 & 0 & 0 & 0 & 0 & 0 & 3 \\
0 & 0 & 0 & 0 & 0 & 0 & 0 & 0 & 0 & 0 & 0 & 0 & 0 & 0 & 0 & 0 & 0 & 0 & 0 & 0 \\
3 & 0 & 0 & 0 & 0 & 0 & 0 & 0 & 0 & 0 & 0 & 0 & 0 & 0 & 0 & 0 & 0 & 0 & 0 & 0 \\
0 & 0 & 0 & 0 & 0 & 0 & 0 & 0 & 0 & 0 & 0 & 0 & 0 & 0 & 0 & 0 & 0 & 0 & 0 & 0 \\
0 & 0 & 0 & 0 & 0 & 0 & 0 & 0 & 0 & 0 & 0 & 0 & 0 & 0 & 0 & 0 & 0 & 0 & 2 & 0 \\
0 & 0 & 0 & 0 & 0 & 0 & 0 & 0 & 0 & 0 & 0 & 0 & 0 & 0 & 0 & 0 & 0 & 0 & 3 & 0 \\
0 & 0 & 0 & 0 & 0 & 0 & 0 & 0 & 0 & 0 & 0 & 3 & 0 & 0 & 0 & 0 & 0 & 2 & 0 & 0 \\
0 & 0 & 0 & 0 & 0 & 0 & 0 & 0 & 0 & 0 & 0 & 0 & 0 & 0 & 0 & 0 & 0 & 0 & 0 & 0
\end{pmatrix}.$$

Using the method designed, after 500 iteration, the optimal solution to this service granularity problem is got which is shown in Table 124.1.

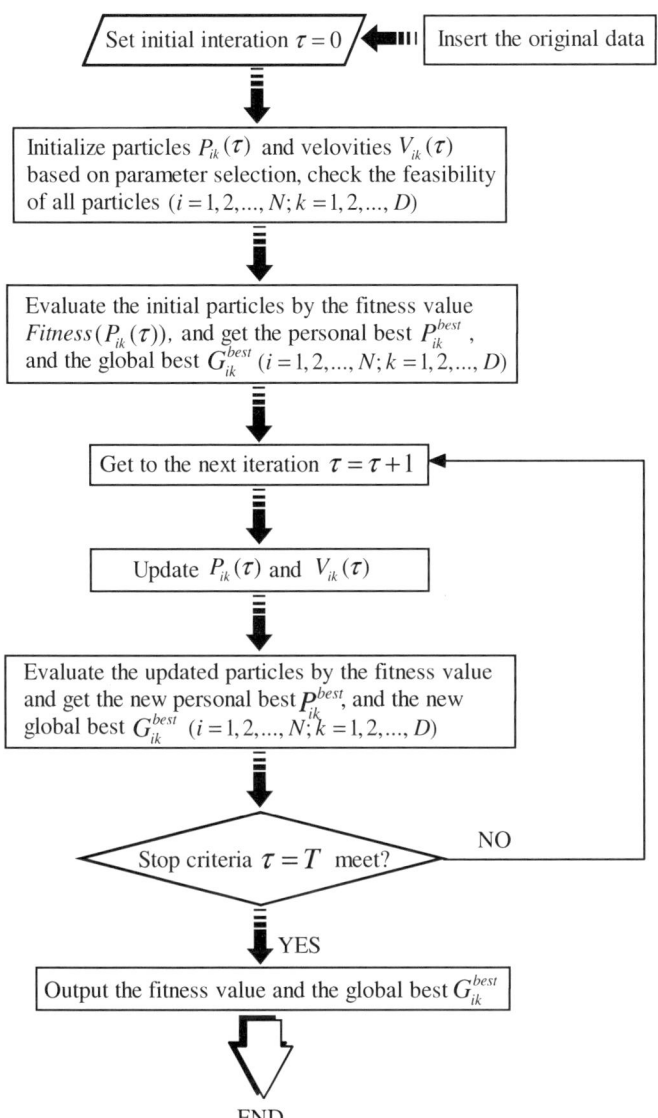

Fig. 124.3 Framework of algorithm

Table 124.1 The solution of proposed method and algorithm

The first service module	1	7	0	0	0	0	0	0	0	0	0	0
The second service module	2	17	9	3	16	12	20	10	19	8	18	15
The third service module	4	5	6	11	0	0	0	0	0	0	0	0
The fourth service module	13	14	0	0	0	0	0	0	0	0	0	0

124.6 Conclusion

This paper researches into the relationship between elements of services, establishes the weighted directed graph to describe the logic of these relationships. Following this, this paper discusses the strategies of service granularity design with comprehensive considerations of coupling degree and cohesive degree. A mathematical model which maximize the coupling degree and minimize the cohesive degree is set up. Then this paper designs multi-objective particle swarm optimization to solve this problem. The approach is applied to a real realistic corporation to illustrate the effectiveness of the proposed model and algorithm. The result shows the effectiveness of the proposed service granularity model and algorithm.

References

1. Chen H, Fang D, Zhao WD (2009) The relationship of business service granularity and process in SOA. Comput Eng Appl 45(27):7–10 (In Chinese)
2. Chen TY, Chen YM et al (2010) A fuzzy trust evaluation method for knowledge sharing in virtual enterprises. Comput Ind Eng 59:853–864
3. Clerc M, Kennedy J (2002) The particle swarm-explosion, stability, and convergence in a multidimensional complex space. IEEE Trans Evol Comput 6(1):58–73
4. Coello CC, Lechuga MS (2002) MOPSO: A proposal for multiple objective particle swarm optimization. In: Proceedings of the IEEE congress on evolutionary computation, pp 1051–1056
5. Coello CC, Pulido G, Lechuga MS (2004) Handling multiple objectives with particle swarm optimization. IEEE Trans Evol Comput 8(3):256–279
6. Eberhart RC, Shi Y (2000) Comparing inertia weights and constriction factors in particle swarm optimization. In: Proceedings of the IEEE congress on evolutionary computation, San Diego, USA
7. Ernest MJ, Nisavic M (2005) Adding value to the IT organization with the component business model. IBM Syst J 46(3):387–403
8. Eberhart RC, Shi Y (2001) Tracking and optimizing dynamic systems with particle swarms. In: Proceedings of IEEE congress on evolutionary computation (CEC 2001), pp 94–97
9. Fang D, Liu J, Zhao WD (2009) A service design method oriented to the flow of business. Comput Int Manuf 5:874–883 (In Chinese)
10. Gordijn J, Yu E, Raadt B (2006) E-service design using i* and e/sup 3/value modeling. IEEE Softw 23(3):26–33
11. Hamta N, Fatemi Ghomi SMT et al (2013) A hybrid PSO algorithm for a multi-objective assembly line balancing problem with flexible operation times, sequence-dependent setup times and learning effect. Int J Prod Econ 141:99–111
12. Hirano H, Yoshikawa T (2012) A study on two-step search using global-best in PSO for multi-objective optimization problems. In: Soft computing and intelligent systems (SCIS) and 13th international symposium on advanced intelligent systems (ISIS), pp 1894–1897
13. Huang VL, Suganthan PN, Liang JJ (2006) Comprehensive learning particle swarm optimizer for solving multiobjective optimization problems. Int J Intell Syst 21:209–226
14. Kashan A, Karimi B (2009) A discrete particle swarm optimization algorithm for scheduling parallel machines. Comput Industial Eng 56:216–223
15. Kennedy J, Eberhart R (1995) Particle swarm optimization. In: Proceedings of the IEEE conference on neural networks, IEEE service center, piscataway, pp 1942–1948

16. Ling S, Iu H et al (2008) Hybrid particle swarm optimization with wavelet mutation and its industrial applications. IEEE Trans Syst Man Cybern-Part B: Cybern 38:743–763
17. Reyes-Sierra M, Coello CC (2006) Multi-objective particle swarm optimizers: a survey of the state-of-the-art. Int J Comput Int Res 2:287–308
18. Robinson J, Sinton S, Rahmat-Samii Y (2002) Particle swarm, genetic algorithm, and their hybrids: Optimization of a profiled corrugated horn antenna. In: IEEE Antennas and propagation society international symposium and URSI national radio science meeting, San Antonio, pp 168–175
19. Sha DY, Hsu CY (2006) A hybrid particle swarm optimization for job shop scheduling problem. Comput Ind Eng 51:791–808
20. Van den Bergh F, Engelbrecht AP (2010) A convergence proof for the particle swarm optimiser. Fundam Informaticae 105:341–374
21. Xia W, Wu Z (2005) An effective hybrid optimization approach for multi-objective flexible job-shop scheduling problems. Comput Ind Eng 48:409–425

Chapter 125
A Fuzzy Synthetic Performance Evaluation Model for the Strategic Alliance of Technological Innovation in China

Cheng Xiang, Jingxia Fan and Gang Xiang

Abstract Based on realizing the performances achieved by the strategic alliances of technological innovation in China, this paper developed guiding thoughts of all-round evaluation and principles, and then established a fuzzy synthetic performance evaluation model. Experts consulting and fuzzy synthetic evaluating method were used to building the criteria system, weight system and measurement standards as well as the calculating tools. This model has been applied to evaluating performances achieved by a strategic alliance of technological innovation centered in an innovative enterprise in Yunnan, China. Responding analysis and suggestions were proposed for the alliance and relevant government office.

Keywords Strategic alliance of technological innovation · Performance · Fuzzy synthetic evaluation · Model · Innovative enterprise

125.1 Introduction

Strategic Alliance of Technological Innovation (SATI) is a new developing organization mode of strategic co-operating alliance which is consisted of enterprises, research institutes and universities. It is oriented to market, tilted by interests [2]. In order to develop the innovative enterprises, the SATI's establishment has been entered in "The Whole Implementing Plan of the Nation's Technological Innovation Program" by the Ministry of Science and Technology of PRC and other departments

C. Xiang
School of Life Science and Technology, Kunming University of Science and Technology, Kunming 650093, People's Republic of China

J. Fan · G. Xiang (✉)
School of Management and Economics, Kunming University of Science and Technology, Kunming 650093, People's Republic of China
e-mail: gang_xiang@aliyun.com

J. Xu et al. (eds.), *Proceedings of the Eighth International Conference on Management Science and Engineering Management*, Advances in Intelligent Systems and Computing 281, DOI: 10.1007/978-3-642-55122-2_125, © Springer-Verlag Berlin Heidelberg 2014

[3]. Since 2010, a large number of SATIs have been organized in China and achieved effects. Some of them are especially centered in the innovative enterprises.

Since then, it has become an important issue to make performance evaluation to the SATI scientifically, especially for the SATI centered in the innovative enterprise. It will promote the SATI's building and encourage the innovative enterprise to realize sustainable innovation and development by using the guiding specialty of performance evaluation in both theoretical research and management practice [1, 4, 7–9]. Nowadays, it could be found that there were some researches on performance evaluation of the SATI which directly evaluated the resulting fruits or the realized extent of the achievement objectives [1, 2, 7]. Obviously, it would be right to directly evaluate the resulting fruits as the target performance for some SATIs outside of China which have successfully implemented for long period and achieved many resulting fruits. However, this mode may be unsuitable for evaluating the performance of most Chinese SATIs since they have been just organized, operated in a short period and achieved a little of resulting fruits.

So, it is necessary to research on performance evaluation of the SATI in China further comprehensively and scientifically according to the real situation of the SATI in China.

Supported by NSF of China, the Science and Technology Department of Yunnan Province and some innovative enterprises, based upon the realize for the organizing, operating and resulting performances produced in the real implementing process of SATI in China, this paper developed guiding thoughts and principles of the evaluation. Then, by using experts consulting and fuzzy synthetic evaluating methods [5], a fuzzy synthetic evaluation model for evaluating the performance of the SATI in China was established. Finally, the evaluation model was applied in a case that the performance evaluation for a real SATI centered in an innovative enterprise was provided and the conclusion was made.

125.2 Basic Guiding Thoughts and Principles

In this section, the authors initially founded guiding thoughts and principles for the model of the performance evaluation of SATI in China.

1. Basic Guiding Thoughts of the Performance Evaluation
Based on the observation of the SATI in China, it was realized that the organization performance, operation performance and maybe some resulting performance have been produced with the implementing of the organized SATI. The three parts of performance formed the main body of the whole performance of the SATI in which the resulting performance was the core, the operation performance was a critical one and the organization performance was the base. None of the three parts was dispensable. So, according to the fundamental goal of the SATI: "Organizing, Operating and Achieving good fruits" declared by the Ministry of Science and Technology of PRC [3], our basic guiding thoughts were established as: "Towards the

organizing, operating and resulting performances, evaluate the performances of the SATI in all-round".

2. Evaluation Principles

Based upon above basic guiding thoughts, three critical evaluation principles were provided as follows.

(1) The systematic evaluation principle

According to systems thinking and the basic guiding thoughts, the evaluating model should be a whole system that would be formed of three sub-systems as organizing performance, operating performance and resulting performance. Further, each sub-system formed by some concrete evaluating criteria that would be interacted and relatively independent each other and would be able to measure. This systematic evaluation principle provided a foundation to implement synthetic performance evaluation for the SATI.

(2) The satisfactory stakeholders' principle

The members of the SATI were various. Not only was an entity (normally, an innovative enterprise) as the center, but also were other important strategic co-operators such as enterprises, research institutes and universities. All the members organized as a group of stakeholders. So, it was necessary to obey the "To Satisfy all Stakeholders" principle of the Contemporary Management Sciences and fully considering the various needs of the performances evaluation criteria from all strategic cooperators in all key points to building a synthetic evaluation model for evaluating the performance of the SATI [6]. For example, in the resulting performance evaluating, as the center of the SATI, an innovative enterprise might trade the developing of core technology and obtaining the economic benefits from the innovation as the most important criteria, but for the cooperators from universities or research institutes, the most interested criteria might be obtaining more research projects and funds, fostering human talents and publishing papers. All above needs would be satisfied in the model, especially in evaluation criteria system designing, weights assignment and measuring standards determining.

(3) The practicability principle

That was, when to design every evaluating criterion and its measuring standard, the designer would be as much as possible to use the real data that were able to get from hysterical or current statistical information system.

125.3 Establishing a Fuzzy Synthetic Effectiveness Evaluation Model

Establishing an Evaluation Criteria System and Model Based on above guiding thoughts and evaluating principles, a performance evaluation criteria system and a fuzzy synthetic performance evaluation model by using experts' consulting and fuzzy synthetic evaluation methods were built as in follows.

Table 125.1 The performance evaluation criteria system of SATI with criteria weights

First class criteria (P_i)	Second class criteria (P_{ij})	Weights (R_{ij})
Organization performance (P_1)	Alliance establishment (P_{11})	$R_{11}(0.12)$
	Alliance's organization establishment (P_{12})	$R_{12}(0.09)$
	Alliance's constitution (P_{13})	$R_{13}(0.09)$
Operation performance (P_2)	Intensity of investment in R&D programs (P_{21})	$R_{21}(0.027)$
	Satisfaction degree of R&D staff (P_{22})	$R_{22}(0.036)$
	Satisfaction degree of facility (P_{23})	$R_{23}(0.027)$
	Number of innovational Projects (P_{24})	$R_{24}(0.06)$
	Technique level of innovation projects (P_{25})	$R_{25}(0.036)$
	Combination and utilizing the resource of alliance (P_{26})	$R_{26}(0.036)$
	Operation of the alliance organization (P_{27})	$R_{27}(0.027)$
	Ratio of encouraged R&D staff (P_{28})	$R_{28}(0.027)$
Resulting performance (P_3)	Number of completed innovational projects (P_{29})	$R_{31}(0.16)$
	Number of alliance's patents (P_{30})	$R_{32}(0.12)$
	Times to set the industry standards (P_{31})	$R_{33}(0.12)$

125.3.1 Building the Evaluation Criteria System

According to the basic guiding thoughts and principles mentioned above, an evaluation criteria system for performance evaluation of the SATI was built, including 3 first class criteria which involved organization performance, operation performance and resulting performance and 14 second class ones (Table 125.1). This system was built as a result of well discussion and communication with managers in relevant government offices and the innovative enterprises according to the realistic situation of the SATI in China, especially in Yunnan Province.

125.3.2 Establishing an Performance Evaluation Model

Based on the built evaluation criteria system, the authors established an evaluation model for performance evaluation of the SATI on following three steps.

Step 1 Building the weights matrix of the evaluation criteria. The weights of the first class criteria were expressed as $R_i (i = 1, 2, 3)$, and $\sum_{i=1}^{3} R_i = 1$. And the weights of the second class criteria were set as: $R_{ij} (i = 1, 2, 3; j = 1, 2, \cdots, 8)$, and $\sum_{i=1, j=1}^{i=3, j=8} R_i = 1$.

According to experts' opinions, survey results and the recognition of the relative importance of the organization performance, operation performance and resulting performance in the synthetic performance evaluation of a SATI, the weights of the three first class criteria were assigned as 0.3, 0.3, and 0.4. Furthermore, the weights of 14 second class criteria were assigned as in Table 125.1.

So, the weights matrix of the evaluation criteria was founded as follows,

$$
R_i = \begin{bmatrix} R_1 \\ R_2 \\ R_3 \end{bmatrix} = \begin{bmatrix} 0.3 \\ 0.4 \\ 0.5 \end{bmatrix}, \quad R_{ij} = \begin{bmatrix} R_{11} & R_{12} & R_{13} & 0 & 0 & 0 & 0 \\ R_{21} & R_{22} & R_{23} & R_{24} & R_{25} & R_{26} & R_{27} & R_{28} \\ R_{31} & R_{32} & R_{33} & 0 & 0 & 0 & 0 \end{bmatrix}
$$

$$
= \begin{bmatrix} 0.12 & 0.09 & 0.09 & 0 & 0 & 0 & 0 \\ 0.027 & 0.036 & 0.027 & 0.06 & 0.036 & 0.036 & 0.027 & 0.027 \\ 0.16 & 0.12 & 0.12 & 0 & 0 & 0 & 0 \end{bmatrix}.
$$

Step 2 Establishing the measurement standards of the evaluation criteria

Since the performance evaluation for the SATI was of intensely fuzzy characteristic, experts could be invited to measure with fuzzy language.

The fuzzy language ranks were excellent, good, medium, less-well and poor, which corresponded to the fuzzy value set as $(\tilde{9}, \tilde{7}, \tilde{5}, \tilde{3}, \tilde{1})$ and the fuzzy intervals set as $\{[8, 10], [6, 8], [4, 6], [2, 4], [0, 2]\}$ (Table 125.2).

Hence, the set of fuzzy measurement standards $(\tilde{P}_0)$ could be expressed as:

$$
\tilde{P}_0 = \begin{bmatrix} \text{excellent} \\ \text{good} \\ \text{medium} \\ \text{less-well} \\ \text{poor} \end{bmatrix} = \begin{bmatrix} \tilde{9} \\ \tilde{7} \\ \tilde{5} \\ \tilde{3} \\ \tilde{1} \end{bmatrix} = \begin{bmatrix} [8, 10] \\ [6, 8] \\ [4, 6] \\ [2, 4] \\ [0, 2] \end{bmatrix}. \qquad (125.1)
$$

Step 3 Establishing the fuzzy synthetic performance evaluation model

Utilizing the evaluation criteria, weights and fuzzy measurement standards listed in Table 125.2, experts were invited to measure the performance of a specific SATI. Then according to the experts' judgments, a mathematics model was established with a fuzzy synthetic method.

In details, if experts were invited to measure the performance of SATI-M by using the evaluation criteria and the fuzzy language ranks of the measurement standards listed in Table 125.2, the distribution set of the performance ranks given by all experts was $A_l (l = 1, 2, 3, 4, 5)$, obviously $\sum_{l=1}^{5} A_l = A$. Here, A meant the quantity of experts invited. Then, the distribution ratio set of the performance ranks was obtained as $a_l (l = 1, 2, 3, 4, 5)$, $\sum_{l=1}^{5} a_l = a$. And the distribution ratio set of the performance ranks of the 14 second class criteria was $a_{lk} (l = 1, 2, 3, 4, 5; k = 1, 2, \cdots, 14)$.

Therefore, the fuzzy average value of each second class criterion given by experts' evaluation could be calculated as $\tilde{P}_{ij} = a_{lk} \cdot \tilde{P}_0$, $(l = 1, 2, 3, 4, 5; k = 1, 2, \cdots, 14)$. So, the matrix of the average fuzzy values of the second class criteria given by experts' evaluation was:

Table 125.2 Fuzzy measurement standards of the performance evaluation criteria system of SATI

Fuzzy measurement standards

Measurement rank		Excellent $\tilde{9}$	Good $\tilde{7}$	Medium $\tilde{5}$	Less well $\tilde{3}$	Poor $\tilde{1}$
Value		$\tilde{9}$	$\tilde{7}$	$\tilde{5}$	$\tilde{3}$	$\tilde{1}$
Interval		[8 ~ 10]	[6 ~ 8]	[4 ~ 6]	[2 ~ 4]	[0 ~ 2]
Second class criteria (P_{ij})	P_{11}	Multiple projects and contract;	Be of multiple programs and programs	Be of contract; without programs	Be of common goals	None
	P_{12}	Org. and Res. are explicit and members. are on duty	Org. and Res. are explicit	Org. is explicit; Res.is obscure	Org. and res. is obscure	None
	P_{13}	Be of a Cons. containing rules of sharing Res., rights and profits	Be of a Cons. rules of sharing Res., rights and profits	Be of a simple Cons.	Be of a Cons. without sharing Res., rights and profits	None
	P_{21}	≥3.0%	2.0 ~ 3.0%	1.0 ~ 2.0%	0.5 ~ 1.5%	<0.5%
	P_{22}	Completely satisfied	80% satisfied	60% satisfied	40% satisfied	≤20% satisfied
	P_{23}	Completely satisfied	80% satisfied	60% satisfied	40% satisfied	≤20% satisfied
	P_{24}	≥5	3 ~ 4	2	1	0
	P_{25}	International advanced	National advanced	Regional advanced	At the average level of the region	lower than the region level
	P_{26}	Combined well, utilized well	Combined well, utilized less well	Combined not enough, utilized less well	Parts of resource wasted	A lot of resource wasted
	P_{27}	Communicating frequently	Communicating occasionally	Communicating rarely	Without communication	Conflict
	P_{28}	20 ~ 30%	10 ~ 20%	5 ~ 10	0 ~ 5%	0%
	P_{29}	>3	3	2	1	0
	P_{30}	Approved more than 2	Approved 2	Applied 2	Applied 1	0
	P_{31}	>2	2	1	1	0

$$P_{ij} = \begin{bmatrix} \tilde{P}_{11} & \tilde{P}_{12} & \tilde{P}_{13} & 0 & 0 & 0 & 0 \\ \tilde{P}_{21} & \tilde{P}_{22} & \tilde{P}_{23} & \tilde{P}_{24} & \tilde{P}_{25} & \tilde{P}_{26} & \tilde{P}_{27} & \tilde{P}_{28} \\ \tilde{P}_{31} & \tilde{P}_{32} & \tilde{P}_{33} & 0 & 0 & 0 & 0 \end{bmatrix}. \tag{125.2}$$

The value of experts' fuzzy synthetic performance evaluation of the SATI-M was calculated as $\tilde{P} = \sum_{i=1, j=1}^{i=3, j=8} \tilde{P}_{ij} \cdot R_{jt}$, or

$$P_{ij} = \begin{bmatrix} \tilde{P}_{11} & \tilde{P}_{12} & \tilde{P}_{13} & 0 & 0 & 0 & 0 \\ \tilde{P}_{21} & \tilde{P}_{22} & \tilde{P}_{23} & \tilde{P}_{24} & \tilde{P}_{25} & \tilde{P}_{26} & \tilde{P}_{27} & \tilde{P}_{28} \\ \tilde{P}_{31} & \tilde{P}_{32} & \tilde{P}_{33} & 0 & 0 & 0 & 0 \end{bmatrix} \cdot \begin{bmatrix} R_{11} & R_{21} & R_{31} \\ R_{12} & R_{22} & R_{32} \\ R_{13} & R_{23} & R_{33} \\ 0 & R_{24} & 0 \\ 0 & R_{25} & 0 \\ 0 & R_{26} & 0 \\ 0 & R_{27} & 0 \\ 0 & R_{28} & 0 \end{bmatrix}. \tag{125.3}$$

125.3.3 The Application of the Model and Analyzing the Evaluated Results

Above evaluation criteria system, weights, measurement standards, experts consulting and fuzzy synthetic evaluation methods had formed a fuzzy synthetic performance evaluation model that could be able to use to evaluate an exact SATI in China and to obtain quantitative evaluation results.

Furthermore, according to the fuzzy measurement standards ($\tilde{P}_0$), one could make further important analysis to the average values of experts' fuzzy performance valuation of the SATI in the second class evaluation criteria sector.

Finally, using the fuzzy measurement standards taking a reversal judgment on the fuzzy value of the synthetic performance evaluation for this SATI calculated above, the final judgment conclusion for this SATI could be clearly made in language such as "Excellent, Good, Medium, Less-well and Poor". This judgment conclusion could give useful reference for managing the SATI.

125.4 Case Study

By using above model, the authors had finished a case study to evaluate the performance for the Strategic Alliance of Technological Innovation in Precious Metal Industry centered in Company G.

1. Background

The SATI in Precious Metal Industry is centered in Company G., a leading company of the precious metal industry and an excellent innovative enterprise in Yunnan Province. The SATI has established in February 2012. It includes 32 members that involving 12 enterprises, 12 universities and 8 academic institutions. The organization of the SATI contained a council of the SATI, a committee of experts and a secretariat. In the SATI, Company G is the core member who undertakes the task to lead the advanced techniques, boom the industry and achieve sustainable innovation. The aim of the SATI was utilizing the obtained merits and the foundation to promote the development of the precious metal industry through accomplishing every member's task. Since 2012, the SATI had implemented 3 cooperated projects of technological innovation and achieved effects.

2. The Performance Evaluation of the SATI in Precious Metal Industry

In order to evaluate the performance of the SATI in Precious Metal Industry centered in Company G, five experts were invited to judge the SATI's performance by using the evaluation criteria system and measurement standards given in Table 125.2. The results were listed in Table 125.3.

Through data normalization of the distribution of the experts' evaluation set $A_l (l = 1, 2, 3, 4, 5)$, the ratio set of the distribution of the experts' evaluation about the secondary evaluation criteria, a_l, $(l = 1, 2, 3, 4, 5)$, $\sum_{l=1}^{5} a_l = 1$, were obtained and shown at the column of Ratio of the distribution of experts' evaluation in Table 125.4.

The set a_{lk}, $(l = 1, 2, 3, 4, 5; k = 1, 2, \ldots, 14)$ could be built.

According to: $\tilde{P}_{ij} = a_{lk} \cdot \tilde{P}_0$, $(l = 1, 2, 3, 4, 5; k = 1, 2, \cdots, 14)$ the average values of experts fuzzy evaluation were calculated on each second class criterion and listed in the column of Average values of experts' fuzzy evaluation ($\tilde{P}_{ij}$) in Table 125.4.

Then, the matrix of the average values of experts' fuzzy evaluation on all the second class evaluation criteria were built as:

$$
P_{ij} = \begin{bmatrix}
\tilde{P}_{11} & \tilde{P}_{12} & \tilde{P}_{13} & 0 & 0 & 0 & 0 \\
\tilde{P}_{21} & \tilde{P}_{22} & \tilde{P}_{23} & \tilde{P}_{24} & \tilde{P}_{25} & \tilde{P}_{26} & \tilde{P}_{27} & \tilde{P}_{28} \\
\tilde{P}_{31} & \tilde{P}_{32} & \tilde{P}_{33} & 0 & 0 & 0 & 0
\end{bmatrix} = \begin{bmatrix}
8.\tilde{6} & 8.\tilde{2} & 7.\tilde{4} & 0 & 0 & 0 & 0 \\
7.\tilde{4} & 7.\tilde{8} & 7.\tilde{8} & 7.\tilde{0} & 7.\tilde{4} & 7.\tilde{0} & 7.\tilde{0} & 6.\tilde{6} \\
6.\tilde{6} & 7.\tilde{4} & 7.\tilde{4} & 0 & 0 & 0 & 0
\end{bmatrix}.
$$

Finally, the value of the fuzzy synthetic performance evaluation for this SATI was calculated as:

$$
P_{ij} = \begin{bmatrix}
\tilde{P}_{11} & \tilde{P}_{12} & \tilde{P}_{13} & 0 & 0 & 0 & 0 \\
\tilde{P}_{21} & \tilde{P}_{22} & \tilde{P}_{23} & \tilde{P}_{24} & \tilde{P}_{25} & \tilde{P}_{26} & \tilde{P}_{27} & \tilde{P}_{28} \\
\tilde{P}_{31} & \tilde{P}_{32} & \tilde{P}_{33} & 0 & 0 & 0 & 0
\end{bmatrix} \cdot \begin{bmatrix}
R_{11} & R_{21} & R_{31} \\
R_{12} & R_{22} & R_{32} \\
R_{13} & R_{23} & R_{33} \\
0 & R_{24} & 0 \\
0 & R_{25} & 0 \\
0 & R_{26} & 0 \\
0 & R_{27} & 0 \\
0 & R_{28} & 0
\end{bmatrix} = 7.\tilde{3}.
$$

Table 125.3 Performance evaluation of the SATI in Precious Metal Industry (With A_i)

First class criteria (P_i)	Second class criteria P_{ij}	Distribution of experts' evaluation ($\sum_{i=1}^{5} A_i = A$)				
		Excellent (A_1)	Good (A_2)	Medium (A_3)	Less well (A_4)	Poor (A_5)
Organization performance (P_1)	Alliance establishment (P_{11})	4	1	0	0	0
	Alliance's organization establishment (P_{12})	3	2	0	0	0
	Alliance's constitution (P_{13})	2	2	1	0	0
Operation performance (P_2)	Intensity of investment in R&D programs (P_{21})	1	4	0	0	0
	Satisfaction degree of R&D staff (P_{22})	2	3	0	0	0
	Satisfaction degree of facility (P_{23})	2	3	0	0	0
	Number of innovational Projects (P_{24})	0	5	0	0	0
	Technique level of innovation projects (P_{25})	3	0	2	0	0
	Combination and utilizing the resource of alliance (P_{26})	1	3	1	0	0
	Operation of the alliance organization (P_{27})	1	3	1	0	0
	Ratio of encouraged R&D staff (P_{28})	0	4	1	0	0
Resulting performance (P_3)	Number of completed innovational projects (P_{29})	0	4	1	0	0
	Number of alliance's patents (P_{30})	2	2	1	0	0
	Times to set the industry standards (P_{31})	3	0	2	0	0

Table 125.4 Performance evaluation of the SATI in precious metal industry (with A_l) and $\tilde{P}_{ij}$

Second class criteria P_{ij}	Ratio[a]					Aver[b]
	Excellent (a_1)	Good (a_2)	Medium (a_3)	Less well (a_4)	Poor (a_5)	
Alliance establishment (P_{11})	0.8	0.2	0	0	0	$\tilde{8.6}$
Alliance's organization establishment (P_{12})	0.6	0.4	0	0	0	$\tilde{8.2}$
Alliance's constitution (P_{13})	0.4	0.4	0.2	0	0	$\tilde{7.4}$
Intensity of investment in R&D programs (P_{21})	0.2	0.8	0	0	0	$\tilde{7.4}$
Satisfaction degree of R&D staff (P_{22})	0.4	0.6	0	0	0	$\tilde{7.8}$
Satisfaction degree of facility (P_{23})	0.4	0.6	0	0	0	$\tilde{7.8}$
Number of innovational Projects (P_{24})	0	1	0	0	0	$\tilde{7.0}$
Technique level of innovation projects (P_{25})	0.6	0	0.4	0	0	$\tilde{7.4}$
Combination and utilizing the resource of alliance (P_{26})	0.2	0.6	0.2	0	0	$\tilde{7.0}$
Operation of the alliance organization (P_{27})	0.2	0.6	0.2	0	0	$\tilde{7.0}$
Ratio of encouraged R&D staff (P_{28})	0	0.8	0.2	0	0	$\tilde{6.6}$
Number of completed innovational projects (P_{29})	0	0.8	0.2	0	0	$\tilde{6.6}$
Number of alliance's patents (P_{30})	0.4	0.4	0.2	0	0	$\tilde{7.4}$
Times to set the industry standards (P_{31})	0.6	0	0.4	0	0	$\tilde{7.4}$

[a]Ratio of distribution of experts' evaluation ($\sum_{i=1}^{5} A_i = A$)
[b]Average values of experts' fuzzy evaluation $\tilde{P}_{ij}$

3. Evaluated Results Analyzing

According to the fuzzy measurement standards ($\tilde{P}_0$), making further analysis to the evaluation results of the SATI, especially to the average values of experts' fuzzy performance valuation in the second class evaluation criteria sector, some major results were indicated as follows.

Among the 14 second class evaluation criteria, the average fuzzy experts' evaluation values were higher than $\tilde{8}$ on two criteria, $\tilde{7}$–$\tilde{8}$ on ten and $\tilde{6}$–$\tilde{7}$ on two.

(1) The performances evaluated on criteria of Alliance establishment and Alliance's organization Establishment which were belonging to the first class criterion of

organization performance, obtained the highest average values of experts' evaluation at $8.\tilde{6}$ and $8.\tilde{2}$, near the excellent rank. It indicated that this SATI was effective in organization. This implied that the SATI centered in an innovative enterprise would be remarkable in organization performance.

(2) The average fuzzy values of experts' evaluation of the ten second class criteria that involved operation performance and resulting performance of the SATI were mostly arrived at $\tilde{7}$. This meant they were in good rank.

(3) The performances evaluated on criteria Ratio of encouraged R&D staff, and Number of completed innovational projects, were only above the medium rank with the average values at $6.\tilde{6}$. It suggested that this SATI should take much more encouraging approaches to the R&D staff and enhancing more accomplishing innovation projects further.

Finally, using the fuzzy measurement standards taking a reversal judgment on the fuzzy value of the synthetic performance evaluation calculated above ($\tilde{P} = 7.\tilde{3}$), the final judgment conclusion for this SATI was "Good". This judgment conclusion highlighted that the SATI of the Precious Metal Industry centered in Company G, had achieved rather satisfactorily performance after organized.

This case study had provided valuable guidance and reference for the SATI in Precious Metal Industry centered in Company G to promote further implementation. Moreover, it initially examined the practicability of the built fuzzy synthetic performance evaluation model.

125.5 Conclusions

Here, three conclusions were made in follows.

1. The successful application of the model in this case study indicated that the in all-round evaluation guiding thoughts, evaluating principles, and the evaluation criteria system including three sub-systems of organization performance, operation performance and resulting performance were reasonable. Also, it showed a good practicality of this model. The guiding thoughts, principles, the built model and the analysis results of the case study provided a useful tool and references to promote the SATI, especially the SATI centered in an innovative enterprise, to achieve better performances.

2. Moreover, the studied case implied that the SATI centered in an innovative enterprise would be remarkable in organization performance. This result may provide an important scientific assumption: the SATI centered in an innovative enterprise must be more effective than other SATI not centered in an innovative enterprise. If the assumption is scientifically proved, the result will become strong decision-making support to enhance the manager's confidence to organizing the SATI centered in an innovative rather than in other entities. So, more comparative case studies on the performance evaluation of the real SATI centered in or not in an innovative enterprise, should be performed in the future research.

3. Finally, this model would be suitable for the performance evaluation of the SATI in China that just is in the beginning stage. With the SATI entering a new stage of the process, the weights and the measurement standards of the model should be modified to suit the new situation.

Acknowledgments It is grateful for managers of the Strategic Alliance of Technological Innovation in Precious Metal Industry and the Company G, providing plentiful materials for our case study. Acknowledgement is made to National Natural Science Foundation of China (Grants No. 70862002 and No. 71262016) for support of this research.

References

1. Gao WG (2008) The development and highlights from international industrial strategic alliances of technological innovation. Rep Sci Technol Dev Res Shanghai Sci Dev Res Cent 12:1–8 (In Chinese)
2. Li GW, Li LL (2012) A review of the research on strategic alliance of industrial technology innovation. Sci Technol Prog Policy 11:156–160 (In Chinese)
3. Ministry of Science and Technology of the Peoples Republic of China (2009) Guiding opinions for promoting the construction of the Strategic Alliances of Technological Innovation. http://www.most.gov.cn/tztg/200902/t20090220?67550.htm (In Chinese)
4. Ming-hong S, Mo H (2013) Path selection mechanism of cluster innovation on strategic emerging industries. In: IEEE international conference on management science and engineering (ICMSE), pp 1962–1967
5. Mesiar R, Štěpnička M, Šostak A (2013) Fuzzy sets: Theory and applications. Fuzzy Sets Syst 232:1–2
6. Verbeke A, Tung V (2013) The future of stakeholder management theory: a temporal perspective. J Bus Ethics 112(3):529–543
7. Wu HX (2006) A Study on the performances of multinational corporations' strategic alliances. PhD thesis, Shandong University (In Chinese)
8. Xiao GL, Zhao ZG, Li F (2013) Impact of supporting units and the secretariats on strategic alliance for industrial technology innovation. Stud Dialectics Nat 2:54–58 (In Chinese)
9. Xiang G, Long J (2010) Evaluation innovative enterprise based on the sustainable innovation driving power, capability and benefits. Inq Into Econ Probl 12:122–125 (In Chinese)

Chapter 126
The Evaluation of College Students' Entrepreneurial Competency Based on PCA-ANN Model

Xiaofeng Li and Taojun Feng

Abstract Entrepreneurial competency is the key to decide the success or failure of College students' innovative undertaking. How to measure a college student' entrepreneurial competency is a long lasting difficult problem in the current educational field. In this paper, the index system for the evaluation of college students' entrepreneurial competency is constructed. Based on this, the PCA-ANN model for the evaluation of college students' entrepreneurial competency is established by combining PCA (principal component analysis) and ANN (artificial neural network) method, and an algorithm frame is designed based on PCA-ANN model. This method can not only reduces the dimensions of the data set and decreases the input variables of ANN but also avoids the influence of the coefficient between different indexes, accelerates the convergence and enhance the adaptability of the network. Theoretical analysis and experimental results show the feasibility and validity of the PCA-ANN model. The research work supplies a new way for evaluation of college students' entrepreneurial competency.

Keywords Entrepreneurial competency · Evaluation · Principal component analysis · Artificial neural network

126.1 Introduction

With the rapid development of our economy and the profound change of society, the higher education has entered the stage of popularization. The following with them is great changes of the college graduates' employment situation. In order to ease college students' employment pressure, the central and local government of China has implemented the strategy of increasing employment and promot-

X. Li (✉) · T. Feng
School of Business, Sichuan University, Chengdu 610064, People's Republic of China
e-mail: Lixiaofeng@scu.edu.cn

J. Xu et al. (eds.), *Proceedings of the Eighth International Conference on Management Science and Engineering Management*, Advances in Intelligent Systems and Computing 281, DOI: 10.1007/978-3-642-55122-2_126, © Springer-Verlag Berlin Heidelberg 2014

ing entrepreneurship, and introduced many policy measures which are helpful to college students' entrepreneurship. This has achieved a certain effect [16]. However, according to statistics, China's overall entrepreneurial success rate is 30 %, college students' entrepreneurial success rate is only 2 %, only holding ten percent of successfully entrepreneurial enterprises [8]. Investigating its reason, the shortage of the college students' entrepreneurial competence is an important factor. Evaluating the entrepreneurial competence of college students objectively and accurately not only helps students to know themselves correctly, grasp the opportunity of employment entrepreneurship, make appropriate choices, reduce the opportunity cost of entrepreneurial, also is helpful for colleges and universities to understand the status of the college students' entrepreneurship, and attaches great importance to the cultivation and development of college students' entrepreneurial quality. Therefore, how to combine China's national conditions and construct competence evaluation system of college students' entrepreneurial that is suitable for the social development, has very important theoretical significance and practical value.

At present, the research literature about college students' entrepreneurship is more, but less research about college students' entrepreneurial competence. So, how to scientifically measure college students' entrepreneurial competence has not yet had a set of effective methods [2, 10, 15]. This paper presents the PCA-ANN model of the college students' entrepreneurial competence evaluation by combining the principal component analysis (PCA) and artificial neural network (ANN). The model firstly decreases the dimensions of original sample data by the principal component analysis (PCA), and then selects the main components that retain primary information of original sample data as the input of neural network. So that it not only reduces the input dimension of network and decreases the size of the network training, but also deletes the redundant information, eliminates the correlation among the indexes, improves the efficiency of network training and the college students' entrepreneurial competence evaluation, simplifies complex problems of college students' entrepreneurial competence evaluation and has strong practical operability. At the end of the article, we prove the feasibility and effectiveness of this model through example.

126.2 Establish Evaluation Index System

College students' entrepreneurial competence is an ability for college students to identify and capture business opportunities, and make a creative project into practice so as to realize their self-development and the value of life. Evaluating college students' entrepreneurial competence needs to build an index system reflecting students' entrepreneurial competence. On the basis of this, the mathematical model is set up to evaluate college students' entrepreneurial competence quantitatively and qualitatively.

126.2.1 The Principles for Establishing the Evaluation Index System

There are many factors which influence college students' comprehensive quality and the relationships of them are complex. Therefore, In process of building college students' comprehensive quality evaluation system, we must follow these principles below.

1. Comprehensiveness principle. Evaluation index system should describe the connotation and feature of college students' entrepreneurial competence from different levels and points. So that it could guarantee the evaluation result accurately feeds back college students' entrepreneurial competence and entrepreneurial education's effect.
2. Guiding principle. Evaluation index system should reflect entrepreneurial education's basic requirement on college students, play a significant role in the promoting of college students' entrepreneurial traits and entrepreneurial abilities.
3. Level principle. We should set the index system according to progressive arrangement compose a united whole which has a clear, connected, reasonable structure. And lower levels' specific evaluation index should analyze and explain to upper levels.
4. Independence principle. Evaluation index should be relatively independent, that reflect one respect of student, there is no cross and similar phenomena involved between each index. And try to avoid duplication of information to take advantage of the evaluate effect of different indexes.
5. Practicability principle. We have to proceed from realities when designing the evaluation system, those evaluation indexes can obtain enough information in education practice, can quantified describe the state of evaluation object in education programmers, at the same time, try to simplify evaluation system.

126.2.2 The Way to Establish the Evaluation Index System

According to competency characteristic model [9], college students' entrepreneurial competence is divided into three categories: entrepreneurial potential competence, entrepreneurial skill competency, entrepreneurial knowledge competence. Then, according to the principle of establishing college students' entrepreneurial competence evaluation system, and reference relative literatures at home and abroad [1, 3, 12, 14], we gradually decompose first-class indexes, and initially build the framework of college students' entrepreneurial competence evaluation index system, which include 3 first-class indexes, 31 second-class indexes. Next, we design the questionnaire, asking interviewee (relative field experts, managers) to assess every evaluation index. Combined with expert opinions, the 11 third class indexes are canceled. At last, we build college students' entrepreneurial competence evaluation index system, which are shown in Table 126.1.

Table 126.1 The college students' entrepreneurial competence evaluation index system

First-class index	Second-class index
Entrepreneurial potential competence	Entrepreneurial motivation
	Entrepreneurial quality
	Self-diagnosis
	Values
	Previous experience
Entrepreneurial skill competency	Opportunity recognition ability
	Innovation ability
	Selflearning ability
	Interpersonal skills
	Team cooperation ability
	Strategic planning competency
	Group decision-making ability
	Resource integration capability
	Marketing ability
	Financial management ability
	Psychological adjustment ability
Entrepreneurial knowledge competence	Enterprise management knowledge
	Economic policy knowledge
	Industry background knowledge
	Relevant legal knowledge

126.3 PCA-ANN Model of College Students' Entrepreneurial Competence Evaluation

126.3.1 Overview of PCA Method

Principal Components Analysis, PCA, is proposed in 1901 by Pearson. By using the thought of dimension reduction, PCA converts multiple variables to a few variables, in which each principal component is a linear combination of the original variables and between the principal components are unrelated, thus most of these principal components can reflect the beginning variable information and inclusive information does not overlap each other [11]. With the advantages of geometric principle clear, simple calculation steps, strong commonality etc., PCA method has been widely applied to the fields of the data processing, process monitoring, fault classification and diagnosis etc., and achieves remarkable results.

Suppose there are n samples and each sample has p variables, constituting a $n \times p$ order data matrix,

$$U = \begin{bmatrix} u_{11} & u_{12} & \cdots & u_{1p} \\ u_{21} & u_{22} & \cdots & u_{2p} \\ \vdots & \vdots & \ddots & \vdots \\ u_{n1} & u_{n2} & \cdots & u_{np} \end{bmatrix} == \left(u_1, u_2, \cdots u_p \right), \tag{126.1}$$

where $u_j = (u_{1j}, u_{2j}, \cdots, u_{nj})^T$, $j = 1, 2, \cdots, p$.

Recording $u_1, u_2, \cdots, u_p$ are the original variable indicators, $z_1, z_2, z_3, \cdots, z_m$ ($m \le p$) are new variable indicators, then

$$\begin{cases} z_1 = l_{11}u_1 + l_{12}u_2 + \cdots + l_{1p}u_p \\ z_2 = l_{21}u_1 + l_{22}u_2 + \cdots + l_{2p}u_p \\ \cdots \\ z_m = l_{m1}u_1 + l_{m2}u_2 + \cdots + l_{mp}u_p. \end{cases} \tag{126.2}$$

Requiring the model below should meet the following conditions:

1. z_i and z_j ($i \ne j$; i or $j = 1, 2, \cdots, m$) are irrelevant to each other.
2. The variance of z_1 is greater than the variance of z_2 and the variance of z_3, and so on.
3. $l_{k1}^2 + l_{k2}^2 + \cdots + l_{kp}^2 = 1$, $k = 1, 2, \cdots m$.

So, we call z_1 the first principal component; call z_2 the second principal component, and so on. There is m principal component. Here l_{kp} is called coefficient of main component.

From the above analysis, we know that the essence of the principal component analysis is to determine the load l_{ij} ($i = 1, 2, \cdots, m$; $j = 1, 2, \cdots, p$) of the original variable indicators u_j ($j = 1, 2, \cdots, p$) in the principal component z_i ($i = 1, 2, \cdots, m$). We can prove that they are corresponding eigenvectors of relevant m larger eigenvalues. The detailed calculation steps of principal component analysis can be found in the literature [6].

126.3.2 BP Artificial Network Architecture and Algorithm

A standard back propagation neural network is shown in Fig. 126.1. The first layer consists of n input units. Each of the n input units is connect to each of the r units in the hidden layer. The r output units of the hidden layer are all connected to each of the m unit in the output layer.

BP is a supervised learning algorithm for multilayer networks [4]. The algorithm aims at minimizing the MSE between the actual output of the network and the desired output. Gradient descent search is user in BP. In BP learning, a set of patterns of the form $< x_1, \cdots, x_n, y_1, \cdots, y_m >$, where $x_1, \cdots, x_n$ are the components of the input vector and $y_1, \cdots, y_m$ are the components of the desired output vector, is repeatedly given to the network until the learning of weights converges [5].

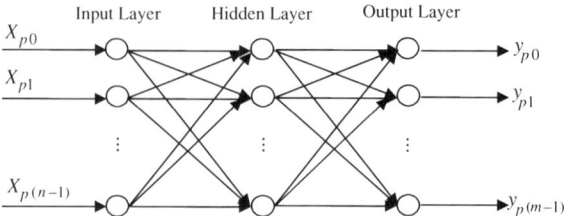

Fig. 126.1 A standard BP artificial neural network

If the BP neural network has N units in each layer, The transfer function is the sigmoid function, $f(x) = \frac{1}{1+e^{-x}}$, the training samples involve M different patterns (X_p, Y_p), $p = 1, 2, \cdots, M$. Corresponding the Input sample P, let net_{pj} represents the input total of unit j, let O_{pj} represents the output value, that is:

$$\text{net}_{pj} = \sum_{j=0}^{N} W_{ji} O_{pj}, \quad O_{pj} = f(\text{net}_{pj}). \tag{126.3}$$

The error between input values and output values is as following:

The revise connection weights of BP neural network are as following:

$$\delta_{pj} = \begin{bmatrix} f(\text{net}_{pj})(d_{pj} - O_{pj}), & \text{corresponding the output units} \\ f(\text{net}_{pj}) \sum \delta_{pk} W_{kj}, & \text{corresponding the input units,} \end{bmatrix} \tag{126.4}$$

where η represents the learning rate, it can increase convergence in speed, α represents the momentum coefficient. The value of α is a constant, it affects the connection weights of next step. Details of the traditional BP artificial neural network algorithm can be found in the original paper by Tian Jingwen and Gao Meiquan [13].

126.3.3 The Basic Principles of PCA-ANN Model Construction

From Table 126.1 we know that many factors affect college students' entrepreneurial competency. Correlation may exist between the data of these indexes, if regarding all of them as artificial neural network input variables will obviously increase the complexity of the network, reduce network performance, greatly increase the calculation of running time, and affect the accuracy of the calculation. PCA provides a good idea to solve this problem.

We can decrease the dimensions of the original sample data, remove the redundant information, simplify the neural network training set, and reduce the complexity and training time of neural network system by PCA. First compress second-class indicators of college students' entrepreneurial competency evaluation by using the method of PCA. Next, use the simplified second-class indicators as input variables of the BP network, and then study by the improved BP algorithm. The idea to do so is that, PCA can start from the relevance of data and found the data pattern to extract data rules, reduce data variable, but does not exist advantage in knowledge inference and prediction. However, neural network's earning ability, reasoning ability and classification ability are strong, also is good at extracting rules and information from large amounts of data, and has a good dynamic prediction function. Therefore, we would organically combine the two methods through learning from each other, in order to improve capabilities of artificial neural network to deal with complex issues, non-linear problems.

126.3.4 Design the PCA-ANN Model

During designing PCA-ANN model of college students' entrepreneurial competency evaluation, we firstly determine the comment level of college students' entrepreneurial competency. Then determine the assignment method of the second-class evaluation indexes. At last, build PCA-ANN model's three layers network structure.

(1) Confirm College Students' Entrepreneurial Competency Comment Level

Let $V = \{v_1, v_2, \cdots, v_n\}$ represent the evaluation levels of college students' entrepreneurial competency. In this paper, set $n = 5$, establish 5 kinds of evaluation levels. Comment set consisting of evaluation levels is $V = \{v_1, v_2, \cdots, v_5\} = \{excellent, fine, medium, ordinary, bad\}$.

(2) Assignment Method to Evaluation Indexes

In order to evaluate college students' entrepreneurial competency objectively and avoid one-sidedness, during quantizing the second-class indexes of college students' entrepreneurial competency evaluation index system; we should combine students' self-assessment, mutual evaluation and teachers' evaluation. At first, organize an assessment team and this group is made up of 3 representatives of classmates, 3 course teachers, 1 counselor, 1 head teacher and 3 relevant field experts. Then, ask the group members to assess every factor index x_k ($k = 1, 2, \cdots, 20$) of evaluation object' entrepreneurial ability. Evaluation standard is as follows.

Suppose one member considers the evaluation index x_i is "bad", then its score will be u_1 ($u_1 \in [0, 1]$); considers the evaluation index x_i is "ordinary", then its score will be vu_1 ($u_1 \in [0, 1]$); and so on.

Suppose the member a ($a = 1, 2, \cdots, 11$) gives vu_a score to index x_i ($i = 1, 2, \cdots, 20$), then the evaluation value xv_i of index x_i is:

$$xv_i = \frac{1}{11} \sum_{a=1}^{11} vu_a. \tag{126.5}$$

(3) PCA-ANN Model Structure

According to the traditional BP neural network structure, we divide PCA-ANN model for college students' entrepreneurial competence evaluation into three layers:

1. Input layer: Firstly, give values to second-class indicators of college students' entrepreneurial competence evaluation index system. Then, compress every the index data and reduce dimension by using the PCA method, extract the principal component as input variables of neural network.
2. Hidden layer: Refer to literature [7] that has mentioned-BP neural network dynamically adjusted learning algorithms. Firstly, set up hidden layer unites large, let network self-regulated learn, and finally get the right size of hidden layer unites.
3. Output layer: College students' entrepreneurial competency is divided to: excellent, fine, medium, ordinary, bad. So, in artificial neural networks, respectively, we use the output vector $(1, 0, 0, 0, 0), (0, 1, 0, 0, 0), (0, 0, 1, 0, 0), (0, 0, 0, 1, 0), (0, 0, 0, 0, 1)$ to describe. Therefore, the neural network's output layer unites are 5.

126.3.5 Basic Algorithm of PCA-ANN Model

Combine PCA with ANN method, we establish college students' entrepreneurial competency PCA-ANN model, the basic algorithm procedures are as follows:

Step 1 According to the college students' entrepreneurial competency comment set, combined with the assignment way of evaluation index give values to second-class indicators of college students' entrepreneurial competency evaluation; then, use PCA method to simplify the index data and extract the principal component as input variables of the neural network.

Step 2 Set neural network output layer unites as 5, and initialize other parameters of network (including a given study accuracy ε, the provisions of the iterative step number M_0, hidden units limit γ, learning parameters η. The initial hidden unites should be appropriate to take a large number).

Step 3 Enter the learning sample, make the sample parameter values into the [0, 1].

Step 4 Random values between $[-1, 1]$ are assigned to the initial weight matrix.

Step 5 Use dynamically adjusted algorithm to train BP neural network, in order to ensure weight matrix between each layers.

Step 6 Judge whether the number of iterations exceed the prescribed number of steps or meet the learning accuracy requirements or not. If yes, terminate the algorithm; if no, return to Step 5 and keep learning;

Step 7 Give values to the entrepreneurial competence evaluation second-class indexes of evaluated object, process the data and make them into [0, 1].

Step 8 Input processed data to the trained BP neural network and calculate the output.

Step 9 According to the output results, combined with college students' entrepreneurial competency evaluation set, make the evaluation of the object's entrepreneurial competency.

Table 126.2 Contribution rate of characteristic value and cumulative contribution rate

Principal component	Eigenvalue	Variance contribution rate (%)	Cumulative contribution rate (%)
1	5.236	36.352	36.352
2	3.267	17.126	53.478
3	2.016	10.875	64.353
4	1.455	9.425	73.778
5	1.23	8.689	82.467
6	0.757	5.632	88.099
7	0.336	4.236	92.335

126.4 Empirical Research

Regarding the class of 2010 undergraduate students in business School of Sichuan University as the research object, we use PCA-ANN model built in this paper to evaluate college students' entrepreneurial competency. We randomly select 37 students from 2010 undergraduate students in business school of Sichuan University, they are: student S_1, student S_2 student S_3, $\cdots$, student S_{37}. Previous 26 students are taken as a training sample of the PCA-ANN model, and the after 11 students as forecast sample.

126.4.1 Simplification of College Students' Entrepreneurial Competency Evaluation Index

Use the method which is introduced in Sect. 126.3.4 of this paper, we get the estimated values of entrepreneurial competency evaluation index x_i ($i = 1, 2, \cdots, 20$) of previous 26 students (student S_1, student S_2, student S_3, $\cdots$, student S_{26}). Then, use the principal component analysis function of SPSS 19.0 statistical software to analyze the sample data, get the correlation coefficient matrix and contribution rates of each characteristic root. As shown in Table 126.2. It can be seen from the table, the contribution rates of previous 7 principal components is up to 92.3 % and take them as input variables of the neural network.

126.4.2 PCA-ANN Model

According to the result of simplification of college students' entrepreneurial competency evaluation index, neural network use 7 input variables (i.e. input layer take 7 units) and the middle hidden layer take 15 units, the output layer for 5 units, the network structure is 7-15-5. Then, initialize the network (take the error limit $\varepsilon = 0.0002$,

Table 126.3 PCA-ANN network inference output

Student	Condition[a]	Sample output	Network inference output	Results[b]	Effect[c]
S_{27}	Fine	$(0, 1, 0, 0, 0)$	$(0.0016, 0.9986, 0.0029, 0.0168, -0.0032)$	Fine	True
S_{28}	Ordinary	$(0, 0, 0, 1, 0)$	$(0.0031, 0.0867, 0.0323, 1.001, 0.0223)$	Ordinary	True
S_{29}	Medium	$(0, 0, 1, 0, 0)$	$(0.0031, -0.0016, 0.9687, 0.0027, 0.0323)$	Medium	True
S_{30}	Excellent	$(1, 0, 0, 0, 0)$	$(0.9897, 0.0112, 0.0852, -0.0596, 0.0029)$	Excellent	True
S_{31}	Ordinary	$(0, 0, 0, 1, 0)$	$(0.0127, 0.0037, 0.0145, 0.9910, -0.0459)$	Ordinary	True
S_{32}	Excellent	$(1, 0, 0, 0, 0)$	$(0.9933, 0.0075, 0.0979, 0.0662, 0.0041)$	Excellent	True
S_{33}	Fine	$(0, 1, 0, 0, 0)$	$(0.0381, 0.9035, 0.0141, 0.1238, 0.0091)$	Fine	True
S_{34}	Bad	$(0, 0, 0, 0, 1)$	$(0.0376, 0.0452, 0.0285, 0.0065, 0.9768)$	Bad	True
S_{35}	Excellent	$(1, 0, 0, 0, 0)$	$(0.7987, 0.9325, 0.0257, 0.0369, 0.0762)$	Fine	False
S_{36}	Bad	$(0, 0, 0, 0, 1)$	$(0.3461, 0.2567, 0.0082, -0.0397, 0.9901)$	Bad	True
S_{37}	Medium	$(0, 0, 1, 0, 0)$	$(0.0651, -0.1233, 0.9826, 0.0752, 0.0629)$	Medium	True

[a] Entrepreneurial competency condition
[b] Network evaluation result
[c] Network prediction effect

learning rate $\eta = 0.5$, iteration steps $M_0 = 20000$). Next, convert principal component analysis of each indicator evaluation data of 26 students (student S_1, student S_2, student S_3, $\cdots$, student S_{26}) to [0, 1] (each estimated value should be divided by 10). Then input processed data as the study sample data to the neural network, train the network by the improved BP algorithm, the network structure is automatically adjusted to 7-9-5 (7 input layer units, 9 hidden layer units, 5 output layer units) after training, at the same time we get optimize network weights matrix. With the trained neural network, we evaluate the entrepreneurial competency of these students (S_{27}, S_{28}, $\cdots$, S_{37}) of 2010 in business school of Sichuan University. The inference results (output) of the network are shown in Table 126.3.

126.4.3 Compared with the Traditional BP Neural Network

Now, we use the traditional BP neural network to establish evaluation model of college students' entrepreneurial competence. 20 factors affect college students' entrepreneurial competency so that the structure of BP network is 20-33-5 (20 input layer units, 33 hidden layer units, 5 output layer units); take the error limit $\varepsilon = 0.0002$, learning rate $\eta = 0.5$, iteration steps $M_0 = 20000$. Next, convert principal component analysis of each indicator evaluation data of 26 students (student S_1, student S_2, student S_3, $\cdots$, student S_{26}) to [0, 1] (each estimated value was be divided by 10). Then input processed data as the study sample data to the neural network, train the network by the improved BP algorithm, the network structure is automatically adjusted to 20-27-5 (20 input layer units, 27 hidden layer units, 5 output layer units) after training, at the same time we get optimize network weights matrix. With the trained neural network, we evaluate the entrepreneurial competency of the students

Table 126.4 Traditional BP neural network inference output

Student	Condition[a]	Sample output	Network inference output	Results[b]	Effect[c]
S_{27}	Fine	(0, 1, 0, 0, 0)	(0.0067, 0.9827, 0.0257, 0.0169, 0.0023)	Fine	True
S_{28}	Ordinary	(0, 0, 0, 1, 0)	(0.042, 0.2452, 0.0441, 0.9673, 0.0062)	Ordinary	True
S_{29}	Medium	(0, 0, 1, 0, 0)	(0.0062, −0.0091, 0.8997, 1.0001, 0.0787)	Ordinary	False
S_{30}	Excellent	(1, 0, 0, 0, 0)	(0.9916, 0.0229, 0.0862, 0.0959, 0.0046)	Excellent	True
S_{31}	Ordinary	(0, 0, 0, 1, 0)	(0.0075, 0.0125, 0.0310, 0.6899, 0.9052)	Bad	False
S_{32}	Excellent	(1, 0, 0, 0, 0)	(0.9910, 0.0076, 0.0925, 0.0265, 0.0121)	Excellent	True
S_{33}	Fine	(0, 1, 0, 0, 0)	(0.0721, 0.9592, 0.0991, 0.0874, 0.0672)	Fine	True
S_{34}	Bad	(0, 0, 0, 0, 1)	(0.0361, 0.0762, 0.0326, 0.0263, 0.9907)	Bad	True
S_{35}	Excellent	(1, 0, 0, 0, 0)	(0.9098, 0.0925, 0.0826, 0.0763, 0.0572)	Fine	False
S_{36}	Bad	(0, 0, 0, 0, 1)	(0.6465, 0.4637, 0.0151, 0.0332, 1.0023)	Bad	True
S_{37}	Medium	(0, 0, 1, 0, 0)	(0.0903, 0.3613, 0.9907, 0.0768, 0.0585)	Medium	True

[a] Entrepreneurial competency condition
[b] Network evaluation result
[c] Network prediction effect

$(S_{27}, S_{28}, \cdots, S_{37})$ of 2010 in business school of Sichuan University. The inference results (output) of the network are shown in Table 126.4.

It can be seen from Tables 126.3 and 126.4, the prediction accuracy of BP neural network is 72.7 %, but PCA-BP model has higher prediction accuracy with 90.9 %. When using PCA-BP model to forecast, input variables of neural network are simplified, the size of network becomes smaller and the needed training time is the shorter than he traditional BP neural network, so as to improve network's operation efficiency and prediction ability.

126.5 Conclusion

There are many factors influencing college students' entrepreneurial competence, and there factors influence each other, information overlapping. So it is difficult to use traditional statistical model to evaluate entrepreneurial competence. In this paper, the evaluation model of college students' entrepreneurial competence based on PCA and ANN method is provided. This model not only can acquire the major feature attributes of college students' entrepreneurial competence, but also can cancel redundancy information and reduce the input variables of neural network. At the same time, it reduces neural network's complicacy and train time, improves neural network learning ability, reasoning ability and classification ability and achieves a dynamic evaluation of college students' entrepreneurial competence. Therefore, the model is scientific and effective, it has considerable practical value.

Acknowledgments This research was supported by the education discipline of National Natural Science Foundation for Young Scientists, China, No. CIA110139.

References

1. Bing G, Hua Y (2009) Comprehensive evaluation model of entrepreneurial team's entrepreneurial competency research in college students' entrepreneurship competition. J Northeast Normal Univ Philos Soc Sci Ed 6:224–227 (In Chinese)
2. Chen W (2012) College students' comprehensive quality state analyze and evaluation research. Jiangsu College, Zhenjiang (In Chinese)
3. Chen Z, Li J (2011) The establishment of college students' entrepreneurial competency mode. Econ Res Guide 16:292–295
4. Ergezinger S, Tomsen E (2001) An accelerated algorithm for multiplayer perceptions optimization layer by laye. IEEE Trans Neural Networks 06:31–42
5. Han L (2002) Theory, design and application about artificial neural network. Chemical Industry Press, Beijing
6. Jolliffe I (1986) Principal component analysis. Springer, New York
7. Li X, Xu J (2004) Improvement the model of dynamically adjusted algorithm train bp neural network. Chin Manage Sci 12(6):68–72 (In Chinese)
8. Liao J (2013) International experience of college students' starting an undertaking and the inspiration to china. Reformation Strategy 29(4):113–116
9. Liu Z (2013) Competency establishment model—Personnel selection and assessment. Science Press, Beijing
10. Lu J, Chen J (2009) An overview of the studies of the status and the thinking of college students' entrepreneurial competence. Educ Manage 7:50–51
11. Pearson K (1901) On lines and planes of closest fit to systems of points in space. Philos Mag 2(6):559–572
12. Song D, Fu B, Tang H (2011) Establishment of entrepreneurial university's entrepreneurial competency evaluation index system. Sci Technol Prog Countermeasures 28(9):116–119
13. Tian J, Gao M (2006) Research and application about artificial neural network algorithm. Beijing Institute of Technology Press, Beijing
14. Zhang Y (2009) Thinking about evaluation system of completing college students' entrepreneurial competence. Sci Technol Inf 15:141–142
15. Zhao S, Wang D (2009) Reflect on college students' comprehensive quality evaluation system. Chang Chun Technician Acad J Soc Sci Version 4:624–626 (In Chinese)
16. Zhu X (2013) An overview of the studies of the status and problems of china's job market for college graduates. J High Educ Manage 7(5):121–124

Chapter 127
Self-Interested Behaviors of Managers About Selective Environmental Information Disclosure and Influencing Factors Researching

Jing Xu

Abstract For the past few years, all sectors of the society are keeping eyes on environmental governance and disclosure, and also a lot of controversy paid more attention to self-interested behaviors of managers in environmental information disclosure field at home and aboard. In addition, the paper discusses the self-interested behaviors of managers and characters embedding on environmental information disclosure places, contents and standards and analyses the related factors affecting self-interested level of managers. The study finds: (1) No matter what the disclosure places, contents and standards are, there are obvious self-interested behaviors of managers in many aspects and the non state-owned listed firms are more significant than the state-owned listed firms. (2) From the angle of corporate governance, both CEO duality and managers' shareholdings in non owned-stated listed firms and institutional investors' shareholdings in owned-stated listed firms will constrict managers' self-interested behaviors. (3) In other related influence factors, firms scale, financial level and cash flow affect self-interested behavior of managers in selective environmental information disclosure.

Keywords Environmental information · Selective disclosure · Self-interested behaviors

127.1 Introduction

The rapid economic development brings out the increasing environmental pollution and ecological crisis. In that case, people begin to rethink the relationship between industrial developments and environmental protection [1]. The sustainable development theory expects the enterprises must protect environment while pursuing their

J. Xu (✉)
Business School, Sichuan University, Chengdu 610064, People's Republic of China
e-mail: xujing@scu.edu.cn

J. Xu et al. (eds.), *Proceedings of the Eighth International Conference on Management Science and Engineering Management*, Advances in Intelligent Systems and Computing 281, DOI: 10.1007/978-3-642-55122-2_127, © Springer-Verlag Berlin Heidelberg 2014

own economical interests [2]. The adequate disclosure is not the only prerequisite of environmental protection. It's also the important tools of supervision of business environmental activities and environmental governance. Also, through the publication, enterprises could maintain their social image.

At present, we couldn't be optimistic to the quality of environmental information disclosure in China. Selective disclosure, question obscureness and other phenomena still exist. Based on the agent theory, owning to the contradiction of shareholders and management, it is possible that the managers would sacrifice interest of shareholders to earn their own. They may choose to show some information selectively and have self-interested behaviors because the reasons that reduction of environmental payment, increment of economical performance and evasion from public supervision. These kinds of activities will impact the rational estimation on listed companies from external investors, and also effect the sustainable development in the future.

Therefore, under the background of voluntary disclosure, do the managers have the motivation and capacity to influence the disclosure pattern and content of environmental information by their authority? Otherwise, if the information has been revealed, are these kinds of information easy to control, or be beneficial to the managers? If lucky enough, do the promotion of management decease the possibility of self-interested behaviors? Based on the discussion of questions above, we do the empirical research on firms in heavy polluted industry of SSE during 2009–2011. Through the model design to evaluate the different situations of environmental information disclosure in several companies, we observe whether there are selective disclosure, self-interested behaviors and the factors that influence the extent of self-interest.

The rest of article is organized as follows. Section 127.2 discusses the theory and formally develops our hypothesis. Section 127.3 describes our research design. Section 127.4 presents the results and analyses their implications. Section 127.5 concludes the paper and gives suggestions.

127.2 Theoretical Analysis and Hypothesis Construction

In fact, since the construction of modern corporation system, the contradiction between managers and shareholders never disappear. They have different interest preference so that the managers use power to chase their personal gain and vise versa. So, their self-interested behaviors not only show in compensation decision, over-investment and duty consumption, they will strongly impact the channels and content of environmental information disclosure. By reason of voluntary information disclosure is revealed according to their self willing and demand [3], it is the important way to raise the transparent level of environmental information and increase the external acceptance on firms' value [4]. The stakeholders are the information receivers. But the managers make the decisions. Therefore, the motivation and quality of environmental information disclosure are affected by their self-interested behaviors directly [5]. On one hand, managers could use their power to change the content of environmental report.

On the other hand, managers could use their power to change the channels and methods of environmental information disclosure. Besides, the performance assessment system for the managers also could affect the quality of information disclosure. We believe, based on the interest contradiction between shareholders and stakeholders, the managers who are selfish have the motivation and authority are willing to influence the content and channels of environmental information disclosure. In this situation, we propose the hypothesis 1:

Hypothesis 1. There are self-interested behaviors of selective disclosure in the environmental information publish.

In terms of disclosure of environmental information for executives selfish behaviors, Cormier and Gordon [6] used electric power enterprises in Canada as the research samples, because nature of enterprises are different, they take various social responsibility. There is prominent gap of information disclosure among them. Compared to the private enterprises, state-owned companies have higher quality of environmental information disclosure.

On the opposite, through the study on the environmental information of IPO around American listed companies, Moreover, there is conflict on the question of strategic committee. Hacksotn and Milne [7] thought the board with strategic committee take more consideration on environmental responsibility and have higher quality of environmental information. However, Cowen et al. [8] found there is no noticeable relationship between strategic committee and level of environmental information.

Based on these theories, we propose the hypothesis 2 in this paper:

Hypothesis 2. If the level of management go up, the extent level of selfish activities will go down.

Indeed, the extent of self-interest will reflect the difference depend on the financial situation, the size of enterprise, and social pressure. Cormier and Gordon [6] thought the financial situation is the key determinant in the decision. Liabilities level has a negative correlation with level of information disclosure. Based on the analysis above, we propose hypothesis 3.

Hypothesis 3. because the reason of difference of financial situation, firm's size and etc, the extent of behaviors of managers' selective disclosure will be different.

127.3 Sample Data and Research Design

1. Sample and Data Resource

In May 2008, Shanghai stock exchange (SSE) issued new regulation <Environmental information disclosure guide of listed companies in SSE>. This regulation encourages more normalization and standardization around listed companies information disclosure. Therefore, we choose some SSE listed companies in heavy pollution industry as the sample and filter on these conditions.

(1) Eliminate the new listed companies after 2008.
(2) Eliminate the ST companies affected by extreme factors.
(3) Eliminate the companies that have a change on CEO position.
(4) Eliminate the companies that have data deficient.

Finally, we get 110 companies and 440 samples. For the record, heavy pollution industry includes 9 types of firms. They are: excavation (B); food, beverages (C0); textiles, clothing, fur (C1); wood, furniture (C2); paper, printing (C3); petroleum, chemical, plastic (C4); metals, non-metal (C6); pharmaceuticals, biological products (C8); electricity, gas and water production and supply industry (D). All the data come from CSMAR database, Wind database and annual reports of the companies.

2. Descriptive Variables

(1) Explained Variables
   We use the related indexes in environmental reports of sample companies to reflect the characters of managers' selective disclosure behaviors. Specifically, they are channels of disclosure, quality of disclosure, standard of disclosure, composite scores. Besides, we will use these methods to test hypothesis 1 that whether the managers have the selfish behaviors. On one side, from perspective of channels of disclosure, quality of disclosure, standard of disclosure, if the channels are "few" rather than "many", if the quality is "qualitative and simple" rather than "quantitative and detailed", if the standard are "low" rather than "high". We can announce that self-interested behaviors are existed. On the other hand, according to the three perspectives, we can construct a comprehensive score index and analyze all the average scores of each companies. If the composite scores are obvious lower than theoretic point, we can estimate there are self-interested behaviors.

(2) Explanatory Variables
   The factors of selective disclosure selfish behaviors of environmental information include corporate governance structure, financial situation, external pressure and other aspects of corporate identity. Inside, corporate governance structure consists of 6 variables. They are CEO duality, MBO, share proportion of institutional investors, share proportion of largest shareholder, nature of stock rights and proportion of independent directors in the board. Other factors include corporate profitability, cash flow, financial leverage, size of company and media attention. The specific definitions are in the Table 127.1 below.

(3) Model Design
   Based on the analyses and related researches above, this paper construct these regression models to verify the effect from corporate governance, financial situation and company characters on managers' self-interested behaviors.

$$Prob(\text{STA}) = \beta_0 + \beta_1 * \text{GOV}_i + \beta_2 * \text{FIN}_i + \beta_3 * \text{SCA} + \beta_4 * \text{MED} + \varepsilon.$$

STA is dummy variable. It gets 1 point if it's higher than composite score of listed companies, otherwise 0. GOV_i is the i-th governance variable of listed companies, including previous six variables. FIN_i represents the i-th listed company's financial position variable. SCA is scale variable and MED is media attention variable.

Table 127.1 Definition of explanatory variables

Sorts of variables	Description of variables	Sign	Definition
Governance structure (GOV)	CEO duality	AUT	It gets 1 point when the CEO is also director of board, otherwise, 0 point
	MBO	MAN	It gets 1 point when the managers hold shares, otherwise, 0 point
	Institutional investors hold shares	INS	Proportion of institutional investors' share-holding/total equity
	Share proportion of largest shareholder	FIR	Percentage of largest investors' share-holding/total equity
	Nature of share rights	EQU	If the controller is government or state-owned enterprise, the variable take the value of 1, otherwise 0
	Proportion of independent shareholders	IND	Proportion of independent directors of the board
Financial situation (FIN)	Return on equity	ROE	Retained profits/the remaining shareholders' equity
	Cash flow	CAS	Cash flow/total asset
	Leverage	LEV	Debt/asset
Firm's characteristic	Size of company	SCA	Logarithm on total asset
Media attention	Media attention	MED	China securities journal, Shanghai securities journal and other online financial medias' reports about sample companies

127.4 Empirical Results and Discussion

1. Characteristic Analysis on Managers' Selective Disclosure and Self-interested Behaviors of Environmental Information

(1) Analysis on Publish Channel

From the amount of media channels, there are fewer channels, more information-hidden phenomenon will appear. So, it's hard to supervise and chase the responsibility for public and media. In Table 127.2, the analytical results on annual reports, social responsible reports, prospectus and environmental reports show that there are 22.7 % of listed companies have shown nothing on environment among these channels. Then, there are 40 companies have one report, which account for 36.4 % in total sample companies. In these firms, 19 of them publish in prospectus, 17 of them publish in social responsibility reports and only two firms present in annual reports and environmental reports respectively. After that, 34 of total sample companies publish two reports, which take over 30.9 % in total. Compared to that, those companies have three reports are 11, which only occupy 10 % in total. In addition, many of

Table 127.2 Analytical results of media channels

Number of reports	Number of company	Proportion (%)
0	25	22.7
1	40	36.4
2	34	30.9
3	11	10
Total	110	100

Table 127.3 Analysis on quality of environmental information disclosure

Quality of disclosure	Numbers of companies	Proportion (%)
Excellent	10	9.1
Good	15	13.6
Normal	57	51.8
Poor	28	25.5
Total	110	100

them choose to represent environmental information on other reports rather than environmental ones. There are just seven companies reveal their information on environmental reports. It implies that there are self-interested behaviors during the decision process of media channels.

(2) Analysis on Disclosure Quality

We have research on the quality of disclosure in social responsibility reports. For the score system, we reckon those companies who both have qualitative and quantitive reports as "excellent", reckon those companies who both have qualitative and quantitive reports but not specific as "good", mark those companies who only have qualitative description as "normal" and mark those companies who have nothing as "poor". In the Table 127.3, there are only ten firms got "excellent", which occupy 9.1 % in total. Those who got "normal" are as many as 57 that accounts for 51.8 %. There are 25.5 % of sample companies got "poor". The data above means managers try to reduce the channels to reveal less information and they obscure the content of reports. When they have to choose between presentations or hiding, many of them choose the simple, qualitative and unspecific ways. Therefore, the self-interested behaviors are remarkably existed.

(3) Analysis on Standard of Disclosure

Based on the <Sustainable development report: 3rd version> of Global Reporting Initiative (GRI), we use these performance index to evaluate standard of disclosure. Results of Table 127.4 show that 28 of them have shown nothing, which agree with Table 127.3. In the 82 companies who present something, there are 16 firms have shown the environmental information in mainbody of reports, appendixes and reports of third party. They account for 14.5 % in total. In the second place, 23 companies have revealed in the mainbody of reports and appendix. They account for 20.9 % in total. Only 39.1 % of sample companies have presented in the mainbody. The number is 43. All the results above show that nearly half of enterprises' managers

Table 127.4 Analysis on the standard of environmental information disclosure in listed companies

Standard of disclosure	Numbers of companies	Proportion (%)
Disclosure on the mainbody, appendix and reports of third party	16	14.5
Disclosure on the mainbody, appendix	23	20.9
Disclosure on the mainbody	43	39.1
Non-disclosure	28	25.5
Total	110	100

Table 127.5 Composite score system of sample companies

Channel of disclosure		Quality of disclosure		Standard of disclosure	
Quantity	Score	Quality	Score	Standard	Score
0	0	Poor	0	Nothing present	0
1	1	Normal	1	Disclosure on the mainbody	1
2	2	Good	2	Disclosure on the mainbody, appendix	2
3	3	Excellent	3	Disclosure on the mainbody, appendix and reports of third party	3

still choose the lowest standard to publish their environmental information ignoring the guide of policy and media. Therefore, we can announce that there are prominent self-interested behaviors in terms of standard of disclosure.

(4) Analysis on Composite Score System

For better judgment and evaluation to the level of managers' selective disclosure, we mark channels of disclosure, standard of disclosure and quality of disclosure separately. And we use composite score to reflect the level of selective disclosure. If the composite score are low, it means the implications of selective disclosure are existed. Table 127.5 displays all the situation of composite score. Third point is the highest points on each term. The lower point indicates that clearer existence of selective disclosure is.

We can see from the Table 127.6, the averages of every item are all less than 2. The average of channels of disclosure is 1.31, which indicates that listed companies publish their reports but control the revealed range. The average of quality of disclosure is 1.06 and it reaches the "normal" exactly. That implies all managers know the vague reports will be susceptible to attach the attention from the supervision. However, the specific reports will lead to external investors doubts. The average of standard of disclosure is 1.65 that is the highest. It means the managers will build a framework of disclosure accordance with the relevant provisions strictly. However managers are unlikely to update and improve. As a result, majority of them refuse the appraisable requests from the third party. On the other hand, the average of composite score is 4.02. It is undoubted that the score is absolute low. This is a further evidence of the self-interested behaviors in the selective disclosure.

Table 127.6 Composite score system of sample companies

	Minimum	Maximum	Average	Variance
Channels of disclosure	0	3	1.31	0.936
Quality of disclosure	0	3	1.06	0.87
Standard of disclosure	0	3	1.65	1.01
Composite score	0	8	4.02	2.563

Table 127.7 Results of T test on groups of the composite score

Variables	STA (high group)	STA (low group)	T test
AUT	0.12	0.27	−2.073**
MAN	0.29	0.19	1.309
INS	10.93	11.95	−0.628
FIR	37.35	31.77	1.776*
EQU	0.71	0.39	3.479***
IND	37.2	35.9	1.114
ROE	7.89	11.7	−0.501
CAS	0.02	0.01	0.892
Lev	0.49	0.46	0.824
SCA	23.01	21.13	3.035***
MED	461	367	1.836*

* Significant at 10 % level
** at 5 % level
*** at 1 % level

2. The Analysis on the Influential Factors of Selective Disclosure and Self-interested Behaviors (1) Grouping analysis in the terms of composite score

We use the average point 4.02 as the standard, whose composite score below the standard will be assigned to group of low points. These managers of companies have more self-interested behaviors. In the meantime, we will assign the companies whose score is higher than the average to the group of high points. These managers of companies have less self-interested behaviors. In terms of corporate governance, financial situation, characters of enterprises and etc., we compare the difference in the two groups. We can see from the Table 127.7, the variables of CEO duality, share proportion of largest shareholders, nature of equity, size of companies and media attention are significantly different. Among these data, index of CEO duality with the group of high points are substantially lower than the counterpart with the group of low points on the level of 5 %. The indexes of share proportion of largest shareholder are remarkably on the high side on the level of 10 %. Furthermore, in terms of the numbers and degree of media attention, the index of group of high points is significantly higher than the index of group of low points on the level of 10 %.

(2) Regression Analysis

Regression Analysis on the Whole Sample: To find the effects from each factor on managers' selective disclosure behaviors, we define the average of composite score as the standard. If the STA is more than the mean point, it gets 1 point, otherwise 0. And also, we use the binomial Logistic analysis to estimate and examine our models.

Table 127.8 Regression analysis on whole samples

Variables	Regressor of corporate governance variable	Regressor of other variables	Regressor of all variables
AUT	−1.174*** 3.852	–	−0.945** 2.889
MAN	1.044** (3.609)	–	0.976** (3.125)
INS	−0.003 (0.015)	–	−0.017 (0.383)
FIR	0.012 (0.684)	–	0.007 (0.19)
EQU	1.227*** (7.333)	–	0.902** (3.352)
IND	0.055 (1.762)	–	0.059 (2.012)
ROE	–	−0.001 (0.869)	−0.001 (0.006)
CAS	–	0.04 (0.354)	0.075 (2.293)
LEV	–	−0.758* (2.368)	−0.65* (2.877)
SCA	–	0.658*** (10.393)	0.504** (5.333)
MED	–	0.000 (0.724)	−0.001 (0.343)
Constant term	−3.224** (3.977)	−14.633*** (11.611)	13.895*** (8.673)
−2 Log liklihood	131.544	130.871	119.654
Nagelkerke R2	0.226	0.233	0.339

It can be seen from chart 10, on one hand, the situation of corporate governance has great influence on managers' self-interested behaviors and selective disclosure. First, the index of CEO duality has a negative correlation with index of composite score on the level of 5%. It means the CEO duality pose a threat to the quality of disclosure. The duty of chairman of the board is decision and supervision. If the manager has two positions, it indicates that he/she is both executive and the supervisor. This situation will definitely weaken the power of board and reduce the function of agent cost, which is why there is a negative correlation. Sora [9] and Forker [10] also explained the theory. Second, the index of MBO has a positive correlation with index of composite score on the level of 5%. It's because that MBO can combine the interests of managers and other shareholders. Therefore, for the common benefit, the managers are more willing to present environmental information and raise the share price. In this case, managers will care more about the external reputation, social responsibility and the sustainable influence they make to the future development. Third, the data shows the index of nature of companies has a positive correlation with index of composite score on the level of 1%. It means that the state-owned companies' environmental reports are better than those in the private enterprises. State-owned companies naturally have more social responsibility than private enterprises. In the aspects of environmental governance and disclosure, according to the Signaling Model of Spence [11], state-owned companies are more prone to send the signal actively for delivering the environmental information. At last, indexes of institutional investors, proportion of largest shareholder and proportion of independent directors don't have a correlation with index of composite score, which means they hardly have a prominent function on disclosure governance of environmental information.

On the other hand, other characters in companies could remarkably impact the level of environmental information disclosure. We find the financial leverage has a negative correlation with composite score on the level of 10 %. It means if firm has more debt, it will have more selective disclosure and managers' self-interested behaviors. The higher leverage indicates more financial risk. There is not spare cash flow to improve environmental performance. Moreover, company size has a positive correlation with composite score on the level of 1 %. That is because bigger companies have more media attention and "butterfly effect" so that they are more cautious about environmental information disclosure. Other factors such as ROE, cash flow and media attention don't have a great influence on composite score among the whole sample companies. To test the results, we group the samples into "state-owned" group and "private" group. We testify the two independent sample groups and the results are shown in Table 127.8.

127.5 Research Conclusion and Limitation

In recent years, managers' self-interested behaviors had been being the hot research area in academia. This paper studies about the managers' self-interested behaviors during environmental information disclosure and do the empirical research with the samples of 2009–2011 heavy polluted listed companies in SSE. From the perspectives of channels of disclosure, quality of disclosure standard of disclosure and the composite score, we deeply studied that if there are managers' selective disclosure and self-interested behaviors. Furthermore, we explore the main factors which affect the self-interested behaviors. The results show that:

1. Managers have obvious self-interested behaviors during the process of environmental information disclosure. There are some phenomena such as few channels of disclosure, the quality of disclosure is qualitative and simple description and inappropriate standard of disclosure.
2. In terms of corporate governance, except the situations that MBO and separated positions of chairman of board and CEO can constrain the managers' self-interested behaviors, other factors hardly have the supervisory and controllable function. Now, the corporate governance structure has limited restraint impact on self-interested behaviors among managers. However, in private companies, the institutional investors are inclined to earn the profit from the share prize fluctuation and they are detrimental to the environmental information disclosure. At that time, they are called transactional institutional investors.

One side, sectors of supervisory should guide and regulate that listed companies should use more quantitive indexes under the voluntary disclosure principles. More importantly, set the minimum standards on channels of disclosure and standards of disclosure. On the other side, improving the function of corporate governance mechanism can promote the supervisory function in the self-interested behaviors.

That kind of activities will restrict and reduce the serious self-interested behaviors and selective disclosure which come from the higher and higher managers' authority.

References

1. Wang X, Xu X, Wang C (2013) Public pressure, social reputation, internal governance and environmental information disclosure. Nankai Bus Rev 2(16):82–91
2. Zhen C, Xiang C (2013) Empirical research on the environmental information disclosure among Chinese listed companies-based on the 170 listed companies in SSE. Sci Technol Prog Policy 6:98–102
3. Meek GK, Roberts CB, Gray SJ (1995) Factors influencing voluntary annual report disclousures by U.S., U.K. and continental european multinational corporations. J Int Bus Stud 26(3):555–572
4. Luo W, Zhu C (1999) Agent cost and voluntary disclosure. Economical Res J 26(3–4):883–917
5. Aboody D, Kasznic R (2000) Stock option awards and the timing of corporate voluntary disclosures. J Acc Econ 29(1):73–100
6. Cormier D, Gordon IM (2001) An examination of social and environmental reporting strategies. Acc Auditing Accountability J 5(14):587–617
7. Hackston D, Milne MJ (1996) Some determinants of social environmental disclosure in New Zealand companies. Acc Auditing Accountability J 1(9):77–108
8. Cowen SS, Ferreri LB, Parker LD (1987) The impact of corporate characteristics on social responsibility disclosure: a typology and frequency-based analysis. Acc Organ Soc 2(12):111–122
9. Sora SA (2004) The ethical dilemma of merging the roles of CEO and chairman of the board. Corp Gov 4(2):64–69
10. Forker JJ (1992) Corporate governance and disclosure quality. Acc Bus Res 86:111–124
11. Spence A (1979) Investment strategy and growth in a new market. J Econ 10(1):1–19

Chapter 128
In Search of Entrepreneurial Opportunities in Digital Economy

Jian Pang, Xinmin Zhu and Jia Liu

Abstract Digital economy has been considered a dominant way to sustain economic growth in developed countries for recent years. However in this field there is a huge gap between technology accumulation and technology application, i.e. technology application far falling short of accumulation. It is this huge gap that creates a lot of entrepreneurial opportunities for Chinese firms. Generally speaking, technology applications firstly depend on discovery of social problems as well as design of their solutions, thus most of entrepreneurial opportunities are social-problem oriented in essence. The authors, taking these tendencies into account in this article, deeply analyze the definition, characteristics and identification of entrepreneurial opportunities in digital economy, compare US Apple Corporate and China Changhong Corporate, and stress necessity of shifting the focus of innovation from technology orientation to a social-problem one. So in the huge markets of China, social problems should always be aimed at first, and customer demands found second, and solutions designed third, and finally key resources integrated for development of these solutions. It is believed that in digital economy this social-problem oriented innovation process can provide more entrepreneurial opportunities as well as more innovative products and services for innovators to construct dominant application paradigms, so to win battles of new technology commercialization.

Keywords Digital economy · Entrepreneurial opportunity · Social-problem orientation · Application paradigm

J. Pang (✉) · X. Zhu · J. Liu
Business School, Sichuan University, Chengdu 610064, People's Republic of China
e-mail: 281880351@qq.com

J. Xu et al. (eds.), *Proceedings of the Eighth International Conference on Management Science and Engineering Management*, Advances in Intelligent Systems and Computing 281, DOI: 10.1007/978-3-642-55122-2_128, © Springer-Verlag Berlin Heidelberg 2014

128.1 Introduction

Identifying and selecting right commercial opportunities is one of the most important ability of a successful entrepreneur [10]. Entrepreneurship is to discover, evaluate and exploit a profitable opportunity. Thus, the entrepreneurship research focuses on discovery and exploitation of opportunities in recent years [13]. Entrepreneurial opportunity is a key concept of entrepreneurship research [3], but the understandings of the nature and definition of entrepreneurship are various. Many scholars do a series of studies on opportunities from the respects of discovery, creation, social cognition and synergy, etc. [8]. Casson considers entrepreneurial opportunity is a status in which new product, service, raw material and organization method are sold with higher price than the production cost [4]. Venkataraman proposes that entrepreneurial opportunities are generated by the interaction of a series of thoughts, concepts and actions, create products and services which do not exist in the current market. He also proposes that two key elements of successful entrepreneurship are the presence of lucrative opportunities and enter prising individuals [13]. Based on Schumpeter innovation theory, Shane et al. offer that opportunities can create new products, new services, new raw materials, new markets and new organization methods via new ways, new results and causal relationships [7]. Short et al. comprehensively considered the opportunities from the angels of discovery and creation theory of opportunities, dynamic environment of opportunities and transformation from thoughts and dreams to opportunities, then defined that opportunities are the thoughts or dreams discoveredor created by firms, which reveal latent profit is existing as time going by analysis [8]. The opportunities are generated from information asymmetry, the external shocks, the change of market supply and demand relations, promotion of production efficiency and informal economy [5].

Plenty of researches on entrepreneurial opportunity have been presented in recent years. These researches focus more on the traditional economics condition than on the emerging information technology, communication technology and digital technology. The current rapid development of information technology, communication technology, and Internet technology is a typical representative, which brings huge external shock to traditional economy and market [9]. These shocks introduce a revolutionary change in many aspects, such as supply and demand relation, production efficiency, resource allocation in the traditional industry and market. As the development of information technology, communication technology and internet technology, especially rapid development of mobile internet technology, all kinds of social problems and demands constantly are emerging. How to match technology and social problem is the key point to exploit entrepreneurial opportunities.

The national strategies for digital economy have been promulgated and implemented by such major economic powers as the United States, EU, and Japan. In these countries the digital economy is regarded as first remedy for economic recovery and sustainable development. Due to the characteristics of the digital economy, the entrepreneurial opportunity is the total solution for the social problem based on the existing technology. And by the comparison of US Apple Corporate and China

Changhong Corporate, we find that there are many characteristics of entrepreneurial opportunity in digital economy, such as social problem orientation, timeliness, dominant paradigm, and winner-take-all. The entrepreneurial opportunity should be discovered and exploited by taking full advantage of the huge market and potential in China. Therefore, we emphasized the enter prises should discover the social problem first, integrate the resources and technologies to establish dominant application paradigm second, and construct the business ecosystem third in order to seize the heights of economic development (Table 128.1).

128.2 The Development of Digital Economy

1. Digital Economy and Its Nature
Based on information and communication technology via internet, mobile communication network and the internet of things etc, digital economy realizes the digitalization of trade, communication and cooperation, and effectively promotes the economic and social development and progress. The development of digital economy will disrupt the law of industries, which will be the engine of economy and society development.

Digital economy is based on information and communication technologies and shows some characteristics.

- Regular upgrade. The pace of the technology development in the digital economy depends on the processing speed of computer chip, which will double in every 18 months.
- Increasing marginal revenue, decreasing marginal cost. Due to regular upgrade of technology, the value to customers is marginal revenue increasing. New technologies cover 50 % market share automatically, so the previous products will be out of market. Correspondingly, the prices of previous ones will cut off 50 %, so marginal cost will be decreasing.
- Network externality. The value of the networks is equal to the square of the number of nodes. The more the computers are in the net, the more the value of the net is.
- Network lock-in effect. When the number of users achieves a certain level, the user will be locked in psychologically and behaviorally, which makes strong path dependence and behavioral inertia to the user.

2. The Development of Digital Economy
In the face of the financial and economic crisis and the sluggish recovery, the world's major economies have been focusing on digital economy and launched the national strategies of digital economy in order to achieve the overall economic recovery driven by digital economy. In June 2006, the Singapore government started a 10-yeariN2015masterplan to build a global intelligent nation [11]. UK developed "Digital Britain" plan in 2009 and announced "Digital Economy Act 2010" in April 2010 [12]. I-Japan Strategy 2015 was launched in 2009 [6]. Australian government started

Table 128.1 Summary of digital economy strategies

UK 2010.4 Digital Economy Act 2010

Goal	Protect stakeholder rights in digital economy; promote the healthy, rapid and orderly development of digital economy, such as music, media and games etc
Action	1. Strengthen government's supervision and management
	2. Strengthen protection on music, media and game etc
	3. Specifically stipulate for broadcast, TV, wireless communication, game analysis and public lending fees etc

Australia 2011.5 National Digital Economy Strategy

Goal	Cover broadband infrastructure, on-line education, government internet, medical services, on-line office etc.; share the convenience of the digital economy to all the society
Action	1. Government increases investment on broadcast, digital TV and other infrastructure to build "E-government"
	2. Use intelligent technology to promote digital business capability and develop on-line content mode
	3. Increase the convenience of experience and participation to people in the digital activities

Singapore 2006.6 iN2015

Goal	Invest 4 Billion SGD in 10 years, build Singapore to a global intelligent nation with information technology everywhere through integration innovation and cooperation of information communication
Action	1. Strengthen the innovation on ICT
	2. Invest on information communication infrastructure
	3. Develop information communication industry with global competitiveness

Japan 2009.7 i-Japan Strategy

Goal	Establish a safe and dynamic digital society, realize the convenient application of information technology, ensure information security and build a new Japan
Action	1. Priority to the digitalization development of electronic government, health care, education and human resources
	2. Promote the industry and local economic recovery and nurture new industries
	3. Vigorously develop digital economic infrastructure

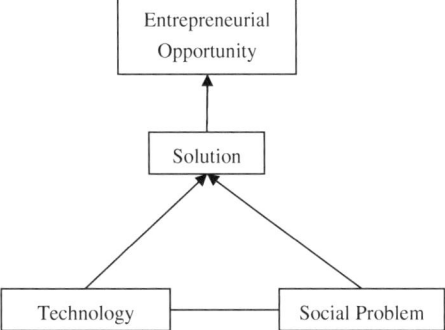

Fig. 128.1 Elements of opportunity in digital economy

the National Digital Economy Strategy in May 2011 [1]. France have been implemented the strategy of the Digital France 2012. The feasibility of Canada 3.0 is under discussion. Obviously, the world's major economies are all focusing on the development of the digital economy with the core of ICT technology and infrastructure, which is the booster of economic development.

128.3 The Entrepreneurial Opportunity in Digital Economy

In the era of digital economy, the technology supply is adequate due to the technology regular upgrade, but the social problem and the rigid demand are in shortage relatively, so discoveries of social problems and applications have become the key to grasp the entrepreneurial opportunity in digital economy. Essentially, entrepreneurial opportunity in digital economy is a kind of opportunity to solve difficult problem, or called the market-oriented opportunity. Only when using the relevant technology can meet the social demand or solve social problems, it will be out of the lab and accepted by the public. In the process of application, technology can realize itself improvement and upgrade, play a role in a wide field, eventually accepted by the mainstream market and achieve the social value and benefit to mankind. Thus, the entrepreneurial opportunity in digital economy is the solution to the social problem by utilizing existing technology (shown in Fig. 128.1).

1. Existing Technology

According to Moore's law, the number of transistors in the chip doubles every 18 months. Correspondingly, chip's processing speed will be doubled, and the cost reduces half. In the era of mobile internet, due to the separation of chip's design and production and the features of SOC mode and Turnkey mode, the updating cycle of mobile intelligent terminal equipment shortens from 18 months to 6–12 months or shorter. The speed is the key of industry development. The application and utilization of technology usually are relatively insufficient or lag, which causes that a lot of technologies can only stay in the stage of theoretical study in the lab, unable to develop and perfect in the practice and application, so the real value of the technology itself is limited. Thus, the existing technology is sufficient to meet the requirement and demand to solve social problem. The support of existing technology makes it possible to exploit and develop the entrepreneurial opportunity.

2. Social Problem

Social problem is the ultimate goal of the entrepreneurial opportunities development and utilization. Information, network and digitalization bring a lot of convenience to people's life, but also inevitably bring about problems and troubles, such as how to manage and filter the huge information, how to protect the information and transaction security and how to protect the copyright of software, music and film etc. These social problems are not only very urgent, but cover wide range of the economic development and everyone's life. Obviously, the development of digital technology and intelligent technology bring entrepreneurial opportunities in many fields, such information communication, music, mechanical processing and transportation etc. During development and upgrading, we will face new problems and challenges, so how to quickly discovery and solve these social problems becomes the key problems which restrict the development of the relevant industries.

3. Solution

The key point of discovering and exploiting opportunity in digital economy is to build up the solutions and implement the effective match between existing technologies and social problems. During above process, suitable technologies for specific social problem can be discovered in the existing technologies. With the aid of integration and optimization of various social resources, the real value of technology can be fully realized, which make the public enjoy the convenience and benefit from technology upgrade and update, and solve the social problems in the economy and society at the same time. These will improve the efficiency of economic and social developments, makes people happier, enable firms to maximize the market share, gain more profit and realize the sustainable development of enter prises.

As to entrepreneurial opportunity, existing technology is the foundation, social problem is the goal, and the solution is the key. Only realizing the perfect combination among the existing technologies, the social problems and solutions, we can truly discovery and develop entrepreneurial opportunities in digital economy.

128.4 The Characteristics of Entrepreneurial Opportunities in Digital Economy

The characters of opportunities in digital economy are shown as below, because there are different laws between the traditional and the digital economic development (Fig. 128.2).

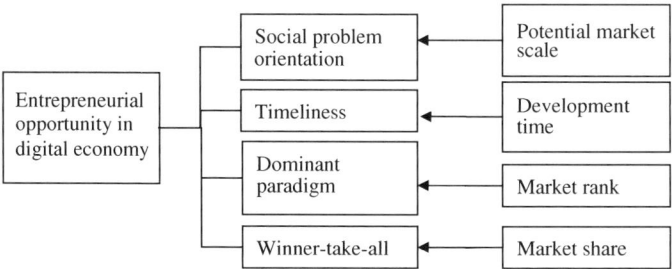

Fig. 128.2   Diagram of characteristics of entrepreneurial opportunity in digital

1. Social Problem Orientation

In the age of digital economy, the technology development is advanced with abundant reserves, so social problems or rigid demand, the premise of technology commercialization and application, are relatively insufficient and lagged. Thus, discovery, identification and creation of social problems are the critical point of entrepreneurial opportunity in digital economy. The chance of entrepreneurial opportunity in digital economy depends on the extensity, urgency and persistence of the social problem. The more urgent, the more extensive and the more rigid the social problems discovered, identified and created, the greater the entrepreneurial opportunity is. Only when the social problems solved by the opportunities are rather urgent in potential market with certain scale, they will be exploited and developed.

In digital economy, compared with abundant technological supply and reserves, the social problems are relatively insufficient. This limits the technology application. From this angle, entrepreneurial opportunities in digital economy are oriented by social problems. Only when the potential market is big enough to exploit, the created value is higher. The potential market scale is an important indicator to identify and exploit the entrepreneurial opportunity. Focusing on the huge potential market of digital music at the beginning of this century, US Apple Corporate introduced a portable music player—iPod and adoptediPod+iTunes mode, which not only solve the copyright issue, but also provide high quality digital musical product to music lovers. Facing huge smart mobile phone market and focusing on main operation system—Nokia Symbian, which can't meet the social problems to realize interaction among users via mobile internet and multiple personalized applications, Apple builds an interaction platform for users, applications and game developers by iPhone+AppStore mode to satisfy the users' requirement of personality and diversification, and achieves the huge sale of iPhone.

2. Timeliness of Entrepreneurial Opportunity

The processing speed of chip will double in every 18 month, and the price of chip cut off 50 % correspondingly. Chip is the heart of computer. And the computer is the foundation and the cell of the digital economics development. The innovation in digital economy, especially technological innovation, with a spontaneous pace which isn't affected by external factors, makes the entrepreneurial opportunity in digital economy with the pace and timeliness. Since the entrepreneurial opportunity in digital economy only exists in a certain stage, once missed "the window of opportunity", the opportunity will be eliminated. The timeliness brings more entrepreneurial opportunities for corporate and individuals in digital economy than ever.

Due to the timeliness of entrepreneurial opportunity in digital economy, the entrepreneur has to shorten the cycle time from discovery to exploitation, eventually launch the products to market. Otherwise, the opportunity would be exploited by others, and the opponents would imitation the solution to social problem. Obviously the development time is the key point to develop entrepreneurial opportunities in digital economy. Whatever the early iPod, or the latest iPhone and iPad, Apple Corporate shorten development cycle and improve launch speed through integrating the global resources. An iPhone is made up of about 500 components provided by over 200 suppliers. Through the seamless control of the whole supply chain and two important processes—design and sales, Apple Corporate launches a new product every year, leads the trend of the smart phone and continuously attracts the attention of the market and consumers.

3. Dominant Paradigm

Due to the nature of regular upgrade, the individuals who control the latest and the most advanced technologies won't own the privilege to identify and discover opportunity in digital economy. The first entrant may not be the final winner. The final winners should make the technological application paradigms to be the dominant ones which will be accepted and utilized by the public [2]. In most cases, the supply of technologies in digital economy exceeds demand, so a large number of technologies, which have been developed, are idle or under used. The main reason is that the relevant social problems to be solved can't be discovered, or is that the bridge does not exist between the existing technologies and social problems, i.e. the lack of dominant paradigm. Above all, the technology dominant application paradigm is the ultimate purpose of the entrepreneurial opportunity discovery and development in digital economy.

The market rank of products or services can reflect the extent of dominant paradigm. The paradigm by market leader, which is accepted by mass, will be the dominant paradigm in the market. Since Windows operation system of Microsoft was launched, it has been the leader in computer operation system market, and become the dominant paradigm of personal computer. In the mobile Internet age, IOS of Apple and Android of Google have been the dominant paradigms of smart phone. But the Android system is the preferred one because of its openness and freeness. Market ranking is an important dictator to measure the extent of the dominant paradigm.

4. Winner-Take-All

The characteristics of digital economy are mentioned above, such as increasing marginal revenue, decreasing marginal cost, development more difficult but duplication easier, the network value is equal to the square of the number of nodes in the net, etc. These characteristics bring about the characteristic of "simultaneous global expansion and winner-take-all" of the new products or services in the digital economy, which is totally different with the characteristic of "gradient transfer" of entrepreneurial opportunity in traditional economy. On the one hand, simultaneous expansion make more individuals to use the products or services, increase the network value, bring the lock-in effect and attract users; on the other hand, the Internet breaks the traditional isolation of time and space, so the solution to the social problems by existing technology can occupy the main market and make profits as much as possible in the whole world. Market share may reflect the value of opportunities. In digital economy, the technologies of Internet, IT and so on realize simultaneous expansion globally, especially those digitalized operation software transmitted via the Internet, such as Windows operation system of Microsoft etc. The higher market share, the higher profits. To keep and increase market share, firms need to keep innovating, establish entrant barrier and constantly left the competitors and imitators behind. The success of Apply Corporate is to keep launching and upgrading new products every year, from iPod, iPhone to iPad. This continuously stimulates the demand of the market, so each launch of the new products will be a hot sale.

128.5 Empirical Analysis: The Success of Apple Products Versus The Failure of Changhong PDP TV

The development of digital economy brings not only challenges but also opportunities to enter prises. To realize sustainable development, enter prises positively seek the opportunities in digital economy and keep exploring and exploiting the rules of opportunities. In the process of discovery and exploration, some enter prises succeeded, such as US Apple Corporate; some ones failed, such as China's Changhong Corporate. The comparison of the sales and sales growth rate of Apple and Changhong from 2006 to 2012 (shown in Fig. 128.3), may further verify the characteristics and rules of entrepreneurial opportunity in digital economy in order to provide strong support and guidance to theoretical research and practice.

1. The Success of Apple Products

With the excellent performance of iPod, iPhone, iPad and relevant services, Apple got over 150 billion sales and 37 billion net profit, and realized the transformation from the computer makers to the consumer electronics supplier. The manufacture of iPod depends on the existing MP3 technique, but the good sale of iPod relies on the iPod+iTunes business model which lock the Soft Ware and Hard Ware mutually. This model focuses on the demands of the music lovers and the recording companies.

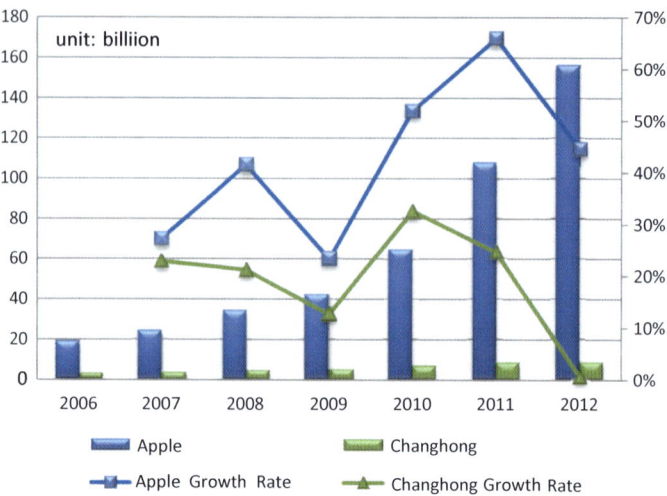

Fig. 128.3 Apple versus changhong sale from 2006 to 2012

The former need legal high-quality digital music, and the later need the copyright protection. Apple Corporate not only succeeds on iPod but also achieve huge profit from digital music sales. Facing the demand of more personalized in the digital economy, Apple integrated the existing technology, built platform, APP Store, launched the iPhone, lead the development of smart phone and made the full-screen smart phone as the industrial dominant paradigm. When iPad launched, it provided better quality office and entertainment experience to a large number of users, and locked the previous users, which increased the network value. Besides these, Apple also used cloud technique, launched the iCloud service to realize files synchronous sharing on different devices.

From iPod, iPhone to iPad, Apple kept discovering and identifying the social problems and demand between the uses and relevant service suppliers, abided by the development rules of product regular upgrade in digital economy, insisted the content to win, launched the new products and new services, seized the dominant application paradigm in the consumer electronics field, stimulated customers' purchasing desire in order to realize development and expansion.

2. The Failure of Changhong PDP TV

As the leader of Chinese color TV, China's Changhong Corporate has been focusing on the development of PDP TV (Plasma Display Panel TV) since 2005. Through many ways such as M&A, independent R&D and alliance, Changhong took five years to build up the whole domestic industrial chain from plasma technological R&D, panel production to TV manufacture and finally realized the mass production of PDP TV in 2010. In the past 5 years, as LCD TV and OLED TV had been developed and matured, many TV companies, such as LG, HITACHI and so on, withdrew the

PDP TV field. This made the market demand of the PDP TV shrunk. Changhong tried to expand market demand through big-size screen, intelligent and 3D technique, but the sales still were poor.

Changhong didn't realize that opportunity in digital economy is social-problem-oriented rather than the one technological-oriented, which made the poor performance of PDP TV. As the technique of LCD and OLED have been matured, they will synchronously extend to the whole world market. This won't leave any time and opportunity of PDP TV to realize gradient transfer. Additionally, the opportunity in digital economy has high timeliness. Once missing the "window of opportunity", the value of opportunity will not exist. At the beginning of this century, Changhong missed the opportunity while the CRT TV is urgent to upgrading and the technique of LCD and OLED is not so matured and stable. So the PDP TV development of Changhong is doomed to failure.

128.6 Conclusion

The digital economy becomes the key field in many major advanced economies. Facing the advanced technology and abundant accumulation, the key point to develop opportunities in digital economy is to discovery, identify and create social problems and rigid demand. The entrepreneurial opportunity in digital economy is a kind of opportunity to solve difficult problems with some characteristics, such as social-problem-oriented, timelessness, predominant of application paradigm, and winner-take-all. Taking the advantages of Chinese economics transformation, high demand of market and so on, firms should aim at our social problems and rigid demand, integrate the whole world technologies and resources, develop application innovation to establish dominant application paradigm in order to realize the sustainable development.

References

1. Australian Government (2013) National digital economy strategy. http://www.nbngovau/digitaleconomystrategy
2. Balachandra R, Goldschmitt M, Friar JH (2004) The evolution of technology generations and associated markets: a double helix model. IEEE Trans Eng Manage 51(1):3–12
3. Busenitz LW, West GP et al (2003) Entrepreneurship research in emergence. J Manage 29(3):285–308
4. Casson M (1982) The entrepreneur: an economic theory, 2nd edn. Edward Elgar, Oxford
5. Eckhardt JT, Shane SA (2003) Opportunities and entrepreneurship. J Manage 29(3):333–349
6. IT Strategy Headquarters (2009) I-Japan strategy 2015. Striving to Create a Citizen-Driven, Reassuring & Vibrant Digital Society
7. Shane S (2012) Delivering on the promise of entrepreneurship as a field of research. Acad Manage Rev 1:10–20

8. Short JC, Ketchen DJ et al (2010) The concept of opportunity in entrepreneurship research. J Manage 36(1):40–65
9. Sirmon DG, Hitt MA, Ireland RD (2007) Managing firm resources in dynamic environments to create value. Acad Manage Rev 32(1):273–292
10. Stevenson HH, Roberts MJ et al (1994) New business ventures and the entrepreneur. Irwin, Homewood
11. Theinfocomm development authority of Singapore, realising the in 2015 vision (2006) Singapore: an intelligent nation, a global city, powered by INFOCOM. http://www.ida.gov.sg
12. UK Government (2010) Digital economy act. http://www.legislationgovuk/ukpga/2010/24/contents
13. Venkataraman S (1997) The distinctive domain of entrepreneurship research: an editor's perspective. In: Katz J, Brockhaus R (eds) Advances in entrepreneurship, firm emergence, and growth, vol 3. JAI Press, Greenwich, pp 119–138

Chapter 129
Designing a Risk Assessment System for China's Third-Party Mobile Payments

Lei Xu and Wuyang Zhuo

Abstract Mobile payments are playing a key role in the rise of the mobile e-commerce industry. Mobile payment security (MPS) has become the largest barrier as customers are unwilling to trust mobile e-commerce without a secure transaction. A comprehensive risk assessment is a critical step in MPS risk early warning, planning and management. It has become necessary to develop dedicated decision support systems (DSS) for the risk management of mobile payments. This paper presents the theoretical principles underlying the design and development of a DSS for MPS risk assessment. This system is capable of addressing multiple qualitative judgments with quantitative support, which is accomplished through the development and use of a framework in a web-based fuzzy decision making environment.

Keywords Third party mobile payments · Security risk assessment · Web-based support system

129.1 Introduction

Customers are more likely to adopt mobile payments if they are confident that the provider has made this service secure by protecting the customers' funds and confidential account information [1]. The overriding reason that 38.2 % of Chinese customers are reluctant to use mobile payments is lack of faith in the security of the mobile payment systems [2]. It is estimated that mobile e-commerce in China will be

L. Xu (✉)
Department of Business Management, Xihua University, Chengdu 610039, People's Republic of China
e-mail: leihsu@163.com

W. Zhuo
School of Economics and Trade, Xihua University, Chengdu 610039, People's Republic of China

J. Xu et al. (eds.), *Proceedings of the Eighth International Conference on Management Science and Engineering Management*, Advances in Intelligent Systems and Computing 281, DOI: 10.1007/978-3-642-55122-2_129, © Springer-Verlag Berlin Heidelberg 2014

entering a crucial development phase in the next 3 years [3], so mobile payment security (MPS) has become the largest barrier as customers are unwilling to trust mobile e-commerce without a guarantee of secure commercial information exchanges and safe electronic financial transactions. Mobile payments require a coordinated and secure exchange of payment information between several unrelated entities, which means that there are several integrated security risks for these stakeholders [4]. Therefore, the payment service providers (PSP), financial institutions (FI), mobile network operators (MNO), mobile device (MD) manufacturers, and content providers (CP) must make robust moves to tackle the MPS challenges to avoid system risk, prevent security breaches and safeguard their assets. However, the stakeholder assets are often not fully protected because of inherent technical and control vulnerabilities [5]. Thus, a comprehensive risk assessment is a critical step in MPS risk early warning, planning and management. Accordingly, it has become necessary to develop dedicated decision support systems (DSS) which can effectively assess the systematic security risks for the risk management of mobile payments.

The Chinese mobile payments industry is now entering a stage of rapid growth after ten years of development driven by the surge in the Chinese mobile internet economy and mobile e-commerce. According to China Internet Network Information Center (CNNIC), the gross merchandise volume (GMV) of China's (third-party payments) TPP reached 1 trillion Yuan in the first quarter of 2013. At the same time, China's mobile payments GMV reached 64.61 billion Yuan in same period, of which mobile TPP had the largest share at 69.2 % [6]. Therefore, in this paper we regard the mobile TPP as the representative mobile payments mode for the performance of the MPS risk assessment. Mobile TPP involves many different stakeholders, including customers and MD, TPP service providers with certificate issuers, PSP as FI, merchants or CP, as well as the MNO [7]. Accordingly, the MPS risk impact domains involved are multi-faceted and encompass all these mobile TPP system elements.

Some previous research has been conducted on MPS issues [8, 9] as well as on systems risk assessment and the relevant DSS [10–13]. However, research so far has been fragmented and somewhat unfocused as much of it has shown a lack of concern for mobile payment risk management and related support systems. In general, security measures from customer viewpoint and technical security of mobile payments are best covered by contemporary research [14, 15]. In this paper, a collaborative web-based decision making system (DSS) is designed, with which MPS are looked at as a holistic and systematic concern, and any risk assessment methods must be able to accommodate this vision. Additionally, the DSS needs to embrace approaches that respond to the interaction and uncertainty in mobile payment systems [16].

129.2 MPS Risk Assessment

1. Methodology

Risk assessment is an integral part of a risk management process designed to provide the appropriate levels of security for mobile payments. In facilitating decision support

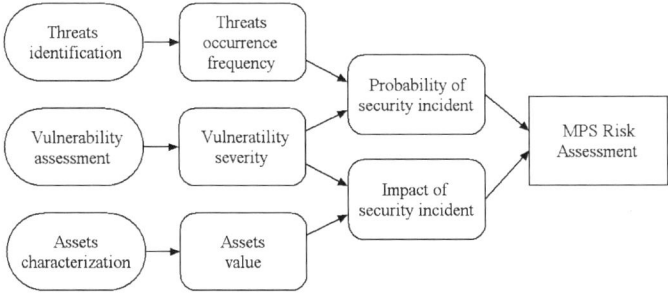

Fig. 129.1 MPS risk assessment conceptual diagram

for MPS risk assessment, the key attributes in the risk assessment process need to be identified. This involves characterizing the critical assets to be protected, identifying the credible threats from the various stakeholders, and assessing the vulnerabilities and risks [17].

Thus, the risk assessment can be essentially broken into three aspects including assets, threats, and vulnerabilities. Associated with the value of the assets, the likelihood of threats, and severity of the vulnerabilities, the risk level of a mobile TPP system can be identified. It can be seen from Fig. 129.1 that the key to an MPS risk assessment is to identify the probability and the potential impact of a security incident. The occurrence probability of a security incident depends on the occurrence frequency of the threat and the severity of the vulnerability. The impact of the security incident depends upon the severity of the vulnerability and the asset values.

The analytical part of the risk assessment process is usually performed qualitatively using the expert judgments. The expected outcome is a qualitative determination of the risk to provide a sound basis for ranking the security risks and supporting the countermeasures priority decision making. Incompleteness and vague information in the assessment process has been the primary influence on risk assessment effectiveness [18]. Therefore, using fuzzy numbers to assess the MPS risk criteria can assist in representing the problem more realistically. Further, the MPS risk assessment aims to position the risk analysis process on a more objective base. However, due to the complex risk assessment environment and increasing information uncertainty, the determination of index weights and the quantitative assessment of qualitative criteria become difficulties [19]. Artificial intelligence techniques, such as fuzzy logic, neural networks, entropy theory, and Bayesian networks, have been widely applied with traditional methods to solve various uncertain decision-making problems for systems security risk assessment [20].

2. Assessment Index System

Risk assessment index system should be established for the MPS risk assessment. The index system is established by decomposing the clusters and subordinate criteria to allow for the recognition of the MPS risk factors. The clusters indicate the guidelines which influence the achievement of the MPS risk assessment and the criteria point out

Table 129.1 Mobile TPP security risk assessment index system

Cluster	Criterion	Risk impact domain
Vulnerability (VB)	Mobile device	Customer
	Database	Financial Institution
	Wireless communication protocol	Mobile network operator
	Mobile terminal operating system	Customer
	Software	TPP service provider
	Operating system and app	Financial Institution
Threats (TR)		
Confidentiality	Transaction information tampering or data missing	Financial Institution
	Transaction information leakage	TPP service provider
Integrity	Data integrity destruction	Financial Institution
Authentication	Legal user masquerading	Mobile network operator
	Unauthorized data modification	TPP service provider
Availability	Service interruption or prohibition	Financial Institution
Non-repudiation	Transaction repudiation	Customer or merchant
Impact on assets (IA)		
Consequence	Payment information lost	TPP service provider
	Transaction process lost	Customer or merchant
	Data leakage	Financial Institution
Severity	Service ability interruption	Financial Institution
	Communication jamming and service delay	Mobile network operator
Recovery cost	Data recovery	Financial Institution
	Service recovery	Customer or merchant

the measures that need to be considered. From the security risk assessment framework and the attribute definitions, the vulnerability (VB), threats (TR) and impact on assets (IA) are determined as the three clusters. Besides these, five properties have always been considered essential for secure transactions; confidentiality, integrity, authentication, availability and non-repudiation [21–23]. Based on the expansion of these five properties, after an analysis of the MPS incidents using historical records and expert knowledge, 20 main incidents are initially determined as the default MPS risk criteria, which are then classified into the three risk clusters to measure the SRV. The default index system, as well as the corresponding stakeholders in the risk impact domains, is shown in Table 129.1.

129.3 System Design

1. System Architecture

The web-based system for MPS risk assessment is a security risk level DSS covering the business process and the responsibilities of all stakeholders from TTP service

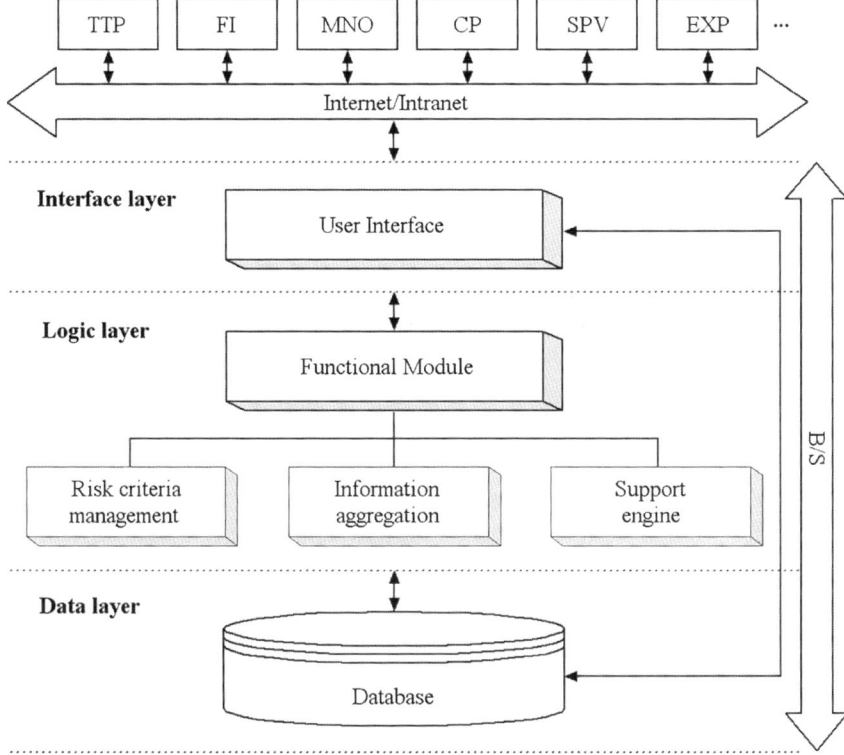

Fig. 129.2 The designed system architecture

providers, FI, CP to MNO, as well as administrators (SPV) and domain experts (EXP). It is necessary to build a browser/server (B/S) model-based network system i.e. an internet/intranet platform because users are in different locations. Conceptually, the system architecture relies on a typical three-tier architecture composed of an interface layer, a data layer and a logic layer, as shown in Fig. 129.2.

This connected functional infrastructure is built upon a presentation-level access (user interface), and is supported by a database for storing the criteria and supporting decision elements. The interface layer allows the users to input and exchange information. The data layer consists of external databases dealing with the real-time uploaded data from multiple sources. The logic layer processes information from the data layer and responds to the users through the interface layer. The logic layer is the core component and is comprised of three major functional modules: risk criteria management, information aggregation and the support engine. The risk criteria management module allows for the assignment of users as criteria experts, provides basic management functions, and allows for knowledge reuse from the previous index system.

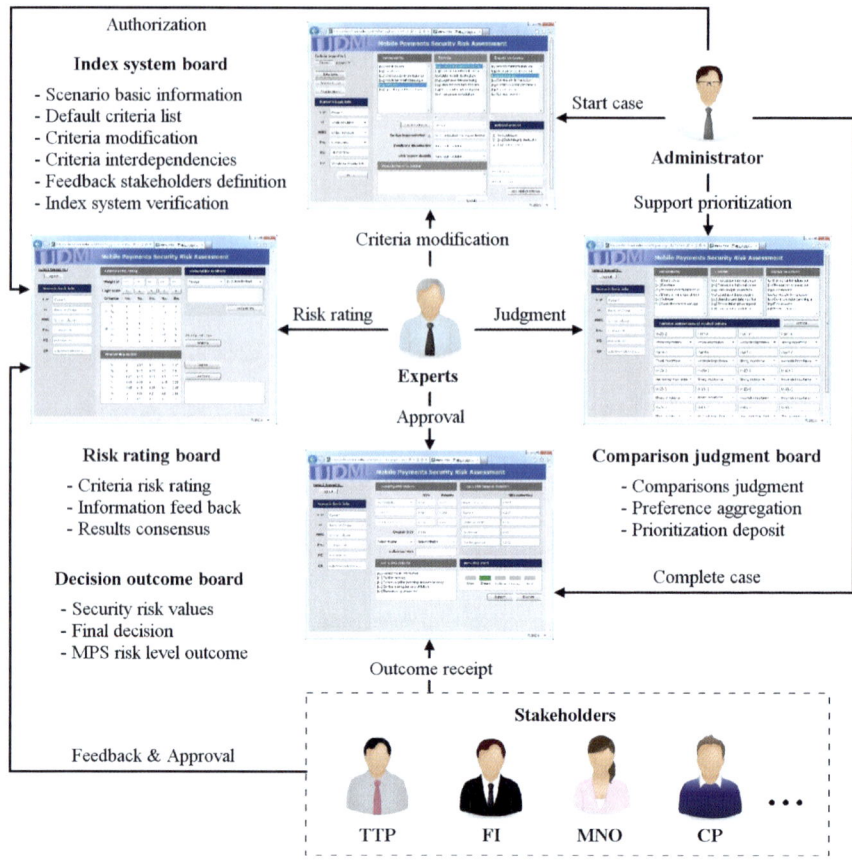

Fig. 129.3 General using procedure of the system

2. Use Procedure

From the general decision making model with respect to security risk management, the overall decision process can be described in four main phases; security problem reorganization with criteria determination, risk analysis based on criteria significance and prioritization identification, and criterion risk rating with information aggregation and risk level decisions. Given the nature of the system architecture defined above, users can gain access to the DSS for the MPS risk assessment on the Internet through Internet Explorer. The user accounts cover the three categories of administrator, experts and stakeholders. The general procedure for using the proposed web-based DSS is shown in Fig. 129.3.

129.4 Conclusion

In this paper, we design a system for MPS risk assessment. This system is capable of addressing multiple qualitative judgments with quantitative support, which is accomplished through the development and use of a framework in a web-based fuzzy decision making environment. The analysis clusters from the VB, TR and IA give a structure that is meaningful to multi-perspective MPS risk management. The system demonstration of the initial prototype was also presented, which showed that the combination of a qualitative fuzzy set and quantitative intelligence techniques results in a valid MPS risk assessment system which can be used to evaluate the relative importance and the risk rating of related criteria. In conclusion, as MPS encompasses an ever-greater scope of organizational relevance and responsibility, it has become necessary to develop decision support methodologies and systems, which are capable of dealing practically with the complex and multifaceted nature of MPS decision making.

Acknowledgments This work is supported by a Program of Natural Science Foundation of Sichuan Provincial Education Department (Grant No. 13212693), the Philosophy and Social Sciences of Sichuan Province [Grant No. SC13JR07], the Scientific Fund of Systems Science and Enterprise Development Research Center of Sichuan (Grant No. Xq13c05) and the Key Scientific Research Fund of Xihua University (Grant No. z1221207).

References

1. Schierz PG, Schilke O, Wirtz BW (2010) Understanding consumer acceptance of mobile payment services: an empirical analysis. Electron Commer Res Appl 9(3):209–216
2. CNNIC (2013) Chinese online payment security status report (in Chinese). http://www.cnnic.net.cn/hlwfzyj/hlwxzbg/dzswbg/201211/P020121121376535616383.pdf. Accessed July 2013
3. Deloitte China (2012) Trends and prospects of mobile payment industry in China 2012–2015: creating innovative models, boosting mobile financial services
4. Robert CD, Matthew WH, Elizabeth AK et al (2012) Mobile payments: an evolving landscape. Supervisory Insights 9(2):3–11
5. Dahlberg T, Mallat N, Ondrus J et al (2008) Past, present and future of mobile payments research: a literature review. Electron Commer Res Appl 7(2):165–181
6. iResearch (2013) Mobile payment heats up the third party payment market. http://english.iresearch.com.cn/views/4942.html. Accessed Jul 2013
7. Au YA, Kauffman RJ (2008) The economics of mobile payments: understanding stakeholder issues for an emerging financial technology application. Electron Commer Res Appl 7(2):141–164
8. Kadhiwal S, Zulfiquar AUS (2007) Analysis of mobile payment security measures and different standards. Comput Fraud Secur 6:12–16
9. Lin P, Chen HY, Fang Y et al (2008) A secure mobile electronic payment architecture platform for wireless mobile networks. IEEE Trans Wirel Commun 7(7):2705–2713
10. Alonso S, Herrera-Viedma E, Chiclana F et al (2010) A web based consensus support system for group decision making problems and incomplete preferences. Inf Sci 180(23):4477–4495
11. El-Gayar OF, Fritz BD (2010) A web-based multi-perspective decision support system for information security planning. Decis Support Syst 50(1):43–54

12. Feng N, Li M (2011) An information systems security risk assessment model under uncertain environment. Appl Soft Comput 11(7):4332–4340
13. Rees LP, Deane JK, Rakes TR et al (2011) Decision support for cybersecurity risk planning. Decis Support Syst 51(3):493–505
14. Deans PC (2005) E-commerce and M-commerce technologies. IGI Global, Pennsylvania
15. Zhou T (2013) An empirical examination of continuance intention of mobile payment services. Decis Support Syst 54(2):1085–1091
16. Pérez IJ, Cabrerizo FJ, Herrera-Viedma E (2010) A mobile decision support system for dynamic group decision-making problems. IEEE Trans Syst Man Cybern Part A Syst Hum 40(6):1244–1256
17. Bajpai S, Sachdeva A, Gupta J (2010) Security risk assessment: applying the concepts of fuzzy logic. J Hazard Mater 173(1):258–264
18. Li W, Zhou J, Xie K et al (2008) Power system risk assessment using a hybrid method of fuzzy set and monte carlo simulation. IEEE Trans Power Syst 23(2):336–343
19. Haimes YY (2005) Risk modeling, assessment, and management, vol 40. Wiley, New York
20. Flanagan DP, Harrison PL (2012) Contemporary intellectual assessment: theories, tests, and issues. Guilford Press, New York
21. Karnouskos S (2004) Mobile payment: a journey through existing procedures and standardization initiatives. IEEE Commun Surv Tutorials 6(4):44–66
22. Marianne C (2012) Evolving mobile landscape challenges and opportunities. In: Technology, compliance and risk management forum. Banker Association, New York
23. Varshney U (2002) Mobile payments. Computer 35(12):120–121

Chapter 130
A Method of Modeling and Evaluation of the Customer Satisfaction

Bo Zhang and Xin Liu

Abstract The customer satisfaction for employment of graduates is an essential evaluation, which has got more and more attention and gradually become a focus research in management science and engineering. This paper designs an evaluation architecture of customer satisfaction for employment of graduates based on the criteria layer analysis, and the hierarchical evaluation model and fuzzy integrated evaluation method is proposed based on the hierarchical analysis and fuzzy comprehensive evaluation. Experimental sampling data analysis results show that the model proposed and its evaluation method are scientifically and rationally.

Keywords Customer satisfaction · Employment of graduates · Evaluation indicator · Hierarchical evaluation model · Fuzzy integrated evaluation · Experimental sampling

130.1 Introduction

The customer satisfaction for employment of graduates is an essential quality evaluation, which pays for attention and has gradually become a research focus in the management science and engineering area. The social evaluation of university student's employment supports the development of our society and orients to improve education quality [12]. A several qualitative appraisal, such as excellent, good and pass, has been used in such research area in China currently, which has a great blindness and randomness consequently [7–9]. How to use scientific and

B. Zhang
Enrollment and Employment Office, Sichuan University, Chengdu 610065,
People's Republic of China

X. Liu (✉)
Business School, Sichuan University, Chengdu 610065, People's Republic of China
e-mail: liuxin67@sina.com

J. Xu et al. (eds.), *Proceedings of the Eighth International Conference on Management Science and Engineering Management*, Advances in Intelligent Systems and Computing 281, DOI: 10.1007/978-3-642-55122-2_130, © Springer-Verlag Berlin Heidelberg 2014

rational method research on the customer satisfaction has attracted much interest of scholars [1–6, 10, 11, 13, 14].

The evaluation is so complicated due to the factors of the customer satisfaction which has a big quantity and different degree of influence. The papers [4] and [13] have applied the fuzzy mathematics theory to study the relevant indicators of the evaluation of the degree of customer satisfaction in order to propose a set of comparable and objective customer satisfaction evaluation indicator system and fuzzy integrated evaluation model, which can achieve qualitative and quantitative analysis. Because of high cost of the questionnaire, this evaluation method has a poor practical usage.

This paper design an evaluation indicator architecture of customer satisfaction based on the criteria layer decomposition and is organized as follows. First, the hierarchical evaluation model and fuzzy integrated evaluation method of customer satisfaction is proposed based on the hierarchical analysis and fuzzy comprehensive evaluation. Subsequently, a type of random sample questionnaire is applied. Finally, a quantitative result of customer satisfaction has achieved by the statistical analysis of questionnaire data.

130.2 Indicators and Evaluation Model of Customer Satisfaction

130.2.1 Evaluation Indicators of Customer Satisfaction

The speciality of employment involves: the property and the scale of the employed enterprise are different, cause to the difference of quality between employment. The weights of the employ requirement varies with the property and the stage of development of the enterprise.

Therefore, the research of this paper to evaluate the satisfaction of the employment include the overall target, the layer of the standard and the layer of the quota, shown in Table 130.1.

To obtain the information of the satisfaction of employment would be an open question, which involves getting the qualification of the satisfaction of employment from human resources experts and employing enterprise and so on, then analyzing and collecting the information, filtering the evaluation quota of the satisfaction of employment.

130.2.2 The Satisfaction of Evaluating Model and Method

In connection with the quote of the satisfaction, combine analytic hierarchy and fuzzy comprehensive evaluation together.

Table 130.1 The index factors of the satisfaction of the employment

The overall target	The layer of the standard	The layer of the quota
The satisfaction of the employment	The quality of knowledge	Professional knowledge
		The level of foreign language
		The level of computer qualification
	The quality of ability	Innovation
		Learning
		Research
		Practical ability
		Thinking of logic
		Literal expression
	The quality of humanity	Interpersonal interaction
		Management of organization
		Teamwork
		Humanistic knowledge
	The quality of profession	Responsibility
		Professional dedication
		Enthusiasm of work
		Enterprising spirit
		Ideological and moral cultivation
	Psychological quality	Social adaptability
		Flexible reaction
		Will and quality

1. The hierarchical fuzzy evaluation model of customer satisfaction

First, derive the relative significance of each indicator from the degree of importance, build the judge matrix, and determine the weight of evaluation factors. Then, establish the hierarchical factor set, which determines the evaluation matrix, and use the weighted average model to confirm fuzzy evaluation vectors of set. Finally, divide the quantitative value of customer satisfaction from the judgment criteria (Fuzzy comprehensive evaluation value).

2. The fuzzy integrated evaluation method of customer satisfaction

This evaluation method involves following steps:
Step 1 Build the judgment matrix and combine the weight for computation. This process calculates the order of importance of factors between some elements in last level and this level, which is called hierarchical single sorting. The relative weighted vectors means the order is represented by relative size. To improve the evaluation of scientific credibility and feasibility, a questionnaire is passed out to some human resources experts who have extensive experiences in assessment and evaluation to establish judgment matrix. Specific methods are shown as follows:

• Determine the judgment matrix C in criteria level which has 5 levels, and determine comparison matrix between criteria level and solution level, and then have a consistency check. Matrix Ckn, Cab, Chu, Cpr, Cps are Compared and recorded as program levels "knowledge quality" indicators, "competency" indicators, "human quality" indicators, "professional quality" indicators of "psychological qualities" index contrast matrix;

- Calculate the composition weights and suppose uc is weight vector between criteria level and level. The weight vector of "knowledge quality", "competency quality", "human quality", "career quality" and "mental quality" in solution levels are $uckn$, $ucab$, $uchu$, $ucpr$, $ucps$, respectively.

Step 2 Build the factor set of the satisfaction hierarchically and the relative judgment matrix.

First level: $U = u_1, u_2, u_3, u_4, u_5$, which is the comprehensive evaluation of graduate satisfaction. Each of factors in U is composed of some sub-factors. Second level: Second layer: u_1 is the set of knowledge quality, u_2 is the set of competency quality, u_3 is the set of human quality, u_4 is the set of career quality, and u_5 is the set of mental quality.

The fuzzy evaluation matrix can be calculated by its fuzzy evaluation table of indicator system. Denote the level of fuzzy evaluation (very dissatisfied, somewhat dissatisfied, neither satisfied nor dissatisfied, somewhat satisfied, very satisfied) as $(v_1, v_2, v_3, v_4, v_5)$. Therefore, the membership of single factor x of fuzzy evaluation set for the evaluation class v_i which is ratio between the number of people and the total number of samples when the indicator x choose ith evaluation level. Taking the research costs and the practicality of the evaluation methods into account, a random sample survey is conducted by the employers, and data were obtained from employers recruiting at the school the previous year comprehensive satisfaction of graduates. That satisfaction evaluation is a reflection of the average of the evaluation of multiple evaluation object.

For simplicity, this paper has an assumption of 5 evaluation scales which is uniformly distributed in $[0, 1]$, specifically $[0, 0.2)$, $[0.2, 0.4)$, $[0.4, 0.6)$, $[0.6, 0.8)$, $[0.8, 1]$, corresponding to 5 evaluation interval spaces (evaluation interval spaces can be modified according to the different evaluation objects). Assuming satisfaction levels to meet the normal distribution:

$$p_{ji} = \frac{\int_{0.2 \times (i-1)}^{0.2 \times i} N_j dx}{\int_0^1 N_j dx}, \quad i, j = 1, 2, \cdots, 5,$$

where p_{ji} represents the probability of the factor x that makes the comprehensive evaluation grade j to i. So the degree of membership r of evaluation level v_i of the single factor x can be calculated as:

$$r = \frac{\sum_{j=1}^{5} p_{ji} \times R_j}{R},$$

where R represents the number of graduates whose evaluation grade is j of the single factor x. R is the sum of all samples who is evaluated, and R is equal to the sum of the sample survey on employment in last year; therefore, $R = R_1 + R_2 + R_3 + R_4 + R_5$.

Step 3 Calculate the fuzzy evaluation with the weighted average model

bkn, ab, bhu, bpr, bpc denote the set of fuzzy comprehensive evaluation of indicators of knowledge quality, ability and quality, humane quality, professional quality and psychological quality. B represents fuzzy comprehensive evaluation of Index layer versus guidelines layer. Then

$$\begin{cases} bkn = uckn \times R_1 \\ bab = ucab \times R_2 \\ bhu = uchu \times R_3 \\ bpr = ucpr \times R_4 \\ bps = ucps \times R_5. \end{cases}$$

That is to say, $B = uc \times R$.

Denoting $h = [h(v_1), h(v_2), h(v_3), h(v_4), h(v_5)]$ as fuzzy evaluation grade vector, fuzzy evaluation value of various levels can be calculated as follows: $P = h \cdot v^T$, where, v is a vector corresponding fuzzy set of evaluation.

Step 4 The correction of Satisfaction factor

Considering the importance of each factor as not the same, the key factors for improvement to be generally determined according to the following method:

- The evaluation, which has not reached the general satisfaction level, can be considered as the key factors for improvement;
- We can calculate the corresponding relative satisfaction of the satisfaction close items, particularly the ratio of satisfaction and weights. Relative satisfaction contains the satisfaction and weights, and it is proportional to the satisfaction, inversely proportional to the weights. If one indicator's satisfaction degree is low and the weight is high, the relative satisfaction degree is low, so the factor of low relative satisfaction degree is the key factor that needs to improve.

130.3 The Analysis

Through the random sample questionnaire, we get the questionnaire data statistics table which, includes 74 educational institutions, 161 employments in the past year according to classifying the enterprises into the type of education and the type of research.

We choose the judgment matrices as follows:

$$C = \begin{bmatrix} 1 & 1 & 5 & 3 & 4 \\ 1 & 1 & 6 & 5 & 4 \\ 1/5 & 1/6 & 1 & 1/2 & 1 \\ 1/3 & 1/5 & 2 & 1 & 2 \\ 1/4 & 1/4 & 1 & 1/2 & 1 \end{bmatrix}, \quad Ckn = \begin{bmatrix} 1 & 6 & 7 \\ 1/6 & 1 & 2 \\ 1/7 & 1/2 & 1 \end{bmatrix},$$

Table 130.2 Evaluation rank probability distribution

P_{ij}	Very displeasure	Displeasure	Common	Pleasure	Very pleasure
Very displeasure	0.8114	0.1870	0.0016	0.0000	0.0000
Displeasure	0.1575	0.6836	0.1575	0.0014	0.0000
Common	0.0013	0.1573	0.6827	0.1573	0.0013
Pleasure	0.0000	0.0014	0.1575	0.6836	0.1575
Very pleasure	0.0000	0.0000	0.0016	0.1870	0.8114

$$Cab = \begin{bmatrix} 1 & 2 & 3 & 6 & 7 & 8 \\ 1/2 & 1 & 3 & 5 & 6 & 8 \\ 1/3 & 1/3 & 1 & 4 & 5 & 7 \\ 1/6 & 1/5 & 1/4 & 1 & 2 & 5 \\ 1/7 & 1/6 & 1/5 & 1/2 & 1 & 4 \\ 1/8 & 1/8 & 1/7 & 1/5 & 1/4 & 1 \end{bmatrix},$$

$$Chu = \begin{bmatrix} 1 & 1/3 & 1/5 & 5 \\ 3 & 1 & 1/2 & 8 \\ 5 & 2 & 1 & 9 \\ 1/5 & 1/8 & 1/9 & 1 \end{bmatrix}, \quad Cpr = \begin{bmatrix} 1 & 1 & 3 & 2 & 2 \\ 1 & 1 & 4 & 1 & 3 \\ 1/3 & 1/4 & 1 & 1/2 & 1 \\ 1/2 & 1 & 2 & 1 & 3 \\ 1/2 & 1/3 & 1 & 1/3 & 1 \end{bmatrix},$$

$$Cps = \begin{bmatrix} 1 & 1/3 & 1/2 \\ 3 & 1 & 4 \\ 2 & 1/4 & 1 \end{bmatrix}.$$

We can check the consistency and get relative weight vector using MATLAB. We suppose the probability distribution function of the employment satisfaction degree evaluation rank v_i satisfies normal distribution $N(\mu_i, \sigma_i^2)$, $i = 1, \cdots, 5$. Based on the distinction of each evaluation rank, we can value the mean $\mu_1 = 0.1$, $\mu_2 = 0.1$, $\mu_3 = 0.1$, $\mu_4 = 0.1$, $\mu_5 = 0.1$. The actual analysis values the data. Through calculating, we can come up with Table 130.2. The fuzzy evaluation grade vector is below: $h = (h(v_1), h(v_2), h(v_3), h(v_4), h(v_5))^T = (0.1, 0.3, 0.5, 0.7, 0.9)^T$.

So the knowledge quality indicator u_1, ability quality indicator u_2, humanistic quality indicator u_3 and professional quality indicator u_4 at all levels can be calculated. The value of respective fuzzy feature evaluation is $P_1 = 0.6111$, $P_2 = 0.6331$, $P_3 = 0.5802$, $P_4 = 0.5521$, $P_5 = 0.5713$. The fuzzy comprehensive evaluation value of the overall objective is $P = 0.6078$, and the satisfaction interval is $[0.6, 0.8)$, which overall assessment of job satisfaction is relatively satisfaction.

130.4 Conclusion

Through collection of easy sample survey data, this paper builds a rational, quantitative fuzzy comprehensive evaluation model which is easy to implement, and combines the analytic hierarchy process and fuzzy comprehensive evaluation, and pays attention on the enterprises scale of employment user, differences in the type of talent needed, etc. On the statistical analysis of survey data, it can obtain comprehensive quantitative evaluation of job satisfaction.

References

1. Bin L (2009) Performing evaluation system on the quality of college students' employment based on job satisfaction. J Yanshan Univ (Philos Soc Sci Edn) 1:038
2. Deng YK (2004a) Analysis of pros and cons on building job satisfaction evaluation model-the eecond study assessed patterns of employment satisfaction. Guangxi Soc Sci 10:178–179 (In Chinese)
3. Deng YK (2004b) Design of job satisfaction index system and fuzzy comprehensive evaluation model-the eecond study assessed patterns of employment satisfaction. Guangxi Soc Sci 11:1102–1109 (In Chinese)
4. Fornell C (1992) A national customer satisfaction barometer: the Swedish experience. J Mark 56:6–21
5. Hu ZH, Zhou ZW, Li LB (2008) The quality and importance of university graduates based on the employer. J Kaili Univ 26:73–75 (In Chinese)
6. Huang CJ (2004) Countermeasure on employment of college students. J Xi'an Agric Univ 39(2):124–129 (In Chinese)
7. Liu CH (2000) Exploration on establish of college graduates quality inspection system. China High Educ Eval 2:87–93 (In Chinese)
8. Luo JJ, Luo ZY (2001) Comprehensive ability of university graduate: evaluation from viewpoint of employment. China Popul Sci 5:552–558 (In Chinese)
9. Lv YL (2000) Employer's requirements for quality of graduates. SEN Univ Forum 8:814–820 (In Chinese)
10. Rust RT, Zahorik AJ, Keiningham TL (1995) Return on quality (ROQ): making service quality financially accountable. J Mark 59:58–70
11. Wang PL, Yan JZ (1997) Analysis and reflections on social evaluation results of the quality of graduates. Educ Res 4:523–529 (In Chinese)
12. Yang HJ, Han DH (2006) Exploration of employment satisfaction evaluation model. Kunming Univ Technol (Soc Sci Edn) 8:715–718 (In Chinese)
13. Zhou JM (2003) Preliminary analysis of social evaluation of seven-year clinical quality graduates. High Educ Proj Res 2:128–133 (In Chinese)
14. Zhu AS (2009) Employer survey report of graduate. Sci Technol Innovation Herald 14:1320–1324 (In Chinese)

Chapter 131
Analysis and Comparison of Life Cycle Assessment and Carbon Footprint Software

Marta Ormazabal, Carmen Jaca and Rogério Puga-Leal

Abstract Environmental management is receiving increased attention by researchers, policy makers and companies as today's world is facing major environmental problems such as global warming and waste. As a consequence, companies are progressively improving their environmental practices and behaviors with the aim of reaching more advanced stages in their environmental management. Specifically, those companies most committed to environmental issues have started to introduce the concept of life-cycle and footprint thinking in their organizations. Nevertheless, there are currently many different software tools and companies may not know which one is most suitable. Consequently, the main objective of this research is to study the different software applications to help companies choose the most suitable one in each case. To achieve this objective, the paper provides an in depth review of the literature on life-cycle and footprint thinking. As a result, more than 20 software tools have been analysed and compared. It is important to highlight that the appropriateness of one software program relative to another may very much depend on the user's scope or objective, as the database of each program could be different.

Keywords LCA · Carbon footprint · Software tool · Environment

M. Ormazabal · C. Jaca (✉)
Department of Industrial Management, School of Industrial Engineers, University of Navarra,
P º Manuel Lardizabal 13, 20018 San Sebastian, Spain
e-mail: cjaca@tecnun.es

R. Puga-Leal
UNIDEMI, Departamento de Engenharia Mecânica e Industrial, Faculdade de Ciências
e Tecnologia, FCT, Universidade Nova de Lisboa, 2829-516 Caparica, Portugal

J. Xu et al. (eds.), *Proceedings of the Eighth International Conference on Management
Science and Engineering Management*, Advances in Intelligent Systems and Computing 281,
DOI: 10.1007/978-3-642-55122-2_131, © Springer-Verlag Berlin Heidelberg 2014

131.1 Introduction

As Cramer [9] stated, environmental management "involves the study of all technical and organizational activities aimed at reducing the environmental impact caused
by a company's business operations". Environmental management is an important
issue in today's organizations [19]; it has been only a few decades since companies
realized the importance of environmental management within their organizations [7].
Because of society's growth in awareness about the environment following a number of disasters, environmental regulations that have pushed companies to improve
their environmental management have been enacted [3, 7, 8, 26]. Another reason
for socially responsible business practices has been concern about the environmental effects that their activities have, in part because of the deterioration caused by
past activities [1, 7]. Consequently, there is also an increasing demand for products
and services that minimize environmental impact [22]. This environmental pressure
from the various stakeholder has had a great influence on companies' environmental behavior [5, 14], and the companies have been able to respond and improve
their environmental practices thanks to the different tools that are available and the
standards that companies are following to improve their environmental management
[2, 4]. It is important to highlight that not all the companies have reached the same
stage of maturity in terms of environmental management [16]. In the academic literature there have been several classifications of the different stages that companies
might be in. Ormazabal et al. [21] proposed an environmental management maturity model of six maturity stages ranging from low maturity to high maturity: legal
requirements, responsibility assignment and training, systematization, ecological and
economic benefits (ECO^2), eco-innovative products and services, and leading green
company. Many companies reach the systematization stage thanks to environmental
management certifications such as ISO 14000 [20], but only a few move on to the
second half of the model. As a consequence, this paper is focused on this second half
and more specifically on some of the software tools that are becoming widely used
by companies to assess a product's life cycle and analyses their carbon footprint [10,
17]. As there are many software applications available, the aim of this paper is to
present them in order to help companies decide which software tool is most suitable
for them. The following section of this paper: Sect. 131.2 introduces the methodology used in this research. Sections 131.3 and 131.4 present the available Life Cycle
Assessment (LCA) and Carbon Footprint (CF) tools. Finally, the main conclusions
of this study are discussed in Sect. 131.5.

131.2 Methodology

Calculating any environmental impact involves a complex process. For this reason
there is a wide variety of software tools on the market that are designed to undertake an LCA, a CF analysis or determine other environmental indicators. The main
elements of these tools are the database used for calculations and the methodology

applied. Both these elements are related, and in most cases, the available software applications are suitable for working with one or more specific database. In other cases, a particular database is incorporated into the software. In terms of the databases, they need to be well defined and regularly updated given that technological advances cause the premature aging of the validity of existing data. From the point of view of environmental impact measurement, two parameters are critical [23, 24]:

- The volume, quality, accuracy and relevance of data available to the user in the software, and
- The software package's user-friendliness.

This paper reviews different software applications available for calculating LCA and CF according to the criteria established by different environmental organisms and academic authors. Applications and databases associated with LCA and CF are in a very dynamic situation, as there are tools that appear, disappear or change continuously. For this reason, the proposed methodology for this paper is as follows:

- Review the tools available in the market, selected from the information and recommendations included on web sites on environmental management such as Ihobe and the Joint Research Center's Institute for environment and Sustainability [11].
- Select the most interesting tools and analyse them.

The main methods for assessing the software detailed in the paper were reviewing the current literature on the available databases and software and trying the software.

131.3 LCA Software

Life cycle assessment (LCA) is a methodological framework for estimating and assessing the environmental impacts attributable to the life cycle of a product, including the sourcing of raw materials, manufacturing, distribution, transportation, and end-of-life disposal. According to many authors, LCA can be separated into several interrelated stages [12, 13]. The stages are: goal definition, inventory, impact assessment and interpretation for improvement. The first stage, goal definition, defines the purpose, goal and boundaries of the assessment. The second stage involves labor-intensive data collection and the calculation of the emissions and burdens associated with every unit process system related to the product. The results of this stage take the form of a life cycle inventory. The impact assessment evaluates the potential and actual environmental impacts. The final stage of LCA is the improvement assessment, where the changes that are needed to bring about environmental improvements are evaluated and reviewed.

The basic function of any LCA software package is to determine energy and mass balances on an item or model and allocate emissions, energy uses, etc. in order to facilitate the calculations associated with the inventory and impact assessment stages. In order to review some of the most common tools available to help companies to calculate LCA, we developed a list of the best known software packages for life cycle

Table 131.1 Software tools for LCA calculation

Tool	Developer	Approach	Web
AIST-LCA 4	National Institute of Advanced Industrial Science and Technology (AIST), Japan	Generic	www.aist-riss.jp/main/
Athena	Athena Sustainable Materials Institute, Canada	Building and construction	www.athenasmi.org/
BEES 4.0	National Institute of Standards and Technology, USA	Building materials	www.nist.gov/el/economics/BEESSoftware.cfm
CMLCA 4.2	Leiden University, Institute of Environmental Sciences (CML), Holland	Generic	www.cml.leiden.edu/software/
E^{3}DATABASE	Ludwig-Bölkow-Systemtechnik GmbH, Germany	Generic	www.e3database.com/
EARTHSTER 2 TURBO	GreenDelta GmbH	Generic	www.greendelta.com/
ECO-BAT 4.0	Haute Ecole d'Ingénierie et de Gestion du Canton de Vaud, Switzerland	Building and construction	www.eco-bat.ch/
GaBi	Pe-International	Generic	www.gabi-software.com/
GEMIS	Oeko Institut, Germany	Generic	www.gemis.de
LEGEP	LEGEP Software GmbH, Germany	Generic and Building	www.legep.de/?lang=en
OpenLCA	GreenDelta GmbH	Generic	www.openlca.org/
REGIS	Sinum AG-EcoPerformance Systems	Generic	www.sinum.com/
SABENTO	ifu Hamburg GmbH, Germany	Chemical	www.sabento.com
SIMAPRO	PRé-Consultants	Generic	www.pre-sustainability.com/
SULCA 4.2	VTT Technical Research Centre of Finland	Generic and forest	www.vtt.fi/index.jsp
TEAM	Ecobilan Pricewaterhouse Coopers	Generic	www.ecobilan.pwc.fr/en/boite-a-outils/team.jhtml
TESPI	ENEA, Italy	Generic	www.elca.enea.it/
UMBERTO	ifu Hamburg GMBH	Generic	www.umberto.de/en/
USES-LCA 2.0	Netherlands Center For Environmental Modeling	Terrestrial, freshwater, and marine ecosystems	www.cem-nl.eu/useslca.html

analysis. The Table 131.1 was built from information available at the Ihobe [15] and Building Ecology webpages [6].

The most popular or commonly used software packages were selected from Table 131.1 in order to have a preliminary set of tools. SimaPro, GaBi, Umberto, Earthster 2 Turbo and OpenLCA were selected because they offered a variety of characteristics.

Of the five tools, SimaPro and GaBi, which have been in the market for more than 15 years, are the most commonly used overall. Both have closed code and high license pricing.

Another widespread business tool is Umberto, developed by ifu Hamburg. This software is more difficult to use than SimaPro and GaBi, and its price is similar. It doesn't bring significant innovations to the previous software.

There is a new type of LCA software tool called Software as a Service (SaaS), where instead of purchasing the software license the company pays for the service. This an interesting option because it allows users to reduce the costs associated with LCA calculations. However, in the business world companies are reluctant to use an outside service to handle critical information like environmental indicators. An example of SaaS software is Earthster 2 Turbo. Earthster 2 Turbo was launched in collaboration with three of the leading environmental companies: Seventh Generation, Tetra Pak and Walmart. The objective of this software was to develop an accessible tool for making updated data available to designers, manufacturers, suppliers and environmental experts. Earthster 2 Turbo allows users to share information, to update databases and to improve the update process. This tool has also the possibility of integrating social impacts into the calculation. Some of the social impacts than can be considered are, for example, workers' rights, impact on local inhabitants, effects of producing goods and services in global or local supply chains, etc. The user can fit the social risks and impacts to adjust the analysis to the particularities of each place.

Finally, we analyzed the freeware package Open LCA in order to include open and free code software in the study. Open LCA is a widely known software tool that is easy to handle and it allows the user to calculate all the stages associated to LCA. In addition, this software was created and developed by GreenDelta, with the support of PE International (makers of GaBi), PRé Consultants (creators of SimaPro) and UNEP (United Nations Environment Programme). Another advantage of this tool is that it offers users the possibility of working with different databases, such as those used by GaBi, and others. Initially Open LCA was designed for calculating the environmental impact of products and processes, but it can now add economic aspects.

In what follows, three of the selected software packages are analyzed, taking different aspects into account. The software packages chosen for analysis are: SimaPro, GaBi and OpenLCA.

The aspects analyzed are as follows:

Graphic interface

This is considered to be an essential feature, since it facilitates the user's ability to operate the data and results.

All three tools allow users to create the process for calculating LCA in an easy way. All of them allow the life cycle to be represented in a diagram chart, in which processes are linked by flow-connections. The three tools present Sankey diagrams, which is a great help for visualizing the contribution of each process to the total impact. SimaPro and OpenLCA allow graphical display forms of the processes, which generates a greater impact.

Of the three software packages analyzed, GaBi is that one offers the most intuitive and clear graphical interface for modeling.

Uncertainty analysis

The three software packages include uncertainty analysis through Monte Carlo analysis. This is also the most common method of analysis among LCA software tools.

Comparison of results

The three software packages allow comparisons between product and processes to be made. GaBi allows two results with different parameter values to be compared for example, two possible transport means for the same plan. In contrast, the other two software tools are able to compare the results between different products.

The Open LCA software allows comparisons to be made more easily than the others, and it offers more visual results, showing the list of all impacts.

Results

The three tools provide the minimum requirements, which are the Life Cycle Inventory Analysis and the Life Cycle Inventory. The obtained results will differ depending on the data used (databases) and the definition of the processes that have been included.

SimaPro offers additional output calculating the total output of the results, weighting environmental impacts. GaBi is probably the most dynamic software since it allows changes to be made to the parameters in order to visualize the impact immediately.

Reporting

The results are usually presented in reports because they are often used to communicate to others in the company, or a third party. The three analyzed software packages enable results to be exported, both graphically and in tables to text editors such as Word or Excel. GaBi goes a step further and offers a basic text editor that is built into the tool itself, which allows editing and the addition of tables or graphs, which are automatically modified if inventory changes.

The following table presents the aspects mentioned (Table 131.2).

However, as previously mentioned, the appropriateness of the software depends on the necessities of the user.

Table 131.2 Characteristics of the selected software tools for LCA calculation

ASPECT	GaBi	SimaPro	OpenLCA
Graphic interface	Intuitive flowpath	Sankey diagram	Flowpath and Sankey diagram
Uncertainty Analysis	Monte Carlo	Monte Carlo, others	Monte Carlo
Comparison of results	Yes	Yes	Yes
DataBase	GaBi-4500, datasets, ecoin-vent, U.S. LCA (NREL)	EcoInvent, U.S. Input/output, U.S. LCI, Dutch Input/output, Swiss Input/output, LCA food, industry data v2, Japanese Input Output, IVAM	No
Information exchange	Yes, between database users or GaBi users	Yes, between SimaPro and EcoSpold users	Yes, between Ecospold or ILCD
Reports	Self editor for reporting or exportation to excel	Graphical reporting of results or exportation to excel	Graphical reporting, custom tables or exportation to excel

131.4 Carbon Footprint Software

In recent decades, society has shown an interest in climate change and global warming issues, and as a consequence carbon footprint calculations have been in strong demand in recent years [18, 27, 29]. There have been many definitions of carbon footprint and suggestions about how it should be calculated [27]. Tierney [25] describes a carbon footprint as a way for an individual or company to determine the impact of its activities on the environment. It might also be seen as the amount of greenhouse gas (GHG) emissions that are directly or indirectly caused by an activity or by the life cycle of a product [28]. Wiedmann and Minx [29] suggested that the term carbon footprint should only be used for analyses that include carbon emissions.

It might seem that LCA and CF are quite similar. The main difference is that CF is based on a single indicator, while LCA focuses on multiple indicators [27]. The environmental vision of estimating the carbon footprint can cause problems when interpreting the results by omitting other environmental impacts. We might study a system that does not have special environmental problems in terms of greenhouse gas emissions, and yet there may be other environmental impacts that are not considered in the carbon footprint analysis.

There are many software tools for estimating a company's carbon footprint, though they differ in terms of implementation, data and objectives. These tools might focus on:

Table 131.3 Software tools for CF calculation

Tool	Subclass				Web
	a	b	c	d	
Sofi and GaBi (PE International)	X	X	X	X	www.pe-international.com
Autodesk				X	www.labs.autodesk.com
Carbon Footprint	X	X	X	X	www.carbonfootprint.com
IHS Energy and Carbon Solution	X	X	X	X	www.ihs.com
Intelex	X	X			www.intelex.com

a Organization, b Projects, c Activities, d Products

Table 131.4 Sectorial software tools for CF calculation

Tool	Sector	Area of emission				Web
		a	b	c	d	
Center for Clean Products and Clean Technologies	Electric-Electronic (computer)				X	www.epeat.net
Cement CO_2 Protocol	Building	X	X	X		www.wbcsdcement.org
Build carbon Neutral	Building		X	X	X	www.buildcarbonneutral.org
Superfos	Packaging				X	www.superfos.com

a Organization, b Projects, c Activities, d Products

- Organizations, by considering the emissions associated with the activities of the company.
- Projects, by considering the emissions associated with the projects undertaken by the organizations or by final users.
- Activities, considering the emissions associated with the final user actions.
- Products, if we consider the emissions associated with the product life cycle.

The following sections describe the most important software tools, which have been divided into two groups: general and sectorial.

General Software

There are some software tools that are independent of a company's sector. The Sofi and GaBi software from the consulting firm PE international, Autodesk, Carbon Footprint, HIS Energy and Carbon Solution and Intelex are some of the general software for estimating a company's carbon footprint. The following table compares them, taking into account our measurement (Tables 131.3 and 131.4).

Sofi and GaBi are based on the LCA methodology. Autodesk Inventor helps users make decisions about selecting materials that reduce the environmental impact caused by a product. It offers large material inventories, where it is possible to add new materials and new properties to the materials. Carbon Footprint includes a series of tools for the carbon footprint estimation both at the business and domestic levels.

The IHS Energy and Carbon Solution is an integrated platform that advises different business activities on carbon footprint data management and communication associated with products, supply chain and different locations within the company. The Intelex software analyses the carbon footprint emissions at the organization level.

Sectorial Software

Some tools have been developed with a specific focus, meeting the needs of a particular industrial sector and thus allowing a better approximation.

The tool from the Center for Clean Products and Clean Technologies aims to advise companies on quantifying the carbon footprint in electronic equipment. Currently, the tool is designed to evaluate computers. The Cement CO_2 tool is for the building sector and it is based on the IPCC (Intergovernmental Panel on Climate Change) guides. Build Carbon Neutral is focused on the CO_2 emissions associated with the building sector. Superfos is a European manufacturer of packages, so it has designed a tool for making precise carbon footprint calculations related to packaging.

131.5 Conclusions

First of all, it is very important to recognize the difference between LCA and CF. While the issue of carbon footprint is becoming increasingly important given the problem of global warming problem, we need to take into account the fact that a smaller carbon footprint is not always synonymous with superior environmental performance. For this reason it is highly recommended that carbon footprint studies be carried out with other global tools, such as LCA.

This paper classifies the different LCA and CF software tools available on the market. Despite there being many tools, each of them has a different purpose or might differ from other tools in some aspect. Consequently, it is important to know what the user is looking for and then determine which tool best covers his/her requirements. For this reason, this paper presents the main characteristics of the different software tools and the links to their corresponding webpages. In this way, this paper will help users to know where to start when they are planning an environmental analysis.

References

1. Angell LC, Klassen RD (1999) Integrating environmental issues into the mainstream: an agenda for research in operations management. J Oper Manag 17(5):575–598
2. Ardente F, Beccali G et al (2006) Poems: a case study of an italian wine-producing firm. Environ Manag 38(3):350–364
3. Bansal P, Roth K (2000) Why companies go green: a model of ecological responsiveness. Acad Manag J 43(4):717–736
4. Boiral O, Gendron Y (2011) Sustainable development and certification practices: lessons learned and prospects. Bus Strategy Environ 20(5):331–347

5. Bremmers H, Omta O et al (2007) Do stakeholder groups influence environmental management system development in the dutch agri-food sector? Bus Strategy Environ 16(3):214–231
6. BuildingEcology (2014) Life cycle assessment software, tools and databases. http://www.buildingecology.com/sustainability/life-cycle-assessment/life-cycle-assessment-software
7. Claver E, López M et al (2007) Environmental management and firm performance: a case study. J Environ Manag 84(4):606–619
8. Collins E, Roper J, Lawrence S (2010) Sustainability practices: trends in new zealand businesses. Bus Strategy Environ 19(8):479–494
9. Cramer J (1998) Environmental management: from 'fit' to 'stretch'. Bus Strategy Environ 7(3):162–172
10. Emilsson S, Hjelm O (2002) Implementation of standardised environmental management systems in swedish local authorities: reasons, expectations and some outcomes. Environ Sci Policy 5(6):443–448
11. EuropeanCommission (2014) Our thinking: life cycle approach. European platform on life cycle assessment. http://eplca.jrc.ec.europa.eu/
12. Fava JA (1993) A conceptual framework for life-cycle impact assessment. Society of environmental toxicology and chemistry and SETAC foundation for environmental education
13. Goenaga MO (2013) EMM model. Environmental management maturity model for industrial companies. Ph.D. thesis, Universidad De Navarra
14. González-Benito J, González-Benito Ó (2006) A review of determinant factors of environmental proactivity. Bus Strategy Environ 15(2):87–102
15. Ihobe (2009) Análisis de ciclo de vida y huella de carbono, dos maneras de medir el impacto ambiental de un producto. INFOR
16. Jabbour CJC (2013) Environmental training and environmental management maturity of brazilian companies with ISO 14001: empirical evidence. J Clean Prod.10.1016/j.jclepro.2013.10.039
17. Knight P, Jenkins JO (2009) Adopting and applying eco-design techniques: a practitioners perspective. J Clean Prod 17(5):549–558
18. Lee K (2011) Integrating carbon footprint into supply chain management: the case of hyundai motor company (HMC) in the automobile industry. J Clean Prod 19(11):1216–1223
19. Lorente JC, Jiménez JDB (2010) Un análisis de las dimensiones de la gestión ambiental en los servicios hoteleros. Dirección y Organización (30)
20. Ormazabal M, Sarriegi JM (2012) Environmental management evolution: empirical evidence from spain and italy. Bus Strategy Environ 23(2):73–88
21. Ormazabal M, Sarriegi JM et al (2013) Evolutionary pathways of environmental management in UK companies. Corp Soc Responsib Environ Manag.10.1002/csr.1341
22. Park J, Seo K (2006) A knowledge-based approximate life cycle assessment system for evaluating environmental impacts of product design alternatives in a collaborative design environment. Adv Eng Inform 20(2):147–154
23. Rebitzer G, Ekvall T et al (2004) Life cycle assessment. Part 1. Framework, goal and scope definition, inventory analysis, and applications. Environ Int 30(5):701–720
24. Rice G, Clift R, Burns R (1997) Comparison of currently available European LCA software. Int J Life Cycle Assess 2(1):53–59
25. Tierney J (2008) Are we ready to track carbon footprints? Cell 510:435–5189
26. Tsai M, Chuang L et al (2012) The effects assessment of firm environmental strategy and customer environmental conscious on green product development. Environ Monit Assess 184(7):4435–4447
27. Weidema BP, Thrane M et al (2008) Carbon footprint. J Ind Ecol 12(1):3–6
28. Wiedmann T (2009) Editorial: carbon footprint and input-output analysis—an introduction. Econ Syst Res 21:175–186
29. Wiedmann T, Minx J (2007) A definition of 'carbon footprint'. Ecol Econ Res Trends 1:1–11

Chapter 132
The Affecting Factors of Inflows of International Hot Money to China

Xiaojuan Chen, Liuliu Kong, Zhaorui Zhang and Xiang Zhou

Abstract Using a nonlinear Markov regime switching VAR model and impulse response functions, this paper tested monthly data from January 2000 to December 2012 to analyze the relationship between hot money inflow and the affecting factors in different regimes. The empirical results show that the actual real estate price is the strongest factor on the international hot money, and actual stock price and the forward exchange rate fluctuation are the second important factors, no matter the world economic and financial situation is under a stable or a turbulent period.

Keywords International hot money · Economic globalization · MS-VAR model

132.1 Introduction

Although China has adopted the relatively strict policy on capital flowing since reform and opening, a lot of signs reveal that international hot money inflowed and outflowed on a large scale by passing the control policies. During Asian financial crisis broke out in 1997, large amounts of hot money flowed out of China. After joining the world trade organization in 2001, China's economy integrated into the global economy more closely, which made the government relax the capital account controls gradually. In 2002, the error and omission items in Balance of International Payments of China turned to positive for the first time after 12-year consecutive negative, which meant that the direction of international hot money shifted from outflow to inflow. And this trend became stronger since China adopted the new foreign exchange policy in 2005. While it reversed since the global financial crisis broke out in 2007, which strengthen the de-leveraging trend of global financial institutions. After the crisis,

X. Chen · L. Kong (✉) · Z. Zhang · X. Zhou
Business School, University of Shanghai for Science and Technology,
Shanghai 200093, People's Republic of China
e-mail: kongliuliu@usst.edu.cn

J. Xu et al. (eds.), *Proceedings of the Eighth International Conference on Management Science and Engineering Management*, Advances in Intelligent Systems and Computing 281, DOI: 10.1007/978-3-642-55122-2_132, © Springer-Verlag Berlin Heidelberg 2014

the U.S. Federal Reserve launched rounds of QE policy, causing a large number of international hot money flooding into the emerging-market countries rapidly for the relatively higher investment returns.

It had practical significance for China's financial system to maintain the security as well as the stability of asset price to find and analyze the factors that affect international hot money to flow in and out of China.

132.2 Literature Review

Domestic and foreign researches on determinants of international hot money flows are mainly as follows.

Fernandez-Arias [2] used panel data from 1988 to 1995 of 13 developing countries to do empirical analysis, showing that the influence of interest rate changes was greater than that of the fundamental conditions of countries. The empirical research of Reinhart et al. [1] once again proved that relative low interest rate in developed countries is the determinant of capital flowing into developing countries. Martin and Morrison [7] discussed the reasons why international hot money flowed into China, and its channel as well as the impact in details. They insisted that the key factors that led the international hot money flowed into China were the relative interest spreads and the expectations of forward appreciation of RMB. Kim and Yang [4] empirically investigated the effects of capital inflows on asset prices by employing a panel VAR model, suggesting that capital inflows indeed have contributed to asset price appreciation in the region, although capital inflow shocks explained a relatively small part of asset price fluctuations. Jiang [3] showed how to restraint the inflow of international hot money and excessive liquidity effectively in theory. Li et al. [5] found that quantitative easing monetary policy that the Federal Reserve engaged in may become "hot money" that disrupts the capital market and intensifies macroeconomic instability.

By constructing a nonlinear MS-VAR model, Zhu and Liu [9] tested monthly data from 2005 to 2009 for empirical analysis and found that there is a mutually reinforceing relationship among inflow of hot money, appreciation of RMB, and rising of China's real estate price. Zhao and Su [8] adopted a quadruple arbitrage and margin cointegration model to demonstrate that the expectation of RMB appreciation does not significantly influence the hot money inflow no matter in the long run or the short run. It is the rise of real estate price and the interest gap conspicuously drive the hot money inflow.

Combining with the Markov Regime Switching Vector Autoregression model, this paper tested monthly data from January 2000 to December 2012 to carry on the research about the relationships between international hot money and factors to affect it.

132.3 Empirical Testing and Analysis

132.3.1 Data Sources

1. Actual interest difference between China and American

This paper adopts the indirect method to estimate the scale of hot money inflow, as Liu [6] did.

Hot money scale (hm) = Position foreign exchange Purchase increment (FP) − Foreign direct investment increment (FDI) − Trade surplus (TS).

The FP data of this article were from the People's Bank of China website, in addition, FDI and TS data were from RESSET financial databases.

2. RMB NDF forward exchange rate

We use e to represent 1-year forward period of RMB NDF in market in Hong Kong.

The data were from WIND financial database.

3. Actual interest difference between China and American

Let 1-year benchmark deposit rate (r_d) to represent domestic interest rate and the U.S. federal funds rate (r_f) to represent U.S. interest rate. Both (r_d) and (r_f) are the real interest rate after excluding inflation, thus actual interest difference between China and U.S. $r = r_d - r_f$).

The data were from WIND financial databases and Department of Labor of the United States.

4. Domestic capital market price

rsh represents capital market price and it is the closing price of Shanghai Stock Exchange Composite Index excluded inflation.

The data were from RESSET financial databases.

5. Domestic housing market price

Housing market prices (hp) = National commercial housing monthly sales / National commercial housing monthly sales area. And the real estate price is rhp after excluding inflation.

The data were from the WIND financial database.

In this paper, all variables are monthly data. All variables were made seasonal adjustment by Census-x12 (additive form) method. $lrsh$ and $lrhp$ that are logarithms of rsh and rhp are used to overcome heteroscedasticity. OX-MSVAR Software was used to analyze the data.

Table 132.1 Unit root tests

Variable	ADF test		
	T-statistic	p	(c, t, k)
hm	-3.846	0.0002^*	$(0, 0, 1)$
e	-2.264	0.4508	$(c, t, 0)$
Δ	-12.420	0.0000^*	$(0, 0, 0)$
r	-1.673	0.0891^{**}	$(0, 0, 3)$
$lrhp$	-5.307	0.0001^*	$(c, t, 1)$
$lrsh$	-2.699	0.0766^{**}	$(c, 0, 4)$

Note Δ firsr-order difference; c intercept; t trend; k lag order; * 1 % significance level; ** 10 % significance level

132.3.2 Model Estimates

1. Time series ADF test

   As shown in Table 132.1, the forward exchange rate is non-stationary at a significant level of 10 %, but the first-order differential form is stationary at a significant level of 1 %. The other variables are stationary at a significant level of 10 %.

2. The selection of MS-VAR model

   This paper constructs a MS-VAR model including hm, Δ, r, $lrhp$ and $lrsh$. As the equation's mean, intercept, coefficient and variance changed with regimes, different MS-VAR models can be obtained: MSM-VAR, MSI-VAR, MSA-VAR and MSH-VAR. If mean and variance are changed with regimes, the model is MSMH-VAR; If intercept and variance changed with regimes: the model is MSIH-VAR; and so on. Before a model form is chosen, the number of regimes and lag orders should be determined. As hm can be defined two states of low volatility and high volatility, Δ can be appreciated or depreciated, r may be positive or negative, $lrhp$ may be low growth or high growth and $lrsh$ may be also defined as two cases, so the model be assumed with two regimes. And according to FPE, AIC, HQ criterions, lag order 2 is chosen. Therefore, MS(2)-VAR(2) model is chosen.

   Table 132.2 shows that, the HQ and SC value of MSH(2)-VAR(2) is superior to other models, but the LL and AIC value of MSIH(2)-VAR(2) is better than that of MSH(2)-VAR(2). In addition, the value of LR linearity test of MSIH(2)-VAR(2) is 184.3884, which is more significant than MSH(2)-VAR(2) (its value is 171.8563), besides the P value of chi-square statistic is less than 5 %, so MSIH(2)-VAR(2) is the most suitable model.

3. Model estimate results

   By comparing the intercept, standard deviation and mean in each equation under different regimes of Table 132.3 and Fig. 132.1, it can be concluded that the world economic and financial situation is relatively stable in regime 1. in that regime, hot money inflow is in a state of low volatility and quantity, the margin of expectations of forward appreciation of RMB is low, the real interest difference between China and U.S. fluctuates small, real estate prices fluctuate largely and volatility of stock

Table 132.2 The selection of optimal model

Model class		LL	AIC	HQ	SC
Linear systems	VAR(2)	−391.1508	6.0281	6.5931	7.4146
	MSM(2)-VAR(2)	−435.4788	6.6991	7.3186	8.2242
	MSI(2)-VAR(2)	−384.2257	6.0291	6.6486	7.5542
	MSA(2)-VAR(2)	−1635.9934	22.9803	23.9619	25.3967
	MSH(2)-VAR(2)	−305.2226	5.1271	5.8271	6.8503
Nonlinear systems	MSMA(2)-VAR(2)	−1189.1658	17.2048	18.2266	19.7202
	MSMH(2)-VAR(2)	−368.2845	6.0168	6.7570	7.8390
	MSIA(2)-VAR(2)	−337.1815	6.0677	7.0896	8.5832
	MSIH(2)-VAR(2)	−298.9566	5.1105	5.8508	6.9328
	MSMAH(2)-VAR(2)	−1189.1658	17.4009	18.5434	20.2134
	MSIAH(2)-VAR(2)	−275.3263	5.4552	6.5978	8.2678
MSH(2)-VAR(2)	LR linearity test:171.8563		Chi(15) = [0.0000], Chi(17) = [0.0000]		
MSIH(2)-VAR(2)	LR linearity test:184.3884		Chi(20) = [0.0000], Chi(22) = [0.0000]		

Table 132.3 MSIH(2)-VAR(2) Model

	State	hm	de	r	$lrhp$	$lrsh$
Intercept	regime1	−424.885814	0.137367	−0.094168	0.343845	−0.296026
		[−2.8136]*	[1.3311]	[−0.2986]	[3.7862]*	[−2.5692]*
	regime2	−429.312209	0.129332	−0.035972	0.369900	−0.314647
		[−2.5762]*	[1.1777]	[−0.1077]	[3.8710]*	[−2.5452]*
Standard	regime1	61.563578	0.054544	0.167454	0.064951	0.064868
deviation	regime2	291.255078	0.099842	0.370399	0.034535	0.078511
Variable	regime1	55.87962	−0.012070	−0.512469	3.321788	2.823651
mean	regime2	118.1506	−0.019079	1.455	3.571216	3.03427

Note number in [] represents t value, * indicates significant at a level of 1 %

price is low. On the contrary, the world economic and financial situation is in the turbulent period in regime 2.

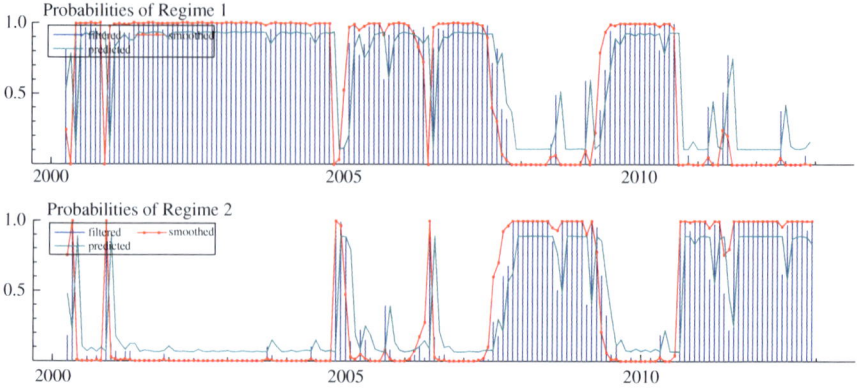

Fig. 132.1 The probability of regime 1 and 2

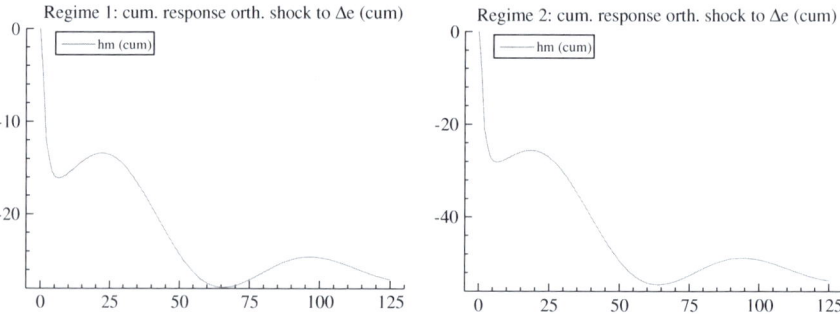

Fig. 132.2 Accumulated response of hm to Δ

132.3.3 MSIH (2)-VAR (2) Model and Impulse Response Functions Analysis

1. The accumulated response of hot money to the shock of forward foreign exchange changes

From Fig. 132.2, one positive standard deviation shock (RMB depreciation) is given. In regime 1, hot money will flow out sharply from China in current period, slow down until period 5, reaching the minimum value −29, and then followed by a slight increase, and finally stops stably at around value −26. In regime 2: the changing trend of accumulated response of hot money is almost the same as regime 1, but its volatility of the accumulated response will be largely increased, then reaching the minimum value −49, and finally stop stably at around value −46. It shows that when the expected depreciation of the RMB against the U.S. dollar, the motivation of arbitrage will drive hot money flee out of China, but the extent of outflows varies in different regimes.

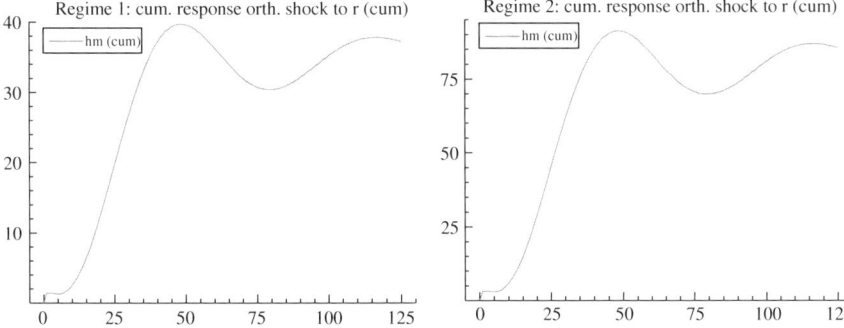

Fig. 132.3 Accumulated response of *hm* to *r*

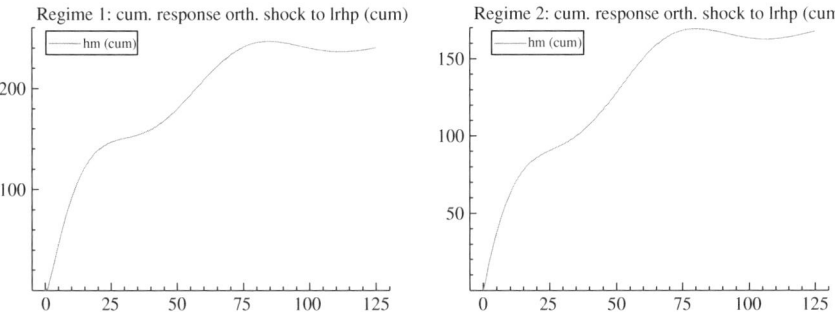

Fig. 132.4 Accumulated response of *hm* to *lrhp*

2. The accumulated response of hot money to the real interest difference

From Fig. 132.3, one standard deviation shock of real interest difference between China and U.S. (China's real interest rate is higher than the U.S.) is given. In regime 1, the accumulated response of hot money will have little change in previous 10 periods, but will flow rapidly into China later, reaching the maximum value 36 at period 50, then falling and raising, and finally stops stably around at value 32. In regime 2, the changing trend of accumulated response of hot money is almost the same as regime 1. But its volatility of the accumulated response will be largely increase, reaching the maximum value 80, and finally stop stably at around value 74. It means that when the real interest difference increases, the motivation of arbitrage will drive hot money flow into China, but the extent of outflows varies in different regimes.

3. The accumulated response of hot money to real estate prices

From Fig. 132.4, one standard deviation positive shock of real estate price is given. In regime 1, hot money will flow rapidly into China in current period, with a slight decline speed after period 25, reaching the maximum value 220 in period 80, and finally stop stably at around period 80. In regime 2, the changing trend of accumulated response of hot money is almost the same as regime 1. But its volatility will be reduce, after reaching the maximum value 150, and finally stop stably at around value 140,

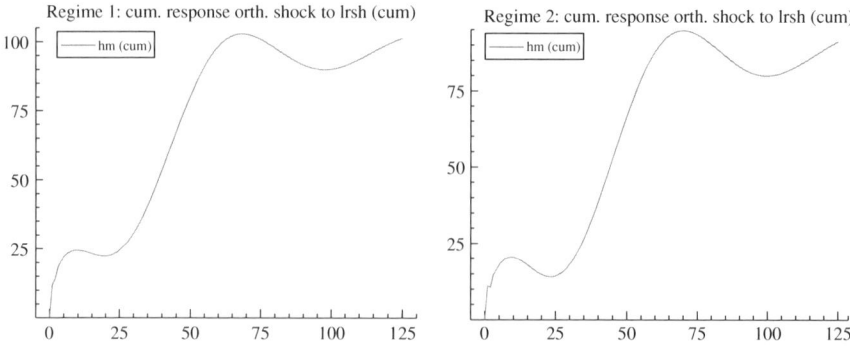

Fig. 132.5 Accumulated response of *hm* to *lrsh*

indicating that when the actual real estate prices rise, the motivation of arbitrage will drive hot money flow into China, and the extent varies in different regimes.

4. The accumulated response of hot money to actual stock prices

From Fig. 132.5, one positive standard deviation shock of actual stock price is given. In regime 1, hot money will flow rapidly into China in current period, with a slight decline in about period 10, then start to flow in increasingly from period 25, and slow down, reaching the maximum value 92 in period 70, then fall and raise, and finally stop at around 80. In regime 2, the changing trend of accumulated response of international hot money is almost the same as regime 1. But its volatility will be reduced, then start to flow in increasingly till reaching the maximum value 85, and finally stop stably at around value 75. It shows that when the actual stock price rises, the motivation of arbitrage in China's capital market will drive hot money flow into China, and the extent of inflows varies in different regimes.

132.4 Conclusions and Policy Recommendations

132.4.1 Conclusions

Based on MS-VAR model, there exists two regimes (two kinds of economic conditions) in the observation period, and the impulse response function showed that each factor has influence on the hot money in both regimes, but the effect is different.

Generally speaking, if the forward exchange rate (direct quotation) rises, RMB depreciation expectation will lead hot money flow out in both regimes. However, the degree of its influence is different. In regime 2, the depreciation expectation of RMB will cause the exodus of hot money, and the influence is as twice as that in regime 1. real interest difference has a positive impact on the hot money flow in, however, it has a lag order10. The influence degree is different in different regimes, and the maximum impact in regime 2 is over twice as that in regime 1. Real estate price has a positive impact on the hot money flow. Different from the former two factors, the influence degree of real estate price on the hot money in regime 1 are much greater than that in regime 2. Actual stock price has a positive impact on hot money, slightly different in different regimes.

In short, no matter the world economic and financial situation is under a stable or turbulent period, real estate price has the greatest influence and the strongest attractiveness on the hot money, following by the actual stock price, then the forward exchange rate. It illustrates the main motivation of the hot money flowing into China is to take the real estate and capital markets revenue, and the following motivation is waiting for arbitrage interest brought by the RMB appreciation expectation.

132.4.2 Policy Recommendations

First, financial authorities of China need to establish a perfect monitoring and early warning mechanisms for short-term hot money flow, increase the punishment on illegal short-term hot money flow and increase the costs of illegal speculative capital transactions or evacuation. Second, the government should stabilize the domestic real estate market and the stock market, strengthen the regulation of large-scale entry of foreign capital in the real estate market and stock market. Third, the government should continue and deepen the market-oriented reforms of foreign exchange rate market to weaken the expected appreciation of the RMB. Fourth, the government should push market reform of interest rate system to reduce the interest difference between home and abroad.

Acknowledgments The research is supported by the key Building Subject of Shanghai Board of Education—"The Operation and Control of Economics System" (J50504) and "the Innovation Fund Project For Graduate Student of Shanghai" (JWCXSL1302)—The impact of international hot money on China's financial markets-based on MS-VAR model.

References

1. Calvo G, Leiderman L, Reinhart C (1994) Inflows of capital to developing countries in the 1990s: causes and effects. IDB Working Paper
2. Fernandez-Arias E (1996) The new wave of private capital inflows: push or pull? J Dev Econ 48(2):389–418
3. Jiang Z (2011) A studying on the status of international hot money and management counter-measures in China. In: Advances in electrical engineering and electrical machines. Springer, pp 739–745
4. Kim S, Yang DY (2011) The impact of capital inflows on asset prices in emerging asian economies: is too much money chasing too little good? Open Econ Rev 22(2):293–315
5. Li J, Liu Y, Ge S (2012) China's alternative remittance system: channels and size of "hot money" flows. Crime, Law and Soc Chang 57(3):221–237

6. Liu Y (2008) Whether overseas "hot money" promote the rise of stock market and housing? Financ Res 10:48–70 (In Chinese)
7. Martin MF, Morrison WM (2008) China's 'hot money' problems. In: Congressional research service reports, DTIC Document
8. Zhao R, Su Z (2012) Does rmb appreciation expectation indeed drive hot money influx to China?—base on quadruple arbitrage and margin cointegration model. Financ Res 6:95–109 (In Chinese)
9. Zhu M, Liu L (2010) Short-run international capital flows, exchange rate and asset prices-an empirical study based on data after exchange rate reform since 2005. Finance Econ 5:5–14 (In Chinese)

Chapter 133
Analysis on Food Quality and Safety Supervision Based on Game Theory

Zongtai Li and Hua Li

Abstract The paper analyzes the regulation of the quality and safety of food based on game theory. The study shows that the situation of food quality and safety remains grim in China, which is mainly due to lack of control and supervision. It is necessary to strength the punishment on manufacturing and selling inferior food behavior, however, it is more important to increase the accountability of the government and enhance the punishment on malpractice of the supervisor. The supervisor should play a proactive role in food regulation, in order to radically reduce the behavior of manufacturing and selling inferior food, and to protect people's lives and improve their health.

Keywords Food quality and safety · Supervision · Game

133.1 Introduction

With development of commercialization and socialization in food economy, the activities of food production, management and consumption become more and more complicated. The food quality safety information turns difficult to be identified. Blind and mercenary producers and operators are apt to succeed with their illegal behaviors. The consumers as the vulnerable side in obtaining information have difficulty in supervising the producers and operators directly. Food safety has become an important public affair, and it is necessary for the government to provide effective supervision and management.

Food quality safety is a tough issue worldwide. The Chinese government has been committing to food quality safety construction. Till now, the relevant laws

Z. Li (✉) · H. Li
College of Economics and Management, Beijing University of Agriculture,
Beijing 102206, People's Republic of China
e-mail: lizongtai@sohu.com

J. Xu et al. (eds.), *Proceedings of the Eighth International Conference on Management Science and Engineering Management*, Advances in Intelligent Systems and Computing 281
DOI: 10.1007/978-3-642-55122-2_133, © Springer-Verlag Berlin Heidelberg 2014

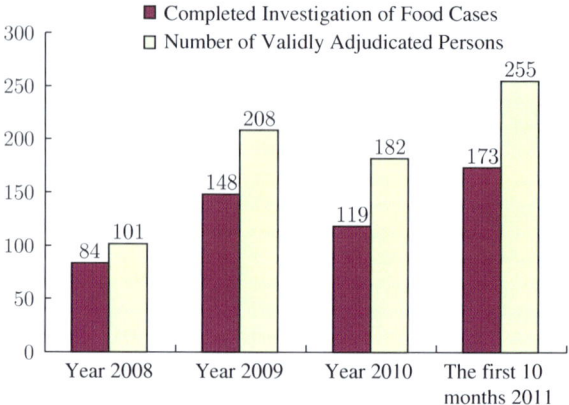

Fig. 133.1 Situation of food safety crime

and regulations system, technical standard system, monitoring system, certification system and risk analyzing system, etc. have been established; the plans of "Ecological Agriculture", "Pollution-free Food" and "Organic Food" have been launched; the system of "From Farmland to Dining-table" has been constructed; in addition, annual special examination and daily routine inspection have been intensified. However, the situation of food quality safety in China is still severe. According to the information from the Supreme People's Court, the number of food safety cases and the number of the adjudicated persons present a progressive increase trend in recent years in China (Fig. 133.1) [1]. The "Melamine Event" of Sanlu Group, the "Dyed Steamed Bun Event" occurring in Shanghai, the "Clenbuterol Issue" of Shineway Group, and the "Hogwash Oil Event" down to date are the typical cases among them. These situations indicate that the government has evidently promoted the striking effort against food safety crimes; however, they also indicate that criminals are motivated by profit, and the criminal activities which threaten food safety are still rampant.

Food quality safety attracted more and more attentions since the 70–80 s of last Century. Nelson [2], Darby and Karni [3], from a consuming perspective, categorized goods in the market into "searched goods (identifiable goods)", "experienced goods" and "credit goods". The American economist, Akerlof [4], analyzed the phenomena and mechanism of market failure in a "lemon market" (the market in which the seller and the buyer have asymmetric information). Chinese scholars performed analysis on and explored reasons for the food quality issue in China based on the above theory: to begin with, it is common that the production of food companies are large in quantity, wide in scale, small in size, as well as scattered and messed, The employees have weak cognition of law-abiding and trustworthiness. Thry may either have fluke mind, or are blinded by the lust for gain, which results in the frequently occurring behaviors of cheating on techniques and materials, and abuse of pesticide, hormone and forbidden additive, etc [5]. Secondly, the modern food production is socialized, and the consumers may still able to judge the characteristics of foods as for the "experienced goods" (such as delicacy degree, juice quality, spiciness, taste, flavour, etc. of products), but it is difficult for the consumers to judge the characteristics of the foods as for the "credit goods" (by the characteristics in the aspects such as whether the food contains antibiotics, hormone, pesticide residue or not and the quantity thereof), which results in dysfunction of market credit mechanism, and failure for the consumers to effectively supervise the producers and operators [6]. Thirdly, the government is weak in supervision and management. Objectively, the professional food supervision and management department in China is merely set on counties or above, where equipments is insufficient, personnel are small in number, and the supervision and management ability is generally weak; Subjectively, some supervision and management departments have sluggish behavior in administration, which results in weak supervision and management frequently; what is more, some supervisors and managers adopt mild attitude of supervision and management towards the behavior of producing and selling the false and inferior, in order to seek department interests or regional interests, which the illegal operators are willing to take risks [7]. Among these reasons, the core is supervision and management. If supervision and management is weak, laws and regulations, technical specifications, operation specifications, and even education and training, etc. cannot be implemented.

In respect to the research on the government supervision of food quality safety, Zhao and Gao [8] analyzed the asymmetric information of the food quality safety management and suggested the necessity of government supervisions; Chen [9] discussed about the relationship between government supervision and food quality safety and suggested relative solutions. These studies are generally qualitatively analysis and introduction of experiences, which are lack of in-depth theoretical analysis on putting forward thoughts about government supervision. The paper provides in-depth analysis on the game playing between government supervision functions and food producer based on Game Theory, which has theoretical and practical significance for understanding the government supervision on the food quality safety.

133.2 Establishment of Food Quality Safety Game Model

1. Characteristics of Relevant Game Subjects

The game theory is a theory which studies the decision-making behavior of the game players during the gaming process, which takes players, strategies, and payoffs as the basic concept, and takes Nash Equilibrium as the basic theory, which provides tools for analyzing the behaviors of each interested party in the real world. The game players here indicate the governmental supervision and management department and the food producers and operators, whose characteristics are assumed as the follows respectively:

(1) The governmental supervision and management department indicates the relevant department which directly executes food quality safety supervision and management. Supervision and management is executed under established

institutional environment and technological environment, wherein the supervisors and managers may have opportunistic behavior, the attitude and action of the supervisors and managers decide the effect of supervision and management, and the two are positively related.

(2) The producers and operators indicate the personnel in the links such as food production, processing, storage and transportation, marketing and selling, etc. Each link has a certain quality safety standard, wherein the producer and operator also have opportunistic behavior. Whether the producer and operator will positively execute the standard or not will affect the food quality safety.

(3) The governmental supervision and management department and the food producers and operators are assumed to be risk neutral, then the expected value is used to judge merits of the decision-making.

2. Establishment of Two-side Game Model

Participants: the governmental supervision and management department, the food producers and operators.

Strategy: the strategy of the governmental supervision and management department is assumed as loose supervision and management against strict supervision and management, wherein the probability of loose supervision and management is P_g, and the probability of strict supervision and management is $1 - P_g$; The strategy of producer and operator is assumed to be producing false and inferior food against producing safe food, wherein the probability of the two strategies are P_t and $1 - P_t$ respectively.

Payoffs: when the governmental supervision and management department adopts strict strategy and performs its duty, there is no gain or loss for it; when it adopts loose supervision and management, it will gain profit R (by eating into the expense of manpower, material resources and financial resources); however, if problems occur, it will be punished by higher authorities for failure to perform its obligation, wherein the loss is assumed to be $(-P)$. if the producer and operator produce and sell safe agricultural products, then whether the governmental supervision and management department adopts strict supervision and management or not, the producers and operators will always gain legal profit V; if the producer and operator select producing and selling false and inferior agricultural products, when the governmental supervision and management department adopts strict supervision and management, the producer and operator will be punished and suffer the loss of producing and selling cost, which are assumed as $(-D)$ in total; when the governmental supervision and management department turns a blind eye and adopts loose supervision and management, the producer and operator will gain illegal profit B, wherein B is greater than V, because the profit and cost differ greatly for counterfeit and shoddy products. Game Analysis Model is shown in Table 133.1.

3. Static Analysis for Food Quality Safety Supervision and Management Gaming

Within a period that is not too long, the payoff parameters in Table 133.1 and the structure thereof will not change. According to Nash Equilibrium Principle in the

Table 133.1 Gaming between food producer and operator and the governmental supervision and management department

Producer and operator	Governmental[a]	
	Strict supervision and management $(1 - P_g)$	Loose supervision and management P_g
False and inferior agricultural products P_t	$-D, 0$	$B, -P$
Safe agricultural products $(1 - P_t)$	$V, 0$	V, R

[a] Governmental supervision and management department

game theory, as the rational economic man, the producer and operator and the governmental supervision and management department will not stick to a certain strategy; but will have scruple of the other party, and try to avoid the other party's knowledge or guess of its selection, and randomly select different strategies. The probability matrix will be formed, which just perfectly fails the other party to take an opportunity by a loophole, so as to avoid loss of itself, that is, to prevent the other party to incline to a certain strategy with pertinence so as to get ascendancy over itself in the game [10]. Specifically, one party will make the expected values of different strategies of the other party to be of no difference.

The probability Pt for the producers and operators to produce and sell false and inferior food shall make the expected payoff of the governmental supervision and management department by loose supervision and management to be equal to the expected profit by strict supervision and management, which can be derived from Table 133.1, $(-P) \times P_t + (1 - P_t) = 0 \times P_t + 0 \times (1 - P_t)$, $P_t = R \div (R + P)$.

This is the optimal P_t. That the probability for the producer and operator to produce and sell false and inferior food is greater than P_t will render loss of the governmental supervision and management department, which will consequentially incur strict supervision and management strategy selected thereby; and that the probability for the producer and operator to produce and sell false and inferior food is smaller than P_t will render loss of its own opportunity to gain a profit greater than V.

Similarly, if the probability P_g of loose supervision and management by the governmental supervision and management department makes the expected profit of producing and selling false and inferior food by the producer and operator to be equal to the expected profit of producing and selling safe food thereby, then $(-D) \times (1 - P_g) + B \times P_g = V \times (1 - P_g) + V \times P_g$, $P_g = (D + V) \div (D + B)$.

This is the optimal P_g. That the probability for the governmental supervision and management department to adopt loose supervision and management is greater than P_g will render problems, which will incur increased possibility of punishment by higher authorities; That the probability for the governmental supervision and management department to adopt loose supervision and management is smaller than P_g will render loss of its own opportunity to gain more R.

It can be seen from these analyses that the optimal strategies of the producer and operator and the governmental supervision and management department are respectively: $P_t = R \div (R + P)$, $P_g = (D + V) \div (D + B)$, wherein the optimal strategy depends on the profit values V, B, R and cost or punishment value D, P of both parties, which can be taken as the controlling factors for the management of food quality safety.

4. Dynamic Analysis for Food Quality Safety Supervision and Management Gaming

For a relatively long period, the profit and loss parameters in Table 133.1 are varied. From the angle of supervision and management, here the effects of variation of the punishment upon the supervision and management department and the punishment value upon the producer and operator on the game equilibrium strategy are discussed [11].

Firstly, the basis on which the governmental supervision and management department selects the strategy is: $W = (-P) \times P_t + R \times (1 - P_t)$, which is the expected profit of the governmental supervision and management department by loose supervision and management, wherein the minimum request is that $W \geq 0$, as shown in Fig. 133.2. In the figure, P_t^* represents the optimal level on which the producer and operator produce and sell false and inferior food. Initially, the probability for the producers and operators to produce and sell false and inferior food cannot be at the optimal level. If $P_t > P_t^*$, i.e. the probability to produce and sell false and inferior food is greater than the optimal level, $W < 0$, which will consequentially incur strict supervision and management by the governmental supervision and management department; If $P_t < P_t^*$, i.e. the probability to produce and sell false and inferior food is lower than the optimal level, $W \geq 0$, on the given premise that the governmental supervision and management department is rational, the governmental supervision and management department will choose loose supervision and management, which will render the producers and operators to increase the probability to produce and sell false and inferior food with profit obtainable, till P_t^* is reached.

If the punishment level ($-P'$ as shown in Fig. 133.2) is increased upon the governmental supervision and management department, and if the probability P_t^* for the producer and operator to produce and sell false and inferior food remain unchanged, the expected profit of the governmental supervision and management department by loose supervision and management in short term is a negative value, then the governmental supervision and management department will certainly select strict supervision and management. Under the strict supervision and management by the governmental supervision and management department, the producers and operators will have to reduce the probability to produce and sell false and inferior agricultural products, so as to reach a new mixed equilibrium P_t^*. With increasing punishment $(-P)$, P_t^* gets smaller and smaller, i.e. the probability for the producers and operators to produce and sell false and inferior food gets smaller and smaller.

Secondly, the basis on which the producer and operator selects the strategy is: $E = (-D) \times (1 - P_g) + B \times P_g$, which is the expected profit of the producer and operator by producing and selling false and inferior food, wherein the minimum request is that $E \geq 0$, as shown in Fig. 133.3. In the figure, P_g^* represents the optimal level on which the governmental supervision and management department select loose supervision and management. When $P_g > P_g^*$, i.e. the probability of loose supervision

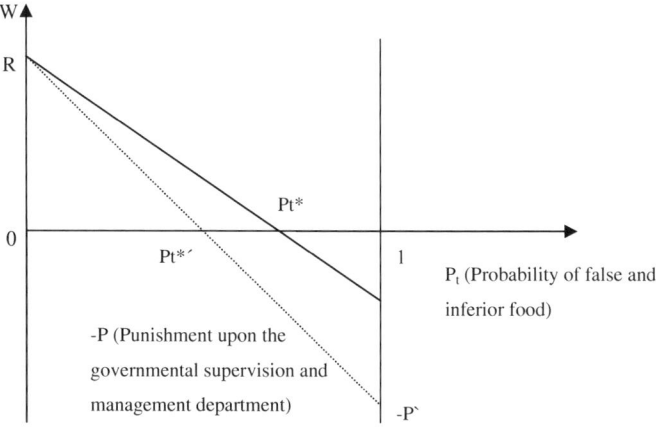

Fig. 133.2 Mixed strategy of producer and operator

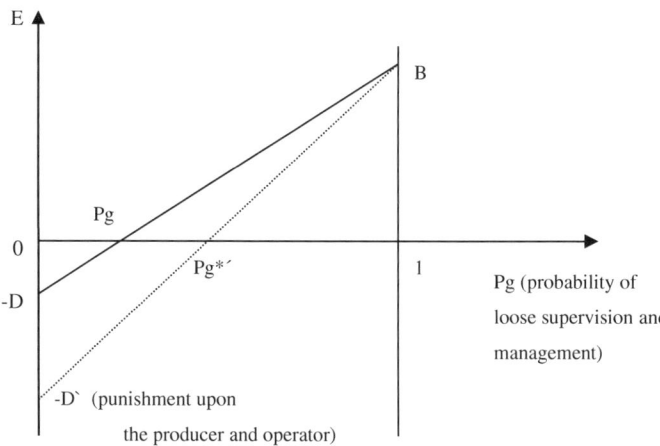

Fig. 133.3 Mixed strategy of governmental supervision and management department

and management by the governmental supervision and management department is greater than the optimal probability level, $E \geq 0$, the rational producer and operator will select producing and selling false and inferior food; When $P_g < P_g^*$, i.e. the probability of loose supervision and management by the governmental supervision and management department is smaller than the optimal probability level, $E < 0$, the producer and operator will select producing and selling safe food; under the circumstances that the producer and operator select produce and sell safe agricultural products, the greater the probability of loose supervision and management by the governmental supervision and management department is, the greater the profit will be till P_g^* is reached.

If the punishment is increased upon the producer and operator ($-D'$ as shown in Fig. 133.3), with the probability P_g^* of loose supervision and management by the governmental supervision and management department unchanged, the value for the producers and operators to select producing and selling false and inferior food is negative, then the producers and operators will choose to produce and sell safe food. After gaming for a certain period of time, the governmental supervision and management department will increase the probability of loose supervision and management, till reaching a new mixed equilibrium P_g^*. With increasing punishment ($-D$), P_g^* is greater and greater, i.e. the probability of loose supervision and management by the governmental supervision and management department is greater and greater.

Thirdly, the probability of producing and selling false and inferior food by the producers and operators and the probability of loose supervision and management by the governmental supervision and management department are analyzed together. It can be seen from the analysis above that, by increasing the punishment level upon the governmental supervision and management department, the probability of producing and selling false and inferior food by the producer and operator will decrease; but by increasing the punishment level upon the producers and operators, the probability of loose supervision and management by the governmental supervision and management department will increase instead. The same measure of punishment renders different influences for both parties.

In practice, the punishment upon the producers and operators by the government of China is not necessary to be mild. The "Food Safety Law" enforced in June, 2009 stipulates that: the highest penalty amounts 10 times of the original cost of the unsafe and inferior food. The laws and regulations issued in September, 2010 even include death penalty as a punishment measure. Why such severe punishment still cannot shock and frighten the dark minded producers and operators? If investigated, all the multiple factors mentioned in the beginning of the article can be deemed as the cause. However, the prominent problem is that the supervision and management department is dependent on high pressure of laws and regulations, and loosens supervision and management, which renders supervision and management not in place and even negligence of duty. As far as merely the year 2011 is concerned, the inspecting authorities in China registered, investigated and disposed 57 criminals accused of delinquency or dereliction, misconduct in food safety management, etc. In the case of the "clenbuterol Issue", disciplinary action of party discipline and government discipline was taken against 74 government employees. This indicates that it is relatively easy to issue laws and regulations, standards and requirements, while it is key to avoid absence of supervision and management. As shown in the analysis, only with strict accountability and increased punishment upon the governmental supervision and management department, may the probability of producing and selling false and inferior food by the producer and operator decrease, and the false and inferior agricultural products reduce from the very fountainhead, so that the life and health of the people can be further improved.

133.3 Conclusion and Suggestion

The food safety situation in China is still severe; therefore, laws and regulations, technical specifications, basic equipments, and education and training, etc. need to be implemented simultaneously, wherein it is the key to implement supervision and management. It is obligatory to intensify the punishment upon the behaviors of producing and selling false and inferior food; however, it is more important to reinforce accountability and punishment upon the governmental supervision and management departments. When the supervisors and managers keep a positive posture of supervision and management all the time, can the false and inferior agricultural products be reduced radically, so as to further improve the life and health of the people. Therefore, the following measures should be taken: firstly, increase the number of organizations and equipments on the basic level to decrease the difficulties in supervision and management; secondly, reinforce personnel training and improve ability of supervision and management; and thirdly, intensify accountability and punishment of loose supervision and management upon the governmental supervision and management departments, so as to increase pressure on nonfeasance.

Acknowledgments Fund Project: The Modern Agricultural Industry Technology System (2013 year)—Beijing Poultry Innovation Team.

References

1. Xing S (2011) Supreme law: criminals in food safety case convicted and given heavier punishment. http://news.qq.com/a/20111125/000075.htm (In Chinese)
2. Nelson P (1970) Information and consumer behavior. J Polit Econ 78:311–329
3. Darby M, Karni E (1973) Free competition and the optimal amount of fraud. J Law Econ 16:67–88
4. Akerlof GA (1970) The market for lemons: quality, uncertainty and the market mechanism. Q J Econ 84:488–500
5. Yang S, Song T (2013) Issues of agro-food quality and security in China: characteristics, cruxes and countermeasures. Res Agric Mod 5:293–297
6. Qiao J, Li B (2008) To explore the causes and countermeasures of food quality safety problem in China. Chin J Anim Sci 8:23–26 (In Chinese)
7. Zhang C (2009) Food quality safety supervision system reconstruction at market failure and government failure. Gansu Soc Sci 2:242–245 (In Chinese)
8. Zhao Y, Gao S (2010) The "lemons" problem existing in agricultural produce market and its countermeasures—a literature review from the perspective of food quality and safety. Chin Agric Sci Bull 10:70–76 (In Chinese)
9. Chen XJ (2011) An analytical framework and supervision system for Chinese government to protect food quality and safety. J Nanjing Norm Univ (Soc Sci) 1:29–36 (In Chinese)
10. Jin W, Xing J (2010) Gambling analysis of agricultural products quality safety detection. Econ Res Guide 5:246–248 (In Chinese)
11. Zhang W (2004) Game theory and information economics, 1st edn. Shanghai People's Publishing House, Shanghai (In Chinese)

Chapter 134
Research on the Design of Smart Portal for Chinese Cities

Yong Huang, Ronghua Zhu and Hao Dong

Abstract Since the 21st century, Chinese cities' rapid development has brought a range of problems such as environment, transport, security, public health management, which has challenged the city management. Through the advanced information techniques such as internet and cloud computing, smart management of the city can be achieved. After the issues mentioned above are analyzed, the definition of smart city is given. Applying the ideas and methods of information system, an urban smart portal which comprises external and internal logic architecture is designed for smart management of Chinese cities.

Keywords City management · Smart city · Smart portal

134.1 Introduction

In 2010, China's GDP surpassed Japan to become the world ranking second largest economy country. Chinese cities have developed so rapidly that a range of problems about city management appeared.

134.1.1 Background

The world's oldest city, Memphis, was built in the lower reaches of the Nile River around 3200 BC. In China, the most ancient city that has a continuous history is

Y. Huang · R. Zhu (✉)
Business School, Sichuan University, Chengdu 610065, People's Republic of China
e-mail: 869079376@qq.com

H. Dong
Nanjing Digital China Ltd., Nanjing 210000, People's Republic of China

J. Xu et al. (eds.), *Proceedings of the Eighth International Conference on Management Science and Engineering Management*, Advances in Intelligent Systems and Computing 281, DOI: 10.1007/978-3-642-55122-2_134, © Springer-Verlag Berlin Heidelberg 2014

Suzhou, which has kept the size of two thousand five hundred years ago, when the king HE LU constructed. There is no way of detailed research about how the city managers managed the city thousands of years ago, but the management of a modern city can be studied. China now has more than 660 cities and Chinese government hopes that China's urbanization rate will reach 51.5 % by the end of 2015. In general if urbanization rate of a country reaches 50 %, it marks the initial formation of the country's urbanization. It continues to be a problem in every city manager closely concerned how to make a city wise and serve well for the public.

At present, many countries have started the construction of smart city, mainly located in the United States, Germany, France, and China, Singapore, Japan. The construction of smart city is at the stage of the limited size and small-scale exploration. July 2009, the IT Strategy Headquarters of Japanese government designed a people-oriented "i-Japan 2015 strategy", which advocated "peace of mind and vibrant digital society", so that the digital information technology could flow into every corner like air and water. The strategy includes three core areas of e-government, health and education. September 2009, the US city of Dubuque and IBM jointly announced that it would build America's first smart city. IBM would use a range of new technologies to rebuild Dubuque City, where all the resources of its fully digital resources would be connected together. Dubuque City could detect, analyze and integrate various data, and intelligently respond to public demand, reduce urban energy consumption and costs, would be more suitable for residential and commercial development. March 2010, the European Commission introduced the "Europe 2020 Strategy", which proposed three key tasks, namely smart growth, sustainable growth and inclusive growth.

China's smart city construction has just started, being in active use of the Internet of Things, cloud computing and other new technologies, promoting smart urban construction. At present, there are three main modes of smart city construction, namely: the Internet of Things industry-driven construction mode, such as Wuxi; construction mode of the information infrastructure as a precursor, such as Wuhan; construction mode of social services and management applications to breakthrough, such as Beijing.

134.1.2 Related Literatures

The topic of city management has been studied long by some scholars. Recently Xuemei proposeda multilayer transfer interpretive structural modeling (ISM) to construct the sustainable resolution mechanism of urban flood disaster [17]. Lixia built a set of evaluation index system to measure competition of the financial industry cluster in different regions in China, the empirical results in the paper provided a theoretical support for local government to take a series of measures to promote the formation of financial agglomeration and establish a regional financial center [19].

Although the topic of city management has been studied for a long time, the discussion of the smart city began in the past ten years. In 1996, A run analyzed the

Singapore case and concluded that the challenge of constructing a smart city was to put IT in the service of humankind instead of using it for the subversion or the destruction of the values and ways of life people held dear [3]. Taewoo suggested strategic principles aligning to the three main dimensions (technology, people, and institutions) of smart city: integration of infrastructures and technology-mediated services, social learning for strengthening human infrastructure, and governance for institutional improvement and citizen engagement [14]. Taewoo viewed the smart city movement as innovation comprised of technology, management and policy; he then discussed inevitable risks from innovation, strategies to innovate while avoiding risks [15]. Hafedhi dentified eight critical factors of smart city initiatives: management and organization, technology, governance, policy context, people and communities, economy, built infrastructure, and natural environment [8]. Hans explored smart cities as environments of open and user-driven innovation for experimenting and validating future internet-enabled services [9]. Jung introduced an integrated road mapping process for services, devices and technologies capable of implementing a smart city development R&D project in Korea and applied a QFD (Quality Function Deployment) method to establish interconnections between services and devices [10]. George used a procedure based on fuzzy logic to propose a model that can estimate "the smart city" [5]. Andrea presented the optimization of a Tubular Permanent Magnet-Linear Generator for energy harvesting from vehicles to grid and developed the optimization process by means of hybrid evolutionary algorithms to reach the best overall system efficiency and the impact on the environment and transportation systems [1]. Luis presented a novel architecture which took advantage of both the critical communications infrastructures already in place and owned by the utilities as well as of the infrastructure belonging to the city municipalities to accelerate efficient provision of existing and new city services [11]. There were several research papers about smart city in IEEE Communications Magazine of 2013 June. Nils presented a theoretical framework for the analysis of platform business models that involved public actors and city governments in particular, in the value network [12]. Zuqing presented d a few energy-saving algorithms to improve the energy efficiency of cable access networks that would play an important role in the ICT sector of future smart cities [20]. Giuseppe proposed a crowd sensing platform with three main original technical aspects: an innovative geo-social model to profile users along different variables; a matching algorithm to autonomously choose people to involve in participity Actions and to quantify the performance of their sensing; and a new Android-based platform to collect sensing data from smart phones [6]. Gang presented six research issues in trace analysis and mining of moving objects in a city and discussed five promising application domains in smart cities [13]. Ignasi proposed a procedure to make smart cities happen based on big data exploitation through the API stores concept and a viable approach to scale business within that ecosystem [16]. Antoni identified a number of privacy breaches that could appear within the context of smart cities, leveraged some concepts of previously defined privacy models and defined the concept of citizens' privacy as a model with five dimensions: identity privacy, query privacy, location privacy, footprint privacy and owner privacy [2].

China's first literature of smart city research is 2003's Dave [4]. Dave introduced Calgary's experiences and methods in a program called smart communities project as an initiative to "connect the Canadians". Guo proposed the method of the smart city evaluation system whose core was the smart city evaluation indexes system, evaluation method and optimization strategy [18]. Guo Xinjian studied the relationship between smarter manufacturing and smarter city and analyzed various requirements of smarter manufacturing in smarter city [7].

The remainder of the article is organized in three sections. The problems mentioned are stated clearly and the causes of the problems are analyzed deeply in Sect. 134.2. In Sect. 134.3 in-depth analysis is made to describe the specific problems of urban management and an urban smart portal including external and internal logical architecture is designed. Section 134.4 concludes.

134.2 Problem Statement

While Cities are growing steadily, cities consume 75 % of worldwide energy production and generate 80 % of CO_2 emissions [10]. The rapid development of urbanization has brought a series of problems such as city environmental pollution, traffic congestion, security management, and public health management.

1. Environmental pollution

   In order to pursue the rapid pace of development, city managers tend to ignore the protection of the environment, carry on a large scale of urban infrastructure projects including buildings, roads. As a result, large number of urban green space was occupied by reinforced concrete high-rise, people begin to worry about PM2.5 average of 24 h in their cities, and in some cities migratory birds rarely return.

2. Traffic congestion

   The wise traffic management is forgotten during the rapid development of the city, so urban traffic carrying capacity limit is usually exceeded. The long-term overload of city transportation network, the private car ownership out of control and non-intelligent urban traffic management make people's travel inconvenient because time is wasted on the road.

3. Security management

   A lot of people now move to cities, lack of effective registration and management brings law and order problems. It is more difficult to find out criminals hiding in crowd, safety monitoring system of city is unable to detect criminal behavior in time,the fact that Indian cities malignant cases occurred in succession in recent years is an example.

4. Public health management

   People always get sick and require good treatment, but difficult and expensive medical treatment get people very uncomfortable.When having a common disease, people used to go to big hospitals and queue for a long time to see doctors, but only got a very short time of diagnosis, while doctors are very idle in other small medical institutions.

The above problems led to a decline in the quality of people's lives, leading to low levels of public dissatisfaction of urban services and city managers.These problems can be solved in the context of smart city construction, applying ideas and methods of information systems to build a universal smart city management structure.

134.3 Approach

Through the construction of urban smart portal, social resources will be integrated, and the efficiency of government public services will be improved.Furthermore, the cost of public access to government services can be reduced, and the quality of community life can be enhanced as well.

Smart city means that the activities and demands of urban residents, enterprises and the government can be perceived and processed intelligently. Smart city utilizes emerging information technologies such as Internet, cloud computing and so on in the area of urban infrastructure, management of environment and community, and municipal management. A city will be built as a new urban ecosystem, which covers urban residents, enterprises and the government, supported by new techniques mentioned before. An urban smart portal is a service system whose resource is integrated, flexible, open and intelligent. It can achieve precise matching and optimal combination among the service, channels and the object, and can improve the public services effectively. The following research is mainly focused on Chinese cities for the reason of convenience.

134.3.1 Requirements Analysis

1. Present Situation Analysis of Urban Public Service

Services associated with urban residents and enterprises can be divided into the government's services, public utility services, business services and information services. Citizens generally get the services through the following channels such as the administrative management service center, government departments hall, the web site of each department, service hot line, and streets and community service, etc.; get the public utility services through the service hall, the sites and the services agent; and get the business and information services from their residential communities, internet, and so on. Four aspects of problems can be found in the services as bellow.

Dispersion of the service

Each department and unit has services related to residents, but these services are distributed on different channels such as the service hall and the websites of every unit.

Relatively traditional service mode

The main external services from departments and units are mainly provided through the service hall, service stations, telephone, website and other channels. Although intelligent terminals, mobile phones and other applications are increasingly common at present, the application of public service on these new technology platforms is rare.

Passive service

Services are provided only when citizens come, call on or submit their application online at present. The transformation from passive service to active service is needed now.

Lack of personalized service

At present, services provided by departments and units are mainly for the masses, but the personalized custom-made reference service is in short.

2. Description of Requirement

Integrate the service resources

The government services information and the service contents of each department can be brought together into one website, which can make it convenient for citizens to get the information.

Construct a unified platform

Through the integration of existing systems, a unified city smart portal can be established, providing multi-channel services such as hot line services, service website, mobile phones, and intelligent terminals. People can obtain services provided by different departments and units only through this platform when putting forward a demand in his virtual life space on the platform.

Expand the service channels

A variety of service channels such as urban smart portal, citizen hotlines, mobile terminal, etc., are required to be built. In this way, we can make full use of different channels and achieve the complementary advantages among them.

Perfect the service system

According to different requirements, we can provide point to point services, diversified services, personalized services on the smart portal.

134.3.2 Design of Urban Smart Portal

Based on the requirement analysis demonstrated in Sect. 134.3.1, the urban smart portal should make it possible for citizens to submit applications, handle administrative matters, inquire work schedule and evaluate the service quality online via the Internet, mobile phones, mobile terminals and other channels. The following parts describe the logical design of the smart portal.

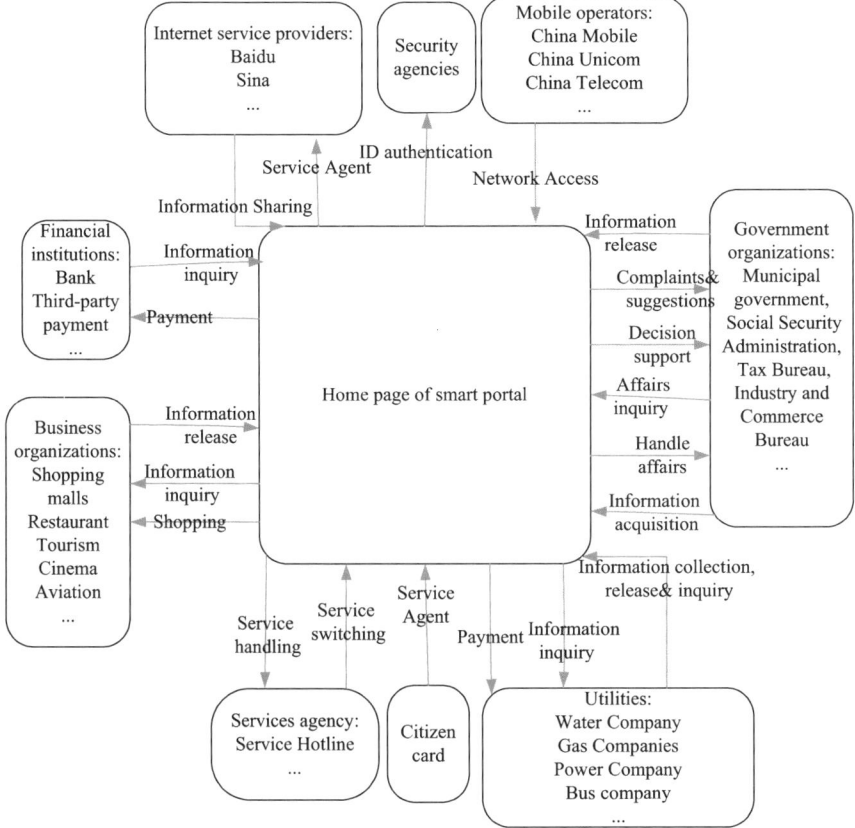

Fig. 134.1   The external logic architecture of urban smart portal

1. External logic architecture of urban smart portal

External logic architecture refers to the relationship between the smart portal and the external entities; the external logical architecture of the urban smart portal is shown in Fig. 134.1. The relationship between urban smart portal and every external entity is described as follows.

As an operational and management unit of urban smart portal, municipal government now can release information such as policies and so on, it can also accept citizens' suggestions and complaints that are proposed through the smart portal. Furthermore, the administrative departments of municipal government can get performance management and decision-supporting provided by the smart portal. As service provider for citizens, other government departments can release information, provide services such as information inquiry and affairs handling online, they can also provide datum of various indicators by data exchange for the smart portal, which will be used for management and decision making.

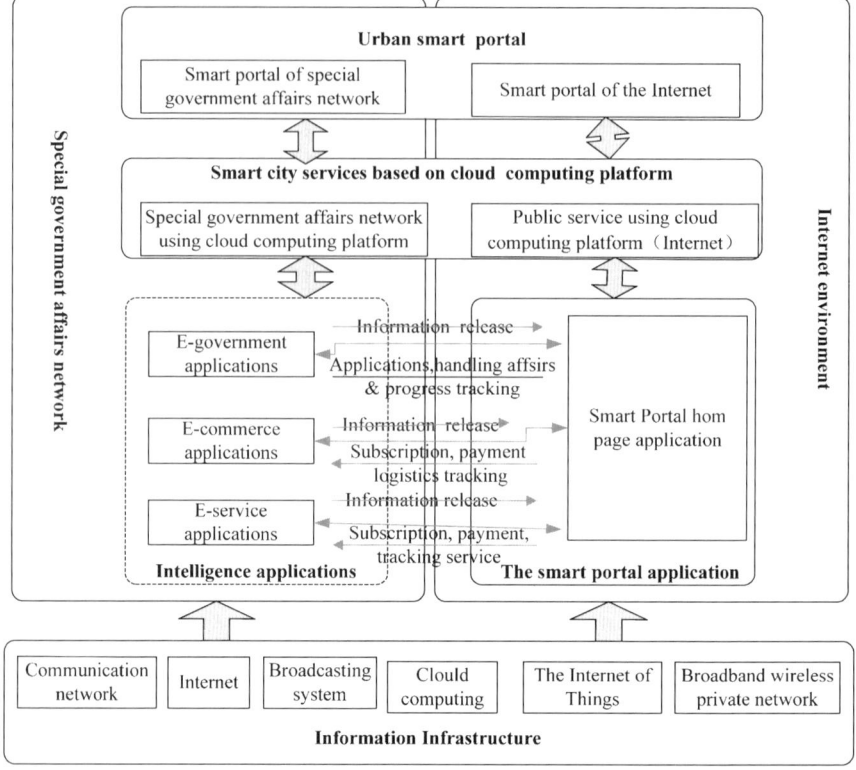

Fig. 134.2 The internal logical architecture of urban smart portal

Utilities release information and provide services such as information inquiry, affairs handling and charging online. Commercial organizations can release information such as special offers and group purchase, and provide online shopping service through the smart portal. Financial institutions mainly provide services such as online payment and information inquiry. Security agencies provide the third-party security certification for the smart portal. Services agency such as hot lines and smart portal can handle and switch each other's services as well as realize information sharing between them. Cooperation with the urban smart portal, the internet service providers can share information and provide service agency access for the smart portal. Mobile operators such as China Mobile, China Unicom, China Telecom provide network access services for the smart portal.

2. Internal logic architecture of urban smart portal

The internal logic architecture of urban smart portal is shown in Fig. 134.2. In the figure we can see that the smart portal can achieve a variety of intelligent applications based on the urban information infrastructure such as the Internet, communications network, cloud computing and so on. The intelligent applications include

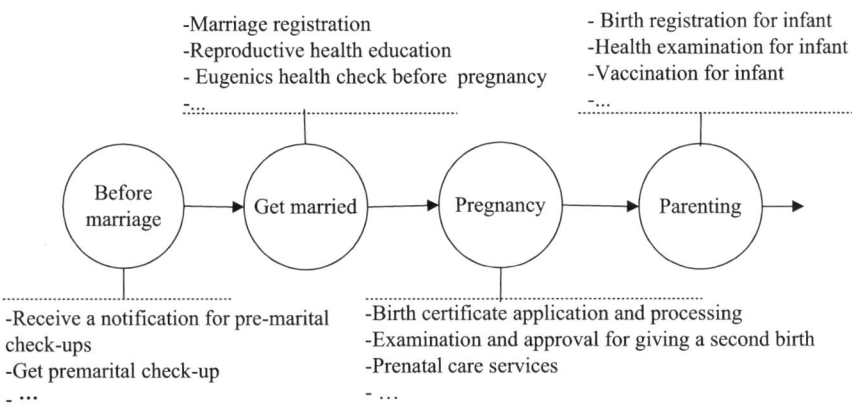

Fig. 134.3 Marriage and childbearing process

e-government, e-commerce and e-services which are associated with citizens is shown in the figure as well. While the smart portal application is mainly refers to the home page application, by which way, citizens can get information and services provided by e-government, e-commerce and so on.

3. Example description

Through the smart portal, citizens can access to various services and handle related matters through a combination of online and offline mode. The marriage and child rearing process with Chinese characteristics is chose to be described, as seen in Fig. 134.3. All procedures involve services of multiple departments, which can be handle din a way of a combination of online and offline. There are many processes in the marriage and childbearing integration services, now we take application process to give a second birth for example, which belongs to the pregnancy procedure, to do the example description.

First, people can get, fill in and submit the application form for giving birth to another child. Then, they submit their IDs, marriage certificate and birth certificate to community workers for auditing, the workers scan and upload the files, and then sign opinions online, submit the audit conclusion to the street office. When the audit is completed, it will be submitted to the Family Planning Commission by the street office for approval. After that, the audit conclusion will be publicized for 10 days by the community office. If there is no disagreement during the time, certificate of permitted birth will be issued to the citizen, who can get the certificate form the community. While during the publication and audit, one who is not qualified for having another child, will be informed by emails or short message service, and get written notification of result on the smart portal. Finally, if there are different opinions on the audit conclusion given by street office or county's family planning office, one can request for reconsideration, and get the result provided by department of a higher rank. The specific process is shown in Fig. 134.4.

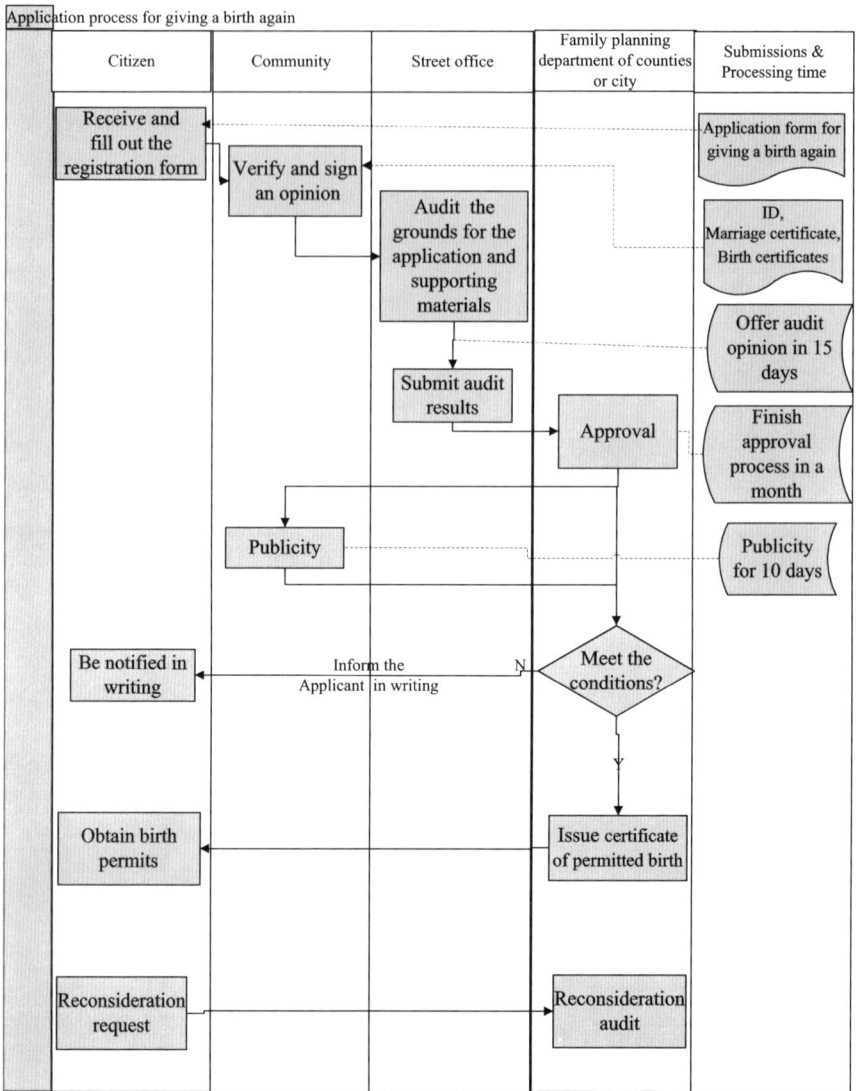

Fig. 134.4 Application process for giving a second birth

134.4 Conclusion

An urban smart portal which consists of external and internal logical architecture is proposed. Most social resources are integrated through the construction of the smart portal. From the smart portal, citizens can get many kinds of services, enterprises can realize online sales of their products and services, and city managers can manage the city efficiently. But in the future, more specific research and construction of the smart portal require participation of related professionals and lots of efforts provided by all components of the society.

References

1. Andrea P, Francesco G et al (2012) Novel speed bumps design and optimization forvehicles' energy recovery in smart cities. Energy 5:4624–4642
2. Antoni MB, Pablo AP, Agusti S (2013), The pursuit of citizens' privacy: A privacy-aware smart city is possible. IEEE Communications magazine june, pp 136–141
3. Arun M (1999) Smart cities: the singapore case. Cities 16:13–18
4. Bronconnier D (2003) Calgary, Canada-a smart community. Technoeconomics Manage Res June 7
5. George CL, Mariacristina R (2012) Definition methodology for the smart cities model. Energy 47:326–332
6. Giuseppe C, Luca F, et al. (2013) Fostering participaction in smart cities: A geo-social crowd sensing platform. IEEE Communications magazine june, pp 112–119
7. Gu X, Dai F et al (2013) Relationship between smarter manufacturing and smarter city. Comput Int Manuf Syst 19:1127–1133
8. Hafedh C, Taewoo N, Shawn W (2012) Understanding smart cities: An integrative framework. In: 45th Hawaii international conference on system sciences, pp 2289–2297
9. Hans S, Nicos K, Marc P (2011) Smart cities and the future internet: towardscooperation frameworks for open innovation. Future Int Assembly LNCS 6656:431–446
10. Jung HL, Robert P, Sang-Ho L (2013) An integrated service-device-technology roadmap for smartcity development. Technol Forecast Soc Change 80:286–306
11. Luis S, Ignacio E et al (2013) Integration of utilities infrastructures in a future internet enabled smart city framework. Sensors 13–14(438–14):465
12. Nils W, Pieter B (2013), Platform business models forsmart cities: From control andvalue to governance and public value. IEEE Communications magazine june, pp 72–79
13. Pan G, Qi G, et al. (2013) Trace analysis and mining forsmart cities: Issues, methods, and applications. IEEE Communications magazine June, pp 120–126
14. Taewoo N, Theresa AP (2011) Conceptualizing smart city with dimensions of technology, people, and institutions. In: 12th annual international conference on digital government research, pp 282–291
15. Taewoo N, Theresa AP (2011) Smart city as urban innovation: Focusing on management, policy, and context. ICEGOV2011, pp 185–194
16. Vilajosana I, Llosa J, et al. (2013) Bootstrapping smart cities through a self-sustainable model based on big data flows. IEEE Communications magazine june, pp 128–134
17. Wang X, Huang R (2013) Research on construction of sustainable resolution mechanism for urban flood disaster based on ISM. In: SixthInternational conferenceon management scienceand engineering management, pp 201–208
18. Xirong XG, Wu X (2013), Research and construction of smart city evaluation system. Comput Eng Sci 167–173
19. Yu L, Yu W, Wen W (2013) The empirical research between the financial industry clusters and regional economicdevelopment. In: Sixth international conference on management scienceand engineering management, pp 329–341
20. Zhu Z, Lu P et al (2013) Energy-efficient wideband cable access networks in future smart cities. IEEE Communications magazine June, pp 94–100

Chapter 135
The Technological Paradigm for Growing Trends of Carbon Utilization

Bobo Zheng, Qiao Zhang and Lu Lv

Abstract The world's energy demand are growing but conventional fossil fuels such as oil, coal and natural gas are drying up at present. Meanwhile renewable energies are becoming more prevalent but are still a long way from being common place in many parts of the world. Under this circumstance, CCUS technologies could play a key role in abating CO_2 emissions immediately. This study propose a concept of the carbon utilization technological paradigm (CUTP) which includes CUTP competition, CUTP diffusion and CUTP shift phases. This paper aims to apply technological paradigm theory to analyze the process of CUTP buildup, and to find out BECCS which is the future trajectory of carbon utilization technologies based on Web of Science, NoteExpress and NodeXl softwares. The regulatory framework in this paper provides scholars with new thoughts for their further studies. Our research also guide and accelerate development of carbon utilization technologies. It draws a conclusion that subsidies and incentives should be provided in specific fields such as capital costs and technological support. Finally, BECCS must be one of the integral energies buildup.

Keywords Technological paradigm · Technological trends · GHG mitigation · Carbon utilization

135.1 Introduction

The growth of global economic and energy consumption continue to soar since the 21st century. Consequently the economic development and social progress are affected by the energies. Although countries in the world have reached an agreement

B. Zheng (✉) · Q. Zhang · L. Lv
Low Carbon Technology and Economy Research Center, Sichuan University,
Chengdu 610064, People's Republic of China
e-mail: bobogeorge@163.com

J. Xu et al. (eds.), *Proceedings of the Eighth International Conference on Management Science and Engineering Management*, Advances in Intelligent Systems and Computing 281, DOI: 10.1007/978-3-642-55122-2_135, © Springer-Verlag Berlin Heidelberg 2014

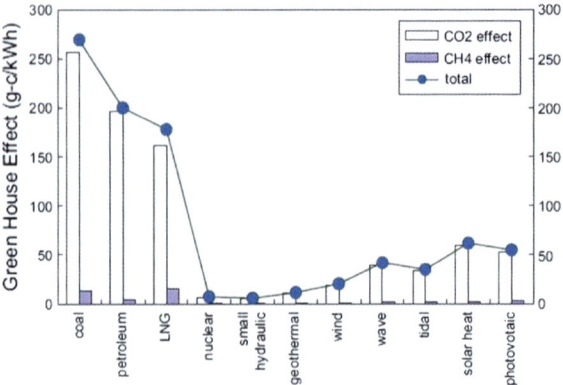

Fig. 135.1 Greenhouse effect by fuel

to build a low-carbon economy, a lot of fossil energy uses, which account for near 80 % energy of the word we used at present, directly lead to more serious environmental pollution and global climate changing as shown in Fig. 135.1 [8, 9]. It is generally accepted that GHG emissions control is the most challenging environmental policy issue worldwide, and carbon dioxide (CO_2) are chief culprits among them. Statistics indicate that CO_2 emissions concentration has increased from a pre-industrial level of 280– 380 ppm at present. As a result, the average global temperature has led to an increase by about 1 °C during the same time period [4]. The International Panel on Climate Change (IPCC) and other similar studies have predicted that the concentration of CO_2 in the atmosphere may reach 570 ppm on present trends [23, 24]. According to studies, energy-related CO_2 emissions would increase from 27.1 Gigatonnes (Gt) of CO_2 in 2008 to 40.4 Gt in 2030 [12]. As a result, it will leads to global average temperature rise of 6 °C or so.

In order to achieve a cost optimized scenario for stabilizing CO_2 emissions at 450 ppm by 2050 suggested by The International Energy Agency (IEA) in 2009 [10], the clean and renewable incorporate energy technologies buildup is necessary, which involves greater energy efficiency, commercialization of renewable energy technologies, nuclear power, and carbon capture storage and utilization (CCUS). Developing and implementing sustainable energy policies is the most obvious path, which would not only provide sufficient energy, but also protect the environment and the integrity of ecosystems. Given the cost and technical feasibility of clean energies such as the solar energy, the wind power and so on, most of these new ones are still more expensive and less deployed than conventional energy sources which can steadily meet the primary world's energy demand now [14]. Conventional fossils fuels including oil, coal and natural gas, on the other hand, remain the most economically available and dominant source of energy. Coal, oil and gas can account for 25, 24 and 21 % of global energy consumption respectively [19]. Under this circumstance, CCUS technologies could play a key role in abating CO_2 emissions immediately. It is accurate to consider CCUS technologies as a chain of technologies created to capture CO_2 from carbon sources, transport it to storage sites and extract it from the atmosphere, making all kinds of production by utilization [11]. Furthermore, CCUS is a promising technology for a reduction in the overall cost of stabilization [1]. IEA's studies strongly agree with this viewpoint, indicating that 19% of total emissions reduction could come from CCUS [10]. Although it is not a real renewable technology, CCUS methodologies especially carbon utilization technologies are important bridges between our current lifestyles and an environmentally friendly ultimate goal.

The paradigm was firstly defined by Kuhn in his masterpiece "structure of the scientific revolution" in 1962 [13]. He believed that paradigm refers to those commonly accepted scientific achievements, and it provides typical questions and answers for communities of practice over a period of time. In 1982, Dosi, a technical innovation economist, combines the concept of technology innovation with paradigm, and puts forward technological paradigm (TP). The TP is a model which can solve economic problems, while the solution is based on the principle of natural science [3]. He also thought that theories of technical change have generally followed two different basic approaches. One theory is "demand-pull" which means that market forces are the determined factor for technical change. The other is "technology-push", which prescribes technology as an independent factor. Any existence has its own gradually evolutionary process, TP build its evolution based on the corresponding driving force following this rule. The two main factors driving the evolution of TP are market demand and dominate technologies competition in the industry respectively. Thus, finding these factors are extremely important through the theoretical framework of TP.

In this paper, we analyze trends of carbon utilization technologies combining TP rules, as a result propose a novel concept of the carbon utilization technological paradigm (CUTP). As mentioned above, under the promotion of market demand and dominate technologies competition, the evolution of TP has specific characteristics at each stage, which is also verified by the analysis of keywords focus trend from network graphs by using NodeXL. Following the TP rules, each evolution of TP can be roughly divided into three stages as shown in Fig. 135.2. After research and analysis from keywords trend, we can draw a conclusion that CUTP is accordance with TP rules. CUTP competition, as a first stage, is competition-oriented. There are some primary technologies at this stage. Under dual action of demand-pull and technology-push, leading technologies appears and prevails in potential commercial market which push and build the CUTP diffusion as the second stage of CUTP. In the end, the technologies become not advanced with the development of the market, which are the characteristics of the third stage. As the market becomes saturated, the original advanced technology can not meet the development needs of the new market. It is the urgent that novel advanced technologies should be invented and commercialization, which is the CUTP shift stage as the third one.

In this article, we firstly presents the method used for literature's keywords analysis in Sect. 135.2 and then draws the conclusion based on the result of keywords focus trend under the guide of theoretical framework of TP. Then the novel concept of CUTP is proposed. Through theoretical framework of paradigm, we can review carbon utilization technologies, and also conclude a methodology for future CUTP

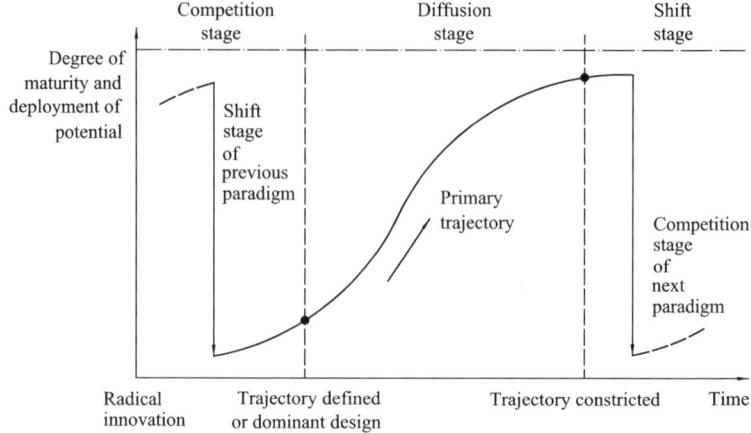

Fig. 135.2 The technological paradigm evolution diagram

development and acquire new ideas of it. So Sect. 135.3 offers an overview of CUTP shift. Evaluation and discussion is presented in Sects. 135.4 and 135.5 finally offers conclusions and recommendations for future research.

135.2 Literature Review

In this section, we look at the past achievements in carbon utilization literature and establish the search strategy and selection criteria.

1. The Data Analysis System Buildup

To find out the keywords trends, the data analysis system (DAS) is composed of the Web of Science database (WoS), NoteExpress and NodeXL. WoS was chosen as the primary database, NoteExpress was selected for the general characteristics analysis, and NodeXL was used as a further analysis tool for the literature. Consequently, a progression step by step meta-synthesis method was developed as the basis for further in-depth research.

Web of Science. As the worlds most trusted citation index covering the leading scholarly literature, Web of Science (WoS) is an online subscription-based scientific citation indexing service maintained by Thomson Reuters that provides a comprehensive citation search. It gives access to multiple databases that reference cross-disciplinary research, which allows for in-depth exploration of specialized subfields within an academic or scientific discipline. Therefore, Web of Science provides researchers, administrators, faculty, and students with quick, powerful access to the world's leading citation databases.

NoteExpress. NoteExpress is the most professional literature retrieval and management system in China. NoteExpress can help us search efficiently and automatically (including from the Internet), download and manage literature and research papers through a variety of ways. The core function of NoteExpress covers all aspects of knowledge management including knowledge acquisition, management, application,and mining, which is the essential tool for academic research and knowledge management.

NodeXL. The tool used for keywords focus trend analysis is NodeXL. NodeXL is a powerful and easy-to-use interactive network visualization and analysis tool that leverages the widely available MS Excel application as the platform for representing generic graph data, performing advanced network analysis and visual exploration of networks. With NodeXL, we can enter a network edge list in a worksheet, click a button and see your graph, all in the familiar environment of the Excel window. This tool generally supports multiple social network data providers that import graph data (nodes and edge lists) into the Excel spreadsheet. Here, we use NodeXL to analyze keywords focus trend.

2. Keywords Focus Trend Analysis

In order to analyze keywords focus trend, we firstly find about 400 pieces of related articles from WoS, which has the same topic about carbon utilization. In this step, after the first search, we will reselect articles to get the target literature through filtering like selecting the research field and literature types etc. Then, we download these 400 pieces of related articles directly to NoteExpress by the link between WoS and NoteExpress. When we finish importing the articles records to NoteExpress and removing duplicates, we can extract every keyword from these articles, and import them to NodeXL grouped by the year, which varies from 1991 to 2013. Therefore, 'year' and 'keyword' as two vertexes form an edge. After calculating and rearranging, we can see the result network in graph, which is shown in Fig. 135.3. It shows that biomass energy are the research focus almost through 1991 to 2013.

In this figure, we can clearly see keywords focus show a trend of each phase with the increment of the year. Below the vertexes of 'year', there are several keywords such as 'biomass energy', 'energy', 'technology', which are the main topic of our research. So, these keywords are the focus throughout the whole years from 1991 to 2013. From the year 1991 to 1999, we can find that the keywords 'nitrogen', 'growth', 'microalgae', 'photosynthesis' and so on are the main research focus because the connecting lines between 'year' and 'keyword' mainly concentrate in the above years. Besides, from the year 2000 to 2009, keywords such as 'carbon', 'ecosystems', 'soil organic carbon', 'carbon sequestration', etc are the main research focus for the same reason talked above. Meanwhile, in the figure, keywords like 'biomass energy' is the research hot spot through 2001–2013.

3. Analysis Results

Above analysis shows that the carbon utilization research focus is generally divided into three stages. we can summarize that nitrogen is the main carbon utilization technology in the first stage. Soil organic carbon and ecosystems are the two main biomass technologies in the second stage, which is also coming from the keywords such as gasification, carbon sequestration etc. Finally, we can predict the potential carbon

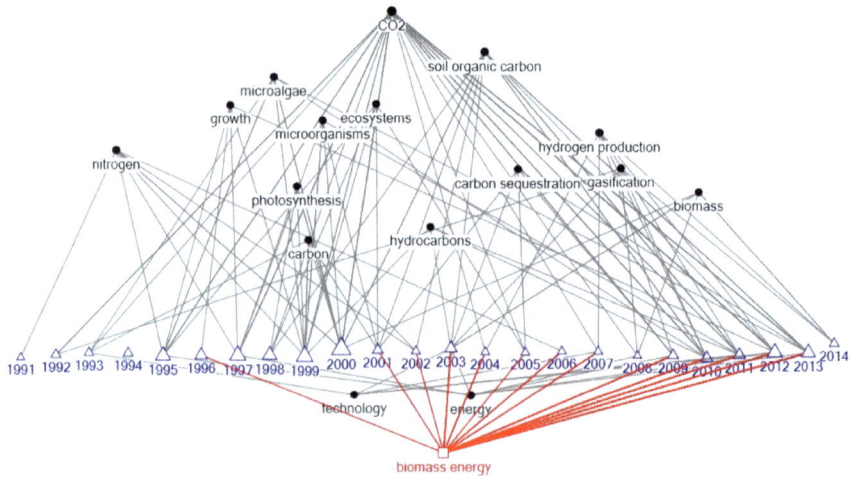

Fig. 135.3 Keywords focus trend of carbon utilization

utilization technology will be biomass energy. Therefore, these three stages are just in accordance with theoretical framework of paradigm. So, we can combine the TP with carbon utilization and propose a novel concept of the carbon utilization technological paradigm (CUTP), which includes CUTP competition, CUTP diffusion, CUTP shift. These stages will be discussed in the below article.

135.3 The Future Trajectories of CUTP

Apart from CUTP competition and diffusion stages, we only focus on the trend trajectories of CUTP shift phase in this paper. Among the technological options for meeting current GHG stabilization targets, it is believed by [5] that biomass energy with carbon capture and storage (BECCS), theoretically, might be essential for achieving stabilization targets of below 450 ppm CO_2. BECCS is able to remove atmospheric CO_2 and to meet the energy demand, which is the delivery of power and heat with net negative emissions [19]. Bioenergy production from agricultural and forestry resources is absolutely necessary, not only to mitigate climate change, but also to ensure the security of energy supply. Application of BECCS is a compromise between fossil fuel alternatives and GHG emissions controlling [21]. As biomass both are an important carbon sink and a substitution for fossil-fuel, it is worthwhile to focus on CO_2 balance in biomass energy systems with CO_2 capture and storage [15].

Rhodes proposed that implementation of gasification, post combustion capture and oxyfuel combustion with CO_2 capture, make it possible to apply CCS to biomass in the production of electricity, hydrogen and liquid biofuels [17]. This can be used in a wide range of fields, including biomass power plants, combined heat and power

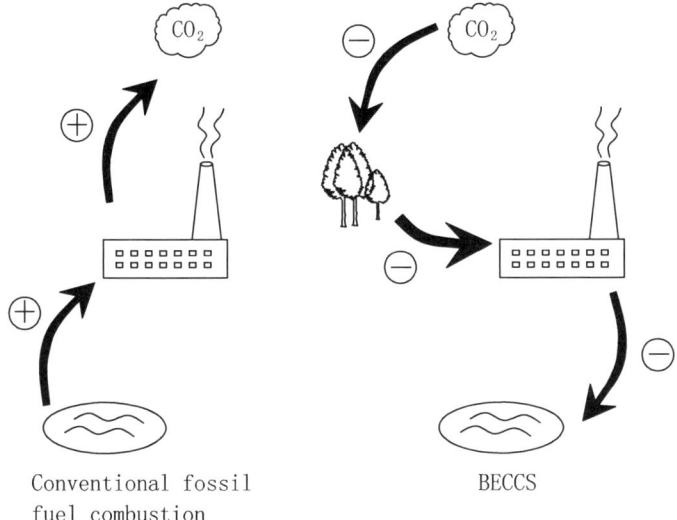

Conventional fossil   BECCS
fuel combustion

Fig. 135.4 Comparison of BECCS and conventional fossil fuel combustion

plants, flue gas streams from the pulp industry and so on. As shown in Fig. 135.4, unlike conventional CCS, which can at most realise zero emissions, BECCS can achieve negative emissions considering the permanent CO_2 geological sequestration and sustainable biomass harvesting practices during its growth via photosynthesis [19]. If the CO_2 emitted within the closed cycle were to be captured and permanently stored, it could achieve net zero or even negative emissions. Considering that biomass is sustainable because CO_2 from biomass combustion or transformation is neutral, BECCS can make a contribution to a net removal of anthropogenic CO_2 emissions worldwide [18].

According to Obersteiner's study, BECCS not only removes carbon from the natural carbon cycle, but also provides low-carbon energy products through the process of combustion and photosynthesis [16]. Because this mythology is likely to substantially reduce the mitigation costs of CO_2 in the atmosphere, it will become one of the future energy mix [18]. However, as BECCS also has region compatibilities, some countries and regions will be more suitable for large BECCS applications than others. For example, Sweden may be well placed an established biomass energy plants in the process industry and access to offshore storage sites, maintaining relatively low power sector emissions [15]. On the contrary, biomass energy resource in UK is relatively small. So BECCS is probably designed to co-firing instead of a dedicated biomass system. This means that establishing smaller scale biofuel plants adjacent to the existing CO_2 transport and storage infrastructure will maximize their performance. In a word, BECCS must be one of the integral energies buildup and a promising trajectory towards the sustainable world in the future. Moreover, from the perspective of the development of technology and cost control, it can also better commercialization in a large scale.

135.4 Evaluation and Discussion

As this article has mentioned, the CUTP evolution is slower than ordinary commercial paradigms. Additional incentive is one of the main reasons, which could be useful for encouraging biological sequestration, capture, or storage. Providing a safe solution for long-term CO_2 storage, BECSS can give leverage to the carbon-reduction potential of the world's biomass resources. The advantage for policymakers of providing the extra incentive at capture is that they can use the same basic administrative infrastructure as for stimulating CCS from fossil fuels. However, some countries may wish to put the BECCS incentive into biological sequestration if this fits better with their biomass policies. BECCS has not been economically viable because environmental taxes only focus in conventional environmental policy instruments, which is a key issue confronting researchers. Moreover, the regulations providing guidance for GHG reduction in the Kyoto framework do not consider BECCS as eligible for the first commitment period of the protocol [7]. As a result, there are still no incentives for governments and sectors to capture CO_2 from biomass. To overcome this barrier, it is strongly recommended that a compromise policy based on a carbon tax and a subsidy on BECCS should be taken into consideration. In the end, a specific subsidy for emissions reduction as the prime mover is an important guarantee to stimulate BECCS of technological development [18].

In a word, in view of energy and the environment, the effective utilization of GHG emissions, particularly CO_2, propose both a serious challenge and a major opportunity for sustainable development. Different fuel, materials even chemicals can be synthesized using CO_2, therefore CO_2 utilization can contribute to enhancing sustainability. In this way, it should be a sustainable way in the long term when renewable energies are used as a basic energy [22]. On the other hand, negative CO_2 emissions through the implementation of BECCS could give leverage to the carbon-reduction potential of the world's biomass resources [15]. Biomass production is still subject to a range of sustainability constraints like fresh water, scarcity of arable land, loss of biodiversity, competition with food production, deforestation and so on [20]. Besides these deficiencies, BECCS offer a promising opportunity for a net removal of CO_2 emissions while meeting energy needs.

135.5 Conclusions

This study aimed to apply TP theory to analyze the CUTP process, to conclude a novel direction form future carbon utilization technologies by Wos, NoteExpress and NodeXL. According to the result analysis, there are nitrogen, growth, microalgae, photosynthesis and so on in the early period of carbon utilization. With the development of economy and society as well as the increasing demand for energy, some new technologies which are more suitable for market revolution is needed, and this kind of technology become the dominant technology in that period. It easily shows that carbon, ecosystems, soil organic carbon, carbon sequestration are the prevail technologies in CUTP diffusion phase. However, some limitations or problems appear with development of potential market and environment requirements, biomass energy will become a research hot issue of carbon utilization through 2001–2013. After primarily analyzing the future routine of CUTP shift stages, it can explain the evolution of CUTP S-curve as shown in Fig. 135.2.

From perspective of the environmental issues, a variety of low-carbon methods are about to take some effects, which include the high efficiency of fossil fuels use or the exploitation of renewable and new energy like solar, wind, biomass etc. Technology policies are one of the options available for the reduction of carbon emissions and the use of energy [6]. Especially CCUS technologies play a vital role in GHG abating. As Bowen's study pointed out, CCUS should be a corporate technology strategy buildup [2]. In this paper, from the different carbon utilization technologies' maturity periods, we analyze the keywords focus trend first by some methods, and then get the result that there are three stages: CUTP competition, CUTP diffusion and CUTP shift, which are in accordance with paradigm greatly. On the other hand, scientists all over the world are making efforts to improve the environment and some achievements have been done for low-carbon economy, there is still no paradigmatic framework as a common guidance for CCUS technologies now. Hence, the progress achieved in the various technologies is at different levels, with some of them overlapping. In this article, We hope that this idea could provide new future research thoughts to scholars all over the world. Further, the specific subsidy incentives for governments and sectors should be considered to allow for acceleration in CUTP development.

At this phase, however, there are still many uncertainties about which technologies could bring about real improvements and which really have no opportunities for reducing CO_2 emissions. Further researches are required to evaluate specific cases selected in this field.

Acknowledgments This research is supported by the Major Bidding Program of National Social Science Foundation of China (No. 12&ZD217).

References

1. Akimoto K, Tomoda T (2006) Costs and technology role for different levels of CO_2 concentration stabilization. In: Avoiding dangerous climate change, Cambridge University Press, pp 355–360
2. Bowen F (2011) Carbon capture and storage as a corporate technology strategy challenge. Energy Policy 39(5):2256–2264
3. Dosi G (1982) Technological paradigms and technological trajectories: a suggested interpretation of the determinants and directions of technical change. Res Policy 11(3):147–162
4. Drage T, Blackman J (2009) Evaluation of activated carbon adsorbents for CO_2 capture in gasification. Energy Fuels 23(5):2790–2796
5. Erlach B, Harder B, Tsatsaronis G (2012) Combined hydrothermal carbonization and gasification of biomass with carbon capture. Energy 45(1):329–338

6. Greening AL, Greene DL, Difiglio C (2000) Energy efficiency and consumption-the rebound effect: a survey. Energy Policy 28(6):389–401
7. Grönkvist S, Möllersten K, Pingoud K (2006) Equal opportunity for avoided CO_2 emissions: a step towards more cost-effective climate change mitigation regimes. Mitig Adapt Strat Glob Change 11(5–6):1083–1096
8. Gurney KR, Mendoza DL (2009) High resolution fossil fuel combustion CO_2 emission fluxes for the United States. Environ Sci Technol 43(14):5535–5541
9. Hoel M, Kverndokk S (1996) Depletion of fossil fuels and the impacts of global warming. Resour Energy Econ 18(2):115–136
10. IEA (2009) IEA CCS Technology Roadmap
11. IPCC (2005) Intergovernmental panel on climate change special report on carbon dioxide capture and storage. IPCC, Cambridge University Press, New York
12. Kaygusuz K (2012) Energy for sustainable development: a case of developing countries. Renew Sustain Energy Rev 16(2):1116–1126
13. Kuhn T (1962) The structure of scientific revolutions. University of Chicago Press, Chicago
14. Lund P (2006) Analysis of energy technology changes and associated costs. Int J Energy Res 30(12):967–984
15. Möllersten K, Yan J, Moreira JR (2003) Potential market niches for biomass energy with CO_2 capture and storage–opportunities for energy supply with negative CO_2 emissions. Biomass Bioenergy 25(3):273–285
16. Obersteiner M, Azar C (2001) Managing climate risk. Science 294(5543):786–787
17. Rhodes JS, Keith DW (2005) Engineering economic analysis of biomass IGCC with carbon capture and storage. Biomass Bioenergy 29(6):440–450
18. Ricci O (2012) Providing adequate economic incentives for bioenergies with CO_2 capture and geological storage. Energy Policy 44:362–373
19. Ricci O, Selosse S (2013) Global and regional potential for bioelectricity with carbon capture and storage. Energy Policy 52:689–698
20. Sachs I (2007) The biofuels controversy. In: United nations conference on trade and development
21. Schmidt J, Leduc S (2011) Cost-effective policy instruments for greenhouse gas emission reduction and fossil fuel substitution through bioenergy production in Austria. Energy Policy 39(6):3261–3280
22. Song C (2006) Global challenges and strategies for control, conversion and utilization of CO_2 for sustainable development involving energy, catalysis, adsorption and chemical processing. Catal Today 115(1):2–32
23. Speight JG (2007) Natural gas: a basic handbook. Gulf Publishing Company, Houston
24. Susan S (2007) Climate change 2007-the physical science basis: working group I contribution to the fourth assessment report of the IPCC, vol 4. Cambridge University Press, Cambridge

Chapter 136
Cultural Determinants on Information Technology

Cheng Luo and Michael Amberg

Abstract The way we develop and implement information technology is affected by our cultural values. A body of research of cultural impacts on IT product design, IT project implementation, IT-enabled process change and IT-enabled organizational change has demonstrated that information technology is deeply dependent on the environmental context and social human factors from a cross-cultural perspective. However, most research were based on Hofstede's unrivaled theory of culture, which were criticized by many IS researchers as too limited and simplistic. To our knowledge, there still lacks a comprehensive investigation which includes all possible determinants of culture that have impacts on information technology. In this study we identify a set of dimensions of culture from the most known culture theories through quantitative and inductive analysis to form a research framework of culture and a classification of cultural impacts on information technology, which provides an agenda for systematically researching the cultural impacts on information technology.

Keywords Cultural determinants · Information technology · Cross-cultural management of IT

136.1 Introduction

Whether information technology is culture-neutral or not has been long argued by researchers and practitioners. A set of studies treats IT as an independent influence on human behavior that exerts unidirectional impacts over humans and organizations,

C. Luo (✉)
Business School, Sichuan University, Chengdu 610064, People's Republic of China
e-mail: 18980007676@189.cn

M. Amberg
Business Information Systems III, University of Erlangen-Nuremberg,
91054 Erlangen, Germany

J. Xu et al. (eds.), *Proceedings of the Eighth International Conference on Management Science and Engineering Management*, Advances in Intelligent Systems and Computing 281, DOI: 10.1007/978-3-642-55122-2_136, © Springer-Verlag Berlin Heidelberg 2014

similar to those operating in nature [1]. Studies of this perspective [2] have focused on IT as "hardware" that humans use in productive activities or daily life, whether industrial devices or information systems.

On the contrary, and more recent, another set of work on information technology has reverted to a "soft" determinism where the force of information technology is moderated by human actors and environmental contexts, especially by cultural differences and structural properties. Indeed, information technologies do work by the same universally applicable scientific laws in every enterprise, in every country, even on the moon. But "the very success of the universalistic philosophy now threatens to become a handicap when applied to interaction between human beings from different cultures" [3]. Plenty of exploratory research works have shed light on the cultural impacts on information technology. Amberg [4] has defined a research model DART which extends Davis's TAM model to comprise influences of environmental contexts on technology acceptance such as cultural differences, economic situations and so on. Sun [5] and He [6] analyzed the cultural impacts on Website Design, whose works were based on Hofstede and Hall's theory about culture. Martinsons [7] has focused on impacts of national culture in business process change and Group Support Systems (GSS). Gallivan [8] has argued that adoption, assimilation and diffusion of information technology are more subject to cultural differences. Trillo [9] considered the cultural diversity in the design and development of user interfaces of software and proposed a Cultural User Interface (CUI). Even standard business software such as ERP software, which is considered "looks and behaves pretty much the same all over the world" [10], is to certain extent subject to culture variables [11]. Recently, Waring [12] intended to add to the debate through a longitudinal case study of an integrated information system implementation undertaken within a large UK university. Skoumpopoulou [13] put forward that trust in the system, the business processes around SITS, the data quality and the people who operate it is extremely important and where lost can impact upon other activities in the organization. And Lee [7] has found that people tend to seek information on their own from direct and formal sources in individualistic cultures, whereas in collectivistic cultures, people rely more on subjective evaluation of an innovation, conveyed from other-like-minded individuals who already have adopted the innovation.

Literature review in information system (IS) field has shown the significant impacts of culture on information technology, ranging from product design to system implementation and deployment. Sensitivity to cultural differences becomes a critical success factor for international IT product design, IT project implementation and IT-enabled organizational change. But scratching the surface of culture can result more complexity, without offering tools to systematically analyze it. Our current knowledge of which, why and how cultural factors affect the planning, design and implementation of information technology products and projects are still limited. To our knowledge, most of the existing researches in IS field are based on Hofstede's unrivaled five dimensions of culture. Several highly regarded MIS researchers have pointed out that studies based on Hofstede's dimensions of national culture have inherent limitations which need to be addressed in future studies. For example, Myers and Tan [14] criticized these studies based on Hofstede's theory "too simplistic" and argued for "a more dynamic view of culture". And this study seeks to advance our theoretical knowledge by systematically analyzing the cultural dimensions defined by Hofstede, Trompenaars, Hall, Schwartz and Kluckhohn, excluding overlapped and dependent dimensions of different authors and identifying independent dimensions of culture. Johannes [15] tried to use Hofstede's cultural dimensions to uncover what commonalities were constructed in the process, and three elements seem to play a role when cultures meet: Reduction of communicative uncertainty, construction of shared meaning, and appropriate use of technology. Esma [16] studied the relations between Hofstede's dimensions, Schwartz's cultural values, Worldwide Governance Indicators and UNPD Human Development Index by correlations, multiple regression analysis, moderator and mediator analyses as well as path analyses.

The theoretical research in this study identified a set of independent cultural variants by searching for equivalences across cross-sectional culture frameworks. This can overcome the problem of simply relying on a limited scope of concepts from one perspective. The identified cultural variables were then used to consider various facets of IT products and IT projects to develop a series of propositions from cross-cultural perspective. Then empirical test will be carried out to validate these propositions. This will be addressed in another paper.

136.2 Culture Literature Review

In order to analyze the cultural impacts on information technology, we must first identify and accept the multiple hidden dimensions of culture. Many anthropologists and psychologists have defined and researched "culture" from different perspectives and accordingly identified many determinants of culture. Hofstede [17] defined culture as "the collective programming of the mind which distinguishes the members of one group or category of people from another". Shalom [18] viewed culture as a latent when presenting his theory of seven cultural value orientations and applying it to understand relations of culture to significant societal phenomena, and he thought the culture can be measured only through its manifestations. A survey of the literature at the time of writing reveals that several cultural frameworks exist. We focus on those advocated by Hofstede, Trompenaars, Hall, Kluckhohn and Schwartz. Table 136.1 provides an overview of the most known cultural frameworks in cultural research literature.

Despite the differences are noticeable, we can still find some common features of culture defined by different authors. Some of the similarities are both in semantics and in nature, for example, Hofstede's individualism/collectivism and Trompenaars's individualism/communitarianism. Both authors define this dimension of culture as the degree to which a society values individual independence versus community membership. Therefore we can consolidate these two dimensions into one. Some features are different in semantics, i.e. Hofstede's power distance and Schwartz's Hierarchy/Egalitarianism. Power distance is defined as the degree to which a society accepts a hierarchical structure and unequal distribution of social power, while

Table 136.1 Overview of cultural literature

Authors	Cultural dimensions
Hofstede	Power distance
	Individualism/collectivism
	Masculinity/femininity
	Uncertainty avoidance
	Long-term orientation/short-term orientation
Trompenaars	Universalism/particularism
	Individualism/communitarianism
	Neutral/emotional
	Specific/diffuse
	Achievement/ascription
	Attitude to time
	Attitude to environment
Hall	Communication context
Kluckhohn	Nature of people
	Persons relationship to other people
	Persons relationship to nature
	Primary mode of activity
	Conception of space
	Persons temporal orientation
Schwartz	Conservatism
	Intellectual autonomy
	Affective autonomy
	Hierarchy
	Egalitarianism
	Mastery
	Harmony

Hierarchy/Egalitarianism also reflects motivational domain of value about the distribution of social power. On account of the similarity in nature we can treat with these two determinants as one dimension when analyzing the cultural impacts on information technology.

Since all five frameworks are describing the same phenomenon, culture, one may argue that the existing difference in definition of culture by different authors is due to semantics and that one should delve into the details of the frameworks to find better matches and correlations. The differences found in these frameworks are partly due to the differences in authors' viewpoints, but also because of the complex and comprehensive nature of culture. It is quite likely that each of the frameworks captures some unique aspects of culture which are not found in the others. It is also quite possible that some characteristics of culture have not yet been found or might escape this classification. To find out new distinctive features of culture exceeds the scope of our work. Our concentration is to analyze the existing dimensions of culture based on the five most known frameworks at hand and to identify independent determinants of culture to form the research base for further analysis of the cultural influence upon information technology. This work is discussed in the next section.

136.3 Cultural Determinants on IT

136.3.1 Research Methodology

There exist two research philosophies in IS research field for a long time: epistemology (what is know to be true) and doxology (what is believed to be true). Accordingly Galliers [19] identified two major research approaches, namely positivist and interpretivist (anti-positivist). Positivists claim that reality is relatively stable and can be observed and described from an objective viewpoint [20] without interfering with the phenomenon being studied. The representative research methodologies of Positivist are laboratory experiments, field experiments, surveys, case studies, theorem proof, forecasting and simulation [19]. Alavi and Carlson [21] also indirectly supported this view, who, in a review of 902 IS research articles, found that all the empirical studies were positivist in approach.

Interpretivists, however, contend that only through the subjective interpretation of reality can the reality be fully understood. They argue that scientists cannot avoid affecting those phenomena they study. The representative research methodologies of Interpretivist are argument, reviews, action research, case studies, futures research and role playing [19].

It is not easy to definitively say which methodology is intrinsically better than any other methodology, many researchers calling for a combination of research methods in order to improve the quality of research [22]. We adopt this point of view and believe that both research methods are valuable if used appropriately, so we include elements of both the positivist and interpretivist approaches in our research.

136.3.2 A Research Framework of Culture

Hofstede and Trompenaars have collected abundant data used for empirical test and measured the cultural difference through theoretical reasoning and statistical analysis. This enables us to analyze and compare cultural variables defined by the two authors not only through inductive reasoning but also through statistical analysis techniques. We conducted the spearman rank correlation analysis to compare dependence of some dimensions defined by Hofstede and Trompenaars, using SPSS as an instrument. The results of the correlation analysis accord well with that of theoretical reasoning and are consequently expected to be convinced. We could not carry out same quantitative analysis of determinants defined by other three authors owing to lack of quantitative data. Research approaches of interpretivist thus play an important role in the analysis of dimensions of other authors, in comparison with those backed up by the availability of empirical data.

Our aim in this study is to develop a framework of culture as a basis to analyze cultural impacts on information technology. Therefore we focus more on human values and behaviors whose impacts are relative directly linked to development and implementation of information technology and organizational process, rather than general beliefs about the way we see the world. Some dimensions, i.e. Kluckhohn's nature of people, which argued that man is a mixture of good and evil, are too subjective and rely more on general beliefs. We can find little relation between this dimension and information technology and consequently exclude it of our framework. Another set of dimensions, such as Trompenaars's neutral/emotional, specific/diffuse, Kluckhohn's conception of space, focus more on individual level rather than on cultural level, although generally claimed to work at the culture level. We do not deny the possible impact of these dimensions on information technology because of their significance in understanding and communicating with people of another nation, yet they are not included in our framework due to lack of common sense in terms of core values at the culture level. Excluding the less influential dimensions, we focus on the rest cultural dimensions.

All five dimensions defined by Hofstede are adopted in our framework not only because of its wide and enhanced authority in the culture research field but also its direct or indirect influence in information technology studies validated by many researchers in IS field [23, 24]. Although these dimensions' power distance and individualism are strongly negatively correlated and represent only one single factor in a confirmatory factor analysis, Hofstede [25] argued that they are conceptual different and were independently developed as indices with respect to extensive literature base. Furthermore, the impacts of these indices on information technology are not the same. Thus we include these dimensions as independent factors in our framework.

Comparing Trompenaars's work with Hofstede's, we found the dimension individualism/communitarianism is very similar to the dimension individualism/collecti-vism of Hofstede, both in semantics and in nature. The further spearman rank correlation analysis also shows significant statistical correlation between these two dimensions. The result of the correlation analysis is demonstrated in Table 136.2 and the original data of the country values based on these two dimensions can be found in [3]. This dimension is consequently excluded from our framework, for it can be substituted by Hofstede's similar dimension of culture.

For another two dimensions of Trompenaars, namely achievement and attitudes to time the result of the correlation analysis showed that attitudes to time is to some extent correlated to the dimension long-term/short-term of Hofstede, and that achievement/ascription is at certain level correlated to the dimension masculinity/femininity of Hofstede. Tables 136.2, 136.3 and 136.4 showed the results of correlation analysis.

The extent of correlation of these dimensions are not as strong as that of individual, which could be attributed to the inadequate number of items in the questionnaire designed by Trompenaars [3] compared to that of Hofstede. Comparing the definition and interpretation of these dimensions, we can find that they are very similar in nature. Attitude to time is interpreted by Trompenaars [3] as the time horizon of person, how man values past, present and future. In Hofstede's work, long-term/short-term is associated with the values of people toward future and tradition. Both dimensions, as well as Kluckhohn's dimension person's temporal orientation, all reflect people's

Table 136.2 Spearman rank correlation analysis of individualism of Hofstede and Trompenaars

			Individualism of Hofstede	Individualism of Trompenaars
Spearman's rho	Individualism of Hofstede	Correlation Coefficient	1	0.613[a]
		Sig. (2-tailed)	–	0.001
		N	26	26
	Individualism of Trompenaars	Correlation Coefficient	0.613[a]	1
		Sig. (2-tailed)	0.001	–
		N	26	26

[a] Correlation is significant at the 0.01 level (2-tailed)

Table 136.3 Spearman rank correlation analysis of Hofstede's long-term/short-term and Trompenaars's attitudes to time

			Long-term/ short-term	Attitudes to time
Spearman's rho	Long-term/ short-term	Correlation coefficient	1	0.634[a]
		Sig. (2-tailed)	–	0.015
		N	14	14
	Attitudes to time	Correlation coefficient	0.634[a]	1
		Sig. (2-tailed)	0.015	–
		N	14	14

[a] Correlation is significant at the 0.05 level (2-tailed)

Table 136.4 Spearman rank correlation analysis of Hofstede's masculinity/femininity and Trompenaars's achievement/ascription

			Achievement/ ascription	Masculinity/ femininity
Spearman's rho	Achievement/ ascription	Correlation Coefficient	1	−0.452[a]
		Sig. (2-tailed)	–	0.04
		N	21	21
	Masculinity/ femininity	Correlation Coefficient	−0.452[a]	1
		Sig. (2-tailed)	0.04	–
		N	21	21

[a] Correlation is significant at the 0.05 level (2-tailed)

values towards future or past (tradition), thus we include only one in our framework. Similarly, masculinity/femininity of Hofstede and achievement/ascription of Trompenaars can all be interpreted to reflect the degree to which a society values achievement in terms of success and status. So we only include Hofstede's dimension masculinity/femininity as a representative of consolidation.

Attitudes to environment of Trompenaars, together with Kluckhohn's person's relationship to nature, all reflect the role people assign to their natural environment. That means dominating the environment or conforming to the environmental change. We argue that this dimension has significant impact on the IT-enabled organizational change and process change, and the correlation analysis also showed that there exists

Table 136.5 Spearman rank correlation analysis of Hofstede's uncertainty avoidance and Trompenaars's universalism/ particularism

			Uncertainty avoidance	Universalism/ particularism
Spearman's rho	Uncertainty avoidance	Correlation coefficient	1	−0.480[a]
		Sig. (2-tailed)	–	0.038
		N	19	19
	Universalism/ particularism	Correlation coefficient	−0.480[a]	1
		Sig. (2-tailed)	0.038	–
		N	19	19

[a] Correlation is significant at the 0.05 level (2-tailed)

no significant correlation to other dimensions in our framework. Hence we include this dimension in our framework as a new variable "attitude to change".

Trompenaars's dimension universalism/particularism reflects people's attitude to rule and exceptional situation. Universalist is rule-based and tends to resist exceptions that might weaken the rule, whereas particularist's judgement varies with the changing environment, even the rule is violated. This corresponds well to Hofstede's dimension uncertainty avoidance, which reflects the degree to which a society creates rules and beliefs to avoid or minimize ambiguous situations. The result of correlation analysis also demonstrates this correlation in Table 136.5. Therefore this dimension is not included in our framework either, represented by Hofstede's alternative dimension.

Hall's dimension communication context and Kluckhohn's dimension person's relationship to other people concern the way people communicate with others or the way in which information is transmitted. Information in low context transactions is more codified in comparison to that in high context transaction, where pre-programmed information with minimal code is transmitted in message. This dimension can to a great extent influence the design of information technology, especially user interface, as we have found in the literature [5, 6, 9, 26]. Thus, it would be useful to integrate this dimension into our framework.

Schwartz's seven dimensions can be summarized in three value dimensions. Conservatism (also called embeddedness) is a value type that emphasizes the maintenance of traditional values or the traditional order. This dimension has similar implication to the time orientation dimension in our framework. Opposed to this value are two distinct autonomy value types, intellectual autonomy and affective autonomy. The two autonomy types both promote individual benefit, rather than group benefit. Schwartz's harmony value type reflects a harmonious relationship with the environment, whereas its opposite value type mastery emphasizes an active mastery of the (social) environment. The implication of these two dimensions is same to our dimension attitude to change. Another summarized dimension can be found with a further two opposing value types: hierarchy versus egalitarianism. We can not find difference in nature between this value type and the dimension power distance. Therefore we argue that all dimensions of Schwartz's work are a summary of other authors

Table 136.6 Cultural
dimensions in IT studies

Cultural dimensions	Original authors
Power distance	Hofstede
Individualism/collectivism	Hofstede, Trompenaars
Masculinity/femininity	Hofstede
Uncertainty avoidance	Hofstede
Time orientation	Hofstede, Trompenaars, Kluckhohn
Attitudes to change	Trompenaars, Kluckhohn, Schwartz
Communication context	Hall, Kluckhohn

and can be interpreted by dimensions in our framework. Consequently none of these dimensions is included in our framework.

Through quantitative and inductive analysis we identify seven dimensions of culture to form a framework for analyzing cultural impacts on information technology. Table 136.6 showed an overview of this framework.

Based on this framework, the next step of our systematical research is to find out which cultural dimensions have what influence upon which facets of information technology. We contend that not all dimensions of culture have impacts on each facet of IT. For example, concerning IT-enabled business process change, it is found that the dimension time orientation can have significant impact on IT-enabled process re-engineering. Short-term oriented cultures will have relatively greater preference for radical IT-enabled process re-engineering than long-term oriented cultures, for this type of initiative can impair the general status of the organisation and its people and thus endanger long-term survival [27]. But we can not find obvious impact of this dimension in the IT product design phase. On the contrary, we find the dimension communication context can to a great extent influence the cross-cultural user interface design, but no significant influence of this dimension can be found in the implementation phase of IT product. Therefore a classification of cultural impacts on different phases of IT product is necessary in terms of systematical analysis. Schwalbe [28] divided the life cycle of IT products into four phases: Concept, Development, Implementation and Close out. We utilize this taxonomy and our framework of culture to classify the cultural impacts on information technology. Table 136.7 provides an overview of our classification.

A subsequent set of propositions are suggested based on this classification, which will be discussed in another paper. Our classification is neither complete nor scientifically confirmed. It needs to be continually complemented and validated. We hope to set an agenda for a systematical analysis of cultural impacts in IS field.

Table 136.7 Classification of cultural impacts

| | Information technology product | | | |
	Concept	Development	Implementation	Close out
Power distance		Process model design	IT-enabled organizational change	Customer acceptance
Individualism /collectivism			Change management	Customer acceptance
Masculinity /femininity			IT-enabled process change	
Uncertainty avoidance			Process change	Customer acceptance
Time orientation	Partner choosing		Process change implementation strategy	Customer acceptance
Attitudes to change			IT-enabled organizational change	
Communication context		Product design: UI, functional module		Customer acceptance

Correlation is significant at the 0.05 level (2-tailed).

136.4 Conclusions

This study is one of our serial research works toward a systematical analysis of cultural implications and impacts on information technology. A body of organizational and IS literature has demonstrated that an awareness of cultural difference is critical to the success of the design of IT product, implementation of IT project and IT-enabled organizational change. We suggested a systematical research toward this issue, which presently lacks in this research field. The first step of this systematical research is to identify all the independent dimensions of culture which directly or indirectly influence different facets of information technology. In this paper we utilize both methodologies in positivism and interpretivism to analyze the most known culture theories of Hofstede, Schwartz, Trompenaars, Hall and Kluckhohn and derive a framework of culture from these authors. We hope that the proposed framework of culture and classified cultural impacts on information technology can shed light on the cultural determinants in IT studies for further research.

Based on our framework of culture, we subsequently propose a set of propositions to provide an agenda for researching the cultural impacts on information technology and IT-enabled process and organizational change. An empirical test will be carried out through questionnaires and interviews in companies in Germany and China from a cross-cultural perspective to validate these propositions. This is beyond the scope of this paper and will be addressed in a future paper.

References

1. Giddens A (1984) The constitution of society: outline of the theory of structure. University of California Press, Berkeley
2. Hiltz SR, Johnson K (1990) User satisfaction with computer-mediated communication systems. Manage Sci 36(6):739–764
3. Trompenaars F, Hampden-Turner C (1997) Riding the waves of culture. Nicholas Brealey Publishing, London
4. Amberg M, Hirschmeier M (2003) Dart—an acceptance model for the analysis and design of innovative technologies. In: Proceedings of the seventh conference on synergetics, cybernetics and informatics
5. Sun H (2001) Building a culturally-component corporate web site: an exploratory study of cultural markers in multilingual web design. In: Proceedings of the 19th annual international conference on computer documentation
6. He S (2001) Interplay of language and culture in global e-commerce: a comparison of five companies' multilingual websites. In: SIGDOC Conference
7. Lee SG (2013) The impact of cultural differences on technology adoption. J World Bus 48:20–29
8. Gallivan JM (2001) Organizational adoption and assimilation of complex technological innovations: development and application of a new framework. ACM SIGMIS Database 32(3):51–85
9. Trillo NG (1999) The cultural component of designing and evaluating international user interfaces. In: Proceedings of the 32nd Hawaii international conference on system sciences
10. Waloszek G (2003) Chi 2003—new horizons, but what are they? http://www.sapdesignguild.org/community/readers/reader_chi2003_gw.asp
11. Mertens P, Ludwig P et al (1997) Mittelwege zwischen individual-und standard software. In: International conference on business information systems—BIS 97, pp 15–44
12. Teresa W (2013) Through the kaleidoscope: perspectives on cultural change within an integrated information systems environment. Int J Inf Manage 32:513–522
13. Waring T, Skoumpopoulou D (2012) Emergent cultural change: unintended consequences of a strategic information technology services implementation in a united kingdom university. Stud High Educ 19:1–17
14. Myers MD, Tan F (2002) Beyond models of national culture in information systems research. J Global Inf Manage 10(1):14–29
15. Cronjé JC (2011) Using hofstede's cultural dimensions to interpret cross-cultural blended teaching and learning. Comput Educ 56:596–603
16. Esma G (2013) How are cultural dimensions and governance quality related to socioeconomic development? J Socio-Econ 47:170–179
17. Hofstede G (2001) Culture's consequences: comparing values, behaviors, institutions, and organizations across nations. Sage Publications, Thousand Oaks
18. Schwartz SH (2013) National culture as value orientations: consequences of value differences and cultural distance. Handb Econ Art Cult 2:547–586
19. Galliers RD (1991) Choosing information systems research approaches. In: Information systems research: issues, methods and practical guidelines. Alfred Waller, Henley-on-Thames
20. Levin WC (1988) Sociological ideas: concepts and applications. Wadsworth Publishing Company, Wadsworth
21. Alavi M, Carlson P (1992) A review of mis research and disciplinary development. J Manage Inf Syst 8(4):45–62
22. Kaplan B, Duchon D (1988) Combining qualitative and quantitative methods in information systems research: a case study. Manage Inf Syst Quart 12(4):571–586
23. Garfield M, Gogan JL (2003) Vertical and horizontal information flows: the case of SARS. In: 11th CCRIS Meeting
24. Zakour AB (2004), Cultural differences and information technology acceptance. In: Proceedings of the 7th annual conference of the Southern association for information systems, pp 156–161

25. Hofstede G (1980) Culture's consequences: international differences in work-Related values. Sage Publications, Beverly Hills
26. Gould EW, Zalcaria N, Yusof SA (2000) Applying culture to website design: a comparison of Malaysian and us websites. In: Proceedings of IEEE professional communication society conference
27. Matinsons M, Hempel PS (1998) Chinese business process reengineering. Int J Inf Manage 18(6):393–407
28. Schwalbe K (2002) Information technology project management. Course Technology, Boston

Chapter 137
The Impact of Consumer Perceived Ethical Value on Trust and Brand Loyalty: Personality as Moderation Variable

Muhammad Kashif Javed, Muhammad Nazam, Jamil Ahmad
and Abid Hussain Nadeem

Abstract The purpose of this study is to examine the significance of perceived ethical value on trust and brand loyalty as well as to examine personality as moderation variable between perceived ethical value and trust on brand loyalty. A theoretical framework with hypothesized relationships is developed and tested in order to answer the research question. To test the research models, Cho-test sub-group moderation variable is use. Hypotheses which were suggested within the framework of the research model were tested with structural equation modeling. In this regard, data were collected for the fast moving consumer goods (FMCG). The results showed that perceived ethical value have positive impact on trust, perceived ethical value and trust have positive impact on brand loyalty. Personality is not moderated the relationship between perceived ethical value and trust to brand loyalty.

Keywords Perceived ethical value · Trust · Brand loyalty · Personality

137.1 Introduction

In recent research work, the importance of ethical corporate brand identity, stakeholder perceptions and outcomes are depicted in a wide array [22]. Also recent research has focused on formulation of ethical perception of consumers [28]. To focus on ethical issues, brands are portraying themselves as fair and socially responsible. Therefore, more brands claim to be ethical, there are brands that use ethical and environment friendly practices in their portfolio (Toyota prius), brands that are using

M. K. Javed (✉) · M. Nazam · J. Ahmad · A. Nadeem
Uncertainty Decision-Making Laboratory, Sichuan University, Chengdu 610064,
People's Republic of China
e-mail: mkjaved3@gmail.com

J. Xu et al. (eds.), *Proceedings of the Eighth International Conference on Management Science and Engineering Management*, Advances in Intelligent Systems and Computing 281, DOI: 10.1007/978-3-642-55122-2_137, © Springer-Verlag Berlin Heidelberg 2014

ethical practices in their supply chain (American apparel), and brand that invest in social causes (Pret a manager offer food to the homeless) [10]. Recent research on corporate branding has specially focused the perceptions that other stake holders have about the corporate brand reputation. Ethicality has become an important element for corporate brands. To promote their brands, an increasing number of companies following the ethical dimension [27].

This is because adopting an unethical behavior harms the reputation within buyers. With all the negative consequences this may signify for firm's retail and financial performance effect by not Purchases again, due to consumer objections, and may need penalty payments [5]. Furthermore, though existing consumer ethics studies have extensive to realizing of consumer faiths, positions, and behavior toward unethical business exercises, research on how consumer ethical perceptions can determine trust, satisfaction, and loyalty is relatively low [32]. Moreover, while consumer demographic and cultural aspects are vital in forming ethical perceptions and afterward reactions.

The call for a better realizing of consumers perceptions and reactions to brand unethical attitude is excusable on four main bases: first, as consumers are main stake holders in the marketing interchange procedure, it is essential to know their ethical perceptions in order to plan efficient business planning and programs [34]; second, brands are more and more trusting on fewer, but more extraordinary relationships with customers, and participation in unethical exercises is very likely to threaten these efforts [9]; third, consumers are probably to take several punitive assesses (boycotting goods), if they sense that their providing organizations are behaving unethically [6]; finally, there is demonstration that a arising number of consumers takes into thoughtfulness corporate social responsibility dissemination (such as ethics) in making their purchasing decisions [7].

Highlights the inconsistency among business and consumer views on ethicality, Brunk [5] distinguishes six areas of perceived ethicality origins. These areas contemplate with how a brand understanding with, employees, environment, consumers, overseas community, local community and business community and economy. Shea [28] highlights that while Brunk's perceived ethicality formulation does a good job about cognitive proportion of consumer behavior toward ethical/unethical attitude on component of the companies, there is a need to question the affective and behavioral proportions as well. We believe that our paper deals this concern by comprising affective and brand loyalty consequences from the consumers view, in increase to brand trust.

Moreover, Brunk [5], congruent with Shea [28], concludes that current research remains unsatisfactory about how powerfully ethical thoughtfulness feature in consumer's purchase conclusions and connect among perceived ethicality and consumer behavior demands further exploration. This study is a step in this way. The dissertation on corporate branding has paid special consideration to the perceptions that multitudinous stake holders have almost a given company and about the corporate brand reliability [1]. According to Morsing [27], morality has become a significant factor for corporate brands, and a flourishing number of companies are applying the ethical proportion as a strategic component in terms of determining and encouraging their brands. This is a key issue as while many investments in CSR and ethical actions are constructed at the corporate level, it is however true that an significant part of the fiscal return might be obtained at the product degree. This study deals this issue by studying the link among consumers perceived ethicality perceptions of a brand and product brand loyalty. Brand loyalty is central as it expedite customer retention proposition by producing an bearing attitudinal bind with customers that is not just compulsive by components such as price or accommodation [8], thus flourishing the anticipation of future re-buy and commendation.

The consideration of brand trust is established on the idea of a consumer brand relationship which is seen as a alternate for human link between companies and customers. Brand trust is one powerful element that effects customer loyalty. Trust turns the most important element in the relationship between a company and its customers and the relationship between a brand with customers [23]. As far as we aware, except recent growth in research under a corporate umbrella, on the impact of consumers ethical perceptions on the brand loyalty, there is no literature on it. This is a main issue because ethical activities and investments in CSR are made at the corporate level. At the product level, an important part of financial return might be obtained however it's true. Also, research on variables that effect brand loyalty have not introduced the type of personality, but personality can determine behavior, including attitudes for the brand. By studying the connection between consumers ethical perceptions of a brand at a corporate level and product brand loyalty, this paper addresses this issue by placing the type of personality as a moderating variable.

137.2 Theoretical Framework and Hypotheses

An ethical brand promotes itself with out harm the public good and has the certain attributes such as integrity, honesty, diversity, responsibility, quality, respect and accountability [19]. Several models have been constructed to explain consumer's ethical decision making, the most attractive being those by Iglesias et al. [19]. In accordance to these models ethical decision making subsist of a fixed classification of stages, which include evaluation, judgment, behavior, intention, behavioral judgment and moral perception. Recent research proves that to be ethical is in the best interest of brands [30].

Brand stake holders and consumers are expecting brands to reflect their ethical concern so they are more demanding [24]. This is because ethical brands are believed by consumers to be highly reputable and brand leaders [14]. Lau and Lee [20] propose that brand trust is the consent of customers in the face of the risks linked with the brand purchased, will allow for a positive and profitable outcomes. There are three components that affect trust in the brand according to Lau and Lee [20] is the brand itself, the producer of the brand and consumers. The third component relates to the three entities that admit the brand relationship with consumers. Customer perceived value is vital of the long-run gains [25]. In other meanings, consumer trust will higher when the perceived value of customers based on what has been obtained and went through by customers based on their forfeits too high or customer perceived value. Customer Perceived Value behaves importantly in building up customer confidence. Something same was proposed by Sirdeshmukh et al. [29]. Similarity among product attributes and consumer perceived value to reduce doubt and help build up trust in an authentic form of expectations versus the possibility of going to another brand.

Particular services such as mobile phone air time, such retention can be measured by the duration of time that the customer has used a service and, by the customers repurchase of the brand [4]. The brand meanings and perceptions elaborate by a consumer not only based on the brand but also on the interactions that the consumer establishes with other stake holder groups and consumers such as the public opinion and brand employees [15].

Morgan and Hunt [26] suggest that opportunistic behavior on factor of a substitute partner is probably to have a negative affect on the trust between the partners. So, at some point, the relationship among a brand and a consumer can be seen as a dyadic substitute relationship that based on perception of reciprocal trust. Nevertheless, the brand perceptions and significance deduced by a consumer not only depend on directly tends with the brand. But as well on the synergy that the consumer establishes with other consumers and stakeholder accumulation, such as the brand employees and public view [15]. An emerging pour of research on service branding and corporate branding [1] also stresses the role of this broad variety of stake holders in the brand building procedures. Altogether, we indicate that the trust among a brand and its consumer, in addition to depending on the consciousness of equity and lack of opportunistic behavior for the partner. Also depends on the consciousness of fair, creditworthy, and accountable behavior of the brand for a bigger audience. Following recommendations by Becker and Becker [3] that query into business ethics is relevant at various levels of analysis-individual, company and social scheme-this research will focus on the perceived ethical value of corporate brands due to the enhancing relevance of multiple stake holders in the brand building process. In business with the former discussion, it is predictable that a corporate brand that is perceived by the consumer to be ethical is as well probably to be perceived as selling products and services that are clean, fair, reliable, and trustworthy. So, the ethical perception of a corporate brand will be shifted into a greater extent of trust for its product brands.

Based on the above discussion, the hypotheses are developed as follows:

Hypotheses 1. Perceived ethical value has positive effect on trust.

All brands are touch points and the brand stake holders are main in the corporate brand building process [19]. Barbalet proposed that emotion makes an important herald of trust, and in fact, trust is regarded as a social emotion [2]. Likewise, Glomb et al. [12] comments that promoting through behavior that helps others has a positive affect on intuitive state, and proposes that humanitarian and courteous conduct is followed by an strengthen positive influence. We do not want to enter into the argue that if an individual 'does good' to coordinate affect; nevertheless, we want to indicate that if an individual does good, he or she senses a positive temperamental state as a consequence. Leading this logic to the brand management field, it is arguable to presume that a brand that is perceived as being ethical (doing good) will arouse positive emotional reactions in its consumers, and this may be pondered as, for example, a higher degree of brand affect. This could responsibility due to a deontological congruous in the ethical behavior of the brand with particular conscience (moral identity) and beliefs; or because of a teleological view that assesses the presence of an comprehensive good that causes a sense of personal satisfaction in affirming a consideration [17]. According to the discussion in this section, suggest that perceived ethicality of a corporate brand will be shifted into a higher brand affect toward its product brand.

Hypotheses 2. Perceived ethical value has positive effect on brand loyalty.

Brand loyalty as a deeply agreed dedication to re-buy or re-sponsor a favored product/service consistently in the future. Thereby, doing insistent same brand set buying, despite situational determines and marketing attempts having the potential to have shifting behavior [2]. Brand wrongful conduct has negative effects for customer-brand relation, one of them being a wrong effect on purchase intention. We indicate that this wrong effect is due to the diminution in trust that arbitrates the relationship between perceived ethicality and brand loyalty [16]. There is a relationship among trust and commitment, both of these act a fundamental role in arising long run relationships [26]. Sirdeshmukh describe two aspects of trust that are gained from front line management and employees policies and patterns, and also mention a convinced relationship between these two aspects and loyalty [29].

Hypotheses 3. Trust has a positive effect on brand loyalty.

It is said that brand trust is an significant mediator component on the customer behaviors earlier and later the purchase of the product; and it consideration long term loyalty and substantiate the relation between two parties [21]. Brand trust can be defined as the disposition of the average consumer to trust on the power of the brand to perform its described functions. Even if there is a difference among brand trust and brand affect when the processes are considered; brand trust is one of the substantial variables that has an significance on brand affect [7].

One of the most important components which directly determine brand loyalty is brand trust. The significance of the trust on loyalty becomes completely relevant and important in case of conclusion taken for altering the brand due to high stage of perceived risk and abstruseness [18]. Trust plays a central role in increasing brand loyalty and also has an significance on the factors such as sustaining market share and price compliance which are related with marketing results [13]. so the Brand trust summarizes customer prospects of brand intentions, in situations affecting perceived risk and indications that the ethical or unethical behavior by the brand will yield a number of responses by consumers.

Hypotheses 4. Personality moderates the relationship between CPE, and trust on brand loyalty.

Personality is a form of features and the way a person acts. According to Suryabrata [31] personality is determined by types can be assorted into two namely extrovert and introvert. Extrovert personality types are persons who have characteristics: many friends, sociable, enjoy parties, need others company, do not like studying and reading

Fig. 137.1 Research model
(proposed model)

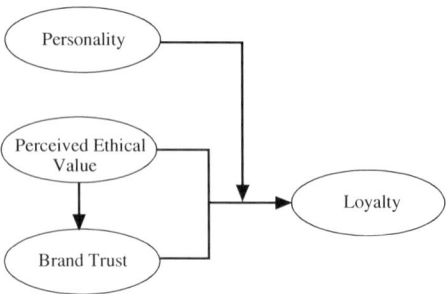

alone. Extrovert peoples mainly determined by the objective world, which is something that exists outdoor of him/herself, drawn out preference, feelings, actions, and
thoughts are determined by the surroundings both socially and environmentally non
social environment. That person is convinced for society, his heart clear and sociable,
relationships with others positively. Those who are introverted types of individuals
have characteristics: do not like to party, does not like to hangout, do not need other
peoples to talk, do not have many friends, and love to read alone. These personality
differences can cause differences in attitudes, including attitude toward the brand
ethicality.

The overall theoretical framework of this study is given in Fig. 137.1.

137.3 Methodology

The sample size is determined according to Hair et al. [33] which says that the size
of a representative sample for analysis using Structural Equation Modeling is the
range 100–200. Thus in this study 140 consumers were sampled, who was taken
with accidental sampling method. The data were collected for the fast moving consumer goods (FMCG), using an online consumer panel. There are various reasons
for choosing this sector. Almost everyone around us needs to fulfill their daily necessities through FMCG and their degree of awareness and interest towards this sector is
comparatively high. As well; today, FMCG is essential for the consumers; due to the
daily usage of such items. Furthermore these products do not only render functional
benefits but also they are able to render psychological benefits, as per requirement.
Therefore; during the choice of these products, the trust for the brand also the effective factors can come to the cutting edge. Between these contexts, it can be said that
the FMCG brands have the potential to enforce both brand trust and brand loyalty
on the consumer trough their ethicality. Testing a model of empirical research using
a chow test regeresi moderation subgroups by using SPSS and Microsoft excel.

137.4 Goodness of Fit Model

The first model declared fit because the first regression equation model of the influence of brand trust on the brand ethical value has $R^2 = 0.199$ with a F Stat. 33.666 and Sig. 0.005. where in the second regression equation model influence brand ethical value and brand trust to brand loyalty has $R^2 = 0.243$ with a value of F Stat. 20.459 and the Sig. 0.000 so the both models declared fit.

137.5 Hypothesis Testing

The influence of brand ethical value to the brand trust based on regression analysis of the first model obtained the results as shown in Table 137.1.

Table 137.1 reports the results of Brand Perceived Ethical Value on Value Trust. Based on the productivity of the t stat 5.87 is greater than t table value 1.976 and Sig. $0.000 < alpha$ 0.05. Then the first hypothesis which says brand Perceived ethical value has positive effect on brand trust is accepted.

Table 137.2 reports the results of Brand Perceived Ethical Value and Brand Trust on Brand loyalty. Based on the productivity of the t stat 5.87 is greater than t table value 1.976 and Sig. $0.004 < alpha$ 0.05. Then the second hypothesis which says brand Perceived ethical value has positive effect on brand loyalty is accepted, and t stat 5.87 is greater than t table value 1.976 and Sig. $0.000 < alpha$ 0.05. Then the third hypothesis which says brand trust has a positive effect on brand loyalty is accepted.

To test the moderating influence of regression analysis is used sub-group with Chow Test [11] with the following formula:

$$F = \frac{(RSSr - RSSur)/k}{(RSSur) - (n1 + n2 - 2k)}, \quad F = \frac{(2148849 - 2086309)/2}{(2086309) - (140 - 4)} = 2.039.$$

According to the results of test chow, calculated F values obtained for 2.039 is less than the value of F table with $df(2; 134)$ of 3.064, it can be purposed that the regression equation between sub-groups of observations introvert and extrovert personality types did not dissent, this depicts that type of personality did not moderate the relationship between brand Perceived ethical value and brand trust to brand loyalty.

Table 137.1 Regression analysis influence of brand ethical value on value trust

	Coefficients	Std. error	Beta	t value	Sig.
Constant	1.430	2.948		0.486	0.638
Brand perceived ethical value	0.587	0.100	0.455	5.870	0.000

Table 137.2 Regression analysis influence of brand perceived ethical value and brand trust on brand loyalty

	Coefficients	Std. error	Beta	t value	Sig.
Constant	2.199	3.383		0.650	0.527
Brand perceived ethical value	0.387	0.138	0.258	2.805	0.004
Brand trust	0.382	0.100	0.327	3.820	0.000

137.6 Conclusion and Recommendation

Most of the results are consequential and are significant with the literature with little exception. Research model is also a good fit of variables. The results showed that perceived ethical value have positive impact on trust and brand loyalty, personality are not moderated the relationship between perceived ethical value and trust to brand loyalty. Recommendations of this study can be characterized as follows: whereas the regression equation model fit, but it's still very low of R^2, either the first regression model and in the second regression model, therefore in future studies will need to add the independent variables as preliminary of brand trust and brand loyalty. The research found that personality type variables are not able to moderate the relationship between brand perceived ethical value and brand trust to brand loyalty. For further studies it is suggested to use environmental consequences as a mediating or moderating variable. Its affect can also be seen through Core brand image. Core brand image and environmental consequences can also be used as sub variables of brand attitude. Further by utilizing improved results. Furthermore this research can also be further initiated on the basis of demographic diversity.

Acknowledgments The authors wish to thank the anonymous referees for their helpful and constructive comments and suggestions. The work is supported by the National Natural Science Foundation of China (Grant No. 71301109), the Western and Frontier Region Project of Humanity and Social Sciences Research, Ministry of Education of China (Grant No. 13XJC630018), the Philosophy and Social Sciences Planning Project of Sichuan province (Grant No. SC12BJ05), and the Initial Funding for Young Teachers of Sichuan University (Grant No. 2013SCU11014).

References

1. Balmer JM, Gray ER (2003) Corporate brands: What are they? What of them? Eur J Mark 37(7/8):972–997
2. Barbalet JM (1996) Social emotions: confidence, trust and loyalty. Int J Sociol Soc Policy 16(9/10):75–96

3. Becker LC, Becker CB (1992) Encyclopedia of ethics. Garland Publishing, NewYork
4. Bhattacharya C (1997) Is your brand's loyalty too much, too little, or just right? Explaining deviations in loyalty from the Dirichlet norm. Int J Res Mark 14(5):421–435
5. Brunk KH (2010) Exploring origins of ethical company/brand perceptions—a consumer perspective of corporate ethics. J Bus Res 63(3):255–262
6. Connolly J, Prothero A (2003) Sustainable consumption: consumption, communities and consumption discourse. J Bus Ethics 6(4):275–291 (In Chinese)
7. Creyer EH (1997) The influence of firm behavior on purchase intention: Do consumers really care about business ethics? J Consum Mark 14(6):421–432
8. Dick AS, Basu K (1994) Customer loyalty: toward an integrated conceptual framework. J Acad Mark Sci 22(2):99–113
9. Geyskens I, Steenkamp JBE, Kumar N (1998) Generalizations about trust in marketing channel relationships using meta-analysis. Int J Res Mark 15(3):223–248
10. Gibbons JD (1973) A question of ethics. Am Stat 27(2):72–76
11. Ghozali I (2006) Aplikasi analisis multivariate dengan program SPSS. Badan Penerbit Universitas Diponegoro, Semarang
12. Glomb TM, Bhave DP et al (2011) Doing good, feeling good: examining the role of organizational citizenship behaviors in changing mood. Pers Psychol 64(1):191–223
13. Gommans M, Krishnan KS, Scheffold KB (2001) From brand loyalty to e-loyalty: a conceptual framework. J Econ Soc Res 3(1):43–58
14. Grisaffe DB, Jaramillo F (2007) Toward higher levels of ethics: preliminary evidence of positive outcomes. J Pers Selling Sales Manag 27(4):355–371
15. Hatch MJ, Schultz M (2010b) Toward a theory of brand co-creation with implications for brand governance. J Brand Manag 17(8):590–604
16. Huber F, Vollhardt K et al (2010) Brand misconduct: consequences on consumer-brand relationships. J Bus Res 63(11):1113–1120
17. Hunt SD, Vitell S (1986) A general theory of marketing ethics. J Macromark 6(1):5–16
18. Ibáñez VA, Hartmann P, Calvo PZ (2006) Antecedents of customer loyalty in residential energy markets: service quality, satisfaction, trust and switching costs. Serv Ind J 26(6):633–650
19. Iglesias O, Sauquet A, Montaña J (2011) The role of corporate culture in relationship marketing. Eur J Mark 45(4):631–650
20. Lau GT, Lee SH (1999) Customer trust in brand loyalty. J Mark Focus Manag 4:341–370 (In Chinese)
21. Liu CT, Guo YM, Lee CH (2011) The effects of relationship quality and switching barriers on customer loyalty. Int J Inf Manag 31(1):71–79
22. Luchs MG, Naylor RW et al (2010) The sustainability liability: potential negative effects of ethicality on product preference. J Mark 74(5):18–31
23. Matzler K, Grabner-Kräuter S, Bidmon S (2008) Risk aversion and brand loyalty: the mediating role of brand trust and brand affect. J Prod Brand Manag 17(3):154–162
24. Maxfield S (2008) Reconciling corporate citizenship and competitive strategy: insights from economic theory. J Bus Ethics 80(2):367–377
25. McDougall GH, Levesque T (2000) Customer satisfaction with services: putting perceived value into the equation. J Serv Mark 14(5):392–410
26. Morgan RM, Hunt SD (1994) The commitment-trust theory of relationship marketing. J Mark 58:20–38
27. Morsing M (2006) Corporate moral branding: Limits to aligning employees. Corp Commun Int J 11(2):97–108
28. Shea LJ (2010) Using consumer perceived ethicality as a guideline for corporate social responsibility strategy: a commentary essay. J Bus Res 63(3):263–264
29. Sirdeshmukh D, Singh J, Sabol B (2002) Consumer trust, value, and loyalty in relational exchanges. J Mark 66:15–37
30. Story J, Hess J (2010) Ethical brand management: customer relationships and ethical duties. J Prod Brand Manag 19(4):240–249
31. Suryabrata S (1988) Psikologi kepribadian. Rajawali, Jakarta

32. Valenzuela LM, Mulki JP, Jaramillo JF (2010) Impact of customer orientation, inducements and ethics on loyalty to the firm: customers' perspective. J Bus Ethics 93(2):277–291
33. Var I (1998) Multivariate data analysis. Vectors 8:6
34. Vitell SJ (2003) Consumer ethics research: review, synthesis and suggestions for the future. J Bus Ethics 43(1–2):33–47

Chapter 138
Equipment Condition Monitoring: The Problematicon Statistical Control Charts

Suzana Paula Gomes Fernando da Silva Lampreia, Valter Martins Vairinhos, José Fernando Gomes Requeijo and Vitor José de Almeid e Sousa Lobo

Abstract Every system is subject, along its life cycle, to several degradation processes that progressively degrade its state and increase its probability of failure (reliability reduction). This is true for the generality of systems—mechanical, electrical, software, human or organizational. If nothing is made—if there is no maintenance—every system will eventually fail. In order to start the condition control and improve its reliability, fixed sensors should be chosen in order to collect data of vibration, oil and water pressure, and temperature and particle size- among others. The actual equipment condition must be known, estimated or predicted from the collected data [1]. Unless the state or condition of system is directly observed, that condition is a latent variable in the sense of statistical theory—a variable not directly observed but with observable effects in the manifest variables, as the just mentioned observed variables associated to sensors. To provide support for the decision maker— at last for critical selected systems—it has been shown that some control statistical techniques are effective to ascertain trends and predict needs of future interventions out of observed data. When data is not appropriate, the combination of statistical techniques with simulations and can be considered. With this work we intend to show that control charts can be decisive as instruments of control of equipment monitoring and functioning [4], although there are some problems when applied to the conditioned maintenance; some of those problems can be overcome using a EWMAQ modified chart, adjusting its parameters [3].

Keywords Condition based maintenance · Control charts · Statistical problems

S. P. G. F. da Silva Lampreia (✉) · V. J. de Almeid e Sousa Lobo · V. M. Vairinhos
CINAV—Mechanical Engineer Department, Naval Academy, Alfeite
2810-405, Almada, Portugal
e-mail: suzanalampreia@gmail.com

V. M. Vairinhos
CENTEC (IST), Rovisco Pais Avenue, 1,1049-001 Lisbon, Portugal

J. F. G. Requeijio
Mechanical and Industrial Department, Science and Technology of the Universidade Nova of Lisbon, 2829-516 Caparica, Portugal

J. Xu et al. (eds.), *Proceedings of the Eighth International Conference on Management Science and Engineering Management*, Advances in Intelligent Systems and Computing 281, DOI: 10.1007/978-3-642-55122-2_138, © Springer-Verlag Berlin Heidelberg 2014

138.1 Introduction

When the decision to apply a condition-based maintenance system is taken, we should define which parameters represent the real machine condition in order to carry out an intervention when the need actually exists, avoiding acting at random or based on wrong information.

In order to follow systems deterioration, sensors can be used to collect the operating parameters needed to perform online monitoring.

Whether we are in an industrial or maritime environment, anavalship for instance,it frequently occurs that large volumes of data are collected without further systematic treatment and analysis. It only makes sense to collect and store data if this data is organized and analysed for the benefit of maintenance teams, translating this information in useful knowledge about the real behaviour and condition of machinery and systems [1] Sometimes, too much information can also bring conflictual management [5]; therefore, a careful previous planing of information really needed is necessary-and stored in databases that collect not all, but only significant data needed for planning and control. Data must serve decision processes perfectly identified and characterized, through statistical treatments that supply results under the form and nature really needed by maintenance organization and management. As an example of such statistical information, the modified control charts can be used—under some constraints—to control equipment [4]. The main objective of are liability study should be to provide information as a basis fordecision.The data and information resulting from the reliability study does not tell us exactly what to do, but must reveal and suggest the direction to be taken, as an indicator [2].

138.2 Modified Control Charts

To minimize some of the problems related to equipment monitoring, given to the data volume, the use of short run control charts can help troubleshoot, even without enough data to provide reliable indicators. It can be a good decision making support to know whether or not take a maintenance action.

The reasons for a shortage of data are:

- Monitoring equipment characteristics;
- Change in operation regime;
- Maintenance strategy adamant;
- An unique maintenance system;
- No unified organization of data collection.

The control charts to be applied for the online equipment monitoring must suffer some changes. In this paper, the short run control charts are developed and applied to individual observation.

Table 138.1 Modified short run control charts to equipment monitoring

	Chart	Limits
Q	$Q_r(X_r) = \Phi^{-1}\left(G_{r-2}\left(\sqrt{\frac{r-1}{r}}\left(\frac{X_r - X_{r-1}}{S_{r-1}}\right)\right)\right)$	$UCL = 3, AL = 2.5, CL_Q = 0$
QM	$Q_r(X_r) = \Phi^{-1}\left(G_{r-2}\left(\sqrt{\frac{r-1}{r}}\left(\frac{X_r - (T_L)_{r-1}}{S_{r-1}}\right)\right)\right)$	$UCL = 3, AL = 2.5, CL_Q = 0$
$CUSUMQM$	$C_r = \max(0, C_{r-1} + (Q_r - k))$	$\alpha = 1\%(ARL = 100)$ to LA
		$\alpha = 0.2\%(ARL = 500)$ to UCL
		Gan Abacus

In the short run control charts the mean of the X variable can't be controlled, so it is transformed to Q variable. In the short run charts the mean control is allowed since the fourth observation.

The statistics for instant $r = 4, 5, \cdots$ represent the modified charts are in Table 138.1.

The distinction in relation to the non-modified Quenseberry charts is that the value of the mean $\overline{X}_{r-1}$ is replaced by $(T_L)_{r-1}$.

$(T_L)_r = (T_L)_N - 3\sigma_{r-1}$, for observation $(r-1)$, where $(T_L)_N$ is the standard vibration limit. $(T_L)_N$ represents the limit defined by the manufacturer or International Standards. X_r is the observation for instant r, $\Phi^{-1}(\cdot)$ the inverse of normal distribution function and $G_\nu(\cdot)$ the T-student distribution function with ν freedom degrees [3].

The mean and the variance for instant r are, respectively: $\overline{X}_r = \frac{1}{r}\sum_{j=1}^{r} X_j$. The mean and the sample variance for time r $(\overline{X}_r$ and $S_r^2)$, can be expressed as a function of X_r and the values at times before $(r-1)$. Since variables $Q(X)$ are $N(0, 1)$ the control charts limits, considering only the maximum values, equal or above zero, are given by $UCL_Q = 3$ and $CL_Q = 0$.

The special charts use early data, so they are more sensitive than the QM charts. The results for this charts become more reliable and consistent. In the case of the modified special chart, the statistic Q is the same calculated in the QM charts.

To the EWMAMQM chart, $\sigma_{\overline{X}} = \sigma_\varepsilon/\sqrt{n}$, $\Delta = \delta\sigma_{\overline{X}}$, $T_L = (T_L)_{Norma} - \Delta_S e \Delta_S = \delta_1\sigma$, being δ_1 a constant, $\sigma_E = \sigma_{\overline{X}}\sqrt{\frac{\lambda}{2-\lambda}}$ the standard deviation.

The analysis of modified Q_X charts aren't like in the non-modified $Q(X)$ chart, so different rules must be defined. The analysis of modified Q_X is next described on the proposed methodology (Fig. 138.1).

138.3 Suggested Methodology

- Select the equipment based on the relevance and importance of its characteristics for the ship.
- Define the parameters to monitor.
- Define the regime in which the measures are going to be taken.

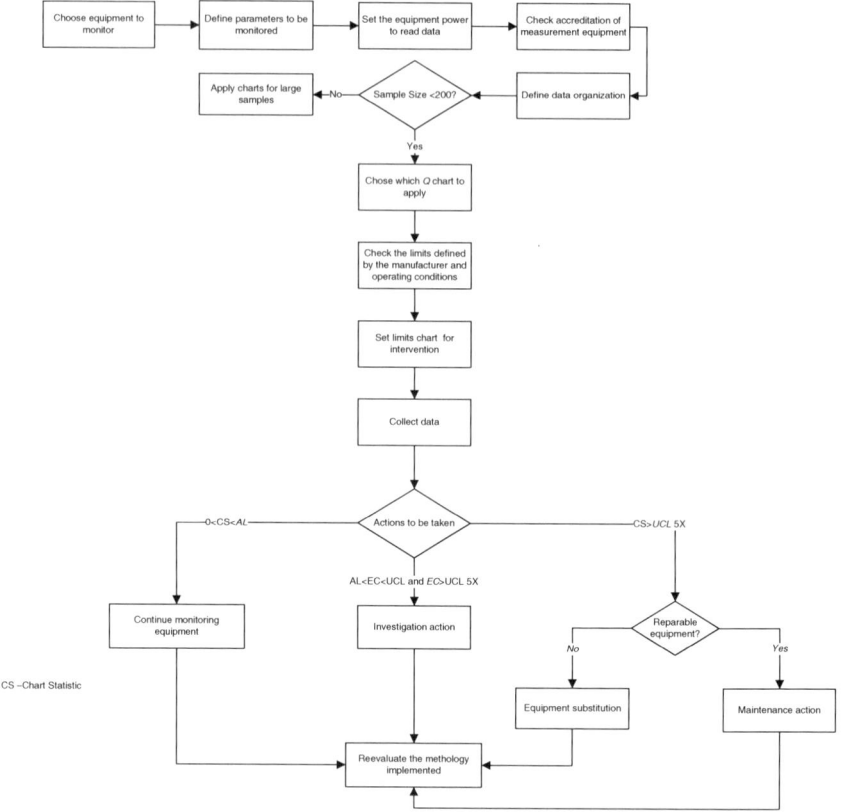

Fig. 138.1 Operational actions

- Check measurement equipment certification. Record also the meteorological conditions.
- Set an organization adequate for the online monitoring (eventual use the MIMOSA normative).
- Specify the sample size (m) for specified statistical procedures. For lack of data, which is the case, use short run control charts.
- Assuming data independence, the short run control chart that best fits the data should be applied. In this case QM, $CUSUMQM$ and $EWMAQM$ are built, based on the collected data.
- Define the limits for intervention:

  – The Upper Control Limit (UCL) is 3 and the Alert Level (AL) is defined by 2.5.
  – The maximum levels are define by the time of functioning and by the environment in which the machine is.
  – The defined rules to act in the equipment, using the Q charts are:

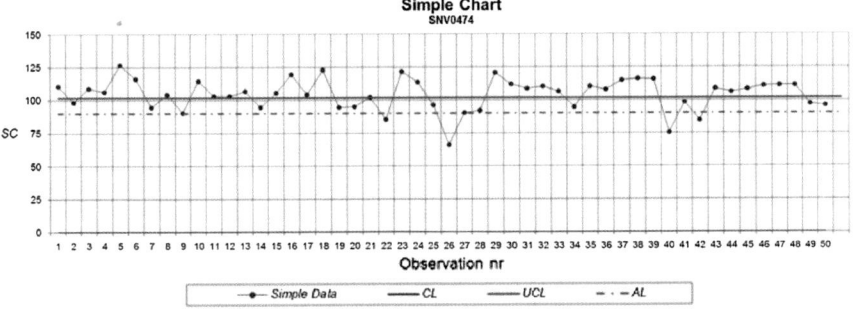

Fig. 138.2 Simple chart for SNV0474 sensor

> Do an eventual anomaly investigation if there is more than 5 consecutive points between the AL and the UCL.
> Proceed to a maintenance action if there is more than 2 consecutive points above UCL.

- Collect Data at the defined points/variables to monitor.
- Equipment parameters are estimated for each instant using univariate sample (data) mean and variance.
- After sufficient data accumulation check if the methodology to be applied should, eventually, be adapted to the new reality.

138.4 Case Study

The case study is based in a Gas Turbine from CODOG propulsion of a ship. When monitoring a ship equipment the meteorological conditions should be collected. In this case, all situations that act as noise factors should be minimized. Readings shall be made after parameters stabilizations, and for each power equipment, an independent study should be done.

The turbine used in the study had an abnormality in the compressor of the gas generator. Therefore, this article presents the results for the sensor that has been shown more sensitive to this anomaly.

The vibration sensors readings, from the referred gas turbine, were taken at 8,000 rpm from the compressor. In this paper, only the SNV0474 vibration sensor are represented, because it best represent the state of the turbine.

1. Traditional Short Run

The vibration data could be monitored online without statistical treatment. But, as can see from Fig. 138.2, the highest values are not highlighted, nor the lower values can be ignored, because they are all above the UCL (Observation nr 10–11).

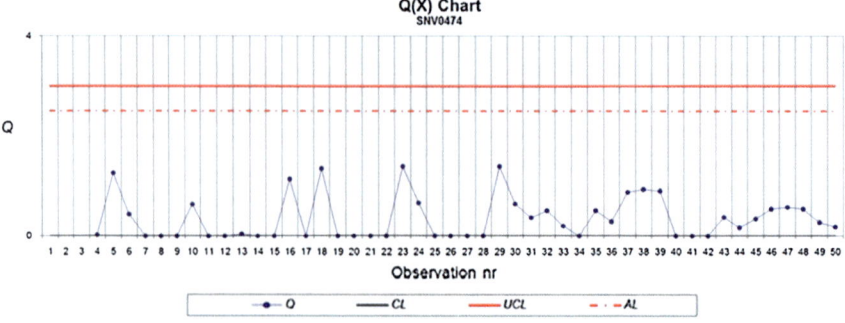

Fig. 138.3 Q chart for SNV0474 sensor

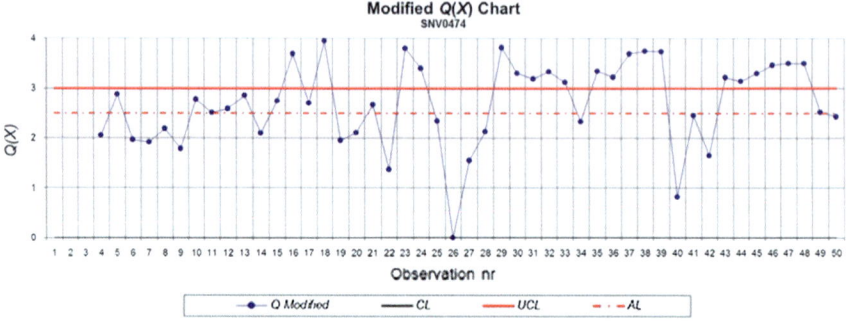

Fig. 138.4 Q chart for SNV0474 sensor

Table 138.2 Limits values for different k for the CUSUMQM chart

		$k = \delta/2$			
		0.25	0.5	0.75	1
ARL	LSC ($\alpha = 0, 2\%$) $- h$	8, 5	5, 1	3, 5	2, 7
	LA ($\alpha = 1\%$) $- h1$	5, 51	3, 5	2, 5	1, 8
		CUSUMQM	CUSUMQM1	CUSUMQM2	CUSUMQM3

Applying the chart without the modification, not replacing the mean with the manufacturer value (TL), observing the Fig. 138.3, we can see that the graph does not represent the equipment real state. No observation is above AL or UCL.

For QM chart (Fig. 138.4) the obtained results show the turbine real state corresponding to the observation time. For the 26 observation there is an abrupt decrease in the vibration values; this means, eventually, that the equipment functioning was not stable or that sea condition wasn't the best.

2. CUSUMQM Chart

Based on the results obtained for QM chart, and since the data is the same, the application of CUSUMQM for SNV0474 is going to be presented (Table 138.2).

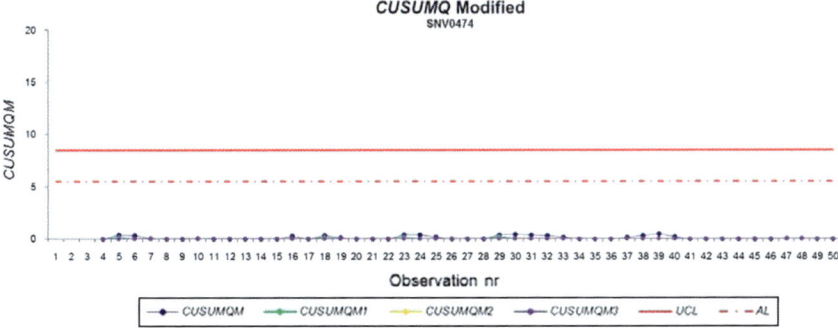

Fig. 138.5 CUSUMQ chart for SNV0474 sensor for different k

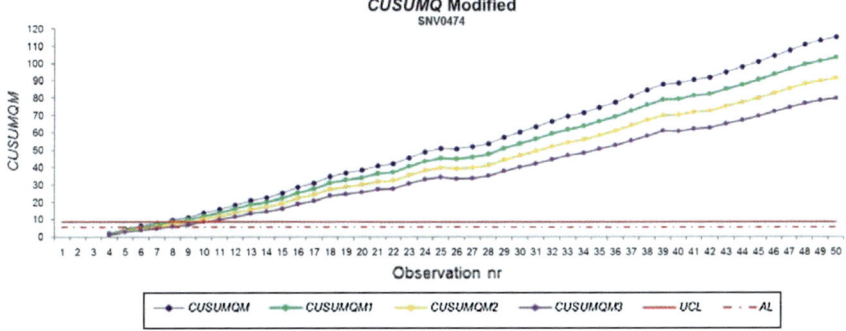

Fig. 138.6 CUSUMQM chart for SNV0474 sensor for different k

It was considered $\alpha = 1\%$ $(ARL = 100)$ to the definition of AL $\alpha = 0,2\%$ $(ARL = 500)$ for the CUSUMQM application.

Also for the CUSUMQM (Fig. 138.5) for different k values, was tested the original chart and the result wasn't satisfactory, and it didn't had the need of any intervention.

The CUSUMQM chart (Fig. 138.6) shows a high sensitivity. We can observe the representation of different k, but it is for $k = 0.25$ where the sensitivity is higher.

3. EWMAQM Chart

For both harts, EWMAQ and EWMAQM the λ and K from Table 138.3.

Representing the chart without modification, we verify that like CUSUMQ, it doesn't show the adequate sensitivity (Fig. 138.7).

The EWMAQM chart (Fig. 138.8) shows an adequate and strong sensitivity relatively to the expected results in applying control charts to the equipment control.

For $\lambda = 0.05$ the results are more linear, but if we want an accurate the study of an elected instant, we should observe the chart with a higher λ values. With this procedure for example in observation 36 we can observe an eventual tendency in the observations.

Table 138.3 EWMAQM chart limits

δ	0.5	1	1.5	2	0.5	1	1.5	2
λ	0.05	0.13	0.25	0.37	0.08	0.18	0.34	0.5
K	2,7	2,9	3	3,04	2,1	2,3	2,5	2,56
	EWMAQM ARL = 500 (LSC)	EWMAQM1	EWMAQM2	EWMAQM3	EWMAQM ARL = 100 (LA)	EWMAQM1	EWMAQM2	EWMAQM3

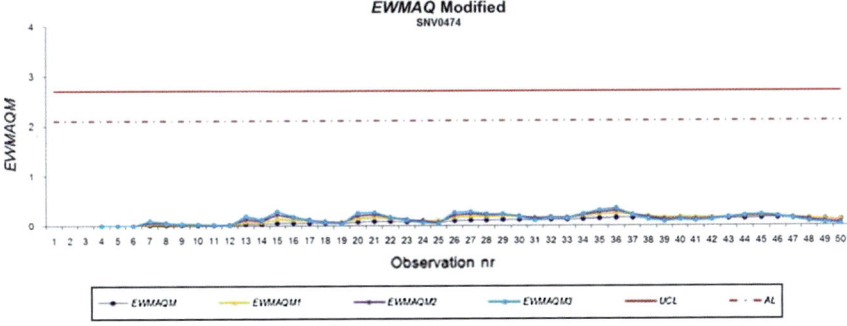

Fig. 138.7 EWMAQ chart for SNV0474 sensor for different λ

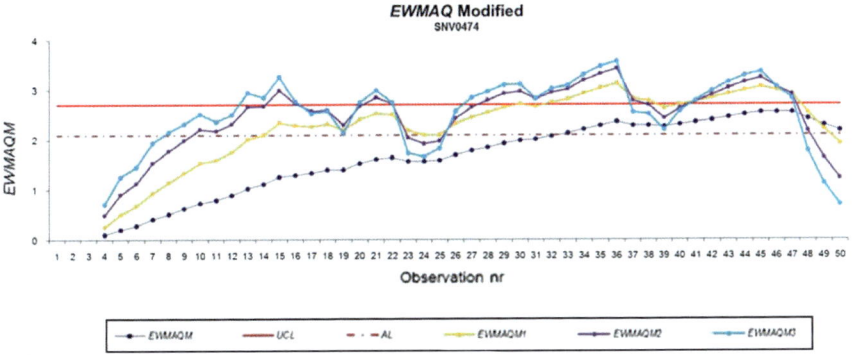

Fig. 138.8 EWMAQM chart for SNV0474 sensor for different λ

4. Problems in Using Control Charts

Instead of the results for EWMAQM charts we must considered that there wasn't any destructives tests. So destructives tests are an important step to certify these charts.

Manufacturer limits can be adjusted according to the operating time of the equipment and according to a risk, which is admissible to take.

Decide a profound intervention based only on the defined rules can conduce to unneeded interventions. Others non-destructives tests must be considered.

We verify that analysing the same data with different charts, the results are modified, and some of it doesn't shows solid results.

The measure instant cannot correspond to equipment stabilization moment.

138.5 Conclusions

- The control charts can be used for equipment online monitoring.
- When there is shortage of data the short run modified control charts should be applied.
- The QM control chart has a good sensitivity.
- CUSUMQM charts shows high sensitivity, and the most reasonable results was for $k = 0.25$.
- The rules for intervention should be defined and decided based on the experience, in the operational needs and on the financial and human resources economy.
- Another important step is to analyse Gan and Crowder abacus, and maybe introduce some adjustments.
- The EWMAQM had shown the most adequate sensitivity for $\delta = 0.5$.
- This charts should be tested in other equipment, other type of charts should be tested.
- To decide over the results other statistic studies should be developed.

Acknowledgments Both Portuguese Naval School and CINAV are kindly acknowledged for the use of the machinery workshop and also for the fruitful collaboration of the Faculty of Science and Technology from the Universidade Nova of Lisbon.

References

1. Alzghoul A, Lofstrand M (2011) Increasing availability of industrial systems through data stream mining. Comput Ind Eng 60(2):195–215
2. Hameed Z, Ahn S, Cho Y (2010) Practical aspects of a condition monitoring system for a wind turbine with emphasis on its design, system architecture, testing and installation. Renew Energy 35(5):879–894
3. Lampreia S, Requeijo J et al (2012) Maritime diesel engine condition accompaniment based on Q, CUSUMQ, EWMAQ e MQ control charts. O Propulsor, pp 29–39
4. Lampreia S, Requeijo J et al (2013) Equipment condition monitoring with an application of mewma control charts and others charts. In: 11th international conference on vibration problems, Lisboa, p 191
5. Vanneste S, Wassenhove L (1995) An integrated and structured approach to improve maintenance. Eur J Oper Res 82(2):241–257

Author Index

J. Xu et al. (eds.), *Proceedings of the Eighth International Conference on Management Science and Engineering Management*, Advances in Intelligent Systems and Computing 281, DOI: 10.1007/978-3-642-55122-2, © Springer-Verlag Berlin Heidelberg 2014

Printed by Printforce, the Netherlands